FRUITS

Tropical and Subtropical

— Volume 1 —

THE EDITORS

Professor T. K. Bose awarded M.Sc. (Ag.), Ph.D. degree from Calcutta University, West Bengal, India. Former Horticulturist and Secretary of the Agri-Horticultural Society of India; Project Coordinator, All India Coordinated Floriculture Improvement Project (ICAR); Professor and Head, Department of Horticulture; Dean, Faculty of Agriculture; Director of Research, Bidhan Chandra Krishi Viswavidyalaya, West Bengal; Member, State Agriculture Commission, West Bengal; Member of Scientific Advisory Committees of ICAR, CSIR (Govt. of India) and Government of West Bengal. Awarded D.Sc. (Honoris Causa) from OUAT, Bhubaneshwar, Odisha and UBKV, Coochbehar, West Bengal, India. Guided around 45 students for their Ph.D. degree, published more than 250 research papers and authored 30 books on horticultural crops.

Dr. V. A. Parthasarathy has about half a century of research experience in Horticulture particularly in the field of genetics and biotechnology. Former Director, Indian Institute of Spices Research, Kerala, India. President of National Academy of Biological Sciences. Edited/authored over 50 books, about 100 review/ invited papers and over 225 research papers.

Professor S. K. Mitra awarded M.Sc. (Ag.), Ph.D. degree from Bidhan Chandra Krishi Viswavidyalaya, West Bengal, India. Former Head, Department of Fruit Science; Dean, Faculty of Horticulture; Dean, Post Graduate Studies, Bidhan Chandra Krishi Viswavisyalaya, West Bengal, India. Guided 20 Ph.D. students, published more than 200 research papers and authored15 books on horticultural crops.

Professor B. Ghosh awarded M.Sc. (Ag.), Ph.D. degree from Bidhan Chandra Krishi Viswavidyalaya, West Bengal, India. Former Head, Department of Fruit Science, Officer in Charge of Mango database Project, DBT, Government of India. Guided 6 Ph.D. students, authored 5 books, published more than 80 research papers.

Professor I. Chakraborty awarded M.Sc. (Ag.) from Calcutta University, Ph.D. from Bidhan Chandra Krishi Viswavidyalaya, West Bengal, India. Former Head, Department of Post Harvest Technology, Bidhan Chandra Krishi Viswavisyalaya, West Bengal, India. Recipient of J.S. Pruthi award conferred by AFSTI, CFTRI, Mysore, India. Guided 7 Ph.D. students, published more than 100 research papers and written 4 books on horticultural crops.

Late Professor D. Sanyal awarded M.Sc. (Ag.), Ph.D. degree from Bidhan Chandra Krishi Viswavidyalaya, West Bengal, India. Associate Professor, NEHU, Nagaland and Professor in the Department of Horticulture, Institute of Agriculture, Visva Bharati, West Bengal. Published more than 100 research papers and written 15 books on horticultural crops including 10 volumes of "Dictionary of Horticulture".

FRUITS
Tropical and Subtropical

— *Volume 1* —

— *Senior Editor* —

V.A. Parthasarathy

ICAR – Indian Institute of Spices Research
32/482 C Narmada Nilayam
Bharathan Bazar, Chelavoor
Calicut 673571, Kerala, India

— *Editor Emeritus* —

T.K. Bose

Faculty of Horticulture
Bidhan Chandra Krishi Viswavidyalaya
West Bengal 741 252, India

— *Edited by* —

S.K. Mitra • B. Ghosh • I. Chakraborty

Faculty of Horticulture
Bidhan Chandra Krishi Viswavidyalaya
West Bengal 741 252, India

D. Sanyal

Palli Siksha Bhavan (Institute of Agriculture)
Sriniketan 731236, Visva Bharati
West Bengal

2022

Daya Publishing House®

A Division of

Astral International Pvt. Ltd.

New Delhi – 110002

Fourth Revised & Illustrated edition 2021

Publisher's Note:

Every possible effort has been made to ensure that the information contained in this book is accurate at the time of going to press, and the publisher and author cannot accept responsibility for any errors or omissions, however caused. No responsibility for loss or damage occasioned to any person acting, or refraining from action, as a result of the material in this publication can be accepted by the editor, the publisher or the author. The Publisher is not associated with any product or vendor mentioned in the book. The contents of this work are intended to further general scientific research, understanding and discussion only. Readers should consult with a specialist where appropriate.

Every effort has been made to trace the owners of copyright material used in this book, if any. The author and the publisher will be grateful for any omission brought to their notice for acknowledgement in the future editions of the book.

Cataloging in Publication Data--DK
Courtesy: D.K. Agencies (P) Ltd. <docinfo@dkagencies.com>

Fruits : tropical and subtropical / senior editor, V.A. Parthasarathy ; editor emeritus, T.K. Bose ; edited by S.K. Mitra, B. Ghosh, I. Chakraborty, D. Sanyal. -- Fourth revised & illustrated edition.
volumes cm
Includes index.
ISBN 9789354616105 (HB)

1. Tropical fruit--India. I. Parthasarathy, V. A., editor.

LCC SB359.F78 2021 | DDC 634.60954 23

Published by : **Daya Publishing House®**
A Division of
Astral International Pvt. Ltd.
– ISO 9001:2015 Certified Company –
4736/23, Ansari Road, Darya Ganj
New Delhi-110 002
Ph. 011-43549197, 23278134
E-mail: info@astralint.com
Website: www.astralint.com

From the Desk of Editor Emeritus

The undersigned had initiated the publications entitled 'Fruits: Tropical and Sub-tropical', 'Vegetable Crops' and 'Commercial Flowers' in one volume each in the 1980's in collaboration with co-editors, humbly expresses from the core of his heart that the updated revised editions of the above mentioned publications with colour illustrations in thirteen volumes which was planned in June, 2020, are being printed from October this year, a rare event in Horticultural Science has been possible by the blessing of Almighty God, with the cooperation of dedicated horticulturist as editors and contributors.

T.K. Bose

Editor Emeritus

From the Desk of Editor Emeritus

Preface to the Fourth Edition from Senior Editor

It is nearly two decades since the third edition hit the stands. Lot of water has flown under the bridge. Many contributors to earlier editions have retired (some deceased as well). We pray the almighty to lay them in rest those who contributed to the earlier editions and since deceased. In India, the last two decades I saw a huge development in horticulture and the country has made huge strides. In terms of number of horticultural crops, the National Agricultural Research System works on over 130 horticultural crops.

India holds an eminent position in cultivation and research on Horticulture. We are the second largest producer of fruits and vegetables in the world leading in the production of mango, banana, coconut, cashew, papaya, pomegranate while our productivity of grapes, banana, cassava, peas, and papaya is the highest in the world. In the research front, we have one of the best networks of research institutes under ICAR. We have 10 Central Institutes, 6 Directorates, 7 NRCs, 13 AICRPs and 6 Network Projects/Outreach programmes. We have 62 Agricultural Universities and Under NARS; we work on about 130 horticultural crops and have developed/released about 1600 varieties of fruits, vegetables, ornamentals, spices, plantation, medicinal and aromatic crops. Our strength in our genetic resources is well conserved with about 73000 accessions being maintained at various gene banks under the control of National Bureau of Plant Genetic Resources, India.

Indian horticulture has great documentations available. Even before Linnaeus developed the hierarchical system of classification of nature, a Portuguese governor of Cochin (Kochi), Van Rheede, documented the horticultural wealth of Malabar region in 12 volumes taking about 3 decades to complete with Indian Botanists.

Van Rheede commenced his work in 1678 and completed it in 1703. History tells us the great history of Indian horticulture.

Research on perennials, particularly, fruit breeding is time consuming and tedious. Contrast to the annuals, this is life time work. We tried to rope in the best in the field to write the eleven chapters with suitable illustrations in this volume. I feel the present volume would be a valuable reference material for post graduate students, research workers, teachers and also the policy makers.

Kolkata **Senior Editor**

Preface to the Fourth Edition from Editors

The first edition of the book entitled 'Fruits: Tropical and Subtropical' published in 1985 was greatly appreciated and proved valuable to the students, teachers, researchers and extension specialists in horticulture in general and fruit science in particular, in different parts of the world. Then the book was revised twice in 1993 and 2002 with the title 'Fruits: Tropical and Subtropical' in one and two volumes, respectively due to its increasing demand. All the previous editions, which contained updated scientific information of all commercially grown fruit crops, were considered as valuable contribution on the subject. After about 18 years of the last revised publication in 2002, voluminous information pertaining to fruit crop improvement, production, protection and post harvest management has been generated necessitating further revision in four volumes.

This particular edition has been enlarged by incorporation of up-to-date information on various aspects of fruit crop production, protection and improvement particularly through biotechnological interventions with meaningful illustrations, and published in four volumes. Efforts have been made to cover as much information as possible. First volume in this edition includes 11 chapters on tropical and subtropical fruit crops covering more than 600 pages. Most of the contributors have taken care to review up-to-date research works and prepared the manuscript. The revision has been made by the editors and in few cases new contributors have been associated with some chapters.

We express our sincere thanks to all contributors who sincerely participated in revision, editing and proof-reading of the manuscript and we appreciate their commendable efforts to bring out this publication. We express our sincere thanks to senior editor Dr. V.A. Parthasarathy for thoroughly editing the manuscript and offering valuable suggestions in strengthening the same.

Apart from own collection of photographs, editors have also taken some illustrations which are in public domain from different websites. Therefore, we acknowledge with enormous thanks to all State Agricultural Universities and Research Institutes under Indian Council of Agricultural Research, New Delhi, India; Public and Private Agricultural Institutes across the world for taking illustration from their websites to make this book a valuable knowledge for the readers across the world. We think images can be powerful teaching tools, as illustrations to in-class lectures, or for studying concepts outside of the classroom. We also think identification of fruit crops, their package of practices, cultivars, utilization, diseases and pests with the help of suitable illustrations in a consolidated manner will make a dent for the students and teachers of fruit science.

The help and cooperation rendered by Dr. Koushik Dutta, Kolkata in preparing manuscript is highly appreciated.

Our thanks are due to Shri Prateek Mittal, Director, Astral International (P) Ltd., New Delhi for his keen interest in bringing out this publication. We are thankful to Shri Debasish Roy and Shri Sukanta Kanji of Durga Reprographics, Kolkata for processing the manuscript.

Kolkata **Editors**

Contents

List of Contributors

Aswath, C.

ICAR-IIHR, Hessaraghatta Lake Post, Bengaluru 560 089, Karnataka

Bhattacharya, R. K

Department of Horticulture, Assam Agricultural University, Jorhat 785013, Assam

Bose, T.K.

Department of Horticulture, Faculty of Agriculture, Bidhan Chandra Krishi Viswavidyalaya, Mohanpur 741252, West Bengal

Chakraborty, Ivi

Department of Post Harvest Technology, Faculty of Horticulture, Bidhan Chandra Krishi Viswavidyalaya, Mohanpur 741252, West Bengal

Ghosh, B.

Department of Fruit Science, Faculty of Horticulture, Bidhan Chandra Krishi Viswavidyalaya, Mohanpur 741252, West Bengal

Hazarika, B.N.

College of Horticulture and Forestry, Central Agricultural University, NH52, Medog, Pasighat 791102, Arunachal Pradesh

Irulappan, I.

Horticulture College and Research Institute, Tamil Nadu Agricultural University, Coimbatore 641003, Tamil Nadu

Jaiswal, Rohit

ICAR-Central Institute for Subtropical Horticulture Rehmankhera, Kakori 226101, Lucknow, Uttar Pradesh

Karale, A.R.

Department of Horticulture, Mahatma Phule Krishi Viswavidyalaya, Rahuri 413722, Maharashtra

Kore, V.N.

Division of Fruit and Horticultural Technology ICAR-IARI, Pusa 110012, New Delhi

Majumder, P.K.

Division of Horticulture, ICAR- IARI, Pusa110012, New Delhi

Mishra, Maneesh

ICAR-Central Institute for Subtropical Horticulture, Rehmankhera, Kakori 226101, Lucknow, Uttar Pradesh

Mitra, S.K.

Department of Fruit Science, Faculty of Horticulture, Bidhan Chandra Krishi Viswavidyalaya, Mohanpur 741252, West Bengal

Narayana Reddy M.A.

Department of Horticulture, University of Agricultural Sciences, Bangalore 560065, Karnataka

Pathhak, P.K.

Murshidabad Krishi Vigyan Kendra Digha, Milebasa, Kalukhali, Bhagwangola, Murshidabad, West Bengal

Parthasarathy, Utpala

ICAR – Indian Institute of Spices Research, 32/482 C Narmada Nilayam, Bharathan Bazar, Chelavoor, Calicut 673571, Kerala

Parthasarathy, V.A.

ICAR – Indian Institute of Spices Research, 32/482 C Narmada Nilayam Bharathan Bazar, Chelavoor, Calicut 673571, Kerala

Rathore, D.S.

CSK Himachal Pradesh Agriculture University, Palampur, Himachal Pradesh

Saha, Tanushree

Department of Fruit Science, Faculty of Horticulture, Bidhan Chandra Krishi Viswavidyalaya, Mohanpur 741252, West Bengal

Sakthivel, T.

ICAR-IIHR, Hessaraghatta Lake Post, Bengaluru 560 089, Karnataka

Senthil Kumar, R.

ICAR-IIHR, Hessaraghatta Lake Post, Bengaluru 560 089, Karnataka

Sharma, D.K.

Division of Horticulture, ICAR- IARI, Pusa110012, New Delhi

Sharma, V.P.

Division of Fruit and Horticultural Technology ICAR-IARI, Pusa 110012, New Delhi

Singh, Sanjay Kumar

ICAR-National Research Centre on Litchi, Musahri, Ramna 842002, Bihar

Vanlalneihi, B.

College of Horticulture and Forestry, Central Agricultural University, NH52, Medog, Pasighat 791102, Arunachal Pradesh

Vinoth, S.

Bioversity International ICAR-IIHR, Hessaraghatta Lake Post, Bengaluru 560 089, Karnataka

①

MANGO

Sanjay Kumar Singh, P.K. Majumder, and D.K. Sharma

1.0 Introduction

No other fruit, except banana, is so closely associated with the history of agriculture, nay, the very history of civilization itself, as the mango is. Records suggest that it has been in cultivation in the Indian subcontinent for well over 4000 years now (De Candolle, 1904) and has been, since time immemorial, the favourite of the kings and commoners alike because of its luscious taste and captivating flavour. In India, it is the part and parcel of life, being connected with all phases of life right from birth to death. Ancient Sanskrit scriptures of various periods abound in rich references to this plant and the vast uses to which its every part is put.

Legend has it that a mango grove was presented by Amradarika to Lord Buddha so that he might use it as a place of repose (Popenoe, 1920). Way back in 327 BC, Alexander the Great too had its view on the Indus Valley when he was there on his conquering spree of the world.

It finds mention in the travelogues of ancient travellers beginning from Hiuen Tsang (632-645 AD) who was the first foreigner to bring mango to the notice of the outside world. Ibn Haukal (902-908 AD) and Ibn Batuta (1325-1349 AD) have written about it (Popenoe, 1920). Amir Khusrau (1330 AD), the great Sufi saint and poet, sung its praise during the time of Mohammed Tughlaq (1325-1351 AD). In relatively recent times, Nicolo Conti (1419-1444), Ludovici de Varthema (1503-1508), Fryer (1673) and Hamilton (1727) have all mentioned about it (Popenoe, 1920). However, the greatest tribute to this esteemed plant was paid by Emperor Akbar (1556-1605) who established the *Lakh Bagh* (a mango orchard having 100,000 plants) in Darbhanga (Bihar) when large orchards of fruit trees were unknown. In *Ain-i-Akbari*, a masterpiece of Persian literature, written by the great chronicler Abul Fazl (1590 AD) during the reign of Emperor Akbar, the cultivars, cultivation and quality of mangoes have been discussed in detail.

2.0 Composition and Uses

The mango, because of its great utility, occupies a pre-eminent place amongst the fruit crops grown in India and is acknowledged as the King of Fruits of this country. Amir Khusrau (1330 AD) has stated (Popenoe, 1920) –

'The mango is the pride of the garden,

The choicest fruit of Hindustan,

Other fruits we are content to eat when ripe,

But the mango is good in all stages of growth'.

2.1.0 Composition

The composition, in general, differs with the cultivar and the stage of maturity. The un-ripe green mangoes are reported to have 90 per cent moisture, 0.7 per cent protein, 0.1 per cent fat, 8.8 per cent carbohydrate, 0.01 per cent calcium, 0.02 per cent phosphorus, 4.5 mg/100 g iron, carotene (as vitamin A- 150 I.U.), 30 g/100 g riboflavin, 3 mg/100 g ascorbic acid (Anon., 1962).

The composition of ripe mango fruits is depicted in Table 1.

Table 1: Composition of Ripe Mango Fruits*

Constituents	*Amount*
Moisture (per cent)	73.0-86.7
Carbohydrate (per cent)	11.6-24.3
Protein (per cent)	0.5-1.0
Fat (per cent)	0.1-0.8
Fibre (per cent)	1.1
Lipid (per cent)	0.8-1.36
TSS (°Brix)	12.0-23.0
Total sugar (per cent)	8.7-17.93
Acidity (per cent)	0.12-0.38
Total phenol (per cent)	1.20-7.83
Vitamin A (µg/100 g -carotene)	6375-20750
Vitamin B1 (mg/100 g)	40
Nicotinic acid (mg/100 g)	0.3
Riboflavin (mg/100 g)	50.0
Ascorbic acid (mg/100 g)	6.8-38.8
N (per cent)	0.46
P (per cent)	0.195
K (per cent)	0.88
Ca (per cent)	0.412
Mg (per cent)	0.082
S (per cent)	0.542
Cu (ppm)	24.0
Zn (ppm)	11.5
Mn (ppm)	10.0
Fe (ppm)	50.0

* Anonymous (1966), Pathak and Sarada (1974), Lakshminarayana (1980), Sanyal *et al.* (1991), Raghupathi and Bhargava (1994), Chandra and Chandra (1997).

The sugars in mango comprise of sucrose, glucose, fructose and maltose (Anon. 1994). Others that have been reported to be present are xylose (Sankar, 1963), arabinose (Wali and Hassan, 1965), sedoheptulose and mannoheptulose (Ogata *et al.*, 1972). The mineral content (per cent DW) of all fruit parts of all cultivars was generally lower at harvesting stage than at half-maturity and the mineral content of peel was generally higher than that of other fruit parts at both stages of maturity (Chattopadhyay and Sarkar, 1992). On the other hand, Raghupathi and Bhargava (1994) obtained maximum N, P, Ca, Mg, Fe, Zn and Cu concentrations in the mesocarp, while S and Mn concentrations were greatest in the endocarp and the kernel respectively. In general, the dorsal part of the fruit was sweeter than the ventral parts, and the basal part was sweeter than the middle or apical parts (Simao and Pimentel -Gomes, 1994).

The soluble protein content was found to decrease up to 44 days after fruit set and increased thereafter until 96 days (Tandon and Kalra, 1983). Elahi and Khan (1973) identified 12 amino acids including the essential ones like alanine, aspertic acid, lysine, leucine, cystine, valine, arginine, phenylalanine and methionine in fruits of 4 mango cultivars grown in Pakistan. These accounted for 10-40 mg/100 g of edible pulp. Shukla *et al.* (1985) noted as many as 13 amino acids during growth and development of Safeda Malihabad mango. The number of amino acids increased from 9 at early stages (1-4 weeks after fruit set) to 13 during the later part of fruit development (10-13 weeks after fruit set). Aspartic acid, glutamic acid, phenylalanine, alanine and histidine were found throughout the development of fruit. Proline and glycine could be observed only during later part of fruit growth, while leucine and threonine only at the initial stage.

The lipid content in peel and pulp of five mango varieties ranged from 0.75 to 1.70 per cent and 0.80 to 1.36 per cent respectively (Pathak and Sarada, 1974). Total kernel lipids in Alphonso mango amounted to 11.6 per cent of the dry kernel and consisted of 96.1 per cent neutral and 3.9 per cent polar lipids which comprised 2.9 per cent glycolipids and 1.0 per cent phospholipids (Hemavathy *et al.*, 1987).

According to Jain (1961), small amount of tannin is also present in the flesh (0.16 per cent) and skin (0.105 per cent) which is responsible for astringency. Mango peel, which amounts to 10-20 per cent of the fruit by weight, is also a good source of pectin (Narasimhachar *et al.*, 1989). From the nutritional point of view the mango is a rich source of vitamin A, almost as rich as butter (Singh, 1960). The carotenoid pigments, ß-carotene (provitamin A) increase with ripening. Also it has a fair amount of vitamin C. The vitamin C content varies with cultivars (Sanyal *et al.*, 1991). Singh and Chadha (1961) reported higher ascorbic acid content in Langra than in Dashehari; lowest being in Fazli. The smaller fruits had higher amount of ascorbic acid content than larger ones (Palaniswamy *et al.*, 1974). The highest concentration of ascorbic acid was observed in fruits just after fruit set and the concentration decreased with fruit development. Twelve weeks after fruit set, the cv. Langra contained the most ascorbic acid, followed by Aswina and Fazli. The ascorbic acid concentration of fruits decreased with storage at room temperature (Gofur *et al.*, 1994). Ghosh (1960) reported the presence of folic acid (vitamin B) in green mangoes to an extent of 3.6µg/100 g. Vitamin B1 (thiamine) and vitamin B2 (riboflavin) range between 35-63 µg and 37-73 µg/100 g fresh weight respectively. A more detailed account of the pre- and post-harvest physiology of mango is given by Krishnamurthy and Subramanyam (1973). Sadhu and Bose (1976, 1982) studied the chemical composition of fruits of 37 mango cultivars and reported marked variations in the constituents of different cultivars.

The aroma and flavour vary widely among mango cultivars and there is no one typical formulation of flavour component for mangoes (Wilson *et al.*, 1990). Monoterpene hydrocarbons represented nearly 49 per cent (w/w) of the total volatile aroma constituents in mango cv. Kensington, with á-terpinolene at 26 per cent being the most abundant single constituent, and 16 esters represented 33 per cent (MacLeod *et al.*, 1988). Fifty components were identified in bambangan (*M. panjang*) fruit volatiles and 63 in binjai (*M. caesia*). Among bambangan fruit volatiles, monoterpene hydrocarbons (91.3 per cent) and esters (7.6 per cent) predominated, with á -pinene (67.2 per cent) and á -phellandrene (11.0 per cent) constituting the two most abundant components. The binjai fruits contained esters (71.3 per cent) and alcohols (23.2 per cent) as the two major chemical classes and ethyl 3-methyl-butanoate (40.0 per cent) as the most abundant constituent (Wong and Siew, 1994).

The seed kernels contain fat (7.5-8.8 per cent), protein (6.1-6.8 per cent), crude fibre (1.3-2.4 per cent) and ash (2.2-2.8 per cent). The main fatty acids of the kernel fat are palmitic (6.9-7.3 per cent), stearic (44.3-44.4 per cent), oleic (38.9-42.1 per cent), and linoleic (4.5-7.4 per cent) acids (Augustin and Ling, 1987). Oils from the roots and flowers have a lower iodine and saponification value, and less unsaponifiable matter than that from the seeds (Khan *et al.*, 1994).

2.2.0 Uses

Young and unripe fruits, because of their acidic taste, are utilized for culinary purposes as well as for preparing pickles, *chutneys* and *amchoor*. Ripe fruits are utilised in preparing squash, nectar, jam, cereal flakes, custard powder and baby food, mango leather (*am-papar*) and toffee. Besides, fruits of some cultivars like Alphonso and Dashehari are sliced and canned for catering to the needs of consumers during the off-season. Alphonso was found to be the most suitable for canned juice preparation due to its very good colour, consistency and flavour, followed by Padiri (Gowda and Ramanjaneya, 1995). Of the 10 mango cultivars evaluated, La Paz, San Felipe, Santa Craz, and Senora were considered suitable to processing for pulp production in view of their high pulp yield and good pulp flavour, colour and aroma (Pinera *et al.*, 1995). The best cultivars for processing were Haden, Haden 2H, Glenn, Joe Welch and Torbet, which had very good organoleptic characteristics (Donadio, 1995). High carotenoid content in Amrapali gives it potential for juice production and blending (Khurdiya and Roy, 1989).

The various plant parts are put to several other uses, *viz. leaves* : tender ones as vegetable in Java and the Philippines, ash of burnt leaves as a household remedy for burns and scalds, fumes from burning leaves for relief from hiccups and affectations of throat, and leaves are masticated to tone up the gums; *flowers* : formerly used in preparing an otto (*am attar*), dried flowers have curative properties for treating diarrhoea and chronic dysentery, and also yield tannin (15 per cent); *bark* : yields mangiferin and tannin (16-20 per cent), useful against diphtheria and rheumatism; the gum-resin exuded by mango stem bark is used in local medicine and the exudation is enhanced by ethephon (Bhatt and Shah, 1985). The wood finds use in furniture, flooring, packing boxes, match boxes and splint, brush backs, boats and oar blades (Anon. 1962). Vinegar of good quality, containing 4.5-5.0 per cent acetic acid and characterized by mild mango flavour, could be obtained from mango peels and stones (Ethiraj and Suresh, 1992). Mango peel is a good source of dietary fibre (Larrauri *et al.*, 1996) and the stone can be an important source of edible fat which has physical characteristics similar to cocoa butter (Narasimha Char and Azeemoddin, 1989).

Uses of mango

Mango fruit with slices

Jam from mango

Jelly from mango

Squash from mango

Uses of mango

Nectar from mango

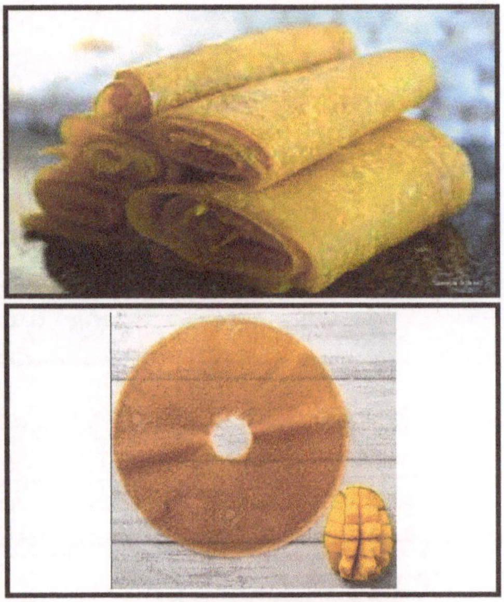

Leather from mango

Ice flakes from mango

Mango flavoured custard powder

Mango based baby food

Toffee from mango

3.0 Origin and Distribution

The genus *Mangifera*, which belongs to the family Anacardiaceae, originated in South-East Asia at an early date. According to Mukherjee (1958), the natural spread of the genus is limited to the Indo-Malaysian region, stretching from India to the Philippines and New Guinea in the east. Evidence based on morphological, phyto-geographical, cytological, anatomical and pollen studies indicates that the genus had its origin in the continental region of (i) Burma (Myanmar), Thailand, Indochina (Cambodia) and (ii) the Malay Peninsula, since these happen to be the main centres of species formation; the Sunda Islands (Java, Sumatra and Borneo), the Philippines and the Celebes-Timor group forming the secondary centres of development. The highest concentration of *Mangifera* species is reported to be in the Malay Peninsula (19), followed by the Sunda Islands (16) and the Eastern Peninsula (14); many of the species being common between them (Mukherjee, 1949a).

The genus, which has been subdivided into sections on the basis of the presence or absence of the disc in the flowers (Mukherjee, 1949a), is reported to contain 41 species in all, but almost all the edible cultivars of mango belong to the single species *Mangifera indica* which originated in the Indian subcontinent. Occurrence of wild forms of *M. indica*, allied species *M. sylvatica* and *M. caloneura*, fossil leaf impre-ssions of *M. pentandra* (a species similar to *indica*), and presence of numerous cultivated and wild varieties in India have been cited as some of the major reasons in favour of *Mangifera indica* having originated in the Indo-Burma region through allopolyploidy, possibly amphidiploidy (Mukherjee, 1958).

The few other species which contribute edible fruits (though of relatively inferior fruit quality) are *M. caesia, M. foetida* and *M. odorata*, which are confined to the Malaysian region.

Mangoes have been cultivated for an estimated 400 years in India and active selection of superior cultivars has taken place (Mukherjee, 1972; Singh, 1978). However, it was virtually unknown to any botanist until 1605 when Carol Clusius first mentioned of it in his writings (Mukherjee, 1949b). The name *Mangifera* was given for the first time by Bontius in 1658 when he referred to this plant as *arbor Mangifera* (the tree producing mango). Later, it was mentioned in the literature as *Mangifera indica, Mangas domestica* or *Mangas sylvatica*. Linnaeus also referred to it as *Mangifera arbor* in 1747, prior to changing the name to its present form (*Mangifera indica*) in 1753, in his much quoted book *Species Plantarum* (Mukherjee, 1949b).

Besides the Indian subcontinent, mango is now found in several countries of the tropical and subtropical world where it was introduced by the Muslim missionaries, Spanish voyagers and Portuguese explorers during the 15th to the 18th centuries. According to Hayes (1957) it was being cultivated at the head of the Persian Gulf by the 16th century.It was introduced to the humid, tropical countries of the Malay Archipelago, and South-East Asia during the 4th and 5th centuries B.C. where it conti-nued to develop and has become naturalized (Mukherjee, 1972). It was introduced in the Philippines after 1600, in the Moluccas in 1665 and in Yemen in the later part of the 18th century (Burns and Prayag, 1921). It is also reported that the mango was being grown in England under glasshouse conditions as early as 1690 and that the trees at Kew were in fruiting during 1818. In Mexico, it was introduced before 1778 by the Spanish travellers from the Philippines. The Portuguese, who carried mango to South Africa in the 16th century from Goa, were responsible for introducing it in Brazil by 1700 (Popenoe, 1920). It was in cultivation in Barbados in 1742 and in Jamaica in 1782. According to Pope (1929) it was introduced in Hawaii between 1800 and 1820. It reached Azores in 1865 and Queensland in 1870 (Burns and Prayag, 1921). According to Knight and Schnell (1994), mango introduction to Florida began in 1861 with the importation of No.11, a polyembryonic, seed propagated cultivar from Cuba. The US Department of Agriculture introduced Mulgoa, an improved cultivar from India, in 1889. Florida has become a contemporary secondary centre of diversity for mango germplasm due to the introduction of large number of diverse genotypes together with the species- breeding and adaptation.

Presently, besides India, it is being cultivated in Pakistan, Bangladesh, Myanmar, Sri Lanka, Thailand, Cambodia, Vietnam, Malaysia, the Philippines, Indonesia, the Fiji Islands, Tropical Australia, Egypt, Israel, Sudan, Somalia, Kenya, Uganda, Tanzania, SouthAfrica, Niger, Nigeria, Zaire, Madagascar, Mauritius, the USA (Florida, Hawaii, Puerto Rico), Venezuela, Mexico, Brazil and the West Indies Islands.

4.0 Taxonomy

The genus *Mangifera* belongs to the order Sapindales in the family Anacardiaceae which is a family of mainly tropical species with 73 genera (c. 850 species), with a few representatives in temperate regions. The other distant relatives of *Mangifera* are cashew (*Anacardium occidentale*), gandaria (*Bouea gandaria*), pistachio (*Pistacia vera*), marula (*Sclerocarya birrea*), ambarella (*Spondias cytherea*), yellow mombin (*Spondias mombin*), red mombin (*Spondias purpurea*), imbu (*Spondias tuberosa*), dragon plums (*Dracontomelum* spp.) kaffir plum (*Harpephyllum caffrum*), *etc*. Apart from edible fruit Anacardiaceous species also yield other valuable products like wood, gums and resins, wax and varnishes and tanning materials. It is also a family well known for the dermal irritation produced by some of its members, including some *Mangifera* spp. whose resinous sap may induce allergic reaction.

5.0 Botany

5.1.0 Tree Shapes, Branching and Longevity

Mango trees, grown from seeds are known as 'seedlings' have a long straight bole. Tree is sympodially branched (Scarrone's model). Grafted trees on the other hand are dwarf with spreading branches. However, the shape of the canopy also depends on the space available for its development. Isolated trees, getting sufficient space for their growth may differ in tree shape with the same variety grown in the orchard. On shallow soils the growth is stunted. Cultivars like Latra and Creeping are spreading in growth habit, thus they can be trained as creepers.

Considerable variation in canopy characteristics of Indian mango cultivars has been observed. Seedling trees live much more than 100 years whereas grafted ones live only 80 years or less. One of the largest trees known is that from Chandigarh (India), with a trunk of 3.5m in diameter, limbs of 75cm diameter, the crown spreading over 2250 m^2 with an annual production of about 16000 fruits in peak years at the age more than 100 years old (Singh, 1960). Seedling tree measuring a spread of 125 ft./42 m and a girth of 25 ft./8 m has been reported to exist in Brazil (Popenoe, 1920).

5.2.0 Tree

Tree is medium to large (10-40 m in height), evergreen with symmetrical, rounded canopy ranging from low and dense to upright and open. Bark is usually dark grey-brown to black, rather smooth, superficially cracked or inconspicuously fissured, peeling off in irregular, rather thick pieces. Exudate of the live bark transparent, a dark yellowish brown, drying brown, consisting of a resin mixed with a gum. The bark contains 78 per cent resin and 15 per cent gum in addition to tannic acid. Terminal bud is small, enveloped by small, lanceolate acute bud scales. Twigs are not very thick, smooth, apically angular, glabrous, glossy and dark green root the tree forms a long unbranched long tap root (up to 6-8 m and more) plus a dense mass of superficial feeder roots. Feeder roots develop at the base of the trunk or slightly deeper; these produce anchor roots, and sometimes a collection of feeder roots develops above the water table. The fibrous root system extends away from the drip line. Effective root system of an 18- year old mango tree may observe a 1.2 m depth with lateral spread as far as 7.5m (Bojappa and Singh, 1974). The leaves are simple, exstipulate, alternately arranged, 15-45 cm in length. The petiole varies in length from 1 to 12 cm, always swollen at the base. It is grooved on the upper side. The phyllotaxy is usually 3/8 but as the leaves are arranged very closely at the tips they appear to be whorled. Leaves are variable in shapes like oval-lanceolate, lanceolate, oblong, linear-oblong, ovate, obovate-lanceolate or roundish-oblong (Singh, 1960). The apex ranges from acuminate to nearly round. The margin is usually entire, sometimes slightly undulated and wavy, rarely twisted or folded. The length and breadth varies from 12 to 45 cm and 2 to 12 cm, respectively, depending on variety and growth. The secondary veins are quite prominent, and in some of the varieties, range from eighteen to thirty pairs. The upper surface is shining and dark green while the lower is glabrous light green. The leaves appear in flushes. They are flaccid and pendulous when young. The colour of young leaves generally vary form variety to variety, generally being tan-red, pink, yellow-brown in colour. As the leaf grows, its colour changes from tan-red to green, passing through many different shades, and become dark green at maturity. The leaves have fibres and crackle when crushed. They strongly smell of turpentine (some cultivars do not smell). The leaves contain a good amount of mangiferin (xanthone). In India, it was obtained as "Indian Yellow" from the cow's urine. Cows were fed exclusively with mango leaves and ultimately excessive feeding on leaves lead to the death of the animal.

5.3.0 Inflorescence

The inflorescence is pseudo-terminal, originating from a bud, together with the new leafy sprout; there are cultivars with lateral inflorescence. The inflorescence is a narrowly to broadly conical panicle up to 45 cm long depending upon cultivar and environmental conditions during its development. It is usually bracteate, but may sometimes be ebracteate. The bract if present is leafy, elliptical and concave. The colour of the panicle may be yellowish-green, light green with crimson patches or with crimson flush on branches. It is generally pubescent but sometimes may be glabrous. The branching of the inflorescence is usually tertiary, rarely quaternary, but the ultimate branching is always cymose (Singh, 1960). The panicle bear 500-6000 flowers of which 1-70 per cent is bisexual, remainder are male depending on the cultivar and temperature during its development.

5.4.0 Flowers

Hermaphrodite and male flowers are produced in the same panicle, usually with a larger number of the later. The size of both male and hermaphrodite flowers varies from 6 to 8 mm in diameter. They are subsessile, rarely pedicellate, and have a sweet smell. Pedicels are very short or missing; they are articulate with a panicle branch of the same diameter, which is often mistaken for the pedicel (Barfod, 1988). The calyx is usually five partite. The lobes are ovate-oblong and concave. The corolla consists of five pale yellow petals (rarely four to eight), which are twice as long as the calyx and contain three to five ridges on the ventral side. The petals are in bud imbricate and slightly contorted. They are thin, yellowish and after expanding horizontally, the upper half rather irregularly and not very pronouncedly reflexed, they are free at their base. The ridges are slightly dark. The upper half and the margin of the petal are white. On fading, the petals become pinkish. Between the corolla and androecium there is an annular, fleshy, and five-lobed disc (Singh, 1960; Kostermans and Bompard, 1993).The androecium consists of stamens and staminodes, altogether five in number, of which usually one, or rarely two, are fertile and the rest are sterile. However, in cultivar Pico, three fertile stamens have been reported. As many as ten stamens, which occur in other members of the genus, may also occasionally be found in the form of primordia only (Maheshwari, 1934). All the stamens are inserted on the inner margin of the disc. The position of the fertile stamen and pistil may be either parallel or oblique to each other (Naik and Gangolly, 1950). The fertile stamens are longer than the staminodes and are nearly equal to the length of the pistil. The colour of the anther is pink, which turns purple at the time of shedding.The ovary is sessile, one-celled, oblique and slightly compressed in its lateral aspect. It is placed on the disc. The ovule is anatropous and pendulous, and shows one-sided growth. The style arises from the edge of the ovary and ends in a simple stigma. Sometimes three carpels may develop in a flower (Singh, 1960). The pollen grains are of variable shapes, with the size varying from 20 to 35 micron. Small amount of pollen is produced in *M. indica*; the grains are sphaeroidal to prolate sphaeroidal, radially symmetrical, subangular in polar view, isopolar, with a few giant triploid ones of up to 50 micron, they are 3-monocolporate, goniotreme, sides convex-subprolate; apertures equidistant and zonal; ecto-aperture (colpus) extends slit-like from pole to pole.

5.5.0 Fruit

The fruit is a more or less compressed, fleshy drupe. It varies considerably in size, shape, colour, presence of fibre, flavour, taste and several other characters. The most characteristic feature of the mango fruit is the formation of a small conical projection developing laterally at the proximal end of the fruit, known as the beak. It may be quite prominent in some, less so in others, while in some varieties it is represented merely by a dot. A wide sinus is always present just above this beak. The pistillate area of the fruit located near the base of the beak is known as the *nak*. The shape of the fruit varies from rounded to ovate-oblong or longish, with the length varying from 2.5 to 30 cm in different varieties. The base may be depressed or elevated or may be intermediate. The skin is gland-dotted and at maturity its colour exhibit different mixtures of green, yellow, and red shades. It may be smooth or rough. The acrid juice, with turpentine like smell, present in the stalk or sometimes in the fruits, is known as *chenp* in Hindi is due to myrcene and ocimene. Its main irritating constituent has been identified as an allergenic urushiol, 5-heptadecenylreorcinol (Anonymous, 2018a).

6.0 Species and Cultivars

6.1.0 Species

It is now apparent on the basis of taxonomic and recent molecular evidence that mango probably evolved within a large area including north western Myanmar, Bangladesh and northeastern India (Mukherjee, 1997). Bompard and Schnell (1997) also suggested that great species diversity of *Mangifera* in Malay Peninsula, Borneo and Sumatra is the available evidence points to a Sudanic origin for the genus. This however must not minimize the particular importance to the region stretching from Myanmar to Indo-China as another centre of diversification, as attested by a range of species belonging the section *Euantherae* (section including *M. calunera, M. cochinchinensis* and *M. pentendra*). Further *M. indica* apparently originated in region on western border of the secondary centre of diversification mentioned above. Truly wild common mango trees have been recorded in Bangladesh (Chittagong Hills), northeastern India (Assam valley) and Myanmar. Owing to its cultivation and dissemination for thousands of years, in India semi-wild trees can be found in the forests throughout the subcontinent.

Indian peninsula is the most important centre of diversity for *M. indica*. Other *Mangifera* species *viz., M. andamanica, M. khasiana, M. sylvatica* and *M. camptosperma* have been reported from India. *Mangifera sylvatica* occurs in the Eastern Himalayas (up to 1300 m altitude) in Sikkim, Darjeeling district, Khasi Hills, Upper Assam and Surma valley; and in the Andaman Islands. Mangifera andamanica, an endemic and rare species is widespread in tropical, wet evergreen forests of Andaman Islands. Mangifera indica is distributed throughout the tropical and subtropical parts of India. Wild forms occur in abundance in forests as well as in cultivated areas. Many wild types occur almost throughout the tropical and subtropical hilly forests and ravines up to 1000 m and in the Andamans and Nicobar Islands. The total number of distinct mango varieties named and maintained in India has been estimated to be over one thousand. About 30 varieties are commercially grown in India.

Some *Mangifera* species has direct role in improvement (Singh *et al.*, 2018)

Mangifera Species	*Special Characteristics*
M. caesia	White pulp, sweet, fragrant
M. decandra	Rootstock for waterlogged conditions
M. gedebe	Rootstock for waterlogged conditions
M. indica var *mekongensis*	Fruit twice a year and hence a good parent in crossing
M. inocarpoides	Rootstock for waterlogged conditions
M. pajang	Fruit can be peeled like banana and could yield new varieties by selection
M. similis	Good parent for breeding freestone mango

Mango is a medium to large (9-31 m) evergreen tree with an open or dense symmertrical canopy, long tap root and dense fibrous roots. Lanceolate shaped leaves are simple and leathery in texture. The inflorescence is a much branched panicle bearing both male and hermaphrodite flowers. The fruit is a fleshy drupe with variable shape (nearly round, oval or ovoid-oblong), size (60 g to 2.3 kg) and colour (Crane and Campbell, 1991; Sukonthasing *et al.*, 1991; Campbell, 1992). Fruits are single seeded which may be either mono or poly-embryonic.

Mangifera indica is an allopolyploid with 2n=4x=40, rest of the other species like *M. sylvatica, M. foetida, M. caesia* and *M. caloneura* are also tetraploid except cv. Vellaikolamban which is an allo-octaploid (i.e. 2n=8x=80).

Mangifera species in Thailand were surveyed during the period 1995-97, and a total of 13 species were identified (Eiadthong *et al.*, 1999). Among these species *M. foetida, M. odorata* and *M. sylvatica* have been cultivated as fruit crop for the local market, as well as *M. indica*. Based on 55 morphological and ecological characters, 13 *Mangifera* species were classified into two major groups; the first group consists of *M. indica, M. sylvatica, M. oblongifolia, M. collina* and *M. cochinchinensis*, and the second group of *M. caloneura, M. pentandra, M. flava, M. macrocarpa, M. gedebe, M. griffithii, M. odorata* and *M. foetida*. All the species in the first group were distributed in a monsoon climate. The other 9 wild species also seemed to have potential value as rootstocks and breeding materials for disease and/or pest resistance or tolerance to critical environmental conditions.

6.2.0 Cultivars

Due to the long history of cultivation in this subcontinent, about a thousand cultivars of mango are known to exist in India. All these have originated as superior chance seedlings arising from natural crossing or gene mutation. These selections were later maintained true-to-type through asexual propagation. Almost all these cultivars are of monoembryonic type.

Mango cultivars can be divided into two groups based on their embryo types: the monoembryonic (Indian) type is predominantly distributed in subtropical areas and susceptible to anthracnose and the polyembryonic (Southeast Asian) type is most common in tropical areas and is tolerant to anthracnose (Lespinasse and Frederic, 1998). The Indian type possesses a zygotic (sexually produced) embryo,

Species of mango (Mangifera)

Mangifera indica

Mangifera foetida

Mangifera quadrifida

Mangifera andamanika

Mangifera griffithii

Mangifera similis

Mangifera kemanga

Species of mango (Mangifera)

Mangifera odorata

Mangifera gadebe

Mangifera pentandra

Mangifera minor

Mangifera cochinchinensis

Mangifera zeylanica

Mangifera altissima

Mangifera sylvatica

and the fruit skin is mainly red, whereas the Southeast Asian type has several nucellar embryos (produced from the mother plant), and the skin is mainly green to yellow (Iyer and Degani, 1997, Viruel *et al.*, 2005).However, the International Check-list of Mango cultivars comprising 793 cultivars from all over the world has been prepared by Pandey (1985). The polyembryonic types, numbering about a dozen, also exist in India, but these are much inferior in fruit size and quality (*viz.* Bappakai, Chandrakaran, Goa, Kurukkan, Olour, Bellary, Kasargod, Mazagaon, Nileswar Dwarf and Salem). These are mostly wild and are confined to the southern states, especially on the west coast. Besides these, some polyembryonic cultivars have been reported from other countries as well, *viz.* Cambodiana, Carabao, Cecil, Higgins, Paho, Peach/Turpentine/Apricot, Pico, Sabre, Saigon, Simmonds, Samini and Strawberry (Singh, 1960).

Detailed descriptions of the numerous edible cultivars of mango have been given in a number of monographs published from time to time (Maries, 1902; Mukherjee, 1948b; Naik and Gangolly, 1950; Singh and Singh, 1956; Gangolly *et al.*, 1957; Maity *et al.*, 1981). However, of these, only a score or so are of commercial significance in the different mango-growing regions in the country. In general, the cultivars are location-specific, and the commercial cultivars of one region do not do so well when grown in other areas.

In the Philippines, 2 cultivars (Carabao and Pico) out of 53 cultivars in the collections are grown commercially (Lazo, 1971). The most widely grown cultivars in Hawaii are Pierie and Haden (Yee, 1976). Tommy Atkins is the most important cultivar in Florida although several cultivars exist in the collection (Malo, 1976). In South Africa, Zill, Kent and Haden are the cultivars recommended for planting (Hobson, 1969). The Common and Kensington are commercially important cultivars in Queensland, Australia (Anon., 1977). The cultivars Fiji, Peach, Jarra, Parrot and Kerosene are commonly grown in Fiji (Iqbal, 1982). According to Beddoe (1994), the most popular cultivars of West Indies are Buxton Spice and Blackman (in Guyana), Starch, Doux Doux, Zabricot, Calabash, Teen and Cream (in Trinidad and Tobago), Ceylon, Peach and Grosbeck (in Grenada), Imperial (in St. Vincent), Pere Louis and Graham (in St. Lucia), Amoury Polly (in St. Kitts Nevis), and Tommy Atkins, Zill, Sensation and Haden (in Jamaica).

A number of mango cultivars are grown in different parts of China. Wang (1999) reported that cultivar Tainong I produces oblate/oval-shaped medium-sized fruits (average weight 221 g) with orange-yellow skin with a faint blush, and orange-yellow, fine, tender flesh with a low fibre content. The flesh is very juicy and aromatic, with a soluble solids content of 20 per cent. Storage and transport quality are good and the trees have some resistance to anthracnose. Matoumang originated as a chance seedling.

The trees are vigorous, precocious and productive, with 10-year-old trees yielding

30-37.5 t/ha. It is suitable for tropical and subtropical climates and red soils. The fruits mature at the end of June to mid-July, and are medium in size, weighing around 200 g. The skin is thin with a yellow or greenish yellow colour, and the flesh is orange-red, fine and tender with a few fibres and an acid-sweet flavour (soluble solids content 16-19 per cent) (Cao and Wang, 1999).

The physico-chemical characteristics of mango fruits vary considerably with genotypes and environmental conditions (Syamal and Mishra, 1987 and 1989b; Sanyal *et al.*, 1991). The 'sucking' mango cultivars are small fruited, highly fibrous types which cannot be sliced easily *e.g.* Gaurjit, Safeda Jauhari, Sukul, Sinduri, Yakuti, Gilas and Mithwa (Rabbani and Singh, 1989). The most productive cultivars in Mexico were Sin Semilla, Diplomatico, Edward, Haden and Manila (early; May to mid June), Tommy Atkins, Springfels, Davis Haden and Fabian (mid-early; mid June to mid July) and Kent and Keitt (late; mid July to mid August). It was also noted that interplanting of Sin Semilla with any variety except Manila reduced its fruit quality, and that Manila and Carabao were slow to enter their productive phase (Santos De La Rosa and Mosqueda Vazquez, 1989). In relation to production and tree growth, the most productive cultivars in Brazil were Parvin, Zill, Dixon, Florigon and Davis Haden. Tree growth varied greatly among cultivars and Parvin was one of the best in terms of fruit quality (Donadio *et al.*, 1994). It is concluded that the cv. Osteen, introduced to Sinaloa, Mexico, has potential as a commercial cultivar (Siller-Cepeda *et al.*, 1994).

Campbell and Campbell (1993) divided the Florida mango cultivars into 5 groups based on production volume. Group 1, with the greatest production, consists of Tommy Atkins, Keitt and Van Dyke. Group 2 contains Palmar and Irwin, closely followed by group 3 with Kent and Golden Nuggets. Group 4 consists of cv. Haden, from older orchards and group 5 is the minor cvs. Brooks, Davis Haden, Edward, Glenn, Fascell, Lippens, Parvin, Sensation, Smith and Zill. Dazhonmang, a chance seedling was selected in Guangdong, China. The fruits have strong aroma and good eating quality with TSS content of 15-18 per cent (Cas and Wang, 1999). Honey Gold cultivar of mango was derived from a cross involving Kensington Pride via selection of highly attractive fruit with brilliant orange-yellow skin, firm but juicy flesh, rich sweet flavour, non-turpentine. Trees are small and produce fruits which weigh an average of 490 g (Anon., 1999).

The popular commercial cultivars of different regions in India are:

Northern region	Dashehari, Langra, Chausa and Bombay Green
Eastern region	Himsagar, Langra, Fazli, Lakshman Bhog, Krishnabhog and Gulabkhas
Western region	Alphonso, Pairi, Kesar, Rajapuri, Mankurad and Jamadar
Southern region	Bangalora, Neelum, Swarnarekha, Pairi (Peter), Banganpalli, Mulgoa and (Alphonso)

The variety *Kerala Dwarf* was found dwarfest with agreeable fruit quality. Hence, this has been used in the breeding programme to impart dwarfing characters to the progeny. *Janardan Pasand, Manjeera, Amrapali, Nileswar Dwarf* are found to be dwarf genotype. The cultivars *Totapuri, Amrapali, Neelum and Banglora* have been used to impart regular bearing in the hybrids. The coloured Florida and Indian cultivars *viz., Eldon, Janardan Pasand, Lal Sundari, Suvarnarekha, Sensation,* and *Vanraj* have been used in hybridization programme for developing the varieties having the red peel colour. In addition to this, *Elaichi* and *Bhadauran, Samar Bahis Rampur, Milan Sahib* has been used in mango breeding programme for imparting resistance to mango malformation. *Bombay Green* possesses resistance for bacterial canker. *Tommy Atkins* is moderately tolerant to anthracnose. *Parish* and *Fairchild* have been reported to resist against it. *Pairi* is highly susceptible for powdery mildew. *Neelum, Zardalu, Banglora, Totapuri Khurd* and *Janardan Pasand* are most tolerant to it and used in hybridization programme (Singh *et al.*, 2018).

Singh *et al.* (2019) found that the Langra (vernacular name *Maldah* in the region) is the predominant cultivar of Bihar (likewise '*Jardalu*' in Bhagalpur Region) and 'Amrapali' in Jharkhand. The unique local varieties which were conserved in the region asare *Jalbanda* (large sized fruit with sweet at mature green stage), *Jalmarai* (uneven shape, but very tasty with hard peel), *Bhastara* (syn. Baramasi, bearing twice in a year) *Sabja* (syn. Bombaiya, an early and sweet even at mature green stage), *Dalma* (syn. Kalkatia Malda, dark greened, large sized), Malda seedling (fruit weighs up to 750 g, superior in taste and sweetness thancompared to grafted one), *Bhont* (thin stoned, papaya shaped and mid maturing), *Sinduria* (earliest maturity), *Raja* (sweet and easily digestible), *Kaveri, Bahadura, Balajee, Kerwa, Tairia* (heavy fruit bearer with regularity), *Baelkhash* (leaves/fruits contains flavour of bael), *Chitranjan* (late maturity, tastes like curd of buffalo milk) *etc.*, and '*Gaurjeet*, (good smelled and highly flavoured) of West Champaran, which has an excellent market in the bordering area of Bihar, Uttar Pradesh and Nepal.

Many Indian mangoes have GI tags (Once the varieties get the GI tag, producers of these mangoes are likely to fetch a good price and farmers of other regions cannot market their fruits by misusing the name. Alphonso of Ratnagiri, Kesar of Gir (Gujarat) and Marathwada, Banganapalli in Andhra Pradesh, Zardalu of Bhagalpur, Appamidi of Karnataka, Himsagar, Laxman Bhog and Fazli of Malda (West Bengal), among others, have attained this distinction.

Salient characteristics of these cultivars are given in the following paragraphs:

Dashehari

This is the most popular cultivar of northern India because of its attractive appearance, excellent taste and pleasing flavour. It is a mid-season cultivar, maturing towards the end of June. The fruits are medium in size (4 to 8 per kg), elliptical-oblong in shape and have an attractive greenish yellow colour. It is a good cropper, though biennial in bearing.

Langra

Some people in the North rank it even higher than Dashehari. It is also a mid-season cultivar. The fruits are large in size (3 to 4 per kg), oblong-oval in shape and have lime-green colour. It has excellent sugar/acid blend and a characteristically pleasant flavour. It is a heavy yielder, especially after the age of 15 years. However, this is also biennial in bearing.

Chausa

This is one of the sweetest mangoes, lacking somewhat in acidity. It is a late cultivar and matures towards the end of July or beginning of August. It starts bearing good crops only after 15-20 years. The fruits are large-sized (3 to 4 per kg), almost oblong in shape (with a characteristic sinus) and bright yellow in colour. Its major drawbacks, besides biennial bearing, are its high susceptibility to mango-malformation and very vigorous growth habit.

Bombay Green

This is the earliest cultivar of northern India, maturing during the first half of June. The fruits are medium in size (4 to 6 per kg), ovate in shape and yellowish green in colour. The taste is good and yield moderate. However, this is also biennial in bearing and highly susceptible to mango-malformation.

Himsagar

This is the most popular commercial cultivar of West Bengal which matures during the second week of June. The fruits are large in size (4 fruits per kg), oval in shape with yellowish green colour and somewhat rough skin. The taste is very sweet and the flavour is characteristically pleasant. This is a heavy yielder though biennial in bearing.

Fazli

This is a very late cultivar (matures late in August) with very large-sized fruits (2 fruits per kg). The tree is very vigorous and a medium to heavy cropper, with biennial bearing habit. The fruit quality is relatively poor and its importance lies in its lateness to mature.

Zardalu

This is considered to be a matchless fruit of Bhagalpur (Bihar). It matures towards the end of June. The fruits are medium in size (5 to 6 per kg), oblong-oval in shape (with a slight sinus) and have attractive apricot-yellow colour. The fruit quality is good with a pleasant flavour. It is biennial in bearing.

Krishnabhog

This is a mid-season cultivar, maturing by middle of July. The fruits are medium to large-sized (3-4 fruits per kg), round in shape and have light yellow colour towards the basal end of the fruit. It is a heavy yielder and biennial in habit.

Gulabkhas

The fruit is favoured for its characteristic rose flavour and very sweet taste. The fruits mature in June and are of medium size (5 to 6 per kg), oblong to oblong-oblique in shape (with a well-marked to shallow sinus). The fruits are amber-yellow in colour, with reddish blush towards the base and the sides. This is heavy yielder, but biennial in bearing.

Alphonso (Badami)

This is one of the finest of Indian mangoes and is rated to be the best by many in home and abroad. However, it is also biennial in bearing. It is very specific in its requirements and does best only on the west coast of Maharashtra (Ratnagiri), although it is grown to some extent in the South too. The fruits are very attractive, large-sized (3 to 4 per kg) and oval in shape (with a prominent ventral shoulder). The fruits have an attractive pinkish blush towards the basal end. The taste is superb, with an excellent sugar/acid blend. The flavour is captivating. Besides being a table cultivar, much in demand, it is also a favoured fruit of the processing industry because it retains its characteristic flavour even during processing. It is a medium bearer.

Mankurad

This variety is of commercial importance in Goa and in the neighbouring Ratnagiri district of Maharashtra. The variety develops black spots on the skin in rainy season. Fruit is medium in size, ovate in shape and yellow in colour. Fruit quality is very good but keeping quality is poor. It is a mid season variety.

Pairi

This is an excellent cultivar of western and southern India. The fruit is very attractive, with crimson tinted shoulders on a yellowish green background. The fruits are medium in size (4 fruits per kg), ovate in shape (with a prominent beak and very slight sinus) and mature around mid-May and June. The bearing is heavy and fruit quality is good. It is biennial in habit.

Kesar

The fruits are medium to large-sized (3 to 4 per kg), oblong in shape with an attractive light apricot-yellow colour. The taste is very good and sugar/acid blend is excellent. It is a moderate cropper but biennial in bearing. It is a famous cultivar of Gujarat.

Rajapuri

This is another famous cultivar of Gujarat. The fruits are very large-sized (2 to 3 per kg), ovate to ovate-oblong in shape and have a bright pinkish blush (on the upper 1/3 of the fruit) on golden yellow background. The beak is very distinct and sinus is almost absent. The bearing is moderate and the fruits mature by the beginning of June. It is a good quality mango but biennial in habit.

Jamadar

This is another good cultivar of Gujarat. The fruits are medium-sized (5 to 6 per kg), ovate-oblique in shape with apricot-yellow colour. It is a medium cropper but biennial in bearing.

Vanraj

It is a highly prized variety of Vadodara district of Gujarat and fetches good returns. Fruit is medium in size, ovate oblong in shape and colour is deep chrome with a blush of jasper red on the shoulders. Fruit quality and keeping quality are good. It is a mid season variety.

Fernandin

This is one of the oldest varieties of Bombay. Some people think that this variety originated in Goa. Fruit size is medium to large, fruit shape is oval to obliquely oval and fruit colour is yellow with a blush of red on shoulders. Fruit quality and keeping quality are medium. It is a late season variety, mostly used for table purpose.

Bangalora

This is one of the most widely cultivated, mid-season cultivars of southern India. It is a heavy yielder and one of the few regular bearing cultivars of mango. The fruits are large-sized (2 to 3 per kg) and very typical in shape, *i.e.* oblong (bottle-necked towards the base) with a prominent sinus and beak. The fruits are attractive and have apricot-yellow colour. The skin is thick and keeping quality very good. The fruit quality is relatively inferior but is preferred by the processing industry because of its dependable regular supply.

Neelum

This is yet another heavy yielding and highly regular bearing commercial cultivar of the South and the fruit quality is relatively more acceptable than Bangalora. The fruits are medium in size (4 to 6 per kg) and the shape is ovate-oblique (roundish). The sinus is somewhat prominent and the beak distinct. The colour is orange-yellow and the taste is somewhat good, with an acidic blend. This cultivar has a wide adaptability and reaches the northern market late in season, *i.e.* end of August to beginning of September. Keeping quality is fairly good.

Swarnarekha

This is one of the few table cultivars of commerce which have highly coloured fruits of attractive pinkish red. This appears early in the northern markets, *i.e.*, about the last week of April. The fruit size is medium (4 to 6 per kg) and the shape is ovate-oblong. The flesh is somewhat fibrous and the taste is good with an acidic blend. The bearing is moderate but biennial.

Banganpalli

This is yet another of the most widely cultivated cultivars of the South which is sent to the northern markets very early in the season (i.e. towards the end of April), even when it has not attained proper maturity. Nevertheless, it is good in taste if tasted towards the later half of May. This is also known as Baneshan in the South and Safeda in the North. The fruits are large-sized (2 to 3 per kg) and the colour is very attractive golden-yellow, with a very smooth skin. The shape is obliquely oval. The bearing is moderate and fairly regular.

Mulgoa

This is an excellent late cultivar of the South but the bearing is rather light and biennial. The fruits size is large (2 to 3 per kg) and shape roundish oblique, with a sunken basal cavity. The taste is very sweet and flavour delightful. Two strains, red and white, are reported.

Fruit quality of some mango cultivars are presented in Table 2.

Table 2: Fruit Quality of some Mango Cultivars

Cultivar	*Total Soluble Solids (°Brix)*	*Total Sugar (per cent)*	*Acidity (per cent)*	*Vitamin C (mg/100 g of pulp)*	*References*
Bombay Green	22.10	16.47	0.380	49.50	
Dashehari	21.79	11.15	0.250	42.50	Singh *et al.* (1985)
Langra	17.10	10.36	0.190	131.75	
Chausa	23.00	16.70	0.158	26.56	
Fazli	16.00	10.67	0.317	11.87	
Alphonso	19.00	15.86	0.340	33.50	Salvi (1983)
Ratna	23.00	16.81	0.260	25.00	
Mallika	24.00	18.60	0.330	19.00	Singh *et al.* (1977)
Amrapali	22.80	17.20	0.120	35.00	Sharma *et al.* (1981)
Bombay Yellow	12.00	8.70	0.320	17.00	Ghosh *et al.* (1985)
Rani Pasand	14.40	11.10	0.160	30.00	
Safdar Pasand	21.60	18.20	0.190	31.20	
Jehanara	16.60	14.81	0.180	13.10	
Khasulkhas	18.00	16.06	0.350	12.50	Sanyal *et al.* (1991)
Krishnabhog	16.60	14.70	0.380	19.30	
Totapuri Red Small	19.40	17.31	0.120	36.20	
Zardalu	15.90	14.19	0.250	6.80	

Pusa Surya

It was developed by IARI, New Delhi in the year 2002. The trees is semi-dwarf and suitable for closer planting (6m x 6m) and fruit ripens by mid-July in northern India, medium to large (260 to 290 g) with attractive apricot-yellow peel colour with pink-blush and pulp with TSS (19.0 °Brix) with long shelf-life (10 to 12 days) at room temperature. Suitable for both domestic and international markets with average yield of 12-15 kg per tree (10th year).

Arka Neelachal Kesari

Developed at ICAR-IIHR (CHES), Bhubaneshwar, Odisha. Extra early (ripening in March), Colour, high yield and fruit fly free in coastal Orissa. TSS 18 °Brix, excellent fruit quality, fruit weight 200-230g, crimson red fruit colour, free from fruit fly and malformation.

Dashehari 51

It is clone of Dashehari developed by ICAR-CISH, Lucknow. A regular bearing and high yielding clone of Dashehari was selected by exploiting clonal variability. Besides bearing regular crop every year, this clone gives 38 per cent more yield than other Dashehari trees.

Paiyur 1

It is clonal selection from Neelum variety of mango from Karukkanchavadi village of Dharmapuri district developed by TNAU Coimbatore (1992). Trees are dwarf, low spreading and suitable for high density planting (400 trees/ha as against the recommended 100 tress/ha). Orchard management practices including harvesting are easy to perform because of short stature of the trees. This is a regular bearer with a mean yield of 22.3 kg per tree (8929 kg/ha) at nine years after planting. Fruits are medium long, ovate with yellow skin. Fruits have excellent taste with good keeping quality. Fruits are medium in size with an average weight of 150g - 200g.

Menaka

It is the selection from Gulabkhash at BAU, Sabour, Bhagalpur, Bihar. It has high yield potential regular bearer, fruit size 15 x 9.2 cm, average fruit weight 300 g, yellow peel colour at ripening, deep yellow pulp colour at ripening, pulp content 75 per cent, TSS 20 °Brix and acidity 0.14 per cent.

Subhash

It is chance seedling selection from Zardalu developed at BAU, Sabour, Bhagalpur, Bihar. Its fruits are similar to Langra in respect to shape and size but quality resembles to Zardalu. Regular bearing, fruit size 11 x 7.7 cm, average fruit weight 220 g, deep yellow peel colour at ripening, yellow pulp colour at ripening, pulp content 66.1 per cent, TSS 24 °Brix and acidity 0.29 per cent.

Niranjan

This variety was developed by Vasantrao Naik Marathwada Krishi Vidyapeeth, Parbhani, Maharashtra. The Fruits are harvested at Dussehra & Diwali, sweet pulp, regular bearer, off season bearer with yield upto 300-500 fruits/tree.

Pant Sinduri

It is clonal selection of Dashehari developed at G. B. Pant University of Agriculture and Technology, Pantnagar, Uttarakhand. Trees are medium in height with round top canopy. Fruit colour is yellow with pink shoulder. Average fruit weight is up to 200g. Fruit pulp is yellow with pleasant aroma. Total soluble solids vary from 16-18 per cent with average yield up to 150 kg per tree. Fruit matures during last week of May to first week of June.

Pant Chandra

It is clonal selection of Dashehari developed at G. B. Pant University of Agriculture and Technology, Pantnagar, Uttarakhand. The plants are tall with erect growth habit. Fruit colour at maturity remains green. It is a mid-season variety. Fruit weight is up to 150g. Fruit pulp is reddish yellow with total soluble solids of 18 per cent and pleasant aroma.

Phule Abhiruchi

It is the selection from GK-PM-5 at Mahatma Phule Krishi Vidyapeeth (MPKV), Rahuri in the year 2014.It is good for pickle purpose.

Cardozo Mankurad

It was selection from Mankurad by ICAR - Central Coastal Agricultural Research Institute, Ela, Old Goa. This germplasm is a promising selection for its attractive fruit colour, bigger fruit size (320g) with better shelf life, higher contents (78.29 per cent) of fibreless pulp, higher fruit yield and regular bearing tendency.

Chhattisgarh Nandiraj

It is the clonal selection of local mango from Nayapara, Jagdalpur developed at IGKV, Raipur. It has attractive shape, appearance and pleasant aroma. The proposed strain has an average productivity of 280 kg/tree and average fruit size 200-250 g with acid blend. It is very much comparable to Deshari and having good keeping quality and suitable for export (TSS is 21.30 Brix and total sugar is 17.87 per cent).

Hybrids of Mango

The hybrids of mango developed in India at various research stations are presented in Table 3.

Table 3: Mango hybrids developed in India*

Centre	*Hybrids*	*Parents*	*Characteristics*
ICAR-IARI, New Delhi	Amrapali	Dashehari x Neelum	Distinctly dwarf, highly regular and precocious. Fruits regular for harvest by the 3rd week of July in northern India. Fruit size small to medium (120 to 160g) and fibreless. It has high TSS (22.8 °Brix) and ß-caratene content (16,830 µg/100g pulp). Suitable for blending mango pulp. Ideal for high density planting (1,600 plants per ha).
	Mallika	Neelum x Dashehari	World's first mango hybrid released for commercial cultivation. Regular in bearing with semi-vigorous growth habit. Size of the fruit is large (307g), good taste (TSS 24 °Brix), fibreless with good flavour. Fruits ready for harvest by the 3rd week of July in northern India. Suitable for processing and export. It has become a major commercial export variety of Karnataka, Andhra Pradesh and West Bengal.
	Pusa Arunima	Amrapali x Sensation	Regular bearer, semi-vigorous and suitable for closer planting (6 m x 6 m). Fruits medium to large in size (230 to 250 g) with attractive red peel and medium TSS (19.5 °Brix). Has long shelf-life (10 to 12 days) with yield upto 15-20 kg per tree in 10th year.

Centre	Hybrids	Parents	Characteristics
	Pusa Prathibha	Amrapali x Sensation	It is a regular bearer, attractive fruit shape, bright red peel and orange pulp. It has 7 to 8 days shelf- life at room temperature after ripening. The plants are semi-vigorous and about 278 plants of this hybrid may be accommodated in a hectare (6 m x 6 m) as against 100 plants of Dushehari (10 m x 10 m). On per plant basis, it yields about 3.0 times higher than Dushehari, which includes 'On' and 'Off' year fruiting. It has good sugar-acid blend and above all, uniformity in fruit size, which is lacking in Amrapali.
	Pusa Shresht	Amrapalli x Sensation	It is unique hybrid having regularity in bearing, attractive elongated shape, red peel and orange pulp. Plants are semi-vigorous and suitable for closer planting (6m x 6m). Fruit medium sized (228g) with attractive red peel colour and higher pulp content (71.9 per cent). The total soluble solids are 20.3 per cent, vitamin C (40.3mg/100g pulp) and ß-carotene content (10,964 ug/100g pulp). It has pleasant flavour with self life (7 to 8 days) at room temperature.
	Pusa Pitamber	Amrapali x Lal Sundari	It is regular bearer, semi-vigorous and suitable for closer planting (278 plants/ha). Fruits have attractive oblong shape, bright yellow peel. It is moderately resistant to mango malformation with fruits weight about 213 g with attractive yellow peel colour with higher juicy pulp (73.6 per cent). It has medium total soluble solids (18.8 per cent), rich in vitamin C (39.8 mg/100 g pulp) and ß-carotene content (11,737 µg/100g pulp).
	Pusa Lalima	Dashehari x Sensation	Regular bearer, semi-vigorous and are suitable for closer planting. Fruits are attractive in shape and have red peel and orange pulp. The fruit (209 g) have attractive red peel colour and higher pulp content (70.1 per cent). Pulp with medium total soluble solids (19.7 per cent), vitamin C (34.7 mg/100 g pulp) and high ß-carotene content (13,028 µg/100 g pulp). It has good flavour with shelf life (5 to 6 days) with yield upto 50-60 kg per tree.
ICAR-CISH, Lucknow, UP	Ambika	Amrapali x Janardhan Pasand	Fruits oblong oval, colour bright yellow with dark red blush, pulp dark yellow, firm with scanty fibre and weighs about 300-350g. TSS is 21°B. It is a regular bearing variety and late in maturity. Yields about 80 kg/plant by about 10 years of planting under improved cultural practices.
	Arunika	Amrapali x Vanraj	The tree produces dwarf and compact canopy. Fruits are smooth, orange yellow with red blush, weigh about 190-210 g, medium sized, ovate oblique, pulp orange yellow, firm with scanty fibre, pulp about 65 per cent, TSS 24.6 °B. The hybrid has potential for both internal and export markets because of its attractive fruit colour. Average fruit yield is about 69 kg/plant at 8 years of planting.
ICAR-IIHR Bengaluru	Arka Aruna	Banganapalli x Alphonso	Dwarf, precocious, medium and regular bearer, fruits large, fibreless, good flavour, free from spongy tissue. Pulp recovery is about 80 per cent. TSS is around 20°Brix. Suitable for backyard planting, and amenable for 5 x 5 m planting, a mid- season variety.

Centre	Hybrids	Parents	Characteristics
	Arka Anmol	Alphonso x Janardhan Pasand	It is a semi-vigorous type, regular bearing and is amenable for high density planting. It is a late season variety, fruits coming to harvest during second to third week of June. Fruits are oblong in shape, medium sized, weighing on an average about 300-330 g. Stalk is inserted vertically. Fruits on maturity will be dull green in colour, attaining uniform golden yellow colour on ripening. Skin is thin and smooth. Pulp is orange coloured, firm and free from fibre and spongy tissue with good sugar-acid blend. TSS is around 19 °Brix. The pulp recovery is about 70-75 per cent. Keeping quality is good.
	Arka Neelkiran	Alphonso x Neelum	It is a late season variety, fruits come to harvest during last week of June. Tree is semi-vigorous in nature. Fruit is elliptic in shape, medium sized weighing on an average about 270-280g. Fruit stalk is inserted vertically. Fruits on ripening get golden yellow blush. Skin is smooth, medium thick. The pulp is deep yellow in colour and firm. It is free from fibre and spongy tissue. Pulp recovery is 70 per cent. Keeping quality is good.
	Arka Puneet	Alphonso x Banganpalli	It is a mid-season variety. Plants are vigorous and regular bearing. Fruit stalk is attached vertically. Fruits are elliptical and on ripening become yellow coloured with red blush. Pulp is firm, orange coloured and is free from fibre and spongy tissue. TSS is around 21°Brix. The pulp recovery is about 65-70 per cent. Keeping quality is good.
	Arka Udaya	Arka Amprapali x Arka Anmol	It is a sweet, high-yielding variety of mango with a long shelf-life. It can stay fresh at room temperature for about 10 days without refrigeration. It is a late-season variety and yields after other mango fruit trees have stopped fruition. The tree bears fruit every year. The yield was high at around eight tonnes. The trees were planted at a density of five by five metres.
	Arka Suprabhath	Double cross mango hybrid Amrapali (Dashehari x Neelam) x Arka Anmol (Alphonso x Janardhan Pasand)	It is a medium vigorous, regular and bunch bearing, high yielding (35-40 kg/plant after 4 years of planting), fruit weight ranged from 240-250g, fruit shape is like Alphonso and has pulp colour of Amrapali, deep orange firm pulp, pulp recovery (>70 per cent), TSS(>22B), acidity (0.12 per cent), carotenoids (8.35mg/100g FW) and flavonoids (9.91 mg/100g FW) and it has got the shelf life of 8-10 days at room temperature.
	H-13-1	Amrapali x Sensation	Fruits attractive, better quality, good keeping quality, suitable for export.
RFRS, Kodur, Andhra Pradesh	Au-Rumani	Rumani x Mulgoa	Prolific, regular bearer, fruits similar to Rumani in appearance, fibreless and good flavour.
	Neeleshan	Neelum x Baneshan	Fruits medium, fibreless, pleasant flavour, regular bearing, good for canning.
	Neelgoa	Neelum x Pera Mulgoa	Regular bearer, fruits akin to Neelum in shape and colour but Mulgoa in size, fibreless, moderately juicy.

Centre	Hybrids	Parents	Characteristics
	Neeluddin	Neelum x Himayuddin	Medium tree, regular bearer, fruits small to medium, fibreless, juicy with characteristic flavor.
	KMH-1	Cheruku-rasam x Khader	Semi dwarf, regular bearing, fruits are fibreless and have high brix and low acidity.
RFRS, Anantha-puram Rajupet, AP	Swarna jahangir	Chinnasu varnarekha x Jehangir	Prolific bearer, fruits medium, fibreless juicy, sweet.
	No.2/2	Neelum x Banganapalli	Regular bearer, good quality fruits.
	No.2/7	Chinnasuvar-narekha x Neelum	Dwarf, regular bearer, good fruit quality.
RFRS, Paria, Gujarat	Neeleshan Gujarat	Neelum x Baneshan	It bears attractive obovoid shaped fruits heavily and regularly. The average weight of fruits is about 318g with higher pulp content of 76 per cent. On ripening the skin become cadmium yellow while the pulp becomes attractive golden yellow in colour. The fruit skin is thin, smooth, moderate adhering to pulp with firm non-fibrous texture. Fruit TSS is 16-17 per cent with moderate acidity 0.21 per cent and total sugars 13.09 per cent. The fruits are suited for table purpose and have very good keeping quality of more than 10 days after ripening.
	Neeleshwari	Neelum x Dashehari	Fruit size is medium weighing about 228g with prominent sinus. The skin color on ripening turns apricot yellow while pulp colour becomes yellow. The thin smooth skin adheres to moderately firm textured flesh. The non-fibrous juicy pulp suited for table as well as juice purpose and has moderate keeping quality. The TSS is more than 19.00 per cent with moderate acidity 0.21 per cent and higher total sugars 13.90 per cent.
	Neelphonso	Neelum x Alphonso	The mango hybrid Neelphonso was released in the year 1986 from AES, NAU, Paria. This hybrid is moderate regular in bearing but have late bearing tendency. The fruits ready for the harvest in July-August and when most of the mango variety will be harvested. The oval oblique shaped fruits of Neelphonso are medium in size weighing about 200g. The skin colour on ripening becomes apricot yellow while the pulp becomes orange yellow. Due to thick, smooth skin the fruits do not get damaged by rain. The fruits are free from spongy tissue disorder and are moderate resistant against fruit fly. The pulp texture firm, non-fibrous, highly suited for table and juice purpose with excellent keeping quality. The TSS is more than 21.5 per cent with lower acidity 0.18 per cent and higher total sugars 14.42 per cent.
Horticul-tural Research Station, Periya-kulam, Tamil Nadu	PKM-1	Chinnasu varnarekha x Neelum	Regular bearer during both on and off years, suitable for long distance transport, fruit weight 250-300 g, long, narrow with prominent beak and yield recorded upto 336 fruit (102.7kg/tree/ha).
	PKM-2	Neelum x Mulgoa	Large size fruits (681.3 g) high pulp content (67.5 per cent), abundant juice, pleasant flavour, good keeping quality with high TSS. The fruit comes to harvest during May.

Centre	Hybrids	Parents	Characteristics
Bihar Agricultural College, Sabour, Bhagalpur, Bihar	Mahmud Bahar	Bombai x Kalapadi	Sturdy and medium tree, regular bearer, fruits small, fibreless, good quality.
	Sundar Langra	Langra x Sundar Pasand	It has high yield potential with quality similar to Langra and regular bearing in nature, fruit size 10.5 x 7.5 cm, average fruit weight 240 g, yellowish red peel colour at ripening, light orange pulp colour at ripening, pulp content 75.8 per cent, TSS 22 °Brix and acidity 0.19 per cent.
	Jawahar	Gulabkhas x Mahmood Bahar	Mid-season variety. Fruits mature from second week of June, precocity and regularity in bearing. The fruits become greenish yellow on ripening. Its fruits are of medium size and average fruit weight is 215g per fruit. Its pulp is light yellow in colour, sweet in taste and pleasant flavoured and remain firm even after ripening. The TSS, acidity and pulp percentage are 22.5, 0.14 and 79.5 per cent respectively.
	Prabha-shanker	Bombai x Kalapadi	It is the most adaptable varieties for Bihar, having high yield potential, medium size tree resembles Bombai in shape, but superior than Bombai in respect to yield and fruit quality parameters. Regular bearing, fruit size 9.2 x 6.9 cm, average fruit weight 192.5 g, greenish yellow colour of peel at ripening, yellowish orange pulp colour at ripening, pulp content 64.1 per cent, TSS 23.0 °Brix and acidity 0.24 per cent. The average yield of 15 years old tree gives 1.0-1.25 quintal fruit.
	Alfazli	Alphonso x Fazli	One of the most adaptable varieties for Bihar, having tall tree size, fruits are bigger in size. Regular bearing, fruit size 12.2 x 7.7 cm, average fruit weight 460 g, yellowish green peel colour at ripening, light yellow pulp colour at ripening, pulp content 79 per cent, TSS 18.5 °Brix and acidity 0.24 per cent.
	Sabari	Gulabkhas x Bombai	Regular bearer, Peel colour: red, Pulp colour: reddish yellow, Avg. Fruit Wt. 153g., Pulp content: 74 per cent, TSS-22.5 per cent.
RFRS Saharanpur, UP	Hybrid-1	Dashehari x Totapuri Hyderabad	High yield and good fruit quality.
	Hybrid-2	Dashehari x Totapuri Hyderabad	High yield and good fruit quality.
	Hybrid-3	Dashehari x Fajarizafrani	High yield and good fruit quality.
RFRS, Sangareddy, AP	A.U. Rumani	Rumani x Mulgoa	Regular and heavy bearer, fruits round, sweet.
	Manjira	Rumani x Neelum	Precocious, regular and prolific bearer, fruits are medium, fibreless, sweet.

Centre	Hybrids	Parents	Characteristics
RFRS, Vengurla, Maharashtra	Ratna	Neelum x Alphonso	Moderately vigorous tree, regular and precocious bearer, large size, ovate shape, slightly elongated fruits, beak absent, greenish orange colour with prominent oil glands, orange coloured flesh excellent taste, free from spongy tissue, excellent for processing as well as table purpose, cluster bearing habits.
	Sindhu	Ratna x Alphonso	Thin stone (6-7 g) and non-viable, higher percentage of pulp, regular bearer, and attractive red blush on fruits, free from spongy tissue, cluster bearing habit. With high yielding (200-250 fruits per trees).
Navsari Agricultural University Navsari, Gujarat	Sonpari	Alphonso x Baneshan	The trees are vigorous in growth, heavy yielder and regular in bearing. The fruits are obliquely oval in shape like Baneshan, big in size weighing 360-550g. The TSS is more than19.5 per cent with lower acidity 0.18 per cent and higher total sugars 14.46 per cent. The keeping quality is very good and fruits remain in good condition for more than 10 days at room temperature. The fruits mature in second week of June.
MPKV Rahuri, Maharashtra	Sai Sugandha	Totapuri x Kesar	Late bearing, table purpose.
BSKKV, Dapoli, Maharashtra	Konkan Ruchi	Neelum x Alphonso	Regular bearer, large fruit size, thick skin, acidic, best for pickle purpose.
	Suvarna	Alphonso x Neelum	It has good qualities of Alphonso and has tendency to bear every year like Neelum, it has cluster bearing habit, high percentage of hermaphrodite flowers in panicle and yield upto 260-280 fruits per tree.
	Konkan Raja	Bangalora x Himayuddin	This variety has better fruit size (616 g) than parents Bangalora (170.3 g) and Himayuddin (350 g). Less sour due to less acidity 0.19 per cent and TSS 8.50 °Brix of unripe fruits. It is best for salad purpose. This variety has high percentage of perfect flowers (28.8 per cent), regular bearer having higher pulp percentage (83 per cent) than the parents Bangalora (72.68 per cent) and Himayuddin (68.07 per cent), early maturing and good attractive shape and colour. The average yield is (65.85 kg/plant).
	Konkan Samrat	Alphonso x Tommy Atkins	Medium fruit size (284.50 g) as compared to the Alphonso (250 g) and Tommy Atkins (484 g). The pulp percentage (73.28) is similar to Alphonso (74.60 per cent) and Tommy Atkins (71.79 per cent). Good T.S.S. (20.03 °Brix). This variety has high percentage of perfect flowers (27.52 per cent). Regular bearing, cluster bearing, spongy tissue free and low fibre. This is second hybrid having Alphonso as female parent and first hybrid of exotic parent. This variety is ideal for table purpose.

ICAR-IARI (Indian Agricultural Research Institute); ICAR-IIHR (Indian Institute of Horticultural Research); ICAR-CISH (Central Institute for Subtropical Horticulture); RFRS – Regional Fruit Research Station,

* Pandey (1993), Chandra and Chandra (1997), Ray (1999), Anon. (2018a).

Cultivars of West Bengal and eastern region of India

Bimli

Bira

Ranipasand

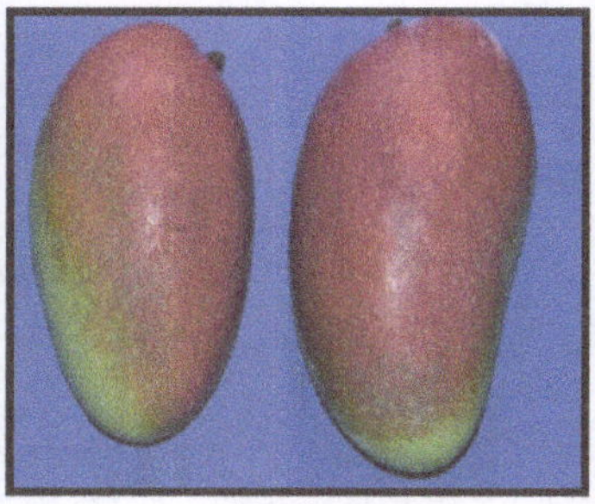

Sinduri

Sarenga

Panja

Bhaduria

Mulayemjam

Kishenbhog

Mango cultivars

Amrapali

Aswina

Champa

Kohitoor

Laxmanbhog

Sarikhas

Baisakhi

Kanchamitha

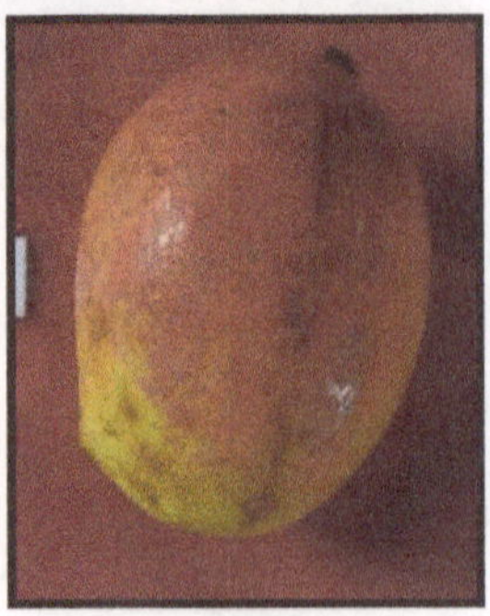

Gulabkhas

Lucknow safeda

Mango cultivars in bearing condition

Cultivars of North India

Langra

Dashehari

Chausa

Cultivars of South India

Banganpali

Swarnarekha

Totapuli red small

Swarnarekha

Rumani

Bangalora

Neelum

Neelashan

Neelgoa

Neeluddin

Swarna jahangir

Manjira

Cultivars of western region of India

Alphanso

Jamadar

Pairi

Kesar

Rajapuri

Mankurad

Exotic cultivars of mango

Tommy Atkins

Edward

Baily Marvel

Harris

Pascual

Siagnom

Philippine

Exotic cultivars of mango

Important mango hybrids produced in India

Pusa Lalima

Pusa Pratibha

Pusa Pitambar

Pusa Surya

Pusa Shreshth

Ratna

Important mango hybrids produced in India

Pusa Arunima

Amrapali

Arka Anmol

Mallika

Arka Neelkiran

Table 4: Important Mango Cultivars in Major Producing Countries*

Country	*Cultivars*
Australia	'Kensington Pride', 'Banana', 'Earlygold', 'Glenn', 'Haden', 'Irwin', 'Keitt', 'Kent', 'Zill'
Bangladesh	'Aswina', 'Fazli', 'Gopal Bhog', 'Himsagar','Khirsapati', 'Langra', 'Kishan Bhog', 'Kohinoor', 'Kua Pahari', 'Mohan Bhog'
Brazil	'Bourbon', 'Carlota', 'Coracao', 'Espada', 'Itamaraca', 'Maco', 'Magoada', 'Rosa', 'Tommy Atkins'
China	'Baiyu', 'Guixiang', 'Huangpi', 'Huangyu', 'Macheco', 'Sannian', 'Yuexi No. 1'
Costa Rica	'Haden', 'Irwin', 'Keitt', 'Mora', 'Tommy Atkins'
Ecuador	'Haden', 'Keitt', 'Kent', 'Tommy Atkins'
Egypt	'Alphonso', 'Bullock's Heart', 'Hindi Be Sennara', 'Langra', 'Mabrouka', 'Pairie', 'Taimour', 'Zebda'
Guatemala	'Haden', 'Kent', 'Tommy Atkins'
Haiti	'Francine', 'Madame Francis'
India	'Alphonso', 'Banganapalli', 'Bombay', 'Bombay Green', 'Chausa', 'Dashehari', 'Fazli', 'Fernandian', 'Himsagar', 'Kesar', 'Kishen Bhog', 'Langra', 'Mallika', 'Mankurad, 'Mulgoa', 'Neelum', 'Pairi', 'Samar Behisht Chausa', 'Suvarnarekha', 'Totapuri', 'Vanraj', 'Zardalu', 'Amrapali', 'Bangalora', 'Gulabkhas'
Indonesia	'Arumanis', 'Dodol', 'Gedong', 'Golek', 'Madu', 'Manalagi', 'Cengkir', 'Wangi'
Israel	'Haden','Tommy Atkins','Keitt', 'Maya', 'Nimrod', 'Kent', 'Palmer'
Kenya	'Boubo','Ngowe','Batawi'
Malaysia	'Arumanis', 'Kuala Selangor 2', 'Golek', 'Apple Rumani', 'Malgoa', 'Apple Mango', 'Maha-65', 'Tok Boon'
Mali	'Amelie','Kent'
Mexico	'Haden', 'Irwin', 'Kent', 'Manila', 'Palmer', 'Sensation', 'Tommy Atkins', 'Van Dyke'
Myanmar	'Aug Din', 'Ma Chit Su', 'Sein Ta Lone', 'Shwe Hin Tha'
Pakistan	'Anwar Ratol', 'Baganapalli', 'Chausa', 'Dashehari', 'Gulab Khas', 'Langra', 'Siroli', 'Sindhri', 'Suvarnarekha', 'Zafran'
Peru	'Haden', 'Keitt', 'Kent', 'Tommy Atkins'
Philippines	'Carabao', 'Manila Super', 'Pico', 'Binoboy', 'Carabao', 'Dudul', 'Pahutan', 'Senora'
Singapore	'Apple Mango', 'Arumanis', 'Golek', 'Kaem Yao', 'Mangga Dadol'
South Africa	'Fascell', 'Haden', 'Keitt', 'Kent', 'Sensation', 'Tommy Atkins', 'Zill,
Sri Lanka	'Karutha Colomban, Willard, Vellai Colomban, Petti amba, Malwana amba, Parrot Mango and Peterpasand, Dapara, Hingurakgoda
Thailand	'Nam Doc Mai', 'Ngar Charn', 'Okrong', 'Rad', 'Choke Anand', 'Kao Keaw', 'Keow Savoey', 'Pimsenmum'
USA	'Keitt', 'Kent', 'Tommy Atkins'
Venezuela	'Haden', 'Keitt', 'Kent', 'Tommy Atkins
Vietnam	'Combodiana'

*Anonymous (2018a).

7.0 Breeding and Improvement

Despite an enormous wealth of mango cultivars available in the country, an ideal cultivar of mango is still lacking. Most of the present-day cultivars appear to have been selected for characters like fruit size and quality, and period of maturity. Some of the more important characters like precocity, dwarfness, prolificity and regularity of bearing, self-fruitfulness and resistance to pests and diseases remained unselected. These characters are now of vital importance for making the best use of our shrinking land resources, reducing the cost of cultivation and for improving the productivity per unit area. However, combining all the desirable characters in a single cultivar is difficult, if not impossible, since the mango is a highly heterozygous crop of suspected allopolyploid origin. An ideal cultivar can, therefore, be built only in stages and above all most of the commercial mango cultivars have a strong biennial bearing tendency.

Thus the main objectives of breeding have been regular bearing and good fruit quality. An ideal mango variety should possess dwarf tree habit; precocity, regular bearing, attractive (golden apricot) colour at ripening, medium size (5 in a kg) and good quality fruit (high pulp: stone ratio, firm and fibreless flesh); freedom from physiological disorders; highly tolerant to various diseases including malformation and pests; and good shipping quality and high keeping quality.

Mango breeding has gained increasing attention in recent years, and conventional breeding methods (cross breeding, mutation breeding, and seedling selection) are mainly used for creating new varieties of mango. However, these methods are time-consuming and difficult to improve upon because of some inherent characteristics of mango (Iyer and Schnell, 2009). Many important economic traits of mango are controlled by a number of loci. Quantitative trait locus (QTL) analysis and marker assisted selection (MAS) can shorten selection times and accelerate the breeding process of new varieties of mango. Genetic maps, especially high-density genetic maps, are particularly valuable tools for QTL mapping and MAS. Four genetic maps of mango have been generated to date.

Although at present, none of the commercial varieties of mango has all these characteristics combined into one, it would be futile to take up all these objectives at a time in any mango breeding programme. Therefore, these characters are to be incorporated at stages and the variety is to be built up gradually. A good mango variety should have a high ratio (3.31 to 4.0) of edible to non edible matters. In fact, this improved mango would be "an ideal cultivar", but is there an ideal or perfect mango cultivar?

The mango varieties must be conserved preferably by *in-situ* conservation strategy or by *on farm* conservation method. Morphological characterization is the first step that should be done before advanced biochemical or molecular studies are carried out. For the management of mango genetic diversity increasing knowledge on the extent and distribution of mango diversity in the target species is a pre-requisite. Determining current mango germplasm conservation efforts (location, collection, conservation and use) is of paramount importance (Singh *et al.*, 2019).

Mango markets show an exigency for high performing cultivar, therefore the plant and fruits should be of following characteristics. The plant should have high productivity (an average of 40 t/ha in planting density of 476 plant/ha), dwarf but spreading canopy with vase type architecture, precocious and regular production, resistant to pests and disease. The fruit characters like weighing 400-500 g, ovate to ovate-elongate shape, peel preferable blush ground color with yellow spots at consumption point, gold-yellow to orange and fiberless pulp, sweet (Brix > 18 per cent), excellent flavor, free of physiological disorders, long shelf-life and juicy pulp with high yield (> 70 per cent of pulp), mainly those fruits to process juice are desirable (Singh *et al.*, 2018).

The major bottlenecks in breeding are long juvenile phase; highly heterozygous and cross pollination; natural crossing is low; complex nature of flower and panicle; excessive fruit dropping; problem of polyembryony which makes identification of hybrid seedling difficult; lack of knowledge of inheritance pattern and tedious hybridization.

It can be minimized by minimizing the high fruit drop, shortening juvenility, polyembryony dilemma for the breeder and asset in rootstock propagation. Isozymes may be used to identify the zygotic seedling from the nucellar one's as the nucellar seedling should have same isozyme alleles as that of the maternal parent (Singh *et al.*, 2018).

Mango was previously propagated through seed and hence old orchards in India were mostly seedling origin. Some seedling progenies gave rise to varieties such as *'Chinnaswarnarekha'* and *'Mundappa'*. The popular, salt tolerant rootstock (13/1) was identified in Israel by this technique. A selection from the open pollinated variety of *'Naomi'* was related as *'Tango'* in Israel which is a mid season variety having tree of medium size, oblong fruit with pointed beak and shallow sinus, of medium size (380 g) with orange skin colour. Most of the Indian mango cultivars do not posses good skin colour. To meet the international demand of good fruit colour, some exotic cultivars of mango had been introduced as *Tommy Atkins, Zillete, Haden, Sensation* and *Julie* from Miami, Florida (USA) and *Amoelie* from Belgium and *Sweet* from Thailand.

Arka Neelachal Kesari variety, a clone of Gulab Khas, has been selected for early maturity in coastal areas of Odisha (Pandey *et al.*, 2013). *Gulab Khas* and *Gulab Khas Red* grown in Bihar and Uttar Pradesh show mid-season fruit maturity there. Superior clones of Dashehari namely, 'Pant Sinduri' and 'Pant Chandra' have been selected by GBPUA&T, Pantnagar, District Udham Singh Nagar of Uttarakhand (India). 'Pant Sinduri' was released in 2004 for commercial cultivation (Pandey, 2015).

Agriculturally important place of Pusa, Bihar has diversity of many commercial cultivars of mango *viz.* Kanchan (*Bathua*), *Malda* (Langra), Sukul, *etc.* grown on large scale, Sipia and Bombai are grown by many farmers, and on smaller scale mango varieties like *Paharpur Sinduria*, Fazli, Sunder Langra, Krishanbhog, *Zarda, Kerwa, Mir Zaffar, Zardalu, Chinia*, Neelam, Dasheri, Mallika, *Kapuria* and Romani are grown. A Studies revealed a clone from *Bhuskaul* community, Pusa, Samstipur,

Bihar [(Clone no. 77 of *Malda* seedling) had almost similar fruit characteristics to *Malda* with very late maturity (August end)] with fruit weight up to 420.0 g with TSS 27.40 °B and having very thin stone. The present study highlights the need for and demands of diversity rich areas of *Pusa* community in Bihar to conserve and protect seedling mangoes for the benefit of posterity with high value traits for future promotion (Singh *et al.*, 2015).

7.1.0 Improved Techniques of Hybridization

In the early days, during 1911-60, large sized muslin bags having 2-3 iron rings were utilized in the hybridization programme (Sen *et al.*, 1946). These were cumbersome in handling since each panicle enclosed in a bag was to be opened and closed everyday during the flowering period for making pollination and fewer panicles could be handled in the short-flowering period. Also, the number of flowers pollinated on a single panicle was rather too high because the same few panicles were utilized for a number of days, the number of flowers pollinated on a single panicle running into hundreds. This appears to be highly unnatural as it is rather impossible to imagine a cultivar of mango which could, on an average, bear more than one fruit per panicle. This technique was wasteful since the outcome was, by any standard, meagre.

In view of these difficulties, improvements were brought about to increase the efficiency of the mango hybridization work. Emphasis was, therefore, laid on utilizing a larger number of panicles and pollinating few flowers per panicle (Mukherjee *et al.*, 1961). The conventional muslin bags have now been replaced by light (100 gauge thickness) small-sized polythene bags which are finely perforated before use. With this technique, the efficiency of mango hybridization has been improved considerably. Since 1961, about 82000 crosses have been made at the IARI and these have yielded 1252 hybrids. This is much larger a population of mango hybrids, even if one pools the number of hybrids raised at all the other mango research centres in India (Singh *et al.*, 1977, 1981).

The self-incompatibility, in some of the most popular cultivars of mango, led to the standardization of the caging technique in which pollination is effected through houseflies, thus doing away with the necessity of hand pollination (Sharma *et al.*, 1972).

A study based on a genetic 'marker' (dominant purple colour of the new leaves and inflorescence, and the dominant beak of the cv. Totapuri Red Small over that of the cv. Dashehari), has shown that the percentage fruit set from the crosses in mango can be doubled simply by doing away with the process of bagging the panicles again after cross-pollination by hand (Singh *et al.*, 1980). The increased fruit set is largely due to preventing the stigmatic and stylar injury to the pollinated flowers, which is otherwise a common occurrence. The time saved by avoiding rebagging of the pollinated panicles can be utilized in handling larger number of panicles in the short span of flowering.

7.2.0 Inter-varietal Hybridization

Mango hybridization work was first started by Burns and Prayag in 1921 at Pune. Subsequently work was taken up in Kodur (A.P.), Saharanpur (U.P.), Patiala

(Punjab), IARI (New Delhi), Sangareddy (A.P.), Periakulam (T.N.), Krishnanagar (W.B.), Paria (Gujarat) and IIHR (Bangalore). Major emphasis in the programme has been on regular and precocious bearing, dwarfness, high percentage of pulp, fibreless flesh, large fruit with red blush, good keeping quality and free from spongy tissue. The commercial variety of the north India was found to be self incompatible. The incompatibility in mango is sporophytic in nature.

The work at Bihar Agriculture College, Sabour yielded two promising hybrids, *viz.* Mahmud Bahar and Prabha Sankar. Same is more or less true for the hybrids evolved at Kodur, *viz.* Neluddin, Swarna Jehangir, Neelgoa and Neeleshan. An intensive mango hybridization programme was initiated at Paria (Gujarat) in 1968. Three mango hybrids namely, Neelphonso (Neelum x Alphonso), Neeleshan Gujarat (Neelum x Baneshan) and Neeleshwari (Neelum x Dashehari) were found promising. The dwarf habit of these hybrids was suited to high density orcharding (Sachan *et al.*, 1985). The progress made so far In India cannot be termed as satisfactory. This applies to the mango breeding work done elsewhere as well (Singh, 1960; Sturrock, 1977).

Today, most of the new Indian hybrids are regular bearing, with good quality fruits (free of spongy tissue) and attractive skin color. All the hybrids have higher pulp yield and possessed lower peel, stone and fibre content.However, it is quite understandable in a crop like mango which is highly heterozygous and endowed with a long juvenile phase.

In such a crop, breeding work is necessarily laborious, and at times even frustrating, and requires sustained effort over a number of years. Some other major contributing factors are lack of well-defined objectives, improper choice of parents, cumbersome and laborious techniques of hybridization and meagre hybrid populations available for selection. This is amply illustrated by the success achieved at the IARI where, taking care of all these points, two improved cultivars (Mallika and Amrapali) could be evolved and released for the growers, although the hybridization work was initiated as late as 1961. Other major landmarks in the field of mango breeding, besides the release of the two cultivars and the discovery of self-incompatibility, are (i) development of improved and more efficient techniques of hybridization, (ii) gaining some knowledge about the inheritance pattern prevailing in mango and (iii) development of sound pre-selection criteria.

Mukherjee *et al.* (1961) was the first to describe intervarietal hybridization with 1.45 per cent success. Pinto and Byrne (1993) suggested an improvement to the technique and increased fruit set to 6.0 per cent, which promotes the enlargement of the hybrid population and shortening of cultivar release. The caging technique and floral induction are essential for the poly-cross method, since there is synchronous flowering of the top-worked cultivars. The discovery of *Caging technique* formed the basis in standardization of improved pollination technique in mango breeding. This technique is involved in the planting of a self incompatible parent along with the male parent in an insect proof cage. Pollination is done by introducing freshly reared houseflies. It has got an advantage of avoiding emasculation and hand pollination. The use of dominant purple colour of new leaves and inflorescence and the dominant beak of the cv. Totapuri Red Small as a genetic marker enhance the efficiency of above technique.

Hybridization technique might be inefficient when used alone; however, it can be highly improved when used together with other technique (s), which is the case of top-working and caging, which is only efficient to use poly-cross method when floral induction is applied in order to synchronize the flowering period of each top-worked cultivar (Pinto *et. al.*, 2004).

Recently, mango cultivars *Pusa Arunima* and *Pusa Surya* released by ICAR-IARI, New Delhi are recommended for commercial cultivation. *Mallika* is recommended for commercial cultivation in south and East Indian region. *Dashehari-51*, a regular bearing superior clone of *Dashehari* with higher yield by about 38 per cent and ripening 15 days later to *Dashehari* has been found suitable for cultivation under Lucknow conditions. *Vashi Badami* has been found free from mango leaf hopper, gall midge and shoot borer and recommended for release for cultivation in Valsad district of Gujarat. *Bangalora, Mallika, Neelum, Suvarnarekha, Van Raj* and *Benishan* have been reported to perform very well under south Telengana conditions. These cultivars can therefore, be grown commercially in this region.

Some varieties developed with other specialties are *Konkan Raja* (Bangalora x Himayuddin) having very large fruit (616g), regularity in bearing, early fruit maturity, high pulp content (83 per cent) and suitability of fruit as salad at fruit maturity stage and *Konkan Ruchi* suitable for making pickle developed at Vengurle, Maharashtra (Pandey, 2015).

7.3.0 Inheritance Pattern in Mango

The fruit size appeared to be governed by additive genes. Red skin colour was reported to be dominant and influenced by duplicate genes (Sharma and Majumder, 1985).

Dwarfism, regular bearing and precocity are governed by recessive gene, while regularity of bearing is linked with precocity (Sharma and Majumder, 1988a). Polyembryony, bunch bearing, presence of beak, biennial bearing & upright habit of the tree are governed by dominant gene. Fruit colour is governed by multiple loci. Fruit size follows transgressive segregation. While malformation and spongy tissue are governed by recessive gene, bacterial canker followed cytoplasmic inheritance. Iyer (1991) studied inheritance of flesh color in Alphonso and Neelum cvs. of mango and found that light yellow color is dominant over orange yellow. There is no maternal effect on juvenility and fertility while there is slight effect of female parent on fruit taste, size, colour and harvest season.

The regular bearing cultivars like Neelum, Bangalora and Totapuri Red Small were found to have the characteristic of putting forth new growth flushes just after harvesting of fruits or even earlier. These newly emerged shoots are able to flower and fruit in the succeeding season. Studies have also established that amongst the regular bearing cultivars, only Neelum happens to be the best combiner. The regularity of bearing of the cvs. Totapuri Red Small and Bangalora appear to be closely linked with their inferior fruit quality (Singh *et al.*, 1981). Likewise, the cv. Bhadauran, which has fruits of poor quality but is the only known cultivar to be free from floral malformation, holds little promise for evolving good quality mango which may also have resistance to malformation.

Self- and cross-pollination effects on mango (*Mangifera indica*) pericarp and seed weights were studied by Dag *et al.* (1999). Selfed and outcrossed offspring were identified using the triosephosphate isomerase (TPI) isoenzyme system. Selfed 'Keitt' progeny showed no significant relationship between TPI isoenzyme genotypes and pericarp and seed weights. No significant differences in pericarp or seed weights were found between outcrossed and selfed fruit for 'Maya', 'Tommy' or 'Kent'. In contrast, pericarp and seed weights from fruit of '6-6' orginating from cross pollination were >50 per cent heavier than self-pollinated fruit.

The studies conducted at the IARI have established that for involving any cultivar in the hybridization programme for transferring a particular desirable character in the hybrids, it must also have at least an acceptable fruit quality.

7.4.0 Pre-selection Criteria

In a long-lived perennial, having a long pre-bearing age, it is advantageous to identify certain characters which may facilitate a quick selection of the desired types amongst the hybrid population. This can save precious time, land and labour. Some pre-selection criteria, which could give a precise indication of the fruit quality, bearing habit and growth habit have now been identified at the IARI. Leaf flavour has direct correlation with the fruit flavour (Majumder *et al.*, 1972b). Emergence of new growth flushes, simultaneously with fruiting or immediately after harvest, is indicative to regularity of bearing (Sharma *et al.*, 1972). Likewise, a lower stomatal density per unit area is indicative to dwarfness (Majumder *et al.*, 1981). Higher phloem to xylem ratio is associated with dwarfing. Genotypes with ratio more than one tend to be least vigorous while those with ratio less than 0.6 are more vigorous (Kurian and Iyer, 1992).

7.5.0 Induction of Mutation

A plant breeder's approach towards finding the solutions for major problems in a crop is largely centered around the creation of a wide genetic variability and its exploitation, besides collection and screening of the existing germplasm. This variability could be inducted in two ways: (i) through conventional hybridization (which has been dealt with earlier) and (ii) by inducing mutations through physical and chemical mutagens (Broertjes and van Harten, 1978).

Although the traditional mango cultivars, individually, do possess one desirable character or the other, cultivars having more of the desirable characters are lacking. Mutation breeding, with a few exceptions in tree crops like apple, apricot, cherry and peach (Broertjes, 1977), has not been given a fair trial by the fruit breeders, in general, and in mango, in particular. In case of mango only a solitary attempt, using a few bud sticks, has been made (Siddiqui *et al.*, 1966) and this too has, probably, not been pursued further.

The main advantages of mutation breeding are, that the basic genotype of the cultivar is usually only slightly altered (as compared to hybridization of two different cultivars), while the improved character(s) is (are) added, and the time required to breed an improved cultivar can be much reduced than when hybridization is employed to achieve the same objectives (Sigurbjornsson, 1970).

The work on induction of mutations was, therefore, initiated at the IARI in 1972 with a view to developing dwarf, regular-bearing, self-fruitful and disease resistant mutants having good fruit quality. This work has led to the standardisation of effective dosages of some of the mutagens and also in the standardization of the propagation technique for the mutagen-treated bud sticks (Sharma *et al.*, 1981). It has been observed that dosages of ã-irradiation beyond 5 kR are lethal to mango. The LD50 for different cultivars ranges between 2 and 4 kR; being 3.9 for Neelum, 2.9 for Dashehari and 2.4 for Mallika. This suggests that the newer a cultivar, the greater is its radiation sensitivity. The effective dosages of EMS (cv. Dashehari) and NMU (cv. Neelum) were found to be 1.50 per cent and 0.05 per cent respectively. Another interesting point that emerged is that the spectrum of changes induced was more or less the same, by both physical and chemical mutagens.

Some very interesting plant types have been obtained in the vM1 generation by stabilising the changes induced through the heading back technique (Sharma *et al.*, 1981). Amongst these are plants with very long and very short extension growths, very thick and very thin shoots, multiple branching shoots, very large and very small lanceolate leaves and small leaves with highly wavy margins, and plants which appear to be compact in growth habit. A few of these induced plant types, both from the cvs. Dashehari and Mallika, appear to be promising from the point of view of dwarfness and, hence, may prove their immense value.Higher TSS and better sugar/acid blend, than the standard Dashehari (control), were also observed in a few plants.

7.6.0 Future Breeding Programme

Despite the advances made in this area, much remains to be done in regard to developing cultivars suited to different regions and differing needs, *viz.* for the export trade and for the processing industry of various products. Improved cultivars, having resistance against some of the more serious diseases and pests, are needed to cut down the extra cost and labour required for checking them through chemical sprays which are, above all, hazardous too.

However, such a breeding programme can be effectively implemented only by wide-ning the germplasm base by collecting all the *Mangifera* spp., cultivars and promising wild types at the major mango research centres. This is essential for studying their characteristics, which could be usefully transferred to the new cultivars through hybridization. However, till now, no attention has been paid to this fundamental aspect and before long many of the species and wild relatives of mango may become extinct in the developing countries of South Asia (where these largely exist in wild state) under the pressure of population explosion and rapid industrialization. One such race has been located by Sharma and Sen Choudhuri (1976) in the forests of southern Tripura. Though its fruits are very small in size, it has the habit of bearing in bunches of up to 50 fruits per panicle. The trees were apparently very healthy and free from malformation. The leaf size is perhaps the largest ever recorded in the genus *Mangifera* and the other leaf characters also do not tally with any of the species or wild types described so far. The north-eastern states, bordering Myanmar and Bangladesh, offer vast scope in this regard since this happens to be the area where the mango originated.

'Creeping', a promising genotype, was identified for introduction of dwarfness in mango. This cultivar has no apical dominance, with around 7 new shoots arising from each main branch (from the lateral buds as well as the tip) and a short internodal length (2 cm), rendering it useful for high density planting. The yield is high and fruit quality is considered good (Iyer and Subramanyam, 1986).

The chromosomal studies conducted thus far in five of the 41 species of *Mangifera* (*viz. M. indica, M. sylvatica, M. caloneura, M. odorata* and *M. zeylanica*) have shown that these species have the same chromosome number (2n=40) and may intercross easily (Mukherjee, 1963). Some of their useful genes can possibly be transferred to mango. The only exception reported in the literature so far is that of the cv. Vellaikollumban which was reported to be a tetraploid, *viz.* 2n=80 (Roy and Visweswariya, 1951). This cultivar was, therefore, studied by Sharma *et al.* (1976) with a view to developing triploids and aneuploid series in mango. However, its cytology revealed it to be a mere diploid, like any other cultivar of mango.

Nevertheless, some scope does appear to exist for developing haploids through cultu-ring of anther, endosperm or meristem, where chromosomal variations are likely to occur. Such studies will lead to a better understanding of the origin and inheritance of characters in mango, besides helping in the development of triploids, trisomics and tetrasomics which may be of immense academic and applied values. Studies of this nature have not been undertaken in tree fruits, excepting guava where useful aneu-ploids have been isolated by crossing triploid with diploid guava (Majumder and Mukherjee, 1972). It was felt that demand for table, sucking and pickles type has been escalated in the region due to decrease in seedlings mango population and awareness about indigestion problems generally experienced after consumption of grafted mangoes (Singh *et al.*, 2015).

Breeding programme need to be intensified with a view to evolving dwarf plant types having regularity and prolificity of bearing, and good fruit quality. This is the only way to meet the twin challenges of population explosion and rapid industrialization which are shrinking our limited land resources. The traditional varieties have sustained us far too long and they can not fit into the modern concept of high density orcharding which alone can increase the productivity per unit area and led to efficient and more economical management of orchards. The combination of mango breeding strategies include the introduction, evaluation and selection of high performance cultivars as well as the use of efficient and cheap breeding methods and the help of molecular markers will assist in obtaining excellent double purpose new cultivars as precocious as possible. A breeding program of this type is one that by its very nature would require many years to complete. It is not a program that one individual could reasonably be expected to complete in his lifetime. Therefore, experiment stations and other research institutions would be the logical organizations to carry out this type of endeavour.

8.0 Propagation

8.1.0 Seed

Seed propagation is not advocated in mango as the seedling-raised plants are never true-to-type. However, seedlings are used as rootstocks. The polyembryonic cultivars can be raised from seed provided the nucellar seedlings are identified properly. In polyembryonic cultivars, the percentage of off-type (zygotic) seedlings ranged from 2 per cent in Sabre to 47 per cent in Florigon. The percentages in Peach, Julie and Kensington were 7 per cent, 9 per cent and 12 per cent respectively (Truscott *et al.*, 1993). The seeds lose their viability very soon. Seed viability of the cvs. Alphonso and Dashehari was significantly affected by storage time and temperature with the best results obtained at 15°C (50 per cent viability at 12 months). Longevity of seeds of both cultivars was extended from 2 weeks to 12 months by pretreatment with 8-hydroxyquinoline sulphate at 15°C (Doijode, 1995). Sensitivity to desiccation decreased after the seeds were treated with ABA (Wang and Fu, 1994).

Seeds are normally sown in well-prepared nursery beds with a spacing of 45 cm between rows and 22 cm between seeds. The highest germination percentage (98.67 per cent) was obtained within 10 days of sowing husked seeds in sawdust and sawdust was found to be better germinating medium than clay soil or peatmoss (Abdel-Galil, 1992). Mango (cv. Xiamao) seeds continued to germinate without supplementary water after removal from their endocarps and vivipary was observed in the seeds of cv. Autumn mango. It was concluded that mango seed recalcitrance may therefore be due to germination of mature seeds on the mother plant (Wang and Fu, 1994). Generally, one-year-old seedlings are used as rootstocks. The most promising cultivars for use as rootstocks were Ceniap-2 (short) and Picode Loro (medium-height), both being polyembryonic (Avilan *et al.*, 1995).

8.2.0 Vegetative Propagation

The methods of vegetative propagation in mango are known in India since very ancient times, as is indicated by early Sanskrit literature (Singh, 1960). However, according to Popenoe (1920), the technique of inarching was first introduced in India by the European Missionaries in Goa.

Methods of vegetative propagation in mango can broadly be divided into 3 categories, *viz.* (a) grafting, (b) layering and (c) cutting.

8.2.1 Grafting

This is the most common method of propagation of mango cultivars and includes inarching and budding. In recent years, more efficient techniques, *viz.* veneer grafting, side grafting and epicotyl or stone grafting have also been standardised.

8.2.1.1 Inarching

It has been a common method of propagating mango cultivars and is still followed by the nursery men all over the country. However, it has the following major disadvantages:

i. It is a cumbersome method since the rootstock seedlings have to be carried to the mother plant for grafting.
ii. It is laborious and time consuming since *machans* (raised structures) are to be fabricated around the mother tree for placing the rootstock seedlings for grafting.
iii. The inarched plants have to be irrigated and looked after for at least 2 to 3 months at various odd places around the tree.
iv. It is uneconomical since only one plant can be obtained by grafting a long scion-shoot.
v. Inarched plants usually get pot-bound and hence mortality on transplanting is rather high, the ideal age of potted seedlings for grafting being about a year.

The highest grafting success rate (92 per cent) in mango cv. Himsagar was obtained with inarching on 30 July (Kumar and Mitra, 1994). Approach grafting has been found to be the most economical method, giving the highest net income (Rs.4.02/graft) and highest survival rate (86.65 per cent) after 6 months (Roy and Hoda, 1996). In an experiment with modified approach method of grafting on mango cv. Banganapalli, Maharana *et al.* (1996) observed that the survival after 60 days in pots increased as the detachment of grafts from mother plant was delayed and it was highest (76.66 per cent) for grafts detached after 50 or 55 days.

8.2.1.2 Veneer Grafting and Side Grafting

These are by far the best methods of propagation of mango since these are easier, more economical, give a higher degree of success and are ideal for establishing *in situ* orchards. Besides, being detached methods of grafting (unlike inarching), these techniques can be successfully utilized for preparing plants anywhere since the scion-sticks can keep well for 5-6 days, during which these can be transported to any place for grafting (Majumder *et al.*, 1972a). Also, these are ideal methods for converting old and inferior seedling trees into productive ones of a desired superior cultivar through top-working. The only difference between these two methods is that in veneer grafting (Mukherjee and Majumder, 1961, 1964) the vertical flap of the rootstock bark is completely removed while grafting, whereas in the other (Kashyap *et al.*, 1972), this flap is retained and tied over the scion. In veneer grafting, only one side of the scion is sliced away in a sloping manner, whereas in side grafting, the scion is sliced on both sides of the lower portion in the form of a wedge.

Proper selection and preparation of scion are of utmost importance in both the methods. The scion should be of matching thickness with the stock, preferably a terminal non-flowered shoot of 3 to 4 months' maturity. Selected scions are defoliated on the mother plant about 7 to 10 days prior to detaching, keeping a part of the petiole intact on the selected terminal shoot. This helps in forcing the buds to swell and in increasing the percentage success in grafting. This method can be successfully employed during March to September in northern India.

Singh *et al.* (1984) recorded 96.6 per cent success when veneer grafting was done in the month of June under Varanasi (UP) conditions while Singh *et al.* (1985) from

Saharanpur (UP) noted the highest percentage of success in the first fortnight of July with three-month-old scions of cv. Dashehari. Bajpai *et al.* (1985) in their experiment with two mango cultivars, Amrapali and Mallika noted higher percentage of success in Mallika but growth of the scions was more vigorous in Amrapali.

Defoliation of scion shoots 10 days prior to detachment markedly increased the scion shoot growth six months after grafting, while storage of scion shoots (for 10 days) resulted in least shoot growth (Dhakal and Hoda, 1986). Reddy and Melanta (1989) made an attempt to correlate between the phenol content of scion shoots of different cultivars and their grafting success. The highest percentage of grafting success was obtained with Dashehari and Totapuri scions with both container and *in situ*-grown rootstocks and lowest in Alphonso. It was interesting to note that the lowest phenol content occurred in Dashehari, while the highest phenol content at all precon-ditioning stages occurred in Alphonso. The phenol content in all scion cultivars, in general, decreased with increasing duration of preconditioning. It is, therefore, suggested that graft union success is related to phenol content in scion. The use of veneer grafting in cv. Himsagar gave 82 per cent success when performed between June and August (Kumar and Mitra, 1994).

8.2.1.3 Cleft Grafting

Yang and Chen (1998) found that cleft grafting was the most suitable for raising young plant material and for topworking. The scion should be more than 5 cm in length with more than 2 buds; after grafting it should be wrapped with polyethylene film to protect against rain. Rootstocks which are weak or too thick are not suitable for grafting and scions that are too tender or too old are also not suitable.

8.2.1.4 Epicotyl/Stone Grafting

This method has been standardized (Mukherjee *et al.*, 1966; Bhan *et al.*, 1969; Majumder and Rathore, 1970) and holds great promise for multiplying plants in larger number and in lesser time. The seeds are placed on a sand bed and covered with 5.0-7.5 cm thick layer of leaf mould for germination. Germinated seedlings of 8 to 15 days are taken out and grafted indoor by beheading the seedlings about 5.0 cm above the stone and inserting the wedge-shaped scions in the vertical split in the beheaded rootstocks. Polythene tape (200 gauge thickness) is used for tying the grafts. Immediately thereafter, the grafts are planted in polythene bags filled with soil and farmyard manure mixture (1:1). The grafts are watered and kept in semi-shade condition to avoid damaging effects of sun and rain.

The successful stone grafts should be shifted to open space when their leaves become pale green in colour and kept in trenches in order to improve their survival (Gunjate, 1985). Reddy and Kohli (1985) obtained higher percentage of success when Totapuri scions were grafted on Alphonso rootstocks than Alphonso scions on Alphonso rootstocks; however, the success was achieved when the grafts were kept inside the mist chamber/glasshouse/thatched house where the temperature and humidity were higher. Dhungana *et al.* (1985) reported that under Kerala conditions, the percentage of success was highest (69.33 per cent) when stone grafting was done during August and was lowest when done in May (20.6 per cent). Experimental results indicated that stone grafting carried out in Jammu, India, from mid-August to mid-September on 10-day-old rootstock seedlings gave 50-55 per cent success and scion shoots of 4 to 4.5-month-old were most suitable (Gupta *et al.*, 1988). Grafting in the warm humid months of June and July gave the highest survival (Gunjate, 1989). Highest percentage of survival (49.50 per cent) was observed when scions were defoliated 10 days prior to stone grafting compared to 15 days prior to defoliation (47.50 per cent) or without any defoliation (33.50 per cent). Treating the scions with IAA at 750 ppm increased the percentage of success and the number of leaves and leaf area/plant (Narwadkar and Anserwadekar, 1985). The highest sprouting and survival were obtained with 8 cm-long scions treated with IAA at 100 ppm, grafted on 6 to 8 cm high rootstocks and covered with transparent bags (Ratan *et al.*, 1987).

In case of epicotyl grafting of mango, there are three main stages in the formation of a graft union (i) callusing stage : extending from 10 to 30 days after grafting, when a mass of callus tissues is formed, (ii) cambial bridge stage : extending from 30 to 60 days after grafting, when cambial continuity between rootstocks and scions is established, (iii) healed union stage : extending from 60 to 120 days after grafting, when vascular tissues are differentiated and complete union between stocks and scions takes place. Callus formation initially starts from the rootstocks, but later callus proliferates from both components (Chakrabarty and Sadhu, 1985).

Variation in the response of mango cultivars and hybrids to epicotyl grafting has been reported. The grafting success ranged from 66.1 per cent in the hybrid Mallika and 64.6 per cent in the cultivar Dashehari to 34 per cent in the hybrid Amrapali (Madalageri *et al.*, 1989). The mango cvs. Alphonso, Pairi and Totapuri, recorded 40.0 per cent, 53.33 per cent and 53.33 per cent grafting success respectively which would be commercially viable due to convenience and ease of stone grafting compared with the traditional method *i.e.* inarching (Patil *et al.*, 1991). The effect of age of seedlings and scions on graft survival was studied by Dhunaga *et al.* (1989). It was observed that the survival rate was in the order of 4 months > 6 months >2 months for scions, and 5 days > 10 days > 15 days for seedlings. On the other hand, Gunjate (1989) reported that the age of rootstock (1-6 weeks) did not significantly affect success.

Healthy vigorously growing seedlings (15-day-old) were transplanted into 18 x 22 cm polybags or in the nursery at 30 x 30 cm. The scions, taken from 5, 15, 25 or 45 year-old mother trees, were grafted on the second or third new flush on the rootstocks. Scion sprouting and survival were best (72 and 70 per cent respectively) when the scions were taken from 25-year old mother trees. The results also indicated that soft wood grafting can be carried out on any growth flush from the rootstock provided that it is coppery-red in colour (Panickar and Desai, 1989). Highest rate of success (86 per cent) of softwood grafting in mango cv. Amrapali was obtained when grafted in July, closely followed by August, September and April. The July grafting also recorded the lowest sprout, however, there was no significant effect of grafting time on diameter of scion shoots (Singh and Sengupta, 1996). A study was undertaken by Radha and Aravindakshan (1998) to evaluate the response of 14 mango cultivars to the epicotyl grafting method of propagation. An average survival of 62.98 per cent was noticed 6 months after grafting. Variation existed in the response of different cultivars to this method of grafting. The best response was noticed in Kalapady (82.1 per cent survival). Mulgoa exhibited the lowest survival (38 per cent).

8.2.2 Budding

This is the most economical method but has not given uniformly good success in different agroclimatic regions, percentage success varying considerably in various areas. This is best suited to areas where general atmospheric humidity remains high. This is certainly not the method for hot and dry regions. The best time of budding is reported to be early spring although in some parts of the country it can be done during the early part of the rainy season, *i.e.* April to June (Singh, 1960). In southern parts of the country, August-September is an ideal period since the weather is relatively cooler also (Gandhi, 1955). The methods most commonly employed are patch, shield and Forkert budding (Singh, 1960). Prasad and Singh (1972) found that patchbudding was superior to shield budding, and budding in May was better than in April or June with both the methods. Cutting back the stocks above the scions after 10 days was better than immediate cutting. The success rate of T-budding in mango cv. Himsagar was only 42.3 per cent when performed between June and August, and then declined dramatically (Kumar and Mitra, 1994).

8.2.3 Layering

This method of propagation, though commercially applicable under high atmospheric humidity in crops like litchi, is of little utility in propagating mango cultivars. Nevertheless, this does hold promise for developing clonal rootstocks for mango. It is practised during the rainy season and leads to own-rooted plants. This is slightly different from rooting of cuttings and the main difference is that the layering is done for indu-cing roots when the shoot is still attached to the mother plant. The method, as practised in mango, is broadly of three types air-layering, pot layering and stooling.

8.2.3.1 Air-layering

It consists of removing a ring of bark of about 2.5 cm width from around the shoot to be rooted and covering it with a rooting medium. Previously, the ringed portion was covered with moistened soil and farmyard manure mixture. Later, sphagnum moss came to be used instead. After covering the ringed portion with a wrapping material (gunny sack piece or a polythene sheet), one had to wait for the appearance of roots, which usually took 2 to 2½ months. However, this method was not found to be applicable commercially because of poor establishment in the nurseries. After the discovery of plant growth regulators (root promoters like IBA and NAA), the percentage success in rooting increased considerably (Thakurta and Dutt, 1941; Singh and Teaotia, 1951; Singh, 1954). Marked improvement in root formation in air layers of important mango cultivars was reported by Sen *et al.* (1961) by using IBA. Subsequently, rooting of air layers was further improved by etiolation treatment followed by application of IBA and NAA (Mukherjee and Bid, 1965). Treating the air layers of cv. Langra with IBA at 0.0735 M increased rooting. Synergistic effect on rooting was also noted when IBA was applied in combination with NAA, chloro-genic acid, pyrogallol or 6-benzylaminopurine (Rajan and Ram, 1985).

In mango cv. Sensation, 33 per cent of the air-layers (ringed just above the base of the subterminal flush) enclosed in black plastic containing sphagnum peat (only ringed part) had rooted, compared with only 7 per cent of those in clear plastic (Oosthuyse, 1991). In cv. Tommy Atkins, treatment with NAA at 2 per cent alone was found effective in increasing rooting, production of fibrous roots and survival of layers in the nursery (Nunez-Elisea *et al.*, 1992). The rooting co-factor activity in mango air-layers increased at callusing and root emergence stages. Treatment with IBA resulted in an early rise in rooting co-factor and a fall in rooting inhibitor level leading to early rooting of IBA-treated mango air-layers (Rajan and Ram, 1989).

8.2.3.2 Pot-layering

This is somewhat more advantageous than the air-layering in the sense that the transplanting shock is much reduced because the shoots to be rooted are bent and buried in a pot containing soil and farmyard manure mixture. Ringing and IBA application are the same as in the earlier type (Singh, 1960).

8.2.3.3 Stooling

This is an improvement over ground-layering (Singh, 1960) and has been standardized for mango (Majumder and Mukherjee, 1961b; Mukherjee and Majumder, 1963). This method is easier and more economical since larger number of rooted shoots can be obtained (4 to 6) from a single plant (stool). However, before establishing the stool-bed, one has to prepare the mother plants through air-layering or through rooting of cuttings. Afterwards these are planted in nursery bed at a spacing of 75x50 cm. After about 2 years' growth, the main stem of the mother plant (stool) is headed back near the ground (about 10 to 12 cm above). This results in the emergence of several shoots on each plant (stool), from just below the cut end. All, but the most vigorous shoots, are ringed in the usual manner and IBA (5000 ppm) in lanolin is applied on the ringed portion for about a week before earthing up. The treated shoots get rooted in about 4 to 6 weeks and are, then, separated from the mother stool. The same plants can be utilized for about 10 years to obtain genetically uniform rooted plants of a given clone (Majumder *et al.*, 1972). Ringing of shoots and treatment with IBA at 5000 ppm resulted in more production of roots in Goa and Moovandan rootstocks compared with Dashehari and Chausa. Such shoots from Rumani, Chausa and ST-9 showed more rooting capacity followed by Goa, Bappakai, Kalapady, Moovandan, Nakkare and Dashehari (Singh and Srivastava, 1982).

In a trial, on effect of growth regulator on stooling of mango cultivars, Sirohi and Ram (1995) obtained a maximum of 70 per cent rooting and 73 per cent survival with IBA (15,000 ppm) + NAA (2,500 ppm) in Bombay Green, 67 per cent rooting and 67 per cent survival with IBA (20,000 ppm) + NAA (10,000 ppm) in Bombay Yellow, 80 per cent rooting and 70 per cent survival with IBA (1500 ppm) + NAA (5,000 ppm) in Amrapali and 57 per cent rooting and 50 per cent survival with IBA (15,000 ppm) + NAA (10,000 ppm) in Sonaton.

8.2.4 Cutting

This method is of great significance for developing clonal rootstocks which are lacking in mango. However, in general, it is rather difficult-to-root hardwood cuttings. This was tried by some early workers but did not yield any encouraging result. It was first attempted by Thakurta and Dutt (1941) using 2 to 3-year-old juvenile plants with 3 per cent IAA in lanolin. Singh and Teaotia (1951) reported that cuttings from bearing trees do not root even with the help of growth regulators.

Mukherjee *et al.* (1967) achieved a major breakthrough by obtaining 50 per cent success from cuttings prepared from 10-year-old trees. This could be possible by invigoration and etiolation treatments. The cuttings made from the basal branches of the trees gave better success than those prepared from the middle or top portions of the plants (Mukherjee *et al.*, 1966).

Intermittent mist has also been found to be very helpful in rooting of hardwood cuttings. This has given a success of as high as 80 per cent when these were ringed and treated with IBA (Sen *et al.*, 1968). The changes in metabolites, levels of endogenous promoters and inhibitors and ethylene during regeneration of roots in mango cuttings taken from seedlings as well as from mature trees by treatment with IBA were reported by Dhua *et al.* (1982). Another simple technique, devised at the Indian Agricultural Research Institute, New Delhi (Reddy and Majumder, 1975), has facilitated rooting of cuttings through bottom heat. The device employed consists of a double tin container, the outer one being 70 cm in height and 46 cm in diameter and an inner jacket having a height of 68 cm and a diameter of 44 cm. The space between inner jacket and outer container is packed with glass wool for heat insulation. Another container which fits into the inner jacket (also constructed from the same material measuring 35 cm in height) and having two electric bulbs of 40 watt each at the bottom end, in between the vacant space in two containers, is fitted into the double container mentioned above. This container is then filled with the rooting media (1 part sphagnum moss: 1 part sand: 1 part grit) and the non leafy hardwood cuttings are planted in it. The temperature is maintained at about 30°C ± 2°C through a thermostat which controls the heating bulbs. The cutting roots in about a month and the success is as high as 97 per cent. Majumder and Prasad (1985) further reported 85.8 per cent rooting was at 35°C basal temperature when combined with wounding and 5000 ppm IBA treatment. When applied through sides, the low concentration of IBA (2500 ppm) increased rooting significantly as compared with higher concentration (5000 ppm). Rooting was the highest (95.6 per cent) when lower concentration of IBA was applied through base and side. The 25°-30 °C bottom-heat temperature was found to be most effective in increasing the percentage of rooting and length of the longest root (Reddy and Singh, 1987; Reuveni and Castoriano, 1993), while rooting was slow at 20°C and poor at 35°C (Reuveni and Castoriano, 1993). Treatment with IBA at 5000 ppm + wounding significantly increased the rooting percentage, number of roots and length of the longest root in cv. Dashehari compared with IBA at 5000 ppm with no wounding (Reddy and Singh, 1987). The juvenile hardwood cuttings taken from 3- and 6-year-old trees gave a significantly higher percentage of rooting, number of roots and length of longest root than cuttings taken from 13-year-old trees (Reddy and Singh, 1987a).

At optimum temperature, the rooting of cultivars was in the order of Turpentine > Gomera > Sabre > 13-1 (Reuveni and Castoriano, 1993).

Pre-treatment of stock plants with cycocel caused marked reduction in the extension of shoot growth resulting in better rooting of cuttings. Rooting was further promoted when IBA was applied to the base of the pre-treated cuttings (Bose and Sadhu, 1985). Pre-treatment of stock plants (10-year-old, cv. Langra) with cycocel (1000 mg/l), CEPA (250 mg/l) and cycocel (1000 mg/l) + CEPA (250 mg/l) improved rooting compared with control, whereas cuttings taken from GA_3 (2000 mg/l) treated plants showed poor rooting. Rooting was further improved when the cuttings were dipped in IBA at 10-2 M and NAA at 2x10-2 M solutions (Sadhu, 1992). Rooting was positively correlated with glucose, fructose and starch concentrations in leaves and stems of cuttings, but appeared to be unaffected by N content and C:N ratio of the stems (Schaesberg and Ludders, 1993). According to Schaesberg *et al.* (1993), under natural light conditions in the mist bed (photon flux density = 100-200 μmol $m^{-2}s^{-1}$), net CO_2 assimilation rates and transpiration rates of both Turpentine (easy-to-root) and 13-1 (difficult-to-root) cuttings dropped after excision from the stock plant. Callus formation and subsequent rooting were accompanied by enhanced CO_2 uptake by the leaves. This was not observed for cuttings that failed to root. Differences in photosynthetic activity of rooted and unrooted cuttings were more pronounced for 13-1 than for Turpentine.

8.2.5 Tissue Culture

Yang and Ludders (1993) standardized the technique of *in vitro* shoot tip culture of different cuttings of mango. Shoot growth was better in G medium (Yang *et al.*, 1984) supplemented with the combination of IBA (0.5 mg/l), BA, zeatin and IAA (each at 1.0 mg/l), 2ip (2 mg/l), casein hydrolysate and glutamine (300 mg/l) and 3 per cent sucrose, than in B-5 basal medium or woody plant medium. Shoot tip initiation and elongation were greater in cv. Turpentine than in the cultivars 13-1 and Gomera and explants collected during May-June produced maximum shoots (Yang and Ludders, 1993). The phenolic exudation from the cut surface of mango explants prevents their successful regeneration *in vitro*. Pretreatment of explants using liquid shaker culture (i.e. in liquid MS medium supplemented with 1 per cent polyvinyl pyrolidone in 250 ml conical flasks on an automated shaker at 75 r.p.m) helped to overcome this problem (Raghuvanshi and Srivastava, 1995).

Litz *et al.* (1982) obtained somatic embryos from nucellar explants of polyembryonic mango cultivars Chino, Heart, Ono and Turpentine on modified MS medium containing 400 mg/l glutamine, 100 mg/l ascorbic acid, 60 g/l sucrose and either 20 per cent coconut water or BA at 4.4 and 8.8 μM. Later it was demonstrated that more efficient recovery of somatic embryos occurred when the nucelli of polyembryonic mangoes were excised and cultured on medium with 1-2 mg/l 2,4-D (Litz *et al.*, 1984). Calli were also induced from nucellar explants excised from fertilized ovules of immature (4.0-4.8 cm long) monoembryonic fruitlets of mango cultivars (Irwin, Ruby and Tommy Atkins). Somatic embryogenesis occurred from callus on modified MS medium and following subculture to a medium lacking growth regulators. The highest percentage of both callus formation from explant cultures and embryoid formation from callus cultures were recorded for Irwin followed by Ruby and Tommy Atkins. Embryogenic cultures of monoembryonic cvs. Lippens (Litz, 1987) and Keitt (Litz and Schaffer, 1987) were subsequently induced. High frequency somatic embryogenesis from explants was achieved in the monoembryonic mango cv. Arka Anmol (Thomas, 1999). For culture initiation, immature fruits of 2-3 cm long, about 30-40 days after pollination were ideal. Culturing half ovules with intact nucellus gave better percentage of callusing and more embryonic cultures than using excised nucellus. Among the basal media, Rugini Olive (RO) medium followed by B5 was superior to the MS medium for culture establi-shment and embryo induction. However, B5 medium was better than the others for embryo conversion and further development. Proliferating embryonic cultures could be maintained in this conversion medium giving a continuous supply of embryos. Embryo maturation was obtained in the presence of 1 mg/l abscisic acid. Germination with shoot and root development was obtained in the presence of 0.01-0.1 mg/l TDZ.

8.2.6 Rootstock

The cultivars of mango, as indicated earlier in this chapter, fall into two broad categories (i) monoembryonic and (ii) polyembryonic. Almost all the commercial cultivars in the world belong to the former category and do not come true-to-type from seed and are, therefore, of necessity to be raised through asexual means. The polyembryonic types are reported to come out true-to-type from seed since the zygotic embryo is crowded out by the surrounding nucellar embryos and gets degenerated, giving rise to nucellar seedlings only (Sachar and Chopra, 1957). However, in the cultivar Pico, both nucellar and zygotic embryos develop almost at the same speed (Juliano, 1937) and in such a case the emerging seedlings may be of both types. The nucellar seedlings amongst these can be identified by uniformity in colour of emerging leaves. Polyembryonic types are, by and large, suited only for developing genetically uniform clonal rootstocks which are presently lacking in mango. For rootstock purposes, the seeds must be sown within a week of extraction from the ripe fruits as described in seed propagation section. The rootstock can also be raised in polybags of 15 x 37.5 cm in size (Nema *et al.*, 1985).

In India and Mexico, monoembryonic seedlings are generally used as rootstocks. Polyembryonic 'Turpentine' seedlings are used as rootstocks in Florida. Either polyembryonic 'Saber' or '13-1' seedlings are used as rootstocks in Israel. In Australia, 'Kensington' seedlings are used as rootstock. Throughout South-East Asia polyembryonic seedlings are used for rootstocks (Reddy and Raj, 2015).

8.2.6.1 Stock-scion Relationship

Since rootstocks for different cultivars have not yet been standardized, it is not possible to indicate the precise relationship. However, in mango, all plants of particular cultivar are known to have the same characteristic canopy shape, despite the rootstocks being of seedling origin and hence highly variable. For instance, trees of Dashehari, Langra or Chausa retain their characteristic canopy shape and can be identified from a distance. Nevertheless, it is quite likely that, as in some other tree fruits, stocks may be having some influence on other characters like vigour, bearing capacity, fruit size and quality. A good beginning was made on

Methods of propagation of mango

Propagation by layering

Propagation by stone grafting

Propagation by stone grafting

Methods of propagation of mango

Grafting knife with detached scions

Propagation by modified cleft grafting - different steps

Methods of propagation of mango

Propagation by inarching

Propagation by chip budding

Micropropagation of Mango

Source: Perez-Hernandez, J.B. and Grajal-Martin, M.J. (2011) *HortScience*, 45, 11.

Irrigation and High Density Planting of Mango

Irrigation by drip method

Irrigation by ring method

High density planting of mango

the standardization of rootstocks at the IARI to achieve a major breakthrough by working out pre-selection criteria for dwarfness so that in future it might become easier to go in for high density orcharding in mango. The criteria which have shown good correlation with dwarfness are – (i) high bark percentage, (ii) small area of xylem vessels and (iii) lower stomatal density (Majumder *et al.*, 1972, 1981). The cultivars Totapuri Red Small and Olour appear to be promi-sing for developing dwarfing clonal rootstock. Studies conducted elsewhere indicate that cultivars Kalapady (Sen, 1939), Olour and Ambalavi (Jauhari *et al.*, 1972), Vellaikollumban (Singh and Singh, 1976), and Belkhas and Parikhas (Mukherjee and Das, 1976) too have potential for imparting dwarfness. Olour has been proved to be a dwarfing and productive rootstock for Himsagar, Langra (Samaddar and Chakrabarti, 1985) and Alphonso (Kohli and Reddy, 1985).

The polyembryonic rootstocks 'Pahutan' and 'Goa' were the best rootstocks for 'Neelum' and 'Olour' and 'Pahutan', the best for 'Baneshan'. 'Eldon' is the best rootstock for 'Parvin' while 'Malda' and 'Eldon' are equally good for 'Tommy Atkins' compared to the control 'Colombo Kidney'. 'Brodie' and 'MYP' were clear examples of rootstocks which impart to 'Kensington Pride' high yield efficiency with smaller size but also others than despite giving a large tree size can give high early yield allowing them to be used for high density plantings (Ex. 'B' and 'Watertank'). The use of the rootstock 'Piva' as interstock has proven very efficient in different countries to reduce tree size. The rootstocks Turpentine' and 'Number 11,' used in Florida, are tolerant of high soil pH while 'Hilacha', is the rootstock more used in Colombia. '13/1' from Israel, 'Gomera 1' ('G1') from the Canary Islands, 'Olour' from India and 'Sukkary' from Egypt are effectively polyembryonic salt tolerant rootstocks. (Sauco, 2016)

A 10-year-study with 3 scion cultivars and 4 rootstocks significantly reduced scion diameter in all cultivars but its effect on the average number of fruits per tree was inconsistent. Eldon was the most effective in reducing canopy volume for Palmer and Irwin while the maximum reduction in canopy volume for Edward was obtained on Julie (Cedeno-Maldonado *et al.*, 1988). On the basis of the girth size, Hussain *et al.* (1989) classified the rootstocks into three grades: Grade I (0.41-0.95 cm), Grade II (0.96-1.20 cm) and Grade III (1.21-1.88 cm). It was observed that grafting success increased with increasing girth of rootstocks. They also noticed that grafting success was greatest with Anwar Rataul for Samar Bahist scions and with Malda for Langra scions. In the *in situ* trial, grafting of cv. Alphonso on Nekkare (polyembryonic) rootstocks gave the highest success with 7-month-old rootstocks, and the lowest with 3-month-old (Reddy and Melanta, 1988). Cumare and Avilan (1994a, b, c) described the characteristics of nine mango cultivars to be used as rootstocks. The cultivars Divine, Tolbert, Tetenene Manzana and Julie had low vigour in vegetative growth and a poor yield (Cumare and Avilan, 1994b). The majority of cultivars reached 100 per cent flowering during the third or fourth week in January with the fruit setting period from the third week in December until the fourth week in June (Cumare and Avilan,1994c).

Promising results have emanated from the use of interstocks in regulating the tree height. According to Perez *et al.* (1988) the cv. Edward produced significantly more fruits than Palmer in the third year, especially with the Eldon interstock, however, in the 6th and 9th years Palmer out-yielded Edward two-fold, especially with the Irwin interstock (in both cases the cv. Eldon was used as rootstocks). Srivastava *et al.* (1989) opined that there was no significant interaction between rootstock and interstock on growth parameters and the polyembryonic Chandrakaran gave more vigorous trees than the monoembryonic ST-9. The combinations of cvs. Manila/Irwin/Irwin and Manila/Esmeralda/Manila scion/interstock/rootstock) were found to be the most compact with relative growth rates of 291.2 and 204.3 mg/g/day respectively, and unit leaf growth rates of 2921.7 and 1968.7 mg/cm^2/ day, respectively (Garcia-Perez *et al.*, 1993). At 9 years, the combinations Manila/ Thomas/Esmeralda and Manila/Irwin/Irwin (scion/interstock/rootstock) reduced both tree height and canopy diameter by 51 per cent and 40 per cent respectively, compared with ungrafted Manila seedlings (Avila-Resendiz *et al.*, 1993).

Experiment with Dashehari scions grafted onto several polyembryonic and monoembryonic rootstocks indicated that rootstocks had some influence on mineral composition of mango trees. The highest levels were obtained with Rumani (N), ST-9 (P and K), Nekkare (Ca, Mg and Mn), Ambalavi (Fe) and Dashehari (Zn and Cu) rootstocks (Thakur *et al.*, 1989). The mango rootstock 13-1 showed relatively high salt tolerance under field conditions (Schmutz and Ludders, 1993).

9.0 Area and Production

Mango (*Mangifera indica*) is the most important fruit of India and is known as "King of fruits". The fruit is cultivated in the largest area *i.e.* 2,312 thousand ha and the production is around 15.03 million tons, contributing 40.48 per cent of the total world production of mango. India rank second among world fruit producing countries despite its lower productivity (11.6) (Table 5). Fruit productivity of India is about half of the productivity of USA (23.3) and Indonesia (22.3), but slightly better than the world average (11.3) and equals that of China (11.6) during 2012-13.

Table 5: Major Fruit Producing Countries (2012-13)*

Country	*Area (ha)*	*Production (MT)*	*Productivity*
China	11834450	137066750	11.6
India	6982015	81285334	11.6
Brazil	2325385	38368678	16.5
USA	1137779	26548859	23.3
Indonesia	796530	17744411	22.3
Philippines	1240370	16370976	13.2
Mexico	1256730	15917806	12.7
Turkey	1102662	14974561	13.6
Spain	1539100	13996447	9.1
Italy	1125593	13889219	12.3
Others	27924521	270594597	9.7
Total	57265135	646757638	11.3

*Anonymous (2018a).

Table 6: Top Ten Mango Producing Countries of the World in 2016*

Country	*Production (Tonnes)*	*Production Share (per cent)*
India	18779000	40.28
China	4771038	10.23
Thailand	3432129	7.36
Mexico	2197313	4.71
Indonesia	2184399	4.68
Pakistan	1606091	3.44
Brazil	1417149	3.04
Egypt	1277008	2.73
Bangladesh	1161685	2.49
Nigeria	917617	1.96
World total	46615462	-

*Anonymous (2016).

Table 7: All India Area, Production and Productivity of Mango in India*

Year	*Area ('000 ha)*	*Production ('000 tonnes)*	*Productivity (t/ha)*
1991-92	1077.6	8715.6	8.1
2001-02	1575.8	10020.2	6.4
2002-03	1623.4	12733.2	7.8
2003-04	1906.7	11490	6
2004-05	1970.4	11829.7	6
2005-06	2080.7	12663.1	6.1
2006-07	2154	13734	6.4
2007-08	2201	13997	6.4
2008-09	2309	12750	5.5
2009-10	2312.3	15026.7	6.5
2010-11	2297	15188	6.6
2011-12	2378.1	16196.4	6.8
2012-13	2500	18002.4	7.2
2013-14	2516	18431.3	7.3
2014-15	2163.5	18527	8.5
2015-16	2208.6	18642.5	8.4
2016-17	2212.2	19506.2	8.8
2017-18	2258.1	21822.3	9.7

*Anonymous (2018b).

The area and production of mango has been almost continuously increasing over the years (Table 2). The area under mango cultivation was 1077.6 thousand ha during 1991-92 which reaches up to 2516 thousand ha in 2013-14 and recorded 2262.8 thousand ha in 2016-17. However, the production has been fluctuating drastically. There was overall increase of 7.41 per cent in productivity of mango from 1991-92 to 2014-15.

The main mango producing states in India are Uttar Pradesh (23.86 per cent), Andhra Pradesh (22.14 per cent), Karnataka (11.71 per cent), Bihar (8.79 per cent), Gujarat (6.00 per cent) and Tamil Nadu (5.09 per cent). Andhra Pradesh was the leading state in area by accounting for 14.72 per cent of total mango area under country in 2016-17 but in 2016-17, total mango production from Uttar Pradesh was 4540.23 thousand MT; while maximum area was reported in Andhra Pradesh (332.97 thousand ha).The productivity of mango was also recorded highest in Uttar Pradesh (17.1 MT/ha) during 2016-17, which is also much higher than the national productivity (8.7 MT/ha) of mango. Andhra Pradesh, Telangana, Bihar, Maharashtra, Gujarat, Tamil Nadu, Odisha, Kerla and West Bengal were the other important states cultivating mango significantly and commercially.

Table 8: State-wise Area ('000 ha), Production ('000 MT) and Productivity (MT/ha) of Mango (Anonymous, 2018b)

States	*2015-16*			*2016-17*			*2017-18*		
	Area ('000 ha)	*Production ('000 tonnes)*	*Productivity (t/ha)*	*Area ('000 ha)*	*Production ('000 tonnes)*	*Productivity (t/ha)*	*Area ('000 ha)*	*Production ('000 tonnes)*	*Productivity (t/ha)*
Uttar Pradesh	263.28	4512.71	17.14	264.94	4341.00	16.38	265.62	4551.83	17.14
Andhra Pradesh	327.31	2803.66	8.57	336.96	4043.47	12.00	363.00	4373.61	12.05
Karnataka	181.7	1725.67	9.50	180.60	1719.73	9.52	183.23	1760.60	9.61
Telangana	194.05	1778.32	9.16	111.65	482.46	4.32	115.99	1080.14	9.31
Bihar	149.14	1464.93	9.82	149.96	1472.38	9.82	149.28	2443.47	16.37
Gujarat	153.18	1241.59	8.11	161.27	1424.87	8.84	162.77	1207.78	7.42
Tamil Nadu	125.98	975.11	7.74	176.42	1282.44	7.27	152.57	1234.00	8.09
West Bengal	96.74	693.39	7.17	99.22	736.90	7.43	103.25	918.35	8.89
Odisha	199.29	778.72	3.91	199.42	817.91	4.10	199.08	805.77	4.05
Maharashtra	162.08	463.17	2.86	157.07	514.87	3.85	166.76	791.36	4.75
Madhya Pradesh	27.89	371.48	13.32	156.84	603.83	13.50	45.52	654.79	14.38
Chhatisgarh	71.52	420.61	5.88	74.17	434.32	5.86	77.03	461.73	5.99

States	2015-16			2016-17			2017-18		
	Area ('000 ha)	Production ('000 tonnes)	Productivity (t/ha)	Area ('000 ha)	Production ('000 tonnes)	Productivity (t/ha)	Area ('000 ha)	Production ('000 tonnes)	Productivity (t/ha)
Jharkhand	50.41	393.67	7.81	50.56	438.54	8.67	54.53	435.86	7.99
Kerala	70.12	382.52	5.46	70.41	388.14	5.51	83.12	439.20	5.28
Uttarakhand	35.91	149.73	4.17	36.42	150.14	4.12	36.48	152.71	4.19
Punjab	6.74	113.5	16.83	6.75	113.69	16.85	6.90	116.52	16.90
Haryana	9.26	89.97	9.72	9.34	96.79	10.37	9.35	98.60	10.54
Rajasthan	5	82.27	16.45	5.16	154.79	29.97	4.97	87.37	17.58
Assam	4.62	46.15	9.99	4.66	47.15	10.12	4.68	48.44	10.34
Tripura	11.75	59.06	5.02	10.64	57.03	5.36	10.33	54.93	5.32
Himachal Pradesh	41.52	37.63	0.91	41.77	48.24	1.16	41.99	31.35	0.75
Jammu & Kashmir	12.67	23.74	1.87	12.74	24.15	1.90	12.96	30.35	2.34
Nagaland	0.57	3.74	6.51	0.64	4.23	6.60	0.64	4.24	6.61
Mizoram	0.87	4.18	4.80	0.89	4.18	4.70	0.91	4.19	4.60
Arunachal Pradesh	0.05	0.03	0.50	-	-	-	-	-	-
Others	6.91	27	3.91	7.40	33.58	4.54	7.17	35.14	4.90
Total	2208.56	18642.53	8.44	2212.24	19506.20	8.82	2258.13	21822.32	9.66

India exports mango to over 40 countries worldwide. The major importing countries of India's Mangoes during the period of 2010-11 were UAE (61.79 per cent), Bangladesh (11.41 per cent), UK (8.92 per cent), Saudi Arabia (3.79 per cent), Kuwait (2.32 per cent), and Bahrain (2.19 per cent) respectively.

Mango production in the country is estimated to lower by 4.36 per cent at 20.44 million tonnes in the crop year 2019-20 ending this month, the latest data from the agriculture ministry. The output of mango, the king of fruits, stood at 21.37 million tonnes during the 2018-19 crop years (July-June).

10.0 Soil and Climate

10.1.0 Soil

The mango can grow well in all types of soil from alluvial to lateritic, except black cotton soils which are considered to be poor. The only pre-requisite is a deep (2 to 2.5 m) and well-drained soil. However, it grows successfully in soft rocky areas of the west coast. The most suitable pH ranges between 5.5 and 7.0 (Whiley, 1984), but it can be grown commercially up to 8.5 with proper nutritional management (Crane and Campbell, 1991). The tree can tolerate drought (Sukonthasing *et al.*, 1991) and short periods of flooding as well (Schaffer *et al.*, 1992). Soils with an appreciable amount of gravel or *kankar* ($CaCO_3$) in the profile too can grow good mangoes provided they are not very alkaline.

For mango, the water table should always be at a depth of 1.80 to 2.40 m.Best quality fruits are produced on soils containing 5 to 10 per cent lime and sufficient quantities of peroxide of iron. Mango is rated as moderately tolerant to salts with 4-6 dsm^{-1}.

Mango is sensitive to saline conditions, producing symptoms like scorching of leaf tips and margins, leaf curling and in severe cases reduced growth, causes abscission of leaves and death of trees (Jindal *et al.*, 1976; Kadman *et al.*, 1976; Gazit and Kadman, 1980). The salt tolerance varies with genotype. The mango rootstocks 13-1 and Turpentine have been found to be salt-tolerant and salt-susceptible respectively (Schmutz and Ludders, 1994). The cv. Dashehari, which had the lowest leaf K: Na ratio was most sensitive to Na scorching (Samra, 1989). Increasing Na^+ concentration (as NaCl) resulted in reduction in vegetative growth and saline conditions significantly reduced transpiration rate (Schmutz and Ludders, 1993). Root respiration of Turpentine was reduced as salinity increased, whereas, root respiration of 13-1 showed little change. Net CO_2 assimilation was reduced in both the cultivars as salinity was increased; thus carbon gain was lower in Turpentine than in 13-1 at high salinity (Schmutz and Ludders, 1994). Schmutz and Ludders (1993) concluded that the intermediate salt tolerance of 13-1 was due to exclusion of Na+ accumulated in the vacuoles of root cells. Inoculation with vesicular arbuscular mycorrhiza (*Gigaspora margarita*) improved the growth and nutrition of mango rootstock cv. Nekkare resulting in greater plant height, stem diameter, shoot biomass and shoot P content (Reddy and Bagyaraj, 1994).

10.2.0 Climate

Although, essentially a tropical fruit, the mango can grow from sea level to an altitude of about 1400 metres provided there is no high humidity, rain or frost during the flowering period. It does well within a temperature ranging from 24-30°C (Corbineau *et al.*, 1986; Whiley *et al.*, 1989), although it can successfully endure even temperatures as high as 48°C during the period of fruit development and maturity (if facilities for irrigation at regular intervals during this period are available). Cold temperatures limit crop production. Exposure to a temperature at or below 0°C for a few hours may severely damage or kill the young trees (Campbell *et al.*, 1977; Young and Sauls, 1981). Unprotected mango trees in Florida have been found to withstand -4°C for a few hours with damage limited to leaves and small branches (Young and Sauls, 1981). Juvenile trees were more susceptible to low temperatures than large ones and were killed after 13 hours at 4° to –6°C (Campbell *et al.*, 1977). Seed germination was rapid between 25-40°C and seedling growth during the first fortnight after germination was optimal at 30°C (Corbineau *et al.*, 1986). The most favourable temperature for the growth of young mango plant was 25°C (Parisot, 1988). From their experiment with container grown trees in controlled temperature,

Whiley *et al.* (1989) found that as temperature increased the mean number of vegetative flushes during a 20 week period increased from 1.02 at 20°/15°C (day/night) to 3.21 at 30°/25°C. Temperature also had an influence on the number of leaves developed on each growth flushes and higher day/night temperature (30°/25°C) resulted in greater number of leaf production per shoot than at 20°/15°C (Whiley *et al.*, 1989). Higher temperatures during fruit development hasten maturity and improve fruit size and quality. However, prolonged exposure of fruits at near maturity to the summer sun when temperatures exceed 35°C may cause sunburning (Schaffer *et al.*, 1994).

The plant grows luxuriantly and yields well from sea level to 1000 m above mean sea level. However, for commercial cultivation of mango crop 600 m altitude is ideal.As the altitude of the place increases over and above 1000 m from MSL, the growth and productivity of the crop are poor. The altitude has a definite role on the time of mango flowering. It has been observed that an increase in every 12 m altitude, flowering is retarded by four days.Similarly for each degree latitude; south or north of the tropics, flowering is delayed by four days.Potted mango cv. Nam Dok Mai trees were maintained at air temperatures of 15°/10° or 30°/20°C (day/night) and 80 per cent RH with soil temperature at 25°C. The greatest decrease in leaf water potential, from -1.0 to -1.5 Mpa, occurred between 06.00 and 08.00 h for both temperature regimes and recovery occurred between 12.00 and 16.00 h. With the 15°/10°C regime, stomatal conductance declined from 1.2 cm/s at 06.00 h to 0.6 cm/s between 12.00 and 16.00 h but recovered to the dawn rates by 18.00 h. At 30°/20°C, the stomatal conductance was lowest at 06.00 h (0.5 cm/s) rising to 2.4 cm/s at 14.00 h and then declining to 1.1 cm/s at 18.00 h. Assimilation of CO_2 increased between 08.00 and 10.00 h for both temperature regimes and declined at 16.00 h, coinciding with a decrease in the photosynthetic photon flux density (Pongsomboon *et al.*, 1992).

Poor mango yields in tropical regions are often attributed to the failure of flower induction, whereas, in subtropical region despite satisfactory blooming, the fruit set is poor. There is a general agreement that day/night temperatures below 20°/15°C trigger flower induction. A strong genotype x environment interaction within various cultivar groups was also reported by Whiley (1993). Mango (cv. Tommy Atkins) trees subjected to cold temperatures (18°/10°C) produced 74 per cent floral, 11 per cent mixed and 15 per cent vegetative buds (Nunez-Elisea *et al.*, 1993).

Low temperatures promoted floral induction of mango, whereas water stress imposed on plants during cool autumn months (night temperature below 15°C) did not increase the proportions of apical buds forming inflorescence (Nunez-Elisea and Davenport, 1994). The proportion of perfect flowers differed little between plants grown in a greenhouse at 20° and 30°C, while water stress slightly lowered the proportion of perfect flowers (Utsunomiya *et al.*, 1995). Temperature and precipitation during the 30 days after anthesis were more important than those during the 10 days before flowering in respect of embryo abortion (Tsai *et al.*, 1996). Cool temperatures (18°/10°C) rather than a short photoperiod caused floral induction, whereas warm temperatures (18°/22°C) rather than a long photoperiod inhibited flowering (Nunez-Elisea and Davenport, 1995). Chen and Zhao (1999) considered temperatures to be a key environmental factor, with low temperatures (19°C/day and 13°C/night) favourable for flower bud differentiation. Singh and Bhargava (2011) has suggested that mango cultivars 'Kesar', 'Amrapali' and 'Rajapuri' (having thicker leaves, high chlorophyll contents, and low *E*) are better performing genotypes under the hot semi-arid environment of Bikaner during the initial year of establishment. Microclimatic factors also exert considerable influences on mango production (Hsu *et al.*, 1995; Chang *et al.*, 1996). It is suggested that rainfall and rainy periods during April and May in Taiwan are key factors affecting mango production (Chang *et al.*, 1996). Increasing the light intensity from 4000 to 30000 lux (at 20 °C) increased the leaf surface area and number of leaves of each morphogenetic unit (Parisot, 1988). Leaf chlorophyll and N contents (DW basis) increased as percentage shading increased, but water use efficiency and specific leaf density were not significantly affected by shading. Light response curves for net CO_2 assimilation had a steeper initial slope for leaves that developed in full sun and 25 per cent shade, than for leaves that developed in 50 and 75 per cent shade. The results indicate greater quantum-use efficiency for leaves developed in high light levels than for those developed in shade (Schaffer and Gaye, 1989).

The annual mean temperature at which mango thrives best is around 26.7 °C. The optimum growth temperature for mango is 23.9° - 26.7 °C. Air temperature of over and above 40.5° C develops spongy tissue in Alphonso mangoes.

The highest yields in mango cv. Haden were obtained under the following conditions: wind speed in December and February being 0.43 and 0.98 km/h respectively; rainfall in October, November and January being 40.10, 15.5 and 3.25 mm respectively (Sergent *et al.*, 1993). Excessive wind has a negative impact on mango production (Schaffer *et al.*, 1994). The amount of rainfall in a given locality is not so important as its intensity and distribution. It can do well in areas having an average rainfall of as low as 25 cm (if irrigation can be provided during the peak requirement period of fruit development) to as high as 250 cm. However, the mango cannot do well in areas which experience frequent rains or high humidity (say, over 80 per cent) during the flowering period. Such conditions are not conducive to a good fruit set and also increase the incidence of serious pests and diseases like powdery mildew and anthracnose. The former is quite common when high humidity is accompanied by low temperatures. These diseases can render the trees completely unproductive, if appropriate and timely measures are not taken to protect the crop. Localities which receive bright sunny days and a relatively low humidity during flowering period are ideal for mango cultivation.

The front injury is exhibited through bark splitting, but this is not visible in the young plants.Oozing of gum from stem bark, death of new shoots, charred appearance of leaves and burning appearance of developing fruit buds are the other symptoms of frost injury. If the temperature is below 1.1° C, the mango plants are adversely affected by frost. A short spell of -3.3° C and consequent longdrawn out cold spell led to the drying out of the young shoots and leaves of mango plants, killing the tree from the top down to a point where the bark was thick and the sap moved very slowly.The young mango trees in vigorous growth may be injured seriously at 0°C. The minimum temperature of -0.6° to 0° C for 1 hr 15 min for 2 consecutive days resulted in appreciable damage to mango trees.

11.0 Cultivation

Once the climatic and soil requirements of mango have been taken into consideration, the next step is to lay out the orchard. The land may be prepared by usual ploughing, harrowing and levelling. There should be a gentle slope to facilitate proper irrigation and prompt drainage to avoid the harmful effects of water stagnation during rains.

11.1.0 System of Planting

The various systems in vogue are the (i) square, (ii) rectangular, (iii) quincunx, (iv) hexagonal and (v) contour (Singh, 1960). Of these, the square system is the most popular in mango. The planting distance will vary with the vigour of the cultivar and the location, ranging between 10 and 12 metres.

The worldwide mango productivity, in general, is poor as compared to other majorfruit crops which are partly due to wider tree spacing. Yields from more closely spaced trees of well managed orchards in Florida and Puerto Rico, where annual or biennial hedging is employed to maintain canopy size, generally exceed 27 t/ha and in recent trials, yields have been increased radically by increasing planting density (Oosthuyse, 1993a). Iyer and Kurian (1992) have emphasized on three approaches in order to achieve more consistent tree growth control : (i) breeding dwarfing rootstocks, (ii) chemical control and (iii) developing new polyembryonic rootstocks. The cv. Amrapali could be planted at a spacing of 2.5 x 2.5 m with a density of 1600 plants per hectare (Majumder and Sharma, 1985). Ram and Sirohi (1985) suggested planting distance of 3.0 x 2.5 m for Dashehari which gave 24.3 to 94.3 quintals yield/ha during 15 to 18 years compared with 1.9 to 3.2 quintals/ha under the normal planting density (12 x 12 m). The cv. Neelum was recommended to the growers in Orissa for high density planting programmes (Ray *et al.*, 1995). In comparison between low (12 x 12 m) and high (10 x 2.5 m) density planting of veneer grafted Dashehari mango, Ram and Sirohi (1989) reported that although there were differences in mean fruit weight, percentage of trees fruiting and fruit yield per tree between the two different plant densities, the differences were inconsistent in different years. Application of paclobutrazol has been reported to reduce the vegetative growth, induce flowering and increase the yield in mango (Kulkarni, 1988; Winston, 1992). An application (soil) rate of 10 g a.i./tree gave the desired effects for at least 2 years in 5-year-old bearing trees whereas in 1- and 2-year-old trees, 1.25 and 2.5 g a.i./trees respectively were safe and effective. Foliar application, on the other hand, failed to exert any influence on shoot growth and flowering (Kulkarni, 1988). Collar treatments of paclobutrazol were significantly more effective than drip line soil drench treatment in reducing the vegetative growth and a dose > 4 ml/tree caused unacceptable compaction of flower panicles. The increase in yield was due to increased fruit number rather than size (Winston, 1992). Foliar application of paclobutrazol (2000 mg/l) in September has been recommended for control of tree size in mango cv. Dashehari (Khader, 1991). In field trials with cv. Alphonso, soil drenching with paclobutrazol at 2.5 g/tree during November and

March in alternate years reduced tree vigour by about 50 per cent compared with control (Iyer and Kurian, 1992). Soil application of paclobutrazol has been found to be most effective in reducing shoot growth rate, internode length, lateral shift growth and increasing flowering. It also increased the leaf N, Ca, Mn, Zn and B contents and decreased P, K and Cu contents (Werner, 1993). Soil application of paclobutrazol (2.5 to 5.0 g/tree) and foliar application of cycocel (8000 mg/l) and alar (3000 mg/l) reduced the tree volume but the effect of alar and cycocel persisted only for one season (Kurian and Iyer, 1993). A high phenol content of the apical bud was associated with the reduction in tree growth caused by paclobutrazol and the phenol content increased with the increasing paclobutrazol concentration (Kurian *et al.*, 1994). Trees treated with paclobutrazol had lower activities of zeatin riboside, zeatin and cytokinin glucoside than control trees (Kurian *et al.*, 1992).

Field experiments conducted at TNAU, Coimbatore, to study the different systems of planting in mango cv. Kalepad revealed that the double hedge row system of planting with a spacing of 10 m between double hedges, 5 m within double hedge and 5 m between plants in double hedge (200 plants ha^{-1}) is the best planting system for obtaining maximum number of fruits and the highest yield per unit area (Jasmine *et al.*, 2019).

11.1.1 High Density Orcharding

High density planting (HDP) (5 x 5 m) or ultra-high density planting (UHDP) system (3 x 2 m) aims to make maximum use of land to achieve high yields in the early periods of orchard along with ease in its management. This system warrants adoption of certain important technologies like formative pruning in the initial years so as to have desirable plant architecture, proper canopy management annually to encourage vegetative growth immediate after harvest, stopping of the vegetative growth during September to favour fruit bud initiation and differentiation (Kumar, 2019).

Recently, experiment conducted by TNAU in collaboration with Jain Irrigation Systems Ltd (JISL), Udumalpet revealed that we can go for still closer spacing of 3 x 2 m called Ultra High Density Planting System (UHDP). The varieties that can be grown successfully under UHDP in different states are shown in Table 2. As the Cultivation Practices are very intensive, high value varieties are recommended. Under UHDP, Mango is planted at 3 m x 2 m which accommodates 674 plants $acre^{-1}$. Kumar (2019) has classified various mango cultvars suitable for HDP.

State	*Varieties*
Andhra Pradesh	Alphonso, Alampur Baneshan, Banganapalli, Totapuri (Bangalora), Mallika
Bihar	Bombai, Himsagar, Langra, Chausa
Goa	Fernandin, Mankurad
Gujarat	Alphonso, Kesar
Karnataka	Alphonso, Bangalora, Neelum, Mallika
Madhya Pradesh	Alphonso, Kesar, Ratna
Tamil Nadu	Alphonso, Banganapalli, Imampasand, Totapuri (Bangalora)
Uttar Pradesh	Bombay Green, Dashehari, Langra, Lucknow Safeda, Mallika, Chausa

ICAR-Central Institute for Subtropical Horticulture, Lucknow, has developed the medium density planting of mango for improving productivity and high economic returns per unit area. This system permits maximum utilization of land, water, nutrients and solar energy. Conventionally, mango is planted at 10 x 10m (100 trees hectare^{-1}) but under medium density planting (MDP) we can accommodate 400 trees per hectare (5 x 5m) and orchardists have to follow canopy management involving training and pruning, are essential aspects right from orchard establishment and continue through out the life of orchard using integration of fertigation technology. The productivity is as high as 14-15 tons could be achieved as against 6-7 tonnes under traditional planting system and fruit drop were reduced to 4.2 per cent.

Benefits of UHDP and associated technologies increase productivity upto 2-3 times, 10 year old UHDP yielded upto 18-20 MT ha^{-1}. It reduces orchard gestation period to three years as compared to conventional orchards and makes orchard crops as profitable as other cash crops (Kumar, 2019).

11.2.0 Preparation of Pits, Planting and Protection of Plants

After marking the places for the plants, pits of 90 x 90 x 90 cm are usually dug out during summer months. This process should preferably be executed, utilising a 'planting board' so that the precise location of the plants in the middle of the pit remains undisturbed. Digging of pits is very essential for somewhat heavy type of soil or soils with a shallow hard pan. While digging, it is necessary to keep the top soil and the subsoil separately in two heaps near each pit for about 2 to 4 weeks. This helps in exposing harmful soil organisms to weathering agencies, providing better aeration in the future rooting zone and in making provision for the nutritional requirements for the healthy development of the plants. However, exposing soils (for long in summer) which have poor organic matter contents is not advisable since it may result in its further losses due to oxidation. While filling back the dug pits, well-decomposed organic matter is mixed with the top soil and the subsoil separately (in equal proportion) and the top soil mixture is filled first at the bottom of the pit. Elaborate manuring of pits has also been suggested by some workers. Burns and Prayag (1921) recommended mixing of 100 lbs (45.4 kg) of well-rotten manure, 5 lb (2.27 kg) of bone meal and 10 lbs (4.54 kg) of wood-ash with the soil of the pit. Allan (1940) recommended mixing of smaller amount, *i.e.* 30 to 40 lbs (13.62-18.16 kg) of manure, 6 lbs (2.72 kg) of bone meal and 6 lb (2.72 kg) of wood-ash in the lower 2 ft (60 cm) of the pit, whereas for the upper 1 ft (30 cm) he suggested mixing of 10 lbs (4.54 kg) of manure, 2 to 3 lbs (908 g to 1.36 kg) of bone meal and 2 lbs (908 g) of neem cake. These workers recommended that the dosages might be increased each year for several years. However, one cannot entirely be guided by such suggestions since these are not based on scientific experiment. In areas where termites are a problem, it is advisable to use chloropyriphos. Planting is done during the rainy season when the soil in the pits has already settled (following a shower or two). While planting, one should be careful that the earth ball does not break and the graft union remains well above the ground level. The planting should preferably be done during cloudy weather and in the evening. The plants should be irrigated immediately after planting. These should be attended to, especially in the initial first week or so to rectify any defect like sinking of soil, leaning of plants, *etc.* In the initial 2-3 years, it is advisable to protect plants against low temperature injury by covering plants with some sort of cover, leaving the eastern side open for entrance of light. Building up slow fires, which emit smoke, and resorting to flood irrigation may also be essential to ward off the ill effects of frost.

11.3.0 Good Agricultural Practices (GAP) in Mango Orchard

GAPs include adoption of improved cultivars; practicing high density planting in cultivars. Focus on regulating flowering and increasing intensity of flowering. There must be pruning of malformed mango shoots at bearing stage and mangment of alternate bearing by reducing fruiting during a year by adoption of pruning and thinning practices. The orchardist has to provide life saving irrigation at the critical stage of fruit growth. Integrated nutrient and water management and management of spongy tissue through sod mulching is essential. Integrated pest and disease management practices and timely control of pests and diseases, especially mango hopper, mango mealy bug and sprays of Bavistin at fruit development stage especially after rains are need of the hour.

11.4.0 Irrigation

The total annual water requirement of mature mango tree is approximately 11976 m^3/ha, while the seasonal water requirement varies from 20 m^3/ha/day in winter, to about 44 m^3/ha/day in summer (Mostert and Wantenaar, 1994). Young trees should be irrigated frequently during lean period for quick growth. However, the amount and timing vary with the climatic conditions (varying seasonal changes) and the physical properties of the soil. For instance, heavier soils will require more water but at longer intervals as compared to lighter soil which will require less water at quicker intervals. The interval between two irrigations may be from 3-4 days in summer and once a fortnight during winter.

In the case of bearing trees, irrigation at regular intervals (10 to 15 days) is an absolute necessity during the fruit development period, beginning from fruit set stage to full development stage. This is helpful in improving fruit size and in reducing fruit drop. Shoot growth, spread and leaf area of mango cv. Dashehari (15-year-old) tree were greater when irrigated at 20 and 40 per cent soil moisture deficit than at 60 per cent and control regimes. The water absorption was maximum during May and June, regardless of the regime, due to the higher temperatures and greater vegetative growth at this time (Chandel and Singh, 1992). The average annual yield in mature mango trees cv. Fascell was maximum with irrigation at –30 k Pa soil moisture tension, excluding winter followed by irrigation applied in winter only *i.e.* in May-July (Wittwer, 1991). Application of irrigation (40 l/tree, twice a week) increased the yield but fruit quality remained unaffected (Chang *et al.*, 1994). However, for obtaining good flowering one must stop irrigation at least 2 to 3 months before the flowering period. Irrigation during this period is likely to promote vegetative growth, which will be detrimental to flowering. Water stress is not a mechanism encouraging the formation of mango flowering but it does affect vegetative growth (Pire and Rojas, 1995). According to Pawar *et al.* (2007), to receive maximum fruit yield with quality fruits, 'Dashehari' mango trees should be provided with 100 per cent followed by 75 per cent fertilizers of recommended doses + 'V' level of irrigation with mulch.

Irrigation at the time of fruit setting (mustard size), marble size stage and fruit development (before maturity) have been identified as the critical stages for optimizing fruit size and quality of mango. The drip irrigation irrespective of fertigation (NPK) and non fertigation gave significantly higher fruit yield (8.13 to 13.56 t ha^{-1}) over basin irrigation (6.40 to 8.34 t ha^{-1}). Fertigation involving NPK at fruit setting and fruit development (marble size) gave maximum fruit yield (13.56 t ha^{-1}). There is a saving of irrigation water of 15.0 to 45.20 per cent and an increase in fruit yield by 19.00 57.50 per cent over basin irrigation. In fertigation, since the nutrients are applied at the time of crop needs, prevents loss of nutrients from leaching beyond the root zone or from accumulating near the surface, thus increasing nutrient use efficiency, quality of fruits and economize fertilizer use by 25 to 50 per cent as compared to normal application. Irrigation to ensure the replenishment of 60 per cent open pan evaporation values gave the highest fruit yields in mango cvs. Dashehari and Langra of 119.15kg, 78.84 kg/tree, respectively. In bearing orchards, three irrigations are recommended after fruit set. First irrigation should be given just after fruit set (mustard size), second at marble size stage and third in the second week of May prior to fruit maturity. It is advisable to irrigate mango trees in basins.

11.5.0 Manuring and Fertilization

This is another vital aspect, which has not merited the requisite attention and the recommendations made in the literature, can, at the most, serve a limited purpose (Sen and Roy, 1945; Sen *et al.*, 1947; Roy *et al.*, 1951; Mallik and De, 1952; Singh, 1962). The current technological status of nutrition in South African mango nurseries is lagging behind that of other subtropical fruits (McKenzie, 1994a). In general, a crop producing 15 t/ha removes 100 kg N, 25 kg P_2O_5 and 100 kg K_2O.

The feeder root density in 40-year-old mango cvs. Fazli, Langra and Himsagar, has been reported to be greater in April than in October and decreased with increasing depth. Maximum feeder root density was observed at the midway point than at the drip-line. Higher yielding trees had higher feeder root density than lower yielding trees (Rao and Mukherjee, 1989a). Older leaves became yellow and exhibited marginal necrosis due to N, P and K deficiency, and brown taints were found on the older leaves of P-deficient plants. Calcium deficiency caused shortening of internodes, while a dark yellow line was observed between necrotic and living tissues in Mg-deficient leaves. Sulphur deficiency resulted in mosaic-like white taints over the entire surface of older leaves. Excess P, K, Ca or Mg caused marked reduction in leaf and stem dry weight (DW) but not whole plant DW. Excess N, P and K were also responsible for chlorosis and necrosis, particularly in older leaves. Excess Ca resulted in bright green leaves with a white, mosaic-like pattern, while the size of new leaves was greatly reduced by excess Mg or S, the latter showing a pinkish coloration (Singh and Saxena, 1994).

Recently, ICAR-CISH Lucknow has obtained best yield by application of 50 kg FYM $tree^{-1}$ and 1.0 kg each of N and K and 500 g P for mango cv. 'Dashehari' and 'Chausa'. These fertilizers should be placed in 30 cm wide and 20 cm deep trenches dug 2.0 m away around the trunk. Since the fertilizers are placed in the feeder root zone, the availability of nutrient improves resulting in increased fruit yield and high NUE. Multi micronutrients deficiencies especially Zn (36.6 98.0 per cent), Mn (15.0 67.6 per cent), B (10.5 42.8 per cent) and Cu (3.0 40.00 per cent) have been observed in the mango orchards in and around Lucknow. Deficiencies could be corrected through foliar spray of $ZnSO_4$ (1 per cent), $MnSO_4$ (0.8 per cent), borax (0.5 per cent) and $CuSO_4$ (0.5 per cent). Alternatively, these can also be applied to soil @ 250 g $ZnSO_4$, 200 g $MnSO_4$, 50 g borax and 150 g $CuSO_4$ along with 50 kg FYM tree 1 in the month of July or August. Foliar sprays are found to be excellent supplements to soil application as they provide rapid absorption of nutrients for immediate utilization in metabolic activities besides ensuring economization of fertilizer use. Since approach is need- based and addresses hidden hunger, it improved size and quality of harvests.

11.5.1 Non-bearing Age

The requirements of a one-year-old plant can reportedly be met by 75 g of nitrogen, 110 g of P_2O_5 and 55 g K_2O (Singh, 1967). Forty to eighty per cent of nitrogen should be in the form of organic manure. These dosages need to be increased by the same amount each year. Another estimate suggests that a one-year-old plant should be supplied with a mixture of 10 kg farmyard manure, 2.5 kg bone meal and 1 kg sulphate of potash (Singh, 1978). These quantities should be increased every year by 5 kg farmyard manure, 0.5 kg bone meal and 0.4 kg sulphate of potash till the age of 10 years. The Central Mango Research Station at Rehmankhera has recommended that during the non-bearing stage, the plants should be supplied with 73 g of nitrogen, 18 g of P_2O_5 and 68 kg of K_2O/tree/year age.

Though a consensus on manurial requirements during the non-bearing age is difficult to reach because of inconsistencies in recommendations, the fact remains that nitrogen application is of vital importance for inducing optimum vegetative growth in this critical phase of the plant. In this connection it would be worthwhile to apply some amount of P and K also, since the former is required to cope up with the high respiration rate and for translocation of carbohydrates and, the latter, is essential for photosynthesis (Singh, 1967). Application of nitrogen increased growth, more especially if P and K were added. An excessive application of nitrogen would lead to K deficiency, unless potash was also supplied to the plant (Roy *et al.*, 1951; Mallik and De, 1952). In young plants, N and K fertilization increased dry matter and uptake of N and K (Kamal *et al.*, 1984). Azotobacter alone or in combination with N promoted vegetative growth of young mango seedlings (Kerni and Gupta, 1986).

11.5.2 Bearing Trees

The recommendations made in the literature are, by and large, centred around the observations made at Sabour by Roy *et al.* (1951) on the basis of pot culture experiments. As per this study, a bearing tree requires 1.6 lbs (726 g) of nitrogen, 0.4 lbs (182 g) of P_2O_5 and 1.5 lbs (671 g) of K_2O. It is also recommended that in the year of heavy fruiting ('on' year) nitrogen application may be doubled. These requirements can be met by applying 200 lbs (90.8 kg) of farmyard manure, 4 lbs (1.82 kg) of castor cake, 10 lbs (4.54 kg) of bone meal, 2 lbs (908 g) of ammonium sulphate and 30 lbs (13.62 kg) of wood ash. Another study, based on loss of nutrients through crop removal in cv. Dashehari, conducted at Saharanpur (Singh, 1962), indicated that applying 20 kg N, P, K and Mg fertilizers (in the ratio of 6:3:10:3) would compensate their losses due to the removal of 1000 kg fruit. However, experience shows that these amounts are rather too low and until more elaborate and systematic investigations are conducted in different regions, it would be better to base one manurial schedule on the work done at Sabour. The amounts of N, P and K absorbed per1000 kg of fruits have been estimated to be 0.98-1.5 kg N, 0.25-0.57 kg P_2O_5 and 1.37-2.19 kg K_2O (Zhaong *et al.*, 1994). Application of 1 kg N, 0.5 kg P_2O_5 and 1 kg K_2O and 3 per cent urea as foliar spray per year after pruning resulted in marked improvement in fruit yield in 60 year old declined Fazli tree (Mallick *et al.*, 1985). In cv. Dashehari, application of 100 g N, 200 g P_2O_5 and 200 g K_2O/tree/ year age was found adequate for optimum yield (Malhi and Nijjar, 1985), while a dose of 100 g N, 100 g P_2O_5 and 200 g K_2O/tree/year age was suggested by Dayal and Misra (1985) for cv. Chausa. Average yield/tree and fruit quality did not differ between N fertilizer applications (Chang *et al.*, 1994). A survey of grower's nutritional practices revealed that orchards with a higher N-status tended to produce green fruits, and those with lower N-status usually resulted in the production of yellow fruits (Smith, 1989; McKenzie, 1994b). Vega and Molina (1999) evaluated the effect of N on cv. Tommy Atkins and used 0, 20, 40, 60 and 80 kg/ha. In the first year 80 kg/ha produced the highest yield (2197 kg/ha), while in the second and third year 60 kg produced yields of 3087 kg and 3965 kg/ha respectively.

According to Kanwar *et al.* (1987), the N rate of 100 g/tree per year of tree age was sufficient for good tree growth and cropping, P had no appreciable effect but K was highly beneficial with regard to yield. Shyamal and Mishra (1989a) reported that application of N:P:K=1:2:1 kg/tree in mango cv. Langra (17-year-old) resulted in the greatest shoot growth, leaf and flower number, fruit set, fruit retention, fruit size, TSS, ascorbic acid and sugar contents, but the lowest total fruit acidity and the shortest duration of flowering. The largest increase in trunk circumference was obtained with 1.0 kg N+0.4 kg+1.0 kg K/tree, while the maximum yield was obtained with 0.5 kg N + 0.4 kg P +1.5 kg K/tree (Suriyapananont, 1992). The best fertilizer combination for mango cv. Dashehari was found to be N:P_2O_5:K_2O::1.5:0.75:1.0 kg/plant/year (Sharma *et al.*, 1993). Higher N and low K (1.5 and 0.5 kg/plant respectively) promoted vegetative growth, whereas higher N and K (both 1.5 kg/ plant) promoted fruiting in mango cv. Fazli (Banik *et al.*, 1997a). Application of K at the increasing rate (up to 1000 g/tree/year) to mango cv. Hindi-Be-Sinnara on a clay-loam soil resulted in increase in fruit number per tree, fruits per panicle and TSS content (Abd-El-Al *et al.*, 1994).

In mango, yield is related to the nutrient status of leaf as well as soil. Leaf nitrogen content before flowering and after harvest, and phosphorus content before flowering were found to be positively correlated with yield. Similarly, available soil nitrogen and phosphorus contents before flowering were also positively correlated with yield. Multiple regression studies indicated that by proper N and P fertilization yield level of mango could be raised (Rao and Mukherjee, 1985a). In cv. Dashehari the maxi-mum fruit production occurred when leaf N at a level of 1.40 to 1.54 per cent was maintained (Chadha *et al.*, 1981). Avilan (1971) reported that N, P, K and Ca contents in cv. Kent were highest before flowering which decreased at fruit formation and again increased during ripening.

Besides regular dose of 10 kg FYM, 100 g nitrogen, 50 g phosphorus and 100 g potassium per year per tree and doses are to be increased till tree to attain 10 years of age, it is desitable to correct micronutrient deficiencies by foliar spray of zinc sulphate (0.5 per cent) and copper sulphate (0.25 per cent) and borax (0.2 per cent), twice after the fruit set at intervals of 15 20 days.

11.5.3 Time of Application

Plants in non-bearing stage can be supplied with fertilizers in several split dosages during summer and rainy seasons. This helps in avoiding losses of nutrients through leaching since the root system of young plants is not so elaborate as to make use of the entire quantity applied at one time.

However, for bearing trees, the time of application is related to the time of production of vegetative flush and that of the fruit bud differentiation (Singh, 1960). The main idea to be kept in mind is the time at which the different major nutrients will be most needed by the trees to show its optimum performance. In this connection, the timing will, therefore, be influenced by the nature of the manures or fertilizers also, *i.e.* whether a given type becomes quickly available to the plants in a short span of time or its availability is slower and spread over a longer duration. For bearing trees, it is advisable to apply full quantity of nitrogenous fertilizer along with half of phosphorus and half of potash just after harvesting the fruits. Remaining quantities of these two fertilizers may be applied during October with the last irrigation. This is also an appropriate time for applying organic manures which release the nutrients slowly.

11.5.4 Mode of Application

A precise knowledge of the rooting pattern and the root system is an essential pre-requisite for scientific application of manures and fertilizers. Systematic studies on this aspect are still lacking. Nevertheless, it is well-known that the mango has a long tap root, sometimes extending down to the depth of water table up to a depth of even 5.5 m (Stephens, 1949). A study conducted in Pakistan showed that the effective root system of an 18-year-old mango tree was confined to a depth of 1.2 m and extended up to 1.8 m from the trunk (Musahib-ud-din, 1960 a,b). However, at Saharanpur, the roots of a 30-year-old seedling tree were found to extend to a distance of 7.5 m on one side of the tree (Singh, 1960).

Studies conducted at the IARI through radioisotope (32P) indicate that the feeding roots of a full grown mango tree are located between 1.20 m and 2.40 m from the trunk (Singh, 1978). Most of the feeding roots were observed at three depths, *i.e.* 30, 60 and 90 cm, the maximum density being within 30 cm. It decreased with depth and laterally from the trunk to the drip line. High yielding plants had significantly greater feeding root densities than low yielding plants at all depths in the drip line (Rao and Mukherjee, 1985b). These observations suggest that for better utilization by the plant, fertilizer applications for such bearing trees should be made only within this area.

With regard to major elements, the highest yields were obtained with soil + foliar applications (Singh and Khan, 1990). Singh *et al.* (1991) obtained highest fruit retention and yield/tree in mango cv. Amrapali, from spraying with 3 per cent urea. However, raising the concentration to 5 per cent caused increase in fruit size and weight, TSS, ascorbic acid, total and reducing sugars.

11.5.5 Micronutrients

Deficiency of micronutrients is, generally, not encountered, although in Florida symptoms of Fe, Mn, Zn, B and Cu deficiencies have been observed in the cultivars Haden and Zill (Smith and Scudder, 1951). Zinc deficiency (in the form of little leaf) was first reported from Florida (Lynch and Ruehle, 1940) and subsequently from Israel also (Oppenheimer and Gazit, 1961). Deficiency of Zn is not reported to affect yield and can be corrected by zinc sulphate sprays. Most of the desirable effects, in respect of vegetative and reproductive growth, were achieved with foliar application of 3000 ppm B (Singh and Dhillon, 1987). Soil applications of B were less effective than foliar sprays (Coetzer *et al.*, 1992) and sometimes were proved to be phytotoxic (Janse van Vuuren *et al.*, 1992). It has been suggested that in cases where B is likely to be deficient, any soil treatment (with low doses) should be carried out at least 3 months before the onset of flowering (Coetzer *et al.*, 1992; Janse van Vuuren *et al.*, 1992). Yield per tree in mango cv. Tommy Atkins could be increased by spraying with B (Solubor at 3 g/l) once when flower panicles began to appear (Coetzer *et al.*, 1991). Combined application of Zn (0.4 per cent) and urea (1 per cent) produced the highest fruit number and yield, whereas the fruit quality was improved by the application of B (0.4 per cent) + urea (1 per cent) to young mango plants (Banik *et al.*, 1997b). Individual fruit weight increased and fruit quality improved but fruit number decreased with increasing trace element application (Singh and Khan, 1990).

11.5.6 Tissue Analysis and Nutrient Requirement

Nitrogen content in leaf of mango declines with age. Chadha *et al.* (1980) suggested relatively stable nutrient contents in leaf between sixth and seventh month, while Young and Koo (1971) reported that the age of the leaf for sampling should be in-between fourth and seventh month. The effect of age of leaf on nutrient composition is given in Table 9.

Table 9: Effect of Leaf Age on Nutrient Composition (Per cent dry matter for major and ppm for trace elements)*

Leaf age (months)	*N*	*P*	*K*	*Ca*	*Mg*	*Mn*	*Cu*	*Zn*	*B*	*Fe*
2-3	1.38	0.133	1.17	1.99	0.220	73	21	56	21	45
10-11	1.20	0.091	0.82	3.44	0.211	141	47	117	34	79
15-16	1.19	0.089	0.83	3.51	0.222	183	34	115	25	78
**	**	**	**	NS	**	**	**	**	**	

** Significant difference, P=1 per cent * Young and Koo (1971).

Kumar (1977), in an experiment in sand culture observed lower leaf nutrient contents in deficient plants than under control (Table 10). In a survey of 30 mango orchards, Samra *et al.* (1978) related yield to leaf nutrient status and considerable variation was recorded in the years of sampling. However, in individual years no relation between yield and nutrient contents could be established.

Table 10: Effect of Nutrient Deficiency on Nutrient Content in Leaves of Mango Grown in Sand Culture (per cent dry matter)*

	Controls	*Slight Deficiency*	*Severe Deficiency*
N	1.88	0.98	0.68
P	0.15	0.08	0.03
K	0.95	0.53	0.24
Ca	2.44	1.50	0.81
Mg	0.32	0.18	0.10
S	0.64	0.52	0.32

* Kumar (1977).

Considerable variations in foliar nutrient element composition among cultivars have been reported. Dashehari was highest in N and Ca, Bombay Green was highest in P, while Malda was highest in Fe and Mn contents. Chausa was highest in K content and Fazli in Mg content (Kumar *et al.*, 1985). Despite wide variations, positive correlations were observed between yield, and leaf and soil N and P, and leaf N, P and K levels decreased from January to September (Rao and Mukherjee, 1989). Chaudhary and Nauriyal (1989) observed that severe Ca deficiency reduced leaf Ca and N, but raised levels of K and B. Severe Mg deficiency reduced leaf Mg but raised levels of K and B. Severe S deficiency caused decline in foliar S, Ca and Mn levels, but increased levels of K, B, Cu and Fe.

A significant correlation was noted between leaf N and K in both bearing and non-bearing shoots and soil N and K contents, while there was no significant correlation between soil P and foliar P (Dhillon *et al.*, 1991). The optimum leaf N concentrations in mango cv. Sensation were 1.40-1.50 per cent and 1.20-1.40 per cent for mature and young trees respectively, whereas the norms for K were 0.85-0.90 per cent (Smith, 1992). Fruit yield and quality of mango (cv. Langra) were not significantly correlated with leaf N, P and K contents. The mineral contents of fruiting shoots were lower than that of vegetative shoots (Dhillon and Malhi, 1993). In Alphonso mango, the norms for classifying the leaf nutrient status were developed using the diagnosis and recommendation integrated system (Raghupathi and Bhargava, 1999). The optimum N ranged from 0.77 to 1.65 per cent, P 22.4 to 33.6 mg per cent, K 0.77 to 1.73 per cent, Ca 0.76 to 1.63 per cent, S 35 to 131 mg per cent, Fe 657 to 961 ppm, Mn 13 to 408 ppm, Zn 7.71 to 18.3 ppm and Cu 14.3 to 17.8 ppm. The plant Mg content was low in all the orchards surveyed. A yield level from 5.4 to 7.4 t/ha was possible for Alphonso mango when soil nutrients were in the optimum range.

11.5.7 Nutritional Disorder

Potassium deficiency led to scorching of leaf margins which starts from tip downwards. Fruit quality is reduced. Trees are easily prone to pest and disease attack. Therefore, apply 1 kg muriate of potash or sulphate of potash along with 2 kg urea and 6 kg super phosphate during July-August in the basin to rectify potassium deficiency. Dropped leaves should also be incorporated along with manures to enrich the soil health and fertility.

During Zinc deficiency, the leaves become small and narrow with leaf margins bent upward or downward. Inter nodal length is reduced drastically and the twig with crowded leaves gives rosette appearance. Pale inferential areas and green veins are typical of zinc deficient leaves. The tree with zinc hunger does not grow well and the yield, size and quality of the fruit are reduced. Small plants with severe zinc deficiency may die. Zinc deficiency is conspicuously seen in alkaline, saline and sandy soils. Spraying zinc sulphate 5 g +10g urea/l water twice at 15 days interval is sufficient to correct this problem.

Due to Iron deficiency, the leaves loose green colour and turn white and this is called "Bleaching".The size of the leaf is reduced and in severe, the leaves dry from tip to downwards. The deficiency is common in soils with high calcium content. Hence, the effect is known as "calcium induced iron chlorosis. Two sprays at fortnight interval with ferrous sulphate 2.5 g per liter are recommended practice.

In copper deficiency, symptoms frequently develop on young trees which generally occur due to heavy nitrogenous fertilization. It may also be accompanied by Zinc deficiency symptoms. The appearance of weak terminal shoots followed by defoliation and die- back of branches, on the top of long drooping or shaped branches of the proceeding cycle of growth usually makes evident that copper is needed. Application/spraying of Copper (250 g/10 years tree) or Copper oxychloride (0.3 per cent) at monthly interval is required.

During boron deficiency, there is cracking of fruits, lusterless leathery leaves with thickened veins are the other associated symptoms. Brown areas in yellow fruit pulp are conspicuous. Application of 250g boron per tree (10-15 year old) with recommended dose of manures during July-August is ideal practice.

In salt injury/toxicity symptoms, the leaves are scorched due to excess salt in soil or irrigation water. The leaves lose their natural colour and turn to bronze colour. Tip burning is also seen in severe cases of salt injury. It is advisable to raise Dhaincha as green manure crop with onset of monsoon in the inter-spaces of the orchard during tree bearing years.Farm yard manure and compost should be applied adequately every year and gypsum filled gunny bag if kept in flowing irrigation water, will reduce salt effect.

11.6.0 Pruning and Training

The mango, being an evergreen plant, hardly needs any pruning. The only pruning that it requires is periodic removal of the dead and diseased branches, as and when one notices them. It was found that in 17-year-old trees of cv. Himayuddin, Rumani, Kalepad and Potalima, when the criss-cross branches were pruned to open the centre and the terminal whorl of shoots were thinned to retain only one or two healthy shoots, gave greater yield than in previous years (Shanmugavelu and Selva-rajan, 1985). Training, however, is an essential practice in the initial 2-3 years. This is done with a view to providing a good framework for the future so that the branches are spaced properly and these do not break with the crop load at the bearing stage. The branches are not encouraged too low on the trunk or too high from the ground level. Fivaz and Stassen (1996a) studied the role of training systems in maintaining high density mango orchards. They observed that the control trees had the highest yields/tree and per hectare but central leader trees had the highest yields/m^3 of tree volume. Projected yield with the central leader system was almost equal to those of controls if the planting density was increased further (from 667 to 1000 trees/ha) for the central leader system. Mango trees are seldom selectively pruned commercially (Young and Sauls, 1981); however, pruning mangoes may result in increased yield and fruit colour (Whiley, 1984). Pruned mango (cv. Tommy Atkins) had approximately one fourth of the centre of the canopy removed during March. Penetration of photosynthetic photon flux was generally greater in canopies of pruned than in non-pruned trees. Total leaf chlorophyll content was greatest for pruned trees during November, but similar for pruned and non-pruned trees during April and July. No significant effect of pruning could be visualized on fruit colour (Schaffer and Gaye, 1989). The circumference of the most recently tipped terminal shoot was more closely related to the number of leaves on the new shoots arising after tipping, than to the total length of those shoots (Oosthuyse, 1993a). The success of pruning has been ascribed to adequate canopy recovery and the general ability of the new shoots produced to initiate inflorescence (Oosthuyse, 1994). There was significant decrease in total shoots per canopy surface after pruning, but 62 days after pruning virtually the difference was no longer evident (Rojas, 1996a). Pruning of mango trees (cv. Tommy Atkins) after harvest caused marked reduction in yield as compared to control trees. However, fruits from pruned trees had a better external colour than fruits from control trees (Fivaz and Stassen, 1996b). Chemical pruning agents, such as cycloheximide (0.25 g/l), dinoseb (0.5 ml/l) and PCP (pentachlorophenol, 5.0 g/l) could be used to destroy apical panicles and in order to induce auxillary panicle development (Shu, 1993, 1994).

The light pruning (30 cm from the apex) in mango cvs. Amrapali, Mallika and Dashehari improved the level of phosphorus but reduced the nitrogen concentration. Severe pruning led to lowest P and Mg content with better S content (Singh *et al.*, 2010). According to Singh *et al.* (2010c) PPO activities were reduced at the FBD stage in 'on'-year shoots. Severely-pruned trees (90 cm from apex) had the highest PPO activities, while the lowest PPO activities were recorded in lightly-pruned trees (30 cm from apex). Shoot bud proline contents were found to be highest in non-pruned trees, and decreased with increasing pruning intensity. Thus moderate pruning can be adopted in high density orchards to obtain sustainable production with improved maintenance of canopy architecture.

In bearing mango trees, not more than 25 per cent biomass should be removed at a time for better canopy management; otherwise it results in excessive vegetative growth. Under high density planting system, remove 10-15 per cent biomass

Nutrient deficiency symptoms in mango

Nitrogen deficiency

Potash deficiency

Phosphorus deficiency

Calcium deficiency

Iron deficiency

Zn deficiency

Boron deficiency

Mg deficiency

Sulphur deficiency

Pruning and training in mango

Pruning and training in young mango orchard

Bush system of training

annually during October-December to increase light penetration inside the canopy. Removal of 10-15 per cent biomass should include criss-cross branches, dead wood and diseased shoots (Kumar, 2019).

11.7.0 Intercropping and Cover-cropping

11.7.1 Intercropping

In tree fruits, it is considered advisable to grow some crops in the initial years of orchard establishment when there is no income from the primary crop. Another advantage is that the vacant land, if left unutilized, will lead to build up of weeds, depletion of nutrients and harbouring of serious pests and diseases. Also, intercropping helps in reducing nutrient losses through leaching. It is, therefore, suggested that to compensate for the early non-productive period intercropping should be practised (Rajput *et al.*, 1989).

An intercrop should, preferably, be some short duration shallow-rooted vegetable crop or even a quick growing fruit plant with a short juvenile period. Amongst vegetable crops, one may profitably go in for onion, tomato, radish, carrot, beans, cauliflower, cabbage and 'palak'. Heavy feeders like colocasia, ginger, turmeric and grain crops should be avoided altogether. Amongst fruit crops (as intercrops) one may grow phalsa, papaya, guava and pineapple in tropical climate and low-chilling peach and strawberry in mild subtropical region.

In general, an intercrop should be planted well away from the mango plants and their requirements for moisture and nutrients be met separately and should not compete with those of mango. With the increasing age of mango plants, the area under inter-crops should be progressively decreased to avoid competition. One must be vigilant to adopt timely and effective control measures against pests and diseases affecting the intercrop. In South Gujarat, on newly planted orchards, intercropping with tomato followed by cluster bean appeared to be more economical with a net cost benefit ratio of 1:1.22 (Bhuva *et al.*, 1985). Intercropping mango orchard with turmeric (April-December) and radish (January-March) or papaya (all year) gave significantly higher yields (Singh and Ram, 1994).

11.7.2 Cover-cropping

Besides the intercrops one may also grow some crops like sunhemp (in light soils) or 'dhaincha' (*Sesbania aculeata*) (for heavy soils) to protect the orchard soil from erosion and also for enriching the soil fertility. These crops are sown with the beginning of the rainy season and are ploughed before the stems become woody (towards the end of monsoon). One can choose any other crop like cowpea, pea, berseem, 'mung' (*Vigna radiata*), 'urid' (*Phaseolus mungo*) and 'guar' (*Cyamopsis tetragonoloba*). Growing of bahia grass (*Paspalum notatum*) cv. A44, in a mango orchard of cv. Irwin on sloping ground as cover crop and cutting twice a year resulted in increasing tree girth by 52 per cent, soil organic matter content by 48 per cent and soil K content by 359 kg/ha, compared with controls (Chang, 1992).

11.8.0 Interculture

Maintenance of good sanitary condition is a must for keeping an orchard in healthy and disease-free condition. In the initial years of establishment, when one has to go in for intercrops or cover-crops, one normally gives preparatory tillage to the plots. However, when the mango trees are full grown, and intercropping is neither feasible nor desirable, it is necessary to plough the soil to a shallow depth at least twice a year, *i.e.* in June and October. Also, the tree basins must be kept free from weed at all times of the year by shallow hoeing and weeding. The best control of weeds in the mango budwood nursery was reported by pre-emergence application of atrazine at 4 kg/ha and post-emergence of gramoxone at 2 l/ha (Pawar *et al.*, 1985).

Grass mulch or sod culture reduced soil temperature and incidence of spongy tissue in ripe fruits (Burondkar *et al.*, 1994). Mulching slightly improved trunk growth in the first 2 years, and mean yield and productivity in the years 4 to 7 (Farre and Hermoso, 1993). Mango plants growing in orchards close to *Leucaena leucocephala*, in general, gave higher growth and yield, whereas reverse effect was noted in the vicinity of *Eucalyptus* sp. (Sharma *et al.*, 1995).

12.0 Flowering, Pollination, Fruit Growth and Development

As indicated earlier, almost all the cultivars of mango are prone to the phenomenon of biennial bearing, and a certain amount of shoot maturity is an essential pre-requisite to their flowering. However, in the case of regular bearing cultivars this is not so and shoots of any size or maturity are capable of flowering and fruiting (Singh, 1959; Singh *et al.*, 1962b). Flowering time is, therefore, closely linked with the time of flower bud differentiation which varies with the cultivars and the area where it is grown.

12.1.0 Flowering

Two types of reproductive shoots typically occur in mango. Generative shoots display only flowers and have floral bracts or non-developed leaves at the base of each lateral inflorescence. The complexes of primary to quaternary branching lateral structures of the inflorescence each terminate with three cymose flowers. The terminal flower opens first, followed by two subtending lateral flowers. These complexes form the lateral inflorescence structures emerging from the central axis of the panicle. Mango flowers are small (5-10mm) in diameter, have a 10 part perianth consisting of four or five sepals and petals that are ovate (Mukherjee and Litz, 2009). Pistils-ovary abortion occurs early in staminate flower development and in perfect flowers, the ovary is superior. The ovule is anatropous and pendulous. Individual flowers are borne collectively on panicles that consist of a central axis that further divides from primary, secondary and tertiatory branches. Panicles develop from dormant apical buds or lateral buds during floral induction. Mango flowers usually open during the night and early morning hours and the flowering duration is usually of short *i.e.* 2 to 3 weeks. (Md Aayesha Jameel *et al.*, 2018)

12.1.1 Flower Bud Differentiation

The flower bud differentiation in India in most cases occurs between October and December (Singh, 1960), although Musahib-ud-din and Dinsa (1946) reported it to occur in August which is rather too early in the light of the present state of knowledge. In the case of *Baramasi* group of cultivars, variously designated as *Baramasi/Dophala/Tephala*, which has an errratic habit of producing some crops (though usually of inferior size and quality) round the year, the flower bud differentiation is not linked with the time of shoot emergence or maturity and the flower bud differentiation usually occurs during May-June and September-October (Singh, 1978). Dormant buds of mango are non-differentiated, consisting of an arrested apical meristem and a set of preformed nodes (Nunez-Elisea *et al.*, 1996). A period of 6 weeks exposure at 15°/10°C (day/night) temperature effectively induced flowering in 94 per cent of the trees (Joubert *et al.*, 1993). Activity of the floral stimulus appears to be promoted by cool temperatures and inhibited by warm ones. Activation of the bud region(s) at near 30°C promoted leaf initiation, whereas activation near 15°C promoted inflorescence morphogenesis (Nunez-Elisea *et al.*, 1996).

It has been reported that IAA and GA contents of leaves were higher at the quiscent and flowering stages than at bud-burst and fruiting, while the kinetin level decreased from the quiscent stage through bud-burst and flowering to fruiting. Starch, ascorbic acid, RNA and DNA contents were higher and protein content was lower at flowering than at quiscent and fruiting stages (Paulas and Shanmugavelu, 1989). A sharp increase in RNA, DNA and soluble protein contents, and a slight decrease in insoluble proteins and amino acids, during October-November, the peak period of fruit bud differentiation, have been observed (Narwadkar and Pandey, 1989). At the onset of fruit bud differentiation, mean RNA content decreased and mean protein content increased, and both then increased during differentiation (Patil *et al.*, 1990). The starch content did not vary during different stages of fruit bud differentiation in the buds which were expected to flower and contents of reducing and non-reducing sugars were lower in those buds during fruit bud differentiation than before its onset (Patil *et al.*, 1991). Gibberellins and cytokinins have been reported to be involved in sexual differentiation of mango (Utsunomiya *et al.*, 1995). Glysine, alanine, serine and threonine levels in fruit buds were higher before fruit bud differentiation and arginine and phenylalanine levels increased in differentiated buds as they developed (Patil *et al.*, 1989).

12.1.2 Time of Flowering

The flowering of mango in India takes place as early as November-December in Rayalasema area of Andhra Pradesh (Gandhi, 1955), February to March in northern India (Singh, 1960) and slightly earlier (January-February) in eastern parts of the country. In Queensland, Australia, flowering occurs between June and October. In Fiji, flowering starts in July and extends through September/October (Iqbal, 1982). The flowering of mango in Egypt occurs during November to January (Nakhla, 1980).

Since the flowering process is entirely dependent on climatic conditions prevailing in an area, at some locations, like Kanyakumari, which have a very specific micro-climate, even cultivars like Neelum, Bangalora and Rumani have a tendency to flower in the off-season and the fruits usually mature during January-February, much earlier than anywhere else in India. Panicle formation and fruit development were observed in June, July and August in mango cv. Haden grafted onto cv. Coquinho in Brazil. Panicle, flower and fruit formation were highest in June than in July or August (Simao *et al.*, 1996).

12.1.3 Duration of Flowering

The flowering period of mango is usually of short duration of 2 to 3 weeks; low temperatures may extend it, whereas higher temperatures may shorten it. The mango tree does not flower simultaneously in all directions and, at least, two distinct flushes are noticed. The panicles located on the eastern and south-eastern aspects of the tree are the first to flower.

12.1.4 Sex-ratio

The mango inflorescence or panicles bear mainly two types of flowers – male and perfect, though neutral flowers are also encountered occasionally. The number of flowers per panicle varies between 1000 and 6000, depending upon the cultivars (Mukherjee, 1953). However, according to Bijhouwer (1937) it ranges from 788 to 9020 in Alphonso and seven cultivars of Java. The percentage of perfect flowers varies between 0.74 in Rumani and 69.8 in Langra (Singh, 1954a). It is reported to be 30.6 per cent in Dashehari (Singh, 1954); 42.9 per cent in Chausa (Randhawa and Damodaran, 1961); 9.20 per cent in Bombay Green and 14.9 per cent in Fazli (Mallik, 1957); 8.40 per cent in Himsagar (Majumder and Mukherjee, 1961a); 16.41-55.7 per cent in Neelum, 6.61-21.0 per cent in Bangalora, 3.67 per cent in Baneshan, 5.67-8.44 per cent in Mulgoa, 32.49 per cent in Pairi (Bhujanga Rao and Rangacharlu, 1958); 55.2 per cent in Carabao and 47.9 per cent in Pico (Pimentel *et al.*, 1984); 75.3 per cent in hybrid Sai-Sugandh (Desai *et al.*, 1994) and 25.6 per cent in Mallika (Narayanaswamy and Thimmaraju, 1990). Varieties with longer panicle produced larger number of flowers, especially male flowers (Thimmappaiah and Suman, 1987). Of the 4 cultivars assessed, Tommy Atkins had the highest percentage of hermaphrodite flower (39 per cent) compared with 7 per cent, 6 per cent and 4 per cent in Kensington, Mulgoa and Lucio-1, respectively (Jasso Argumedo, 1995).

The sex-ratio in different cultivars is greatly influenced by the the environment in which they are grown. Even the same cultivar behaves differently in different localities. This is illustrated by the fact that some of the south Indian cultivars of mango, like Neelum, Baneshan, Allumpur, Janardhan Pasand and Willard when grown under north Indian conditions have significantly lower proportion of perfect flowers than under south Indian conditions (Singh *et al.*, 1965). Polyembryonic cultivars appear more susceptible to poor fruit set than monoembryonic cultivars. Under subtropical conditions, Searle *et al.* (1995) reported that the percentage of fertilized fruitlets per inflorescence, two weeks after anthesis was 54.7 for Kensington (polyembryonic) compared with 93.9 for Irwin (monoembryonic). Some workers have suggested that trees exposed to low temperatures during inflorescence development have reduced numbers of hermaphrodite flowers (Singh *et al.*, 1965; Singh and Dhillon, 1987). Chaikiattiyos *et al.* (1997) found for Nam Dok Mai (polyembryonic) the percentage of hermaphrodite flowers on trees grown in tropical climates of Thailand was significantly greater than on trees grown in the subtropics of Thailand. Sukhvibul *et al.* (1999a) studied the effect of temperature on inflorescence development and sex expression in two monoembryonic (Irwin and Sensation), and two polyembryonic (Nam Dok Mai and Kensington) mango cultivars. Trees were subjected to natural winter temperature to induce flowering prior to transfer into controlled environment glasshouse rooms under day/night temperaure regimes of 15°/5°, 20°/10°, 25°/15° and 30°/20°C for 20 weeks. Inflorescence development did not progress when trees were held at 15°/5°C. Cooler temperature (20°/10°C) delayed the start of anthesis (42.4 days) compared with trees grown at 25°/15°C (23 days) and 30°/20°C (16.1 days). At 20°/10°C, the delay in the start of anthesis was greatest for Sensation (55.5 days) and least for Nam Dok Mai (25.5 days) while at other temperatures there was little difference between cultivars. The distribution of hermaphrodite flowers within the inflorescence was independent of temperature with the highest percentage, found in the apical half of the inflorescence (Table 11). There was an inverse relationship between the length of the anthesis period and temperature, with anthesis occurring over 30 days at 20°/10°C and reducing to fewer than 10 days at 30°/20°C. Temperature also had an inverse effect on the total number of flowers per inflorescence with 619.6 ± 108.0 (mean for all cultivars) at 20°/10°C decreasing to 431.3 ± 80.5 at 30°/20°C. Sukhvibul *et al.* (1999b) also reported that warmer temperatures generally increased the inflorescence size of all cultivars. At 20°/10°C, the pollen viability of Sensation was significantly lower than the other cultivars, but there was no significant difference between cultivars held at 25°/15 and 32°/20°C. Low temperatures caused morphologcal changes in styles, stigmas, ovaries and anther size in all cultivars, and changes were especially pronounced in Kensington.

Table 11: Effect of Temperature during Inflorescence Development on the Percentage of Hermaphrodite Flowers per Inflorescence*

Cultivar	*Day/Night Temperature*		
	20°/10°C	*25°/15°C*	*30°/20°C*
Nam Dok Mai	3.5	24.3	28.0
Kensington	20.1	31.0	25.1
Irwin	46.8	47.6	28.8
Sensation	57.1	34.6	31.6
Mean	31.9 ± 4.7	34.4 ± 6.0	28.4 ± 4.6

*Sukhvibul *et al.* (1999a).

The effects of cytokinin application and night temperature on flower bud differentiation and bisexual flower formation were investigated in Irwin mango (Yamashita and Ogawa, 1998). BA at 50 ppm or paclobutrazol at 50 ppm was applied as a soil drench to young potted trees at the end of September. BA promoted flower bud differentiation while paclobutrazol did not. In a second experiment, young potted trees grown outdoors were transferred into a greenhouse at the end of

December and kept at a night temperature of 30° or 20°C until flowering. The effects of applying a soil drench of 50 ppm BA were examined at a night temperature of 20°C. The percentage of bisexual flowers increased in response to a night temperature of 30°C (particularly in the basal part of panicles) or to a night temeprature of 20°C combined with BA treatment (particularly in the apical section of panicles).

Investigations into the causes of this phenomenon revealed that lower temperatures during the period of panicle development resulted in an increased proportion of male flowers. The early flush, emerging on the eastern and south-eastern sides of the tree was, therefore, found to have a higher sex-ratio (i.e. lower proportion of perfect flowers in comparison to male flowers). On the other hand, late emerged panicles, which experienced relatively warmer environment during their developmental period, had a higher percentage of perfect flowers (Majumder and Mukherjee, 1961a; Singh *et al.*, 1966). Also, it has been observed that the panicles located in the inner portions of the tree had a significantly higher proportion of perfect flowers than the panicles located on the periphery. The proportion of perfect flowers varied in different portions of the same panicles, the highest being in the terminal portion. It was also reported to improve with the age of the plant. The proportion of perfect flowers in a cultivar becomes detrimental to optimum fruit set only when it drops down to as low as 1 per cent or so. However, the situation can be improved significantly by exogenous application of NAA (200 ppm) at the time of flower bud initiation (Singh *et al.*, 1965). The initial fruit set in mango is directly related to the proportion of perfect flowers and it appears to be a genetical characteristic of a cultivar but the final retention depends on a number of factors including climate, soil moisture, diseases and pests. For instance, the initial fruit set in the cv. Langra is very high and impressive, but the ultimate retention per panicle is better in the cv. Dashehari which has a relatively lower proportion of perfect flowers. On the contrary, Simao *et al.* (1996) failed to establish any correlation between the number of perfect flowers and the number of fruit production, but the maximum number of fruits was found in the upper central region of the panicle and on the north-facing side of the tree. The average fruit set in mango hybrid Sai-Sugandh was found to be 20.76 per cent of all flowers and 28.22 per cent for perfect flowers (Desai *et al.*, 1994).

12.1.5 Effect of Cincturing and Growth Substances

Leonardi *et al.* (1999) conducted a series of field experiments to evaluate cincturing and chemical treatments as a means of improving mango productivity. A standard cincture was compared with a modified technique in which twine was tied tightly into a cincture groove (twine). The chemical treatment was based on a morphactin formulation (MF) and was introduced to the trees by either painting directly onto the bark of the tree trunk (MF-paint) or soaking twine in MF before tying it into a trunk cincture (MF-twine). The amount of morphactin applied varied with tree age and was in the range 0-0.06 g active ingredient (a.i.) per tree. Tree responses, measured in terms of vegetative growth, flowering and fruiting, were compared with trees that had been treated with a physical cincture only, or with paclobutrazol (up to 5.0 g a.i.), applied as a collar drench. In young (3- to 8-year-old) trees, twine, MF-twine and MF-paint had a positive effect on flowering and fruiting. These trees had earlier, more intense flowering, produced early (September) maturing fruit (up to about 4-fold increase) and had high fruit production (up to about 2-fold increase in fruit number) compared with controls. In some cases vegetative growth was reduced by 50-60 per cent compared with untreated trees. Twine and MF-twine are favoured over MF-paint because (i) the paint must be applied annually, incurring high labour costs, and (ii) paint treatments carry the risk of overdosing the trees with morphactin. The positive effects of Twine and MF-twine treatments were sustained with large responses in flowering and/or fruiting 2-4 years after application. The responses in fruit production from paclobutrazol, applied at rates based on manufacturer's recommendations were less than with Twine, MF-twine of MF-paint.

12.1.6 Flowering Physiology

Flowering mechanism in mango is a complex and still poorly understood phenomenon. Besides genotypes, it clearly depends on environmental factors. Usually new vegetative flush appears and become mature, on which flowering occurs. After withdrawal of rainfall and dry spell, winter begins which triggers flowering in mango. Flowering usually occurs in three flushes in the season. Growth of mango is not continuous. It occurs as intermittent, short lasting flushes from apical buds and lateral buds. A healthy mango shoot completes four to five flushes per year and new shoots as primary growth arise as laterals from axillary buds around the fruited twigs of the previous year. Extension growth is extension of terminal portion of shoots already produced. Primary growth and extension growth depends on the variety and environment under which they are grown. The vegetative or reproductive fate of the resting buds is not pre-determined at the time of shoot initiation (Ramirez and Davenport, 2010). Mango trees produce basically three types of shoots *viz.* (i) vegetative shoots which bears only leaves, (ii) generative shoots which bears terminal panicles and (iii) mixed shoots produces both leaves and inflorescences within the same internodes. Vegetative shoots undergo distinct changes from early shoot development to maturation of leaves.

Florigenic promoter (FP) is continuously synthesized in mango leaves and its synthesis is governed by temperature. FP is translocated through phloem to apical buds and it was confirmed by girdling experiments. Branches which were girdled to isolate from the rest of tree and defoliated completely produced only vegetative shoots (Ramirez and Davenport, 2010). It is temperature regulated and vegetative promoter (VP) is age dependant. High ratio of FP/VP favours floral induction, low FP/VP favours vegetative growth and intermediate ratios favours mixed shoots. Florigenic promoter is up regulated on exposure to cool temperature (<18 °C) in sub-tropical conditions. In tropics, regardless of temperature, floral induction occurs in terminal stems that have attained sufficient time to rest of at least 4-5 months depending on cultivar. VP is gibberellin or closely associated with gibberellin synthesis pathway. To induce flowering in warm conditions the levels of VP must drop to sufficiently low levels with stem age (4 months) to raise the FP/VP ratio.

The tri-factor hypothesis of flowering in mango proposed by Kulkarni (2004) proposes an interactive role for a putative, cyclically produced floral stimulus in leaves, a floral inhibitor in leaves and fruits, and bud activity during the floral cycle. During dormancy following a vegetative cycle, genetic and environmental factors determine the level of synthesis of the putative floral stimulus. Flowering occurs only if certain correlative factors are present, for example if the receptor bud becomes active. If fruits are or have been recently present on the stem, vegetative growth will result. Presence of the putative floral inhibitor in leaves interferes with expression of the floral stimulus resulting in vegetative growth. The level of the floral stimulus determines the response: high levels give rise to normal panicles, intermediate levels give rise to mixed panicles and low levels result in vegetative growth (Sandip *et al.*, 2015).

According to Davenport (2007), site of florigenic promoter's synthesis is leaves. The main environmental factors influencing mango flowering is temperature and cool temperatures of 15 °C day/10 °C, night induce flowering in subtropical condition (Davenport and Nunez-Elisea, 1997). Very high and very low temperature during flowering is harmful to pollen and tree fails to flower. The main limiting factor of mango tree survival is severe frost.

The shoot initiation followed by floral differentiation of apical bud, and panicle emergence and these developmental events occur in most of the mango cultivars sometimes during October-December under tropical as well as subtropical conditions. The induction of floral bud formation has strong links to prevailing environmental conditions and age of terminal resting shoots (Davenport, 2007) as under tropical locations, the flower induction occurs in response to age of previous year shoot, while cool inductive conditions are vital to floral induction under sub-tropical conditions. It is documented that a low temperature around 15-18°C and 6-8 months old matured shoots have a strong possibility for floral growth initiation (Murti and Upreti, 2000). Ramirez and Davenport (2010) illustrated the basic mechanism underlying mango flowering, involving an interaction between putative temperature dependent florigenic promoter and an age regulated vegetative promoter, with a high ratio in favour of florigenic promoter under low temperature conditions contributing to floral development. It is found that the florigenic promoter that is graft transferable and short-lived, has potential to move long distances during cool conditions and carbohydrates in the donor tissues are anticipated as driving force in the transport of florigenic promoter (Davenport *et al.*, 2006). It is found that the florigenic promoter has strong link to phytohormonal factors, which act in association with carbohydrate production and supply during differentiation for expression of floral responses (Davenport, 2009; Murti and Upreti, 2000).

Flowering has strong links to the phytohormonal synthesis and balance in the developing reproductive organs. Hormonal regulation of shoot initiation and induction events results reproductive shoot formation. During early shoot initiation, inductive signal shifted from floral (F) to Vegetative (V) or V to F, forming F/V or V/F transition shoots by altering temperature. Auxin from leaves and cytokinins from roots both are involved in initiation cycle under inductive temperatures. The minimum inductive period at 10/15 °C (12h light/12h dark) required for complete floral induction and development was found to be 35 days for both cultivars namely *viz.*, Tommy Atkins and Keitt (Yeshitela *et al.*, 2004).

Besides, carbohydrates reserves depicted as key energy producing chemicals also play important role in floral induction process in many crop species. The increased capacity of leaf phloem loading system is contributory to floral induction. The increased assimilate supply to the shoot apex contributed to floral initiation; the strategic role for carbohydrates in the flowering process was established. Thus, the adequate availability or supply of carbohydrate reserves is crucial to floral bud development and floral initiation. A high endogenous ratio of carbon to nitrogen in plants is stimulatory to flowering whereas a low C: N ratio favours vegetative growth (Corbesier *et al.*, 2002). In mango, Singh (1960) demonstrated that initiation of flowering mainly depends on maintenance of higher C/N ratio. This builtup in C: N ratio, vital for floral growth points towards definite increases in carbohydrates. The role of gibberellins in the carbohydrate mobilization is by stimulating their degradation to hexoses (Jacobsen and Chandler, 1987). Thus, a reduction in gibberellins important to induction of flowering could favour carbohydrate accumulation for eliciting of flowering. Davenport (2009) suggested that the maturity of terminal shoots and accumulation of carbohydrates in the shoot apex are in some way linked to the synthesis of the floral stimulus.

In many perennial fruit species including mango, gibberellins have been shown to suppress floral process and its response depends upon concentrations, growth stage, and climatic conditions of the location. In contrast to delaying inflorescence initiation in cool temperatures, GA did not delay vegetative growth during warm temperatures, thereby revealing that GA prevents initiation of reproductive shoots of mango rather than inhibiting floral induction. It is believed that increased gibberellins might enhance or maintain the synthesis and production of other hormones and/ or modify assimilate partitioning to suit for inhibitions in flowering.Concentration of GA increased with in buds with increase in stem age. The endogenous level of gibberellins in mango is high in "off' year shoot-tips than in "on" year shoot-tips, the paclobutrazol inhibit gibberellins in mango buds concomitant with profuse induction in flowering. Auxins influence flowering in many perennial plants but their direct role in mango flowering is still not conclusive. It is believed that these act in association with other phytohormones like ethylene, cytokinin or gibberellins through manipulation in cell elongation process. Cytokinins are important regulator of shoot meristematic activity and their high production in apical meristem during active growth facilitates reproductive morphogenesis. The fact that smudging and external applications of ethrel stimulates mango flowering suggests ethylene plays important role in floral induction. There is higher ethylene concentration in Totapuri and Neelum (regular) at flowering stage compared to juvenile Langra (biennial) during flowering (Murti and Upreti, 1996). Various indirect evidences linked to ethylene production such as extrusion of latex in terminal buds at flower initiation and leaf epinasty near apex during panicle growth also confirm involvement of ethylene in mango flowering process. Besides, KNO_3 has been shown to stimulate flowering under tropical condition in number of mango cultivars. As KNO_3 is suggested to induce ethylene production and efficacy of KNO_3 is suppressed by ethylene biosynthesis inhibitors, the involvement of ethylene appear an important factor in mango flower process. Abscisic acid has an inhibitory effect on cell elongation. It also regulates adaptive stress responses in plants. As stress conditions are required for floral morphogenesis, its increased concentrations are expected to facilitate floral growth though stress adaptive mechanism involving osmotic adjustment and synthesis of stress responsive genes. Chacko (1968) observed that the shoots of 'Dashehari' during 'On' year and 'Totapuri' trees had relatively higher levels of this inhibitors during flower bud initiation than the shoots of 'Dashehari' in 'off' trees indicated that the inhibitors are important in the initiation of flowering in mango. Since the abscisic acid is antagonistic to both gibberellins and auxins, thereby affecting cell elongation and thereby providing conditions for floral bud differentiation. Upreti *et al.* (2013) found profound accumulation of abscisic acid in paclobutrazol is untreated and treated trees of Totapuri during pre-burst stages and concluded its possible significance in creating situation for floral growth in differentiating buds. Singh *et al.* (2010) found that the major nutrients (primary and secondary) reduced during flowering, while during 'on' year the N, P and Ca levels increased. In contrast, Mg and S observed to be higher during vegetative phase in mango cvs. Amrapali, Mallika and Dashehari.

The requirement of high C: N ratio for floral initiation, induction in C: N ratio by paclobutrazol (Upreti *et al.*, 2013) and induction of flowering through girdling (Davenport, 2009) depicts that increase in carbohydrate availability is important attribute to floral initiation in mango. Mean increase in carbohydrate availability and translocation is vital to the floral initiation in mango (Shivu Prasad, 2014). Thus, shoots with higher carbohydrate content are expected to favour flower initiation, provided inductive conditions are prevailed. For this a critical level of carbohydrate is desired to express flowering favourable activity. As reproductive growth is a high energy requiring developmental event, the requirement of high carbohydrate demand at flowering is obvious. The increase in carbohydrates are concomitant with changes in phytohormones and we expect that declining gibberellins levels would favour built of simple carbohydrates, as one of principal effect of gibberellins is to mobilize carbohydrates by stimulating their degradation to simple sugars. Thus, an environment where gibberellins are high, no starch accumulation can take place. However, gibberellins concentration needs to fall below threshold levels to show carbohydrate accumulation tendency and floral initiation. Distinct differences in carbohydrate pattern is seen in growing shoots and with approach of flowering consistently higher production of total sugars and reducing sugars in apical buds of untreated as well as paclobutrazol treated trees was witnessed with peak concentration at bud burst (Shivu Prasad *et al.*, 2014). The increased carbohydrates demand is due to the outcome of enhanced photosynthetic rate as result of increased photosynthetic efficiency required for encouraging reproductive development process. In addition, Davenport (2007) believed occurrence of putative florigenic promoter in mango leaves that is loaded into the phloem and actively translocated by mass flow generated by high concentration of photon assimilates for the induction of floral differentiation. The mass flow of assimilated sugars from the leaves might help in the transport of a florigenic promoter to facilitate production of generative buds.

Flowers and flowering in mango

Male flowers

Male and hermaphrodite flowers

Mango tree in full bloom

Mango panicles

Pollination by bee

Stages of fruit setting and fruit growth

Inflorescence

Pin head stage

Marble stage

Unripe fruits of mango

Phytohormone play a decisive role in floral induction and bud development, therefore changes in IAA, ABA, cytokinin-ZR, DHZR and gibberellins contents of leaves and buds play the main role in flowering. Their concentration varied at different phenological stages in different variety (Singh *et al.*, 2014). The major gibberellins identified in the leaves and buds were recorded high content of GA_3, GA_4, GA_4, GA_4 and GA_9. GA_3 in buds and leaves was high during off-season in the mango. In contrast, GA_4, GA_5 and GA_7 were recorded high during main season in the trees grown, GA_9 was invariably high in the buds and leaves of during the offseason flowering.

Zhang *et al.* (2005) and Davenport *et al.* (2006) isolated a *CONSTANS*-like gene (*MiCOL*) from mango leaf DNA by a combination of genomic walking and PCR methods. *CONSTANS* is a circadian expression gene interacting with the photoperiodic pathway in *Arabidopsis* (Putterill *et al.*, 2004). This gene is central to activation of the FT gene in *Arabidopsis* during long days, but because mango is non-photoperiodic, the role of this gene in mango flowering systems remains unclear. The mango ortholog has 79 per cent, 76 per cent, and 62 per cent homology with two apple *CONSTANS* genes, *MdCOL2* and *MdCOL1*, and the *Arabidopsis CONSTANS* gene (*AtCO*), respectively. Efforts to isolate the FT or homologous gene responsible for synthesis of the protein, FP, however, have, thus far, been unsuccessful.

12.2.0 Pollination and Fruit Set

The mango is a highly cross-pollinated crop. Pollination takes place through insects, though earlier it had been suggested that it can take place through gravity and wind (Maheshwari, 1934). The insects reported to be responsible for effective pollination (Singh, 1954b) are *Musca domestica* (the common housefly), *Melipona* sp. (a small Dipterous fly) and *Syrphidae* sp. (a kind of hoverfly). In nature, more than 50 per cent of the flowers do not receive any pollen grain. The number of pollen grains per pollinated flower, in nature, is also rather too low, less than 3 per flower (Wagle, 1929; Sharma and Singh, 1970a). Some earlier workers had suggested that possibility of self-pollination in some cultivars cannot be ruled out (Wagle, 1929; Bijhouwer, 1937; Dijkman and Soule, 1951). However, bagging of panicles did not result in fruit set (Musahib-ud-din and Dinsa, 1946; Singh, 1954b). Nevertheless, studies have shown that the mango ovary is capable of attaining a substantial increase in size without any stimulus of pollination (Naik and Rao, 1943; Marloth, 1947; Singh, 1954b). Later studies have shown that unpollinated flower of the cv. Dashehari could continue on the panicles up to the 13th day after anthesis, the mean size of the ovary and ovule at that stage being about 2½ times more than that at anthesis (Sharma and Singh, 1970a). Certainly, this cannot be considered as the fruit set stage.

Although some workers (Dijkman and Soule, 1951) had suspected the existence of self-sterility in mango, it was not known until 1962 when Singh, Majumder and Sharma reported its occurrence for the first time in the cultivar Dashehari. Subsequently, detailed studies were conducted in the four popular cultivars of northern India (*viz.* Dashehari, Langra, Chausa and Bombay Green) and these were all found to be self-incompatible (Mukherjee *et al.*, 1968; Sharma and Singh,

1970b). The type of self-incompatibility observed in mango was of an uncommon type (Sharma and Singh, 1970b); thus far reported only in the case of *Theobroma cacao* (Knight and Rogers, 1955; Cope, 1962).

The initial fruit set, following self-pollination in the four cultivars, was rather negli-gible (0 to 1.68 per cent) as compared to that after cross-pollination (6.40 to 23.40 per cent); from the 15th day after pollination the selfed fruitlets were invariably smaller and majority of these dropped down within 4 weeks of pollination. None of the selfed fruitlets reached even the half-grown stage (Sharma and Singh, 1970b). Self-incompatibility was reported in mango cv. Langra and the cv. Dashehari was recommended as pollenizer for Langra (Ram and Sirohi, 1994). On the other hand, Desai *et al.* (1994) reported that self-fertilization in mango hybrid Sai-Sugandh resulted in 32.5 per cent fruit set. Fruit retention was 0.128 per cent of fruits set, with 0.29 fruits retained per panicle.

An analysis of the causes of this phenomenon revealed that the processes culminating in the fertilization of the ovules were the same as after compatible cross-pollination, but in the selfed fruitlets the fertilized ovules got aborted due to abnormalities in the proembryo and endosperm and due to overgrowth of the cells of the nucellus (Sharma and Singh, 1970b). Physiological studies, conducted later at IARI, showed that the level of auxin-like substances was higher in the cross-pollinated fruitlets as compared to that of the self-pollinated ones (Pandey *et al.*, 1972). A study utilizing ^{32}P further indicated that the cross-pollinated fruitlets acted as a stronger physiological sink in which nutrients and other metabolites moved preferentially, as compared to the selfed fruitlets. The levels of RNA and DNA were also found to be higher in the cross-pollinated fruitlets (Pandey *et al.*, 1974b). According to Robbertse *et al.* (1994) the time of pollination and female parent both played an important role in pollination success, although the interaction between female parent and time of pollination had the largest effect. This is a clear indication of differences in receptivity existing among the cultivars. In Tommy Atkins, Sensation and Kent, pollination success rate was much higher in the morning than in the afternoon, while in Isis and Keitt, the reverse was observed. Night temperatures below 10°C resulted in pollen grains with a low viability (<50 per cent). A temperature between 15 and 33°C during the phase from meiosis to the prevacuolate microspore was optimum for pollen development *i.e.* 70-85 per cent pollen viability (Issarakraisila and Considine, 1994). Highest percentage of successfully pollinated stigmas (stigmas with germinating pollen grains) in mango cv. Haden was observed when pollination occurred at 25 and 30°C (Wet *et al.*, 1989). Fruit set at the 15th day ultimately determined yield and was significantly correlated with retention and yield (Thimmappaiah and Suman, 1987). It was observed that CPPU [forchlorfenuron] at 20 to 60 ppm had some effects in the early stage of fruit setting in mango and that application at full bloom showed better results than those at 10 days after inflorescence emergence (Subhadrabandhu and Iamsub, 1996).

12.3.0 Fruit Growth and Development

This aspect has been studied in Dashehari, Langra and Chausa (Singh, 1954b) and the growth curve is of the sigmoid type (Singh, 1978; Kundu and Mitra, 1997).

On the contrary, Chen and Huang (1995) reported that the pattern of fruit growth in mango cv. Zihua was double sigmoid. Four periods of fruit growth could be identified, namely, stage-0: a period of initial limited growth (duration was correlated with the date of anthesis), stage-I: a period of rapid increase (duration-39 days), stage-II : a period of delayed increase (duration-7 days), stage-III : a second period of rapid increase and maturity (duration-49 days). Size increase occurred mainly during stage-I, while DM mostly accumulated in stage-III. The fibrous stony pericarp hardened in stage-II. Development of the fruits of Langra and Dashehari started in the last week of March and was completed by the second week of June. Percentage increase in growth (in terms of length, breadth and thickness) was highest in April, followed by that in May, least growth being in the month of June. However, maximum increase in weight and volume occurred in May, followed by April and June (Saini *et al.*, 1971). The weight of fruit and stone, specific gravity, TSS and total sugar content in Himsagar and Fazli increased throughout fruit development while starch, titratable acidity and ascorbic acid contents increased during the early growth stages but declined on attaining maturity. In both the cultivars maximum increase in weight was observed between 55 and 105 days after fruit set (Ghosh *et al.*, 1985).

The period of rapid growth in mango was directly related to the contents of auxin and gibberellin-like substances in the seeds (Chacko *et al.*, 1970). Failure in fertilization of the ovules or its subsequent post-fertilization abortion (as occurred in incompatible pollination) led to complete cessation of fruit growth and resulted in fruit drop. Natural parthenocarpy is not very common in mango. In Dashehari, natural parthenocarpy has been found to occur to a certain degree. The fruits are usually flat, mostly undersized and seedless and in some cases a rudimentary cotyledon is also present. Thimmappaiah and Singh (1983) reported that in Dashehari, on an average, each tree produced 5.44 per cent parthenocarpic fruits of which 3.94 per cent was completely seedless and 1.43 per cent fruits had rudimentary cotyledons showing sign of degeneration. Low temperature coupled with reduced activity of pollinators, absence of pollinizers and enhanced selfing largely contributed towards parthenocarpy in mango. Exogenous applications of benzyladenine (250 ppm) at anthesis and later a combination of ß-naphthoxyacetic acid (10 ppm) and GA_3 (250 ppm) at fortnightly intervals, resulted in the production of seedless fruits in Dashehari. These fruits, although smaller in size, were superior in fruit quality as compared with normal-seeded fruits (Chacko and Singh, 1969).

12.3.1 Fruit Drop

The natural fruit drop in mango is rather too high; amounting to about 99 per cent at various stages of growth, more especially during the initial four weeks (Singh, 1960). In the cultivars Bombai, Langra and Fazli, about 13-28 per cent of the perfect flowers set fruit and only 0.1-0.25 per cent reach maturity stage (Sen, 1939). Studies conducted by various workers showed that in mango, in general, only about 0.1 per cent of the perfect flowers developed fruits to maturity (Bijhouwer, 1937; Naik and Rao, 1943; Singh, 1954b). Fruit drop was high (96.87 to 99.69 per cent) in all cultivars (Bombay Green, Bombay Yellow, Golapkhas, Himsagar, Langra and Safdar Pasand, with no significant difference between cultivars (Sanyal and Maity, 1989).

Amongst the causes attributed to this phenomenon are lack of pollination, low stigma-tic receptivity, defective perfect flowers (having defective embyro sac development at anthesis), poor pollen transference, occurrence and extent of self-incompatibility, competition between developing fruitlets and drought or lack of irrigation (Singh, 1960; Sharma and Singh, 1970a, 1970b). Some other contributing factors are unfavourable climatic conditions during fruit development period (*viz.* winds and hail-storms), high incidence of serious diseases like powdery mildew and anthracnose and pests like hopper and mealy bug (Hayes, 1957; Singh, 1960). Fruit drop was found to be positively correlated with fruit growth (length and breadth) and the extent of drop was greatest in April (Sanyal and Maity, 1989). Fruits abscised within about 5 weeks after flower shedding and the main wave occurred in the first 2 weeks (Chen and Huang, 1995).

The extent of fruit drop in mango can be reduced significantly by regular irrigation during the fruit development period. Timely and effective control measures against major pests and diseases can also be of great help.

Attempts have also been made to control fruit drop through the application of plant regulators like NAA and 2,4-D. Gokhale and Kanitkar (1951) reported that post-setting drop in Alphonso could be checked by the application of these regulators at 25 ppm concentration. Subsequently, 2,4-D (30 ppm) showed good results in controlling fruit drop in Neelum, without having any adverse effect on fruit size or TSS (Rao and Subba Rao, 1963). Roy *et al.* (1963) also observed some reduction, but at lower concentrations (5 and 10 ppm) in the cultivars Gulabkhas, Himsagar, Langra and Bombai. Gill (1966) tried a number of growth regulators including 2,4-D, NAA and 2,4,5-T on Dashehari, Langra, Chausa and Bombay Green. Of these, 2,4-D appeared to be the best at concentrations lower than 20 ppm about six weeks after fruit set (end of April). Though these treatments resulted in harvesting of 3 to 50 per cent more fruits than the control, its deleterious effects on fruit development were, nevertheless, quite common. In Dashehari, foliar application of urea (2 per cent) proved effective in increasing the harvest as compared with control. Spraying of NAA at 50 ppm (Prakash and Ram, 1985) or 2,4-D at 10 ppm (Singh *et al.*, 1986) effectively controlled fruit drop in Chausa. NAA at 5 ppm induced maximum retention in Langra, while GA_3 at 10 ppm was found effective for Dashehari (Singh *et al.*, 1986). In cv. Langra, 40 ppm NAA was very effective in increasing fruit retention when applied at mustard stage of fruit development (Yadav and Ram, 1986). The increase in fruit retention by using various chemicals suggests that endogenous levels of these chemicals or their analogues were deficient in these stages. Therefore, exogenous application increased the fruit retention by increasing their levels or by antagonising adverse effects of endogenous growth hormones like ethylene and ABA (Singh and Ram, 1983). The interaction of urea (2 or 4 per cent) and NAA (10 or 20 ppm) was significant for fruit retention and physical characters but not for chemical composition (Singh *et al.*, 1994). In cv. Totapuri foliar spray of NAA (10 ppm) + urea (2 per cent) or 2,4,5-T (20 ppm)+urea (2 per cent) and in cv. Suvarnarekha NAA (10-20 ppm) + urea (2 per cent) just before fruit set (when the protruding ovary became visible to the naked eye) were the most effective treatments for increasing fruit retention (Behera *et al.*, 1994).

However, a review of the work done on controlling fruit drop through plant growth regulators revealed that their application did not yield any material gain in fruit retention during the 'on' year, when the crop load was heavy and the drop was also more. During the 'off' year more favourable results were obtained (Chadha, 1963) but these did not seem to be of any practical significance.

12.3.2 Biennial Bearing

This is one of the most burning problems, since it renders mango cultivation less remunerative to the growers. This term is synonymous with 'alternate' bearing which denotes yield variation in alternate years, *i.e.* a year of optimum or heavy fruiting is followed by a year of little or no fruiting. This is an age old phenomenon, common to almost all the cultivars of mango and finds mention even in *Ain-i-Akbari* written by Abul Fazl during the reign of Emperor Akbar. The few cultivars of mango which are exception to this rule are relatively inferior in fruit quality.

The terms 'periodicity of cropping' and 'irregular bearing' are sometimes erroneously used to describe the phenomenon of biennial bearing. These terms do not imply variation in yield in alternate years. The terms 'irregular bearing' and 'periodicity of cropping' imply that cropping does not follow a systematic pattern, *i.e.* an optimum crop is obtained only once in a number of years. Such behaviour is largely due to lack of proper orchard management practices. On the other hand, the alternate or biennial bearing habit is of a genetic nature since this trait can be identified amongst the mango plants right from the initial years of fruiting. This is contrary to the view-point prevalent in the literature (Singh and Khan, 1940; Ghandhi, 1955; Singh, 1960), wherein it is mentioned that the biennial bearing habit sets in only after the age of 10-12 years. The observations made during the last two decades at the IARI on the bearing behaviour of hundreds of hybrid seedlings have established that regular and biennial bearing hybrids can be identified right in the second year of fruiting. The degree of bienniality in the cropping of some south Indian mango cultivars and hybrids was evaluated by Rao *et al.* (1995). The cv. Neelashan (Neelum x Baneshan) produced the highest mean yield and weak bienniality. There was no relationship between yield potential and bienniality. The problem of biennial bearing has been studied in great depth by a number of workers during the last 4 decades or so (Singh, 1960; Singh, 1978). Several aspects like climatological factors, age and size of shoots, C/N ratio and hormonal balance have been investigated into with a view to arriving at a better understanding of the problem and for providing a solution thereof.

The commercial varieties of North India, namely, Dashehari, Langra and Chausa are biennial bearers, while south Indian varieties like Totapuri Red Small, Bangalora, and Neelum are known to be regular bearers. Among the most important commercial varieties of the South, Neelum (Kazaladdu), Banganapalli (Sappatai), Bangalora (Kizhimooku, Totapuri) Kalepad and Senthura (Chinnaswarnarekha) are moderate to heavy bearers and are considered to be fairly regular. Choice varieties such as Alphonso (Gundu), Imam Pasand (Himayuddin), Mulgoa, Peter (Pairi, Nadusalai) *etc.*, are mostly erratic in bearing (Sharma *et al.*, 2017).

12.3.2.1 Climatological Factors

Adverse climatic conditions like rain, high humidity and low temperature sometimes convert an 'on' year into an 'off' year directly or by promoting the incidence of diseases like powdery mildew and anthracnose. Frequent frosts or low temperatures during the flowering period adversely affect the fruit set, thereby turning an 'on' year into a year of low crop or no crop. Climatological factors as such do not form the basic causes of biennial bearing.

12.3.2.2 Age and Size of Shoots

As early as 1935, in the Philippines, Galang and Lazo reported that the cv. Carabao would not differentiate flower buds unless the shoots are of certain length and girth and with a number of leaves of a particular size. It has also been reported by some earlier workers in India (Singh and Khan, 1940; Naik and Rao, 1942; Roy, 1953; Singh, 1960) that of the various vegetative flushes, varying from 3 to 5 in different parts of the country, it is the earliest emerged flush, which shows certain dormant period (without any further extension growth) and is of 8 to 10 months maturity, which is capable of producing flower buds. This necessarily means that a certain amount of physiological maturity is pre-requisite to flowering in the case of cultivars which are prone to this phenomenon.

However, studies conducted at Saharanpur (Singh, 1959) have shown that in the case of the regular bearing cultivar Rumani, even December-emerged shoots are capable of producing flower buds in the following February. This has also been established in another regular bearing cultivar Neelum, wherein shoots emerging in October are in a position to flower by the following March (Singh *et al.*, 1962a). However, this is not the case with biennial bearing cultivars.

In the biennial bearing Dashehari, even when shoots of desired maturity were induced to the extent of 83.0 per cent (following defruiting at 'pea' stage in the last week of March), only 15.2 per cent of these flowered next year (Singh *et al.*, 1968). This clearly suggests that a vast majority of the non-flowering shoots (though of correct maturity) is probably lacking in some vital substance or substances necessary for flower bud formation. The tree, therefore, takes a year to recoup this loss, thus causing biennial bearing habit in the tree.

12.3.2.3 Carbon/Nitrogen Ratio

The C/N ratio, in relation to flowering in mango (cv. Langra) was first studied by Naik and Shaw (1937), but no definite conclusion could be arrived at. Subsequently, Sen (1943a) reported that irregular bearing in mango was caused by nutritional deficiency, especially that of nitrogen. He further reported that a proportionate increase in nitrogen led to vegetative growth whereas its proportionate decrease favoured flowering (Sen, 1946).

Studies on this aspect have since been conducted by a number of workers (Mallik, 1953; Singh, 1960; Sen *et al.*, 1963), and it is indicated that higher starch reserves, total carbohydrates and C/N ratio favour flower bud formation in most of the cultivars studied except Baramasi and regular bearing cultivars.

Studies on the nitrogenous constituents in the stems and leaves of different cultivars of mango indicated presence of higher amounts in the trees which were expected to flower, although no correlation between flower bud initiation and total nitrogen content could be derived (Singh, 1960; Sen *et al.*, 1963; Chacko, 1968).

The available evidence suggests that in fruit plants, nitrogen and carbohydrate reserves play important roles in flower bud initiation even if these do not form the primary cause of the phenomenon of biennial bearing (Singh, 1978). It is quite probable that an accumulation of these compounds may create favourable conditions for the synthesis and action of the substances responsible for flowering.

There is a variation in flowering ability among the terminal branches within a mango tree. The extent of variations in net photosynthesis rate (P_N), transpiration rate, stomatal conductance (gs) and mesophyll efficiency (mesophyll capacity to fix CO_2) were more pronounced in trees that showed irregular bearing (cv. Langra) than in those showing regular bearing (cv. Rumani). In regular bearing trees, the narrow range of variation in P_N was related to a higher mesophyll capacity, which in turn might have resulted in higher carbon-build up in productive branches that could lead to regular flowering (Shivashankara and Mathai, 1995). Leaf relative water content (RWC) varied significantly with cultivars but not with time. The regular bearing cvs. Malviya Bhog and Amrapali had significantly higher RWC and lower water saturation deficit in September than in May (Devi *et al.*, 1996). According to Narwadkar and Pandey (1989a) translocation of P is involved in controlling alternate bearing in mango.

12.3.2.4 Hormonal Balance

The problem of biennial bearing in mango also appeared to be closely associated with the fruit development process and the inhibitory influence of the developing fruits on vegetative growth (Singh, 1960; Singh *et al.*, 1968). Studies conducted on the physiological aspects of the flowering process in mango showed that higher levels of auxin-like substances and an inhibitor (similar to ABA) and lower levels of gibberellin (GA_3) like substances were vital for a floriferous shoot in mango (Chacko, 1968). Choudhuri and Rudra (1971) found an inverse relationship between the level of endogenous inhibitor in the shoot and vegetative growth and suggested that higher inhibitor content promoted flowering in mango. Further studies are needed to bring about an ideal combination of these substances in precise amounts and in correct sequence. However, it seems rather impossible in a crop like mango, which is rather unmanageable in comparison with annuals. A flowering stimulus was thought to be synthesised in the leaves and defoliation caused rapid loss of the induced floral stimulus, followed by rapid resynthesis with the new vegetative growth (Nunez-Elisea and Davenport, 1989). The higher IAA, ABA and lower GAs were found in 'on' year of mango shoots.

Shoot-tip pruning removes the source of apical dominance and stimulate axillary bud development. The control trees had the lowest IAA and the highest GA-like substances while lowest level of GA-like substances was in moderately pruned trees. The higher level of ABA is conducive to the process of flowering in mango and its manipulation by spray of synthetic substance or by pruning at appropriate time may be helpful to induced flowering during off season (Singh *et al.*, 2009b).

The distinct difference in the bearing behaviour of the regular and biennial bearing cultivars leaves hardly any doubt that the problem is an inherent one and a lasting solution can only be obtained through genetic engineering.

12.3.2.5 Control of Biennial Bearing Problem

The biennial bearing problem has been a major bottleneck in the expansion of the mango industry. The growers have to split the profits of an 'on' year into two, in order to take care of the 'off' year. Also, because of glut in the 'on' year, the price offered to the orchardist is less remunerative. During an 'off' year or a year of low production, the consumer too has to pay an exorbitant price for the fruit. Obviously, this problem has, therefore, posed a major challenge to the horticultural scientists during the last seven decades. Consequently, a number of remedial measures have been proposed from time to time to overcome this limitation in mango production. These include proper upkeep and maintenance of the orchards, deblossoming, smudging and crop regulation through chemicals, pruning and growing of regular bearing cultivars.

Significant earliness in flowering (85.4 day) and advancement in harvesting (82 day) is achieved with the application of PBZ on 15th May. PBZ application on 15th June resulted early flowering (56 day) and harvesting (69 days) than the control (Burondkar *et al.*, 2014). The findings of study indicate huge potential for realizing about 5-6 times higher returns from "Alphonso" produced in February-March as compared to May harvest.

Regular use of paclobutrazol may have detrimental effects on longevity of crop; therefore, its integration through inorganic/organic nutrition seems to be essential for getting sustainable production of mango. Small amount of paclobutrazol residue remains in the soil for 2-3 years, therefore, varietals response to paclobutrazol persistence in relation to flowering and fruiting is also need to be studied.

Motaium *et al.* (2019) observed the average yield in the off year is 47 kg/tree at 250 mg L^{-1} boron but 9 kg/tree at 0.0 mg L^{-1} boron rate (the control treatment). The interaction treatment of 250 mg L^{-1} boron + 1500g nitrogen/tree is the best treatment as it resulted in the highest values for all the tested parameters. The average yield of this treatment is 53.5 kg/tree.

12.3.2.6 Proper Upkeep and Maintenance of Orchards

A proper cultural schedule is of paramount importance for maintaining fruit trees in healthy and disease-free condition, and thereby, obtaining better plant performance. Wagle (1931) reported that proper manuring of mango orchards (10 to 12-year-old) leads to the solution of this problem. A number of other cultural practices, like root pruning, application of salt and giving incision to the bark were advocated by Balakrishnamurti and Jogiraju (1932). Necessity for regular ploughing thrice a year (in autumn, winter and beginning of monsoon), liberal manuring with farmyard manure or compost, irrigation after fruit set, ringing of branches in high rainfall areas and regions of late rains and keeping the orchards open has also been emphasised by Sen (1943a) for overcoming this problem.

However, these treatments have not been found to be of any help (Cheema *et al.*, 1954; Gandhi, 1955; Singh, 1961; Singh, 1978). Experience shows that proper maintenance of trees may help in reducing erratic or irregular bearing but it cannot induce regularity of bearing in biennial cultivars.

12.3.2.7 Deblossoming

This practice has been recommended by some earlier workers with a view to obtaining some crop every year by reducing the crop load in the 'on' year (Singh and Khan, 1939, 1940; Sen, 1943b). According to these workers the shoots deblossomed in the 'on' year put forth panicles in the following year which would otherwise be an 'off' year.

Singh (1961) reported that response to deblossoming depends on the cultivar, Dashehari responding somewhat better than Langra. Studies conducted at the IARI showed that deblossoming was helpful only at the full development stage of the panicles and not earlier (Singh *et al.*, 1974). Whole tree deblossoming in the main fruiting season in March, after completion of flowering, resulted in a marked increase in off-season flowering and fruiting in the extended flowering cultivars Bobbili punasa, Neelum and Royal Special, whereas in the single flowering cultivar Banganapalli only vegetative shoots emerged from deblossomed shoots (Kulkarni and Rameshwar, 1989).

12.3.2.8 Smudging and Chemical Regulation

Induction of flowering in mango through 'smudging' (building up slow fires, emitting smoke) is an age-old practice in the Philippines (Gonzalez, 1923; Wester, 1924; Galang and Agati, 1936). Some attribute the flowering to heat whereas the others hold CO_2 responsible for this. However, these have not been found to be helpful in India (Cheema *et al.*, 1954). It has been concluded that growth regulators have the ability to break the cycle of alternate bearing in mangoes (Rath and Rajput, 1990). Ethrel (2-chloroethanephosphonic acid) was also tried to induce flowering in mango during 'off' years and some success was claimed in this regard (Chacko *et al.*, 1972; Pandey *et al.*, 1973a; Mallick, 1985a; Dhatt and Dhillon, 1985; Rath and Rajput, 1990). However, large scale trials conducted at the Central Institute of Subtropical Horticulture (Lucknow) and in some private orchards at Faridabad and Khurja in Uttar Pradesh did not yield favourable results and efficacy of this chemical is doubtful. At the most, it can merely hasten flowering by a few days in an 'on' year. Spraying with 1 per cent to 2 per cent KNO_3 (Rojas and Leal, 1993; Sharma *et al.*, 1993; Ferrari and Sergent, 1996a) and 6 per cent to 18 per cent calcium nitrate (Rojas, 1996 a, b) was found to promote flowering and fruit set in mango. More recently promising results were reported on the induction of flowering in mango with paclobutrazol (Winston, 1992; Miao *et al.*, 1994; Mossak, 1997). All paclobutrazol application rates (2.5 to 15.0 g a.i./tree) significantly increased the percentage of reproductive shoots compared with control and also extended the flowering period (Ferrari and Sergent, 1996b).

The use of 10 g paclobutrazol/tree had a dual effect, as it reduced tree size and induce early flowering. It was suggested that under rainfed conditions, a single application of 10 g paclobutrazol/tree should be followed by further application every two years (Salazar-Garcia and Vazquez-Valdivia, 1997). Yadav and Singh (1998) found that paclobutrazol at 4 g per tree increased the number of fruits and yield and produced high quality ripened fruits when applied as a soil drench, in the first year after treatment. However, the effects were off by the third year, indicating that paclobutrazol needs to be applied annually to increase mango fruit yields. Rao and Srihari (1998) summarized the results of investigation on biennial bearing in Alphonso and suggested methods successful in producing fruits in the expected 'off' year : (i) application of 1 per cent each of urea and KH_2PO_4 after harvest of the 'on' year crop, which favoured development of new fruiting shoots which flowered in the 'off' year; (ii) application of foliar sprays of 100 ppm TIBA or 2000 ppm paclobutrazol or soil application of 10 g paclobutrazol which promoted flowering directly on fruited shoots in the 'off' year by-passing the need for the vegetative phase; (iii) applying NAA to thin flowers which promoted moderate cropping in the 'off' year and (iv) pruning of fruited shoots which promoted lateral flowering in the 'off' year. Among the different treatments with paclobutrazol, Subhadrabandhu *et al.* (1999) concluded that soil drenching (8 g/tree) was most effective in suppressing shoot elongation. Flowering ratio in the trees subjected to all the paclobutrazol treatments was higher than that of the control, but fruit set was only slightly increased by these treatments. Paclobutrazol remained in the soil for up to 11 months when it was applied to soil and for 3 months when applied by foliar spray without soil cover. No residues were detected in the soil following trunk injection, and when paclobutrazol was applied by foliar spray and the soil was covered with a plastic sheet. Levels of paclobutrazol residues were high in leaves subjected to foliar spraying, but not following soil application and trunk injection. In all treatments, no chemical residues were detected in mature fruits.

Shinde *et al.* (2000) standardized the dose and time of paclobutrazol application on tree canopy diameter basis for regular cropping in Alphonso mango. They reported that application of paclobutrazol at 0.75 g/m was well as 1.25 g/m on crown diameter basis, applied during 90 days and 120 days before bud break, significantly increased flowering (68.48 to 80.60 per cent) and fruit yield (92.82 to 117.87 kg/tree) as compared to control (30.03 kg/tree) (Table 12).

Madhava Rao (1971) reported that pruning can be helpful in overcoming the problem of irregular bearing in mango. This type of pruning was recommended for operning the centre of the trees by topping off or thinning of branches which helped in reducing irregular bearing. Saidhu *et al.* (1983) stated that pruning and thinning had no significant effect on the time of floral bud formation. However, pruning and thinning enhanced the leaf ethylene level and the percentage of flowering in cvs. Mulgoa, Neelum and Bangalora. However, these recommendations cannot be helpful in biennial cultivars which do not have the genetic potential to bear regularly. This practice can be of help in permitting light to old and neglected orchards and in improving their erratic fruiting.

Table 12: Effect of Paclobutrazol on Flowering and Fruiting Behaviour of Mango cv. Alphonso*

Treatment Dose of chemical (a.i./m)	*Days before Bud Break*	*Per cent Flowering Shoots*	*No. of Fruits per Tree*	*Yield per Tree (kg)*
Control		33.30	118.6	30.03
0.75 g	60	51.35	178.6	44.28
0.75 g	90	68.48	380.1	92.82
0.75 g	120	73.96	398.9	96.21
1.25 g	60	61.36	298.8	69.15
1.25 g	90	77.93	319.7	93.76
1.25 g	120	80.60	491.9	117.87
C.D. at 5 per cent		4.05	212.0	51.84

*Shinde *et al.* (2000).

12.3.2.9 Pruning and Canopy Managment

Canopy management ensures well structured tree which carry a heavy crop, facilitate spraying operations and ensure the fruit exposed to plenty of light leading to a good blush on the fruit at maturity.

In young trees, it is essential to develop a strong trunk and the trees should be allowed to grow over 1 metre height from ground (single stem), initially. They are then cut back to a height of 0.6 to 0.7 m during October-November to induce primary branches (make a smooth cut with sharp secateur). Heading back results in the formation of new primary branches (3-7) during March-April. Prune the excess branches and allow 3 to 4 in all the directions. The site of first cut is important for the development of a strong frame. Mangoes grow in flushes; each flush is delineated by a concentrated whorl of leaves on the stem. This is referred to as a "ring of buds", as a bud capable of forming into a branch which is situated at the base of each leaf. If the cut is made above this "ring of buds" the resulting regrowth will be a feather duster effect of seven or more shoots developing. Prune primary branches at 60-70 cm height to induce new secondary shoots (7-10 month after the first cut preferably during October-November). These would need to be thinned out to three or four shoots per primary branch if a good tree structure is to be attained. Tertiary branches (2 to 3) can be obtained by pruning the secondary branches at 60-70 cm height.

In bearing teees, first pruning should be done immediately after harvest and should be completed by the end of June/July. It includes skirting (removal of low hanging branches, opening up of branches inside the tree which cross over or clutter up the centre of the tree restricting the penetration of sprays).

In bearing mango trees, not more than 25 per cent biomass should be removed at a time; otherwise it results in excessive vegetative growth. Second pruning may be taken up in the middle of December where tip pruning is a useful practice and the young flushes are cut back to mature wood; the resulting flush may result into a floral one.

The Total chlorophyll and chlorophyll *b* contents are enhanced as a result of moderate pruning. Moderate pruning (60 cm from the shoot tip) in high-density mango orchards can therefore be adopted to restore optimum vegetative growth and fruiting (Singh *et al.*, 2010c).

Pruning the flowered/vegetative shoots 10 cm above the node during June-July (immediately after harvest) followed by soil drenching with paculobutrazol 1 g *a.i* per sq.m during August -September enhanced flowering, fruit set and quality of fruit in Banganapalli mango under irrigated condition in the regular season.

Similarly, pruning 10 cm of terminal shoots after emergence of new growth (floral or vegetative growth) during December–January followed by application of paculobutrazol at 0.75 gm *a.i.* per tree of 9 year old during March–April resulted in good off-season crop in cv Neelum (Balamohan *et al.*, 2015).

The need based fertilizer application based on the soil test report has to be followed and spraying of Potassium nitrate (2 per cent) + NAA (40 ppm) or Potassium dihydrogen phosphate (1 per cent) + Potassium nitrate (1 per cent) during October can assure flowering. The spray of 0.5 per cent urea, if the trees do not flower up to January and in the irrigated orchards, soil drenching of Paclobutrazol at 1 ml/m^2 of canopy area is necessary for regular bearing.

Under high density orchard, the pruning must be completed as soon as possible after harvest preferably before second week of June. Tertiary branches are to be headed back in such a way that the plant height can be maintained at 1.5 to 2 m and having 10 to 15 tertiary shoots per tree. Excess tertiary shoots are to be thinned out to avoid overcrowding. About one month after pruning, thinning of newly emerged shoots is essential to avoid excess shoots and overcrowding. On each tertiary shoot, 3 to 4 new shoots are to be allowed so that 40 to 60 panicles can be obtained in each tree. Dried panicle and dried shoots/branches must be removed at the time of pruning.

12.3.2.10 Rejuvenation

Old trees often produce fruits and difficult to harvest due to their size and harbour pest and diseases and it is difficult to reach the whole tree while spraying to control them. It is possible to cut these trees back to a more manageable size, but depending on the severity of the pruning, one may lose two to three crops. Rejuvenation improves production, fruit size and quality once trees recover. But, rejuvenation of mango trees under rainfed condition is not recommended since the tree may not recover after pruning. There are four models developed for rejuvenation in mango (Balamohan *et al.*, 2016) which is as follows:

Model I

The tall central trunks are cut back to about 3-4 m height from the ground. The actual site to cut back at a point where there are side branches. Here, one half of the tree is cut (through chain saw) back and the remaining trunks and leaves will help protect the stump from sunburn.The freshly exposed trunks and branches to the sun should be painted with white water-based paint diluted three or four times to prevent sunburn and bark damage due to borer. Select the most vigorous shoots

Rejuvenation of old mango orchard to improve production

Methods of rejuvenation

Training of branch after rejuvenation

Intercropping in mango orchard

Intercropping with pineapple

Intercropping with cabbage

Intercropping with brinjal

when emerged, paced evenly around the stump and if possible at differing heights. Remove all the unwanted shoots. This process should be repeated until the selected shoots begin to dominate and take over. Then the rest of the tree can be cut back and the operations may be repeated. By this method, half the yield will be realized throughout the process till it regains the vigour.

Model II

Old, unthrifty trees can often be rejuvenated by skeletonising the tree, *i.e.* cutting back the branches of the tree till only the basic frame is left. Moderate skeletonising would entail cutting back the large branches to healthy wood, while at the same time maintaining the basic structure of the tree. Particularly large unthrifty trees would benefit from a more severe skeletonising, where all the main branches are cut back to the main trunk only leaving about a metre or so of branch. The whole trunk and remaining branches should be painted with diluted paint to prevent sunburn and borer attack. In both these cases there will be a proliferation of sucker growth.

Model III

In old and dense mango orchards, light interception and photosynthetic potentials of trees is reduced resulting poor yield. The branches existing on main trunk are considered as first order branches, the branches existing on first order branch are called second order branches, the branches existing on second order branch are called third order branches, similarly fourth and fifth order. At IIHR, Bangalore, pruning third order branches 30cm from point of origin recorded the maximum yield in Alphonso (86.3 kg/tree). Trial conducted at CISH revealed that pruning second order branches recorded maximum pooled fruit yield of twelve years after pruning (57.99 kg/tree) of Dashehari.

Model IV

To rejuvenate 40-50 years old mango trees, the main branches are pruned at a height of 5 m from the ground level during December. About 3-4 diverging branches are kept for developing healthy umbrella-shaped canopy and rest is removed from the base. Pruned surfaces are smeared with copper oxychloride paste immediately after pruning to check the microbial infection. Pruned trees are kept under intensive care and management. Profuse shoots emerged from prune branches from April onwards. Only 8-10 outward growing well spaced healthy shoots are retained per branch and the rest are removed. Plant-protection measures are seriously adopted especially against stem-borer, leaf cutting weevil and anthracnose. Because pruned trees come into flowering and fruiting after 2 years of pruning, growers ended up with loss by missing crops for 2 year, hence, it is essential to undertake the pruning work in alternate rows. With alternate row pruning, availability of light to unpruned trees in two adjacent rows is greatly improved and their fruiting increased by 2-3 times. Thus, enhanced production from unpruned trees can be compensated.

12.3.2.11 Growing Regular Bearing Cultivars

Choice in this regard is rather limited since only a few of the hundreds of mango cultivars are regular in bearing and these are grown commercially only in the South. Although these are relatively inferior in fruit quality and late in maturity under North Indian conditions, these can be tried to obtain regular cropping in areas where the local cultivars are all biennial. Besides the traditional regular bearing cultivars (*viz.* Bangalora, Neelum, Totapuri Red Small, Rumani), two improved cultivars developed at the IARI (*viz.* Mallika and Amrapali) should also be tried in different regions (Table 13).

Table 13: Fruit Quality of Mallika* and Amrapali along with their Parents**

Characters	*Cultivars*		*Parents*	
	Mallika (Neelum x Dashehari)	*Amrapali (Dashehari x Neelum)*	*Neelum*	*Dashehari*
Average fruit weight (g)	307.00	143.00	120.00	155.00
Pulp (per cent)	74.90	74.80	59.10	68.00
Total soluble solids (°Brix)	24.00	22.80	18.00	21.00
Total sugars (per cent)	18.60	17.20	16.40	15.20
Acidity (per cent citric)	0.33	0.12	0.28	0.19
Ascorbic acid (mg/100 g)	19.00	35.00	31.00	37.00
Total carotenoid pigments	10,392	16,830	5,275	7,452

(expressed as ß-carotene; µg/100 g).

*Singh *et al.* (1977); ** Sharma *et al.* (1981).

Amrapali is dwarf, regular bearer and responds very well to pruning, thus goes well under high density planting. About 1,600 plants can be accommodated in one hectare area, while Mallika in semi-vigorous can be planted at 6-8 m spacing. Amrapali is highly precocious and commercial harvest can be taken after three years of planting. The fruits are sweet, fibreless, with an attractive orange red pulp colour. It is quite suitable for overseas markets. Besides it has excellent processing traits. Amrapali pulp has the highest betacarotenoids contents and can be used for blending pulp of white pulp coloured varieties. Both the varieties are late in ripening thus extending the mango season for almost two extra months. The major states growing commercially these two varities are Odisha, Jharkhand, West Bengal, Gujarat, Karnataka, Madhya Pradesh, Chhattisgarh, Uttar Pradesh, Gujarat, Uttrakhand *etc.*

The cultivar Mallika is semi-vigorous in habit and has a strong tendency to bear every year, besides having fruits of larger size (about 300 g/fruit), have good pulp content (72 per cent) and super fruit quality (Singh *et al.*, 1977). The cultivar Amrapali is precocious, distinctly dwarf, highly regular and prolific in bearing and has good fruit quality (Sharma *et al.*, 1981). This is a cultivar with a future since it has been found to fit into the concept of high density orcharding which will lead to higher plant density per unit area, better management and manifold increase in yield. Another regular bearing cv. Ratna was released from Konkan Agricultural University, Dapoli by hybridization between Neelum and Alphonso (Salvi, 1983). The tree is moderately vigorous attaining a spread of 5.70 m and height of 6.70 m at the age of 10 years. It flowers during December to February, precocious and starts bearing fruits from the fourth year. Fruits start maturing in May and harvesting is completed by the first week of June. The fruits are ovate-oblique and slightly elongate in shape and are greenish orange with prominent oil glands. The flesh is firm and fibreless.Ripe fruits can be kept in good condition for about a week and are resistant to spongy tissues. The physico-chemical characters of the fruits of this hybrid and its parents are presented in Table 14.

Table 14: Fruit Quality of Ratna along with its Parents*

Characters	*Parents*		*Ratna*
	Neelum	*Alphonso*	
Average fruit weight (g)	225.00	250.00	315.00
Average pulp (per cent)	73.90	74.60	78.62
Total soluble solid (°Brix)	17.50	19.00	23.00
Total sugar (per cent)	12.07	15.86	16.81
Acidity (per cent)	0.27	0.34	0.26
Vitamin C (mg/100 g)	22.50	33.50	25.00]

*Salvi (1983).

13.0 Climate Change and its Probable Effects on Mango Production

Temperature, light and CO_2 concentration are important factors affecting photosynthesis, and an expected increase of the level of these factors would have a positive effect on photosynthesis. But higher temperature and CO_2 concentration have counterbalanced effects because they increase both photosynthetic assimilation and respiratory losses. Extreme temperatures (> 45°C) or high levels of light intensity provoke damages to the photosynthesis machinery.

Although mango tree is adapted to dry environments experiencing water stress and high evaporative demand, the expected increase of drought and vapour pressure deficit (VPD) would have a negative effect on photosynthesis because of the rapid stomatal closure of the mango tree when the climatic demand increases. Flooding is expected to increase in regions with more intense and frequent rainfalls, and to decrease in regions becoming drier. In both cases, the effect on mango photosynthesis will probably be negative. Mango is moderately flood-tolerant and flooding leads to a rapid decrease of transpiration, stomatal conductance and maximal photosynthetic assimilation. The decrease of flooding occurrence corresponds to the increase of drought and is dealt with the above method.

Provided nutrient and water are not limited, the increasing temperature will lead to more rapid growth rhythm (shorter time-lag between successive vegetative flushes) and development of growth units and leaves, and to the delay or the avoidance of growth cessation at the beginning of winter. Whiley *et al.* (1989) showed that the mean leaf size of trees grown at 30/25°C (day/night temperature) was 300 per cent larger than that of trees growing at 20/15°C. More dry matter would be allocated to the roots under increasing CO_2 concentration. Drought and flooding have both negative effects on the vegetative development of mango trees as they reduce tree growth. If mango trees can survive without rain or irrigation for more than 8 months, flooded trees can perish within few days.

The floral induction of mango tree is mainly driven by cool temperatures. Consequently, increasing temperatures would have a negative effect on floral induction. But in regions with particularly cool temperature during flowering, increasing temperatures would have a positive effect on pollen viability and fruit set. Temperature also has a negative effect on inflorescence size (Dambreville *et al.*, 2013) and on the number of flowers per inflorescence (Sukhvibul *et al.*, 1999). Floral induction also requires the exposure of mature leaves to light, and higher levels of light intensity could have a positive effect on mango flowering. Drought and higher VPD would have a negative effect on fruit set and retention. Drought could also have an indirect positive effect on floral induction by promoting early growth cessation and vegetative rest required for floral induction.

Increasing temperatures would have a positive effect on mango fruit growth which would be more rapid. In Australia, the estimated duration of mango fruit development decreased by 12-16 days (7-8 per cent) as a consequence of the 1.5°C increase of winter temperatures over the last 45 years (Olesen, 2011). High temperature could also have a positive effect on fruit quality because of stress-induced synthesis of secondary compounds, some of them being of nutritional value. Higher light intensity enhances skin color for colored cultivars (Léchaudel *et al.*, 2013). Since excessive light could be a stress factor, it could, as high temperature, enhance fruit quality. It could also have a positive effect on fruit size through improved photosynthesis (Urban *et al.*, 2003). Higher CO_2 concentration enhances photosynthesis and could therefore have a positive effect on fruit quality. Fruit grown at higher CO_2 concentration have a larger fruit dry mass, mainly due to an increase of the pulp dry mass. The effects of drought on fruit quality are positive and negative and well-known in non-irrigated orchards. Drought reduces fruit size (Spreer *et al.*, 2009) and increases fruit quality, in particular, the dry matter content and sugar concentration.

South Asia region extends from 5° to 30° North latitude and from 64 to 100° East longitude and includes India, the first mango producing country, and Sri Lanka. The predicted climate for the end of the 21st century, with respect to the mean climate of the last 20 years of the 20th century, will be warmer and wetter, with a mean annual increase of 3.3°C for temperature and of 11 per cent for precipitation. The temperature and precipitation increase will not be uniform among seasons, with the largest temperature increase during the flowering season (+3.6°C) and the lowest during the harvest and vegetative growth season (+2.7°C). Precipitation is expected to be lower during flowering (-5 per cent) and higher during the other seasons, in particular during the end of vegetative growth and vegetative rest (+15 per cent). The frequency of extremely warm season and extremely wet season are high, in particular for warm season. More frequent and intense hurricanes are also expected (Normand *et al.*, 2015).

For instance, an increase in temperature during coldest month has made Mango cultivation possible in the valley areas of Himachal Pradesh. Rains during flowering are harmful to Mango causing total failure sometimes. Matching crop varieties to climate is an important activity for profitable mango production in traditional and newer areas.

The warmer climate during flowering and the warmer and wetter climate during the season of vegetative rest will probably lead to a lower floral induction. On the opposite, the hot and wet climate during fruit growth and vegetative growth will promote good fruit growth and important vegetative growth after harvest. But fruit quality could be reduced and pests and diseases problems could be accentuated. Damages by hurricanes and flooding will increase in coastal areas. The mango cultivation in South Asia would be a shift of mango cultivation toward the North or in altitude to find cooler places allowing good floral induction. Mango orchards would also probably be established on the slopes rather than in the valleys or lowlands to avoid the risk of flooding. Irrigation could become facultative because of higher precipitation during fruit growth and vegetative development. Pruning will probably require more labor because of the warmer and wetter climate promoting vegetative growth.

The specific and urgent research at different levels is needed like more knowledge on stress physiology, in particular for high temperature, water stress, light stress and salinity. The genetic improvement of mango plant material with respect to climate change is a promising way of adaptation in climate resilient agriculture. Besides yield and fruit quality, cultivar selection should account for tolerance to drought, high temperature and high level of light. The ability to induce flowering at high temperature should also be an objective of selection. Rootstock selection should aim at developing tolerance to drought and to salinity, and/or at identifying low vigor rootstock with a strong root system able to pick up water in a large volume of soil and to resist to strong wind. Early phenotyping methods must be developed to speed up cultivar and rootstock breeding and selection. Specific orchard management techniques should be generated, like the improvement of irrigation techniques to increase water efficiency, the management of tree canopy and flowering, or the designing of orchards to create a favorable microclimate around the trees.

Finally, a mango crop model appears as an interesting tool to assess more precisely the effects of climate change on mango production, and to propose simulation-based ways for mango cultivation to cope with climate change. The development of an effective and reliable model would probably be long and should gather the efforts of different research teams (Normand *et al.*, 2015).

14.0 Reasons of Low Productivity in Mango and Ways to Address it

There are several reasons that can be attributed for low productivity. Predominant existing orchards are of seedling progenies Even now, in many parts of the country, senile unproductive orchards of seedling origin are being maintained. These orchards with unmanageable canopy neither produce fruits nor the quality.

Besides, these orchards are acting as sources of pest and disease. There are more than 1000 mango varieties exist in India. Most of these varieties are happened to either poor yielders or alternate bearers. The new plantations are now established through mango grafts only. In generally mango orchards used to take 4 to 7 years to reach commercial proportion due to several factors including soil, climatic and cultural methods. Most of the choice varieties are either alternate bearers or poor yielders. Evergreen trees bound to grow continuously due to abundant sunlight throughout the year. Dominant vegetative phase can affect the reproductive phase especially under tropical regions, if not regulated properly. Conventional spacing (wider spacing) is followed in many parts of the country The lands which were not suitable for agricultural crops were used for planting of perennial trees with the view that the natural resources could not support the agricultural crops any more. Hence, the fruit trees were spaced widely resulting into poor land use efficiency and yield. Several pest and diseases are affecting the mango all through the year. The yield loss due to pest and diseases is accounted to more than 15 to 20 per cent in a normal year and even to the extent of 40 to 60 per cent in severe infection. Pests like stem borer and bark borer are some of those causing severe damage to the tree. Spongy tissue in Alphonso is a typical example for physiological disorder in mango which affects both the domestic and international trading. Similarly, flower malformation, black tip and alternate bearing, are the other disorders associated with mango cultivation.

Orchardists can adopt high density or ultra high density orcharding, they can practice micro irrigation and fertigation system. Ensuring better canopy management with maechanised crop cultivation can enhance productivity. By just doing severe (90 cm from apex) and moderate (60 cm from apex) pruning in mango, net photosynthetic rate was much higher than in unpruned trees. The net photosynthetic rate, transpiration rates and leaf relative water content are generally higher in regular bearing cultivars ('Mallika' and 'Amrapali') than the biennial bearer ('Dashehari') (Singh *et al.*, 2009). Pre and post-harvest handling including bagging of fruits, harvesting with pedicel, desapping, plastic crates *etc*, top working and rejuvenation, and use of growth regulators on flower induction and setting are other focused area lareas which needs attention.

15.0 Pests, Diseases, Parasites and Physiological Disorders

15.1.0 Pests

Though the list of pests attacking mango is formidable, only two of these are considered to be devastating, *viz.*, mango hopper and mealy bug.

15.1.1 Mango Hopper (*Amritodus atkinsoni* and *Idioscopus niveosparsus*)

This is the most damaging insect during the flowering season. Until February, the hoppers remain in hiding under the bark crevices and become active during flowering period. Both adults and the nymphs are damaging, especially the latter which have a higher rate of feeding. They suck the sap from tender shoots and panicles. The panicles wither away and the fruit set is adversely affected. Injury caused by oviposition further aggravates the situation. They give out honey dew which develops a sooty mould on the leaves and panicles. Its incidence is very high in overcrowded and neglected orchards.

To manage this menace avoid dense plantings, prune overcrowded overlapping branches after rainy season. Orchards should be kept clean by regular ploughing and removal of weeds. Conservation of bio control agents like predator, *Mallada boninensis*, *Chrysopa lacciperda*, egg parasite, *Polynema* spp. *Gonatocerus* sp. *Tetrastichus* sp. and fungus, *Verticillium lecanii* is essential. Spraying of 0.2 per cent nimbecidin or azadirachtin @2m/l at initial stage of hopper population is recommended. Spray Lambda cyhalothrin 5 per cent EC@ 0.5 ml or imidacloprid 200 SL @ 0.25 ml/l or monocrotophos (0.05 per cent). First spray should be done at early stage of panicle formation if hopper population, is more than 5-10 panicle, second spray at full length stage of panicle and the third spray after fruit setting (at pea size). Chemical spray is to be minimized and should be need based. A rational rotation of insecticide is desirable to counteract the tendency of pest to develop field resistance.

Synthetic pyrethroids, *viz.* permethrin, fenvalerate (at 20, 30 and 40 ppm) or cypermethrin (at 10, 20 and 30 ppm) when sprayed after flowering were found effective in minimising the hopper population (Shah *et al.*, 1985). Spraying of Nimbecidine (0.2 per cent) is effective at initial stage of hoppers management.

15.1.2 Mealy Bug (*Drosicha mangiferae*; *Rastrococcus invadens* and *R. iceryoides*)

The female lays eggs during May under soil clods, around the tree trunk, up to a depth of 5-15 cm. The nymphs emerge in December-January and start climbing up the tree where they congregate together and suck juice from young shoots, panicles and flower pedicels. The affected parts dry up and yield is reduced substantially.

The females can be identified by their flat shape, covered with white mealy powder. Once the pest manages to reach the top of the plant, its control becomes rather difficult. Damage from the mealy bugs can be avoided if the eggs are destroyed by digging around the trunk during hot months. Sticky bands (grease and coal tar in the ratio of 1:2, resin and castor oil in the ratio 4:5) or slippery bands of alkathene, 30 to 45 cm wide, should be applied around the tree trunk about 30-45 cm above the ground level during December. The alkathene bands (20 cm wide, 200 gauze) were found to be the most effective at preventing insects reaching the tree canopies, while Namher bands (made from castor oil, sulphuric acid, axle grease, resin and glycerine and painted in a band of 20 cm wide on the trunk) could be recommended for use only if alkathene sheets were not available (Chandra *et al.*, 1991). However, before applying these bands, care should be taken to plaster all the bark crevices with mud so that the nymphs may not climb up from underneath such bands. Akathene bands need to be wiped clean at least once a week and the nymphs on the lower edges may be collected and destroyed in kerosene oil or in any strong insecticide. In case the nymphs have climbed up the tree, these should be controlled at the earliest by spraying *Beauveria bassiana* product (2g/litre) or 5 per cent NSKE in last week of January around tree trunk. Biological control agents, *Gyranusoidea tebygi* (Bokonon-Ganta and Neuenschwander, 1995) and *Cryptolaemus montrouzieri* (Mani *et al.*, 1995) have been reported to reduce the infestations of *R.*

invadens and *R. iceryoides* respectively. Releasing 10-15 grubs of cocinellid predator, *C. montrozieri* per tree is also helps to control this menace. Apart from *B. bassiana*, coccinellid beetles (predator), *Menochillus sexmaculatus*, *Rodolia fumida* and *Suminus renardi* are natural bio-control agents of this pest.

15.1.3 Stem Borer (*Batocera rufomaculata; B. rubus*)

This pest tunnels through the main trunk or its branches, weakens the plant, and in extreme cases the plant may die. Its presence can be identified by dry hard balls (small-sized) of excreta, emerging from the tunnelled portion. The control lies in cutting and destroying affected branches with grubs and pupae, cleaning the tunnels with a hard wire, pouring kerosene oil, creosote, petrol, crude oil or formalin, and subsequently, closing the entrance of the tunnel with mud after plugging it with cotton wool soaked in any emulsion of monocrotophos (0.05 per cent) or kerosene or petrol in each hole and plug them with mud. Use of green muscardine fungi, *Metarhizium anisopliae* or *Beauveria bassianna* is encouraged. The pest can be kept under check by maintaining the orchard clean and applying propanophos (0.05 per cent) or Imidacloprid (0.005 per cent) or DDVP (0.05 per cent) in hole of insect and plugging with mud.

15.1.4 Shoot Borer (*Chlumetia transversa*)

The damage is caused by caterpillars which enter the young shoots from the terminal end and bore down to a depth of 8-10 cm. Larvae of this insect bore in to the young shoot which results in its drying. The infestation of larvae has also been noticed on inflorescence stalk. Eggs are laid in tender leaves and after hatching larvae enter in mid rib and afterwards in growing shoot from tip downwards. There are four overlapping generation of this pest in a year. The insect can be controlled by Clipping and destruction of infested shoots and spraying (2-3) of Carbaryl (0.2 per cent) or Propanophos (0.05 per cent) or Imidacloprid (0.005 per cent) at fortnightly intervals.

15.1.5 Bark-eating Caterpillar (*Inderbella* sp.)

The caterpillar bores into the bark and at times makes a tunnel through the stem or the branch. Its incidence can be identified by dark brownish excreta appearing in the form of a ribbon covering the affected patches of bark. Larvae of this pest feed on the tree bark and spin brown silken web which consist of its excreta and wood pieces and make tunnels in the stem and branches. Larvae generally feed from April to December and have only one generation in a year. It can be controlled by cleaning the affected portion and killing the caterpillar mechanically by inserting a metallic wire in the tunnel and then injecting kerosene oil, petrol or ethylene glycol and kerosene oil mixture in the ratio of 1:3. The entrance of the tunnel should be plugged with cotton and plastered immediately. The population of this pest can be reduced by keeping the orchard clean and removal of webs from tree trunks and filling the insect holes with emulsion of Monocrotophos (0.05 per cent) or DVVP (0.05 per cent) and plugging with mud.

Pests of mango

Hopper infestation in mango

Mealy bug infestation

Thrips infestation

Fruit fly infestation

Pests of mango

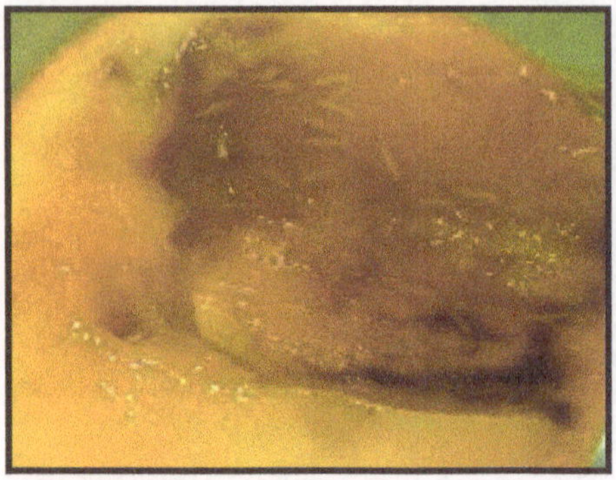

Fruit fly maggot

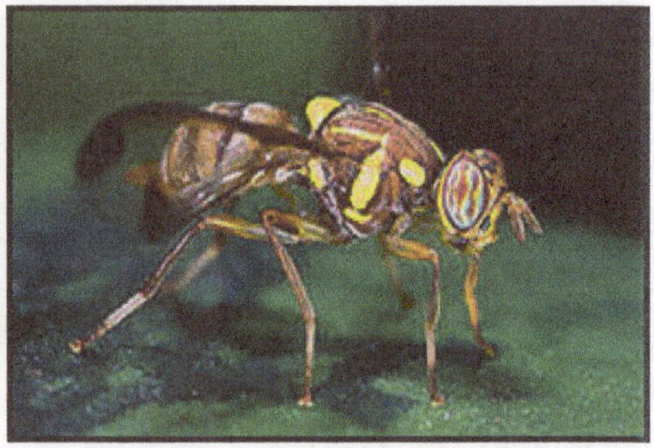

Adult fly

Leaf gall psyllid infestation

Pests of mango

Shoot gall psyllid (Apsylla cistellata) infestation

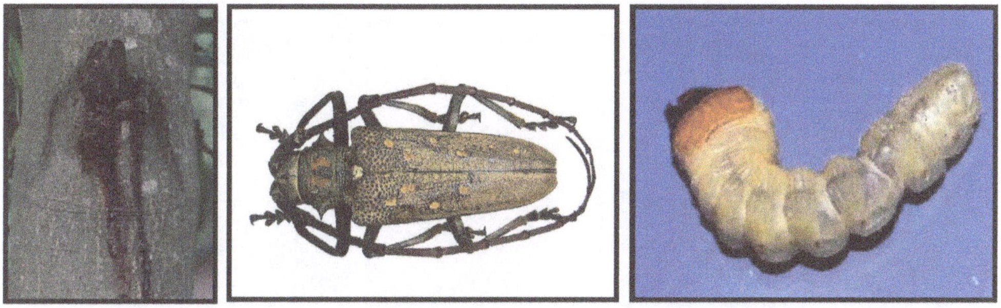

Trunk borer in mango. adult (left) and larva (right)

Shoot borer in mango

Bark eating caterpillar in mango

Pests of mango

Leaf cutting weevil

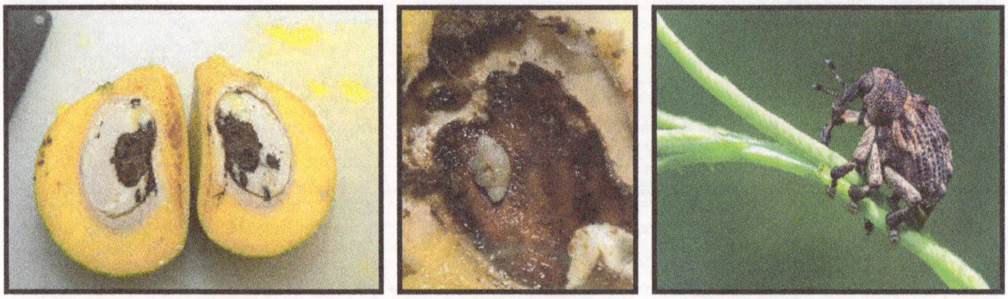

Nut weevil

Red ants

Mite infestation

15.1.6 Leaf-cutting Weevil [*Deporaus (Eugnamptus) marginatus*]

This cuts the newly emerged leaves at the base of the lamina in the form of a clear and sharp cut. The shoots, thus defoliated, lose vigour and get weakened. Its damage in the nursery can be avoided by applying chlorpyriphos to the newly emerged shoots. On grown-up plants, the damage can be checked by malathion spray (0.1 per cent), which should be repeated, if necessary, during the period of vegetative flush.

15.1.7 Shoot-gall Maker (*Apsylla cistellata*)

The damage through this insect is widely prevalent in the northern region, especially in the Terai area of Uttar Pradesh. This minute insect lays its eggs in rows of two, on the underside of the leaves of new flush (March-April) along the mid-rib. The eggs hatch in about 5 to 6 months. The nymphs, on emergence, enter the axillary and terminal buds and turn them into hard conical galls through their secretion. This generally affects the growth of the shoots and the inflorescences. However, there is hardly any effective and efficient control against this pest. The galls are usually seen during September-October. New mango orchard in humid region needs to be discouraged. The pest can be managed by avoidance of new planting in humid region, removal and destruction of infested plant parts and use of Monocrotophos (0.05 per cent) or Imidacloprid (0.005 per cent) or Propanophos (0.05 per cent) at fortnightly intervals.

15.1.8 Leaf-gall Maker (*Procontarinia matteiana*; *Amraemyia* sp.)

Its incidence is characterized by the presence, on the leaves, of round and raised galls of various colours (*viz.* grey, green and brown) depending upon the species. The eggs are laid on the underside of the leaves during March, July and October. The maggots, on hatching, bore into the leaf tissues giving rise to galls on the upper surface of the leaves. The leaves are rendered useless by continuous draining of the sap by the larvae feeding inside the gall. No definite control measure has been evolved as yet. However, 4 sprays of tar oil (2 to 3 per cent) during the egg-laying period at weekly intervals can be of some help in reducing the incidence. Endrin (0.04 per cent) sprays are also effective against this pest. The cvs. Tommy Atkins and Heidi are very susceptible, while the cvs. Sensation, Irwin, Keitt and Kent appear to be resistant to galling (Schoeman, 1995).

15.1.9 Fruit-fly (*Dacus* spp.; *Anastrepha obliqua; Bactrocera tryoni*)

This is a serious pest as the affected fruits become unfit for consumption because of the maggots feeding in the flesh. The fruit-fly lays its eggs in clusters of 150-200 under the skin of the fruit just before ripening. The affected fruits begin to rot and drop down.

The control lies in prompt collection and destruction of the damaged fruits in hot water or by burying them deep in the soil. Poisoned baits placed in wide-mouthed containers @ 10 per hectare are helpful in checking the incidence of fruit-fly. The formulations reported to be effective are (i) sodium fluorosilicate (1 part) and molasses or syrup (6 parts) in water (100 parts), (ii) lead arsenate (1 part) and molasses (24 parts) in water (100 parts) and (iii) 100 ml emulsion of methyl eugenol (0.1 per cent) and Malathion (0.1 per cent). Hanging of methyl eugenol wooden block traps soaked in ethanol, methyl eugenol and malathion (6:4:1) during fruiting period from April to August @10 traps/ha tie them tightly of 3-5 feet above ground level is highly effective to this meance. The flies can also be controlled by bait sprays of carbaryl (0.2 per cent) + protein hydrolysate or molasses (0.1 per cent), beginning at egg-laying stage. Addition of 10 ml methyl eugenol in place of molasses is also recommended. The use of dipterex [trichlorofen] and Malathion, alone or in combination was found to reduce pest numbers (Soto Manitiu *et al.*, 1987). Hot water treatment of fruit at 48 ± 1°C for 60 min or heating Kensington mangoes by means of a hot air circulatory system at > 95 per cent RH to a core temperature of 47 °C which was then held for 15 min was an effective disinfestation treatment (Heather *et al.*, 1997). Three weeks before harvesting, spraying of Deltamethrin 2.8 EC @ 0.5 ml/l +Azadirachtin (3000 ppm) or 2 ml/l is also effective and managing fruit flies and reduces anthrcnose disease and prevents late fruit fall.

15.1.10 Fruit Borer (*Deanolis albizonalis*)

This is the major pest in Odisha, West Bengal and Coastal Andhra Pradesh. Pest is active from January to May. Adults lay eggs on fruits. After hatching larvae bore into fruits. Fully grown caterpillars (25 mm) have red bands on body alternating with white bands caterpillars bore into the fruit at the bottom (beak region) and feed inside reaching Kernels. Entrance hole is plugged with excreta. Affected fruits rot and fall prematurely. Collection of fruits and dead wood after fruit harvest and destroy all fallen fruits. It is advisable to spray Monocrotophos 1.5 ml/L at marble size onwards and repeat with Deltamethrin 28 EC @ 1ml/L after two weeks in case of heavy infection and no spray should be given in fortnight before harvest

15.1.11 Stone or Nut Weevil [*Cryptorrhynchus* (Sternochetus) *mangiferae*; *C. gravis*]

This is mosly prevalent in the southern parts of the country and is reported to be specific to sweet cultivars of mango. Grabs of *Cryptorrhynchus magniferae* (more commonly occurring) damage both the pulp and the cotyledons of the stone whereas those of *C. gravis* develop in the pulp and eat only the fibre of the stone. The eggs are laid in partly developed fruits. The grabs travel through the pulp and enter the seeds where these pupate and the adults come out piercing through the stone and the pulp. General cleanliness in the orchard and destroying the adults in the bark crevices and holes, during August, has been reported to be helpful. Infested bark should be washed with kerosene emulsion.

Spraying of Acephate 75 SP @ 1.5 g/l when fruits are of lime size (2.5-4 cm diameter) followed by Deltamethrin 2.8 EC @ 1ml/l after two or three weeks is chemical method of control. Vapour heat treatment of fruits and irradiation of fruits with 0.25-0.75 KGY to control this weevil is suggested by many researchers. The natural enemies recorded on *S. gravis* include a mite *Rhizoglyphus* sp, ants (Camponatus sp., *Monomorium* sp. and *oecophylla smaragdina*) and fungus *Aspergillus* sp, *Beauveria bassiana* was found to be pathogenic on mango weevil.

15.1.12 Inflorescence Midge

In recent years, the inflorescence midge (Erosomyia *indica*) became serious in certain mango growing pockets particularly in Uttar Pradesh, affecting both inflorescence and small fruits. It affects the crop at 3 stages, *i.e.*, at floral bud burst, fruit set and tender leaves particularly encircling the inflorescence. The first phase is more damaging as the entire inflorescence destroyed before flowering and fruit set. The inflorescence show stunting growth and its axis has curve at the entry point of the larvae and ultimately die before fruit set.

Its attack on inflorescence could be recognized by presence of tiny black spots. Apart from inflorescence midge, 2 other gall midges, *Dasineura amramanjarae* and *Procystiphora mangiferae* have also been found infesting mango inflorescence. The larvae of *D. amramanjarae* are red, while *P. mangiferae* are orange in colour. Accordingly, *D. amramanjarae* infected parts are red whereas *P. mangiferae* are swollen and bigger (bud) in comparison to normal. The pest can be managed by ploughing of orchards and spraying of Propanophos (0.05 per cent) or Imidacloprid (0.005 per cent) at bud burst stage of inflorescence.

15.1.13 Red Ants (*Oecophylla smargdina*)

These are widely prevalent in mango-growing areas all over the world and cause damage to the mango trees indirectly, *i.e.* they make the new leaves useless for photosynthesis since these are moulded together in the form of a nest by the silken substance excreted by larvae. Also, these nests serve as distribution centres for various kinds of scales and mealy bugs, both living and dead, on which they feed. The red ants can be effectively controlled by destroying the nests or by putting in lead arsenate in the nests (@ 1/2 teaspoonful/nest). Spraying any contact insecticides, Dimethoate 1.5 ml/l after disturbing the nest.

15.1.14 Termites

Termites are white in colour, shy to light and remain underground. They feed on root or move upward making the tunnels. They construct mud galleries on tree trunk and under the protection of these galleries; they feed on the bark of the trunks. For the management of termites, chlorpyriphos 20 EC or its dust formulation @ 25 kg/ha in soil around infested plants can be used. Orchards should be kept clean and free of all refuse or unwanted vegetation. All the dead and decaying wood, whether on the tree or on the ground should be regularly removed. Application of finely ground *mahua* cake followed by irrigation helps to drive away the termites.

15.1.15 Mite [*Eriophyes mangiferae (Aceria mangiferae)*]

Mite has been reported to attack mango trees causing distortion and curling of leaves. In general, the infestation is higher in May than in April. A combination of pruning the infested twigs of mango trees and 2 dicofol sprays at 7-day-intervals gave good control of mite (Pena and Ferragut, 1994).

15.1.16 Leaf Webber (*Orthaga euadrusalis*)

Initially caterpillars feed on leaf surface gregariously by scrapping. Later they make web of tender shoots and leaves together and feed within. Several caterpillars may be found in a single webbed up cluster of leaves. It is recommended that pruning and destruction of infested shoots during April to May, ploughing of orchards and loosening of soil around the trees in January and spraying (2-3) with Carbaryl (0.2 per cent) or Propanophos (0.05 per cent) or Imidacloprid (0.005 per cent) at 15-days interval reduces the population of this pest.

15.2.0 Diseases

Amongst the diseases most damaging to mango are the powdery mildew (*Oidium mangiferae*) and anthracnose (*Colletotrichum gloesporioides, Glomerella cingulata*); the damage caused by other diseases like stem-end rot (*Diplodia* sp.), sooty mould (*Microxyphium columnatum, Leptoxyphium fumago* and *Tripospermum myrti*), pink disease (*Pellicularia salmonicolor, Erythricium salmonicolor, Corticum salmonicolor*) and blight (*Macrophoma mangiferae*) are serious only in certain areas. Bacterial spot (*Xanthomonas campestris* pv. *Mangiferae indica, Pseudomonas mangiferae*) is also common in humid areas, especially in the North.

Fungal Diseases

15.2.1 Powdery Mildew (*Oidium mangiferae*)

This fungal disease is widely prevalent in all mango-growing regions and can even completely destroy the crop. The damage caused by powdery mildew is often under-estimated, as disease outbreaks occur early during the flowering stage (Schoeman, 1966). Its incidence is favoured by high humidity accompanied by cloudy weather and low night temperatures during the period between panicle development and fruit set. The temperature range of 11-14°C (minimum) and 27-31°C (maximum) along with moderate relative humidity (64-72 per cent) was most congenial for the development of powdery mildew on mango (Gupta, 1985). The fungus attacks young tissues of all parts of the inflorescences, leaves and fruit. Initially, small isolated patches of powdery white mycelium develop on affected organs. These may later coalesce to cover larger areas of the plant (Schoeman, 1996).

The disease can be managed by pruning of malformed panicles and need based spray of fungicides at different stages starting with Wettable Sulphur (80WP) @ 0.2 per cent at the panicle size of 7.50 -10.00 cm. Next spray of Dinocap (48EC) or Hexaconazole (4WP) + Zineb (68WP) or Hexaconazole (5SL) @ 0.1 per cent should be done 15-20 days after of first spray. Wettable Sulphur can also be repeated for second spray. The first spray should be given at pre-bloom stage, as a preventive step, in areas more prone to the incidence of this disease. If necessary, this may have to be followed by two more applications at full bloom and fruit set stages (at fortnightly intervals). In years of persistent and severe infection the interval between sprays is to be reduced to a week and even additional sprays may be required if there is rain immediately after spraying. Joshi and Chauhan (1985b) suggested alternate spraying of tridemorph (calixin) 400 ppm and dinocap (karathane) 1500 ppm for effective control of powdery mildew.

15.2.2 Anthracnose (*Colletotrichum gloesporioides, Glomerella cingulata*)

This fungal disease is also of wide occurrence, more especially in humid and high rainfall areas. The leaves, shoots, inflorescence and the fruits are all affected by it. The characteristic symptom is the appearance of black necrotic areas on the affected parts. The affected young shoots finally show die-back symptoms. The diseased young fruitlets drop down and if the fungus attacks the fruits at maturity, the disease is enhanced by storage. As the fungus survives on dead or dried twigs, these should be pruned and burnt at the earliest.

Infection on blossom could be reduced effectively by 2 sprays of Carbendazim (0.1 per cent) at 15 day intervals. Its foliar infection can be managed by spray of Copper oxychloride or Copper hydroxide (77WP) @ 0.3 per cent, while latent infection of the pathogen on fruits could be reduced by pre-harvest sprays of Thiophanate methyl (70WP) or Carbendazim (50WP) @ 0.1 per cent. Post-harvest infection of this pathogen can be managed by post-harvest dip of fruits either with hot water alone (52 ± 1°C) or hot water in combination of fungicides, Thiophanate methyl or Carbendazim (0.05 per cent) for ten minutes. Covering of fruits on tree, 15-days prior to harvest with paper bags was also found effective in management of its post-harvest phase. The disease can also be controlled by spraying Bordeaux mixture (3:3:50), blitox or phytolan (0.3 per cent), bavistin (0.1 per cent) or benlate (0.1 per cent) thrice a year, *i.e.*February, April and September. Treated fruits could be stored at 12°C + 5 days at ambient temperature for ripening. The cvs. Tommy Atkins and Keitt were found to be resistant to anthracnose (Paez Redondo, 1995). Tiwari and Singh (1999) reported that cultivars Bambaiya, Bombay Green and Dilpasand were moderately resistant to anthracnose.

Pre-harvest sprays of hexaconazole (0.01 per cent) or Carbendazim (0.1 per cent) at 15 days interval should be done in such a way that the last spray falls 15 days prior to harvest. Hot water treatment at 52 °C for 30 minutes and covering the fruits on tree, 15 days prior to harvest with news or brown paper bags as well as using bio control agent like *Streptosporangium pseudovulgare* is recommended.

15.2.3 Die Back (*Lasiodiplodia theobromae*)

The pathogen causing dieback, tip dieback, graft union blight, twig blight, seedling rot, wood stain, stem-end rot, black root rot, fruit rot, dry rot, brown rot of panicle *etc*. The disease is most conspicuous during October November. It is characterized by drying back of twigs from top downwards, particularly in older trees followed by drying of leaves which gives an appearance of fire scorch. Internal browning in wood tissue is observed when it is slit open along with the long axis. Cracks appear on branches and gum exudes before they die out. When graft union of nursery plant is affected, it usually dies.

There must be diseses free scion wood selection and prune any infected portion, followed by spraying/or pasting of copper oxychloride or pasting with cow dung at the cut ends. Pruning (10-20 cm below dried portion) should be done in such a way that some healthy portion is also removed, to ensure complete eradication of pathogen (3 "below the infection site). The affected branches should be collected and burnt. The cut ends should be pasted with copper oxychloride (50WP) @ 5.0 per cent and/or practicing two sprays of copper oxychloride (50WP) or copper hydroxide (77WP) @ 0.3 per cent at 15 days interval.

15.2.4 Sooty Mould (*Capnodium mangiferae*)

The disease is common in areas where 'hopper' build-up is intense. The fungus develops on the honey dew secreted by the hoppers on the leaves, twigs and inflorescence. The damage caused is due to restricted photosynthetic activity of the affected leaves. The best control was observed with a spray of wettasul [sulphur] + metacid [parathion methyl] + gum acacia. Spraying of starch @2 per cent is very effective. Spray wettable sulphur + monocrotophos + gum acacia (0.2+0.05+0.3 per cent) or Indian oil formulation No. 1 & 2 (4 per cent) at 15 days interval (Om Prakash, 1991). Application of pesticides should cover both sides of leaves.

15.2.5 Pink Disease (*Erythricium salmonicolor*)

This is a serious disease in southern parts of the country, more especially in Kerala where inoculum is present on the plant and humid conditions favour its further spread. Initially, the fungus affects the bark of the young branches (below or above the forks or crotches). Portions of the bark die and peel off from the wood, resulting in some sort of ringing of the branch. The terminal portions of the branch, above the damaged area, die along with the foliage. The disease continues to develop beyond the areas affected initially and pink coloured encrustations with irregular fine cracks are formed. The infection may spread to woody tissues also, ultimately killing them too. An effective control can be obtained by pruning the affected branches at least 30 cm below the infected area at the initial stage. The cut ends of the branches may, then, be protected with a weak fungicidal paint like Bordeaux paste (1:1:30).

15.2.6 Blight (*Drechslera hawaiiensis*)

The fungus survives in the infected bark of the stems and twigs of young mango plants, in blighted leaves and as dormant mycelium in the wood. Blight of young mango plants could be successfully controlled by 4 sprays of 0.3 per cent captaf [captan], 1 per cent Bordeaux mixture, captafol (0.2 per cent), 0.1 per cent carbendazim (as bavistin) and 0.1 per cent thiophanate-methyl (on topsin-M) starting from the first week of July and repeating at fortnightly intervals (Verma and Singh, 1996).

15.2.7 Blossom Blight (*Alternaria* spp., *Colletotrichum gloeosporioides*)

Dark brown to black spots and lesions develop on affected inflorescence. The lesions coalesce after increase in size and affected flowers dry. Under severe infection, whole inflorescence may dry up without fruit set. The disease development is favoured by rains and high relative humidity. Disease can be managed by the spray of carbendazim (12WP) + mancozeb (64WP) formulation @ 0.2 per cent on appearance of first symptoms and repetition of spray at 10 to 15 days interval.

15.2.8 Phoma Blight (*Phoma glomerata*)

The disease is noticed on matured/old leaves only. Fully developed spots are characterized by dark margin and dull grey necrotic centre. In severe cases, spots coalesce to form patches, which result in withering and defoliation of infected leaves. To control this disease, spray copper oxychloride (0.3 per cent) and ensure balanced nutrition which provides resistance to phoma blight.

15.2.9 Gummosis [*Lasiodiplodia theobromae* (Syn. *Botryodiplodia theobromae* Pat.)] [*Physalospora rhodina* and its perfect stage. *Botryodiplodia theobromae* Pat.]

The disease is characterized by the presence of profuse oozing of gum on the surface of the affected wood, bark of the trunk and also on larger braches but more common on the cracked branches. In severe cases, droplets of gum trickle down on stem, bark turn dark brown with longitudinal cracks, rots completely and the tree dries up because of cracking, rotting and girdling effects.

The diseased bark/portion should be removed or cleaned and pasted with Bordeaux paste or copper oxychloride paste or cow dung paste. Application of Copper sulphate 500 gm/tree (depending upon the age of the tree) in soil around the tree trunk is recommended. Gummosis is very less in the orchards receiving regular copper oxychloride sprays for control of leaf spot diseases.

15.2.10 Wilt

It is caused by *Ceratocystis fimbriata* Ellis and Halsted. Symptom of sudden wilt of mango appears as drooping of leaves and severe oozing of gum from main trunk and branches. Sometimes branches of trees wilt one by one and whole tree dries within a year period. Management of wilt can be achieved by soil drench with thiophanate methyl (70WP) @ 0.1 per cent solution @ 100-150 l/tree followed by spray with propiconazole (25EC) @ 0.1 per cent solution. Pruning of affected branches should also be done followed by cut ends pasting with copper oxychloride (50WP) @ 5 per cent. Proper care of trees should be taken for nutrition and irrigation for quick recovery and resistance to infection.

Bacterial Disease

15.2.11 Bacterial Canker or Bacterial Spot (*Botryosphaeria ribis, Fusicoccum* sp., *Hypoxylon serpens*)

The earliest symptom of this disease, on the leaves and the fruits, is the appearance of small dark green water-soaked spots which finally assume the shape of raised black spots. These areas on the fruits develop longitudinal cracks and gum starts oozing out from the splits. Seriously affected fruits drop down and the yield may adversely be affected. Affected fruits are unattractive and unmarketable. This disease is widely prevalent and the infection increases with recurrent rainy weather. However, all the cultivars are not affected equally. The cultivars Bangalora and Neelum are more commonly affected, especially in the North. Rarely, this is encountered in Dashehari also. The cultivar *Mallika* has the drawback, especially when the fruit maturity period happens to be humid and rainy. In dry season the incidence is not noticed. Regular inspection of orchards, sanitation and seedling certification are recommended as preventive measures. Raising healthy seedlings (root stock) and wind-breaks helps in reducing brushing/wounding and thus reduces chance of infection. Three sprays of Streptocycline (200 ppm) at 10 days intervals reduce fruit infection and in severe infection, spraying of Streptocycline (300 ppm) or copper oxychloride (0.3 per cent) is more effective. An antagonistic phytoplane bacterium, *Bacillus coagulans*, is found effective. Bordeaux mixture (4:4:50) may be applied at fortnightly intervals from the first appearance of the disease on the leaves. The sprays are to be repeated immediately after a rain. Prakash and Raoof (1985) suggested spraying of streptomycin sulphate (0.025 per cent) or aureofungin (0.025 per cent). The incidence of bacterial leaf spot was effectively controlled by spraying 2 per cent agromycin-100 (streptomycin) followed by cupravit [copper oxychloride] (Sheikh *et al.*, 1995). Use of healthy stones for root stock, 3 sprayings of Streptocycline (200ppm) or Copper oxychloride (50WP) @ 0.3 per cent alone or its combination is also useful.

15.2.12 Black Banded [*Rhinocladium corticolum* Massee, perfect stage *Peziotrichum corticolum*]

The disease is noticed on the midribs/veins of the leaves, twigs and branches as black velvety raise fungal outgrowth in the form of spots which gradually increase in size and encircle the trunk limbs, branches and twigs. The incidence of disease is very low on the main branches. The disease occasionally spread on the leaves and cause loss. Gunny rubbing on twigs/branches to remove the black growth and spraying of Bordeaux mixture or copper oxychloride (0.3 per cent) can be taken as remedial mesaures.

15.2.13 Ganoderma Root-rot (*Ganoderma lucidum*)

The leaves of affected tree area are lusterless and sparse. Diseased tree wilts and dies. Infected roots are very light in weight and get easily crumbled and powdered with fingers. The fruiting bodies of the fungus (brackets) appear at the base of the tree in rainy season. It is better to expose the roots and drenching with dinocap @ 2 ml/l, covering with soil and incorporation of green leaves in the tree basin.

15.2.14 Red Rust (*Cephaleuros virescens* Kunze)

The presence of the rusty red fructification of the alga on the surface of the leaves, veins, petiole and young twigs and fruits are main symptoms. Initially the spots are greenish grey in colour and velvety in texture which finally turn into reddish brown in colour. If vigour of plant is maintained by balanced nutrients, the disease is less. Spraying Bordeaux mixture (5:5:50) or copper oxychloride (0.3 per cent) and on onset of rain, it is desired to spray fungicide twice during the month of July/August at 15 days intervals.

15.2.15 Lichens [*Strigula elegans*]

Lichens are found on full grown trees of mango, mainly on trunks, branches and twigs in the areas of high humidity, heavy rainfall and poorly managed orchards. It is seen in the form of whitish, pinkish, superficial patches of different shapes on the main trunk, branches, leaves and twigs of the trees. Field sanitation and balanced nutrition, gunny rubbing followed by spraying with commercial caustic soda (1.0 per cent) on trunk and branches are major managment protocol.

Diseases of mango

Anthracnose symptoms in fruit, panicle rachis, leaf and panicle

Powdery mildew infection

Diseases of mango

Bacterial canker infection

Stem end rot infection

Sooty mould infection

Pink disease in mango

15.3.0 Parasites

Uncared mango orchards are commonly infested with several species of the Loranthaceae family, which damage from 60 to 90 per cent of the mango trees in the North (Singh, 1960). The most destructive parasite in this category is *Dendrophthoe falcata* (earlier known as *Loranthus longiflorus*) followed by *Macrosolen cochinchinensis, Helicanthes elasticus* and *Elytranthe capitellata.*

These parasites result in retardation of the growth and yield of the trees and degradation in fruit quality. In more serious cases, the plant may even die. The only dependable method of controlling these parasites is cutting off the infected branches much below the affected area to eliminate all vestiges of the haustorial system from the host plants. Other methods, utilizing chemicals and oil emulsions, have also tried to control the parasites but the results are not consistent.

15.4.0 Physiological Disorders

15.4.1 Black Tip

This is widely prevalent in Punjab, Uttar Pradesh, Bihar and West Bengal. There is development of a small etiolated area at the distal end of the fruit, which gradually spreads and turns nearly black and covers the tips completely. Black-tipped fruits fetch very low market price (Chandra and Yamdagni, 1984).

This condition is caused by the fumes emanating from the brick kilns located near the mango orchards. The distance to which these can cause damage is reported to be about 700 yards (630 metres), although damage has also been noticed up to a distance of 1 mile (1.6 km). This disorder is quite common if the fruiting orchards fall in the direction of the wind from the brick kiln side.

Gases like carbon dioxide, sulphur dioxide and acetylene which constitute the brick-kiln fume are reported to cause the symptoms, and its incidence on fruits is inversely proportional to the distance of the orchard from the brick kiln (Ranjan and Jha, 1940; Sen, 1943b). Experimental evidence proved that fluorine in fumes emitted from adjacent brick kilns was the direct causative factor of this disorder (Zhang *et al.*, 1995). Varietal differences, in regard to susceptibility, have also been reported; the cultivars having large number of lenticels per fruit being more susceptible (Sen, 1943b).

The disorder can be avoided by allowing brick kilns only at a distance of at least 1.6 km in the east and west and 0.8 km in the north and south of the orchard, taking into consideration the usual wind direction during the fruiting season. The chimney height should also be increased to at least 15 to 18 metres to minimise the damaging effects of the smoke in the orchard. Amongst the chemicals tried to prevent the occurrence of this disorder are borax (0.6 per cent) and caustic soda (0.8 per cent) which have given good control (Das Gupta and Sen, 1958; Nauriyal *et al.*, 1972; Chandra and Yamdagni, 1984). These should, preferably be sprayed thrice, *i.e.* before flowering, during flowering and at the fruit set stage. Both borax and washing soda are basic in nature and they neutralize the acidic fumes from brick kilns.

15.4.2 Leaf Scorch

The characteristic symptom is akin to that of potash deficiency, *i.e.* scorching of old leaves at the tips and margins. This condition is quite common during the winter months when the entire old foliage in some regions appears to be affected by this symptom. The leaves fall down and the tree vigour and yield are reduced.

Studies conducted on this disorder have shown that it is caused by an excess of chloride ions which render potash unavailable (Pandey *et al.*, 1971). This disorder is common in saline soils or where brackish water is available for irrigation or where muriate of potash is used as a fertilizer to meet the potash requirements of the plants.

This disorder can be checked effectively by collecting and burning the fallen leaves and using potassium sulphate instead of muriate of potash. Acute condition can be cured by 4-5 foliar applications of potassium sulphate (5 per cent) on newly emerged flushes at fortnightly intervals.

15.4.3 Internal Necrosis

It is mainly due to boron deficiency. The water soaked grayish spots develop on the lower side of the fruit and later the spots enlarge and develop into dark brown necrotic area. The internal tissue starts disintegrating. The pericarp and mesocarp is disintegrated exposing the flesh. Yellow coloured droplets also come out and such affected fruits drop easily. Spay borax (1 per cent) at pea stage followed by two more sprays at 15 days interval and application of 250 g boron per tree (10-15 year old) around the tree basin is recommended.

15.4.4 Spongy Tissue

This disorder is common in the fruits of Alphonso, where it is observed to an extent of as high as 30 per cent. In this disorder a non-edible, sour, yellowish and sponge-like patch with or without air pocket develops in the mesocarp of the fruit during ripening. The symptom at harvest is localized predominantly at the distal end of the fruit, whereas post-harvest exposure of fruits to the sun around midday causes the maximum occurrence of spongy tissue in the middle part of the fruit (Katrodia and Rane, 1989). This malady has become more complicated since the affected fruit presents healthy external appearance. Its symptoms are not apparent at the time of fruit development, picking or when ripe. Affected tissue is visible only when the ripe fruit is cut (Joshi and Roy, 1985). The fruits have a bad odour and become unpalatable. This disorder has adversely affected the export market of this delicious cultivar.

The causes leading to this phenomenon are still being investigated into and to-date no definite conclusion has been possible. However, it is reported that harvesting the fruits at full maturity increases the incidence of the disorder, whereas on harvesting at ¾ maturity stage, the incidence is negligible (Galan *et al.*, 1984). Involvement of calcium is also indicated for its incidence (Gunjate *et al.*, 1979). The disorder is reported to occur due to the heat which is emitted by soil as convective flux. Thus, spongy tissue in Alphonso fruit of mango seems to be a physiological disorder in which the fruit pulp remains unripe because of unhydrolysed starch due to histological and biochemical disturbance caused by heat in the pulp of a mature fruit at the pre- and post-harvest stage (Katrodia *et al.*, 1985; Katrodia, 1989). Joshi and Limaye (1986) reported that percentage breakdown due to spongy tissue in Alphonso mango increased with an increase in fruit weight from 150-200 g (26.67 per cent) to 350 g (60 per cent).

The disorder appeared to be caused by convective heat rising from the soil rather than by nutrient imbalance or pathogens (Katrodia, 1989; Katrodia and Rane, 1989). It was alleviated by mulching or sod culture with *Eragrostis* (*Spartina*) *cynosuroides* (Katrodia, 1989; Katrodia and Sheth, 1989) that reduced the ground temperature by 20°C compared with clean cultivation, thus reducing convected heat (Katrodia and Sheth, 1989). The hybrid cvs. Ratna, Arka Aruna, and Arka Punit have been reported to be free from spongy tissue. The relationship of leaf and fruit transpiration rates with the incidence of spongy tissue in mango cultivars Dashehari (not susceptible to spongy tissue) and Alphonso (susceptible to spongy tissue) was investigated by Shivashankara and Mathai (1999) The significant and negative relationship observed between the fruit transpiration rate and spongy tissue suggests that the lower fruit transpiration rates in cv. Alphonso are a varietal specific trait, which results in slower movement of water and minerals to the fruits from soil leading to the development of spongy tissue.

15.4.5 Soft Nose

Also known as red nose, tip pulp, insidious fruit rot and yeasty fruit rot and more prevalent in late maturing Neelam and Mallika varieties. The typical symptom is breakdown of the flesh towards the apex of the fruit before ripening. Mesocarp cells in fruits of the cultivars Kent and Beverly showed marked cell separation and cell wall disintegration (Burdon *et al.*, 1992). Soft nose has long been cited as a calcium deficiency disorder (Burdon and Moore, 1991). The cv. Kent was more susceptible (90 per cent of fruits affected) compared with Beverly (30 per cent) (Burdon *et al.*, 1992). The cultivar susceptibility clearly plays a significant role which cannot be accounted for in terms of Ca content or distribution (Burdon *et al.*, 1991). The tmely harvesting of fruit and proper nutrition to the plant is essential to inhibit this malady.

15.4.6 Fruit Clustering

The formation of several fruit lets at the tip of panicle and fruitlets are darker green in colour and their shape is slightly curved than the normal fruits. There are the main characteristics of this disorder. These fruits generally hang for more time compared to some normal fruits, which subsequently drop due to other fruit drop reasons. However these fruitlets do not grow more and later drop. The fruits do not have formation of seeds.

During flower opening stage spray of insecticides and fungicides should be avoided and ensure population of pollinators during flowering season and if possible pollinizing cultivars should be planted in the orchard.

15.4.7 Woody Stem Gall

Woody galls of 10-15 inches diameter are formed on limbs and branches. The galls are abundant on Chinnasuvarnarekha, Langra and moderate in Neelam mango.The removal of galls using saw and applying Bordeaux paste to cut surface can control this problem.

15.5.0 Mango Malformation

The disorder was first noticed by Maries (Watt, 1891) in the Indian subcontinent towards the end of the last century. Now it is known to occur in the Middle East, Africa, Israel, Central America, Mexico and the USA (Varma *et al.*, 1974).

During the last two decades or so it has assumed an alarming magnitude in the northern India where it is threatening the very existence of the mango industry. It is of wide occurrence in Punjab, Delhi and western Uttar Pradesh and its incidence is also noticed to some extent in Gujarat and Maharashtra. It is negligible in Bihar, West Bengal and Orissa. The southern mango growing regions, beyond Hyderabad, are virtually free from this malady. Mango malformation is of two types: vegetative and floral; the former being more common on the nursery seedlings and young plants and, the latter, on trees at the bearing stage. It is the floral malformation which directly affects the productivity.

The incidence of malformation varies from year to year, cultivar to cultivar and also with age of trees. The cvs. Bombay Green, Dashehari, Lucknow Safeda and Chousa showed a high percentage of malformed panicles (10.8-24.2 per cent) with Baramasi being the least (0.32-1.92 per cent) affected (Ram *et al.*, 1990). Singh and Singh (1996) observed the highest incidence of this disorder in Khasulkhas (34.51 per cent), followed by Gulabkhas (29.58 per cent), Dashehari (20.5 per cent) and Chausa (16.39 per cent), while the percentage of malformation was significantly lower in Safeda Malihabad, Rataul, Bombay Green and Langra, averaging 8.19 to 11.65 per cent. The cv. Bhadauran was free from malformation. The rate of floral malformation was highest in the American cv. Kensington (19.2 per cent), followed by Mallika (12.3 per cent), and was lowest in Dashehari (4.6 per cent) (Yadava and Singh, 1995). Average percentage of malformation was higher in the older trees than in the younger ones (Singh and Singh, 1996). On the contrary Ram *et al.* (1990) reported that young trees (6- to 15-year-old) were more affected than older trees (16- to >26-year-old). The site of primary infection causing malformation is probably the meristematic region (Usha *et al.*, 1994).

This malady has attracted the attention of many workers and various causes have been attributed to this from time to time, *viz.* cultural practices (Watt, 1891), nutritional practices (Tripathi, 1955; Prasad *et al.*, 1965), mites (Narasimhan, 1954, 1959; Narayanan and Ghai, 1961) and viruses (Sattar, 1946; Puttarudriah and Channa Basavanna, 1961), and fungi (Summanwari *et al.*, 1966). However, the picture is still confusing and a general consensus on the precise causal organism/substances is yet to be reached. Studies conducted at the Division of Horticulture of the IARI, since 1966, have revealed that the floral malformation is broadly of three types, *i.e.* light, medium and heavy. The heavy type is the most damaging because of its failure to set any fruit. Such panicles are characterised by a condensed mass of flower buds which are largely male (Majumder and Sinha, 1972a). These panicles show suppression of apical dominance and have a tendency to hang on the trees round the year and, in this process, these rob the trees of essential nutrients. In

Physiological disorders of mango

Black tip

Soft nose

Leaf scorching

Plant parasites attack in mango

Malformation in mango

Vegetative and floral malformation (A & B - vegetative, C - Healthy & malformed panicle, D - malformed panicle)

Source: Crespo Palomo, María, Arrebola, Eva, Cazorla, Francisco M., Maymon, Marcel, Freeman, Stanley, Aoki, Takayuki, O'Donnell, K., Tores, J. And Vicente, Antonio. (2015) *Plant Dis.*, Doi: 100. 150727061835002. 10.1094/PDIS-02-15-0153-RE.

the light and medium types, flowers of both the sexes are found on panicles and some fruit sets can also be seen, though it is rare. In such cases, it is not uncommon to find a part of the panicle normal, whereas the other portions are malformed. Also, occasionally healthy and malformed flowers are encountered side by side even on the same panicle which is, otherwise, apparently quite normal looking. A better understanding of this phenomenon was reached at the IARI during the 1968 flowering season (February-March) when cv. Neelum showed an incidence of as high as 59 per cent. Since these panicles got damaged due to severe frost, the trees flowered again in June, when the incidence of floral malformation was found to be as low as 4.5 per cent. The maximum temperature during the development of March-flush of panicles was found to range from 21 to 30°C whereas at the time of June-flush it was as high as 39°C. This observation was later confirmed by creating warmer conditions during March, similar to June (42 to 45°C), in a polythene tent enclosing a plant of Neelum. The enclosed tree had 60 per cent floral malformation in the previous year but its incidence under the tent was totally absent (Majumder and Sinha, 1972b). These studies lead to the conclusion that an inverse correlation exists between temperature and the incidence of malformation. Observations at the IARI, in general, have shown that the incidence of floral malformation in the early flush of panicles is much higher than in the late-emerged panicles. Later studies have shown that deblossoming of the early flush of panicles at the bud-burst stage (when the panicles are about 1½ to 2 cm long) can shift the flowering to a later date, when the temperature is high. Panicles emerging after deblossoming, during the same fruiting season, showed a significantly lower incidence of malformation and a higher fruit set (Singh *et al.*, 1974). The increased fruit set was possibly due to an improvement in the sex ratio due to high temperature (Singh *et al.*, 1966) and significant reduction in the number of heavily malformed panicles of the hanging type as well as the others. An analysis of the healthy and malformed panicles in 4 cultivars have shown that the former have higher levels of RNA, DNA, soluble proteins and total nitrogen than the latter (Pandey *et al.*, 1975). However, the concentration of all the fractions of carbo-hydrates (*viz.* reducing, non-reducing, polysaccharides and total carbohydrates) was higher in the shoots showing vegetative malformation. On the contrary, the nitrogen fractions were always less in such shoots (Pandey *et al.*, 1973b). Leaves on twigs with malformed flowers generally had a lower rate of photosynthesis than leaves on twigs with healthy flowers (3.98 and 5.23 $\mu mol\ m^{-2}s^{-1}$ respectively), lower concentrations of sugars, starch and chlorophyll, and a higher rate of respiration (2.86 and 2.08 $\mu mol\ m^{-2}s^{-1}$ respectively (Yadava and Singh, 1995). Detailed chromatography and bioassay studies indicated higher levels of all the four fractions of auxin (*viz.* free neutral, free acidic, bound neutral and bound acidic) in healthy flower buds, as compared to malformed ones (Pandey *et al.*, 1974a). The levels of inhibitors showed a reverse trend. Tissue from malformed inflorescences contained higher levels of endogenous cytokinins than that from healthy inflorescences (Nicholson and Staden, 1988). Higher endogenous levels of ethylene and lower levels of IAA have been detected in tissues in all stages of malformed panicle development than in healthy ones. It was concluded that ethylene plays an important role in causing floral malformation in mango (Singh and Dhillon, 1989a, b). Analysis of stem tissue showed that malformed mango seedlings had lower contents of IAA and GA_3 and higher content of zeatin, abscisic acid and ethylene than healthy seedlings (Singh and Dhillon, 1989c). It appears that malformin-like substances are involved in this disorder (Singh and Dhillon, 1989d), the action of which may be ethylene-mediated (Hansraj and Ram, 1996). Floral malformation in mango is also associated with higher level of ABA, GA and zeatin but with low level of IAA (Singh, 2000). While pruning, it is useful to consider the activity of the growth promoters (gibberellins and cytokinins) and inhibitors (auxins and abscisic acid). According to Singh *et al.* (2010b), the moderate pruning (60 cm from the apex) drastically reduced the floral malformed panicles (9.40, 11.21) than the control (51.23, 41.39) trees in mango. The control (un-pruned) trees in all cultivars showed the higher malformation incidence, while lower was registered after pruning.

15.5.1 Improving the Productivity of Trees Affected by Malformation

Studies conducted at the IARI have established that the productivity of the trees, rendered unproductive by malformation, can be improved by a single spray of NAA (200 ppm) in the first week of October and deblossoming once at the bud-burst stage (Majumder *et al.*, 1970; Majumder *et al.*, 1976). Pal *et al.* (1985) obtained beneficial effects of deblossoming and treatment with cycloheximide at 500 ppm. However, partial deblossoming alone did not improve the productivity of trees showing floral malformation (Singh and Dhillon, 1988). Application of anti-malformins controlled malformations by counteracting malformin production in the panicles. The malformed panicles of Dashehari treated with 2240 ppm glutathione, 2110 ppm ascorbic acid, 140 ppm $K_2S_2O_5$ and 6000 ppm $AgNO_3$ inhibited production of malformations and fruited normally (Bist and Ram, 1985). Paclobutrazol (as cultar) applied either as a foliar spray or as a soil drench at 10-60 g/tree prior to flower bud differentiation (first week of October) reduced the incidence of floral malformation and enhanced yields and fruit quality compared with control (Singh and Dhillon, 1992). Single foliar spray of cobalt sulphate (1000 ppm) prior to flower bud differentiation successfully reduced floral malformation (Singh and Singh, 1993).

16.0 Harvesting and Yield

16.1.0 Stage of Maturity

The stage at which the fruits should be harvested has an important bearing on ripening and fruit quality.Unless the fruits are harvested at the correct stage, when they have attained full size, the characteristic taste and flavour of the cultivar cannot develop. The various criteria recommended for judging maturity are – (i) slight colour development on the shoulders (Cheema and Dani, 1934; Wardlaw and Leonard, 1936), (ii) when one or two ripe fruits fall from the plant naturally, and (iii) when the specific gravity of fruits ranges between 1.01 and 1.02 (Singh, 1960). The first two methods are not of much help since these are not representative of the fruit maturity of the entire tree and the fruits harvested do not ripen uniformly. The last method (Harkness and Cobin, 1951; Mukherjee, 1960), based on specific gravity, is more dependable. For this, fruit samples from various directions of the

Bagging for mango fruits for export promotion

Methods of harvesting mango

Hand picking of mango fruits

Methods of Harvesting Mango

Mechanical harvesting of mango

tree are taken and dropped in a bucket of water, the dipped fruits being indicative of the correct maturity.

Number of days taken by the fruit to mature depends on the cultivar and the climatic conditions and, hence, cannot serve as a guide. However, in general, fruits mature between 90 and 120 days from the fruit set stage. Mature fruits ripen faster than both half-mature and immature fruits when ripening is not induced artificially (Medlicott *et al.*, 1988). Baez-Sanudo *et al.* (1993) opined that it is possible to prolong the marketing season by harvesting fruits earlier than normal. Lee *et al.* (1998) collected fruits of Chiin Hwang mango periodically from 26 June (93 days after anthesis) to 24 August (152 days after anthesis) to determine fruit weight, volume and specific gravity as well as the relationship between degree of maturity, sugar content and the extent of mesocarp disorder. Results indicated fruit weight and volume increased as fruits developed. The specific graity was closely related to fruit maturity, and the former tended to exceed 1.00 by 121 days after flowering. Those fruits picked on day 93 had a soluble solids content of 11.2 per cent and reached 15.2 per cent by 152 days after flowering. The fruit disorder worsened at later stage of ripening; it reached 60 per cent by 135 days after flowering, as compared with 15 per cent on day 121, and a close relationship was found between specific gravity and the extent of the fruit disorder or sugar content. It is suggested that fruits should be harvested on day 120 at a specific gravity index of 1.01 in order to maintain fruit quality, yield and to solve the problem of uneven ripening. As yet, no single criterion for judging maturity correctly has been worked out (Rao *et al.*, 1970, 1972; Hulme, 1971; Krishnamurthy and Subramanyam, 1973). This aspect is of vital concern and needs further investigations.

16.2.0 Method of Harvesting

The fruits are harvested by shaking the branches which results in internal breakdown of the flesh on falling and spoils the appearance also. In the other common method, the fruit picker climbs up the tree with a collecting bag on his shoulder and the fruits plucked are placed in the bag. This method, though more satisfactory, is rather impracticable because of the large size of the tree. Besides, there is danger of breaking the branches also. The most satisfactory method available at the moment is the one which makes use of a bamboo hand tool, called mango picker. This consists of a long bamboo pole fitted with a cutting shear at the distal end, and under which a fruit collecting net is tied. However, harvesting of fruits from tall trees is problematic. Necessity is, therefore, felt for developing dwarf cultivars through breeding.

A vibratory mango harvesting system has been designed for harvesting mangoes by the limb shake-catch method. For obtaining maximum fruit removal with minimum tree damage, the shaker should be operated in the range of 7.5-10.2 cm stroke and 800-700 rev/min (Parameswarakumar and Gupta, 1988). Considerable damage is caused to the skin of mango fruits by contact with caustic latex exuded from the cut or broken pedicel at the time of harvesting. It has been concluded that the primary cause of sapburn is the entry of volatile sap components through the lenticels, resulting in tissue damage and enzymatic browning (Loveys *et al.*, 1992).

Thai mango cultivars (Nam Dok Mai, Nang Klang Wun, Tong Dum and Kaew Sawoey) were less susceptible to sap injury (sapburn) than the Australian cv. Kensington (O'Hare, 1994). The factors affecting the sap injury depend on total sap flow, sap oil content and lenticel distribution. The injury could be largely eliminated by removing the stalks and allowing the fruits to desap in a solution of 1 per cent (w/v) calcium hydroxide (pH 12.5), while a lime solution (pH also 12.5) was only slightly less effective (O'Hare and Prasad, 1992). Placing fruits in cells of specially designed perforated plastic trays resulted in lower percentage of damaged fruits (D'Angles and Parodi, 1993).

16.3.0 Yield

It is a highly variable factor depending upon the cultivars and age of the plants, climatic conditions, incidence of pests and diseases and, above all, on the upkeep of the orchards.

The grafted plants of almost all the cultivars start bearing at about the age of 4 or 5 years. The monoembryonic seedlings take a longer time to come to fruiting, ranging between 8 and 10 years, whereas the polyembryonic types come to bearing earlier. The grafted plants of regular bearing cultivars are precocious and start fruiting in the third year. The seedling trees are very vigorous in growth habit, relatively long-lived and heavier in bearing. The fruit size and quality are highly variable and no two seedlings are alike in regard to these characters. Nevertheless, it is not uncommon to find them in good bearing even up to the age of 300 years (Popenoe, 1920). One seedling tree, named 'Chhappar' in district Ambala is reported to have a trunk circumference of 32 ft (9.75 m) and an area of 2700 sq yards (2257.2 sq m) (Randhawa, 1949).

The yield is reported to be 450 maunds (16875 kg approx.). However, because of the serious disadvantages of seedling trees their cultivation is not advocated anymore. The grafted plants, which are relatively short-lived and less vigorous, bear fruits of highly uniform size and good quality. These start bearing at the age of 5 years (15-20 fruits) and the optimum crop starts from the 10th year when each tree yields about 400 to 600 fruits. The yield continues to increase up to the age of 40 years (2500 fruits), after which it starts declining (Singh, 1960). However, in certain grafted culti-vars, like Langra and Chausa, the full bearing potential is realised much later (15-20 years) than in Dashehari (10 years). On the other hand, some grafted cultivars are characteristically shy in bearing and their yield cannot be considered to be optimum or economical. These were probably selected for their fruit quality alone, *viz.* Himayuddin, Jehangir, Mulgoa, Allumpur Beneshan. The yield behaviour of the ever-bearing types like Baramasi, Dophala or Tephala is highly erratic and such cultivars can hardly fit into the commercial concept of orcharding.

17.0 Post-harvest Management

17.1.0 Ripening

During ripening, mangoes show a decrease in acidity and an increase in sugars (Tripathi, 1980; Morga *et al.,* 1979). The predominant acids are citric with lesser amount of succinic and malic (Shashirekha and Patwardhan, 1976) and tartaric with small amount of citric (Sarkar and Mushi, 1981). Reducing sugars have been found to increase (Modi and Reddy, 1967) or remain constant (Krishnamurthy *et al.*, 1960). Glucose, fructose and sucrose have been reported to be in similar concentrations in ripe mangoes (Shashirekha and Patwardhan, 1976). Krishnamurthy *et al.* (1960) reported sucrose to be the major constituent of sugars during ripening. Meldicott and Thompson (1985) found that ripening of Keitt mangoes was associated with a loss in firmness, peel chlorophyll and pulp acidity, with increasing soluble solids and total sugar (Table 13). The major sugars identified were glucose, fructose and sucrose. All the three sugars increased during ripening, sucrose was found to be in the highest concentration throughout, with fructose the predominant reducing sugar. The loss in acidity was due to decreasing titratable acidity and increasing pH values. The neutral sugars in cell wall, particularly arabinosyl, rhamnosyl and galactosyl residues decreased and polygalacturonase activity increased during ripening. It appears that enzymatic and/or non-enzymatic processes, in addition to polygalacturonaseactivity, are involved in the extensive softening of mango fruits (Mitcham and McDonald, 1992). The close correlation between changes in ß-galactosidase activity, tissue softness, and increased pectin solubility and degradation suggested that ß-galactosidase might play an important role in cell wall pectin modification and softening of mango fruit during ripening (Ali *et al.*, 1995). There was an apparent overall loss of galactosyl and deoxyhexosyl residues during ripening, the latter indicating degradation of the pectin component of the wall (Muda *et al.*, 1995). In an experiment in central Taiwan, Chiin Hwang mango fruits were harvested 93 days after full bloom and were placed at 25°C or 38°C for ripening (Lee *et al.*, 1998). In another experiment fruits were gathered 102 days after anthesis and were stored at 28°C or 38°C, to compare the ripening process under different temperatures. Results indicated Chiin Hwang fruits were unable to complete the ripening process in up to 4 days at 25°C, while ripening was accelerated at 38°C. Two days after treatment (38°C), soluble solids increased to 11.2 per cent, firmness and starch content decreased gradually as total soluble sugar concentration and alpha-amylase activity rose. Total soluble solids stood at 13 per cent or more. There was an increase in total soluble sugars corresponding to the enhanced activity of alpha-amylase at 38°C, but relatively less significant activity was detected at 28. Those ripened at 38°C also possessed orange yellow skin but with green patches and some black spots appeared 8 days after treatment. It is concluded that ripening at 28°C could retain a better post-harvest quality than ripening at 25°C or 38°C. Usually, the fruits ripen in about 5 days under tropical conditions and become over ripe in 7 to 8 days. The usual practice in vogue is to place the fruits in layers, one above the other, with a straw padding (wheat or paddy) in between the two layers. This is usually done in closed but well-ventilated rooms or godowns. Singh and Mathur (1952) reported that a temperature of 19.4-21.1°C, during ripening, was congenial to development of high TSS, reduced acidity and in improving ascorbic acid retention. Recommended optimum post-harvest ripening temperature varied between cultivars from different regions. Thomas (1975) stated that storage temperature below 25°C adversely affected the development of typical aroma, flavour and carotenoid forma-tion of Alphonso mango during ripening.

Temperature between 21°C and 24°C was recommended for ripening of mature Florida mango (Hatton *et al.*, 1965). Kalra and Tandon (1984) reported that fruits of Dashehari mango when harvested 90 days after fruit set and were held at 25°C or 30°C or at ambient temperature (33.5-35°C) for 3 to 12 days, ripening occurred after 6, 8 and 10 days at 30°C, 33.5-35°C and 25°C respectively. Fruits ripened at 0°C, however, had poor organoleptic quality. In cv. Mallika dipping of fruits in ethrel at 500 ppm in hot water (54°C ± 1°C) for 5 minutes and subsequent storage in incubators maintained at 35°C and 80-85 per cent RH showed uniform ripening after 5 days which also improved fruit quality without any adverse organoleptic character (Kalra and Tandon, 1983).

Table 15: Compositional Changes during Ripening of Keitt Mangoes at 22°C*

Storage Time (Days)	*Soluble Solids (Per cent)*	*Total Sugar (Per cent)*	*Glucose: Fructose Ratio*	*Total Sugar: Soluble Solids Ratio*	*Titratable Acidity (m Eq-1 100 g)*	*pH*	*Citric : Malic Ratio*	*Sugar : Acid Ratio*
0	4.9±0.3	3.51±0.5	0.59	0.71	20.2±4.2	3.1±0.1	4.03	0.71
2	9.5±0.4	8.29±0.6	0.56	0.87	11.9±3.0	3.7±0.2	1.56	0.70
4	11.2±1.2	10.12±1.2	0.64	0.908.5±0.6	3.9±0.3	1.75	1.19	
7	11.0±0.7	10.29±0.8	0.54	0.945.3±0.5	4.1±0.1	6.56	1.93	
9	10.6±1.1	8.40±0.7	0.57	0.79	4.0±0.7	4.3±0.2	5.80	2.09

*Meldicott and Thompson (1985).

In an investigation on regulation of cell wall softening enzymes, Reddy and Srivastava (1999) determined the polygalacturonase (PG) activity in Amrapali and Dashehari mangoes under vacuum infiltration with $CaCl_2$. Enzyme activity in untreated fruits was low at the mature green stage but increased gradually to the full ripe stage and later declined in both cultivars. Vacuum infiltration with 1 per cent, 2 per cent, 3 per cent or 4 per cent $CaCl_2$ at 300 mm Hg inhibited PG activity during ripening. Ethylene (1 ppm) treatment enhanced AG activity many times over the control at ambient temperature (25 ±1°C), while at 10°C PG activity declined significantly in ethylene-treated and also in control fruits. This suggests that temperature plays a significant role in the activity of softening enzymes. Various kinds of ripening materials in the form of leaves and sawdust have been tried for ripening mangoes, *viz.* leaves of Albizia lebbeck, Butea frondosa, mango, banana and sugarcane (Singh, 1978). Storage in sawdust, Albizia lebbeck leaves and mango leaves resulted in better ripening and fruit quality.

Other methods of ripening are all artificial, utilising special chemicals or exposure to gases like acetylene and ethylene. Post-harvest treatment with ethephon at 500 mgl/l and 1000 mg/l in hot (54°C ± 10°C for 5 min) or cold (24-28°C for 5 min) water with subsequent holding conditions at 24-28°C showed accelerated fruit ripening, and improved surface colour was achieved by using ethephon in hot water compared with untreated control or treatment of the chemical in cold water (Lakshminarayana *et al.*, 1975). These treatments improved fruit colour but the fruit quality is impaired because of the general tendency of ripening immature fruits to catch in early market and to make more profit. In general, it may be stated that the fruits ripened under room condition (open) contained more ascorbic acid whereas the carbide-treated fruits had the least amount at all stages of ripening (Singh, 1978). Ripening can also be induced by acetylene gas and there is an interaction between gas concentration and exposure time taken to initiate ripening. The 0.8 ml/l acetylene treatment required 24 h to initiate full ripening while 8 h were required with 1.6 ml/l acetylene and 1.0 ml/l ethylene (Medlicott *et al.*, 1990). Calcium carbide (2 g/kg fruit) treatment accelerated ripening and resulted in higher percentage of weight loss during storage (Abbas *et al.*, 1989). The storage life of mango fruits can be extended if the ripening process is delayed. It has been observed that mangoes can be stored successfully for 20 days if they are dipped in 200 mg GA_3/l + 2.5 per cent Vapour Gard (di-l-p-menthene) and stored at 15°C (Khader, 1992). Acetaldehyde was capable of inhibiting the activity of ACC oxidase directly, or alternatively of preventing an increase in the enzyme, thereby providing a possible mechanism for retarding mango fruit ripening (Burdon *et al.*, 1996). The relationship between softening, pectic substances, polygalacturonase (EC 3.2.1.15), pectinesterase (EC 3.1.1.11) and á-galactosidase (EC 3.2.1.23) during ripening of heated (38°C) and non-heated (25°C) mango cv. Nam Dokmai fruit was investigated by Ketsa *et al.* (1998). Fruit softening was accompanied by a progressive increase in water-soluble and ammonium oxalate-soluble pectin and a progressive decrease in alkali-soluble pectin. Polygalacturonase and á-galactosidase activities increased while pectinesterase activity decreased continuously during softening. Heating fruits for 3 days at 38°C resulted in an increase in softening, ammonium oxalate-soluble pectin and activity of á-galactosidase, and a decrease in alkali-soluble pectin and activities of polygalacturonase and pectinesterase, while heating had no pronounced effect on water-soluble pectin. Softening of ripening mango fruits was more closely related to changes in á-galactosidase activity than to polygalacturonase and pectinesterase activities.

Barcelon *et al.* (2000) used an X-ray computed tomographic (CT) technique to analyse the internal changes associated with the ripening process of mango. It was found that CT number, moisture content, titratable acidity decreased with ripening time while soluble solids and pH increased. Changes in gene expression during ripening were studied in mango fruits (Chaimanee *et al.*, 1999). Total soluble proteins were extracted at various stages of ripening, and analyzed by one- and two-dimensional gel electrophoresis. A considerable increase of 64, 42, 39 and 26 kDa polypeptides was observed.

Similarly, total RNA and poly $(A)^+$ mRNA were isolated and subjected to *in vitro* translation. The amount of total RNA increased during the ripening process, especially from the unripe to ripe stage. Analysis of translated products by SDS gel electrophoresis and autoradiography showed the accumulation of 67, 61 and 48 kDa polypeptides. The results indicated that the ripening process in mango is associated with the expression of specific genes.

17.2.0 Storage, Processing and Export

The mango is a climacteric fruit and unless the fruits are stored properly, one cannot be sure of the condition in which the fruits will reach the desired market. Proper storage is absolutely essential during the year of glut. Also, the processing units cannot utilise the entire produce at a time and proper storage conditions become vital. In general, green but mature fruits store better than those harvested ripe from the trees. Fruits (cv. Himsagar) harvested with stalks were less susceptible to storage decay than fruits harvested without stalks (Haque and Dhua, 1993). Experimental results indicate that low temperature prolongs the storage life of different cultivars from 4 to 7 weeks (Table 16).

Table 16: Storage of Mango Fruits at Low Temperature*

Cultivars	*Temperature (°C)*	*Relative Humidity (per cent)*	*Duration (weeks)*
Alphonso	7-9	90	7
Bangalora, Safeda	5.5-7	90	7
Neelum, Raspuri	5.5-9	90	5-6
Haden, Keitt	12-14	90	2
Julie	11-12	90	2
Zill	10	90	3

*Snowdon (1990).

However, very low temperature may cause chilling injury. Chilling injury (skin staining and browning) in mango cv. Zihua fruits was observed after 15 days at 2°C and after 34 days at 5°C but was not observed in fruits stored at 8°C (Ji *et al.*, 1994; Wang *et al.*, 1995). It was suggested that the increase in activity of peroxidase and cellulase enzymes was part of the chilling enzyme syndrome that developed during storage of mango fruits at chilling temperatures (Zauberman *et al.*, 1988). Another simple method advocated for enhancing storage life and for delaying ripening was that of wax coating (Bose and Basu, 1953). The fruits were treated with an aqueous wax emulsion made out of an indigenous raw material (Srivastava, 1967). Treatment with wax emulsion helps to reduce transportation and post-transportation losses in mangoes (Sharma and Kapur, 1967). This helps in improving the appearance and in prolonging the storage-life. According to Garg *et al.* (1971) fruits stored at 29-35°C, treated with 6 per cent waxsol-w-emulsion and pre-packaging (1 kg lots in 200 gauge polythene bags with 0.6 per cent ventilation) or waxed and stored at 5.5-2.7°C, resulted in reduction of weight loss and spoilage with prolonged shelf-life. Dipping Alphonso fruits in warm 25°C ± 1°C solution of carbendazim 300 ppm, captafol 1500 ppm or benomyl 300 ppm for 10 minutes could check the post-harvest rot. The treated fruits could be stored for more than 3 weeks with no infection or post-harvest rot (Joshi and Chauhan, 1985a). In another experiment mango fruits harvested at the green stage of ripeness were immersed or not in hot water (as a quarantine treatment), then treated with Primafresh 50E wax. The use of wax perceptibly reduced fruit dehydration and allowed fruits to be stored for at least 21 days at ambient temperature and humidity, maintaining quality and acceptability, whereas unwaxed fruits suffered perceptible quality deterioration (Alache and Munoz, 1998). Attempts have also been made to increase storage-life by irradiation (Dharkar *et al.*, 1966; Ahmed *et al.*, 1972), controlled atmosphere (Hatton and Reeder, 1966) and freezing (Srivastava, 1967). Hypobaric storage is a promising method to extend the shelf-life of mango. It consists of placing the fruits in a flowing stream of air, substantially saturated with humidity at a reduced pressure (50-150 mm Hg absolute) and controlled temperature (120°C). Due to lowered partial pressure of oxygen, respiration rates are reduced. Volatiles, including ethylene, are effectively removed. Both of the above factors contribute to increase the shelf-life of fruits and by using this method, mango fruits could be stored for a long period (Rameshwar, 1985). Quality afterstorage (4 weeks) was best in fruits hydro-cooled to 15°C, dipped in wax (0.7-1.0 per cent) and stored at 13°C at 60 or 100 mm Hg (Ilangantileke and Salokhe, 1989). Effect of storage temperature on fruit quality was studied by Ueda *et al.* (1999). Fully mature mango (cv. Irwin) fruits (mature red stage), harvested from trees grown in a plastic house in Japan, were stored at : 25°C for 3 days (1-a); 25°C for 3 days + 4°C for 3 weeks (1-b); 25°C for 3 days + 4°C for 6 weeks (1-c); 4°C for 3 weeks (2-a); 4°C for 3 weeks + 25°C for 3 days (2-b); 4°C for 6 weeks (2-c); or 4°C for 6 weeks + 25°C for 3 days (2-d). Each sample was analyzed for Hunter's colour, flesh firmness, respiratory rate, contents of polysaccharides, free sugars and organic acids, and organoleptic traits. In general, L and b values of the peel decreased slightly in general, but the a value increased. The a value of flesh from 1-c fruits only increased slightly, but a values in the other samples did not increase. In all samples, flesh firmness decreased. A decline in respiratory rate was observed for 1-b, 1-c, 2-a and 2-c fruits. In all samples during storage, the concentration of AIS [alcohol-insoluble solids] and starch decreased about by half, while the concentration of pectin decreased gradually. The concentration of sucrose (dominant sugar) decreased gradually. Glucose concentration in 1-a, 1-b and 1-c samples decreased, but increased slightly in 2-a, 2-b, 2-c, 2-d fruits. In all samples, the concentration of citric acid tended to decrease while that of malic acid increased. Fruit weight loss increased with longer storage periods.

The organoleptic ratings in descending order were 1-a > 1-b > unstored fruits > 2-a> 2-b > 2-d more than 2-c> 1-c. Mango fruits could be stored for ~ 3 weeks at 4°C. Mango cv. Kensington fruits could be stored at 13°C for 33 days under 4 per cent CO_2 + 2-4 per cent O_2 (McLauchlan and Baker, 1994). The best conditions for CA storage of mango fruits (cv. Rad) were 4 per cent CO_2 + 6 per cent O_2 followed by ripening at 13°C (Noomhorm and Tiasuwan, 1995). It seems that there exists a relationship between temperature and CO_2 on chilling symptoms. Enhanced CO_2 atmospheres (5 and 10 per cent) alleviated chilling symptoms (O'Hare and Prasad, 1993). Symptoms of CO_2 injury appeared in mango exposed to 50 and 70 per cent CO_2 (Bender *et al.*, 1994). Fruits stored in polypropylene bags (0.031 mm thick) containing either air or 5 per cent CO_2 + 5 per cent O_2, or stored in a low pressure chamber (260 Torr), showed delayed ripening but underwent fermentation within a few weeks at 13°C (Sornsrivichai *et al.*, 1989). The longest storage life occurred in fruits kept in perforated polypropylene bags with and without ethylene absorbent (21 and 23 days respectively), while control fruits had a storage life of < 5 days.

Perforated polypropylene bags contained 12.38-14.94 per cent CO_2 and 8.91-10.72 per cent O_2 during the storage period (Koolpluksee *et al.*, 1993). A suitable method of storing immature green mangoes in a fresh state for processing was reported by Wickham and Majeed (1999). Fruits were stored at 10, 20 or 30°C either in sealed low density polyethylene (LDPE) bags or paper bags or they were waxed before storage. Storage above 20°C proved unsuitable for fruits held longer than one week. Sixty per cent of fruits stored at 10°C in LDPE bags showed no evidence of pathological infection. Development of severe chilling injury limited the use of waxed and control fruits beyond 14 days at 10°C. There were no visible symptoms of chilling injury in fruits stored at 10°C in LDPE bags. The majority of fruits stored in LDPE bags at 10°C maintained good quality characteristics with acceptable flavour and minimal changes in chemical composition up to 21 days.

17.2.1 Mango Export Requirements

As defined by the APEDA, the mangoes shall be whole, firm, sound and fresh in appearance.; the produce affected by rotting or deterioration such as to make it unfit for consumption is excluded; clean, practically free of any visible foreign matter; free of black necrotic stains or trails;free of marked bruising; free of abnormal external moisture, excluding condensation following removal from cold storage;free from damage caused by low/or high temperature; free of any foreign smell and/ or taste; free of damage caused by pests and sufficiently developed and display satisfactory ripeness (Rajan *et al.*, 2015).

When a peduncle is present, it shall not be any longer than 1.0 cm and shall comply with the residue levels of heavy metals, pesticides and other food safety parameters as laid down by the Codex Alimentarius Commission for exports.

Grade	*Requirements*
Extra class	Mangoes must be of superior quality. They must be characteristic of the variety. They must be free of defects, with the exception of very slight superficial defects, provided these do not affect the general appearances of the produce, the quality, the keeping quality and presentation in the package.
Class I	Mangoes must be of good quality. They must be characteristic of the variety. Mangoes may have following slight defects, provided these do not affect the general appearance of the produce, the quality, the keeping quality and presentation in the package. *Defects:* slight defects in shape; slight skin defects due to rubbing or sunburn, suberized stains due to resin exudation (elongated trails included) and healed bruises not exceeding 2,3,4,5 sq. cm. for size groups A, B, C, D, respectively.
Class II	This grade includes mangoes which do not qualify for inclusion in the higher grades, but satisfy the minimum requirements. Mangoes may have following defects, provided they retain their essential characteristics as regards to the quality, keeping quality and presentation.*Defects:* slight defects in shape, slight skin defects due to rubbing or sunburn, suberized stains due to resin exudation (elongated trails included) and healed bruises not exceeding 4,5,6,7 sq. cm. for size groups A, B, C, D, respectively.

In Class I and Class II, scattered suberized rusty lenticels, as well as yellowing of green varieties due to exposure to direct sunlight, not exceeding 40 per cent of the surface and not showing any signs of necrosis are allowed.

Promotion of exports of agricultural products like mangoes is a continuous process. India has filed Market Access Requests for mangoes with several countries like Argentina, Chile, Ecuador, Israel, Malawi, Peru, Russia, Serbia, Uruguay and Uzbekistan. Under the newly introduced Agriculture Export Policy, clusters have been identified in the States of Maharashtra, Gujarat, Uttar Pradesh, Telangana and Andhra Pradesh for export-oriented production of mangoes. MEIS of 7 per cent is available on the export of mangoes. APEDA also provides assistance to Mango exporters under various components of its export promotion scheme.

Table 17: India's Exports of Mangoes during the Last Three Years [Quantity (in MT); Value (in Million USD)]

Commodity Description	*2015-16*		*2106-17*		*2017-18*		*2018-19**	
	Quantity	*Value*	*Quantity*	*Value*	*Quantity*	*Value*	*Quantity*	*Value*
Mangoes Fresh	36779.26	50.10	52761.00	66.46	49671.32	59.45	43063.70	53.80
Mangoes Sliced Dried	3644.09	6.94	376.78	2.40	569.05	3.05	1035.56	3.39
Mango Pulp	128866.01	121.29	130886.07	126.14	110923.73	104.54	72187.96	64.12
Total'	169289.36	178.33	184023.85	195.00	161164.10	167.04	116287.22	121.31

Figures of 2018-19 are provisional only.

Table 18: Export of Mangoes (Fresh/Dried) from India - Country-wise (Quantity in MT; Value in Rs. in Lakhs)*

Country	*2015-16*		*2016-17*		*2017-18*	
	Quantity	*Value*	*Quantity*	*Value*	*Quantity*	*Value*
United Arab Emirates	19974.29	19199.34	28483.16	24744.65	23542.53	18458.37
United Kingdom	1496.28	3205.72	3030.79	4956.51	3728.45	4798.33
Saudi Arabia	1399.91	1675.19	2371.99	2445.85	2670.50	2198.71
Qatar	1016.25	1023.29	2254.19	2147.25	2321.89	1981.35
United States	266.67	734.30	615.53	1548.63	800.63	1772.68
Kuwait	748.35	1298.33	1100.19	1911.26	1300.31	1630.70
Oman	426.84	412.50	897.88	822.57	2230.53	1587.93
Nepal	8704.85	1733.36	9415.38	1603.54	7878.09	1512.12
Bahrain	747.79	633.01	1086.00	978.53	1288.28	908.00
Singapore	597.61	640.31	840.69	878.08	840.62	818.62
Others	1400.41	1508.56	2665.19	2329.16	2578.63	2567.21
Total	36779.25	32063.91	52760.99	44366.03	49180.46	38234.02

Source: APEDA (2018).

17.2.2 Reduction of Chilling Injury

Pesis *et al.* (2000) reported that storage of mango (cultivars Tommy Atkins and Keitt) fruits at 12°C caused slight chilling injury (CI) symptoms on the fruit peel, expressed as red spots around the lenticels (lenticel spotting). A modified atmosphere (5 per cent CO_2 and 10 per cent O_2) was created in 4-kg film-lined cartons by using microperforated poly-ethylene (PE) or Xtend film (XF). For Keitt fruits, a similar atmosphere was also applied using controlled atmosphere chambers. After 3 weeks of storage at 12°C plus one week at 20°C, both modified and controlled atmosphere treatments were effective in reducing Cl.

The most effective reduction was found in fruits packed in the XF film. A second advantage of using XF film was the reduction in the level of sap inside the package due to the lower relative humidity in the XF film (~ ~90 per cent) compared with that of PE packaging (~ ~99 per cent). Exposure of mango (cv. Tommy Atkins) fruits to methyl jasmonate (MJ) vapours (10-4 M) for 24 h at 25°C reduced chilling injury during subsequent storage for 21 days at 7°C and after 5 days of shelf life at 20°C. The chilling tolerance induced by MJ was positively correlated with the reduction in the percentage ion leakage of mango tissue. The overall quality of MJ-treated fruits was also better than that of control fruits. MJ treatment increased the total soluble solids but did not affect titratable acidity or pH (Gonzalez-Aguilar *et al.*, 2000).

17.2.3 Packaging and Transport

The harvested fruits should first be graded according to size and appearance, although this is not the usual practice. Singh *et al.* (1989) classified the Dashehari mango fruits into three categories according to fruit weight. Grade-A fruits (average fruit weight of 210 g) were mostly floaters and Grade-C (average fruit weight 131 g) were sinkers, while Grade-B (167 g) fruits were in between. Ripe fruits and damaged fruits of relatively poor quality are usually retained for local markets and better types are packed for distant places.

Proper packaging is an essential pre-requisite for maintaining good appearance and quality of the produce on reaching the marketing centres. The most common practice in western India is to pack the fruits in bamboo baskets of 50 to 100 fruit capacity. This is followed in the district Malda (West Bengal) also. Sometimes straw is used as a cushioning material to avoid bruising injury to the fruits but this has not been found to be satisfactory (Contractor, 1951). Fruits of some hardy cultivars like Bangalora are often loaded directly into the railway wagons without any sort of packaging and these reach far off places in good condition (Cheema *et al.*, 1954). However, one should bear in mind that fruits for distant markets are to be packaged when full grown but unripe. Quite often, a small packet of calcium carbide is also placed in the box while packing the fruits.

Besides bamboo baskets, wooden crates of various dimensions are also utilised in packaging mango (Singh, 1960). The dimensions of the crates or boxes vary with the fruits of a cultivar to be packed. The wooden packing cases are, however, costlier and uneconomical. Emphasis is, therefore, being laid on utilizing corrugated cardboard boxes which are lighter in weight and can be recycled. For

Storage and packaging of mango

Cold store

Mango storage in amnbient condition

Packaging in double foam net

Packaging and Marketing of Mango

Packaging in corrugated fibre board box

Packaging in wooden box

Transport systems for marketing of mango

export purpose, the fruits are wrapped in tissue paper and cushioning is provided by paper shavings or wood wool. The dimensions of different types of packing boxes are given in Table 19.

Table 19: Dimension of Packing Boxes for Mango Fruits*

Packing Boxes	*Length (cm)*	*Width (cm)*	*Height (Per cent of width)*	*Internal Volume (L)*	*Load Capacity (kg)*	*Mode of Transport*
Three-piece fibreboard	30-35	28-35	30-65	9-25	10	Truck, rail box with wooden frame and air
Corrugated fibreboard	40-45	25-40	30-50	10-30	12	Truck, rail, bliss box, sea and air
Hinged lid box	30-45	20-30	25-30	5-10	6	Air

*Thompson (1996).

In SriLanka, the shallow slatted crates have been found to be an economically accep-table method of transportation for the traders (Wijeratnam *et al.*, 1996).

17.2.4 Processing of Mango

In general, the raw fruits are used for preparing pickle, chutney, powder, flakes, and drinks and the ripe fruits besides being consumed as dessert, are also processed in to various value added products such as juice, nectar, squash, jam *etc.* The mango seed, often called as kernel, contains 8-10 per cent fat, which can be used in soap making and also serves a good substitute for cocoa in confectionary. Even roasted mango kernel is relished by many people.

Mango fruits are utilized at every stage of their growth for the development of several value added products such as *chutney*, mango powder (*aamchoor*) and leather, commonly called as '*Aam papad*' in India.

Variety	*Chief Processing Characteristics*
Alphonso	Good for pulping and canning
Bangalora	Used for processing
Baneshan	Good for canning
Ramkela	Excellent for pickling
Totapuri	Excellent for processing
Kesar	Ideal for pulping and juice concentrates
Chausa	Good for processing purpose
Amrapali	Excellent for pulping
Mallika	Good for processing purpose

Some Processing Type Mango Varieties

Raw mango fruits can be used in a number of ways. In the beginning, mangoes are commonly used for sweet or sour *chutney*. Mangoes can be processed into purée from whole or peeled fruit. Because of the time and cost of peeling, this step is best avoided but with some varieties. Mango oil, also called as mango kernel fat, or mango butter, is extracted from dried kernels of ripe mango fruits by hydraulic press, or by solvent extraction.

18.0 Biotechnology

The mango improvement work by *in vitro* systems was mainly initiated in the early 1980s, and some progress in somatic embryogenesis, *in vitro* selection, transformations imparting disease resistance, control of ripening *etc.* has been made in the last 20 years (Mitra, 2001). The achievements in many of the aspects so far are not very satisfactory, but there is enormous scope to improve the most important tropical fruit by the application of biotechnology. The salient results of the investigations on biotechnology in mango are mentioned below.

18.1.0 Tissue Culture

Certain aspects of tissue culture pertaining to propagation *in vitro* have been discussed under propagation.

18.1.1 Somatic Embryogenesis

Histological analysis revealed the nucellar origin as well as the variable location of adventive embryos within two micropylar portions of the ovules. Embryogenic callus was initiated from the micropylar and chalazal portions of excised nucelli. In suspension culture, proliferation occurred via secondary budding from the epidermis of preformed somatic embryos. These somatic embryos exhibited numerous developmental abnormalities; normal embryo development occurred on solid medium. Embryogenic efficiency of callus was improved by transfer to solid medium containing glutamine, 2,4-D and kinetin. Germination of somatic embryos occurred on medium where sucrose concentration was subsequently lowered and the synthetic growth regulators were replaced by 20 per cent liquid coconut endosperm and 0.025 per cent casein hydrolysate (DeWald, 1987). DeWald *et al.* (1989a) demonstrated that the entire tissues can produce embryogenic cultures on a modified liquid B5 basal medium. They suggested subculture from liquid to solid medium when the somatic embryos reach the heart stage of development. Multiple somatic embryos normally developed from each mass while the smallest embryogenic masses rarely produced individual somatic embryos. By sequential transfer to a medium containing 20 per cent (v/v) coconut water and reduced sucrose concentrations, maturation of embryos and germination were achieved (DeWald *et al.*, 1989b).

Litz *et al.* (1998) studied the efficacy of different genotypes on induction of embryo-genic cultures. In their experiment, nucellus was removed from immature seeds of four mango genotypes (polyembryonic Hindi and Nam Doc Mai and monoembryonic Lippens and Tommy Atkins), and cultured on different media.

Induction of embryogenic competence was mediated by genotype, explanting 2,4-D and the presence of a highly embryogenic nurse culture, although there was considerable difference in genotype response, Hindi had the greatest embryogenic potential, followed by Lippens, Tommy Atkins and Nam Doc Mai. Induction of embryogenic cultures of all genotypes at low frequency was found to occur as results of explanting excised nucellus onto no growth regulator medium. The most effective treatment for inducing embryogenic culture was 2,4-D + embryogenic Parris nurse culture, except for Tommy Atkins, in which the treatment with 2,4-D alone was most effective. Cotyledonary-stage somatic embryos (3-5 mm in length) originating from nucellar explants of cv. Amrapali were encapsulated individually in 2 per cent alginate gel (Ara *et al.*, 1999). The encapsulated somatic embryos (ESEs) germinated successfully on 0.6 per cent agar-gelled medium containing B5 macrosalts (half-strength), MS microsalts (full- strength), 3 per cent sucrose and 2.9 μM gibberellic acid. The percentage of germination of ESEs was higher than that of naked somatic embryos of the same size on the same medium. The germinability of ESEs was increased (73.16 ± 7.08 per cent) when the medium was supplemented with full-strength B5 macrosalts. Of the germinating ESEs, 45.83 ± 3.40 per cent developed into plantlets.

18.1.2 Somaclonal Variation

Anthracnose caused by *Colletotrichum gloesporioides* is considered as a widely occurring disease of mango. The fungus infects blossoms, leaves and fruits which is, however, more acute under favourable conditions (high humidity and high rainfall). The presence of latent infection could be a serious problem during post-harvest stages.

Litz *et al.* (1991) and Mathews and Litz (1992) demonstrated the potentiality of *in vitro* selection of mango showing enhanced resistance to *C. gloesporioides*. Szabados *et al.* (1987) suggested using the culture filtrate of *C. gloesporioides* to select stylosanthes cells for resistance to anthracnose *in vitro*. Jayasankar (1995) believed that the host resistance response may be linked with the production of hydrolases. Jayasankar and Litz (1998) selected embryogenic nucellar culture of two polyembryonic cultivars (Hindi and Carabao) for resistance to the culture filtrate and phytotoxin of a virulant strain of *C. gloesporioides* that was isolated from leaves. The cultures were recurrently selected either with progressively increasing concentrations of culture filtrate or by continuous challenge with the same concentration of either culture filtrate phytotoxin. Mycelium growth was inhibited when the pathogen was co-cultured with the selected, resistant embryogenic cultures.

18.2.0 Biochemical and Molecular Markers

In the last 10 to 15 years, various molecular techniques have been successfully applied in determining the genotypic profiles of individuals and/or populations of numerous wild and cultivated *Mangifera* species, moreover DNA markers, such as restriction fragment length polymorphism (RFLP), amplified fragment length polymorphism (AFLP) and randomly amplified polymorphic DNA (RAPD), offer significant advantages in comparison with morphological and isozyme markers because they can be readily obtained in large numbers, can provide greater discrimination of cultivars, and are unaffected by environmental factors. In this context, RAPD has been employed extensively not only for the determination of genetic variability within plant taxonomic groups but also as an auxiliary tool in breeding programs and in obtaining genetic maps.

Degani *et al.* (1990) characterized 41 mango cultivars derived from self-pollinated and open-pollinated trees using isozymes. They identified 6 loci with 17 allelomorphs and determined the outcross origin of some mango cultivars. Schnell and Knight (1992) successfully differentiated zygotic from nucellar seedlings using different biochemical markers. Degani *et al.* (1992) demonstrated that there were two distinct zones of PGI activity in mango (PG11 and PG12) and suggested that four alleles controlled the PG12 banding. Truscott *et al.* (1993) characterized 88 mango cultivars on the basis of isocitrate dehydrogenase, cytosol aminopeptidase, glucose-6-phosphate isomerase, phosphoglucomutase and triosephosphate isomerase. The outcross origin of some cultivars was supported by the zymograms. The reported parentage of some other cultivars did not conform to their banding patterns.

Eiadthong *et al.* (1999) examined twenty two mango (*Mangifera indica*) cultivars, including 13 Thai, 4 Florida, 3 Indian, one Indonesian and one Philippine cultivar, for genetic variation using SSR-anchored primers of 15-18 oligonucleotides. Of the 40 primers screened, 7 gave reproducible, polymorphic DNA amplification patterns and were selected to construct a DNA fingerprinting table to distinguish the mango genotypes. Banding patterns obtained using these 7 primers allowed each cultivar to be distinguished from the others, indicating that polymerase chain reaction using SSR-anchored primers was an efficient method for cultivar identification. Thai mango cultivars Nang Klangwan and Nong Saeng were very far distant of the genetic relationship from the other cultivars. The remaining 11 Thai cultivars were classified into 3 groups based on the dendrogram. Seven cultivars were placed in the same group as two Florida cultivars (Brooks and Edward), Philippine cv. Carabao and Indonesian cv. Arumanis. Four other Thai cultivars were scattered into two groups, each group contained Indian cultivars.

A wide genetic diversity exists in the mango cultivars in India. Many commercial cultivars originated mainly from this subcontinent. Ravishankar *et al.* (1999) selected 18 commercial mango cultivars, traditionally grown in western, southern, northern and eastern parts of India, to assess genetic relatedness. Total genomic DNA was extracted and subjected to RAPD analysis using 30 arbitrary 10-mer primers. Of these, 27 primers amplified mango genomic DNA. None of these primers produced unique band pattern for each cultivar. Cluster analysis clearly showed two groups - the first consisting of western, northern and eastern mango cultivars and the second group consisting of southern cultivars. It appears that the majority of mango cultivars originated from a local mango genepool and was domesticated later.

Kashkush *et al.* (2001) have used AFLP for both the identification of mango cultivars and the generation of a preliminary genetic map. Six AFLP primer combinations produced 204 clear bands in 16 mango cultivars and six rootstocks. The Band-Sharing between cultivars and rootstocks was on the average 81 per cent. The probability of obtaining a similar pattern for two different mango cultivars and rootstocks was found to be: 6 x 10-3 and 2 x 10-3, respectively. They have constructed a preliminary genetic map by genotyping the progeny of the cross: 'Keitt' x'Tommy-Atkins'. The map is made by 34 AFLP bands, consisting of 13 linkage groups and covering about 160 cM (Lavi *et al.*, 2004).

Schnell *et al.* (2006) used 25 SSRs to estimate the genetic diversity among 203 mangos and five accessions of related species cultivated and maintained in Miami, Florida. The 25 SSRs revealed an average of 6.96 alleles per locus and an average polymorphic information content value of 0.552 among the mango accessions. The 25 SSRs they used for the 208 accessions were moderately polymorphic, suggesting that this number is adequate for studies of genetic diversity and parentage. Hirano *et al.* (2010) used 11 SSRs to conduct a phylogeographical analysis of Myanmar mango landraces, which they compared with accessions from Florida, India, and Southeast Asia. Tsai *et al.* (2013) used 37 SSRs to identify cultivars and estimate their genetic relationships of 22 Taiwanese cultivars. Recently, 176 novel SSR markers were newly identified originated from 'Irwin' by using the Roche 454 GS FLX+ platform in Japan. The start codon targeted (SCoT) markers have also proved useful in evaluating the relationship among the Xiang Ya Mango type cultivars in China. SCoT markers are useful for the confirmation of identical genotypes and clones with different names.

SCoT and other dominant markers are useful for germplasm collection management, and they are technically simple and inexpensive to implement.

Recently, Ravishankar *et al.* (2015) using 14 carefully selected SSR markers on a comprehensive set of 367 Indian mango cultivars identified two main sub-populations representing mango cultivars from the North-East and South-West regions of India.

Significant progress has been made in the area of whole genome sequencing of mango in recent past. A group led by Professor N.K. Singh identified 1.67 million high quality SNPs by sequencing double digested restriction site associated DNA (ddRAD) from 84 diverse mango cultivars from different regions of India and abroad and a database of SNPs was created. They have developed large number of genic simple sequence repeat (SSR) and single nucleotide polymorphism (SNP) markers from RNA sequence data and generated 60 and 58 million short sequence reads of pooled cDNA libraries of RNA from the leaves of mango varieties 'Neelam' and 'Dashehari' using SOLiD sequencing technology and assembled these into 27,528 and 20,771 transcriptome shotgun assembly (TSA) contigs, respectively (Singh *et al.*, 2016).

Mango has a relatively small genome size of 439 Mb that should have made it relatively easy to sequence and assemble the genome using high throughout second generation sequencing technologies. However, high heterozygosity is a real challenge in achieving a high quality reference genome assembly of mango. The first draft genome assembly of mango cultivar 'Amrapali' using Illumina MiSeq overlapping paired-end reads at San Diego PAG meeting (Singh *et al.*, 2014), but the assembled genome size of 492 Mbp in 211,141 contigs was unexpectedly higher than the actual genome size, indicating redundancy in the contig assembly due to high heterozygosity.

18.3.0 Genetic Transformation

Genetic engineering, an alternative approach for improving mango production, emerged just over a decade ago, and offers a sustainable way for addressing certain key breeding objectives. The primary components of mango genetic engineering include: 1) efficient somatic embryogenesis and plant recovery from elite (i.e., nucellar) material; 2) induction of random mutations in embryogenic cultures and challenging for resistance to a specific selective agent; and 3) transformation with a gene that mediates a horticultural trait. Transfer of discrete genetic information between species can be accomplished using cell culture and molecular techniques. Certain genes are cloned in a genetically engineered bacterial plasmid that also contains a promoter, *e.g.*, 35S from CaMV, and selectable marker genes, *e.g.*, neophosphate transferase (NPT II) and a reporter or scorable gene, *i.e.*, â-glucuronidase (GUS).

Non-oncogenic strains of *Agrobacterium tumefaciens* carrying the marker gene *nptII*, which confer resistance to kanamycin, were used for transformation studies of mango somatic embryos. Growth of proembryos in suspension culture was arrested at 12.5 mg/l kanamycin, whereas more mature somatic embryos grown on solid medium were inhibited by 200 mg/l kanamycin (Mathews and Litz, 1990). Somatic embryos of Keitt were successfully transformed with *A. tumefaciens* (C 58) harbouring the plasmid pGV 3850:1103, which confers resistance to kanamycin. Transformed somatic embryos were developed to maturity and germinated (Litz *et al.*, 1990).

The histochemical stain for the presence of GUS confirms the expression of transgenes. Delivery of genes into plant cells is accomplished in two ways: biolistics and *Agrobacterium*-mediated transfer. The latter technique has been utilized for mango transformation (Litz, 2004).

18.4.0 Linkage Mapping

Genetic maps are particularly important and valuable tools for quantitative trait locus (QTL) mapping and marker assisted selection (MAS) of plant with desirable traits. Linking and mapping important mango traits with molecular markers will improve the efficiency of mango breeding. One of the traits of mango that is very distinct is polyembryony in which multiple apomictic embryos develop from the maternal nucellar tissues around the fertilized egg in addition to a single zygotic embryo.

The first linkage mapping study in mango was conducted by Kashkush *et al.* (2001). In this study, a linkage map based on the progeny of a cross between *'Keitt'* and *'Tommy Atkins'* was constructed using 34 AFLP markers, which were mapped in 13 linkage groups that covered 161.5 cM. A second linkage map was constructed by Fang *et al.* (2003) using 60 progeny of a cross between *'Keitt'* and *'Tommy Atkins'*. In this map, 39 loci were mapped in 14 linkage groups. Using whole-genome sequencing technology, several hundred SSR markers have been developed and are now being validated in mapping populations (Arias *et al.*, 2012), which will enable us to develop a high-density genetic map and to identify trait linked markers for mango.

F_1 plants from a cross between *Mangifera indica* L. "Jin-Hwang" and *M. indica* L. "Irwin" and their parent plants were subjected to high-throughput sequencing and specific-locus amplified fragment (SLAF) library construction. After preprocessing, 66.02 Gb of raw data containing 330.64 M reads were obtained. A total of 318,414 SLAFs were detected, of which 156,368 were polymorphic. Finally, 6594 SLAFs were organized into a linkage map consisting of 20 linkage groups (LGs). The total length of the map was 3148.28 cM and the average distance between adjacent markers was 0.48 cM. This map could be considered, to our knowledge, the first high-density genetic map of mango, and might form the basis for fine QTL mapping and MAS of mango (Luo *et al.*, 2016).

18.5.0 Cryopreservation and *In vitro* Conservation

Mango cryopreservation is a useful technique for mango breeding programs. Stored pollen is important for developing breeding programs because it can be effective for cross-pollination breeding works. Mango pollen can be used after two to five years of cryopreservation. Chaudhury *et al.* (2010) employed pollen cryopreservation at between -170 °C to -180 °C over four years to achieve long term storage in several mango cultivars.The enlargement and germination of nucellar embryos were inhibited for as long as 4 weeks after subculture from ABA containing medium. Mannitol at concentration between 7.5 per cent and 12.5 per cent inhibited nucellar embryo development presumably due to osmotic effects; however, there was no residual effect after subculture of somatic embryos onto medium without mannitol. Temperature between 22.5°C and 37.5°C stimulated embryo development, whereas lower temperature (7.5°C and 15°C) delayed germination. There was no germination at 1 month after somatic embryos, pulsed for 8 weeks at 7.5°C, were transferred to 22.5°C; however, after 2 months 86 per cent of these somatic embryos germinated (Pliego-Alfaro *et al.*, 1996).

18.6.0 Genomics

Research on tree crops is expensive, slow, and has often been focused on limits on production in horticultural species. Recently, the development of genomic and transcriptomic tools has contributed to a better understanding of the metabolic and molecular processes involved in floral biology. Therefore, this complex phenomenon could be solved by transcriptome analysis resulted into numerous differentially expressed genes (DEGs), allowing the partial identification of mechanisms that convert 'on' into 'off' buds (Sharma *et al.*, 2015). RNA-seq can be used to look for allele specific expression in heterozygous organisms. This is best done with hybrids between very divergent genotypes. This data can be utilized to infer the presence of cis-regulatory variation that influences gene expression levels, to find out regulatory regions and their function in fruit crops. Several candidate genes would be identified in future studies whose differential expression can be correlated with growth habit and architectural variation in another perennial fruit crops. A detailed study is required to coordinate the regulation among chromatin structure, transcription, regulatory genes 'on' and 'off' mechanism and alternative splicing during vegetative to flowering and fruiting transition. Transcriptome profiling is needed for both types of mRNAs and small regulatory RNAs to study the regular and alternate bearing behavior of perennial fruit crops. Fundamental to the task of analyzing gene expression data is the need to select those genes whose patterns of expression are related to a specific phenotype of interest (Sharma *et al.*, 2017).

19.0 References

Abbas, T., Iqbal, J. and Khan, M.D. (1989) *J. Agril. Res.* (Lahore), **27**: 121-126.

Abd-El-Al, A.A., El-Demerdash, A.M. and Ebd-El-Kader, A.M.M. (1994) *Ann. Agric. Sci., Moshtohor*, **32**: 2029-2038.

Abdel-Galil, H.A. (1992) *Assiut J. Agril. Sci.*, **23** : 137-151.

Ahmed, M., Naqvi, M.H., Hussain, A., Mohyuddin, M., Sattar, A. and Ali, M. (1972) *Pakistan. J. Sci. Ind. Res.*, **15** : 314-19.

Alache, G.J. and Munoz, A.C. (1998) *IDESIA*, **15** : 15-19.

Ali, Z.M., Armugam, S. and Lazan, H. (1995) *Pytochemistry*, **38** : 1109-1114.

Allan, R.G. (1940) *Bull. U.P. Dept. Agric. Fruit Ser.*, **13** : 45.

Anonymous (1962) *Wealth of India*, Raw Materials, C.S.I.R., New Delhi.

Anonymous (1966) *Special Report Series*, No.4, ICMR, New Delhi.

Anonymous (1977) *Res. Rep. Queensland Dept. Agric.*, pp. 12, 127.

Anonymous (2016) *FAO 2016 Production Year Book*, Vol.45, FAO of the United Nationas, Rome, 2016.

Anonymous (1994) *Global Aam's race, Times of India*, New Delhi, June 18, 1994.

Anonymous (1997) *Selected Indicators of Food and Agriculture Development in Asia-Pacific Region*, 1986-96, RAP Publication, FAO of the United Nations, Bangkok, 1997/23.

Anonymous (1999) *Plant Varieties J.*, **12** : 36.

Anonymous (2018a) *https: //mangifera.res.in/partner.php*

Anonymous (2018b) Horticultural statistics at a glance 2018, DAC, Ministry of Ag. & farmers' welfare, GOI

Ara, H., Jaiswal, U. and Jaiswal, V.S. (1999) *Plant Cell Rep.*, **19** : 166-170.

Arias, R.S., J.W. Borrone, C.L. Tondo, D.N. Kuhn, B.M. Irish and R.J. Schnell (2012) In: Shunell, R.J. and P.M. Priyadarshan (eds.) *Genomics of Tree Crops*. Springer, New York, pp. 209–239.

Augustin, M.A. and Ling, E.T. (1987) *Pertanika*, **10** : 53-59.

Avilan, R.L. (1971) *Agon. Trop.*, **21** : 3-10.

Avilan, R.L. and Rengifo, A.C. (1992) *FONAIAP Divulga*, **9** : 32-35.

Avilan, R.L., Rodriguez, M. and Ruiz, J. (1995) *Agronomia Trop.* (Maracay), **45** : 445-456.

Avila-Resendiz, C., Garcia-Perez, E., Matheis-Toledano, L. and Mosqueda-Vazquez, R. (1993) *Acta Horticulturae*, **341** : 281-287.

Baez-Sanudo, R., Siller-Cepeda, J., Bringas, T.E. and Baez, S.M. (1993) *Proc. Interamerican Soc. Trop.Hort.* (Ed. R.J. Campbell), **37**: 148-154.

Bajpai, P.N., Yati, V., Singh, A.R. and Chaturvedi, O.P. (1985) *2nd Int. Symp. on Mango*. India. Abst. No. 3.22, p. 26.

Balakrishnamurti, D. and Jogiraju, G. (1932) *Madras Dept. Agril. Bull.*, No. 24.

Balamohan, T.M., Mekala, P., Rajadurai, S., Priyadharshin, G., Prakash, S., Soorianathasundaram, K., and Kumar, N. (2016) Department of Fruit Crops, HCRI, TNAU, Coimbatore, pp. 28.

Banik, B.C., Sen, S.K. and Bose, T.K. (1997a) *Environ. Ecol.*, **15** : 122-125.

Banik, B.C., Sen, S.K. and Bose, T.K. (1997b) *Environ. Ecol.*, **15** : 269-271.

Banik, B.C., Mitra, S.K., Sen, S.K. and Bose, T.K. (1997c) *Indian Agric.*, **41** : 187-192.

Barcelon, *e.g.*, Tojo, S. and Watanabe, K. (2000) *Intern. Agric. Engineering J.*, **9** : 73-80.

Barfod, A. (1988) *Nord. J. Bot.* **8**: 3–11.

Barwal, R.N. (1995) *Insect Environ.*, **1** : 16.

Beddoe, T.W. (1994) *Trop.Fruit Newsl.*, **10** : 3-4.

Behera, T.K., Rath, S. and Dora, D.K (1994) *Orissa J. Agril. Res.*,**7** : 41-44.

Bender, R.J., Brecht, J.K. and Campbell, C.A. (1994) *Proc. Fla. Sta. Hort. Soc.*, **107** : 274-277.

Bhan, K.C., Samaddar, H.N. and Yadav, P.S. (1969) *Trop. Agric., Trinidad*, **45** : 247-53.

Bhatt, J.R. and Shah, J.J.(1985) *Indian J. Exp. Biol.*, **23** : 330-339.

Bhujanga Rao, C. and Rangacharlu, V.S. (1958) *Indian J. Hort.*, **15** : 173-83.

Bhuva, H.P., Katrodia, J.S., Patel, H.L. and Chundawat, B.S. (1985) *2nd Int. Symp. on Mango*, India, Abst. No. 4.15, p. 35.

Bijhouwer, A.P.C. (1937) *Flowering and Fruiting Habits of the Mango*, H. Veenman and Zonen, Wageningen.

Bist, L.D. and Ram, S. (1985) *2nd Int. Symp. on Mango*, India, Abst. No.9.12, p.98.

Bokonon-Ganta, A.H. and Neuenschwander, P. (1995) *Biocontrol Sci. Tech.*, **5** : 95-107.

Bompad JM. and Schnell RJ. (1997) In: *Mango: botany, production and uses*. pp. 21–47.

Bose, A.N. and Basu, G.(1953) *Sci. Cult.*, **19** : 263.

Bose, S. and Sadhu, M.K. (1985) *2nd Int. Symp. on Mango*, India, Abst. No.3.24, p. 27.

Bojappa KM, Singh RN (1974) *Indian J. Agric. Sci.*, **44**: 32-35

Broertjes, C. (1977) *Acta Horticulturae*, **75**: 19-26.

Broertjes, C. and Van Harten, A.M. (1978) In: *Application of Mutation Breeding Methods in the Improvement of Vegetatively Propagated Crops – An interpretative literature review*, Elsevier Scientific Publishing Co., Amsterdam, p.360.

Burdon, J., Dori, S., Marinansky, R. and Pesis, E. (1996) *Post-harvest Biol. Tech.*, **8** : 153-161.

Burdon, J.N. and Moore, K.G. (1991) *Planter*, **67** : 477-481.

Burdon, J.N., Moore, K.G. and Wainwright, H. (1991) *Scientia Hort.*, **48** : 329-336.

Burdon, J.N., Moore, K.G. and Wainwright, H. (1992) *Acta Horticulturae*, **296** : 15-22.

Burns, W. and Prayag, S.H. (1921) *Bombay Dept. Agric. Bull.*, No. 103.

Burondkar, M.M., Gujate, R.T. and Govekar, M.A. (1994) *J. Maharashtra Agril. Univ.*, **19** : 408-409.

Campbell, C.W., Knight, R.J.Jr. and Zareski, N.L. (1977) *Proc. Fla. Sta. Hort. Soc.*, **90** : 254.

Campbell, R.J. (1992) In : *A Guide to Mangoes in Florida* (Ed. R.J. Campbell), Fairchild Tropical Gardens, Miami, Florida, p. 24.

Campbell, R.J. and Campbell, C.W. (1993) *Acta Horticulturae*, **341** : 54-59.

Cao Xiaoan and Wang Hongxi (1999) *South China Fruits*, **28** : 33.

Corbesier L, Bernier G. Perilleux, C (2002) *Plant Cell Physiol.*, **43**: 684-688.

Corbesier L, Vincent C, Jang S, Fornara F, Fan Q, Searle I, Giakountis A, Farrona S, Gissot L, Turnbull CGN, Coupland G. (2007) *Science*, **316**: 1030-1033

Cedeno-Maldonado, A., Perez, A. and Reyes-Soto, I. (1988) *J. Agril. Univ., Puerto Rico*, **72** : 1-8.

Chacko, E.K. (1968) *Ph.D. Thesis* submitted to the P.G. School, IARI, New Delhi.

Chacko, E.K. and Singh, R.N. (1969) *HortScience*, **4** : 121-123.

Chacko, E.K., Kachru, R.B. and Singh, R.N. (1970) *J. Hort. Sci.*, **45** : 341-349.

Chacko, E.K., Kohli, R.R. and Randhawa, G.S. (1972) *Curr. Sci.*, **41** : 50.

Chadha, K.L. (1963) *Punjab Hort. J.*, **3** : 214-220.

Chadha, K.L. and Pal, R.N. (1993) *Acta Horticulturae*, **341** : 42-54.

Chadha, K.L., Samra, J.S. and Thakur, R.S. (1980) *Scientia Hort.*, **13** : 323-329.

Chadha, K.L., Thakur, R.S., Rajput, M.S. and Samra, J.S. (1981) *Indian J. Hort.*, **38** : 153-156.

Chaimanee, P., Suntornwat, O., Lerrtwikool, N. and Bungaruang, L. (1999) *J. Biochem. Mol. Biol. and Biophysics*, **3** : 75-79.

Chakarabarty, U. and Sadhu, M.K. (1985) *2nd Int. Symp. on Mango, India*. Abst.No. 3.8,p.20.

Chandel, J.S. and Singh, R. (1992) *Acta Horticulturae*, **321** : 561-565.

Chandra, A. and Yamdagni, R. (1984) *Punjab Hort. J.*, **24** : 17-18.

Chandra, A. and Chandra, A. (1997) *Production and Post-harvest Technology of Fruits*, NBS Pub. & Distributors, Bikaner, India.

Chandra, A., Bhati, D.P.S. and Singh, K.M. (1991) *Indian J. Entomology*, **53** : 76-79.

Chang, J.W., Lin, H.L., Yen, C.R. and Lee, K.C. (1994) *Proc. Symp. on Practical Aspects of Some Economically Imp. Fruit Trees in Taiwan* (Eds. H.S. Lin, and L.R. Chang), No. 33 : 21-29.

Chang, J.W., Chang, H.M., Chao, C.N. and Wang, H.L. (1996) *Chinese J. Agrometeorology*, **3** : 151-157.

Chang, M.T. (1992) *Acta Horticulturae*, **292** : 113-120.

Chattopadhyay, P.K. and Sarkar, S.K. (1992) *The Hort. J.*, **5** : 25-30.

Chaudhary, S.K. and Nauriyal, J.P. (1989) *Acta Horticulturae*, **231** : 296-300

Chaudhury, R., Malik, S.K., Rajan, S. (2010) *CryoLetters*, **31**: 268–278.

Cheema, G.S. and Dani, P.G. (1934) *Bombay Dept. Agric. Bull.*, No. 170.

Cheema, G.S., Bhat, S.S. and Naik, K.C. (1954) *Commercial Fruits of India*, Macmillan, London.

Chen, H.B. and Huang, M.Y. (1995) *J. South China Agril. Univ.*, **16** : 73-77.

Chen JieZhong and Zhao Hong Ye (1999) *South China Fruits*, **28** : 34-35.

Choudhuri, J.M. and Rudra, P. (1971) *Indian Agric.*, **15** : 127-135.

Coetzer, L.A., Robbertse, P.J. and Janse van Vuuren, B.P.H. (1992) *Yearb. South Afr. Mango Growers' Assoc.*, **12** : 48-51.

Coetzer, L.A., Robertse, P.J. and Wet, E.De (1991) *Yearb. South Afr. Mango Growers' Assoc.*, **11** : 29-31.

Colyn, J. (1993) *Acta Horticulturae*, **341** : 60-68.

Contractor, J. (1951) *Mango Studies, Fla. Mango Forum*, p. 127.

Cope, F.W. (1962) *Heredity*, **17** : 156-182.

Corbineau, F., Kante, M. and Come, D. (1986) *Tree Physiol.*, **1** : 151.

Crane, J.H. and Campbell, C.W. (1991) *The Mango*, Fla. Co-op. Extn. Ser., IFAS, Univ. Florida, Fact Sheet FC-2.

Cumare, S.J.A. and Avilan, R.L.A. (1994a) *Agronomia Trop.*, **44** : 373-391.

Cumare, S.J.A. and Avilan, R.L.A. (1994b) *Agronomia Trop.*, **44** : 393-416.

Cumare, S.J.A. and Avilan, R.L.A. (1994c) *Agronomia Trop.*, **44** : 413-439.

D'Angles, R.E. and Parodi, M.G. (1993) *Proc. Interamerican Soc. Trop. Hort.* (Ed. R.J. Campbell), **37**: 155-160.

Dag, A., Gazit, S., Eisenstein, D., El-Batsri, R. and Degani, C. (1999) *Sci. Hort.*, **82** : 325-329.

Dambreville, A., Normand, F. and Lauri, P-E. 2013. *Functional Plant Biology*, **40**: 280-291.

Das Gupta, S.N. and Sen, C. (1958) *Curr. Sci.*, **27** : 446-447.

Davenport TL (2000) *Hort Technol.* **10**: 733-739.

Davenport TL (2003) *Hort Science* **38**: 1331-1335

Davenport TL (2006) *Hort Science* **41**: 544-548

Davenport TL. (2009) In: (Ed. R.E. Litz). The Mango: Botany Production and Uses, 2nd edition. CAB International, Wallingford, UK, 2009, 97-169.

Davenport T. L. (2007) *Brazilian J. Plant. Physiol.* **19**(4): 363-376.

Davenport T L and Nunez-Elisea (1997) In: Mango: botany, production and uses. Wallingford: CAB International, 69-146

Davenport TL, Zhang T and Ying Z (2006) In: Proceedings of the 33rd Annual Meeting of the Plant Growth Regulation Society of America. Quebec City, pp. 109-110

Dayal, K. and Misra, S.P. (1985) *2nd Int. Symp. on Mango*, India, Abst. No. 4.3, p. 30.

De Candolle (1904) *Origin of Cultivated Plants*, Kegan Paul, London.

Degani, C., El Batsri, R. and Gazit, S. (1990) *J. Amer. Soc. Hort. Sci.*, **115** : 844-847.

Degani, C., Cohen, M., El-Batsri, R. and Gazit, S. (1992) *HortScience*, **27** : 252-254.

Desai, U.T., Masalkar, S.D., Choudhari, S.M., Kale, P.N. and Nagre, P.K. (1994) *Recent Hort.*, **1** : 11-13.

Devi, T.M., Pandey, S., Patel, D.P. and Tyagi, D.N. (1996) *Indian J. Pl. Physiol.*, **1** : 65-67.

DeWald, S.G. (1987) *Ph. D. Thesis*, Univ. Florida, USA.

DeWald, S.G. (1988) *Dissertation Abst. International, B* (Sci. & Engineering), **48** : 2165B-2166B.

DeWald, S.G., Litz, R.E. and Moore, G.A. (1989a) *J. Amer. Soc. Hort. Sci.*, **114** : 712-716.

DeWald, S.G., Litz, R.E. and Moore, G.A. (1989b) *J. Amer. Soc. Hort. Sci.*, **114** : 837-841.

Dhakal, B.R. and Hoda, M.N. (1986) *South Indian Hort.*, **34** : 184-186.

Dharkar, S.N., Savagaon, D.A., Srirangarajan, A.N. and Sreenivasan, A. (1966) *J. Food Sci.*, **31** : 863-869.

Dhatt, A.S. and Dhillon, B.S. (1985) *2nd Int. Symp. on Mango*, Abst.No. 5.11, p. 47.

Dhillon, W.S. and Malhi, C.S. (1993) *Punjab Hort. J.*, **33** : 34-39.

Dhillon, W.S., Mathi, C.S., Singh, R. and Dhatt, A.S. (1991) *Indian J. Hort.*, **48** : 108-111.

Dhua, R.S., Mitra, S.K., Sen, S.K. and Bose, T.K. (1982) *Abst. 21st International Hort. Cong.*, Hamburg, p. 1334.

Dhunaga, D.B., Aravindakshan, M. and Gopikumar, K. (1989) *Acta Horticulturae*, **231** : 170-174.

Dhungana, D.E., Aravindakshan, M. and Gopi Kumar, K. (1985) *2nd Int. Symp. on Mango*, India, Abst. No. 3.3, p.18.

Dijkman, N.J. and Soule, M.J. (1951) *Proc. Fla. Sta. Hort. Soc.*, **64** : 257-262.

Doijode, S.D. (1995) *Plant Genetic Resources Newsl.*, **104** : 24-25.

Donadio, L.C. (1995) *Acta Horticulturae*, **379**: 137-140.

Donadio, L.C., Soares, N.B. and Sempionato, O. (1994) *Proc. Interamerican Soc. Trop. Hort.*, **38** : 32-36.

Eiadthong, W., Yonemori, K., Sugiura, A., Utsunomiya, N. and Subhadrabaandhu, S. (1999) *Scientia Hort.*, **82** : 57-66.

Eiadthong, W., Yonemori, K., Sugiura, A., Utsunomiya, N. and Suranant, S. (1999) *Japanese J. Trop. Agric.*, **43** : 76-83.

El-Motaium, R.A., Shaban, A. E. A. Badawy, S. H. and Ibrahim A. S. A. (2019) *J. Plant Nutrition*, **42**(20): 2861-2872.

Elahi, M. and Khan, N. (1973) *J. Agric. Food Chem.*, **21** : 229-231.

Ethiraj, S. and Suresh, E.R. (1992) *J. Food Sci. Tech.*, **29** : 48-50.

Fang, J., D. Liu and Z. Ma (2003) Mol. Plant Breed. **1**: 313–319.

Farre, J.M. and Hermoso, J.M. (1993) *Acta Horticulturae*, **341** : 295-302.

Ferrari, F.D. and Sergent, A.E. (1996a) *Revista de la Facultad de Agronomia, Universidad Central de Venezuela*, **22** : 1-8.

Ferrari, F.D. and Sergent, A.E. (1996b) *Revista de la Facultad de Agronomia, Universidad Central de Venezuela*, **22** : 9-17.

Fivaz, J. and Stassen, P.J.C. (1996a) *Yearb. South Afr. Mango Growers' Assoc.*, **16** : 32-35.

Fivaz, J. and Stassen, P.J.C. (1996b) *Yearb. South Afr. Mango Growers' Assoc.*, **16** : 49-54.

Galan, S.V., Fernandez, G.D. and Calvo, R. (1984) *Proc. Fla. Sta. Hort. Soc.*, **97** : 358-360.

Galang, F.G. and Lazo, F.D. (1935) *Phlipp. J. Agric.*, **6** : 129-139.

Galang, G.A. and Agati, J.A. (1936) *Philipp. J. Agric.*, **7** : 245-259.

Gandhi, S.R. (1955) *Farm Bull.*, No.6, ICAR, New Delhi.

Gangolly, S.R., Singh, Ranjit, Katyal, S.L. and Singh, Daljit (1957) *The Mango*, ICAR, New Delhi.

Garcia-Perez, E., Becerril-Roman, A.E., Avila-Resendiz, C., Matheis-Toledano, L. and Mosqueda-Vazquez, R. (1993) *Acta Horticulturae*, **341** : 256-263.

Garg, R.C., Srivastava, R.K., Ram, H.B. and Verma, R.A. (1971) *Progressive Hort.*, **3** : 49-67.

Ghosh, S. (1960) *Sci. and Cult.*, **26** : 287-288.

Ghosh, S.K., Dhua, R.S. and Mitra, S.K. (1985) *Indian Fd. Packer*, **39** : 46-50.

Gill, A.P.S. (1966) *Ph.D. Thesis* submitted to the P.G. School, IARI, Delhi.

Gofur, M.A., Shafique, M.Z., Helali, O.H., Ibrahim, M., Rahman, M.M. and Hakim, A. (1994) *Bangladesh J. Scientific & Industrial Res.*, **29** : 163-171.

Gokhale, A.V. and Kanithkar, U.K. (1951) *Proc. Indian Sci. Cong.*, **38** : 151.

Gonzalez, L.G. (1923) *Philipp. Agric.*, **12** : 15-28.

Gonzalez-Aguilar, G.A., Fortiz, J., Cruz, R., Baez, R. and Wang, C.Y. (2000) *J. Agric. Food Chemistry*, **48** : 515-519.

Gowda, I.N.D. and Ramanjaneya, K.H. (1995) *J. Food Sci. Tech.*, **32** : 323-325.

Gozit, S. and Kadman, A. (1980) *HortScience*, **15** : 669.

Gunjate, R.T. (1985) *2nd Int. Symp. on Mango*, India, Abst. No.3.1, p. 17.

Gunjate, R.T. (1989) *Acta Horticulturae*, **231** : 164-167.

Gunjate, R.T., Tare, S.J., Rangwala, A.D. and Limaya, C.P. (1979) *Indian J. Hort.*, **37** : 140.

Gupta, J.H. (1985) *2nd Int. Symp. on Mango*, India, Abst. No.6.7, p.62.

Gupta, O.P., Jawanda, J.S. and Sharma, K.C. (1988) *Indian J. Hort.*, **45** : 268-270.

Hansraj, D. and Ram, S. (1996) *Recent Hort.*, **3** : 8-13.

Haque, R. and Dhua, R.S. (1993) *Environment & Ecology*, **11** : 333-336.

Harkness, R.W. and Cobin, M. (1951) *Fla. Mango Forum*, pp. 141-146.

Hatton, T.T. and Reeder, W.F. (1966) *Proc. Carib. Reg. Amer. Soc. Hort. Sci.*, **10** : 114-119.

Hatton, T.T., Reeder, W.F. and Campbell, C.W. (1965) *USDA. Marketing Res. Report.*, No.725, p.9.

Hayes, W.B. (1957) *Fruit Growing in India*, Kitabistan, Allahabad.

Heather, N.W., Corcoran, R.J. and Kopittke, R.A. (1997) *Post-harvest Biol. Tech.*, **10** : 99-105.

Hemavathy, J., Prabhakar, J.V. and Sen, D.P. (1987) *J. Food Sci.*, **52** : 833-834.

Hirano, R., O.T. Htun and K.N. Watanabe (2010) *Genome*, **53**: 321–330

Hobson, L. (1969) *South Afr. Citrus J.*, **42** : 15-17.

Hsu, S.H., Hsieh, S.C., Tang, C., Chen, T.J., Wu, C.C. and Lin, P.H. (1995) *Chinese J. Agrometeorology*, **2** : 169-177.

Hulme, A.C. (1971) *The Biochemistry of Fruits and their Products*, 2, Academic Press, London, pp. 233-254.

Hussain, C.M., Khan, K.M. and Nasir, M.A. (1989) *J. Agril. Res.* (Lahore), **27** : 289-198.

Ilangantileke, S. and Salokhe, V. (1989) *Proc. 5th International CA Res. Conf.*, Washington, USA, Vol. 2, pp.103-117.

Iqbal, M. (1982) *Fiji Agric. J.*, **44** : 21-26.

Issarakraisila, M. and Considine, J.A. (1994) *Ann. Bot.*, **73** : 231-240.

Iyer, C.P.A., 1991.*Acta Hort.*, **291**: 109–32

Iyer, C.P.A. and Kurian, R.M. (1992) *Acta Horticulturae*, **321** : 425-436.

Iyer, C.P.A. and Subramanyam, M.D. (1986) *Indian J. Hort.*, **43** : 221-223.

Iyer, C.P.A. and C. Degani (1997) *In:* Litz, R.E. (ed.) The Mango, Botany, Production and Uses. CAB International, New York, pp. 49–68.

Iyer, C. P. A., and Schnell, R. J. (2009) *In: "Breeding and genetics*, (*The Mango: Botany, Production and Uses*), ed R. E. Litz (London: British Library), 67–96.

Jain, N.L. (1961) *Rev. Food Techol.*, **3** : 131-162.

Janse, van Vuuren,B.P.H., Robbertse, P.J., Coetzer, L.A. and Hudson-Lamb, D.C. (1992) *Yearb. South Afr. Mango Growers' Assoc.*, **12** : 14-19.

Jasmine J., Nainar A. P. and Balasubramanyan S. (2009) In: National Seminar on "Production, Post-harvest Technology and Marketing of Mango, HCRI, Periyakulam (TNAU), pp. 74-75.

Jasso Argumedo, J. (1995) *Agricultura Tecnica en Mexico*, **21** : 29-39.

Jauhari, O.S., Teaotia, S.S. and Upadhyay, S.K. (1972) *Acta Horticulturae*, **24** : 107-109.

Jawanda, J.S. and Singh, K.K. (1963) *Punjab Hort. J.*, **3** : 148-153.

Jayasankar, S. (1995) *Ph. D. Thesis*, Univ. of Florida, USA.

Jayasankar, S. and Litz, R.E. (1998) *Theor. Appl. Genet.*, **96** : 6-7.

Ji, Z.L., Zhang, Z.Q., Wang, Y. and Li, X.P. (1994) *Acta Horticulturae Sinica*, **21** : 111-116.

Jindal, P.C., Singh, J.P. and Gupta, O.P. (1976) *Progressive Hort.*, **8** : 65.

Joshi, G.D. and Limaye, V.P. (1986) *J. Maharashtra Agril. Univ.*, **11** : 104.

Joshi, G.D. and Roy, S.K. (1985) *Indian Hort.*, **29** : 21-22.

Joshi, H.U. and Chauhan, H.L. (1985a) *2nd Int. Symp. on Mango*, India, Abst. No. 6.10, p. 63.

Joshi, H.U. and Chauhan, H.L. (1985b) *2nd Int. Symp. on Mango*, India, Abst. No. 8.13, p. 83-84.

Joubert, J.P., Robbertse, P.J., Coetzer, L.A. and Wishart, D.L. (1993) *Yearb. South Afr. Mango Growers' Assoc.*, 13 : 27-33.

Juliano, J.B. (1937) *Philipp. Agric.*, **25** : 749.

Kadman, A., Gazit, S. and Ziv, G. (1976) *Acta Horticulturae*, **57** : 81.

Kalra, S.K. and Tandon, D.K. (1983) *Indian J. Hort.*, **40** : 155-159.

Kalra, S.K. and Tandon, D.K. (1984) *Indian J. Hort.*, **41** : 177-181.

Kamal, A.J.K., Wong, B.F. and Yaacob, O. (1984) *Proc. 6th Int. Colloquium on Plant Nutriton*, Montpellier, France, pp. 993-1001.

Kanwar, J.S., Nijjar, G.S. and Kahlon, G.S. (1987) *Jour. Res., PAU*, **24** : 411-422.

Kashyap, R., Jyotishi, R.P. and Sharma, A.B. (1972) *Acta Horticulturae*, **24** : 99-100.

Kashkush, K., F. Jinggui, E. Tomer, J. Hillel and U. Lavi (2001) *Euphytica*, **122**: 129–136

Katrodia, J.S. (1989) *Acta Horticulturae*, **231** : 814-826.

Katrodia, J.S. and Rane, D.A. (1989) *Acta Horticulturae*, **231** : 873-877.

Katrodia, J.S. and Seth, I.K. (1989) *Acta Horticutlurae*, **231** : 827-834.

Katrodia, J.S., Rane, D.A. and Salunkhe, D.K. (1985) *2nd Int. Symp. on Mango*, India, Abst. No. 96, p. 95.

Kerni, P.N. and Gupta, A. (1986) *Res. Dev. Reptr.*, **3** : 77-79.

Ketsa, S., Chidtragool, S., Klein, J.D. and Lurie, S. (1998) *J. Pl. Physiol.*, 153 : 457-461.

Khader, S.E.S.A. (1992) *J. Hort. Sci.*, **67** : 855-60.

Khader, S.E.S.A.(1991) *Indian J.Hort.*, **48** : 112-115.

Khan, M.A., Nizami, S.S., Khan, M.N.I. and Azeem, S.W. (1994) *Pakistan J. Sci.& Ind. Res.*, **37** : 213-114.

Khurdiya, D.S. and Roy, S.K. (1989) *Acta Horticulturae*, **231** : 709-714.

Knight, R. and Rogers, H.H. (1955) *Heredity*, **9** : 69-77.

Knight, R.J.Jr. and Schnell, R.J. (1994) *Economic Bot.*, **48** : 139-145.

Kohli, R.R. and Reddy, Y.T.N. (1985) *2nd Int. Symp. on Mango*, India, Abst. No. 3.15, p. 23.

Koolpluksee, M., Ketsa, S. and Subhadrabandhu, S. (1993) *Kasetsart J. Natural Sci.*, **27** : 115-124.

Krishnamurthy, G.V., Jain, N.L. and Bhatia, B.S. (1960) *Food Sci.*, **9** : 227-279

Krishnamurthy, S. and Subramanyam, H. (1973) *Tropical Sci.*, **15** : 167-193

Kulkarni, V.J. (1988) *J. Hort. Sci.*, **63** : 557-566.

Kulkarni, V.J. and Rameshwar, A. (1989) *Scientia Hort.*, **39** : 143-148.

Kulkarni VJ (2004) *Acta Hort.* **645** : 61-70.

Kumar, N. (2019) *Advanced Ag. Res. & Tech. J.* **3**(1): 47-53

Kumar, R., Ahlawat, V.P. and Yamdagni, R. (1985) *Progressive Hort.*, **17** : 87-90.

Kumar, S. (1977) *Indian J. Hort.*, **34** : 100-106.

Kumar, V.S. and Mitra, S.K. (1994) *The Hort. J.*, **7** : 71-73.

Kundu, S. and Mitra, S. (1997) *Environment & Ecology*, **15** : 325-328.

Kurian, R.M. and Iyer, C.P.A. (1992) *Scientia Hort.* **50** : 245-248.

Kurian, R.M. and Iyer, C.P.A. (1993) *J. Hort. Sci.*, **68** : 349-354.

Kurian, R.M., Murti, G.S.R. and Iyer, C.P.A. (1992) *Gartenbauwissenchaft*, **57** : 84-87.

Kurian, R.M., Iyer, C.P.A. and Murti, G.S.R. (1994) *Gartenbauwissenschaft*, **59** : 268-270.

Lakshminarayana, S. (1980) In : *Tropical and Subtropical Fruits – Composition, Properties and Uses* (Eds. S. Nagy and P.E. Shaw). AVI Westport, CT, p. 184.

Lakshminarayana, S., Subbaih Shetty, M. and Krishnaprasad, C.A. (1975) *Tropical Sci.*, **17** : 95-101.

Larrauri, J.A., Ruperez, P., Borroto, B. and Saura-Calixto, F. (1996) *Lebensmittel-Wissenschaft & Technologie*, **29** : 729-733.

Lavi, U., K. Kashkush, D. Sa'ada, H. Shats, U. Ravid and E. Tomer (2004) *Acta Hort.* **645**: 51–59

Lazo, F.D. (1971) *Philipp. J. Plant Indust.*, **35** : 41-48.

Léchaudel, M., Lopez-Lauri, F., Vidal, V., Sallanon, H. and Joas, J. 2013. *J. Plant Physiol.* **170** : 567-578.

Lee, S.R., Lin, H.L., Shiesh, C.C. and Lee, K.C. (1998) *J. Chinese Soc. Hort. Sci.*, **44** : 126-136.

Lee, S.R., Lin, H.L., Shiesh, C.C. and Lee, K.C. (1998) *J. Chinese Soc. Hort. Sci.*, **44** : 138-143.

Leonardi, J., Blaikie, S.J., Muller, W.J., Steele Scoti, N. and Chacko, E.K. (1999) *Austalian J. Exp. Agric.*, **39** : 761-770.

Lespinasse, Y. and B. Frederic, (1998) Breeding for fruits. (http: /wchr.agrsci.unibo.it/wc2/lespinas.html).

Litz, R.E. (1987) In : *Plant Tissue and Cell Culture* (Eds. C.E. Green, A.A. Somers, W.P. Hackett and D.D. Biesboer), Alan R. Liss, New York, pp. 407-418.

Litz, Rechard. E. (2004) Biotechnology and Mango Improvement. *Acta Hort.* **645**: 85092

Litz, R.E. and Schaffer, B. (1987) *J. Plant Physiol.*, **128** : 251-258.

Litz, R.E. and Yurgalevitch, C. (1997) *Plant Cell Tissue Organ Cult.*, **51** : 171-176.

Litz, R.E., Knight, R.K. and Gazit, S. (1982) *Plant Cell Rep.*, **1** : 264-266.

Litz, R.E., Knight, R.K. and Gazit, S. (1984) *Scientia Hort.*, **22** : 233-240.

Litz, R.E., Mathews, H., Bharathan, N. and Narayanan, K.R. (1990) *Proc. VII Intl. Congr. Plant Tissue and Cell Cult.*, Amsterdam, p. 67.

Litz, R.E., Hendrix, R.C., Moon, P.A. and Chavez, V.M. (1998) *Plant Cell Tissue Organ Cult.*, **53** : 13-18.

Litz, R.E., Mathews, V.H., Hendrix, R.C. and Yurgalevitch, C. (1991) *Acta Horticuturae*, **291** : 133-140.

Loveys, B.R., Robinson, S.P., Brophy, J.J. and Chacko, E.K. (1992) *Australian J. Plant Physiol.*, **19** : 449-457.

Luo C, Shu B, Yao Q, Wu H, Xu W and Wang S (2016) *Front. Plant Sci.* **7** : 1310.

Lynch, S.J. and Ruehle, G.D. (1940) *Proc. Fla. Sta. Hort. Soc.*, **53** : 167-169.

MacLeod, A.J., MacLeod, G. and Snyder, C.H. (1988) *Phytochemistry*, **27** : 2189-2193.

Madalageri, M.B., Hulamani, N.C. and Patil, V.R. (1989) *Progressive Hort.*, **21** : 173-175.

Madhava Rao, V.N. (1971) *Andhra Agric. J.*, **18** : 242-245.

Maharana, T., Lenka, P.C., Sahu, B.C. and Tripathy, P. (1996) *Orissa J. Hort.*, **24** : 85-88.

Maheshwari, P. (1934) *Curr. Sci.*, **10** : 97-98.

Maiti, S.C., Sen, S.K. and Bose, T.K. (1981) *Tech. Bull.* No.1, Dept. of Hort. BCKV, Kalyani.

Majumder, P.K. and Mukherjee, S.K. (1961a) *Indian J. Hort.*, **18** : 12-19.

Majumder, P.K. and Mukherjee, S.K. (1961b) *Indian J. Hort.*, **18** : 167-178.

Majumder, P.K. and Mukherjee, S.K. (1972) *Nucleus*, **14** : 119-131.

Majumder, P.K. and Prsad, J. (1985) *2nd Int. Symp. on Mango*, India. Abst. No. 3.9, p. 20.

Majumder, P.K. and Rathore, D.S. (1970) *Curr. Sci.*, **38** : 142.

Majumder, P.K. and Sharma, D.K. (1985) *2nd Int. Symp. on Mango*, India. Abst. No. 4, p. 29.

Majumder, P.K. and Sinha, G.C. (1972a) *Acta Horticultuare*, **24** : 221-223.

Majumder, P.K. and Sinha, G.C. (1972b) *Acta Horticulturae*, **24** : 230-234.

Majumder, P.K., Sinha, G.C. and Singh, R.N. (1970) *Indian J. Hort.*, **27** : 130-131.

Majumder, P.K., Chakladar, B.P. and Mukherjee, S.K. (1972a) *Acta Horticulturae*, **24** : 101-116.

Majumder, P.K., Mukherjee, S.K. and Rathore, D.S. (1972b) *Acta Horticulturae*, **24** : 72-73.

Majumder, P.K., Sharma, D.K. and Singh, R.N. (1981) *Nat. Symp. Trop. and Subtrop. Fruit Crops*, Bangalore, p. 3.

Majumder, P.K., Sharma, D.K., Singh, M.P. and Singh, R.N. (1976) *Indian Hort.*, **21** : 16-18.

Malhi, C.S. and Nijjar, G.S. (1985) *2nd Int. Symp. on Mango*, India, Abst. No. 4, p. 29.

Mallick, S.K., Mitra, S.K., Banik, B.C., Maity, S.C. and Sen, S.K. (1985) *2nd Int. Symp. on Mango*, India, Abst. No. 4.19, p. 37.

Mallik,P.C. (1953) *Proc. Bihar Acad. Agri. Sci.*, **2** : 141-143.

Mallik, P.C. (1957) *Indian J. Hort.*, **14** : 1-23.

Mallik, P.C. and De, B.N. (1952) *Indian J. Agric. Sci*, **22** : 151-166.

Malo, S.E. (1976) *Acta Horticulturae*, **57** : 274.

Mani, M., Krishnamoorthy, A. and Pattar, G.L. (1995) *Pest Management in Horticultural Ecosystems*, **1** : 15-20.

Maries, C. (1902) *J. Royal Hort. Soc.*, **26** : 755-770.

Marloth, R.H. (1947) *Fmg. S. Afr.*, **22** : 457-63.

Mathews, H. and Litz, R.E. (1990) *HortScience*, **25** : 965-966.

Mathews, H. and Litz, R.E. (1992) In : *Biotechnology of Perennial Fruit Crops* (Eds. F.A. Hammerschlag and R.E. Litz), CAB International, UK, pp. 433-448.

McKenzie, C.B. (1994a) *Yearb. South Afr. Mango Growers' Assoc.*, **14** : 11-13.

McKenzie, C.B. (1994b) *Yearb. South Afr. Mango Growers' Assoc.*, **14** : 26-28.

McLauchlan, R.L. and Barker, L.R. (1994) *ACIAR Proc.*, Australia, **58** : 41-44.

Md Aayesha Jameel, SM Rajesh Naik, C Madhumathi, D Srinivasa Reddy and KT Venkataraman. *J. Pharmacognosy Phytoch*. P 2018; **7**(6): 2375-2382

Medlicott, A.P. and Thompson, A.K. (1985) *J. Sci. Food Agric.*, **36** : 561-566.

Medlicott, A.P., N'Diaye, M. and Sigrist, J.M.M. (1990) *J. Amer. Soc. Hort. Sci.*, **115** : 426-430.

Medlicott, A.P., Reynolds, S.B., New, S.W. and Thompson, A.K. (1988) *Trop. Agric.*, **65** : 153-157.

Miao, P.S., Wang, X.H., Guo, Z.C. and Wu, Q.X. (1994) *Guangdong J. Agric. Sci.*, **4** : 29-30.

Mitcham, E.J. and McDonald, R.E. (1992) *J. Amer. Soc. Hort. Sci.*, **117** : 919-924.

Mitra, S.K. (2001) In : *Biotechnology of Horticultural Crops*, Vol.1 (Eds. V.A. Parthasarathy, T.K. Bose, P. Das, S.K. Mitra and S. Mohandas), Naya Prokash, Calcutta, pp. 119-160.

Modi, V.V. and Reddy, V.V.R. (1967) *Indian J. Exp. Biol.*, **5** : 233-235.

Morga, N.S., Lustre, A.O., Tunac, M.M., Balogot, A.H. and Soriano, M.R. (1979) *Food Chem.*, **4** : 225-234.

Mossak, I.I. (1997) *Trop. Fruits Newsl.*, **23** : 3-5.

Mostert, P.G. and Wantenaar, L. (1994) *Yearb. South Afr. Mango Growers' Assoc.*, **14** : 21-23.

Muda, P., Seymour, G.B., Errington, N. and Tucker, G.A. (1995) *Carbohydrate polymers*, **26** : 255-260.

Mukherjee, P.K. (1960) *Ann. Rep.Hort. Res. Inst.*, Saharanpur, pp. 78-85.

Mukherjee, S.K. (1948a) *Bull. Botanical Society of Bengal*, **2** : 101-133.

Mukherjee, S.K. (1949a) *Lloydia*, **12**: 73-136.

Mukherjee, S.K. (1949b) *D.Sc. Thesis* submitted to the Calcutta University, Calcutta (India).

Mukherjee, S.K. (1953) *Econ. Bot.*, **7**: 130-162.

Mukherjee, S.K. (1958) *Indian J. Hort.*, **16**: 129-134.

Mukherjee, S.K. (1963) *Punjab Hort. J.*, **3** : 107-115.

Mukherjee, S.K. (1972) *Econ. Bot.*, **26**: 260.

Mukherjee, S.K. and Bid, N.N. (1965) *Indian J. Agri. Sci.*, **35** : 309-314.

Mukherjee, S.K. and Das, Dora (1976) *Progessive Hort.*, **8** : 5-11.

Mukherjee, S.K. and Majumder, P.K. (1961) *Indian Hort.*, **18** : 102-104.

Mukherjee, S.K. and Majumder, P.K. (1963) *Indian J. Hort.*, **20** : 202-207

Mukherjee, S.K. and Majumder, P.K. (1964) *Indian J. Hort.*, **21** : 46-50.

Mukherjee, S.K., Majumder, P.K. and Chatterjee, S.S. (1961) *Indian J. Hort.*, **18** : 302-304.

Mukherjee, S.K., Majumder, P.K. and Goswami, A.M. (1966b) *Sci. and Cult.*, **32** : 377-378.

Mukherjee, S.K., Majumder, P.K. and Rathore, D.S. (1966a) *Ann. Rep.*, IARI, New Delhi.

Mukherjee, S.K., Majumder, P.K., Bid, N.N. and Goswami, A.M. (1967) *J. Hort. Sci.*, **42** : 83-87.

Mukherjee, S.K., Singh, R.N., Majumder, P.K. and Sharma, D.K. (1968) *Euphytica*, **17** : 462-467.

Murti GSR and Upreti KK. (1996) *Plant Physiol. & Biochem*, **25**(2): 167-171.

Murti GSR, and Upreti KK. (2000) In: *Advances in Plant Physiology*, (ed. A. Hemantaranjan), Scientific Publishers, Jodhpur (India). **3**: 109-148.

Musahib-ud-din (1960a) *Punjab Fruit J.*, **23** : 35-42.

Musahib-ud-din (1960b) *Punjab Fruit J.*, **23** : 113-136.

Musahib-ud-din and Dinsa, H.S. (1946) *Punjab Fruit J.*, **10** : 35-42.

Naik, K.C. (1948) *Indian J. Agric. Sci.*, **18** : 35-41.

Naik, K.C. and Gangolly, S.R. (1950) *A Monograph on Classification and Nomenclature of South Indian Mangoes*, Supdt., Govt. Press, Madras.

Naik, K.C. and Rao, M.M. (1942) *Madras Agric. J.*, **30** : 365-374.

Naik, K.C. and Rao, M.M. (1943) *Indian J. Hort.*, **1** : 107-119.

Naik, K.C. and Shaw, R. (1937) *Bihar and Orissa Dept. Agric. Rep.* (1935-36), p. 87.

Nakhla, F.D. (1980) *M.Sc. Thesis,* Zagazig University.

Narasimha Char, B.L. and Azeemoddin, G. (1989) *Acta Horticulturae*, **231** : 744-748.

Narasimha Char, B.L., Azeemoddin, G., Atchyuta, R.D. and Thirumala, R.S.D. (1989) *Acta Horticulturae*, **231** : 749-750.

Narasimhan, M.J. (1954) *Curr. Sci.*, **23** : 297-298.

Narasimhan, M.J. (1959) *Curr. Sci.*, **28** : 254-255.

Narayanan, E.S. and Ghai, S. (1961) *Proc. 48th Indian Sci. Cong.*, Roorkee, Part III, p. 502.

Narwadkar, P.R. and Anserwadekar, K.M. (1985) *2nd Int. Symp. on Mango*, India, Abst. No. 3.4, p. 18.

Narwadkar, P.R. and Pandey, R.M. (1989a) *Acta Horticulturae*, **231** : 399-404.

Narwadkar, P.R. and Pandey, R.M. (1989b) *Acta Horticutlurae*, **231** : 440-447.

Nauriyal, J.P., Chadha, K.L. and Rajput, M.S. (1972) *Acta Horticulturae*, **24** : 215-216.

Nema, B.K., Jain, A.K., Nema, M.K., Nair, P.K.R. and Srivastava, S.R. (1985) *2nd Int. Symp. on Mango*, India, Abst.No. 3.25, p. 27.

Nicholson, R.I.D. and van Staden, J. (1988) *J. Pl. Physiol.*, **132** : 720-724.

Noomhorm, A. and Tiasuwan, N. (1995) *J. Food Process. Preserv.*, **19** : 271-281.

Normand, F., Lauri, P.-.E. and Legave, J.-.M. (2015) *Acta Horticulture*. **1075**: 21-31

Nunez-Elisea, R. and Davenport, T.L. (1989) *Proc. Plant Growth Regulator Soc. Amer.*, pp. 245-47.

Nunez-Elisea, R. and Davenport, T.L. (1994) *Scientia Hort.*, **58** : 57-66.

Nunez-Elisea, R. and Davenport, T.L. (1995) *Scientia Hort.*, **62** : 63-73.

Nunez-Elisea, R., Davenport, T.L. and Caldeira, M.L. (1993) *Acta Horticulturae*, **341** : 192-198.

Nunez-Elisea, R., Davenport, T.L. and Caldeira, M.L. (1996) *J. Hort. Sci.*, **71** : 25-39.

Nunez-Elisea, R., Caldeira, M.L., Ferreira, W. and Davenport, T.L. (1992) *HortScience*, **27** : 926.

O'Hare, T.J. (1994) *ACIAR Proc.*, Australia, **58** : 21-24.

O'Hare, T.J. and Prasad, A. (1992) *Acta Horticulturae*, **321** : 372-381.

O'Hare, T.J. and Prasad, A. (1993) *Acta Horticulturae*, **343** : 244-250.

Ogata, J.N., Kawani, Y., Bevenue, A. and Casarett, L.J. (1972) *J. Agric. Food Chem.*, **20** : 113-115.

Olesen, T.(2011) New Zealand *J. Crop Hort. Science* **39**(2): 119-129

Om Prakash (1991) *International J. Trop. Plant Dis.*, **9** : 277-280.

Oosthuyse, S.A. (1991) *Yearb. South Afr. Mango Growers' Assoc.*, **11** : 57-58.

Oosthuyse, S.A. (1993a) *Yearb. South Afr. Mango Growers' Assoc.*, **13** : 34-39.

Oosthuyse, S.A. (1993b) *Yearb. South Afr. Mango Growers' Assoc.*, **13** : 45-53.

Oosthuyse, S.A. (1994) *Yearb. South Afr. Mango Growers' Assoc.*, **14** : 1-6.

Oppenheimer, C. and Gazit, S. (1961) *Hort. Advance*, **5** : 1-12.

Paez Redondo, A.R. (1995) *Fitopatologia Colombiana*, **19** : 25-30.

Pal, R.N., Chadha, K.L. and Srivastava, K.C. (1985) *2nd Int. Symp. on Mango*, India, Abst. No. 9.11, p. 97.

Palaniswamy, K.P., Muthukrishnan, C.R. and Shanmugavelu, K.G. (1974) *Indian Fd. Packer*, **28** : 12-19.

Pandey, R.M. (1993) *Indian Hort.*, **38** : 4-5.

Pandey, R.M., Rao, M.M. and Singh, R.N. (1975) *Sci. and Cult.*, **41** : 385-395.

Pandey, R.M., Rathore, D.S. and Singh, R.N. (1972) *Incompatibility Newsl.*, **1** : 22-24.

Pandey, R.M., Rathore, D.S. and Singh, R.N. (1974b) *Curr. Sci*, **43** : 694-695.

Pandey, R.M., Singh, R.N. and Sinha, G.C. (1973a) *Sci. and Cult.*, **39** : 148-150.

Pandey, R.M., Bojappa, K.M., Rathore, D.S. and Singh, R.N. (1974a) *Incompatibility Newsl.*, **4** : 83-84.

Pandey, R.M., Singh, G.C., Majumder, P.K. and Singh, R.N. (1971) *Curr. Sci.*, **40** : 356-357.

Pandey, R.M., Sinha, G.C., Singh, R.N. and Majumder, P.K. (1973b) *Indian J.Hort.*, **30** : 475-480.

Pandey, S.N. (1985) *2nd Int. Symp. on Mango*, India, Abst., No. 2.16.

Pandey, S.N. (2015) *Progressive Horticulture*, **47**(1), 20-38

Panickar, P. and Desai, A.G. (1989) *Progressive Hort.*, **21**: 141-144.

Parameswarakumar, M. and Gupta, C.P. (1988) *Paper, Amer. Soc. Agril. Eng.*, pp. 19.

Parisot, E. (1988) *Fruits*, **43** : 293-312.

Pathak, S.R. and Sarada, J.R. (1974) *Curr. Sci.*, **43** : 716.

Patil, A.A., Vadigeri, B.G. and Nalawadi, U.G. (1991b) *Curr. Res.* – UAS, Bangalore, **20** : 135-136.

Patil, P.B., Rao, M.M., Basarkar, P.W., Janardhan, K.V., Srinivasan, C.N. and Nalawadi, U.G. (1989) *Acta Horticulturae*, **231** : 405-411.

Patil, P.B., Rao, M.M., Basarkar, P.W., Srinivasan, C.N. and Nalawadi, U.G. (1991a) *Karnataka J. Agril. Sci.*, **4** : 17-23.

Patil, P.B., Rao, M.M., Janardhan, K.V., Srinivasan, C.N. and Nalawadi, U.G. (1990) *Karnataka J. Agric. Sci.*, **3** : 208-212.

Paulas, D. and Shanmugavelu, K.G. (1989) *Acta Horticulturae*, **231** : 394-398.

Panwar, R., Singh, Sanjay Kr., Singh, C P and Singh, P K (2007) *Indian J. Ag. Sc.* **77** (*10*): 680-4

Pawar, S.S., Gunjate, R.T. and Lad, B.L. (1985) *Pesticides*, **19** : 32-35.

Pena, M.A. and Ferragut, F. (1994) *Boletin de Sanidad Vegetal, Plagas*, **20** : 605-609.

Perez, A., Cedeno-Maldonado, A., Reyes-Soto, I. and Lopez, J. (1988) *J. Agril. Univ. Puerto Rico.*, **72** : 501-508.

Pesis, E., Aharoni, D., Aharon, Z., Ben-Arie, R., Aharoni, N. and Fuchs, Y. (2000) *Post-harvest Biol. Technol.*, **19** : 93-101.

Pimentel, R.B., Coronel, R.E. and Espino, R.C. (1984) *Philipp. J. Crop. Sci.*, **9** : 47-51.

Pinera, R.M., Casals, C., Batista, A.R. and Cerezal, P. (1995) *Alimentaria*, **32** : 29-31.

Pinto, A.C. de Q. and Byrne, D.H. (1993) *Acta Hort.*, **341**: 98-106

Pinto, A.C.Q., Andrade, S. R. M., Ramos, V. H. V. and Cordeiro, M.C.R. (2004) *Acta Hort.*, **645**.

Pire, R. and Rojas, E. (1995) *Proc. Interamerican Soc. Trop. Hort.*, **39** : 145-150.

Pliego-Alfaro, F., Monsalud, M.J.R., Litz, R.E., Gray, D.J. and Moon, P.A. (1996) *Plant Cell Tissue and Organ Culture*, **44** : 63-70.

Pongsomboon, W., Whiley, A.W., Stephenson, R.A. and Subhadrabandhu, S. (1992) *Acta Horticulturae*, **321** : 472-481.

Pope, W.T. (1929) *Hawaii Agri. Exp. Sta. Bull.*, No. 58.

Popenoe, W. (1920) *Manual of Tropical and Subtropical Fruits*, MacMillan, New York.

Prakash, Om and Raoof, M.A. (1985) *2nd Int. Symp. on Mango*, India, Abst. No. **6**.2, p. 59.

Prakash, S. and Ram, S. (1985) *2nd Int. Symp. on Mango*, India, Abst. No.5.28, p. 54.

Prasad, A. and Singh, R.D.(1972) *Punjab Hort. J.*, **13** : 30-34.

Prasad, A., Singh, H. and Shukla, T.N. (1965) *Indian J. Hort.*, **22** : 254-265.

Puttarudriah, M. and Channa Basavanna, G.P. (1961) *Curr. Sci.*, **30** : 114-115.

Rabbani, A. and Singh, I.S. (1989) *Acta Horticulturae*, **231** : 715-720.

Radha, T. and Aravindakshan, K. (1998) *The Hort. J.*, **11** : 25-31.

Raghupathi, H.B. and Bhargava, B.S. (1994) *South Indian Hort.*, **42** : 259-267.

Raghupathi, H.B. and Bhargava, B.S. (1999) *Indian J. Agric. Sci.*, **69** : 648-650.

Raghuvanshi, S.S. and Srivastava, A. (1995) *Plant Cell, Tissue Organ Culture*, **41** : 83-85.

Rajan, S. and Ram, S. (1985) *2nd Int. Symp. on Mango*, India, Abst. No. 3.8, p.20.

Rajan, S. and Ram, S. (1989) *Acta Horticulturae*, **231** : 192-197.

Rajput, M.S., Biswas, P.P. and Joshi, O.P. (1989) *J. Hort. Sci.*, **18** : 184-191.

Ram, R.B., Singh, B.P. and Singh, S.P. (1990) *The Hort. J.*, **3** : 31-36.

Ram, S. and Sirohi, S.C. (1985) *2nd Int. Symp. on Mango*, India, Abst. No. 4.21, p. 38.

Ram, S. and Sirohi, S.C. (1989) *Acta Horticulturae*, **231** : 339-344.

Ram, S. and Sirohi, S.C. (1994) *Recent Hort.*, **1** : 14-17.

Ramirez F. and Davenport T. L. (2010) *Scientia Horticulture*, **126** : 65-72

Rameshwar, A. (1985) *2nd Int. Symp. on Mango*, India, Abst. No. 8.28, p. 89.

Randhawa, G.S. and Damodaran, V.K, (1961) *Indian J. Hort.*, **18** : 29-64.

Randhawa, M.S. (1949) *Indian Fmg.*, **10** : 257-258.

Ranjan, S. and Jha, V.R. (1940) *Proc. Indian Acad. Sci.*, **11** : 267-268.

Rao, D.P. and Mukherjee, S.K. (1985a) *2nd Int. Symp. on Mango*, India, Abst. No.4.7, p.31

Rao, D.P. and Mukherjee, S.K. (1985b) *2nd Int. Symp. on Mango*, India, Abst. No. 4.17, p.36.

Rao, D.P. and Mukherjee, S.K. (1989a) *Acta Horticulturae*, **231** : 286-295.

Rao, D.P. and Mukherjee, S.K. (1989b) *Acta Horticulturae*, **231** : 325-334.

Rao, M.M. and Sriari, D. (1998) *J. Maharashtra Agric. Univ.*, **23** : 19-20.

Rao, M.M., Reddy, B.S., Hussain, S.A. and Ravishankar, H. (1995) *Adv. Agril. Res. India*, **4** : 78-86.

Rao, P.V.S., Giridhar, N., Prasad, P.S.R.K. and Rao, G.N. (1970) *Indian J. Hort.*, **27** : 117-122.

Rao, P.V.S., Giridhar, N., Prasad, P.S.R.K. and Rao, G.N. (1972) *Indian J. Hort.*, **29** : 126-134.

Rao, S.N. and Subba Rao, Ch. (1963) *Punjab Hort. J.*, **3** : 205-208.

Ratan, J., Arvindakshan, M. and Gopikumar, K. (1987) *South Inian Hort.*, **35**: 192-198.

Rath, S. and Rajput, C.B.S. (1990) *Orissa J. Agril. Res.*, **2**: 232-235.

Ravishankar, K.V., Lalitha, A. and Dinesh, M.R. (1999) *J. Hort. Sci. Biotech.*, **75** : 198-201.

Ravishankar, K.V.; Bommisetty, P.; Bajpai, A.; Srivastava, N.; Mani, B.H.; Vasugi, C.; Rajan, S.; Dinesh, M.R. (2015) *Trees*, **29**: 775-783.

Ray, D.P., Das, B.K., Das, T.K. and Jena, L. (1995) *Orissa J. Hort.*, **23** : 77-82.

Ray, P.K. (1999) In : *Tropical Horticuture*, Vol. I. (Eds. T.K. Bose, S.K. Mitra, A.A. Farooqi, M.K. Sadhu), Naya Prokash, Calcutta, India, pp. 102-177.

Reddy, B. and Bagyaraj, D.J. (1994) *Scientia Hort.*, **59** : 69-73.

Reddy, Y.T.N. and Vineeth Raj, A.V. (2015) Acta Hortic. **1066** : 99-108

Reddy, C.V. and Melanta, K.R. (1988) *South Indian Hort.*, **36** : 143-145.

Reddy, C.V. and Melanta, K.R. (1989) *Mysore J. Agric. Sci.*, **23** : 341-143.

Reddy, K.M. and Singh, R.N. (1987a) *Jour. Res. APAU*, **15** : 24-30.

Reddy, K.M. and Singh, R.N. (1987b) *Jour. Res. APAU*, **15** : 129-135.

Reddy, Y.N. and Majumder, P.K. (1975) *Curr. Sci.*, **44** : 444-445.

Reddy, Y.T.N. and Kohli, R.R. (1985) *2nd Int. Symp. on Mango*, India, Abst. No. 3.2, p.17.

Reddy, Y.V. and Srivastava, G.C. (1999) *J. Plant Physiol.*, **4** : 194-196.

Reuveni, O. and Castoriano, M. (1993) *Acta Horticulturae*, **341** : 288-294.

Robbertse, P.J., Coetzer, L.A., Tomer, E. and Smith, M.F. (1994) *Yearb. South Afr. Mango Growers' Assoc.*, **14** : 18-20.

Rojas, E. (1996a) *Revista de la Facultad de Agronomia, Universidad Central de Venezuela*, **22** : 47-56.

Rojas, E. (1996b) *Revista de la Facultad de Agronomia, Universidad Central de Venezuela*, **22** : 37-45.

Rojas, E. and Leal, F. (1993) *Proc. Intermerican Soc. Trop. Hort.*, **37** : 142-147.

Roy, B. and Visweswariya, S.S. (1951) *Review of Work* (1948-51), Maharashtra Association for the Cultivation of Science, Poona.

Roy, R.S. (1953) *Indian J.Hort.*, **10** : 157-160.

Roy, R.S., Chhonkar, V.S. and Prasad, S.N. (1963) *Punjab Hort.J.*, **3** : 209-213.

Roy, R.S., Mallik, P.C. and De, B.N. (1951) *Proc. Amer. Soc. Hort. Sci.*, **57** : 9-16.

Roy, R.S., Mallik, P.C. and Sinha, R.P. (1956) *Proc. Amer. Soc. Hort. Sci.*, **68** : 259-264.

Roy, S. and Hoda, M.N. (1996) *The Hort. J.*, **9** : 27-31.

Sachan, S.C.R., Katrodia, J.S., Chundawat, B.S. and Patel, M.N. (1985) *2nd Int. Symp. on Mango*, India, Abst. No. 2.8, p.9.

Sachar, R.C. and Chopra, R.N. (1957) *Indian J. Agric. Sci.*, **27** : 219-228.

Sadhu, M.K. (1992) *Acta Horticulturae,* **321** : 356-365.

Sadhu, M.K. and Bose, T.K. (1976) *Indian Fd. Packer,* **30** : 24-32.

Sadhu, M.K. and Bose, T.K. (1982) *Indian Agric.,* **26** : 243-256.

Saidhu, T., Madhava Rao, V.N. and Santhanakrishnan, P. (1983) *Indian J.Hort.,* **40** : 139-145.

Saini, S.S., Singh, R.N. and Paliwal, G.S. (1971) *Indian J. Hort.,* **28** : 247-256.

Salazar-Garcia, S. and Vazquez-Valdivia, V. (1997) *J. Hort. Sci.,* **72** : 339-345.

Salvi, P.V. (1983) *Indian Hort.,* **27** : 13.

Samaddar, H.N. and Chakarabarti, U. (1985) *2nd Int. Symp. on Mango,* India, Abst. No. 3.14, p.22.

Samra, J.S. (1989) *Acta Horticulturae,* **231** : 306-311.

Samra, J.S., Thakur, R.S. and Chadha, K.L. (1978) *Scientia Hort.,* **8** : 349-355.

Sandip, M., AN Makwana, AV Barad and BD Nawade (2015) *International J. App. Res.* **1**(11): 1008-1012

Sankar, K.P. (1963) *Sci. and Cult.,* **29** : 51.

Santos De La Rosa, F. De Los and Mosqueda Vazquez, R. (1989) *Revista Chapingo,* **13-16**(62-63) : 63-68.

Sanyal, D. and Maity, S.C. (1989) *Progressive Hort.,* **21** : 300-304.

Sanyal, D., Roychoudhury, R., Dhua, R.S. and Mitra, S.K. (1991) *Indian Fd. Packer,* **45** : 29-33.

Sarkar, S.C. and Mushi, A.A.A. (1981) *Bangladesh J. Agric. Sci.,* **8** : 69-75.

Sattar, A. (1946) *Punjab Fruit J.,* **10** : 56-58.

Sauco, Víctor Galán (2016.) Mango Rootstocks. Literature Review and Interviews. Tropical Fruit Consultant, USA

Schaesberg, N. von, Ebert, G. and Ludders, P. (1993) *Angewandte Botanik,* **67** : 14-16.

Schaesberg, N. von, Ebert, G. and Ludders, P. (1993) *Gartenbauwissenschaft,* **58** : 280-285.

Schaffer, B. and Gaye, G.O. (1989) *HortScience,* **24** : 507-509.

Schaffer, B., Andersen, P.C. and Ploetz, R.C. (1992) *Hort. Rev.,* **13** : 257.

Schaffer, B., Whiley, W.A. and Crane, J.H. (1994) In : *Handbook of Environmental Physiology of Fruit Crops,* Vol. II *Sub-tropical and Tropical Crops* (Eds. B. Schaffer and P.C. Andersen),CRC Press Inc.,Boca Raton, Florida, pp.165-97.

Schmutz, U. and Ludders, P. (1993) *Acta Horticutlurae,* **341** : 160-167.

Schmutz, U. and Ludders, P. (1994) *Angewandte Botanik Berichte,* **5** : 189-193.

Schnell, R.J. and Knight, R.J.Jr. (1992) *HortScience,* **27** : 174-176.

Schnell, R.J., Brown, J.S., Olano, C.T., Meerow, A.W., Campbell, R.J. and Kuhn, D.N. (2006) *J Am Soc Hortic Sci.,* **131**(2), 214–224

Schoeman, A.S. (1995) *Yearb. South Afr. Mango Growers' Assoc.*, **15** : 97-98.

Schoeman, M. (1996) *Inligtingsbulletin – Instituut vir Tropiese en Subtropiese Gewasse*, **286** : 5-7.

Searle, C., Whiley, A.W., Simpson, D.R. and Saranah, J.B. (1995) *Proc. Marketing Seminar and Production Workshop*, Department of Primary Industries, Brisbane, pp. 127-135.

Sen, P.K. (1939) *Annual report, Fruit Research Station*, Sabour (Bihar), India.

Sen, P.K. (1943a) *Indian J. Hort.*, **1** : 48-71.

Sen, P.K. (1943b) *Indian J. Agric. Sci.*, **13** : 300-305.

Sen, P.K. (1946) *Punjab Fruit J.*, **10** : 31-34.

Sen, P.K. and Roy, K. (1945) *Proc. 32nd Indian Science Congress*, Part III, p. 23.

Sen, P.K., Mallik, P.C. and Ganguly, B.D. (1946) *Indian J. Hort.*, **4** : 4-15.

Sen, P.K., Roy, P.K. and De, B.N. (1947) *Indian J. Hort.*, **35** : 34-44.

Sen, P.K., Bose, T.K. and Sashibhusan, T. (1961) *Indian Agric.*, **5** : 167-172.

Sen P.K., Sen, S.K. and Guha, Dipika (1963) *Indian Agric.*, **7** : 133-138.

Sen, P.K., Basu, R.N, Bose, T.K. and Roychowdhury, N. (1968) *Curr. Sci.*, **37** : 144-145.

Sergent, E., Casanova, E. and Leal, F. (1993) *Revista de la Facultad de Agronomia, Universidad del Zulia*, **10** : 287-296.

Shah, A.H., Jhala, R.C., Patel, Z.P. and Patel, R.L. (1985) *2nd Int. Symp. on Mango*, India, Abst. No. 7.12, p. 72.

Shanmugavelu, K.G. and Selvarajan, M. (1985) *2nd Int. Symp. on Mango*, India, Abst. No. 4.18, p. 37.

Sharma, D.K. and Majumder, P.K. (1985) *2nd Int. Symp. on Mango*, India, Abst. No. 2.9, p.10.

Sharma, D.K. and Sen Choudhuri, S. (1976) *Curr. Sci.*, **45** : 305-306.

Sharma, D.K. and Singh, R.N. (1970a) *Indian J. Hort.*, **27** : 1-15.

Sharma, D.K. and Singh, R.N. (1970b) *Hort. Res.*, **10** : 108-118.

Sharma, D.K. and P.K. Majumder, (1988a). *Acta Horticulturae*, **231**: 106–11.

Sharma, D.K., Majumder, P.K. and Kaicker, U.S. (1976) *Curr. Sci.*, **45** : 675.

Sharma, D.K., Majumder, P.K. and Singh, R.N. (1972) *Proc. Symp. Recent Advances in Horticulture*, UP Inst. of Agric. Sci., Kanpur, pp. 66-68.

Sharma, D.K., Singh, R.N. and Majumder, P.K. (1981) *Nat. Symp. Trop. and Subtrop. Fruit Crops*, Bangalore, p.4.

Sharma, G.D. and Kapur, B.S. (1967) *Indian Fd. Packer*, **21** : 34-36.

Sharma, K., Sharma, I.M. and Sharma, N.K. (1995) *Indian Forester*, **121** : 122-125.

Sharma, R.C., Mahajan, B.V.C. and Azad, A.S. (1993) *The Hort. J.*, **6** : 83-87.

Sharma N., Singh SK, Singh NK, Srivastav M, Singh BP, *et al.* (2015) *Transcriptomics* **3**: 110.

Sharma, N., Singh, S K., Prakash,J., Srivastav,M., Mahto, A K., and Singh, N K (2017) *Curr. Trends Biomedical Eng & Biosci.* **7**(2): 555707.

Shashirekha, M.S. and Patwardhan, M.V. (1976) *Lebensmittelwiss. Technol.*, **9** : 369-370.

Sheikh, K.M., Bajwa, M.N., Rashid, A. and Mahmood, T. (1995) *Pakistan J. Phytopath.*, **7** : 29-32.

Shivashankara, K.S. and Mathai, C.K. (1995) *Photosynthetica*, **31** : 135-140.

Shinde, A.K., Waghmare, G.M., Wagh, R.G. and Burondkar, M.M. (2000) *Indian J. Pl. Physiol.*, **5** : 82-84.

Shivashankara, K.S. and Mathai, C.K. (1999) *Scientia Hort.*, **82** : 317-323.

Shu, Z.H. (1993) *Acta Horticulturae*, **341** : 199-205.

Shu, Z.H. (1994) *J. Agril. Res. China*, **43** : 63-68.

Shukla, H.S., Bajpai, P.N. and Prasad, A. (1985) *2nd Int. Symp. on Mango*, India, Abst. No. 5.31, p. 56.

Siddiqui, S.H., Mujeeb, K.A. and Vasti, S.M. (1966) *Proc. Agric. Symp., Atomic Energy Centre*, Dhaka, pp. 34-37.

Sigurbjornsson, B. (1970) *Technical Report Series No. 119*, IAFA, Vienna, pp. 1M7.

Siller-Cepeda, J., Muy-Rangel, D., Araiza, E., Baez, M., Rodriguez, J., Baez, R. and Ireta, A. (1994) *Proc. Interamerican Soc. Trop. Hort.*, **38** : 37-42.

Simao, S. and Pimentel-Gomes, F. (1994) *Revista de Agricultura* (Piracicaba), **69** : 3-10.

Simao, S., Doni, M.E. and Barbin, D. (1996) *Revista de Agricultura* (Piracicaba), **71** : 347-359.

Singh, Anju *et al.* (2014) *In:* International Plant & Animal Genome Conference XXIII / January 10-14, 2014 - San Diego, CA, USA.

Singh, A.K., Singh, B.P., Sharma, R. and Singh, R.N. (1985) *Punjab Hort. J.*, **25** : 1-4.

Singh, A.R. and Singh, N.D. (1996) *Recent Hort.*, **3** : 4-7

Singh, B., Singh, D.S. and Pathak, R.A. (1985) *2nd Int. Symp. on Mango*, India, Abst. No. 3.23, p. 26.

Singh, B.P., Kalra, S.K. and Tandon, D.K. (1989) *Indian J.Hort.*, **46** : 154.

Singh, I.P. and Ram, S. (1994) *Recent Hort.*, **1** : 23-29.

Singh, I.P., Singh, B.P. and Singh, S.P. (1986) *Haryana J. Hort. Sci.*, **15** : 29-32.

Singh, J.N., Rajput, C.B.S. and Prakash, S. (1991) *Haryana J. Hort. Sci.*, **20** : 35-38.

Singh, J.N., Singh, D.K. and Chakravarty, D. (1994) *Orissa J. Hort.*, **22** : 26-30.

Singh, K. and Khan, A. (1990) *Progressive Hort.*, **22** : 44-50.

Singh, K.K. (1967) *The Mango : A Handbook*, ICAR, New Delhi, pp. 70-98.

Singh, K.K. and Chadha, K.L. (1961) *Punjab Hort. J.*, **1** : 171-179.

Singh, K.K. and Mathur, P.B. (1952) *CFTRI, Mysore, Bull.*, No. 2 : 14.

Singh, L. and Khan, A.A. (1939) *Indian J. Agric. Sci.*, **9** : 835-867.

Singh, L. and Khan, A.A. (1940) *Indian Fmg.*, **1** : 380-383.

Singh, L.B. (1954) *Proc. Amer. Soc. Hort. Sci.*, **63** : 128-130.

Singh, L.B. (1960) *The Mango : Botany, Cultivation and Utilisation*, Leonard Hill, London.

Singh, L.B. (1961) *Hort. Advances*, **5** : 17-24.

Singh, L.B. and Singh, R.N. (1956) *A monograph on the Mangoes of Uttar Pradesh* (2 vols), Supdt., Printing, Lucknow.

Singh, M., Chaudhary, A.S. and Prasad, M. (1986) *Haryana J. Hort. Sci.*, **15** : 221-223.

Singh, M.P. and Saxena, P.K. (1994) *Recent Hort.*, **1** : 30-35.

Singh, M.P. (1962) *Indian J. Hort.*, **19** : 103-10.

Singh, N.K. *et al.* (2016) *Indian J. History of Science*, 51.2.2: 355-368

Singh, N.P. and Srivastava, R.P. (1982) *Progressive Hort.*, **14** : 117-120.

Singh, R.N. (1954a) *Science*, **119** : 389.

Singh, R.N. (1954b) *Indian J. Hort.*, **11** : 69-88.

Singh, R.N. (1957) *Hort. Advance*, **1** : 23-33.

Singh, R.N. (1959) *Hort. Advance*, **3** : 28-49.

Singh, R.N. (1960) *Hort. Advance*, **4** : 48-60.

Singh, R.N.(1961) *Indian J. Hort.*, **18** : 1-11.

Singh, R.N. (1978) *Mango*, low-priced Book Series No.3., ICAR, New Delhi.

Singh, R.N., Majumder, P.K. and Sharma, D.K. (1962a) *Sci. and Cult.*, **28** : 484-485.

Singh, R.N., Majumder, P.K. and Sharma, D.K. (1962b) *Curr. Sci.*, **31** : 209.

Singh, R.N., Majumder, P.K. and Sharma, D.K. (1965) *Trop. Agri.(Trinidad)*, **12** : 170-174.

Singh, R.N., Majumder, P.K. and Sharma, D.K. (1966) *Proc. Amer. Soc. Hort. Sci.*, **89** : 228-230.

Singh, R.N., Majumder, P.K. and Sharma, D.K. (1977) In : *Fruit Breeding in India* (Ed. G.S. Nijjar), Oxford & IBH Publishing Co., New Delhi, pp. 25-30.

Singh, R.N., Majumder, P.K and Sharma, D.K. (1981) *Nat. Symp. Trop. and Subtrop. Fruit Crops*, Bangalore, p. 3.

Singh, R.N., Majumder, P.K and Sharma, D.K., Singha, G.C. and Bose, P.C.(1974) *Scientia Hort.*, **2** : 299-401.

Singh, R.N., Majumder, P.K and Sharma, D.K. and Rathore, D.S. (1968) *Indian J. Hort.*, **25** : 64-65.

Singh, R.N., Sharma, D.K. and Majumder, P.K (1980) *Scientia Hort.*, **12** : 299-301.

Singh, R.S. and Ram, S. (1983) *Indian J. Hort.*, **40**: 188-194.

Singh, R.N., Rao, O.P. and Singh, G. (1984) *Progressive Hort.*, **16** : 161-115.

Singh, S. and Sengupta, B.N. (1996) *The Hort. J.*, **9**: 13-16.

Singh, Sanjay Kumar and Bhargava, R. (2011) *J. Trop. Ag.* **49**(1-2) : 104-106

Singh, Sanjay Kumar, Singh, S.K., Sharma, R.R. and Srivastav, M. (2009a) *Indian J. Agri. Sc.* **79**(8): 632–5

Singh, Sanjay Kumar, Singh, S.K. and Sharma, R.R. (2009b) *Indian J. Plant Physiol.*, **14**(4): 392-396

Singh, Sanjay Kumar, Singh, S.K., Sharma, R.R., Srivastav, M. and V.B. Patel (2010a) *Indian J. Hort.* **67**(1): 16-20.

Singh, Sanjay Kumar, Singh, S.K., Sharma, R.R., and V.B. Patel (2010b) *Indian J. Hort.* **67**(Special Issue): 84-89.

Singh, Sanjay Kumar, Singh, S.K. and Sharma, R. R. (2010c) *J. Hort. Sci. & Biotech.* **85**(6): 483–490.

Singh, Sanjay Kumar, Singh, Awtar, Nath Vishal, Parthasarathy, V A, Sthapit B, Rajan, S and Vinoth, S. (2015) *Indian J. Plant Genet. Resour.* **28**(1): 123-131.

Singh, Sanjay Kumar, Rajan, Shailendra and M R Dinesh (2018) *In:* Breeding of Horticultural Crops Vol. 2 -Part: Tropical Fruits. (Eds. Parthasarthy *et al.*,), ISBN 81-7019- (India), *Today & Tomorrow's Printers and Publishers,* New Delhi - 110 002, India. p307-337

Singh, Sanjay Kumar, Nath, Vishal, Rajan, S. and Pandey, S.D. (2019) *Indian J. Plant Genet. Resour.* **32**(2): 200-206.

Singh, S.N. (1963) *Punjab Hort. J.*, **3**: 116-123.

Singh, S.N. and Teaotia, S.S. (1951) *Sci. and Cult.*, **17** : 207-210.

Singh, U.R. (1960) *Hort. Advance*, **4** : 142-154.

Singh, U.R. and Singh, A.P. (1976) *Progressive Hort.*, **8** : 13-19.

Singh, V.K., Saxena, Pooja, Rajan, S., Upreti, K.K. and Ravishankar, H. (2014) In: *National Seminar-cum-Workshop on "Physiology of Flowering in Perennial Fruit Crops"* ICAR-CISH, Lucknow, pp.29.

Singh, Z. (2000) *Acta Hort.* **525**: 229-236

Singh, Z. and Dhillon, B.S. (1987) *Trop. Agric.*, **64** : 305-308.

Singh, Z. and Dhillon, B.S. (1988) *Trop. Agric.*, **65** : 295-296.

Singh, Z. and Dhillon, B.S. (1989a) *Acta Horticulturae*, **239** : 367-70.

Singh, Z. and Dhillon, B.S. (1989b) *Acta Horticulturae*, **239** : 371-374

Singh, Z. and Dhillon, B.S. (1989c) *J. Phytopath.*, **125** : 193-197.

Singh, Z. and Dhillon, B.S. (1989d) *J. Phytopath.*, **125** : 117-123.

Singh, Z. and Dhillon, B.S. (1992) *Acta Horticulturae*, **296** : 51-54.

Singh, Z. and Singh, L. (1993) *J. Hort. Sci.*, **68** : 535-540.

Sirohi, S.C. and Ram, S. (1995) *Recent Hort.*, **2** : 16-23.

Smith, B.L. (1989) *Inf. Bll., Citrus & Subtrop. Fr. Res. Inst., S.Afr.*, **197** : 4.

Smith, B.L. (1992) *Yearb. South Afr. Mango Growers' Assoc.*, **12** : 54-56.

Smith, P.F. and Scudder, G.K. (1951) *Proc. Fla. Sta. Hort. Soc.*, **64** : 243-248.

Snowdon, A.L. (1990) *A colour Atlas of Post-harvest Diseases and Disorders of Fruits and Vegetables.* Vol.I, *General Introduction and Fruits*, Wolfe Scientific Ltd., London, p. 302.

Sornsrivichai, J., Anusadorn, P., Oogaki, C. and Gemma, H. (1989) *Japanese J. Trop. Agric.*, **33** : 6-17.

Soto-Manitiu, J., Jiron, L. and Hernandez, R. (1987) *Turrialba*, **37** : 245-251.

Spreer, W., Ongprasert, S., Hegele, M., Wünsche, J.N. and Müller, J. (2009) *Agric. Water Manage.*, **96**: 574-584.

Srivastava, H.C. (1967) *The Mango : A Handbook*, ICAR, New Delhi, pp. 99-149.

Srivastava, K.C., Chadha, K.L., Singh, N.P., Sinha, G., Rajput, M.S., Lal, B. and Suman, C.L. (1989) *Acta Horti.*, **231** : 239-248.

Stephens, S.E.(1949) *Qd. Agric. J.*, **68** : 71-81.

Sturrock, T.T. (1977) In: *Fruit Breeding in India*. (Ed. G.S. Nijjar), Oxford & IBH Publishing Company, New Delhi, pp. 21-24.

Subhadrabandhu, S. and Iamsub, K. (1996) *Kasetsart J. Nat. Sci.*, **30** : 285-292.

Subhadrabandhu, S., Iamsub, K. and Kataoka, I. (1999) *Japanese J. Trop. Agric.*, **43** : 249-253.

Subrahamanyam, K.V. (1990) *Acta Horticulturae*, **269** : 89-95.

Sukhvibul, N., Whiley, A.W., Smith, M.K., Hetherington, S.E. and Vithanage, V. (1999a) *J. Hort. Sci. Biotech.*, **74** : 64-68.

Sukhvibul, N., Whiley, A.W., Smith, M.K., Hetherington, S.E. and Vithanage, V. (1999b) *Scientia Hort.*, **82** : 67-84.

Sukonthasing, S., Wongrakpanich, M. and Verheij, E.W.M. (1991) In : *Plant Resources of Southeast Asia. 2. Edible Fruits and Nuts*, (Eds. E.W.M. Verheij and R.E Coronel), Pudoc-DLO, Wageningen, The Netherlands, p. 221.

Summanwari, A.S., Raychaudhuri, S.P. and Phatak, S.C. (1966) *Indian Phytopath.*, **19** : 227-228.

Suriyapananont, V. (1992) *Acta Horticulturae*, **321** : 529-534.

Swamy, G.S., Rao, S.N. and Tatachari, Y. (1970) *South Indian Hort.*, **18** : 81-84.

Syamal, M.M. and Mishra, K.A. (1987) *Indian J. Hort.*, **44** : 194-196.

Syamal, M.M. and Mishra, K.A. (1989a) *Acta Horticulturae*, **231** : 276-281.

Syamal, M.M. and Mishra, K.A. (1989b) *Acta Horticulturae*, **231**: 149-151.

Szabados, L., Roca, W.M., Tabares, E., Nunez, V., Soto, M.E. and Laignelet, A. (1987) *Proc. Intl. Congr. Plant Tissue Cult. Tropical Species*, Bagota, p. 10.

Tandon, D.K. and Katra, S.K. (1983) *Plant Physiol. Biochem.*, **10** : 113.

Thakur, R.S., Srivastava, R.P., Chadha, K.L. and Singh, P. (1989) *Acta Horticulturae*, **231** : 232-238.

Thakurta, A.G. and Dutt, B.K. (1941) *Curr. Sci.*, **10** : 297.

Thimmappaiah and Singh, H. (1983) *Indian J. Hort.*, **40** : 195-198.

Thimmappaiah and Suman, C.L. (1987) *Punjab Hort. J.*, **27** : 8-11.

Thomas, P. (1975) *J. Food. Sci.*, **40** : 704-706.

Thomas, P. (1999) *J. Hort. Sci. Biotech.*, **74** : 135-139.

Thompson, A.K. (1996) *Post-harvest Technology of Fruit and Vegetables*. Blackwell Science Ltd., London.

Tiwari, R.K.S. and Singh, J. (1999) *Scientific Hort.*, **6** : 1-4.

Tripathi, J.S. (1980) *Curr. Agric.*, **4** : 277-278

Tripathi, R.D. (1955) *Indian J. Hort.*, **12** : 173-179.

Truscott, M. (1992) *Inligtingsbulletin-Instituut vir Tropiese en Subtropese Gewasse*, **237** : 17-18.

Truscott, M., Human, C. and Visser, G.J. (1993) *J. South Afr. Soc. Hort. Sci.*, **3** : 106-107.

Tsai, J.C., Yao, M.H., Chi, K.S. and Liang, L.S. (1996) *J. Agril. Res., China*, **45** : 186-194.

Ueda, M, Sasaki, K., Utsunomiya, N., Inaba, K. and Shimabayashi, Y. (1999) *J. Japanese Soc. Food Sci. Tech.*, **46** : 16-23.

Upreti KK, Reddy YTN, Shivu P, Bindu GV, Jayaram HL, Rajan S (2013) *Scientia Hort.*, **150**: 414-418.

Urban, L., Le Roux, X., Sinoquet, H., Jaffuel, S. and Jannoyer, M. (2003) *Tree Physiol.*, **23**: 289-300.

Usha, K., Gambhir, P.N., Sharma, H.C., Goswami, A.M. and Singh, B. (1994) *Scientia Hort.*, **59** : 291-295.

Utsunomiya, N., Sogawa, K. and Sasaki, K. (1995) *Acta Horticulturae*, **394**: 183-190.

Varma, A., Lele, V.C. and Goswami, B.K. (1974) In : *Current Trends in Plant Pathology* (Eds. S.P. Rouchoudhury and J.F. Varma), Botany Dept., Lucknow University.

Vega, E. and Molina, E. (1999) *Agronomia Costarricense*, **23** : 37-44.

Verma, K.S. and Singh, J. (1996) *Indian J. Mycology & Plant Path.*, **26** : 75-78.

Viruel, M.A., P. Escribano, M. Barbieri, M. Ferri and J.I. Hormaza (2005) *Mol. Breed.*, **15**: 383–393

Wagle, P.V. (1929) *Agric. J. India*, **24** : 259-263.

Wagle, P.V. (1931) *Agriculture and Livestock, India*, **1** : 286-290.

Wali, Y.A. and Hassan, Y.M. (1965) *Proc. Amer. Soc. Hort. Sci.*, **87** : 264-269.

Wang Caifa (1999) *China Fruits*, **1** : 58.

Wang, D., Sondahl, M.R. and Evans, D.A. (1986) *Intl. Congr. Plant Tissue Cult.*, 6 Meet, p. 339.

Wang, X.F. and Fu, J.R. (1994) *J. South China Agril. Univ.*, **15:** 88-92.

Wang, Y., Li, X.P. and Ji, Z.L. (1995) *Acta Horticulturae Sinica*, **22** : 235-239.

Wardlaw, C.W. and Leonard, E.R. (1936) *Mem. Low. Temp. Res. Sta., Trin.*, **3** : 47.

Watt, G. (1891) *Dictionary of the Economic Products of India*, **5** : 149.

Wei MingQing (1998) *China Fruits*, **4** : 54.

Werner, H. (1993) *Acta Horticulturae*, **341** : 225-231.

Wester, P.J. (1924) *Philipp. Agric. Rev.*, **17** : 293.

Wet, E. De, Robbertse, P.J. and Groeneveld, H.T. (1989) *South Afr. J. Plant Soil*, **6** : 228-234.

Whiley, A.W. (1984) In : *Tropical Tree Fruits for Australia* (compiled by P.E. Page), Qd. Dept. Prim. Indus. Inform. Series Q183018, 25.

Whiley, A.W. (1993) *Acta Horticulturae*, **341** : 168-176.

Whiley, A.W., Rasmussen, T.S., Saranath, J.B. and Wolstenholme, B.N. (1989) *J. Hort. Sci.*, **64** : 753-765

Wickham, L.D. and Majeed Mohammed (1999) *J. Food Quality*, **22** : 31-40.

Wijeratnam, R.S.W., Altes, F.W.K. and Kulathunge, G. (1996) *Trop. Sci.*, **36** : 68-70.

Wilson, C.W., III, Shaw, P.E. and Knight, R.J.Jr. (1990) *J. Agric. Food Chem.*, **38** : 1556-1559.

Winston, E.C. (1992) *Australian J. Expt. Agric.*, **32** : 97-104.

Wittwer, A.N. (1991) *Yearb. South Afr. Mango Grower' Assoc.*, **11** : 6-8.

Wong, K.C. and Siew, S.S. (1994) *Flavour & Fragrance J.*, **9** : 173-178.

Yadav, S.S. and Ram, S. (1986) *Abst. 22nd Int. Hort. Congress*, California, p. 689.

Yadava, R.B.R. and Singh, V.K. (1995) *Indian J. Pl. Physiol.*, **38** : 328-330.

Yadava, R.B.R. and Singh, V.K. (1998) *Indian J. Pl. Physiol.*, **3** : 166-167.

Yamashita, K. and Ogawa, S. (1998) *Japanese J. Trop. Agric.*, **42** : 153-158.

Yang, J. P. and Chen, X. M. (1998) *South China Fruits*, **27** : 29-31.

Yang, Z.H. and Ludders, P. (1993) *Acta Horticulturae*, **341** : 240-247.

Yang, Z.H., Hu, N.Y. and Lu, G.M. (1984) *Acta Univ. Sept. Occ. Agri.*, **1** : 13-19.

Yee, W.Y.J. (1976) *Fruit Var. J.*, **30** : 107-108.

Yeshitela T, Robbertse PJ, Stassen PJC (2004) *New Zealand J. Crop Hort. Sci.*, **32:** 281-293

Young, T.W. and Koo, R.C.J. (1971) *Proc. Fla. Sta. Hort. Soc.*, **84** : 298-303.

Young, T.W. and Sauls, J.W. (1981) *The Mango Inustry in Florida*, Fla. Co-op. Extn. Serv. Bull. 189, Univ. Florida, Gainesville, Florida.

Zauberman, G., Fuchs, Y., Rot, I. and Wexler, A. (1988) *HortScience*, **23** : 732-733.

Zhang, C.L., Huang, H.B. and Kuang, Y.H. (1995) *Scientia Hort.*, **64** : 49-54.

Zhang T, Ying Z, Davenport TL (2005) *Plant Biology* p.151.

Zhaong, H.Q., Luo, J.T., Lu, D. and Yuan, J. (1994) *Guangdong Agric. Sci.*, **5** : 19-22.

②

GUAVA

Maneesh Mishra, S. K. Mitra and T. K. Bose

1.0 Introduction

Guava (*Psidium guajava* L.) belongs to family Myrtaceae and believed to be originated from Southern Mexico or Central America (Morton, 1987) and spread to tropical and subtropical areas of the world. *Psidium* is large genera consisting as many as 150 species listed in Kew Index (Pommer and Murakami, 2009) However, only few species are of importance. Guava is nutritionally rich, exquisite and remunerative crop. It excels over many other fruit crops in terms of productivity, hardiness, adaptability and vitamin C content. It can be cultivated in range of soils and climatic conditions and are less affected by extremes of temperature, hot winds, scanty rainfall, saline soils, water logging *etc*. The major problems facing guava industry in different part of world includes severity of wilt disease, high seed content of diploid commercial varieties and poor yield with small, irregular and mis-shapen fruits of triploid seedless varieties (Negi and Rajan, 2007). Guava is gaining popularity globally under wide range of soil and ecosystems. Changing consumers' preference, availability of exotic varieties across the countries, wilt and other biotic stress and fast depleting genetic resources has become important both for conservation of genetic resources and their use for developing new varieties.

2.0 Composition and Uses

2.1.0 Composition

Guava is a rich source of ascorbic acid and pectin. The ripe fruits contain moisture (77.9-86.9 per cent), dry matter (12.3-26.3 per cent), ash (0.51-1.02 per cent), crude fat (0.10-0.70 per cent), crude protein (0.82-1.45 per cent) and crude fibre (2.0-7.2 per cent). The physico-chemical composition of guava fruits varies widely with cultivars, stage of maturity and season (Das *et al.*, 1995; Kundu *et al.*, 1995; Ghosh and Chattopadhyay, 1996). The total soluble solids content in fruit varies from 8.2 to 10.5 per cent Brix (Mitra, 1983; Kundu *et al.*, 1995). The total sugar content ranges between 4.9 and 10.1 per cent (Table 1). Fructose(59 per cent), glucose (36 per cent) and sucrose (5 per cent) are the predominant sugars in ripe guava fruits (Chan and Kwok, 1975). Fructose is the principal sugar in green ripe fruits while sucrose is the main one in fully ripe fruits (Arenas de Moreno *et al.*, 1995). Fruits are fair sources of vitamin A (about 250 IU/100 g) and contain appreciable quantities of thiamine, niacin and riboflavin (Table 1). The ascorbic acid content ranges from 75 to 260 mg/100 g, which varies with cultivar, season, location and stage of maturity (Singh, 1988; Das *et al.*, 1995; Ghosh and Chattopadhyay, 1996). Pink-fleshed cultivars are poorer in ascorbic acid content than the white fleshed ones (Mitra *et al.*, 1984b). In general, the fruits harvested during winter season are superior in quality to rainy season fruits (Ghosh and Chattopadhyay, 1996). It contains considerable amounts of Ca, P and Fe, however, 80 per cent of the iron remains in seed and is not utilisable (Miller and Bazore, 1945). Guava fruits are a good source of pectin which ranges between 0.5 and 1.8 per cent (Adsule and Kadam, 1995). The pectin content also varies with cultivars and stage of maturity and there is no specific trend in pectin content between pink and white fleshed cultivars/hybrids (Baramanray *et al.*, 1995).

Citric and malic are the predominant organic acids. Besides that, glycolic, tartaric and lactic acids are also present (Adsule and Kadam, 1995). Several volatile compounds have been reported to be responsible for the characteristic flavour of guava, that include hydrocarbons, alcohols, and carbonyls. The decrease in astringence with the advancement of maturity is ascribed to polymerization of leucoanthocyanidins. The pink flesh colour of some cultivars is due to the presence of lycopene (Adsule and Kadam, 1995).

Table 1: Composition of Guava Fruits*

Constituents	*Range*
Moisture (per cent)	77.9-86.9
Dry matter (per cent)	12.3-26.3
Ash (per cent)	0.51-1.02
Crude fat (per cent)	0.10-0.70
Crude protein (per cent)	0.82-1.45
Crude fibre (g/100 g pulp)	2.0-7.2
Sugars	
Reducing (per cent)	2.4-5.2
Non-reducing (per cent)	2.5-3.8
Total (per cent)	4.9-10.1
Acidity (per cent)	0.22-0.39
Ascorbic acid (mg/100 g)	75.2-234.3
Thiamine (mg/100 g)	0.03-0.07
Riboflavin (mg/100 g)	0.02-0.04
Niacin (mg/100 g)	0.20-2.32
Calcium (mg/100 g)	10.0-30.0
Phosphorus (mg/100 g)	22.5-40.0
Iron (mg/100 g)	0.60-1.39

* Wilson (1980), Singh (1988), Das *et al.* (1995), Ghosh and Chattopadhyay (1996).

2.2.0 Uses

Guava fruit is relished when mature or ripe and freshly plucked from the tree. Excellent salad and pudding are prepared from the shell of the ripe fruit. It can be preserved by canning as halves or quarters, with or without seed core (shells). The cv. Allahabad seedless white guavas have been reported to be more suitable for canning as halves (Siddappa, 1982). The guava jelly is well-known to all with an attractive purplish-red colour, pleasant taste and aroma. The common sour wild guava makes the best jelly. High quality nectar can be prepared from guava and nectar prepared from the hybrids was rated better than that of the commercial cultivars (Baramanray *et al.*, 1995). Puree made from cv. EEA 18-40 was the best in flavour, colour and aroma (Pinera *et al.*, 1997). Ripe fruits are also used for manufacturing of ice cream, *sherbet*, cheese and toffee. Guava slices can be dehydrated by air drying

Uses of guava

Guava fruit with slices

Guava fruit with slices

Guava used as salad

Guava used as pudding

Uses of guava

Guava canning

Guava jelly

Guava jam

Guava nectar

or osmotic dehydration and the dehydrated slices are pulverised to obtain guava fruit powder which is a good source of vitamin C. It freezes exceptionally well and the frozen product is practically indistinguishable from fresh fruit (Ruehle, 1948). Guava jamand juice concentrate are also liked for their characteristic taste. Two types of wines, *viz.* guava juice wine and guava pulp wine are also prepared from guava fruits (Bardiya *et al.*, 1974).The seeds yield 5 to 13 per cent oil which is rich in essential fatty acid and can be used in salad dressing (Adsule and Kadam, 1995). In some countries the leaves are used for curing diarrhoea, and also for dyeing and tanning.

Processing of guavas into various products results into wastes including seeds, stone cells and fibrous tissues from the skin generated especially during pulping. These make about 25 per cent and are suitable for processing into animal feeds and other products (El Boushy and van der Poel AFB, 2000). Guava wastes have been found to contain high levels of crude fiber (as much as 61 per cent), significant quantities of ether extracts (mainly oleic and linoleic acids) and 1,336 kcal/kg - 1,808 kcal/kg metabolizable energy values (Omayio, ***et al.***, 2019).

3.0 Origin and Distribution

The common guava originated, along with a number of other fruits, in tropical America and seems to have been growing from Mexico to Peru. The trees were domesticated more than 2,000 years ago. It was spread rapidly throughout the world's tropics by the Spanish and Portuguese soon after the discovery of the New World (Cobley and Steele, 1976). It is now cultivated in most tropical and subtropical regions of the world (Chadha and Pandey, 1986). It is grown in Ceylon from sea level to an elevation of 5000 feet (1515 m) and throughout Burma. At present the major guava producing countries are Southern Asian countries, the Hawaiian Islands, Cuba and India. It is believed to be introduced in India at a very early date, as it is mentioned by Bruton who was in India early in the 17th century.

4.0 Species and Cultivars

4.1.0 Species

All the cultivars of guava belong to the single species *P. guajava* L. Initially, the two species *pyriferum* and *pomiferum* mentioned by Linnaeus are nothing but trees with pear shaped and round shaped fruits. Subsequently, other species were recognized and documented. The wild species are of considerable importance in breeding programs. About 153 genotypes including some *Psidum* spp. are available in the Indian collections and are being maintained at several centres in the country in field gene banks. Some of the wild relatives of *Psidium* are mentioned below.

- ***P. cattleianum***: It is a small tree or shrub with smooth bark. Leaves are obovate elliptic and glabrous. Fruits are obovate to roundish with purplish red or light yellow in colour. Skin is thine; pulp is soft having numerous seeds. Flavour is sweet with good aroma.

- ✰ ***P. molle* Bertd:** It is a shrub or small tree. The leaves are obovate. The lower surface of leaves is reddish velvety. Fruits are pale yellow in colour when fully ripe. It has white flesh with numerous seeds. The flavour is acidic. It is wilt resistant and has potential for rootstock breeding
- ✰ ***P. friedrichsthalianum* Niedenzu:** It is a tall tree of about 7-10m high. The branches are thin and smooth. Leaves are oval or oblong, glossy above and pubescent below. Fruits are globosely, small and sour.
- ✰ ***P. guineese***: It is also known as Brazilian guava. It is a shrub or small tree. Fruits are big in size having flavour, thick flesh and few seeds.
- ✰ ***P. montanum* Swartz**: It is found generally in mountains of Jamaica. The branchlets are four-angled; leaves oblong to oval, glabrous.
- ✰ ***P. cujavilis*:** Growth characters and flowering habit of the plants is just like *P. guineense*. The size of fruit is small to medium, average weight is 30-50 g, and sour in taste.
- ✰ ***P. pumilum*:** Tree is like pyramidal in shape, leaves are light in colour, small in size, non-pubescent, having 13-17 pairs of primary veins. There are seven petals of cream colour, stamens are 252-327 in number, small stigma with medium size of ovary having 4-5 chambers.

4.2.0 Cultivars

The genus *Psidium* consists of around 150 species among which *P. guajava* is the extensively cultivated species. Central and South Eastern Brazil consists of largest diversity of guava (Pommer and Murakami, 2009). Guava diversity also exists in northern South America, West Indies, Cuba and India. Many researchers have given deep insight on genetic diversity of guava (Soubhe Sobrinho, 1951; Kwee and Chang, 1990; Subramanyam and Iyer, 1993; GonzageNeto, 1999; Pereira and Nachitagal, 2002; Negi and Rajan, 2007). Several countries have reported conservation of guava diversity in field gene bank. Pakistan has collected one hundred thirty two germplsam of guava which have been grouped in three different clusters (Mehmood *et al.*, 2014). Morphologogical characterization of 395 accessions of guava and a catalogue with more than 70 descriptors have been developed in Cuba (Rodriguez *et al.*, 2010). Noqueira *et al.* (2014) evaluated the genetic diversity of 66 guava trees collected from southern region of Espírito Santo and from Caparaó, MG, Brazil. A total of 28 morphological descriptors (11 quantitative and 17 multicategorical) and 18 microsatellite markers were used. Only one guava variety Ben-dov is extensively cultivated in Israel. However, Israel has introduced several Brazilian, Thai and Mexican varieties in their field gene bank Zipori *et al.* (2010). Hernandez-Delgado *et al.* (2007) reported 48 germplasm of guava in Mexico. Pathak and Ojha (1988) described genetic resources available in India. Around 70 accessions of guava are maintained and ICAR-IIHR, Bengaluru (www.iihr.res.in), 150 at ICAR-CISH, Lucknow (www.cish.res.in). Guava germplasm has been conserved at several places in India. However, it's difficult to predict exact extent of diversity available in the country. The main varieties grown in different parts of the world has been listed in Table 2.

Distribution of guava in India and in the world

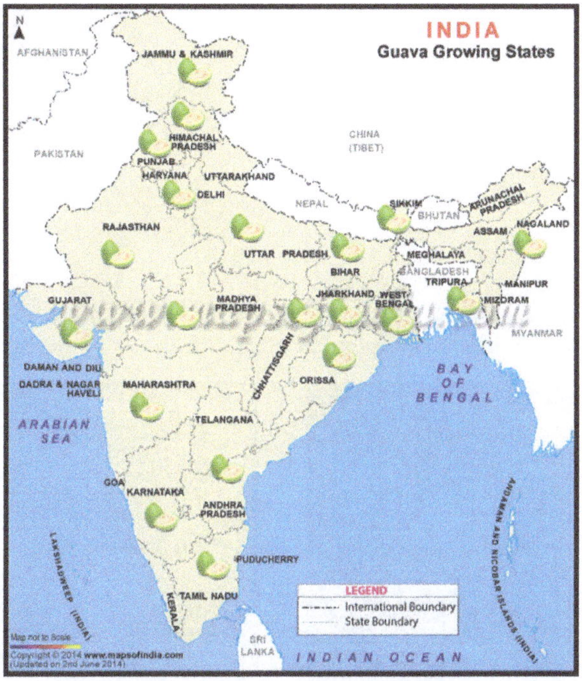

Guava growing states in India

Major guava growing countries in the world

Guava species

Yellow fruits of Psidium cattleianum

Red fruits of Psidium cattleianum

Psidium cattleianum

Psidium littorale (Peruvian guava)

Psidium coriacium

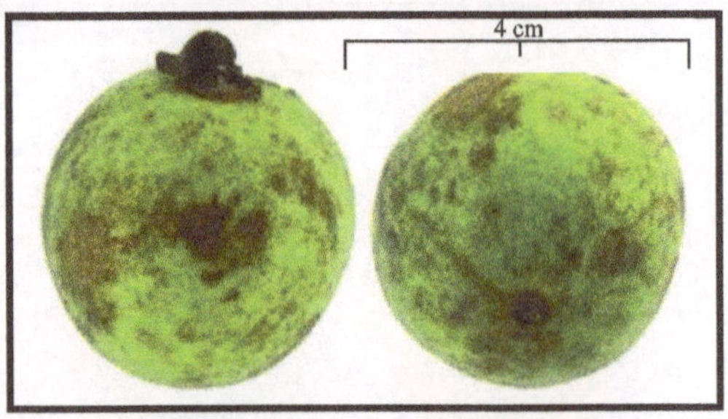

Psidium friedrichsthalianum

Psidium molle

Cultivars of guava

Chittidar

Sardar (L-49)

Allahabad safeda

Kazi

Sweta

Baruipur local

Dudh khaja

Khaja

Lalit

CISH-1

Hybrids of guava

Arka Kiran

Arka Rashmi

Hisar Surkha

Punjab Pink

Propagation of guava

Guava propagation by cutting

Table 2: Major Guava Cultivars in the World

Country	*Variety*
Australia	Beaumont, Sardar, Ka Hua Kula
Bangladesh	Swarupkathi, Mukundapuri, Kanchannagar, Kazi
Brazil	Paluma, Rica, Pedro Sato, Kumagai, Sassaoka, Ogawa, Yamamoto, XXI Century Colombia Puerto Rico, Rojo Africano, Extranjero, Trujillo Costa Rica Tai-kuo-bar
Cuba	Enana Roja Cubana, EEA 1–23, Suprema roja and Belic L-207
Egypt	Bassateen El Sabahia, Bassateen Edûna, Allahabad Safeda
Hong Kong	Pink, Jambu Kapri Putih, Maha 65, Bentong Seedless
India	Allahabad Safeda, Shweta, Dhawal, Apple Colour, Sardar, Karela, Seedless, Red Fleshed, Lalit, Lalima, Benarasi, Sardar, Chittidar, Harijha, Arka Mridula, Arka Amulya, Baruipur
South Africa	Fan Retief, Frank Malherbe Taiwan Tai-kuo-bar, Thailand Glom Sali, Glom Toon Klau, Khao Boon Soom, Vietnam Xa ly nghe, Ruot hong da lang, Xa ly don
Taiwan	Pear Mexico, Media, China, Regional de Calvillo, China, la Labor, Acaponeta, Coyame Puerto Rico, Corozal Mixta, Corriente, Seedling 57-6-79
USA (Hawaii)	Beaumont, Pink Acid, Ka Hua Kula

5.0 Breeding and Improvement

5.1.0 Cytology

Psidium is multi species genus of tropical shrubs and small trees in the myrtle family Myrtaceae. Guava has recently been subject of renewed interest because of acknowledged presence of phenols, carotenoids, antioxidant and flavonoids, giving them relatively high dietary antioxidant value among plant foods. They are therefore considered super fruits, being rich in vitamins A, C and omega-3 & 6 polyunsaturated fatty and especially high levels of dietary fiber. The genus *Psidium* belongs to subfamily Myrtoideae and has a basic chromosome number of X= 11. Most of the commercial guava varieties are reported to be diploids (Negi and Rajan, 2007) with a chromosome number 2n = 22 except the seedless types, which are triploids (Majumder and Singh, 1974). Cytological studies made on structure and behaviour in different varieties of *P. guajava* by many workers indicated that the meiosis was normal with the formation of 11 bivalents at diakinesis and normal distribution of the chromosomes at the later stages (Raman *et al.*, 1969). Pollen grain is triangular with maximum dimension of 27.1µm, exine thickness being 1.3 µm. Polyploidy is frequent in *Psidium*, with somatic chromosome number 2n=33, 44, 55, 66, 77 and 88 (Atchison, 1947; Bolkhovskikh *et al.*, 1969; Goldblatt and Johnson, 1990; Moore, 1977). Guerra (1988) attributed fertility restoration to interspecies hybridization which is further corroborated by existence of intermediate characters between closely related species in Neotropical Myrtaceae (Costa *et al.*, 2008; Costa and Forni-Martins, 2006). Breeding behaviour of aneuploids such as trisomic, tetrasomic, and higher aneuploids has been studied and poor crossability aneuploids and diploids was observed. Triploidy as the main cause for seedlessness was established by studies of D' Cruz and Rao (1962), Sharma and Majumder (1957) and Seth (1959) with aneuploid types: 2n =18, 20, 32, 23, 21 and 30 with varied pollen and seed fertility. In comparison, diploids also exhibit differences for seed content, with high pollen fertility, however it is clear that level of seedlessness obtained in triploids is seldom reached in diploids. Recent study on *Psidium* biosystematics utilized chromosome counts and 2C DNA content (De Souza *et al.*, 2015), the chromosome count ranged from $2n = 18$ (*P. guajava-Purple Guava*) to $2n = 82$ (*P. cattleyanum*). The interspecific variation of nuclear DNA amount presented a coherent and proportional relationship with the ploidy level and was important indicators of graft compatibility. A high level of pollen stainability (> 80 per cent) was seen in all Brazilian guava species with about 50 per cent polyploidy, indicating the importance of polyploidy in the evolution of this group (Rebeiro and Forni-Martins, 2014).

5.2.0 Selection

Genetic variability available in seedling populations has been globally utilized for evolving new guava varieties. Most of the varieties are the results of selection made from seedlings. Few decades back, asexual methods such as grafting were not in vogue and most of the orchards were established using seedling plants. Heterogeneous seedlings provide a good opportunity for selection of promising types. In India, improvement of guava by selection was initiated during 1907 at Ganeshkhind Fruit Experiment Station, Pune. Seeds of varieties grown at different places were used for growing seedlings (Phadnis, 1970). About 600 seedlings were raised and evaluated for fruit and yield characters (Cheema and Deshmukh, 1927). A seedling from open-pollinated seedlings of Allahabad Safeda collected from Lucknow was selected and released as Lucknow-49 (Cheema *et al.*, 1954). The variety after evaluation at Saharanpur (Singh, 1953) and Kodur (Rangacharlu, 1954), was recommended for commercial cultivation and renamed as Sardar (Phadnis, 1970). A promising selection, S-1 was made at Fruit Research Station, Saharanpur having good fruit shape, few seeds, sweet taste and high yield (Singh, 1953, 1959). Three promising selections, *viz.*, ABD 3, BHR 3 and BHR 5 were made from Marathwada (Thonte and Chakrawar, 1981). In Karnataka, from guava plantations around Navalur, 16 good performing seedlings were selected from variety Navalur based on fruit quality, yield and plant characters (Hulamani *et al.*, 1981). Three Allahabad Safeda (AS 1, AS 2 and AS 3) and 2 seedling selections (FS 1 and FS 2) from Faizabad were made (Pathak and Dwivedi, 1988). From open-pollinated 200 seedlings of Allahabad Safeda variety collected from Uttar Pradesh, a dwarf and give high yielding seedling selection was made at Indian Institute of Horticultural Research (IIHR), Bangalore named as Selection-8 (Iyer and Subramanyam, 1988), which was released as Arka Mridula. The fruits are of medium size with white pulp, few soft seeds, excellent quality in terms of sugar content and TSS and good shelf- life (Anon., 1996). An extensive improvement programme for selection of promising types from half-sib progeny of red coloured guava at Central Institute for Subtropical Horticulture (CISH), Lucknow resulted in development of commercial varieties like Lalit and Shweta which are being cultivated in wide agroclimatic conditions due to their high yield, good fruit quality, responsiveness to pruning and wider adaptability. Selection and evaluation studies conducted for two decades on population of 631 open-pollinated seedlings developed from Apple Colour guava collected from diversity rich area of Allahabad and Kaushambhi districts led to development of six selections, *viz.*, CISH-G-1, CISH-G-2, CISH-G-3, CISH-G-4, CISH-G-5 and CISH-G-6 on the basis of fruit yield and quality (Rajan *et al.*, 1996; Negi and Rajan, 2007). CISH-G-3 with attractive saffron yellow coloured fruits with red blush, pink pulp, good blend of sugar and acid was released as Lalit for commercial cultivation. It is suitable for both table and processing purposes. Another selection CISH-G-4 with attractive cream coloured fruits, pink blush, soft seeds and high yield was released as Shweta which is suitable for North India. Three red coloured selections, CISH-G-1, CISH-G-5 and CISH-G-6 were identified and evaluated (Rajan *et al.*, 1996) for two decades and CISH-G-5 was released as Lalima in 2015. Lalima has attractive crimson colour fruits, higher proportion of coloured fruit and good yield with higher shelf life (fruit weight 190 g and TSS 13.7° B) (Anon., 2016). CISH, Lucknow during 1995-2000 surveyed for different variants of Allahabad Safeda and developed half sib population of 267 seedlings out of which CISH-G-35 was selected for heavy bearing, round, smooth, medium to large fruits (200-250 g) attractive fruits. Mature fruits develop light yellow colour on ripening. Pulp is white, sweet in taste having muskiness, TSS 13.4 °B, seeds are soft, moderate in number. CISH-G-35 was evaluated for a decade and named as Dhawal, released in 2015 for commercial cultivation (Anon., 2016). G. Vilas Pasand, a selection made by a farmer near Malihabad (Lucknow) is very productive throughout the year. Fruits are very large (> 350 g), round to ovoid, skin texture is course to smooth, fruit skin pale yellow, colour of pulp is creamy, less seeds. Variants of Apple Colour guava seedlings are being grown as Surka by the farmers in Allahabad and Kaushambi districts. GI has been provided to the Surkha guava grown in certain pocket of Allahabad. A selection from Apple Colour variety was released as Sangam with pink pulp and peel. Pant Prabhat is a promising guava selection made at Govind Ballabh Pant University of Agriculture and Technology, Pantnagar (Uttarakhand). It is heavy yielder with small soft seeded round fruits. Dhareedar a seedling selection has been identified from old seedling orchard of Fruit Research Station, Kuthulia, Rewa. Fruits are medium to large with soft and sweet pulp. In India, several varieties have been identified by the farmers from the seedling populations being grown in different parts of the country (Table 3). Many seedling selection have been reported but without any area under their cultivation. Variation in guava seedlings at the Subtropical Experiment Station, Homestead, Florida, USA resulted in identification of 3 horticultural varieties, namely, Supreme, Red Indian and Ruby (Ruehl, 1946). In Brazil, Canizares (1981) studied over 3000 seedlings obtained by open-pollination of about 60 promising clones. Five plants had fruits weighing more than 400 g each. In South Africa, 6 Nelspriut guava selections were made and compared with a commercial cultivar Fan Retief. Fruit weight was highest in Selection No. 6 and ascorbic acid content in Selection No. 1 (Rensburg and Preeze, 1985). In Cuba, Enana Roja Cubana, N6, Suprema Roja and Belic L-207 are the selections resulted from screening of seedling population (Rodríguez *et al.*, 2010). Varieties with attractive red colour on fruit fetches better price in the market, the fruit quality and shelf life is longer in such varieties than the others. Genotype x environment interaction has been noticed in red peel guava varieties with respect to anthocyanin development in the fruit peel. These varieties require cool nights during fruit maturation period. In Uttar Pradesh, India, the areas suitable for production of red colour guava were mapped with extracting climatic information from the areas of successful production like Allahabad, Fathepur, Raebareli and Lucknow districts. An analysis of climatic data shows that areas where minimum temperature is 8-10°C or even cooler are suitable for the production of red coloured guavas (Rajan, 2008, 2013). Dessert quality of guava fruit is considerably reduced by high seed content. Number of seeds in guava is associated with different seed and fruit characters (Rajan *et al.*, 2008). Variability in seed related characters, like number of seeds/fruit, fruit: seed weight proportion, seed weight/fruit, seeds/100g pulp, 100-seed weight and seed content in 69 accessions collected from diverse geographical regions indicate that a wide range of variation for seed traits exists among varieties (Rajan *et al.*, 2007). Fruit: seed weight proportion and 100-seed weight are two important traits to be kept in mind while making selection for soft and less seeded varieties. Selections from open pollinated seedlings indicate that half-sib progeny may provide ample opportunity for selecting genotypes with suitable seed characters (Rajan *et al.*, 2007b). Correlation and path coefficient analysis revealed that seed weight per fruit, number of seeds per 100 pulp and 100-seed weight were deciding factors for realizing improvement for number of seeds per fruit (Rajan *et al.*, 2008). This indicated the importance of small seeded genotypes in selection of less seeded varieties. The high genetic coefficient of variation and heritability estimates associated with greater genetic advance as percent of mean were recorded for pulp: seed weight ratio, 100-seed weight and number of seeds per fruit, which indicated that these three characters had additive gene effect and therefore effective selections can be made for these characters (Rajan *et al.*, 2005b).

Table 3: Guava Cultivars Grown in India

State	*Cultivars Grown*
Andhra Pradesh	Allahabad Safeda, Sardar, Anakapalli, Hafshi, Safed Jam, Arka Mridula, Lalit
Assam	Am Sophri, Madhuri Am, Safrior Payere
Bihar	Allahabad Safeda, Apple Colour, Harijha, Sardar, Lalit, Shweta
Gujarat	Nagpur Seedless, Dharwar, Dholka, Kothrud, Sardar, Nasik, Sindh, Lalit, Shweta
Jharkhad	Sardar, Allahabad Safeda, Lalit
Karnataka	Allahabad Safeda, Sardar, Arka Mridula, Arka Amulya, Bangalore, Dharwar, Navalur
Madhya Pradesh	Sardar, Allahabad Safeda, Gwalior-27, Hafshi, Seedless
Maharashtra	Sardar, Nagpur seedless, Dharwar, Dholka, Nasik, Sindh, Lalit, Shweta
North-eastern	
States	Allahabad Safeda, Sardar, Red Fleshed, Lalit.
Punjab	Shweta, Sardar, Allahabad Safeda, Lalit
Tamil Nadu	Anakapalli, Banarasi, Bangalore, Chittidar, Hafshi, Nagpur Seedless, Smooth Green
Uttar Pradesh	Sardar, Allahabad Safeda, Lalit, Shweta Apple Colour, Chittidar, Red Fleshed, Allahabad Surkha, Sardar, Mirzapuri Seedless, Surkha
Uttarakhand	Sardar, Allahabad Safeda, Lalit, Shweta, Pant Prabhat
West Bengal	Baruipur, Sardar, Allahabad Safeda, Dudhe Khaja, Gole Khaja, Kabli, Harijha, Sardar

5.3.0 Hybridization

Hybridization is underway at several research institutes for developing new varieties in guava. However, very few hybrids are available across the globe. Reports are available on several hybrid population but without much outcome. In India, at Fruit Research Station, Anantharajupet (Andhra Pradesh), two hybrids (H1 and H6) were found promising with regard to fruit quality and precocity of bearing (Rama Rao and Dayanand, 1977). At the Horticultural Experiment and Training Centre, Basti, Uttar Pradesh, 55 hybrids were obtained using several parents, *viz.*, Allahabad Safeda, Seedless, Lucknow-49, Patillo, Apple Colour, Kothrud and Red Fleshed (Anon., 1982), however, none of the hybrids were promising (Chadha, 1998). Intervarietal hybridization carried out at the Fruit Research Station, Sangareddy, Andhra Pradesh, resulted in selection of two superior hybrids, Safed Jam and Kohir Safeda. Safed Jam is a hybrid between Allahabad Safeda and Kohir (a local collection from Andhra- Karnataka region). It is similar to Allahabad Safeda in growth habit and fruit quality. The fruits are bigger in size with good quality and few soft seeds. Kohir Safeda is from a cross of selected line of Kohir x Allahabad Safeda. Tree is vigorous and the fruits are larger with white flesh and few soft seeds. These hybrids have been released for commercial cultivation, particularly in semi-arid tropical areas of Telangana and Rayalseema (Mitra and Bose, 1985; Shanmugavelu *et al.*, 1987). Involving the cultivars Allahabad Safeda, Seedless, Red Fleshed, Chittidar, Apple Colour, Lucknow-49 and Banaras, 600 F_1 hybrids were raised from intervarietal hybridization at the Indian Institute of Horticultural Research, Bangalore, Karnataka. Arka Amulya, a promising hybrid from the cross between Seedless x Allahabad Safeda was released. It is a high yielder with medium fruit size, white pulp and few soft seeds (Subramanyam and Iyer, 1988; Anon., 1996). Two varieties Arka Kiran and Arka Rashmi were released from ICAR-IIHR, Bangaluru as a result of intensive selection from the hybrid population of Kamsari x Purple Local. Both the varieties are pink pulp and suitable for processing (Dinesh and Vasugi, 2010). At ICAR-CISH, Lucknow, hybridization using parental combinations, *viz.*, Lalit x Purple Guava, Shweta x Purple Guava, CISH-G-1 x Purple Guava and reciprocal crosses a hybrid population (>1500) is being screened for developing guava varieties with high lycopene content, soft and less seeded, suitable for table purpose and processing. Hybrids showed a wide range of variation in fruit weight (60-350 g), TSS (8-14°Brix) and seed hardness (8-23 kg/cm^2) indicating the possibility of evolving varieties with high fruit weight and TSS with low seed hardness. Six hybrids have been developed with red peel and pink pulp colour. At Rajendra Agricultural University, Sabour, Bihar, 210 F1 hybrid seedlings were raised from various intervarietal crosses. A hybrid from a cross Apple Colour x Sardar had maximum fruit weight, vitamin C and pectin content (Singh and Hoda, 1994). Two hybrids, Hisar Safeda (Allahabad Safeda x Seedless) and Hisar Surkha (Apple Colour x Banarsi Surkha) were released by Chaudhary Charan Singh Haryana Agricultural University. Hisar Safeda has round fruits with smooth surface, creamy yellow skin, creamy white pulp, soft seeds, less in number and high TSS (13.4 per cent). Hisar Surkha has round fruits with smooth surface, yellow skin with red dots under low temperature, small to medium size, pulp pink, seed number medium and high TSS. At Taiwan Agricultural Research Institute, Koshinag, Wan and Leu

(1999) selected the crosses between R4 x Pepia and R1 x Pepia resistant against *Myxosporium* wilt in guava. Peipa is a local variety of Taiwan, which is resistant to wilt (Leu and Kao, 1979). Enana Roja Cubana, N6, Suprema Roja and Belic L-207 were used for intraspeciûc crosses development showing variability in yield potential and fruit quality. This will contribute to increase number of cultivars for commercial purposes (Rodríguez *et al.*, 2010). Rajan *et al.* (2007) studies the nature and magnitude of genetic divergence for fruit weight and six seed characters and reported 6 groups of varieties and suggested for developing desired genotypes with better seed characteristics, population can be raised by hybridizing varieties from distantly related groups (Table 4).

Table 4: Important Guava Hybrids Developed in India

Name of Hybrids	*Parentage*
Safed Jam	Allahabad Safeda x Kohir
Kohir Safeda	Kohir x Allahabad Safeda
Arka Amulya	Seedless x Allahabad Safeda
Hisar Safeda	Allahabad Safeda x Seedless
Hisar Surkha	Apple Colour x Banarasi Surkha
Arka Kiran	Kamsari x Purple Local
Arka Rashmi	Kamsari x Purple Local

Physico-chemical composition of fruits of some cultivars is given in Table 5.

Table 5: Physico-chemical Composition of Fruits of some Guava Cultivars*

Cultivars	*Length of Fruits (cm)*	*Diameter of Fruits (cm)*	*Fruit Weight (g)*	*Total Soluble Solids (°Brix)*	*Total Sugar (Per cent fresh weight)*	*Ascorbic Acid (mg/100 g of pulp)*
Allahabad Safeda	5.4-6.4	6.2-6.9	86.5-160.6	8.4-10.5	6.8-8.5	106.0-187.9
Apple Colour	5.0-5.6	5.0-6.2	80.2-120.2	7.8-9.2	6.2-7.3	116.2-194.6
Banarasi	5.1-6.2	5.8-6.4	92.8-132.6	9.2-9.3	6.8-7.7	93.0-172.6
Baruipur	5.6-6.3	6.0-6.2	88.6-125.6	8.8-9.2	6.8-7.6	103.9-182.2
Behat Coconut	5.8-6.1	5.4-6.0	82.2-156.8	9.0-9.4	6.8-7.7	88.8-168.0
Chittidar	5.6-6.6	5.8-7.2	95.6-148.6	8.2-9.0	6.4-7.9	96.4-182.8
Harijha	5.5-6.4	5.8-6.8	81.8-135.2	8.6-9.2	6.3-7.9	92.3-164.6
Kerala	-	4.5-5.3	76.0-117.0	9.1-9.8	-	84.0-185.0
Lucknow-49	5.8-6.6	6.4-6.5	95.8-145.0	9.2-10.0	6.8-8.8	132.5-216.0
Pear Shaped	5.4-6.2	5.8-6.0	78.8-130.8	8.8-9.6	6.7-7.5	75.2-157.4
Red Fleshed	5.2-6.3	5.3-6.0	91.2-135.6	8.2-8.8	6.2-7.6	62.5-144.6
Seedless	4.2-5.0	4.8-5.3	70.8-86.8	8.4-9.2	6.7-7.4	83.3-168.2
Supreme	-	4.5-5.3	63.0-82.0	9.1-9.9	-	98.0-177.0

* Mitra *et al.* (1983), Kundu *et al.* (1995).

6.0 Propagation

Guava is propagated from seeds and also by vegetative methods.

6.1.0 Seed Propagation

Seedling trees bear fruits of variable size and quality, but such trees are generally long-lived. Normally, fully matured seeds of the current season are used for sowing. The seedlings can be raised in the nursery or in polythene bags. The seedlings raised in polythene bags can be transplanted in the field when they are six to eight weeks' old. Guava seeds lose viability within a short period after extraction. The usual practice is to sow the seeds immediately after extraction. Extension of storage life and invigoration has been possible only to a limited extent through the use of different chemicals. But only meagre information regarding seed treatment for maintenance of vigour and viability is available. Haq *et al.* (1973) reported that soaking in cold water had beneficial effect, but treatment with hot water was harmful. Seeds could be stored for 104 days, when they were pre-soaked in ferulic acid at (10-3 M) concentration. Potassium nitrate at 1 per cent and para-hydroxybenzoic acid (10-3 M) were also effective in retaining seed viability of guava (Ghosh, 1978). Seed coating, seed dormancy and tanning can cause poor, irregular and late germination (Ali *et al.*, 2007). Germination of seeds was improved to 90 per cent, with seed-treatment with 10 per cent HCl for 12 hours (Usman *et al.*, 2012, Butt *et al.*, 2013). Kalyani *et al.* (2014) reported that GA_3 has improved seed germination among various growth regulators as compared to other treatments, such as the application of warm water and thiourea. GA3 helps in early germination due to it reacts on the embryo and causes denova synthesis of hydrolysing enzymes particularly amylase and protease and this hydrolysed food is utilized for the growth of embryo. Manoj *et al.* (2013) observed that the guava seeds scarified with 10 per cent hydrochloric acid for 2 minutes were higher than the control with the highest percentage of seed germination, the seed strength index and the survival rate of the seedlings. Rodriguez *et al.* (1983) obtained more than 90 per cent germination by soaking the seeds in water for 4 days followed by treatment with GA3 at 1000 ppm for 24 hours. Soaking of seeds in GA3 (3000 ppm) solution resulted in maximum germination (83.2 per cent) after six weeks and better seedling growth (Chandra and Govind, 1990). A combination of scarifying in sand (for 15 minutes) and germination in sand gave the best total germination (98.15 per cent). Germination generally occurred between 8 to 11 days after sowing (Tavares *et al.*, 1995). Martinez-De-Lara *et al.* (2004) observed, seed multiplication will result in genetic heterogeneity in orchards and plants in the same orchard. Therefore, seed propagation in commercial orchards to increase productivity is not recommended (Pereira, 1990). Chemical analyses of guava seeds indicate 16 per cent of crude protein, 18 per cent of crude oil, 20 per cent of crude fiber and 44 per cent of carbohydrates, based on dry weight. However, the oil of the seeds has 77 per cent of linoleic acid and may have application in production of edible oils and inks (Paull; Bittembender, 2006).

6.2.0 Vegetative Propagation

Guava trees propagated vegetatively are uniform and have a shorter juvenile period, compared to plants multiplied by seed. Scion shoots must be obtained from identified mother plants for vegetative propagation (Singh *et al.*, 2007). Guava can be successfully propagated by cuttings, air-layering, grafting and budding.However, wedge grafting is commercially adopted by nurserymen in the country.

6.2.1 Cutting

The guava tree can be propagated by root or shoot cuttings (herbaceous, semi-hardwood or hardwood). The propagation via stem cuttings was not very significant in older days. Recently, this method has become popular. September to March is the appropriate time for the preparation of cuttings (southern hemisphere), when the plants are in full growth. One can use stem of all kinds to make cuttings and have young leaves that grow actively. Cuttings must have three internodes and four nodes. Due to the length of the internodes, the size of the cuttings can be reduced to ensure that only one node will be immersed in rooting media. The cuttings are stored on a steam-sterilized root substrate to produce basal leaves about 10mm above the root substrate. Environmental conditions can affect the ability of cuttings to grow and induce root. The most common external factors are light, seasons, temperature, humidity, moisture level of the cutting medium and rooting (Hartman and Kester., 2002). Blommaert (1958) observed that softwood cuttings of two local commercial guava cultivars gave 75 to 90 per cent rooting in sand media under intermittent mist when treated with IBA. Teaotia and Pandey (1961) reported that both NAA and IIA at 50 ppm and 100 ppm encouraged rooting of semi-hard wood etiolated cuttings. According to Jolioceur (1962) optimum rooting of 44 per cent was obtained after one month under mist, treated with IBA at 0.08 per cent. Sen *et al.* (1967) also recorded 100 per cent rooting in the leafy cuttings of guava under intermittent mist treatment with IBA. Bhandari and Mukherjee (1969) observed that green-wood cuttings taken from young seedlings and about 10-year-old grafts of Allahabad Safeda gave 100 to 90 per cent rooting respectively, when etiolated and ringed for 39 days and treated with 5000 ppm of IBA at the time of planting. Dhua *et al.* (1982) reported 93.3 per cent rooting in semi-hard wood cutting under mist with p-hydroxybenzoic acid (200 ppm) and IBA (5000 ppm) treatment as quick dip. Soft-wood cuttings with 2 nodes and 4 leaves rooted better than semi-hard wood cuttings when treated with NAA at 2000 ppm (Pereira *et al.*, 1983). Evans (1992) and Singh (2018) reported that the best time to make cuts in the field probably at the beginning of the rainy season. Although guava is hard-to-root, however, results indicated that it can successfully be propagated from cuttings under mist. Abdullah *et al.* (2009) reported that cuttings had 60 per cent to 70.9 per cent survival rooting rates in the mist-free propagator when treated with a concentration of 4000ppm IBA. In the absence of treatment with Indole3-Butyric acid, the cuttings had not induced root. The diameter of the maximum average stem, the number of leaves, the number of roots per plant, the length of the root and the root weight observed in the IBA treated cuts at 1000ppm.The root cuts can give good rooting and survival when the cuts are made in July (Singh *et al.*, 2017). Singh (2017, 2018) found that the application of auxin improves the histological features such as the formation

Propagation of guava

Guava propagation by air layering

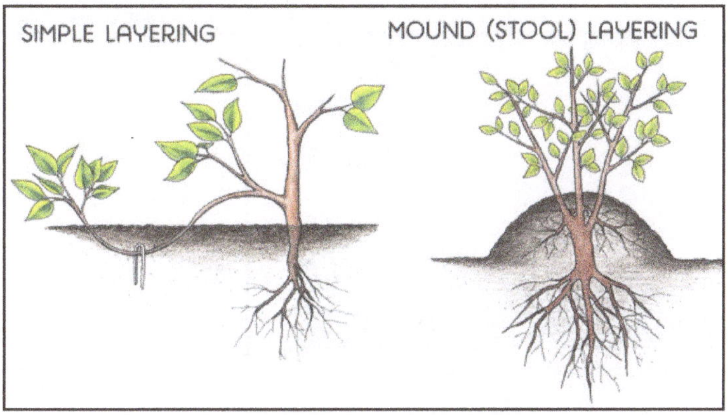

Guava propagation by stooling

Veneer grafting steps in guava

Propagation of guava

Cleft grafting steps in guava

Successful cleft grafted saplings of guava

Patch budding in guava

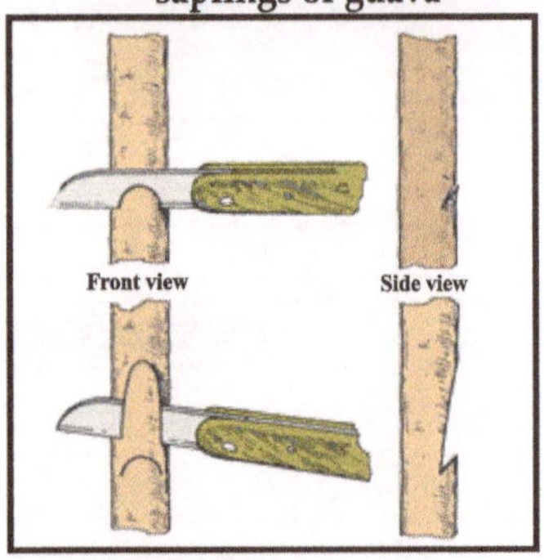

Chip budding in guava

Micropropagation in guava

In vitro proliferation of guava. (A) stock guava plants in the greenhouse; (B) nodal section became browning after sterilization; (C) new shoots break out from healthy nodal sections; (D) shoots proliferated; (E) elongated shoots; (F) rooted shoots by medium method (medium with IBA); (G) rooted shoots by dipping method; (H) guava plantlets acclimatized into the soil for 2 weeks; (I) guava plantlets acclimatized into the soil for 10 weeks [Liu and Yang (2011) *Intl. J. Plant Biol.*, 2:e2].

of callus and tissues and the differentiation of vascular tissue. Zamir *et al.* (2017) found that auxin treatment favored rooting in both cuttings. The maximum number of roots per cutting was between 11 and 17.3 for semi-hardwoods, while that for softwood was between 15.6 and 27.6. Whatever the nature of cuttings, IAA and NAA seems to be the most effective in promoting roots. In contrast, IBA was more effective on softwood cuttings only. However, the ultimate survival of plants is the key to such studies. Throughout the experiment, a maximum survival rate of 28 per cent was found in softwood cuttings treated with IAA, *i.e.* 100 mg per 100 g of talc. The mixture of vermiculite, perlite, sand and peat in the rooting medium, increases rooting with or without using auxins such as IBA and NAA (Gautam *et al.*, 2010, Cheng *et al.*, 2011).

Two-node cuttings of guava cv. Paluma, taken from shoots with fruits, rooted best under intermittent mist when treated with IBA at 200 ppm for 14 hours in darkness (Pereira *et al.*, 1991). Leafy semi-hardwood cuttings produced maximum percentage of rooting when preconditioned with etiolation (10 days) + 10 days in sunlight and treated with IBA at 3000 ppm, followed by girdling (20 days before taking cuttings) + IBA at 3000 ppm (Mitra, 1996). Kareem *et al.* (2016) observed that the IBA (4000ppm) showed maximum results in terms of shooting percent, roots number per cutting, normal root length respectively. The role of leaves in the production and translocation of dry matter (and hence root development) was greater than that of the stem (Pereira *et al.*, 1991). Season has a strong influence on adventitious root formation. Guava tip cuttings (10-12 cm long) with at least 4 leaves from current season's growth, dipped (basal 1-2 cm) in paclobutrazol for 24 hours gave maximum rooting percentage, root length and number of roots per cuttings during mid June and end of September, while there was no rooting response between the end of December and mid March (Rahman *et al.*, 1991).

6.2.2 Air-layering

In guava, two types of layering are commonly practiced *viz.*, air layering and sound/stool layering. Air-layering is one of the most important commercial methods in practice for propogation of guava. Shoots selected for air-layering should be 1 cm in diameter and preferably from previous year's growth. A ring of bark of about 3 cm long is removed and covered with sphagnum moss, previously soaked in water and wrapped with polythene film. The ends of the film are carefully tied and left for rooting. It takes 30-40 days for rooting in the monsoon. Rainy season was found more favourable than spring for air-layering. Bhatt and Chundawat (1982) reported that ringing of branches followed by etiolation by wrapping black polythene on ringed portion before the application of rooting media proved effective in the regeneration of roots. Said and Inayatullah (1966) reported that a media consisting of soil, sand and leaf mould in the proportion of 1:1:1 resulted in better rooting in the layers. Singh *et al.* (2007) reported that one year old guava cv. "Lucknow-49" shoots treated 6000 ppm IBA with soil: sand: poulty manure rooting media produced the highest percentage (76.75 per cent) of survival of the plants after 60days.

Air-layering carried out on 10 July and treated with IBA (10,000 ppm) resulted in the highest percentage of success and in number, length, diameter and weight of roots (Sharma *et al.*, 1991). Highest percentage of survival was obtained with 75 per cent defoliation of the layer shoots after separation from mother plants followed by complete defoliation (Singh *et al.*, 1992). Maurya *et al.* (2012) reported that the air layering with soil and dung + sphagnum + 6000 mg IBA/l showed early emergence of the roots, increase in the number of primary roots, secondary roots, the maximum root length, fresh root and dry root weight compared to control. Maximum survival rate (90.67), the length shoots (7.93 cm) and the number of shoots (18.33) at 60 days of air layers in the polyethylene bag was observed. Manga and Jholgiker (2017) reported that the minimum days for the root emergence, the maximum number of roots, the highest rooting and the survival rate were found in layers prepared in August were applied with IBA-4000 ppm. Immediate application of NAA at 9000 ppm after ringing gave the highest number of primary and secondary roots and heaviest root fresh and dry weight as compared to application after 10 or 20 days later (Patel and Pasaliya, 1995). A combination of different types of auxins seems to have stronger influence on root regeneration than individual treatment. The best plant growth regulator (PGR) treatment was found to be IBA + NAA (Singh *et al.*, 1995; Patel *et al.*, 1996) and the best concentrations of PGR were 3000 ppm (Patel *et al.*, 1996) and/or 10,000 ppm (Singh *et al.*, 1995). Black polyethylene wrapper was better than white wrapper in respect of rooting and growth of air layers (Singh *et al.*, 1995; Patel *et al.*, 1996). Multiple air-layering (2 or 3/stem) is feasible for increasing the efficiency of guava propagation (Mishra and Singh, 1995). Shoots could be layered inside polybags and the best result was obtained in combination with 2-cut-technique of detachment of rooted layers from the mother plants (Verma *et al.*, 1996).

6.2.3 Stooling

Majumdar and Mukherjee (1968) reported a method in which air-layers were grown for 3-5 years and then cut back and allowed to shoot. These shoots were ringed and applied with IBA in lanolin and earthed up to induce rooting; when rooted, the shoots were separated and planted in nursery beds or pots. Stooling twice in a year was found successful thereby producing more number of stool layers. The plants raised from stool made more growth than seedlings during the first year (Rathore, 1984).

With etiolation, 100 per cent rooting of stools of guava cv. Harijha occurred in all the IBAtreatments and in the control (no growth regulator), but with localized (25 mm-wide black adhesive tape applied to the shoots 30 cm from the apex) or no etiolation, rooting only occurred with IBA at 5000 ppm or with IBA at 3000 ppm plus PHB. IBA (5000 ppm) + PHB (200 ppm) resulted in the maximum number of roots in all etiolation treatments (Dutta and Mitra, 1991). Treatment with IBA + NAA (each at 7500 ppm) induced the best rooting in *Psidium chinensis* and *P. cujavillis* (*P. guajava*) stools and promoted establishment (Saroj and Pathak, 1994). Lal *et al.* (2007) found that IBA (7500 ppm) of treatment has a maximum rooting rate (96.67 per cent), average number of roots per shoot (46.93) average root length (8.45 cm) and survival (75 per cent) after transplantation in the field. The highest rooting percentage and rate of survival were obtained with IBA (4500 ppm) + NAA (400 ppm) applied in lanolin paste to the rings of the shoots arising from the stools (Chaudharu *et al.*,1994). Rymbai and Reddy (2010) found that IBA 7000 ppm was found to be significantly effective in inducing rooting, promoting root characters, facilitating the initial sprouting and the maximum number of leaves in stooled shoots. Early sprouting (7.90 days), maximum rooting rate (90.73 per cent), maximum number of primary roots (15.72), maximum number of secondary roots (30.82), more leaves in 45 days after transferring into bags of polyethylene (8.94) were recorded in stool layering (Patil *et al.,* 2016). Studies on biochemical and anatomical factors associated with root initiation in *Psidium* species propagated by stooling revealed that there was a great variation in rooting response. A high concentration of total phenols and high C:N ratio were associated with a high rate of root formation, whereas a high protein concentration had adetrimental effect on rooting. *P. chinensis* was an easy-to-root species, while *P. cujavillis* (*P. guajava*), *P. molle* and *P. cattleianum* were difficult-to-root species in which rooting could be improved by exogenous application of plant growth regulators (7500 ppm IBA + 7500 ppm NAA). Long sclerenchyma cells with little space between them (continuous ring like structure), was observed to be the main cause of shy rooting in *P. friedrichsthalianum,* even after treatment with auxins (Pathak and Saroj, 1995).

6.2.4 Grafting

Inarching is another important method of propagating guava which gives high success rate but it is more laborious as compared to cutting or layering. The maximum rate of graft sprouting, the number of sprouted shoots occurred in grafted plants under shade net condition as compared to grafted plants under polyhouse condition and also the maximum percentage of graft sprouting, fewer days taken by 50 per cent of the sprouting of grafts and the number of days taken for sprouting was also recorded in grafted plants in January (Vanaja *et al.,* 2017). Sweeti *et al.* (2016) reported that the maximum percentage of sprouting (44.76 per cent) was recorded on March 5th and the success rate of the transplant (69.08 per cent) and the maximum length of sprouts (14.00 cm) was observed in grafted plants on 20th February after 90 days. Nelson (1954) reported the success of veneer grafting and suggested that the scion should be selected from terminal growth flushes in which the stem was green and quadran-gular with well-developed axillary buds. Veneer grafting with green-wood section was recommended for top working (Tamburo *et al.,* 1955). They also recorded 73 per cent success when scions were wrapped with vinyl film. Besides these, side grafting method in which the vertical flap of rootstock is retained and tied over the scion during operation is also now in practice. Gotur *et al.* (2017) reported that the wedge grafting in August gave better results in the poly house (69.88 per cent) condition and open field condition (67.12 per cent). The success of grafts in August could be attributed to the optimal temperature and high humidity that prevails during this period, which resulted in the successful bonding of the layers of cambium of stocks and scion, the first formation of calluses and the beginning of the subsequent growth. A high success rate and graft survival was observed in 35 per cent of shaded houses (68.80 and 87.19 per cent respectively), followed by a 50 per cent shading (58.00 and 79.13 per cent, respectively) (Manga and Jholgiker, 2017). Similar results were also obtained by other guava workers (Mahendra *et al.,* 2015; Syamal *et al.,*2012; Rani *et al.,* 2015; Beera *et al.,* 2013; Nanditha *et al.,* 2017; Singh *et al.,* 2018; Shashi *et al.,*2012; Gurjar *et al.,* 2012).

Wedge grafting technology is commercially being employed for propagation of guava. The technique has been developed by ICAR-Central Institute for Subtropical Horticulture, Lucknow. More than 5000 guava nurseries have adopted this technique in the country.

6.2.5 Budding

The rootstock is prepared by removing side growth up to 20-25 cm from the base. Different budding techniques like Forkert, shield, patch, chip, *etc.*, have been tried in guava with different degrees of success. Patch budding on stump shoots (1.2 to 1.9 cm in diameter) proved successful in guava (Dedolph and Bowers, 1960). Jaffco (1970) developed 'green chip' method of budding on guava seedling (5 mm in diameter). He suggested that this method can also be adopted successfully for top working on older trees. Srivastava (1962) compared different methods of budding in guava and concluded that plants propagated by Forkert method grew better than those by patch or shield budding. Forkert method of budding was found to give consistently better result than shield or patch budding (Chandra, 1965). The relative efficancy of 2 methods of budding (patch and modified Forkert) and the best time for budding were investigated in guava cv. Sardar by Singh and Pandey (1998). Sohnika *et al.* (2015) reported that the patch budding from August 15th to 21st showed the highest success rate (92.07 per cent) observed after 90days of guava propagation. Bhatt *et al.*(2013) reported that the patch budding in mid-June showed a better response to the number of sprouting (7.49), percentage of survival (73.33 per cent), average length of the shoot (50.27 cm), average leaf length in new growth (6.67 cm), average leaf width (3.71 cm) and leaf area (53.86 cm^2). Patch budding was more popular method to propagate for Allahabad Safeda among budding (Bhatt *et al.*, 2013). The patch method was more successful than the modified Forkert method. The best time for budding was July followed by August. Poor success was observed in February, March, April, May and June. Buds sprouted the quickest on plants budded in June and July. Poor-performing guava trees can be converted to improved one by on-site top-working. Cutting back of the branches of poor-performing trees and grafting of more productive cv. Beaumont patch buds onto the new shoots gave 95 per cent topworking success rate (Lyannaz, 1994).

6.2.6 Rootstock

In some fruit plants, rootstocks exert a great influence on vigour, cold resistance, fruitfulness, fruit quality, mineral composition of leaves and disease tolerance of the scion, but very little information is available on guava (Sankar, 1967). At the Central Institute for Subtropical Horticulture, Lucknow, Uttar Pradesh, interspecific hybridization was attempted between *P. molle* and *P. guajava and a hybrid rootstock CISH-WRS-1 has been developed* (Fig. 3). The interspecific hybrid rootstock has been found resistant to guava wilt and it is graft compatible with commercial varieties of *P. guajava* (Rajan *et al.*, 2005). Several species of *Psidium* such as *P. cujavillis, P. molle, P. cattleianum, P. guineense* can be suitably used as rootstock for guava. The Chinese guava (*Psidium friedrichsthalianum*) has been found to be resistant to wilt disease and a compatible rootstock. The dwarfing effect of the Chinese rootstock could be used in commercial planting (Edward and Sankar, 1964). The trees grafted on *P. pumilum* had a dwarfing effect and *P. cujavillis* produced the largest but non-uniform and rough skinned fruits. Fruits from trees on Florida seedling stock had the highest total acidity (Teaotia and Phogat, 1971). The Chinese guava was also found to be resistant to *Meloidogyne incognita* and suitable as a rootstock in nematode affected areas (Fernadezdiaz-Silveira, 1975). Singh *et al.* (1976) observed that Allahabad Safeda guava, when grafted on *P. pumilum,* the sugar content of fruits increased considerably while the ascorbic acid content increased by grafting on *P. cujavillis* stock. The yield response was, however, more by using *P. cattleianum* as rootstock for Allahabad Safeda. A pathogenic green algae, *Cephalsuros* sp. was found to affect seriously the leaves and fruits of guava in southern Florida. Marlatt and Campbell (1981) reported that Patillo and Blitch had a low disease incidence, Ruby ´ Supreme 6-29 was moderately susceptible while Webber ´ Supreme and Ruby ´ Supreme 10-30 were most susceptible to the disease.

A potentially dwarf aneuploid No. 82 rootstock of guava had been identified at the Indian Agricultural Research Institute (IARI), New Delhi, through a selection made out of 48 different aneuploid seedling rootstocks (Sharma, 1982). It was observed that aneuploid No. 82 induced substantial dwarfing of Allahabad Safeda in terms of plant height, plant spread and tree volume. It is reported to have strong potentiality of being used as a dwarfing rootstock on commercial scale for increasing the production and profitability of guava orchards (Sharma *et al.*, 1992). In trials with 3 six-month-old seedling stocks (P1 and P2 from *P. guajava* and P3 from *P. friedrichsthalianum*) and 6 budwood sources of *P. guajava* were used. Each terminal scion had at least 3 pairs of buds. All the *P. guajava* scions were cleft grafted on each one of the 3 stocks. The variables under study were : percentage of successful unions or 'takes' (TP), percentage of takes with new shoots (TSP) and percentage of takes with leaves on (LTP). The results showed that all scions and stocks had the same TP and LTP values 22 days after grafting. Among the different scions, there were no significant differences, but there were significant variation between root-stock P3 and the 2 others. There was also a significant difference between the interaction of budwood x rootstock. The rootstocks frequently produce shoots, particularly at the early stage in budded or grafted plants and they should be removed.

7.0 Area and Production

The important guava producing nations are India, Brazil, Mexico, Thailand, USA (Hawaii, Florida), Australia, the Philippines, China, Indonesia, Cuba, Java, Malaysia, Bangladesh, Sri-Lanka, Myanmar and some African countries (Singh and Rana, 1996). Guava occupied 264850 hectares with a production of 4053510 tonnes in India (Anon., 2018). Though it is successfully grown all over the country, the most important guava-growing states are Uttar Pradesh (49530 hectares), Madhya Pradesh (35080 hectares), Bihar (27610 hectares), and Chhattisgarh (21890 hectares). Uttar Pradesh is by far the most important guava-producing state of India, and Allahabad has the reputation of growing the best guava in the country as well as in the world. The District Budaun has got the largest area (4820 hectares) followed by Kasganj (4110 hectares), Aligarh (4020 hectares) of Uttar Pradesh; Bhavnagar 3960 (hectares) of Gujrat; and Solapur (3.550 hectares) of Maharashtra under guava cultivation (Anon., 2018).

8.0 Soil and Climate

8.1.0 Soil

Guava trees are very hardy and can thrive on all types of soil from alluvial to lateritic. It can, however, be grown on heavier but well-drained soil. Guava has been reported to be very (Ogden *et al.*, 1981) to moderately (Jawanda, 1961; Schaffer *et al.*, 1992) flood tolerant and can thrive on poorly drained land where most other fruit trees will not grow (Morton, 1987). The best soils are deep, friable and well-drained. The root distribution studies by the direct excavation method revealed the maximum concen-tration of roots between 0-20 cm soil depth indicating that the active absorption was close to the surface of the soil (Hegde and Tiwari 1981), which leads to suggest that at least the top soil should be rich. The soil rejuvenation with organic sources are essentially required in subtropical regions with low organic matter content and inherent low fertility, in order to improve the soil fertility and sustainability of guava orchards. Application of 10 kg vermicompost should be considered for integration in the nutrient management strategy (Adak *et al.*, 2014).

The acceptable pH ranges between 4.5 and 8.2 (Samson, 1980) or 5.5 and 7.0 (Maxwell *et al.*, 1984). Guava is one of the few tropical fruits that has some degree of tolerance to salinity, which varies from fairly tolerant (Malo and Campbell, 1986), to somewhat resistant (Morton, 1987), to poorly tolerant (Ogden *et al.*, 1981). It performs successfully in soil saturation extract as high as an electrical conductivity (ECe) of 8 to 9 ds.m^{-2} (Mehta *et al.*, 1988). However, Patil *et al.* (1984) reported a threshold ECe of 4.7 ds.m^{-2}, and decrease in growth of almost 10 per cent per 1 ds.m^{-2} above this threshold. It was observed that Red Fleshed guava is more salt tolerant than white guava (Ali-Dinar *et al.*, 1999). Total N, P, K and Na concentrations in the dried leaves of guava seedlings and cv. Sardar, and the total chlorophyll content decreased with increasing salinity level from EC values of 4.5 to 12mmhos/cm while the Ca, Mg and Cl concentrations showed an opposite trend (Singh and Pathak, 1994). Substantial rise in the Na and Cl contents of leaves was noticed with increasing salinity (Makhija *et al.*, 1980; Dhankar *et al.*, 1981; Haggag and Maksoud, 1996) that caused leaf injury and reduced shoot growth, which might be attributed to Na and Cl leaf toxicity and the imbalance of other nutrients at higher levels of salinity (Marler, 1994). All levels of salinity inhibited and delayed seed germination and seedling emergence (Maas, 1984). It appears that guava is more sensitive to salinity stress during germination and emergence than during later growth (Marler, 1994). Fruit weight, size and ascorbic acid content decreased significantly at higher salinity level (Haggag and Maksoud, 1996).

8.2.0 Climate

Guava is successfully grown under tropical and subtropical climate. In areas having distinct winter season, the yield tends to increase and quality improves. It can grow from sea level to an altitude of about 5000 ft (1515 m).

Among the tropical and subtropical fruits, guava is one of the more tolerant to environmental stress (Marler, 1994). The optimum temperature lies between 23° to 28°C (Samson, 1980). Flower abortion occurs in higher elevations of Hawaii where winter temperature goes as low as 7°C (Shigeura and Bullock, 1983). High temperature during summer coupled with low humidity has been reported to reduce fruit set and increase fruit drop (Chadha and Pandey, 1986). Although guava can tolerate some cold, but exposure to freezing temperature may cause shoot dieback (Anon., 1986) and the trees may be killed at –2° to –3°C (Maxwell *et al.*, 1984; Malo and Cambell, 1986). However, Cattleya guava can reportedly survive in temperatures below –5°C (Morton, 1987).

Guava is fairly tolerant to drought (Ogden *et al.*, 1981). The young plants are susceptible to drought and cold conditions. Optimum annual rainfall varies from 100 to 200 cm (Samson, 1980). Light saturation of net photosynthesis is quite high for healthy guava leaves, with saturation not occurring even at a photosynthetic photon flux of approximately 925 m mol. $m^{-2}s^{-1}$ (Walker *et al.*, 1979). Trees are tolerant to be blown over by storm, although defoliation occurs during strong winds. However, growth is sensitive to wind loads (Shigeura and Bullock, 1983).

9.0 Cultivation

9.1.0 Planting

Before planting, the field should be deeply ploughed, harrowed and levelled. The pits of about 1 metre cube should be dug before the monsoon at appropriate distance in the square system of planting. After 15 to 20 days, the pits should be filled with 25 to 30 kg of decomposed farmyard manure mixed with surface soil. Planting usually starts by the onset of monsoon. The soil should be well pressed to keep the plants firmly in position.

9.1.1 Plant Density

Traditionally guava is planted at a spacing of 5m ´ 5m or 6m ´ 6m accommodating 278 to 400 plants per hectare in a square system of planting. The distance between trees has a profound influence on growth, yield, fruit quality and on nutritional levels of leaves.

In guava, the total number of fruits per plant and the number of fruits of different grades were not much influenced at the early stage by plant density, but high density planting had an adverse effect on the quality of fruits. The contents of total soluble solids, total sugar, reducing and non-reducing sugars and ascorbic acid were significantly reduced, while acidity increased with higher plant density (Gaikward *et al.*, 1981). Yadav *et al.* (1981) observed that the volume, spread and number of branches per tree decreased with increasing plant density, while the height of plant increased with the increase in plant population.

Experimental results showed that higher plant densities decreased the fruit weight and size but the yield per unit area increased considerably (Mitra *et al.*, 1984a; Kundu *et al.*, 1993). Guava cv. Allahabad Safeda, planted at 6m ´ 2m spacing and managed by hedge-row system produced the highest yield per hectare with better quality fruits. However, per tree yield was reduced as compared to 6m ´ 6m spacing (Chundawat *et al.*, 1992). Feungchan *et al.* (1992) reported that guava seedlings planted at a spacing of 100 ´ 200 cm produced the greatest shoot number, weight and diameter, while a spacing of 50 ´ 100 cm produced the greatest number of shoots per rai (1 rai = 1600 m^2). In cv. Allahabad Safeda budded on Lucknow-49 seedling rootstock, the maximum yield per tree per year both by number and weight of fruits was obtained with spacing of 6 ´ 6 m (Table 6). However, maximum yield per hectare was obtained from trees spaced at 2 ´ 4 m followed by those at 4 ´ 6 m (Kalra *et al.*, 1994). Spacing did not significantly affect fruit set and flower/fruit drop in guava cv. Sardar. However, trees at 2 ´ 2 m had a lower yield per tree than those at 8´8 m but produced a 10-fold higher yield per hectare (Shant Lal *et al.*, 1996).

In an experiment having different plant densities, it was revealed that the tree height was maximum (5.76 m), trunk circumference was also higher (0.86 m) and canopeis were shorter at the planting distance of 3.0 x 1.5 m (2222 trees ha^{-1}). Yield of per unit area was more (7.24 kg) at the planting distance of 1.5 x 3.0 m. Photosynthetically active radiation (PAR) was found less in closely spaced trees than medium and low ones. Overall, better light penetration was observed in the trees planted at 6.0 x 6.0 and 3.0 x 6.0 m than the other distances at NS/EW canopy edge, inside tree centre, centre between tree in the rows and centre between rows (Singh *et al.*,2007).

Table 6: Number of Fruits and Yield of guava cv. Allahabad Safeda under different Spacing*

Spacing (m)	*Number of Fruits per Tree*		*Yield per Tree (kg)*		*Yield per Hectare (q)*	
	Rainy Season	*Winter Season*	*Rainy Season*	*Winter Season*	*Rainy Season*	*Winter Season*
S1 (2 2)	27	21	3.22	4.01	80.50	100.37
S2 (2 4)	113	198	13.48	24.45	168.56	305.65
S3 (2 6)	131	225	15.02	26.29	125.16	219.03
S4 (4 4)	237	331	28.89	40.87	180.60	255.45
S5 (4 6)	364	503	46.20	63.24	192.20	263.09
S6 (6 6)	408	561	49.43	74.18	136.94	205.48
C.D. at 5 per cent	21.00	5.50	4.31	1.50	17.56	7.52

*Kalra et al. (1994).

9.2.0 Irrigation

Guava can withstand prolonged period of drought (Shigeura and Bullock, 1983). However, the drought tolerance capacity varies with cultivars. The metabolic activity of the plants was better maintained, even at low relative water content (RWC), in cv. Sardar than cv. Allahabad Safeda (Singh *et al.*, 1997). Practically, guava trees are grown as a rainfed crop (Singh, 1969). However, the young trees should be irrigated at regular intervals during lean periods (Hayes, 1970). In the early stage, plants require 8 to 10 irrigations a year, while full grown bearing trees require watering during April to June at bi-weekly intervals to secure higher fruit-set and to reduce fruit drop. Irrigation during winter has been found to be effective in reducing fruit drop and improving fruit size of winter crop.

Cultivation of guava

High density planting of guava

Training and pruning in guava

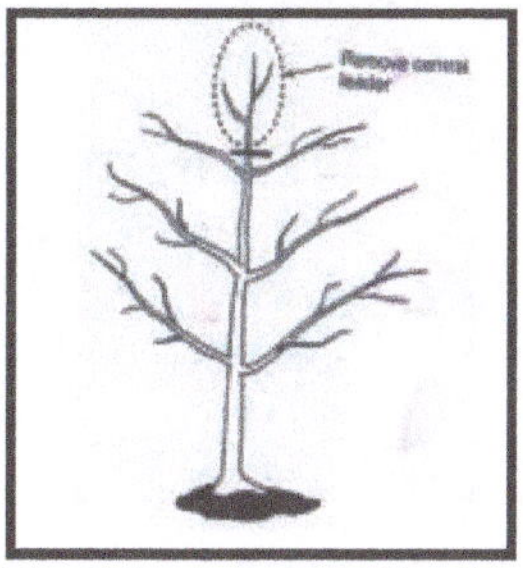

Modified central leader

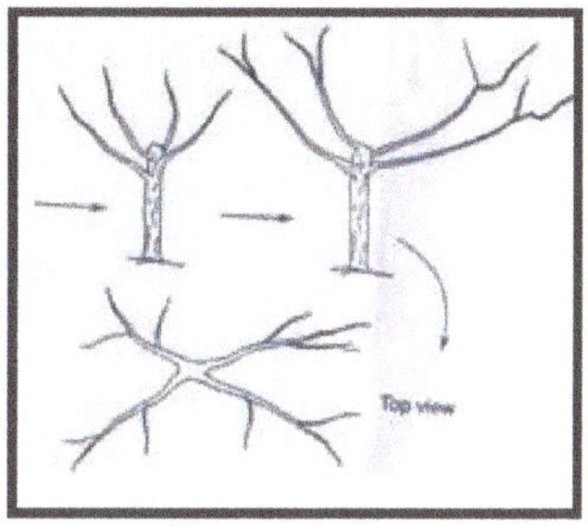

Open centre

Pruning

Tip pruning of guava

30 cm pruning

Headed back of guava tree and canopy development

Black polythene mulching in guava

Guava intercropped with melon

Flower abscission was observed when RWC reached 60.20 per cent in cv. Allahabad Safeda and 53.40 per cent in cv. Sardar and water-stressed condition significantly decreased yields in both cultivars during the rainy season (Singh *et al.*, 1997). The best effect on growth, flowering and fruiting (earliest flowering and fruit maturity, largest fruits and highest yield per hectare) in cv. Sardar was recorded with an irrigation treatment of 60 CPE to a depth of 75 cm (Lal, 1996). The cropping pattern in guava (to reduce rainy season yields) can be regulated to some extent by means of water stress (with holding irrigation) (Singh *et al.*, 1997).

9.3.0 Manuring and Fertilization

The results of the experiments on the effect of manures and fertilizers on the yield of guava clearly indicate that guava is very responsive to the application of inorganic fertilizers along with organic manures. The necessity of manuring guava trees for increasing the fruit production was emphasized by Naik (1949) and Hayes (1970), because the fruits are borne on current season's growth and manuring encourages vegetative growth and fruiting. Sankar (1966), Singh and Krishnamurthy (1967) and Singh and Singh (1970) reported that fertilization of guava not only increased the yield but also improved the quality.

In a field experiment in Sao Paulo, Brazil, Natale *et al.* (1994a) observed that yield increased with N dose, reaching maximum at the highest in the third year. The optimum rate of N, in economic terms, applied to the tree base in August, October, November and December, was 131 kg and 199 kg/ha respectively, in the 2nd and 3rd year of the trial (Natale *et al.*, 1996a). The beneficial effects of P on growth, yield and quality of guava were also demonstrated by Singh and Rajput (1977) and Kumar *et al.* (1995). Application of P_2O_5 at 500g tree^{-1} year^{-1} improved the fruit quality but net return per tree was better with 600g tree^{-1} year^{-1} (Kumar *et al.*, 1995). Potassium application significantly increased fruit yield (Natale *et al.*, 1996b). Tree growth, fruit weight and yield significantly increased at K rate up to 400g K_2O tree^{-1}, while TSS, ascorbic acid, reducing sugars and pectin contents of fruit juice increased significantly with higher rate K_2O application up to 500 g tree^{-1}. Economic analysis showed that 682g K_2O tree^{-1} year^{-1} produced the maximum fruit yield and net return tree^{-1} (Kumar *et al.*, 1996). Mitra and Bose (1985) and Mitra (1987) suggested to apply 260 g N, 320 g P_2O_5 and 260 g K_2O plant-1 year^{-1} in two equal split doses once in the month of January and another in the month of August in the alluvial plains of West Bengal, India. Azhakiamanavalan and Khader (1981) recorded the maximum yield in trees receiving 1.0 kg each of N, P and K in Tamil Nadu. Chonkar and Singh (1981) reported that spraying of 3 per cent urea along with calcium phosphate and muriate of potash each at 1 per cent significantly increased the yield of guava. Wagh and Mahajan (1985) obtained the highest yield (46 kg/tree) from 5-year-old Sardar guava by application of 600 g N, and 300 g each of P_2O_5 and K_2O along with 25 kg of FYM as basal dressing. Mustaq *et al.* (2019) recorded that application of NPK@ 235:118:120 g per plant as per the RDF for HDP along with 25 kg FYM resulted in taller plants, plant spread E-W as well as N-S directions, canopy volume and leaf area index. Higher fruits/plant, average fruit weight, fruit diameter, fruit length, fruit yield per plant and fruit yield (48.58 t/ha). Application of 50 per cent of HDP recommendation along with foliar application of urea @ 1 per cent at 2, 3 and 4 months after pruning is effective. Walling and Sanyal (1995) obtained marked variation in yield and fruit quality in cv. Allahabad Safeda following different levels of N and K application (Table 4). Significant responses were noted in number of fruits, fruit size and fruit weight per plant with the level of N applied, but no significant variations were observed to different levels of K application or N-K combinations (Teran *et al.*, 1996). Kumar *et al.* (2009) concluded that application of 600g nitrogen, 300g phosphorus and 600g potassium per plant in two split doses were found to be optimum in increasing fruit yield and quality of guava var. Allahabad Safeda. Further, with the use of optimum dose of fertilizers, an yield enhancement of approximately 44.44 per cent (Table 7).

Table 7: Effect of NPK Levels on Vegetative Growth of Guava var. Allahabad Safeda*

Treatment	*Plant Height (cm)*	*Stem Diameter (cm)*	*Canopy Spread (cm)*	*Fruit Number (Nos/tree)*	*Fruit Yield (kg/tmree)*
N1 -300g	275.85	9.23	349.90	340.49	35.45
N2- 600g	281.35	9.85	370.51	380.55	38.12
N3- 900g	285.16	10.41	388.13	406.86	42.61
P1-300g	270.21	9.65	348.71	355.74	34.78
P2-600g	277.98	9.86	375.11	396.11	36.97
K1-300g	269.91	9.04	351.85	342.76	36.90
K2-600g	276.32	9.94	371.39	377.62	41.75
K3-900g	280.14	10.57	389.28	412.42	45.45

*Kumar *et al.* (2009).

Increase in N and K rates up to 250g and 200g tree^{-1}, respectively resulted in increase in yield (Walling and Sanyal, 1995) (Table 8). Fruit TSS and total sugar contents were increased with high N and K application rates, but were not influenced by changes in P rate (Ghosh, 1991). In a field experiment the effects of N (0, 460 or 920 g/2 tree unit, applied as urea) and K (0, 300 or 600 g K_2O_2 tree unit, applied as muriate of potash) on yield and yield attributes of 7-year-old guava cv. Gwalior-27 were studied by Tomar *et al.* (1998). N and K applications improved values for the yield attributes and yield/plant over the control. N and K applications also increased the TSS, reducing sugars, acidity, and pulp:seed ratio. N rate had no significant effect on total sugar content while K application increased the amount of sugar in fruits.

Foliar application of 4 to 6 per cent urea in January and July significantly increased the growth, flowering, yield and improved the fruit quality of cv. Safeda (Singh, 1985). Foliar spray with 3 per cent N, 1 per cent P, and 1 per cent K in combination caused marked increase in shoot growth, fruit set, fruit retention and yield (Sharma and Sharma, 1992). Spraying of urea (4 per cent) alone or in combination with single superphosphate (2 per cent) and GA3 (100 ppm) significantly increased the yield plant^{-1}, but the combined spray failed to exert any influence on TSS and acidity of rainy season fruits though there was a significant improvement in ascorbic acid content (Singh *et al.*, 1992b; Singh and Singh, 1995). However, the quality of winter season fruits was generally improved by the combined spray (Singh *et al.*,

Table 8: Effect of different Levels of Nitrogen and Potassium on Yield and Fruit Quality of Allahabad Safeda Guava*

Treatment	Yield (kg tree^{-1})	Weight of Fruit (g)	TSS (°Brix)	Acidity (Per cent)	Ascorbic Acid (mg/100g pulp)
N1 (125g tree^{-1})	65.9	149.2	8.2	0.15	129.2
N2 (250g tree^{-1})	82.6	179.4	8.2	0.16	168.2
N3 (375g tree^{-1})	38.8	140.4	8.3	0.12	141.5
LSD (P=0.05)	24.2	37.0	NS	NS	16.2
K1 (100g tree^{-1})	54.9	158.6	8.1	0.15	164.8
K2 (200g tree^{-1})	72.5	169.1	8.5	0.16	140.8
K3 (300g tree^{-1})	59.9	141.3	8.3	0.12	133.3
LSD (P=0.05)	NS	NS	NS	NS	16.2

* Walling and Sanyal (1995).

1990). Yield and quality were improved significantly by calcium and potassium compounds and calcium nitrate (1.5 per cent) proved more beneficial (Brahmachari *et al.*, 1997. NPK 1000 grams fertilizers application along with constant doses of 3 kg gypsum and 40 kg FYM is more suitable and economical dose for arid region (Arshad, 2015). Dhomane (2013) reported that integration of organic manures and inorganic fertilizers were more effective in increasing yield and quality of guava trees than the inorganic fertilizers alone. Among the various combinations, 75 per cent of nitrogen through urea along with 25 per cent nitrogen through neem cake was found effective in respect of yield parameters like weight of fruit (243.80 g), yield per tree (58.1 kg) and yield per hectare (23.26 tonnes).

9.3.1 Micronutrients

Guava sometimes suffers from deficiency of zinc. The usual symptoms of zinc deficiency are interveinal chlorosis, sparsity of foliage, reduced leaf size and less fruit production. Spraying the trees with 1 lb (0.45 kg) zinc sulphate and 0.7 lb (0.34 kg) slaked lime dissolved in 16 gallons (72.736 litres) corrects the deficiency and improves growth and yield (Singh, 1969). Oosthuizen (1985) suggested spraying of zinc oxide at 200 g per 100 litres of water in the spring to control Zn deficiency. The number of sprays to be done is dependent on the severity and extent of the deficiency. Pre-flowering spray with 0.4 per cent boric acid and 0.3 per cent zinc sulphate increased the yield and fruit size (Rajput and Chand, 1976). Copper deficient plants exhibited chlorosis and marginal necrosis of young leaves, brown pigmentation at both sides of the midrib and dieback of the main and axillary shoots (Agarwala *et al.*, 1991). Spra-ying of copper sulphate at 0.2 to 0.4 per cent also proved effective in increasing the growth and yield of guava (Arora and Singh, 1971).

Uptake of Fe, Zn and Mn has been reported to increase with increasing S application rate up to 4 kg fedden-1 (1 fedden = 0.42 ha), but further increase in S dose resulted in decreased uptake of these microelements (Azzazy *et al.*, 1994).

9.3.2 Tissue Analysis

Leaf analysis has proved to be a dependable diagnostic tool for determination of nutritional status of perennial fruit crops. Rodriguez *et al.* (1984) reported that in the leaves of guava rootstock Cotorrera, the N, P and K contents decreased and Ca and Na contents increased from the apex to the base of the shoots and these nutrients were higher in non-fruiting than in fruiting shoots. The position of the leaf half way on the fruiting shoots was recommended for leaf nutrient analyses. In alluvial soils of tropical climate, the leaf N content of guava cv. Lucknow-49 increased from March to August, which afterwards decreased and again increased from December to January (Table 9). The variations in P and K contents were, however, not so pronounced. The foliar N and K contents also showed variations among the different directions of the plant, the P and K contents were found higher when the leaves were sampled from 1-2 m height of the plant while N content was reduced more from 2-3 m height. The new emerging leaves contained higher N, P and K than the older ones (Sanyal and Mitra, 1990). The leaf N content was higher in guava cv. Rica during flowering than at the beginning of fruiting and sampling during flowering gave optimum correlation with the N rate applied, fruit production and leaf N : K ratio (Natale *et al.*, 1994b). It has been con-cluded that a minimum of 30 leaves are required for the determination of macro and micronutrients in guava (Dahiya and Joon, 1995). Significant variations in foliar Fe and Mn concentrations were noted at different seasons of the year, while Cu and Zn concentrations were not affected by season.The fluctuations in micronutrient concentrations were smallest between November and December, indicating that this is the best time for sampling (Chetri *et al.*, 1996). Treatment with higher rates of N, P and K increased the N, P,

Table 9: Variation in Leaf Nutrient Content in different Months (Per cent dry weight)*

Month	*Nitrogen (Per cent)*	*Phosphorus (Per cent)*	*Potassium (Per cent)*
January	1.92	0.45	1.54
February	1.67	0.46	1.58
March	1.58	0.48	1.46
April	1.62	0.48	1.57
May	1.70	0.52	1.55
June	1.80	0.52	1.54
July	1.86	0.46	1.49
August	1.88	0.44	1.58
September	1.82	0.42	1.42
October	1.78	0.45	1.40
November	1.64	0.45	1.40
December	1.80	0.45	1.56
Mean	1.76	0.46	1.50
C.D. at 5 per cent	0.116	0.01	0.019

*Sanyal and Mitra (1990).

and K status of leaves and stems (Mitra, 1983; Ghosh, 1991); a remarkable reduction in K concentration was, however, recorded under N deficiency (Mitra, 1983). In the rainy season, leaf N level of 2.20 per cent was obtained by the application of 400 g N tree-, which resulted in the longest terminal shoot and highest fruit yield, whereas in winter leaf N level of 1.95 per cent, obtained by the same dose resulted in the highest yield of better quality fruits (Koj Tassar *et al.*, 1989).

9.4.0 Training and Pruning

Training of guava trees has been found to improve yield and fruit quality. The beneficial effects of training were reported from the investigation at the Government Fruit Research Station, Basti (Jauhari and Singh, 1973). The main objective of training guava plants is to provide a strong framework and scaffold of branches suitable for bearing a heavy remunerative crop without damaging the branches. The system of training followed by them was open centre, in which the plants were headed back and four primary shoots were retained for initial framework which was subsequently pruned by cutting one-third to half of their length after 3 months. After making the initial framework, the two side shoots were permitted to grow initially and after 3 to 4 years subsequent doubling of selected branches was continued.

The other system was delayed open centre, in which two tiers of framework were prepared and the centre was kept open at a height of 140 cm. Four side shoots were retained in this system also for initial framework, but only one-third of their growth was pruned. For obtaining the second tier, bud growing just below the cut of the central leader was retained. It was observed that training reduced the tree areas up to 11.1 per cent in open centre and 27.6 per cent in delayed open centre. This gave an opportunity to increase the number of plants and subsequently the yield per unit area in the ratio of reduction in tree growth.

In Maharashtra, the system of training guava plants is somewhat peculiar. The bran-ches are bent downwards and tied to each other. The dormant buds are forced to grow and thereby the yield is increased. The method was not encouraged, because the initial increase in yield was found to reduce considerably in the second year (Gadgil and Gadgil, 1933). In Kodur, the trees trained as cordons proved effective and yielded good crop (Naik, 1949).

As the flowers and fruits are borne on current season's growth, a light annual pruning is considered necessary to encourage new shoots after the harvest. All dead, diseased, crowded growths and suckers coming up from the base and sides of the framework should be pruned back annually. Gonzalez and Sourd (1982), however, opined that pruning lowered yields and ten months after pruning, the crown diameter of pruned trees were similar to that of unpruned ones, but pruned trees were signi-ficantly shorter. The intensity of pruning has profound influence on flowering and fruiting of guava. Pruning (removal of terminal 15 and 30 cm of branches) adversely affected flower production and reduced fruit yield branch-1 (Sheikh and Hullamani 1993). However, severe pruning (30 cm) had a beneficial effect on fruit weight and volume (Sheikh and Hullamani, 1993, 1994). The TSS content increased with severe pruning but total sugar showed no difference and the genotype CIW 2 of Navalur guava selection responded well to pruning (Sheikh and Hullamani, 1994). Pruning at an intensity of 25 per cent in February could regulate fruit yield without affecting fruit quality under high density planting (Chandra and Govind, 1995). Pruning has been reported to influence the cropping pattern and with the increasing pruning intensity the rainy season yield decreased. One leaf-pair pruning proved superior to the other pruning treatments for flower bud initiation and fruit yield of the winter crop (Shant Lal *et al.*, 1996).

Complete removal of leaves along with decapitation of shoots promoted flower bud differentiation (FBD) while ringing with partial or complete defoliation and decapi-tation of shoots did not promote FBD. Decapitation of leafy shoots also promoted FBD, increasing from 44 to 63 per cent and 37 to 54 per cent in Sardar and Allahabad Safeda culti-vars respectively (Singh *et al.*, 1999). Pruning of shoot and spraying of hydrogen cyanide (2 per cent) produced the highest yield in Redcreole type guava (Quijada *et al.*,1999). Pruning was able to accelerate the appearance of flowers and increase the number of generative shoots, the number of flowers per tree, the amount of fruit harvested. Increased flowering response due to pruning is supported by the rate of stomatal conduction, the number of stomata is higher than without pruning. The pruning treatment can accelerate the time, the flower appears 10 days faster than without trimming. Fruit quality such as total dissolved solids, fruit acidity, and fruit diameter was not affected by pruning treatment (Widyastuti, 2019). Training pruning significantly influenced yield and quality attributes of guava. Half shoot pruning in April results in equal yield in rainy and winter season crop during both the year. Lowest yield recorded in unpruned control in winter season crop. However, highest total yield per year was recorded in unpruned control during both the year with inferior fruit quality. Winter season crop obtained with increased TSS, ascorbic acid and total sugars (Shah *et al.*, 2018)

9.5.0 Mulching

Mulching, in general, helps in conservation of moisture in soil by reducing the soil temperature and evaporational loss of moisture. Mulching can be done with black polythene or any organic materials. Mulching with rice husk, rice straw and water hyacinth (*Eichhornia crassipes*) reduced the weed population and increased the yield in guava cvs. Allahabad Safeda, Banarasi and Seedless in comparison to control (Borthakur and Bhattacharyya, 1993).

9.6.0 Intercropping

In the early stages of establishment of guava orchard till the bearing, the interspace can be economically utilized by growing suitable intercrops. Intrcropping with vegetables can fetch better returns to the growers beside generating more employment (Hugar *et al.*, 1991). In a crop combination involving several plantation crops, vegetables and leguminous crops like cocoa, pineapple, peas, cowpea, gram, beans are considered as safe intercrops. However, when the trees are full grown, intercropping is neither feasible nor desirable.

Training and pruning in guava

Headed back of guava tree and canopy development

Bending shoots of guava trees to promote flowering

(Department of Fruits and Orchard management, BCKV, 2016)

Flowering and fruiting of guava

Different stages of flowering

Fruit growth and development of guava

Source: Ramírez, Fernando and Kallarackal, Jose (2018) *Sci. Hort.*, 232: 184-190.

Guava trees in fruit bearing stage

10.0 Flowering, Fruit-set, Fruit Growth and Development

10.1.0 Flowering

In Punjab and other parts of northern India guava flowers twice a year, first in April-May for rainy season crop and then in August-September for winter season crop (Gupta and Nijjar, 1978). In Maharashtra and Tamil Nadu, there is a third crop with flowers appearing in October (Hayes, 1970). In West Bengal, two important seasons of blooming were found by Mitra (1983), once in April-May and again in September-October.

The guava bears solitary flower, or in cymes of two or three flowers, on the current season's growth in the axils of the leaves. Normally, the bearing twigs grow a few centimetres long, putting forth four to five pairs of leaves. Meanwhile, if the flower happens to set, terminal bud ceases to grow till the next growing season (Dasrathy, 1951). Gardener *et al.* (1952) reported that the flowering bud is a mixed type, and the flowering shoot bears the flowers laterally. Both terminal as well as lateral flower bearing shoots was observed by Sehgal and Singh (1967). It was further observed that axillary buds are not produced all over the shoot and may appear scattered (Srivastava, 1962a). The flowering period varies from 25 to 45 days depending on the cultivar, season and region of growing.

Studies on floral biology of several cultivars of guava revealed that the peak period of anther dehiscence was between 05.30 and 07.30 h except in Seedless (peak period between 07.30 and 08.30 h). Maximum number of flowers opened between 05.00 and 07.00 h, except in Seedless (peak period between 07.00 and 08.00 h). Pollen viability was 50.37 per cent for Seedless, but in other cultivars, it varied between 86.24 and 94.32 per cent (Kundu and Mitra, 1994).

10.1.1 Regulation of Flowering

The fruits produced in the rainy season are rather insipid and watery and do not keep well. In Uttar Pradesh, the winter crop is the one ordinarily desired as it is not only larger in size, but of much better quality. Smith (1934) stated that the total quantity of fruits borne in a year is greater if only the winter crop is taken. Rathore (1972) carried out extensive investigation on the cropping pattern of guava in which (i) only rainy season crop was taken by removing flowers for all other crops, (ii) winter season crop was taken by removing flowers for all other crops, and (iii) all crops which appeared during the year were taken as such without any intervention. Results of the experiments showed that shoots made more growth in the rainy season cropping pattern than (i) in winter (ii) or whole year cropping pattern. The shoots grew faster in the first four months, slowed down during the next three months and finally started drying out at the age of one year or initiated new growth which was mostly lateral. More lateral shoots dried in whole year cropping pattern than in other seasons (Rathore, 1978). The percentage of fruit set and mature fruits were found to be the highest in the winter crop followed by the whole year and rainy season crop (Rathore and Singh, 1974). Singh *et al.* (1996) reported that the control trees (from which no flowers were removed) produced the highest yield, followed by only winter and rainy season crops, but the largest and best quality (in terms of TSS and ascorbic acid content) fruits were obtained in the winter. In the Deccan, where the climate is mild, it has been a popular practice to obtain flowering in a desired season. The practice is commonly known as *bahar* treatment and it is achieved by root exposure and root pruning or exposure to hot sun before the onset of monsoon. Root pruning, however, has a harmful effect on the longevity of trees. In order to have a good winter harvest, deblossoming of rainy season crop and root pruning have been suggested even in Uttar Pradesh where root pruning is not desirable. For obtaining winter crop in Maharashtra, Cheema *et al.* (1954) recommended withholding of water and removing the earth from around the upper roots by June 10, and again covering it with soil and manure mixture. Two light irrigations were also suggested before a normally heavy one, when the rains did not start.

Growth regulators have been found very effective in thinning of flowers and manipulating the cropping season. The effective chemicals so far tried are napthaleneacetic acid (NAA), napthalene acetamide (NAD), and 2, 4-dichlorophenoxy acetic acid (2, 4-D). Teaotia and Pandey (1970) observed that spraying with NAA at 100 ppm reduced the rainy season crop considerably, but the best result was obtained by hand thinning of summer flowers, which reduced the rainy season crop by 81 per cent and increased the following winter crop by 18.4 per cent. Rathore (1975) reported that NAA at 80 and 100 ppm greatly reduced the fruit-set of guava when sprayed in the month of April. Chundawat *et al.* (1975) observed that NAA at 100, 200 and 400ppm, applied to the guava cv. Banarasi Surakha for flower thinning, resulted in deblossoming by 25.51 to 82 per cent and recommended it as an alternative for the conventional method of witholding water before flowering. Among the different fruit thinning chemicals used, Kumar and Hoda (1977) suggested NAD at 50 ppm and 2, 4-D at 30 ppm as the most effective chemicals for deblossoming of summer flowers. By deblo-ssoming the summer flowers, an increase in yield by 200-300 per cent was obtained and NAD at 50 mg/l was more effective than NAA or 2, 4-D (Mitra *et al.*, 1981, 1995; Singh *et al.*, 1993). Treatment with ethrel (1ml l-1) resulted in the highest ethylene forming enzyme activity in leaves and maximum number of flowers and fruits (Castelan-Estrada and Becerril Roman, 1994). The flowering was greatly enhanced by spraying with urea at 25 per cent and ethrel at 2000 ppm (Chandra and Govind, 1994). Brahmachari *et al.* (1996) reported that CCC at 500 ppm induced the earliest flowering and highest number of flowers.

Now-a-days, defoliation has been used to induce new shoots and flowers. In a fruit management programme, Kobayshi (1986) working with Beaumont guava in Hawaii noted maximum number of new lateral shoot production by defoliation with ethepon at 1200 ppm and maximum shoot length with ethepon at 600 ppm. Spraying of urea (10 to 15 per cent) at full bloom caused significant defoliation and deblossoming of the rainy season crop and increased the yield in winter (Dwivedi *et al.*, 1990; Singh *et al.*, 1992; Gaur, 1996; Singh *et al.*, 1996). Further increase in urea concentration increased the percentage of flower/fruit drop, reduced yield and impaired the fruit quality (Dwivedi, *et al.*, 1990; Singh *et al.*, 1996). Exogenous application of flower thinning agents in April resulted in abscission of rainy season flowers and enhanced the winter as well as cumulative yield as compared to control (Table 10). Manual deblossoming was effective in increasing the winter yield, but laborious and uneconomical (Mitra *et al.*, 1995).

Table 10: Effect of Deblossoming on Yield and Quality of Guava Fruit cv. Harijha* (Age of the plant 4 years)

Treatment	*Yield (kg/plant)*		*TSS (°Brix)*		*Ascorbic Acid (mg/100 g)*	
	Rainy	*Winter*	*Rainy*	*Winter*	*Rainy*	*Winter*
Control	25.2	11.8	8.0	11.8	122	187
Manual	-	41.9	-	11.6	-	222
deblossoming						
NAA 50 mg/l	22.2	17.8	8.5	11.0	126	209
NAA 100 mg/l	210	32.2	9.1	10.8	116	253
2,4-D 50 mg/l	17.3	20.0	9.0	10.8	135	195
2,4-D 100 mg/l	12.0	32.4	9.0	11.0	169	220
DNOC 5 mg/l	22.0	24.2	9.3	10.8	133	216
DNOC 10 mg/l	22.1	23.8	9.3	11.4	138	247
Urea 10 per cent	10.4	37.2	8.6	12.0	172	230
C.D. at 5 per cent	6.81	5.99	0.84	NS	26.16	NS

* Mitra *et al.* (1995).

10.2.0 Fruit Set

The initial fruit set in nature is quite high in case of guava and about 80-86 per cent of the flowers set fruits. But due to severe fruit drop, only 34 to 56 per cent of fruits reach maturity. In cultivar like Seedless, the final retention is as low as 6 per cent (Sehgal, 1961). The fruit drop may be due to different physiological and environmental factors. Endogenous auxin, gibberellin and cytokinin contents in ovaries are associated with fruit set in guava (Nagar and Rao, 1986).

The formation of fruit is first noticed after 12 days from flowering (Dasrathy, 1951). Spraying of GA3 at 15 or 30 ppm in the month of January proved effective in increasing fruit retention and subsequently the yield (Rajput *et al.*, 1977). About 90 per cent fruit retention was also recorded in trees treated with GA3 at 200 ppm (Sundarajan *et al.*, 1969). Foliar application of CCC (500 ppm) and PCPA (50 ppm), twice (one before flowering and again a month after fruit set) significantly enhanced fruit set in Sardar guava (Brahmachari *et al.*, 1995). Fruit retention, size, weight and volume were highest with double spraying (before flowering and after fruit set) of GA3 at 100 ppm and 2,4, 5-T at 25 ppm (Hoque and Irabagon, 1994; Brahmachari *et al.*, 1995,1996).

10.3.0 Fruit Growth and Development

The growth of guava fruits followed a double sigmoid curve *i.e.* two periods of rapid growth with a period of relatively slow growth in between (Nandi, 1983).

The weight of fruit and fruit size gradually increased at the first phase, then slowed down and finally increased till maturity. The duration of all the three growth periods of the fruit appeared to be inversely proportional to prevailing temperatures. The better quality of winter season fruit was correlated with lower temperatures during the winter season (Rathore, 1976). The flesh colour of some of the guava cultivars changed from white to pink and fruit acidity, TSS, sugar and ascorbic acid contents increased very slowly at the initial stages of development, then increased rapidly and reached the maximum value at full ripe stage (Le-Riche, 1952). Mukherjee and Dutta (1967) reported that guava cultivars, *viz.*, Safeda, Pyriform and L-49 took approximately 137, 110 and 106 to 138 days, respectively, to reach maturity. In Australia, the cvs. Ka Hua Kula, GAA11-56 CSIRO, GA11–56T3, GA11-56T7, GA11-56RITI and GA11-56R5T2 took 14 weeks to reach maturity when defoliation was done by urea at 250 g/l (Menzel and Paxton, 1986), while in South Africa, it required 22 weeks from full bloom to maturity (Welgemoed and Preez, 1986).

Kundu *et al.* (1998) classified the fruit growth into three distinct stages, *viz.* stage-I, in which fruit growth was rapid for the first 50 days of fruit set; stage II, in between 50 and 90 days of fruit set, when the growth was very slow and stage III, in which an exponential increase in the rate of fruit growth was observed between 90 and 120 days. The TSS-acid ratio was found to increase initially, thereafter it decresed and increased again at the later stages of fruit growth and development.

The presence of tartaric acid at flower bud stage and citric acid during the later stages of development of fruits in Red Fleshed cultivar was reported by Doraipandian and Muthukrishnan (1973). All acids were abundant at the initial stage of development in seedless cultivars whereas in seeded cultivars it was at the later stages. Ripe guava fruits had a higher content of total acidity, total soluble solid, total sugar, carotenoids, protein and fat, while they had a low content of ash, fibre and total solids compared to mature fruits. Over ripe fruits contained high ascorbic acid, moisture and carotenoids. Ripe fruits had the highest content of ester, carbonyl compounds, oxygenated terpenes and limonene that reveals the fruit aroma. Forty components were identified in guava fruits representing different groups of organic compounds, 12 alde-hydes, 13 esters, 10 alcohols, 3 ketones and 2 hydrocarbons. The components responsible for the flavour are the esters which have the highest concentration (44.94 per cent) in ripe fruits and the lowest in mature ones (33.68 per cent). A marked variation in alcohol concentration was observed as it reached 27.33 per cent in the ripe fruits while in over ripe ones it decreased to 8.95 per cent (Ashour *et al.*, 1986). Itoo *et al.* (1980) from Japan reported that ascorbic acid content was highest in mature stage which was mostly present in the peel. El-Buluk *et al.* (1995) studied the biochemical and physical changes in four guava cvs. Ganib, Pakistani, Shambati and Shendi. It was observed that texture and protein contents decreased gradually while the moisture and water-soluble pectin increased during fruit development in all cultivars. Alcohol-insoluble solids (AIS) declined gradually during the initial growing period followed by a rapid decline during later stages of development for all cultivars. Softness and yellowness of fruit were associated with lower protein and AIS and higher moisture and appreciable amounts of water-soluble pectin contents. Selvaraj *et al.* (1999) studied the physico-chemical changes associated with growth and ripening of Allahabad Safeda and Sardar guava. They observed that on an average, the fruits of Allahabad Safeda and Sardar took 165 and 155 days respectively from flowering to ripening. Both the cultivars followed a double sigmoid pattern of growth. The fruit density, total acidity, skin chlorophyll, skin carotenoids, vitamin A, total pectins, crude protein, phosphorus, potassium, calcium, magnesium and sodium content decreased; pulp pH, total tannins, dry matter and iron contents showed marginal changes and TSS, vitamin C, sucrose, glucose and fructose contents increased during fruit maturation and ripening. The fruits of Allahabad Safeda had glucose : fructose ratio below 1 and those of Sardar above 1 during maturation and ripening.

Parthenocarpy is not a common feature in guava even in commercial seedless cultivar (Dasrathy, 1951). However, Balasubrahmanyam and Rangaswamy (1959) obtained parthenocarpic fruit set in the cv. Allahabad Round by self-pollination which they ascribed to be due to the stimulation provided by pollen hormones. In parthenocarpic guava indole-like compounds were isolated by Rangaswami and Kaliperumal (1961). Application of GA3 at 1000 to 8000 ppm in lanolin paste was found to be effective in inducing parthenocarpy (Rao and Rao, 1960; Teaotia *et al.*, 1961). Parthenocarpic fruits produced by treatment with GA3 were oblong in shape and showed 6 to 8 prominent ridges on the fruit surface. The fruits were richer in ascorbic acid content (Rao and Rao, 1960). The seedless nature of Seedless cultivar was believed to be due to auto-ploidy and not as a result of parthenocarpic fruit development (Dastrathy, 1951).

11.0 Pests, Diseases and Disorder

11.1.0 Pests

11.1.1 Fruit Fly

Oriental fruit fly (*Dacus dorsalis*) is common in India. The other fruit flies are Mediterranean fruit fly (*Ceratitis capitata*), Natal fruit fly (*Ceratitis rosa*), melon fly (*Dacus cucurbitace*) and Caribbean fruit fly (*Anastrepha striata*).

During the monsoon, the adults lay eggs on the surface of the fruit. On hatching the maggots enter into the fruit and in most cases fruit drop occurs. Spraying of malathion, demicron, *etc.*, and burning of the infected fruits minimise the incidence. Plastic bucket trap baited with methyl eugenol (4 ml/trap on a cotton wick) and placed on guava trees 5 feet above the ground at a density of 1 per acre caught significant number of male fruit fly (Marwat *et al.*, 1992; Stark and Vargas, 1992). The attractiveness of the trap (white and yellow) enhanced the trap efficiency which is due primarily to intensity of reflected light (Stark and Vargas, 1992). The effect of methyl eugenol baited with dichlorvos for the control of *Dacus dorsalis* in guava orchards was studied by Makhmoor and Singh (1998). Five different concentrations were tested. The pest control was highest with 1 per cent methyl eugenol. *B. dorsalis* population peaked during 20 July to August 3, which almost coincided with fruit ripening. There was overall 51 per cent reduction in fruit damage where methyl eugenol traps were used. In the orchards where the traps were used, 16 per cent of the fruits were damaged on 4th September, compared with 82 per cent in the controls. Highest trap catch (170.66) was obtained in the 6th week (20-27th July) when maximum and minimum temperatures were 36.6°C and 25.5°C, and relative humidity 90.3 and 57 per cent respectively.

11.1.2 Mealy Bug

Guava has been reported to be attacked by mango mealy bug (*Drosicha mangiferae*) and citrus mealy bug (*Planococcus citri*). The tiny small bugs suck sap from young leaves, twigs and flowers. The affected parts dry up and the yield is considerably reduced. Banding the tree trunks with 30 cm wide, 400 gauge polythene combined with grease on 5 cm area on its lower end and plugging the space between trunk and the band significantly controlled the pest (Dalaya *et al.*, 1983). Spraying of several insecticides *viz.* chlorpyrifos (0.08 per cent), acephate (0.08 per cent), cypermethrin (0.02 per cent), methyl demeton (0.08 per cent), monocrotophos (0.08 per cent), quinalphos (0.08 per cent), dimethoate (0.08 per cent), fenvalerate (0.02 per cent), carbaryl (0.2 per cent), phosphamidon (0.08 per cent) and phenthoate (0.08 per cent) have been reported to be equally effective in controlling larvae of mango mealy bug infesting guavas (Ghule and Dhumal, 1992). The encyrtid parasitoid (*Leptomastix dactylopii*) could completely suppress the citrus mealy bug in guava orchards (Mani, 1994). Neem oil and Pongamia oil (both at 4 per cent) are recommended for the control of citrus mealy bug on guava (Hussain *et al.*, 1996).

11.1.3 Scale Insect

The guava scale, (*Chloropulvinaria psidii*) is one of the serious pests in Maharashtra. These green sticky insects are found on the leaves, shoots and fruits. Spraying twice with fish oil, rosin soap, 500 g in 36 litres of water, or with crude oil emulsion, 7.8 kg in 450 litres of water has been suggested to minimise the pest incidence. Dimethoate at 100 ml/100 l water is also effective. Pruning of affected branches is recommended for checking their damage.

11.1.4 Bark Eating Caterpillar (*Indarbela quadrinotata; I. tetraonis*)

Larvae on hatching eat up the bark and bore into the trunk. Infestation is more in old and neglected orchards. Seedless and Apple cultivars are more susceptible than Allahabad Safeda, Strawberry and Cattleya guava (Butani, 1974). The infestation can be reduced by keeping the orchard clean. The larvae can be killed by inserting wire in the holes.

11.1.5 Thrips (*Rhipiphorothrips cruentatus*)

Thrips cause damage to the fruits during March, August and September (Dahiya *et al.*, 1995) and they can be controlled by spraying of malathion at 1 ml/l.Beside these major pests, the pomegranate butterfly (*Virachola isocrates*) has been reported to damage by boring into fruits. The infestation began at the marble stage of the fruits and the peak infestation was recorded in mid-August to September (Biswas *et al.*, 1995). Spraying with carbaryl (0.2 per cent) at 10 to 12 days intervals reduced the population of pomegranate butterfly. The cv. Sindh was found to be the most resistant (Yadav and Pandey, 1994), while the cv. Swarupkathi was least susceptible and cv. Kazi piara was highly susceptible to this pest (Biswas *et al.*, 1995).

Pests of guava

Fruit fly infestation in guava

Mealy bug infestation on fruits and leaf in guava

Scale insect infestation in guava

Pests of guava

Bark eating caterpillar, adult (left) and larvae (right) of guava

Thrips infestation in guava

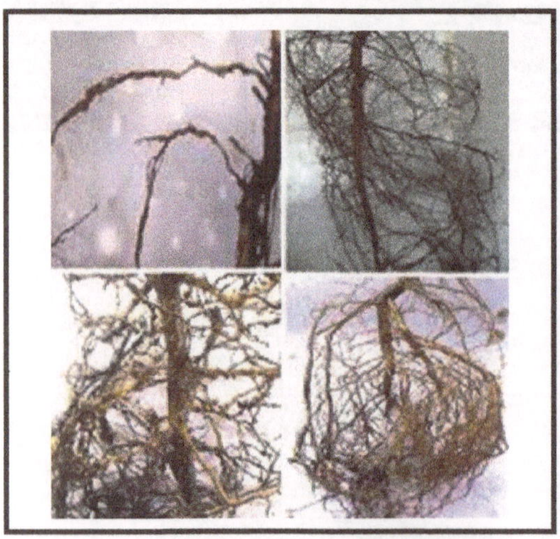

Nematode infestation in guava

11.1.6 Nematodes (*Meloidogyne mayaguensis*)

Severe root-knot nematode symptoms were observed in commercial guava plantations at Nelspruit, Mpumalanga, South Africa (Willers, 1997b). The affected trees were stunted with poor vigour and unthrifty appearance. The yield was reduced and the tree gradually declined. Application of cadusafos, a nematicide (100 g/kg) @ 10 g/m^2 in the drip zone provided effective protection against root-knot nematode (Willers, 1997a and b).

11.2.0 Diseases

Amongst the diseases, most damaging to guava are the wilt and anthracnose; other important diseases prevalent in certain regions are stem canker, cercospora leaf spot, seedling blight, *etc.*

11.2.1 Wilt

This fungal disease is known to occur in Uttar Pradesh, West Bengal, Bihar, Rajasthan and Madhya Pradesh. The actual cause of this disease is yet to be discovered. The pathogens, *viz., Fusarium oxysporium* f. sp. *psidii, Fusarium solani, Macrophomina phaseoli, Rhizoctonia bataticola* and *Cephalosporium* sp. may incite the disease (Misra, 1996). Das Gupta and Rai (1947) recorded the disease in the severe form the orchards of Lucknow for the first time India. Dey (1948) reported it from Allahabad, Kanpur Guava trees suffered serious losses in 11 districts of UP estimated that guava wilt spread rapidly to cover about 20,000 m^2 area in U.P. Mathur (1956) reported upto 30 per cent trees affected with wilt in Allahabad, Farrukhabad, Unnao (15-30 per cent), Kanpur and Jaunpur (5-15 per cent) and less than 5 per cent in Gorakhpur, Ballia, Hardoi, Barabanki and Varanasi. Edward and Srivastava (1957) reported wilt as the most serious disease threatening guava cultivation in UP. Later, it was also reported from western parts of UP (Singh and Lal, 1953), Varanasi (Pandey and Dwivedi, 1985), Kaimganj (Farrukhabad), Bithoor (Kanpur), Ganga Ghat (Unnao), Abbubakarpur (Allahabad), Lucknow, Bichpuri (Agra), Sasni (Aligarh) (Misra and Prakash, 1990; Misra, 1987). After the first few reports from U.P., in West Bengal the disease spread in the Gangetic alluvium of Baruipur area in the district of 24 Parganas and in the laterite zone of Jhargram and Midnapur (Chattopadhyay and Sengupta, 1955). Chattopadhyay and Bhattacharya (1968) reported the disease from Kashakul, Bankura. The disease was also been reported from Haryana (Suhag, 1976; Mehta, 1987), and Punjab (Bhargava *et al.,* 2003). It is charcterized by the yellowing of leaves followed by drying of leaves and twigs from the tip and complete wilting of trees within 10 to 15 days. The leaves drop down prematurely and the fruit size became smaller. Chakraborty and Singh (1989) classified the symptoms into two categories: a) slow, b) sudden. *M. phaseolina* (dominant in alkaline soil) was associated with slow wilt while, *F. oxysporium* f. sp. *psidii* (dominant in acid soil) was associated with sudden wilt. The disease occurs more severely in alkaline soils.

Although several pathogens have been reported for the cause of wilt in guava by different workers but *Fusarium oxysporum f.* sp. *psidii* and *Fusarium solani* were found two most important pathogen associated with this disease (Misra, 2006; Misra and

Gupta, 2007; Gupta, 2010). Therefore, in present review the wilt disease of guava has been described in detailed with special reference to *Fusarium* sp. infection in India.

The infection can effectively be minimised by drenching the soil with brassicol and spraying the plant with bavistin (0.1 per cent) at an interval of 15 days at the early stage of infection. Jain (1956) suggested to inject the healthy plants with 0.1 per cent 8-quinolinol sulphate. Soil solarization using a 30 mm transparent polyethylene sheet controlled *F. oxysporium* f. sp. *psidii*, *F. solani* and *R. solani* on guava (Dwivedi, 1993). A local variety of Taiwan, Peipa has been reported to be resistant and *P. friedrichsthalianum* has been recommended as possible rootstock (Leu and Kao, 1979). Cultivars such as Banarasi, Dholka Sind, Nasik and Supreme are resistant to wilt (Pathak, 1980).

Prasad *et al.* (2003) and Misra *et al.* (2004) reported intercropping with turmeric or Marigold to check the wilting of guava. These cultural practices are useful and should be adopted to escape wilt. Considering the complexity of the problem integrated eco-friendly approach for the control of guava wilt was suggested by Misra *et al.* (2003b, 2004) and Misra (2005) utilizing bioagent *Aspergillus niger* strain AN17, using resistant root stock (*P. molle* x *P. guajava*), intercropping of guava with turmeric or Marigold as well proper cultural practices and can be integrated to minimize losses due to the disease. Bioagents like *Aspergillus niger*, *Trichoderma* sp. *Penicillium citrinum* and some bio-dynamic antagonists have shown their effectiveness towards the control of wilt pathogens of guava (Singh *et al.*, 2003; Misra, 2006; Gupta *et al.*, 2009a). When these fungi were tested for the control of wilt pathogen in laboratory conditions, these were found quite effective (Misra *et al.*, 2004). When relative growth of the three bioagents was studied, it was found that *Aspergillus niger* was fastest growing and most effective (Misra and Prasad, 2003). These can be grown easily on any substrate like maize/bajra seeds *etc.* and can also be multiplied on cheap substrates like Sacchrum sp. (grass) and dry and green leaves of *Psidium guajava* (Shukla *et al.*, 2003). It was also found that at village level these bioagent can be multiplied in earthen pots (Misra and Prasad, 2004).

11.2.2 Anthracnose

In the western districts and the terai region of Uttar Pradesh anthracnose in guava is reported to be a serious disease. The affected plants begin to die back from the top of the branch, while shoots, leaves and fruits are readily affected. The growing tips gradually turn dark brown and the black necrotic areas extend backward causing die back (Pathak, 1980). *Pestalotiopsis psidii*, *Colletotrichum gloeosporioides* (*Glomerella cingulata*) and *Botryodiplodia theobromae* were isolated from infected fruits and pathogenicity was confirmed (Hossain and Meah, 1992). The disease spreads in cool as well as in hot dry weather.

The growth of the causal organisms was observed under different temperatures. *Botryodiplodia* sp. was the most fast growing fungus at all temperatures within the identified fungi. Temperatures 28° and 28-30°C were found the most suited for growth of *P. psiddii and C. gloeosporioides*, respectively.The isolation from diseased guava fruit (*Pesidium guajava* L.) and pathogenicity test in this study showed that *Colletotrichum gloeosporioides* is the pathogen responsible for guava anthracnose in

Diseases of guava

Guava canker infection in fruits

Phytophthora fruit rot in guava

Wilt affected guava trees

Anthracnose infection in guava fruit

Cercospora leaf spot in guava fruits and leaves

Rust infection in guava fruit

Fruit begging for protection against fruit fly to obtain quality fruits

Fruit begging for protection against fruit fly to obtain quality fruits

Harvesting of guava

Egypt. (El-Helaly *et al.*, 1963, 1966; Aly *et al.*, 1972; Abdel Aal, 1980; El-Hoseiny, 1990).The disease can be controlled by spraying the trees with copper oxychloride, cuprous oxide, difolatan or dithane Z-78. Application of topsin M (thiophanate-methyl) to plants provided the most significant reduction in fruit infection and disease severity (Hossain and Meah, 1992).

11.2.3 Stem Canker

The infected plants show cracks and lesions on the stem and the affected twigs wilt due to collapse of the stem tissue. Brown spots appear on the calyx end of the fruit. The stem canker is caused by *Physalospora psidii*. Pruning of infected shoots and subsequent protection of wound with blitox or Bordeaux paste are suggested to check the infestation.

11.2.4 Phytophthora Fruit Rot

This disease is caused by *Phytophthora nicotianae* var. *parasitica*. It is a major disease in northern India, particularly in Haryana and Punjab. White cottony mycelial growth occurs at the stylar end of the fruit with initiation of ripening process and covers the entire fruit within 3-4 days. Humid condition is highly congenial for rapid spread of the disease. The disease may occur on unripe fruits and the affected fruits drop down. Metalaxyl (ridomil 25 WP) and copper oxychloride (blitox) gave cent per cent control of the pathogen (Seema *et al.*, 1995).

11.2.5 Grey Blight and Fruit Canker

Pestalotia psidii is the causal organism of this disease. Dark brown corky lesions or cankerous spots with raised margins and depressed centre appear on the fruits. The optimum temperature for the growth and sporulation was 25 to 30°C. Carbendazim, flusilazole and imazalil were very effective in inhibiting the mycelial growth and spore germination (Tsay, 1991).

11.2.6 Cercospora Leaf Spot

Water-soaked patches under the leaf caused by *Cercospora sawadae* are the characteristic symptoms of this disease. The infection can be reduced by spraying lime sulphur (1:30) or 0.3 per cent copper oxychloride (Bose and Muller, 1967).

11.3.0 Disorder

The guava trees growing in the vicinity of factories producing sulphur products are prone to disorder. The leaves develop discoloration and necrosis of the tips and margins. The distal end of fruits become black resembling black tip necrosis of mango caused by SO_2 gases emanating from brick kilns (Verma *et al.*, 1991).

12.0 Harvesting and Yield

Seedling guava trees require 4 to 5 years to bear, while grafted, budded or layered plants start bearing at the age of 2 to 3 years. Guava fruits mature between 88 days(wild guava, *P. guineense*) and 121 days (cv. Chittidar) from fruit set (Singh and Sharma, 1996). Fruits are recommended to harvest between 106 and 121 days after fruit set to ensure higher amount of total sugars and appreciable amount of minerals (El-buluk *et al.*, 1996). The fruits turn greenish yellow with the advancement of maturity. The TSS-acid ratio of 35.8 (Allahabad Safeda), 26.39 (Apple Colour), 28.13 (Chittidar), 34.25 (Lucknow-49) and 34.58 (Seedless) could be used as maturity index (Kundu *et al.*, 1998). The guava fruit should be picked immediately when it is mature, because it cannot be retained on the tree in ripe stage as occasionally ripe fruits are liable to be damaged by birds. Individual hand picking at regular intervals is suggested to avoid possible damage to fruits and plants. Harvesting by shaking the tree is discouraged as the fruits are very delicate at the ripe stage. Mature or half-ripe fruits are mostly liked for consumption.

The yield varies in different cultivars and with the care and management of orchard, age of the plant and season of cropping. The average yield per tree estimated to be 90 kg from the seedling tree and 350 kg from the grafted plants (Singh, 1980). Rao (1946) estimated a yield of 2.2 tonnes per acre, while in West Bengal, Chattopadhyay and Sengupta (1955) estimated fruit yield of 4.5 tonnes per acre. Yield records at Allahabad showed 55-60 kg from a 3-year-old grafted L-49 guava. In trials with 3-year-old grafted plants of different cultivars, Mitra *et al.* (1981) found that all the cultivars produced greater yield in the rainy season compared with that obtained in winter (Table 11). The cultivars L-49, Allahabad Safeda, and Chittidar were superior in yield compared with other cultivars studied.

Table 11: Fruit Yield of different Guava Cultivars (3-year-old grafted plants)*

Varieties	*Yield per Plant (kg)*		*Yield/Plant per Year (kg)*	*Yield/ha (q) (Plant density 278/ha)*
	Rainy Season	*Winter Season*		
Allahabad Safeda	16.0	12.0	28.0	77.84
Apple Colour	15.2	10.2	25.4	70.61
Banarasi	14.8	10.6	25.4	70.61
Baruipur	16.8	8.7	25.5	70.89
Behat Coconut	13.2	9.3	22.5	62.55
Chittidar	16.2	14.2	30.4	86.51
Harijha	12.8	10.1	22.9	56.63
Lucknow-49	21.1	18.1	39.2	108.98
Pear Shaped	11.3	7.8	19.1	53.09
Red Fleshed	15.0	10.8	25.8	71.72
Seedless	3.2	2.2	5.4	15.01

* Mitra *et al.* (1981)

13.0 Post-harvest Management

13.1.0 Packaging and Transport

After harvesting the fruits are packed in wooden boxes or bamboo baskets and transported to the market or processing units. Fruits should never be packed in hessian bags as it may cause considerable damage to the fruits even in short-

Ripening and packaging of guava

Mature fruits of guava

Fresh harvest of guava

Ripening of guava

Foam net packaging of guava

Selling of guava packed in foam net in multistore

Marketing of guava

Farmers selling guava in floating market at Barishal district, Bangaldesh

Packaging of guava fruits for distant market

distance transport. For long-duration transport, a layer of cushion material (dry grass, paddy straw, paper-wool *etc.*) should be used at the bottom of the container. Refrigerated transport can reduce the spoilage of fruits. Siddiqui *et al.* (1991) stated that fruits positioned in the natural posture with the pedicel end vertically upward during transport showed better keeping quality than fruits kept in the reverse or horizontal position. Individual wrapped fruits showed less water loss and decay than fruits packed without wrapping.

13.2.0 Ripening

Guava is a climacteric fruit, showing a typical increase in respiration and ethylene production during ripening (Brown and Wills, 1983). The colour of the skin changes from green to yellow or red during ripening while the colour of the flesh may change from white to creamy white, yellowish pink, deep pink or salmon red (Wilson, 1980). The flesh colour at ripening may be influenced by the presence of carotenoids, lycopene and b-carotene (Wilberg and Rodriguez-Amaya, 1995). The loss of tissue firmness during ripening is accompanied by a decrease in the level of total pectin (Lazan and Ali, 1989). The activities of the softening enzymes polygalacturonase, pectinesterase, b-galactosidase and cellulase showed an increase with ripening (Chin *et al.*, 1994). Sugar and total ash contents increase rapidly during maturation and ripening (El-Buluk *et al.*, 1996). Ripening was delayed in mature-green fruits stored at 10°C. Ripening can be induced by exposing the fruits in ethylene. Treating fruits for 24 h in ethylene atmosphere (100 ml/l) at 20°C resulted in a significant increase in the rate of skin-yellowing and softening of immature-green fruits, whereas ethylene-treated mature-green and quarter-yellow fruits did not exhibit any difference from control (Reyes and Pauli, 1995). A strong aroma develops in fruit during the process of ripening which is due to volatile compound b-carrophyllene (McLeod and de Troconis, 1982). With the advancement of maturity and ripening the production of isobutanol, butanol and sesquiterpenes decreased while ethyl acetate, ethyl caproate, ethyl caprylate and cis-hexenyl acetate increased (Askar *et al.*, 1986). Potassium iodide (5000 ppm) was the best for ripening as indicated by higher yellowing index, better appearance, firmness and organoleptic quality and reduced weight loss during storage (Singh *et al.*, 1996).

13.3.0 Storage

The shelf-life of guava fruit at ambient condition is relatively short due to rapid development of fungal rots. However, the fruits may be stored for a few days to adjust the market demand. Chundawat *et al.* (1976) reported that all the cultivars except Allahabad Safeda could be stored for 2 days at room temperature. Fruits can be stored for 2 to 4 weeks at relatively low temperature coupled with high humid conditions (Table 12). However, chilling injury may occur at 0°C (Wills *et al.*, 1983). Singh *et al.* (1976) stored guava successfully up to 6 days in perforated polythene bags and wooden boxes without rotting and much weight loss. Das and Acharya (1969) observed that fruits with 2 or 3 dips in 6 per cent wax or one dip in 9 per cent wax stored well for one month. Mature green fruits were treated with different concentrations of 2,4-D, 2,4,5-T or GA3 at 100 and 200 ppm or MH at 500 and 1000 ppm. Both ripening and weight loss were accelerated with 2, 4-D and 2,4,5-T and retarded by MH and GA3 treatment (Saha, 1971). Ghosh (1980) reported that fruits of guava cv. L-49 could be stored well up to one month at 10°C. At room temperature, fruits treated with aureofungin and ethylene chlorohydrin plus calcium carbonate could be stored for 5 days without spoilage and up to 7 days with only 25 per cent spoilage.

Table 12: Recommended Storage Conditions of Guava

Temperature (°C)	*Relative Humidity (Per cent)*	*Duration (Weeks)*	*References*
5-10	90	2-3	Hardenburg *et al.* (1990)
100	90	3	Mercantilia (1989)
10	85-90	2-3	Anon. (1991)

The storage of fruits in low density polyethylene bag with celite-permanganate mixture showed minimum physiological loss in weight, slower ripening during storage and the fruit quality was comparable (Dutta *et al.*, 1991). In general, fruits with higher specific gravity can be stored longer than fruits with low specific gravity (Balakrishnan *et al.*, 1994). Preharvest sprays of calcium nitrate (1.5 per cent) and Alar (750 ppm) were effective for minimizing physiological weight loss during storage and maintaining the eating quality and marketability of fruits for more than 8 days (Chandra *et al.*, 1994). High incidence of fruit browning after harvest was generally associated with low TSS and ascorbic acid contents and low acidity. Dipping of the fruits in $CaCl_2$ (3 per cent) and GA_3 (100 ppm) reduced skin browning and polyphenol oxidase (catechol oxidase) activity compard with control (Nawar and Ezz, 1994). Wrapping the fruits in newspaper was found to enhance the shelf life of fruits by about 3 days (Siddiqui and Gupta, 1997). Coating of mature-green fruits in cellulose or carnauba-based emulsions delayed colour development and suppressed increase in the level of total soluble solids (McGuire and Hallman, 1995). Controlled atmosphere storage using high CO_2 (10 per cent) concentration also delayed ripening but showed physiological injuries in fruit (Pal and Buescher, 1993).

14.0 Biotechnology

14.1.0 Tissue Culture

Attempts on propagating guava through tissue culture gave varying degree of success. In a trial on new sterilizing techniques for *in vitro* culture of guava, Broodrijk (1989) obtained 70 per cent success with minimal browning when explants were put in growing medium after preliminary treatment. Fitchet-Purnell (1989) was able to establish shoot tips and nodal sections of the cv. Fam Relif in *in vitro*. The oxidative browning which caused the explant's death was successfully prevented by placing the rinsed explants in petri dishes to dry in air flow for 30-45 minutes, so that the surface was moist when the explants were incubated. The stem and buds remained green and all buds subsequently developed. Prakash (1992) and Prakash and Tiwari (1993, 1996) reported *in vitro* clonal propagation from apical portion of shoot collected from new flush of 12-year-old guava plants. The highest frequency of shoot induction was observed from nodal explants cultured on MS medium supplemented with benzylaminopurine (BAP) at 1 mg/l and polyvinyl polypyrrolidone (PVPP) at 250 mg/l (Siddiqui and Farooq, 1997). Mishra *et al.* (2005) described a complete protocol for micropropagation of guava. Recently, Mishra *et al.* (2019) and Sharma *et al.* (2018) worked out micropropagation of wilt resistant rootstock of guava.

14.2.0 Genetic Transformation

Recombinant DNA technology can facilitate precision breeding in guava particularly for developing wilt resistant scion variety having enhanced shelf life. Chandra and Mishra (2007) highlighted the significance of biotechnological interventions for improvement of guava. A robust *in vitro* regeneration system is prerequisite for developing a gene delivering system. Many workers have reported *in vitro* regeneration through enhanced axillary branching (Amin and Jaiswal, 1988; Jaiswal and Amin, 1987; Mishra *et al.*, 2007) or somatic embryogenesis (Chandra *et al.*, 2004; Chandra *et al.*, 2007). Recently, Mishra *et al.* (2014) reported genetic transformation system in guava using *in vitro* grown shoot tip explant co-cultivated with *Agrobacterium tumefaciens* strain LBA4404 harbouring binary vector pIIHR-JBMch with *endochitinase* and *nptII* genes. The highest transformation efficiency was achieved by wounding explants with tungsten particles (0.5 µm) through particle acceleration system, followed by infection for 45 minutes with *A. tumefaciens*, grown overnight with 100 µM acetosyringone, corresponding to OD600=0.5 followed by co-cultivation for 72 hours under dark condition on co-cultivation medium (MS+100 µM acetosyringone+100 mg/l L-Cystein). Putative transformed explants regenerated shoots on selection medium stressed with 200 mg/l kanamycin for 12 weeks. Molecular analysis of putative transformants by PCR confirmed the integration of *endochitinase* and *nptII* gene in the plant nuclear genome. The histopathological analysis indicated the presence of mycelium in vascular bundles. However, none of the plants showed symptoms of wilt disease during the investigation. *In vitro* pathogen inhibition assay and subsequently spore germination assay signified that the crude leaf extract of transformed plants inhibited the germination of fungal conidia. The leaf tissue studied for expression of *endochitinase* revealed that two individual transformation events manifested very high activity of N-acetyl D-glucosamine (NAG) (0.741 and 0.738 µM/min/µg of protein, respectively) which clearly indicated that transgenic lines could not develop any symptoms of wilt disease due to over expression of *endochitinase* (Mishra *et al.*, 2015).

14.2.1 Gene Cloning

The progress in the knowledge of the mechanisms and functions of genes in plants, using molecular biology techniques, contribute to increase our understanding of the role played by them and their products in plant development. The guava fruit is characterized by a high content of vitamin C, fibre and â-carotenes, all of which are desirable traits for genetic improvement programs. The enzyme hydroperoxide lyase (HPL) is widely distributed in plants and is involved in the biosynthesis of volatile aldehydes and alcohols that are important constituents in the characteristic flavour of fruits, vegetables, and leaves. Valecillos and Fermin (2010) confirmed guava HPL enzyme to share homology with cytochrome P450, placing it in the cytochrome P45074B subfamily as CYP74B5. This enzyme showed a high similarity to known HPL from other plant species even though the catalytic mechanism is still poorly understood. The same group have cloned and sequenced the HPL gene which was reported to be polymorphic, as observed in a subset from a 100 guava germplasm accessions screened from different geographical locations. Effects of HPL gene silencing in guava transformants were also attempted by them by sub-cloning a guava HPL gene into a plant transformation vector, and a transformation protocol was performed (Valecillos and Fermin, 2010).

14.3.0 Molecular Mapping

Molecular crops are starting point for map based cloning and gene isolation. Lepitre and his group (2010) have generated an integrated parental linkage map of guava (*Psidium guajava* L., 2n = 22) based on 1103 segregating AFLP from 119 AFLP primer combinations and 171 SSR marker loci with 258 allelic fragments for the mapping population MP1 derived from a cross between two heterozygous guava cultivars (Enana x N6). The integrated linkage map of MP1 included 578 markers (452 AFLPs, 126 SSRs) distributed on 11 linkage groups was of 2179 cM length with an average linkage group of 198 cM. In addition, a total of 126 so-called RF0 markers and 146 so-called associated markers were determined. This represents a decisive step towards a saturated genetic reference map of *P. guajava* in support of accelerated breeding in guava by marker-assisted selection. QTL analysis have been performed in three guava mapping populations (Enana x N6; MP1, Enana x Suprema Roja; MP2 and Enana times; Belic L-207; MP3) based on the interval mapping method (Ritter, 2010a). Sixteen different traits related to leaf and fruit morphology, fruit characteristics, fruit quality, plant height and yield were recorded in the parents and progeny genotypes of the three mapping progenies. Sufficient variation was available for all characters allowing an efficient QTL analysis. The coefficients of variation (CV) depended strongly on the particular trait under evaluation and ranged from 5.7 to 65.2 per cent. Several significant correlations were detected within leaf, fruit and yield traits which were generally similar in the different progenies. QTLs were detected for all traits under study. A total of 75, 56 and 59 QTLs were detected in populations MP1, MP2 and MP3, respectively. Individual QTLs explained between 2.6 and 31.6 per cent of the variance. QTLs were integrated in the corresponding linkage maps desplaying different traits QTLs being co-located in these maps, reflecting the observed correlations between characters. Some of the QTLs are also co-located or closely linked to SSR markers, which will allow an efficient selection for marker assisted breeding in different genetic backgrounds. The use of RAPD markers by Dominguez-Alvarez *et al.* (2010) helped in elucidation of morphological and molecular variability of guava cultivar Media China seeds subjected to growth conditions of low concentrations of Fe and Zn. The response of plants to nutritional stress and 49 RAPD loci were associated through an analysis of grouping using the NTSYS and DAR win programs. The groups of plants without symptoms were associated to the presence of RAPD loci A20-4, A20-2, and A20-1, with a molecular weight of 0.85, 1.4 and 1.6 kb and a frequency of 74 per cent. Due to the association of more than one locus with the characterized absence of symptoms, it is inferred that this characteristic is polygenic, the combination of the three mentioned loci being necessary to define the groups without symptoms for deficiency of Fe and Zn. The results suggest that RAPD markers are useful to detect genetic variability and that the changes that occur at the level of the genome not always correspond to the changes in the plant phenotype. The found variability has potential to aid in the selection of suitable genotypes as rootstocks in guava, because iron and zinc ions are the most restrictive elements in soils of alkaline pH.

15.0 References

Abdullah,T.A., Aziz, M.A., Rashid, A.A., Saleh, G., Elhory, S.M.A., Panjaitan, S.B., Hailmi, M.S. and Muzamil, M.N. (2009) *Pak. J. Biotech.*, **6**(1/2): 21-26.

Abdel aal, A. F.(1980) Orchard of evergreen fruits (Cairo, Egypt: Dar El Maref), p. 365.

Ali, N., Mulwa, R.M.S., Norton, M.A. and Skirvin, R.M. (2007) *Plant Cell Tiss. Organ Cult.*, **91**(3): 295-298.

Adak, T., Kumar, K., Singha, A., Shukla, S. K., and Singh, V. K. (2014) *The J. Animal Plant Sci.*, **24** (4): 1157-1165.

Adsule, R.N. and Kadam, S.S. (1995) In: *Handbook of Fruit Science and Technology-Production, Composition, Storage and Processing* (Eds. D.K. Salunkhe and S.S. Kadam), Marcel Dekker Inc., New York, pp. 419-433.

Ahmed, S. (1961) *Agric. Pakistan*, **12** : 745-61.

Ali-Dinar, H.M., Ebert, G. and Ludders, P. (1999) *Gartenbauwissenschaft*, **64** : 54-59.

Aly, M. D. H., Morssy, A. A., EL-Sherief, M. A., Diab, M. M. and Abo el ghar, A. I. (1972) A List of Agricultural Researches Done in Egypt Since 1900 ± 1970: *Plant Disease* (Cairo, Egypt: NIDC), p. 489.

Amin, M.N. and Jaiswal, V.S. (1988) *Sci. Hort.*, **36**: 89-95.

Anonymous (1986) In : *Food and Fruit-bearing Forest Species 3*, FAO Forestry Paper 44/3, Chap. 64.

Anonymous (1982) *Report, Fruit Res. Workshop on Subtropical and Temperate Fruits*, Nagpur, p. 119.

Anonymous, (1996) Research Programs and Progress, Indian Institute of Horticultural Research, Bangalore, pp. 10-11.

Anonymous (1996-97) *Annual Report*. CISH, Lucknow, pp. 10-11.

Anonymous (2016) *Annual Report*, CISH, Lucknow, pp. 37-38.

Anonymous (2018) *Horticultural Statistics at a Glance*, Ministry of Agriculture & Farmers' Welfare,

Horticulture Statistics Division, Govt. of India.

Arenas De Moreno, L., Mann, M., Castro De Rincon, C. and Sandoval, L. (1995) *Revista de la Facultad de Agronomia*, Universidad del Zulia, **12** : 467-483.

Arora, J.S. and Singh, J.R. (1971) *Indian J. Hort.*, **28** : 108-13.

Arshad, I. (2015) *Int. Res. J. Plant and Crop Sci.*, **2** (1): 19-23.

Ashour, F.M., Nour El-Din, H., Askar, A. and Zeddan, O. A. (1986) *22nd Int. Hort. Congress.*, California, Abst. No. 1036.

Askar, A., El-Nemr, S.E. and Bassionnyi, S.S. (1986) *Alimenta*, **25** : 162-167.

Atchison, E. (1947) *Amer. J. Bot.*, **34**: 159–164.

Azhakiamanavalan, R.S. and Khader, J.B.M. Md. A. (1981) *Nat. Symp. Trop and Subtrop. Fruit Crops*, Bangalore, p. 60.

Azzazy, M.A., Maksoud, M.A. and Haggag, L.F. (1994) *Annals Agric. Sci.*, **39** : 731-738.

Bakshi, J.C. and Randhawa, M.S. (1967) *Indian Hort.*, **11**: 3-4.

Bajpai, A., Chandra, R. Misra, M, and Tiwari, R.K. (2007) *Acta Hort.*, **735**: 145-153

Bajpai, A., Chandra, R., Rajan, S. and Srivastava, N. (2008) *Indian J. Genet.*, **68** (4): 441-445.

Bajpai, A., Srivastava, N., Chandra R. and Rajan S. (2011) In: *Molecular Markers and Marker Assisted Selection—Fruit Crops, Plantation Crops and Spices* (Volume III) (Eds. H.P. Singh, V.A. Parthasarathy, and K.Nirmal Babu) *Advances in Horticulture Biotechnology*, Westville Publishing House, New Delhi. pp. 79-90.

Balakrishnan, K., Singaravelu, M., Palanisamy, V., Thandapani, V. and Arumugam, R. (1994) *South Indian Hort.*, **42** : 271-273.

Balasubramanyam, V.R. (1959) *Ind. J. Hort.*, **16:** 69-75

Balasubrahmanyam,V.R. and Rangaswamy,G.(1959) *Curr. Sci.*, **28** : 10.

Baramanray, A., Gupta, O.P. and Dhawan, S.S. (1995) *Haryana J. Hort. Sci.*, **24** : 102-109.

Barbosa, J.T., Manica, I., Pinheiro, R.V.P., Conde, A.R. and Anderson, O. (1985) *Fruits*, **40** : 485-489.

Bardiya, M.C., Kundy, B.S. and Tauro, P. (1974) *Haryana J. Hort. Sci.*, **3** : 140.

Bihari, M., and Suryanarayan (2011) *Indian J. Agric. Sci.*, **81**(2): 107-110.

Bhargava, A.K., Sobti A.K. and Ghasolia, R.P. (2003) *J. Phytological Res.*, **16**: 81-84.

Bhatt, R.K. and Chundawat, B.S. (1982) *Nat. Sem. Pl. Propagation*, Calcutta, p. 53.

Bhatt, BB., Tomar, Y.K. and Rawat, S.S. (2013) *Prog. Hort.*, **45**(2) : 278-280.

Biale, J.B. and Barcus, D.E. (1970) *Trop. Sci.*, **12** : 93-104.

Biswas, G.C., Ahmad, M., Karim, M.A. and Mian, M.Y. (1995) *Bangladesh J. Entomol.*, **5** : 33-39.

Braganza, M.A. (1990) *M.Sc. (Ag) Thesis submitted to University of Agricultural Sciences*, Bangalore, India.

Brar, H. S. and Bal, J. S. (2003) *Ann. Agric. Bio. Res.*, **8**(2): 213-217.

Bolkhovskikh, Z., Grif, V., Matvejeva, T. and Zakharyeva, O. (1969) *J. Phytol. Res.*, **16**: 81-84.

Blommaert, K. L. J. (1958) *Fmg. S. Afr.*, **34** : 10-11.

Borthakur, P.K. and Bhattacharyya, R.K. (1993) *Proc. Indian Soc. Weed. Sci. Intern. Symp.*, Hisar, India, **3** : 219-221.

Bose, S.K. and Muller, E. (1967) *Indian Phytopath.*, **20** : 124-138.

Brahmachari, V.S., Kumar, N., and Kumar, R. (1997) *Haryana J. Hort. Sci.*, **26** : 169-173.

Brahmachari, V.S., Mandal, A.K., Kumar, R. and Rani, R. (1995) *Recent Hort.*, **2** : 127-131.

Brahmachari,V.S., Mandal,A.K., Kumar.R. and Rani, R. (1996) *The Hort. J.*, **9** : 1-7.

Broodrijk, M. (1989) *Vrugie*, No. 202, Nelspruit, South Africa.

Brown, B.I. and Wills, R.B.H. (1983) *Sci. Hort.*, **19** : 237-243.

Butani, D.K. (1974) *Pesticides*, **8** : 126-130.

Canizares, Z. J. (1981) *Ciencia y Tecnica en la Agricultura, Citricos y Otros Frutales* **4** (3-4): 7-21.

Chattopadhyay, S.B. and Bhattacharjya, S.K. (1967) *Bull. Bot. Soc. Beng.*, **21**: 107-112.

Chadha, K. L. (1998) *Indian J. Hort.*, **55**(4): 265-296

Chandra, R. and Mishra, M. (2007) *Acta Hort.*, **735**: 117-126

Chandra, R., Bajpai, A., Gupta, S. and Tiwari, R.K. (2004) *Indian J. Biotech.*, **3**: 246-248

Chandra, R., Mishra, M., Abida, M. and Sigh, D.B. (2007) *Acta Hort.*, **735**: 133-138

Cheema, G. S. and Deshmukh, G. B. (1927) *Bull. Department of Agriculture*, Bombay No. 148.

Cheema, G.S., Bhat, S.S. and Naik, K.C. (1954) In: *Commercial Fruits of India*, MacMillan & Co., New York, USA.

Coser, S.M., da Silva Ferreira, M.F., Ferreira, A., Mitre, L.K., Carvalho, C.R. and Clarindo, W.R. (2012) *Sci Hort.*, **148**: 223–229.

Costa, I.R. and Forni-Martins, E.R. (2006) *Caryologia*, Firenze, **59**: 7–13.

Costa, I.R., Dornelas, M.C. and Forni-Martins, E.R. (2008) *Plant Syst Evol.*, **276**: 209–217.

Campbell, C.W. (1963) *Proc. Fla. St. Hort. Soc.*, **76** : 363-365.

Castelan-Estrada, M. and Becerril Roman, A.E. (1994) *Proc. Inter american Soc. Tropical Hort.*, **38** : 152-56.

Chadha, K.L. and Pandey, R.M. (1986) In : *CRC Handbook of Flowering*. Vol.V (Ed. A.H. Halevy), CRC Press, Inc. Boca Raton, p. 287.

Chadha, K.L., Singh, H. and Tandon, D.K. (1981) *Abst. Nat. Symp. Trop. & Subtrop. Fruit Crops*, Bangalore, p. 17.

Chain, L.H., Ali, Z.M. and Lazan, H. (1994) *Proc. Malayasian Biochem. Soc. Conf.*, **19** : 147-50.

Chakraborty, D.K. and Singh, R.N. (1989) *Indian Phytopath.*, **42** : 310.

Chan, H.T. and Kwok, S.C.M. (1975) *J. Food Sci.*, **40** : 419.

Chandra, D. (1965) *Allahabad Fmr.*, **39** : 137-9.

Chandra, R. and Govind, S. (1990) *Progressive Hort.*, **22** : 40-43.

Chandra, R. and Govind, S. (1994) *Indian J. Hort.*, **51** : 340-345.

Chandra, R. and Govind, S. (1995) *Trop. Agric.*, **72** : 110-113.

Chandra, R., Govind, S. and Basuchaudhuri, P. (1994) *Indian J. Hill Farming*, **7** : 51-56.

Chapman, K.R., Paxton, B. and Maggs, D.H. (1986) *Australian J. Expt. Agric.*, **26** : 619-624.

Chattopadhyay, S.B. and Bhattacharjya, S.K. (1968) *Indian J. Agric.*, **38**: 65-72.

Chattopadhyay, S.B. and Bhattacharjya, S.K. (1980) *Indian J. Agric. Sci.*, **38**: 176-183.

Chattopadhyay, S.B. and Sengupta, S.K. (1955) *Indian J. Hortic.*, **12**: 76-79.

Chaudhuri, S.M., Desai, U.T. and Kale, P.N. (1994) *J. Maharashtra Agric. Univ.*, **19** : 458-459.

Cheema, G.S. and Deshmukh, G.B. (1927) *Bombay Department of Agriculture, Bullletin* No. 148.

Cheema, G.S., Bhat, S.S. and Naik, K.C. (1954) *Commercial Fruits of India with Special Reference to Western India.* Macmillan & Co., pp. 253-254.

Cheng, M., Lee, S. and Teck, C. (2011) In: *Taiwan guava production manual.* Cayo District: Horticulture Crop Training and Demonstration Center, pp. 7-15

Chetri, K., Kar, P.L. and Sanyal, D. (1996) *The Hort. J.*, **9** : 21-26.

Chohan, G. S., Sharma, J.N. and Thatai, S. K. (1978) *Indian J. Hort.*, **35**: 212-14.

Chohan,G.S., Kumar, H. and Vij,V.K. (2000) *Indian J. Hort.*,**57**: 54-58.

Chonkar, V.S. and Singh, P.N. (1981) *Abst. Nat. Symp. Trop. and Subtrop, Fruit Crops,* Bangalore, p. 53.

Chundawat, B.S., Gupta, O.P. and Godara, N.R. (1975) *Haryana J. Hort. Sci.*, **4** : 23-25.

Chundawat, B.S., Kikani, K.P., Verma, L.R. and Jadav, R.G. (1992) *Indian J. Hort.*, **49** : 134-137.

Chundawat, B.S., Singh, J.P., Kainsa, R. and Gupta, O.P. (1976) *Haryana J. Hort. Sci.,* **5** : 130-136.

Cobley, L.S. and Steele, W.M. (1976) *An Introduction to the Botany of Tropical Crops,* Longman, London and New York.

Cole, J.T., Cole J.C. and Conway, K.E. (2005) *J. Appl. Hort.*, **7**: 16-19.

Dahiya, K.K., Lakra, R.K. and Rana, J.S. (1995) *Haryana J. Hort. Sci.*, **24** : 239-241.

Dahiya, S.S. and Joon, M.S. (1995) *Crop Res.*, **9** : 121-122.

Dalaya, Y.P., Rajput, S.G., Mali, A.R. and Mohite, P.B. (1983) *Indian J. Plant Prot.*, **11** : 138-139.

Das, B.C., Chakraborty, A., Chakraborty, P.K., Maiti, A., Mandal, S. and Ghosh, S. (1995) *The Hort. J.*, **8** : 141-146.

Das Gupta, S.N. and Rai, J.N. (1947) *Curr. Sci.*, **16**: 256-258.

Das, R.C. and Acharya, B.N. (1969) *Plant Sci.*, **1** : 233-238.

Dasrathy, T.B. (1951) *Madras Agric. J.*, **38** : 521-527.

Daulta, B.S., Godara, N.R., Bhatia, S.K., Singh, H.K. and Bisla, S.S. (1998) *Indian Hort.*, **43**: 22-23.

DeCruz, R. and Babu Rao, G. (1962) *J. Indian Bot. Soc.*, **41** : 316-321.

Dedolph, R.R. and Bowers, F.A. (1960) *Hawaii Fm. Sci.*, **9** : 4-5.

Dhankar, O.P., Makhijha, M., Chauhan, K.S. and Singhrot, R.S. (1981) *Trans. Indian Soc. Desert Technology*, Univ. Centre Desert Studies, **6** : 33.

D'Cruz, R. and Rao, G.B. (1962) *J. Indian Bot. Sci.*, **41**: 316–321.

Dahiya, K.K., Archak, S. and Karihaloo, J.L. (2002) *Indian J. Plant Genet. Resour.*, **15**(2): 112-115.

De Souza, Gracas, A.D., Resende, L.V., De Lima, I.P., Semen, L.S. M and Techio, V.H. (2015) *Plant Syst Evol.*, **301**: 231–237.

Dinesh, M.R. and Vasugi, C. (2010a) *J. Hort. Sci.*, **5**: 94-108.

Dinesh, M.R. and Vasugi, C. (2010b) *Indian J. Agri. Sci.*, **80**(11): 62-63.

Dhomane, P.A. and Kadam. A.S. (2013) *Sch. J. Agric. Sci.* 3(7): 261-263.

Dhua, R.S., Mitra, S.K. and Sen, S.K. (1982) *Nat. Sem. Pl. Propagation*, Calcutta, p. 36.

Doraipandian, A. and Muthukrishnan, C.R. (1973) *South Indian Hort.*, **20** : 24-28.

Dutta, P. and Mitra, S.K. (1991) *Indian Agric.*, **35** : 101-105.

Dutta, P., Banik, A.K., Raychaudhury.R. and Dhua, R.S. (1991) *Indian J. Hort.*, **48** : 213-216.

Dutta, S. (1953) *Indian J. Hort.*, **19** : 137-139.

Dwivedi, R., Pathak, R.K. and Pandey, S.D. (1990) *Prog. Hort.*, **22** : 134-139.

Dwivedi, S.K. (1993) *Soil Biol. Biochem.*, **25** : 1635-1636.

Edward, J.C. and R.N. Srivastava (1957) *Allahabad Fmr.*, **31**: 144-146.

Edward, J.C. and Shankar, G. (1964) *Allahabad Fmr.*, **38** : 249-250.

El Boushy, A.R.Y. and van der Poel, A.F.B. (2000) In: *Handbook of Poultry Feed from Waste: Processing and Use,* Springer-Verlag, New York.

El-Buluk, R.E., Babikar, E.E. and El-Tinay, A.H. (1995) *Food Chem.*, **54** : 279-282.

El-Buluk, R.E., Babikar, E.E. and El-Tinay, A.H. (1996) *Plant Food Human Nutr.*, **49** : 147-154.

El-helaly, A. F., Ibrahim, I. A., Assawah, M. W., El-Arosi, H. M., Abou-EL-Dahab, M. K., Michail, S. H. and ABD-AL-Rahim, M. A. (1963) *Alexandrian J. of Agri. Res. Bulletin.*, **8** (1) : 180.

El-helaly, A. F., Ibrahim, I. A., Assawah, M. W., El-Arosi, H. M., Abou-EL-Dahab, M. K., Michail, S. H. and ABD-AL-Rahim, M. A., Wasfy, E. H. and El-Goorani, M. A. (1966) *Alexandrian J. of Agri. Res. Bulletin.*, **15** (1): 154.

Edward, J.C. (1961) *Allahabad Fmr.*, **35** : 5-9.

Edward, J.C. and Gaurishanker (1964) *Allahabad Fmr.*, **38** : 249-250.

Edward, J.C. and Srivastava, Q.N. (1957) *Allahabad Fmr.*, **31** : 144-146.

Fernadezdiaz-Silveira, M. (1975) *Agricultura*, **3**: 80-85.

Feungchan, S., Yimsawat, T. and Chindaprasert, S. (1992) *Khon Kaen Agric. J.*, **20**: 249-252.

Gadgil, D.R. and Gadgil, V.R. (1933) *A survey of the Marketing of Fruit in Poona*, Gokhale Inst. Politics and Economics, Publ. No. 3.

Gaikwad, N.R., Yadav, E.D. and Patil, A.V. (1981) *Nat. Symp. Trop. and Subtrop. Fruit Crops*, Bangalore, p. 30.

Gardner, V.R., Bradford, F.C. and Hooker, H.D. (1952) In: *The Fundamentals of Fruit Production*, Mcgraw-Hill, New York.Gaur, G.S., *Recent Hort.*, **3** : 21-23.

Gautam, N.N., Singh, K., Singh, B., Seal, S, Goel, A. and Goel, V.L. (2010) *Australian J. Crop Sci.*, Lismore,**4**(9): p.666.

Gerhardt, L.B. DE. A., Manica, I., Kist, H. and Sieler, R.L. (1997) *Pesquisa Agropecuaria Brasileira*, **32** : 185-192.

Ghosh S.K. (1980) *M.Sc. (Ag.) Thesis* submitted to Bidhan Chandra Krishi Viswavidyalaya, Kalyani.

Ghosh, S.N. (1978) *M.Sc. (Ag.) Thesis* submitted to Bidhan Chandra Krishi Viswavidyalaya, Kalyani.

Ghosh, S.N. (1991) *Indian Agric.*, **35** : 231-238.

Ghosh, S.N. and Chattapadhyay, N. (1996) *The Hort. J.*, **9** : 121-127.

Ghule, B.D. and Dhumal, V.S. (1992) *J. Maharashtra Agric. Univ.*, **17** : 321-322.

Golberg, L. and Levy, L. (1941) *Nature*,**148** : 286.

Gonzalez, G. and Sourd, D. (1982) *Ciencia Y Tecnica en la Agricultura, Citricos Y Ortros Frutales*, **5** : 39-51.

Gopikumar, K., Sulladmath, U.V. and Badiger, M.K. (1985) *Singapore J. Prim. Indus.*, **13** : 112-127.

Gotur, Mutteppa, Sharma, D.K., Chawla, S.L., Joshi, C.J. and Navya, K. (2017) *Plant Archives*. **17**(2): 1283-1287.

Gupta, M.R. and Niggar, G.S. (1978) *Indian J. Hort.*, **35** : 12-16.

Gupta, V.K., Misra, A.K. and Gaur, R.K. (2010) *J. Plant Protec. Res.*, **50**: 4.

Gupta, V.K., Misra, A.K., Gaur, R.K., Jain, P.K., Gaur, D. and Sharma, S. (2010) *Biotechnology*, **9**: 176-195.

Goldblatt, P. and Johnson, D.E. (1990) In: *Monographs in Systematic Botany from the Missouri Botanical Garden*, **30**: 1-243.

Gonzaga Neto, L. (1999) In: Recursos gen´eticos e melhoramento de plantas para o Nordeste brasileiro (on line).(Eds. M.A. Queiroz, C.O. de Goedert, S.R.R.Ramos), Versao 1.0. Petrolina, PE: Embrapa Semi-´Arido/Bras´ýlia: Embrapa Recursos Gen´eticos e Biotecn

Guerra, M. (1988) Introducao a citogenetica. *Guanabara*, Rio de Janeiro

Hafeez-ur-Rahman, Khan, M.A., Khokhar, K.M. and Laghari, M.H. (1991) *Indian J. Agric. Sci.*, **61** : 404-406.

Haggag, L.F. and Maksoud, M.A. (1996) *Egyptian J. Hort.*, **23** : 203-209.

Haq, F., Khan, M.S. and Faridullah, I. (1973) *W. Pakistan J. Agric. Res.*, **11** : 121.

Hardenburg, R.E., Watada, A.E. and Wang, C.Y. (1990) *The Commercial Storage of Fruits, Vegetables and Florist and Nursery Stocks*. USDA, ARS, Agric. Handb. 66, pp. 130.

Hayes, W.B. (1970) *Fruit Growing in India*, Kitabistan, Allahabad.

Hegde, H.G. and Tiwari, J.P. (1981) *Nat. Symp. Trop. and Subtrop. Fruit Crops*, Bangalore, p. 83.

Hoque, M.A. and Irabagon, J.A. (1994) *South Indian Hort.*, **42** : 289-292.

Hossain, M.S. and Meah, M.B. (1992) *Tropical Pest Management*, **38** : 181-185.

Hugar, L.B., Murthy, P.S.S., Umesh, K.B. and Reddy, B.S. (1991) *Agricultural Situation in India*, **46** : 211-214.

Hussain, M.A., Puttaswamy, Viraktamath, C.A. (1996) *Insect Environ.*, **2** : 73-74.

Hernandez-Delgado, S., Padilla-Ramirez, J.S., Nava-Cedello, A. and Mayek-Perez, N. (2007) *Plant Genetic Resources: Charecterization and Utilization*, **5**(3): 131-141.

Hulamani, N. C., Mokashi, A. N., Sulikeri, G. S. and Rao, M. M. (1981) National Symposium on Tropical Subtropical Fruit Crops, Bangalore, p. 18

Ibrahim, S. (1943) *Madras Agric. J.*, **31** : 279-282.

Itoo, S.,Yamaguchi, T., Oohata, T. and Ishihama, K. (1980) *Bull. Kagoshima Univ. Japan*. **30** : 47-54.

Iyer, C. P. A. and Subramanyam, M. D. (1988) IIHR Selection-8, an improved guava cultivar. *South Indian Hort.*, **36**(5): 258-259.

Jaiswal, V.S. and Amin, M.N. (1987) *In vitro* propagation of guava from shoot culture of mature trees. *J. Pl. Physiol.*, **130**: 7-12.

Jaffe, A. (1970) *Pl. Propagator*, **16** : 6.

Jain, S.S. (1966) *Indian J. Hort.*, **13** : 102-104.

Jauhari, O.S. and Singh, R.D. (1973) *Indian J. Hort.*, **30** : 9-10.

Jawanda, J.S. (1961) *Punjab Hort. J.*, **1** : 150.

Kakon, A.J., Rahim, M.A. and Alam, M.S. (2005) *J. Subtropical Agri. Res. Dev.*, **3**(1): 34-38.

Kalra, S.K., Sidhu, P.J.S., Dhaliwal, G.S. and Singh, R. (1994) *Indian J. Hort.*, **51** : 272-274.

Kalyani, M.; Bharad, S.G.; Parameshwar, P. (2014) *Intl. J. Biol. Sci.*, **5**(2): 81-91.

Kareem Abdul, Manan Abdul, Saeed Shafqat, Rehman Shoaib Ur, ShahzadUmbreen and Nafees Muhammad (2016) *J. Agric. Environ. Intl. Develop.* - JAEID, **110**(2): 197-203.

Kobayashi, K.D. (1986) *Abst. 22nd. Int. Hort. Cong.*, California, No. 216.

Koj Tassar, Tiwari, J.P. and Shant Lal (1989) *Prog. Hort.*, **21** : 213-217.

Kundu, S. and Mitra, S. K. (1994) *Crop Res.*, **8**(1): 80-85.

Kwee, L.T. and Chong, K.K. (1990) In: *Guava in Malaysia – Production, Pests and Diseases*, Tropical Press, Kuala Lumpur, pp. 21–51.

Kumar, D., Pandey, V., Anjaneyulu, K. and Vishal Nath (2009) *Indian J. Hort.*, **66**(1): 18-21.

Kumar, R. and Hoda, M.N. (1977) *Indian J. Hort.*, **34** : 13-14.

Kumar, R., Kotur, S.C. and Singh, H.P. (1995) *Indian J. Hort.*, **52**: 254-258.

Kumar, R., Kotur, S.C. and Singh, H.P. (1996) *J. Potassium.* Res., **12** : 59-64.

Kundu, S., and Mitra, S.K. (1994) *Crop Res.*, **8** : 80-85.

Kundu, S., Ghosh, S.N. and Mitra, S.K. (1993) *Indian Agric.*, **37** : 157-62.

Kundu, S., Ghosh, S.N. and Mitra, S.K. (1995) *Indian Fd. Packer*, pp.11-16.

Kundu, S., Mitra, S. and Mitra, S.K. (1998) *The Hort. J.*, **11** : 91-96.

Lepitre, V., Nansot, G., Grangeon, R., Pomies, V., Rivallan, R., Risterucci, A.M., ValdésInfante, J., Rodríguez-Medina, N.N., Muth, J., Boike, J., Prüfer, D., Becker, D., Rohde, W., Ritter E. and Billotte, N. (2010) *Acta Hort.*, **849**: 183-191.

Leu, L.S. and Kao, C.W. (1979) *Plant Dis. Report*, **63**: 1077-1079.

Lal, G. (1996) *Annals of Biol.*, **12** : 238-241.

Lal, S., Tiwari, J.P., Awasthi, P. and Singh, G. (2007) *Acta Hortic.*, **735**: 193-196.

Lazan, H. and Ali, Z.M. (1989) In : *Research Imperative in Life Sciences* (Eds. H. Zakri and A. Latiff), University Kebangsaan Malaysia Press, Bangi, Malaysia, pp. 125-150.

Le-Riche, F.J.H. (1952) *Sci. Bull. Dep. Agric. South Afr.*, No. 286.

Leu, L.S. and Kao, C.W. (1979) *Plant Dis. Reptr.*, **63** : 1077-1079.

Lyannaz, J.P. (1994) *Fruits*, **49** : 353-534.

Maas, E.V. (1984) In : *Handbook of Plant Sciences in Agriculture*. Vol. II, (Ed: B.R. Christie) CRC Press, Inc. Boca Raton, FL.

Macmillan, H.G. (1914) *A Handbook of Tropical Gardening and Planting*, Macmillan Co., New York, pp. 183-184.

Madhavi, D.L., Chand, N., Rajalakshmi, D. and Patwardhan, M.V. (1992) *J. Sci. Food Agric.*, **58** : 29-34.

Mahendra, J., Yogesh, S., Singh, S.S. and Mishra S.P. (2015) *Amer. Int. J. Res. Formal Appl. Naturl. Sci.*, **9**(1): 30-32.

Mahoney, M.J. and Tattar, T.A. (1980) *Plant Dis.*, **64**: 854-856.

Majumder, P.K. and Mukherjee, S.K. (1968) *Indian Hort.*, **12** : 11.

Majumder, P.K. and Mukherjee, S.K. (1972a) *Cytologia*, **37**: 541-548.

Majumder, P.K. and Mukherjee, S.K. (1972b) *Nucleus*, **13**: 42-47.

Makhija, M., Dhankar, O.P. and Singhrot, R.S. (1980) *Haryana J. Hort. Sci.*,**9** : 21.

Malo, S.E. and Cambell, C.W. (1986) *The Guava*. Univ. Fla. Institute of Food and Agricultural Sciences, Fruit Crops Fact Sheet FC-4.

Manga, B. and Jholgiker, P. (2017) *Int. J. Curr. Microbiol. App. Sci.*, **6**(6): 2792-2795.

Mani, M. (1994) *J. Entomological Res.*, **18** : 351-355.

Manoj, B. and Rajesh, K. (2013) *Indian J. Agric. Res.*, **47**(4): 347-352.

Marlatt, R.B. and Campbell, C.W. (1981) *Proc. Fla. St. Hort. Soc.*, **93** : 109-110.

Marler, T.K. (1994) In : *Handbook of Environmental Physiology of Fruit Crops*. Vol. II *Sub-Tropical and Tropical Crops* (Eds. B. Schaffer and P.C. Andersen), CRC Press, Inc., Boca Raton, FL, pp. 213-216.

Martinez-De-Lara, J.; Barrientos-Lara, M.C.; Reyes-De-Anda, A.C.; Hernandezdelgado, S.; Padilla-Ramírez, J.S.; Perez, N.M. (2004) *Revista Fitotecnia Mexicana, Chapingo*, **27** (3): 243-249.

Mathur, R.S. (1956) *Indian J. Hort.*, **13**: 26-29.

Marwat, N.K., Hussain, N. and Khan. A. (1992) *Pakistan J. Zoology*, **24** : 82-84.

Maurya, R.K., Ray, N.R., Chavda, J.C., Chauhan, V.B. and Patil, A.K. (2012) *Asian J. Hort.*, **7**(1): 44-47.

Maxwell, L.S., Krezdon, A.H., Will, A.A. and Golby, E.V. (1984) *Florida Fruits Maxwell*, Tempa, FL.

McLeod, A.J. and de Troconis, N.G. (1982) *Phytochem.*, **21** : 2523-2526.

Mehta, N. (1987) *Plant Dis. Res.*, **2**: 116-119.

Mehta, P.K., Kachroo, A., Kaul, M.K. and Yamdagni, R. (1988) *Agric. Rev.*, **9** : 57.

Menzel, C.M. and Paxton, B.F. (1986) *Australian J. Exp. Agric.*, **26** : 123-128.

Mercantilia (1989) *Guide to Food Transport-Fruit and Vegetables*, Mercantilia Publisher.

Mishra, A.K. (1996) In : *Advances in diseases of Fruit Crops in India* (Ed. S.J. Singh), Kalyani Pub. Ludhiana, pp. 183-190.

Mishra, R.S. and Singh, S.P. (1995) *Horticulturist*, **4** : 32-33.

Misra, A.K. (1987) Studies on Diseases of Fruits Crops. CIHNP, Lucknow, pp: 124-125.

Misra, A.K., Prakash, O. and Sen, B. (2000) *Proceedings of the National Seminar on Hi-tech Horticulture*, June 26-28, Banglore, pp: 149-149.

Misra, A.K. and Prasad, B. (2003) *J. Mycol. Plant Pathol.*, **33**: 494-494.

Misra, A.K., Pandey, B.K. and Prakash Om (2003b) In: *Souvenir and Abstract,Zonal Conference (East Zone)*, Indian Society of Mycology and Plant Pathology and Seminar on Plant Diseases of National Importance with Special Reference to Guava Wilt and Mango Malformation. 4-5 April, 2003, CISH and IISR. Lucknow p. 23.

Misra, A.K. and Prasad, D. (2004) *Proceedings of the Symposium on Recent Advances in Fungal Bioagents and Their Social Benefits*, Sept. 10-10, NBRI, Lucknow, pp. 12-13.

Misra, A. K. (2005) In: *Souvenir of the Ist International Guava Symposium*.December 5-8, 2005. CISH, Lucknow. pp. 75-90.

Majumdar, P. K. and Singh, R.N. (1974) *Curr. Sci.*, **33** : 24-25.

Majumder, P.K. and Mukherjee, S.K. (1972a) *Cytologia*, **37**: 541-548.

Majumder, P.K. and Mukherjee, S.K. (1972b) *Nucleus*, **13**: 42-47.

Mathur, R.S. and Jain, S.S. (1960) *Proc. Nat. Acad. Sci.* B., **30** : 33-36.

Mehmood, A., Jaskani, M.J., Khan, J.A., Ahmad, S., Ahmad, R., Luo, S. and Ahmad, N.M. (2014) *Sci. Hort.*, **172**: 221-232.

Mehmood, A., Jaskani, M.J., Ahmad, S. and Ahmad, R. (2014) *Pak. J. Agri. Sci.*, **50**(4): 591-597.

Milan, A.R. (2007) *Acta Hort.*, **735**: 61-69.

Mishra, M., Chandra, R., Pati, R. and Bajpai, A. (2007) *Acta Hort.*, **735**: 155-160.

Mishra, M., Jalil, S.U., Mishra, R.K. Kumari, S and Pandey, B.K. (2016) *Czeck J. Plant Breed.*, **52** (1): 6-12.

Mishra, M., Jalil, U., Sharma, N. and Hudedamani, U. (2014) *Crop Breed. Biotech.*, **14:** 232-237.

Mitra, S. K. and Bose, T. K. (1985) In: Fruits of India: Tropical and Subropical (ed. T. K.Bose), Naya Prakash, Calcutta, pp. 277-297.

Moore, R. J. (1977) Index to plant chromosome numbers 1967-1971. International Association for Plant Taxonomy.

Morton, J. (1987) In: *Fruits of Warm Climates* (ed. Julia F. Morton), Miami, USA. pp. 356–363.

Mitra, S.K. (1983) *Ph. D. Thesis* submitted to Bidhan Chandra Krishi Viswavidyalaya, Kalyani.

Mitra, S.K. (1987) *J. Potassium Res.*, **3** : 160-163.

Mitra, S.K. (1996) *Proc. Nat. Sem. on Plant Bioregulators in Horticulture*, BCKV, Kalyani, pp. 70-75.

Mitra, S.K. and Bose, T.K. (1985) *South Indian Hort.*, **33** : 286-292.

Mitra, S.K. and Bose, T.K. (1999) In: *Tropical Horticulture*, Vol. I, (Eds. T.K.Bose, S.K.Mitra, A.A.Farooqui and M.K.Sadhu), Naya Prokash, Calcutta, pp. 297-307.

Mitra, S.K., Dhua, R.S. and Ghosh, S.K. (1988) *Intensive Agric.*, **26** : 26-27.

Mitra, S.K., Ghosh, S.K. and Dhua, R.S. (1984b) *Sci. Cult.* **50** : 235-236.

Mitra, S.K., Sanyal, D. and Bose, T.K. (1995) *Proc. Nat. Sem. Advances in Res. and Dev. in Horticulture for Export.* Hisar, Haryana, pp. 171-176.

Mitra, S.K., Maiti, S.C., Sen, S.K. and Bose, T.K. (1983) *South Indian Hort.*, **31** : 62-65.

Mitra, S.K., Sen, S.K., Maiti, S.C. and Bose, T.K. (1981) *Nat. Symp. Trop. and Subtrop. Fruit Crops.* Bangalore, p. 17.

Mitra, S.K., Sen, S.K., Maiti, S.C. and Bose, T.K. (1984a) *Bangladesh Hort.*, **12** : 7-9.

Mohan, C., Jhooty, J.S. and Chand, T. (1986) *Plant Dis. Res.*, **1**: 77-78.

Morton, J.F. (1987) *Fruits of Warm Climates, Creative Resource Systems.* Inc., Winterville, NC.

Mukherjee, S.K. and Dutta, M.N. (1967) *Curr. Sci.*, **36** : 674-675.

Mushtaq, Kurubar, A. R., Umesh, M. R., Patil, S. and Hugar, A. (2019) *J. Appl. Natur. Sci.*, **11**(2) 440-444.

Nagar, P.K., and Rao, T.R. (1986) *Sci. Hort.*, **29** : 139-146.

Naik, K.C. (1949) *South Indian Fruits and Their Culture*, Varadachary and Co., Madras, pp. 448-450.

Nand, D., Shanker, G. and Srivastava, A.K. (1991) *Indian Hort.*, **36** : 4-5.

Nandi, S. (1983) *M. Sc. (Ag.) Thesis* submitted to Bidhan Chandra Krishi Viswavidyalaya, West Bengal, India.

Natale, W., Coutinho, E.L.M., Boaretto, A.E. and Banzatto, D.A. (1994a) *Revista de Agricultura*, **69** : 247-255.

Natale, W., Coutinho, E.L.M., Boaretto, A.E. and Pereira, F.M. (1994b) *Fruits*, **49** : 205-210.

Natale, W., Coutinho, E.L.M., Boaretto, A.E. and Pereira, F.M. (1996a) *Hortic. Basil.*, **14** : 196-198.

Natale, W., Coutinho, E.L.M., Boaretto, A.E., Pereira, F.M., Oioli, A.A.P. and Sales, L. (1996b) *Revista Brasileira de Ciencia do Solo*, **20** : 247-250.

Nathani, S.P. and Srivastava, H.C. (1965) *Allahabad Fmr.*, **41** : 65-68.

Nawar, A. and Ezz, T. (1994) *Alexandria J. Agric. Res.*, **39** : 315-329.

Nakasone, H.Y. and Paull, R.E. (1998) In: *Tropical Fruits* (Eds. H.Y. Nakasone and R.E. Paull,), Wallingford, CAB International. pp. 149–172

Negi, S.S. and Rajan, S. (2007) *Acta Hort.*, **735**: 31-37.

Negi, S.S., Misra, A.K. and Rajan, S. (2001) *Indian J. Hort.*, **58**(1/2): 145-151.

Noqueira, A.M., Ferreira, M.F., Guilhen, J.H. and Ferreira, A. (2014) *Genet. Mol. Res.*, **18**: 13(4): 10657-68

Normand, F. (1994) *Fruits*, **49** : 217-227.

Nelson, R.O. (1954) *Proc. Fla. St. Hort. Soc.*, **67** : 228-251.

Normand, F. (1994) *Fruits*, **49** : 217-27.

Ojha, A.P., Tiwari, J.P. and Mishra, K.K. (1986) *Prog. Hort.*, **8**: 205-06.

Ogden, M.A.H., Jackson, L.K. and Campbell,C.W. (1981) *Proc. Fla. St. Hort. Soc.*, **94** : 222.

Oosthuizen, S. (1985) *Inf. Bull. Citrus and Subtrop. Fr. Res. Inst.*, No. 159, p. 19.

Omayio, D.G., Abong, G.O., Michael, W.O., Gachuiri, C.K. and Mwang'ombe, A.W. (2019) *Curr. Agric. Res. J.*, **7** (3): 318-331.

Pal, R.K. and Buescher, R.W. (1993) *J. Fd. Sci. Technol.*, **16** : 115-116.

Pandey, R.R. and Dwivedi, R.S. (1985) *J. Phytopathol.*, **114**: 243-248.

Patel, N.B. and Pasaliya, Y.M. (1995) *Recent Hort.*, **2** : 33-36.

Patel, R.K., Bose, U.S. and Tripathi, S.K. (1996) *Crop Res.*, **12** : 56-60.

Pathak, R.K. and Saroj, P.L. (1995) *Crop Res.*, **10** : 318-323.

Pathak, V.N. (1980) In: *Diseases of Fruit Crops*, Oxford & IBH Publishing Co., New Delhi,pp. 151-164.

Pathak, R.A. and Dwivedi, R. (1988) Report, Fruit Research Workshop Subtropical and Temperature Fruits. Rajendra Agricultural University, Pusa, pp. 76-77.

Pathak, R.K. and Ojah, C.M. (1993) In: Genetic resources of guava. Advances in Horticulture: Fruit Crops – Vol. 1, New Delhi, India, Malhotra Publishing House, pp. 143–147.

Pereira, F.M. and Nachtigal, J.C. (2002) In: *Melhoramento de Fruteiras Tropicais* (Ed. C. H. Bruckner,), Vicosa, UFV, pp. 267–289.

Phadnis, N.A. (1970) *Indian J. Hort.*, **27**: 99-105.

Pommer, C.V. and Murakami, K.R.N. (2009) In: *Breeding Plantation Tree Crops* (eds. S.M. Jain and S.Priyadarshan), Springer, NY, pp.83-120.

Prakash, D.P., Narayanaswamy, P. and Sondur, S.N. (2002) *J. Hort. Sci. Biotechnol.*, **77**: 287–293.

Patil, P.K., Patil, V.K. and Ghonsikar, C.P. (1984) *Intern. J. Trop. Agric.*, **2** : 337.

Paull, R.E. and Bittenbender, H.C. (2006) In: *The encyclopedia of fruit & Nuts* (Ed. J. Janick), Cambridge: Cambridge University pp.541-549.

Pereira, F.M. (1990) *Acta Hort.*, **275**: 103-109.

Pensburg, N. van. and Preez, R. Du (1985) *Information Bull. Citrus and Subtrop. Fr. Res. Inst.*, No. 153, p. 8.

Pereira, F.M. (1984) *Expresa de Pesquisa Agropecuaria* S.A., **2** : 524-528.

Pereira, F.M., Oioli, A.A.P. and Banzatto, D.A. (1983) *Cientifica*, **11** : 239-244.

Pereira, F.M., Petrechen, E. DE H., Benincasa, M.M.P. and Banzatto, D.A. (1991) *Cientifica*, **19** : 199-206.

Phandis, N.A. (1970) *Indian J. Hort.*, **27** : 99-105.

Pinera, R., Hombre, R.De, Batista, A. and Cerezal, P. (1997) *Alimentaria*, **35** : 19-20.

Popenoe, W. (1920) *Manual of Tropical and Subtropical Fruits*, Macmillan Co., New York.

Prakash, Hari (1992) *Ph.D. Thesis* submitted to G.B. Panth Univ. of Agriculture and Technology, Panthnagar

Prakash, Hari and Tiwari, J.P. (1993) *Intl. Conf. on Biotechnology in Agriculture and Forestry*, New Delhi.

Prakash, Hari and Tiwari, J.P. (1996) *Proc. Nat. Symp. Hort. Biotech.*, Indian Institute of Hort. Research, Bangalore, p. 4.

Rahman, H.U., Khan, M.A., Khokhar, K.M. and Laghari, M.H. (1991) *Indian J. Agric. Sci.*, **61** : 404-406.

Rajput, C.B.S. and Chand, S. (1976) *Bangladesh Hort.*, **3** : 22-27.

Rajput, C.B.S., Singh, S.N. and Singh, N.P. (1977) *Haryana J. Hort. Sci.*, **6** : 117-119.

Rajan, S. (2008) In: *Recent Initiatives in Horticulture* (Eds. K.L. Chadha, A.K. Singh, and V.B. Patel), The Horticulture Society of India, New Delhi, pp. 669-677.

Rajan, S. (2013) In: *Geospatial Technologies for Natural Resources Management* (Eds. S.K. Soam, P.D. Sreekanth, and N.H. Rao), New India Publishing Agency, Pitam Pura, New Delhi, pp. 305–320.

Rajan, S., Kumar, R., Mishra, A. K. and Negi, S. S. (2005a) *Indian J. Plant Genet. Resour.*, **18** (3), p. 301.

Rajan, S., Negi, S.S. and Kumar, R. (1996) In: *Breeding high yielding red coloured guava varieties. 2nd Int. Crop Sci. Cong.*, Abstracts New Delhi. p. 279.

Rajan, S., Negi, S.S. and Ram Kumar (1997) Improvement of guava by selection and hybridization, *Ann. Rep. Central Inst. Subtrop. Hort.*, Lucknow, 1996-97, p.10.

Rajan, S., Yadava, L. P., Kumar, R. and Saxena, S.K. (2005b) *J. Appl. Hort.*, **7**: 52–54.

Rajan, S., Yadava, L.P. and Kumar, R. (2012) *Int. J. Innovative Hort.*, **1**(1): 18-23.

Rajan, S., Yadava, L.P., Kumar, R. and Saxena, S.K. (2005c) *Acta Hort.*, **735**: 109–113.

Rajan, S., Yadava, L.P., Kumar, R. and Saxena, S.K. (2007a) *Indian J. Hort.*, **64**(3) : 290–293.

Rajan, S., Yadava, L.P., Kumar, R. and Saxena, S.K. (2007b) *Indian J. Agri. Sci.*, **77**(12): 828–833.

Rajan, S., Yadava, L.P., Kumar, R., Saxena, S.K. (2008) *Sci. Hort.*, **116** (1): 47-51

Raghava, M. and Tiwari, J.P. (2008) *Indian J. Hort.*, **65** (3): 263-370.

Rama Rao, M. and Dayanand, T. (1977) *Andhra Agric. J.*, **24** (1-2): 53 -54.

Raman, V. S., Alikhan, W. M., Manimekalai, G. and Ramalingam, R. S. (1969) *Madras Agric. J.*, **56**: 255-161.

Rangacharlu, V.S. (1954) *Andhra Agric. J.*, **1**: 105-109.

Rensburg-N-van and Preez-R-du (1985) *Information Bulletin*, Citrus and Subtropical Fruit Research Institute 1985, No.153, 8.

Ribeiro, I.J.A. and Pommer, C.V. (2004) *Acta Hort.*, **632**: 75-78

Ritter, E., Rodríguez-Medina, N.N., Velásquez, B., Rivero, D., Rodríguez, J.A., Martínez, F. and Valdés-Infante, J. (2010) *Acta Hort.*, **849**: 193-202.

Rodríguez-Medina, N. N., Valdés-Infante, J., González, G., Fuentes, V. and Cañizares, J. (2010) *Acta Hort.*, **849**: 341-348.

Risterucci, A.-M., Duval, M.F., Rohde, W. and Billotte, N. (2005) *Mol. Ecol. Notes*, **5**: 745-748

Ruehl, G.W. (1946) *Proc. Fla. St. Hort. Soc.*, pp. 127-131.

Ramansomayazulu, M.V. (1953) *Indian J. Hort.*, **10** : 154-156.

Ramirez Villalobos, M.del C. and Salazar Yamarte, *e.g.* (1998) *Revista de la Facultad de Agronomia, Universidad del Zula*, **15** : 211-221.

Rangaswami, G. and Kaliperumal, T.T. (1961) *Curr. Sci.*, **29** : 485-486.

Rao, S.N. and Rao, Bhakara P. (1960) *Andhra Agr. J.*, **7** : 82-90.

Rao, V.K.R.V. (1946) *The Food Statistics of India*, Govt. of India Press, New Delhi.

Rathore, D.S. (1972) *Ph. D. Thesis* submitted to Agra University, India.

Rathore, D.S. (1975) *Progressive Hort.*, **7** : 63-65.

Rathore, D.S. (1976) *J. Hort. Sci.*, **51** : 41-47.

Rathore, D.S. (1978) *Indian J. Hort.*, **35** : 109-111.

Rathore, D.S. (1984) *Punjab Hort. J.*, **24** : 75-78.

Rathore, D.S. (1998) In : *50 Years of Horicultural Research* (Eds. S.P. Ghosh, P.S Bhatnagar and N.P. Sukumaran), ICAR, New Delhi, pp. 141-149.

Rathore, D.S. and Singh, R.N. (1974) *Indian J. Hort.*, **31** : 331-336.

Rathore, D.S. and Singh, R.N. (1976) *Indian J. Hort.*, **33** : 7-13.

Ray, P.K. (1999) In: *Tropical Horticulture,* vol. I. (Eds : T.K. Bose, S.K. Mitra, A.A. Farooqui and M.K. Sadhu), Naya Prokash, Calcutta, pp. 102-177.

Reyes, M.U. and Pauli, R.E. (1995) *Post-harvest Biol. Tech.*, **6** : 357-365.

Rodriguez, H., Pedrera, B. and Carrera, N. (1984) *Ciencia Tecnica en la Agricultura*, **7** : 83-97.

Rodriguez, N., Gordillo, M. and Caballero, S. (1983) *Ciencia Tecnica en la Agricultura*, **6** : 39-48.

Roy, R.S. and Ahmed, S.F. (1951) *Indian J. Hort.*, **8** : 22-23.

Ruehle, C.D. (1948) *Econ. Bot.*, **2** : 306-325.

Sah, H., Lal, S., Singh, C.P. and Kumar, S. (2018) *Int. J. Pure App. Biosci.*, **6**(3): 468-474.

Saha, A.K. (1971) *Indian J. Hort.*, **28** : 11-15.

Said, M. and Inayatuzzah (1966) *W. Pakistan J. Agric. Res.*, **4** : 149-150.

Samson, J.A. (1980) *Tropical Fruits*, Longman, London.

Sanyal, D. and Mitra, S.K. (1990) *Indian J. Hort.*, **48** : 312-314.

Sanyal, D., Mitra, S.K. and Bose, T.K. (1995) *Nat. Symp. Advances in Res. and Dev. in Horticultural Crops for Export*, HAU, Hisar, Haryana, Abst. No. 23.

Saroj, P.L. and Pathak, R.K. (1994) *Adv. Agricultural Res. in India*, **2** : 74-80.

Schaffer, B., Andersen, P.C. and Ploetz, R.C. (1992) *Hortic. Rev.*, **13** : 257.

SeaLand (1991) *Shipping guide to perishables*. SeaLand Service Inc., P.O. Box 800, Iselim, New Jersey, USA.

Seema, A.A., Rani, P. and Mehrotra, R.S. (1995) *J. Mycopathological Res.*, **33** : 77-83.

Sehgal, O.P. (1961) *M. Sc. (Ag.) Thesis* submitted to IARI, New Delhi.

Sehgal, O.P. and Singh, R. (1965) *Indian J. Hort.*, **22** : 26.

Sehgal, O.P. and Singh, R. (1967) *Indian J. Hort.*, **24** : 118-126.

Selvaraj, Y., Pal, D.K., Edward Raja, M. and Rawal, R.D. (1999) *Indian J. Hort.*, **56** : 10-18.

Sen, P.K., Basu, R.N., Bose, T.K. and Roychaudhuri, N.K. (1967) *Proc. Sem. Pl. Propagation* FAO, New Delhi.

Shah, H., Lal, S., Singh, C.P. and Kumar, S. (2018) *Int. J. Pure App. Biosci.*, **6**(3): 468-474.

Shankar, G. (1966) *Fertilizer News*, **11** : 27-28.

Shankar, G. (1967) *Allahabad Fmr.*, **12** : 259-264.

Shant Lal, Tiwari, J.P. and Misra, K.K. (1996) *Annals Agric. Res.*, **17** : 83-89.

Sharma, D.P. and Sharma, R.G. (1992) *Adv. Plant Sci.*, **5** : 313-315.

Sharma, R.S., Sharma, T.R. and Sharma, R.C. (1991) *Orissa J. Hort.*, **19** : 41-45.

Sharma, Y.K. (1982) *Ph. D. Thesis* submitted to Meerut Univ., Meerut.

Sharma, Y.K., Goswami, A.N. and Sharma, R.R. (1992) *Indian J. Hort.*, **49** : 31-36.

Sheikh, M.K. and Hulamani, N.C. (1993) *Progressive Hort.*, **25** : 157-160.

Sheikh, M.K. and Hulamani, N.C. (1994) *Karnataka J. Agric. Sci.*, **7** : 473-475.

Shigeura, G.T. and Bullock, R.M. (1983) In : *Hawaii-history and production*. Univ. Hawaii, College of Trop. Agric. and Human Resources Extension Series 035.

Shukla S.K., Babita Prasad, Prasad, D. and Misra, A.K. (2003) In: *Souvenir of Zonal Conference (East Zone)*, Indian Society of Mycology and Plant Pathology and Seminar on Plant Diseases of National Importance with Special Reference to Guava Wilt and Mango Malformation, 4-5April, 2003. CISH and IISR, Lucknow. pp 54-55.

Siddappa, G.S. (1982) *Indian Food Ind.*, **1** : 73.

Siddiqui, S. and Gupta, O.P. (1997) *Haryana J. Hort. Sci.*, **26** : 101-104.

Siddiqui, S., Sharma, R.K. and Gupta, O.P. (1991) *HortScience*, **26** : 1295-1297.

Siddiqui, Z.M. and Faroq, S.A. (1996) *Adv. Plant Sci.*, **9** : 155-158.

Singh, A. (1980) *Fruit Physiology and Production*, Kalyani Publisher, pp. 323-326.

Singh, A.K. and Pathak, R.K. (1994) *Orissa J. Hort.*, **22** : 31-35.

Singh, A.K., Singh, G., Pandey, D. and Rajan, S. (1996) *Indian J. Agric. Sci.*, **66** : 348-352.

Singh, B. and S.B. Lal (1953) *Agric. Anim. Husb.*, **3**: 78-79.

Singh, B.P., Singh, G. and Singh, A.K. (1996) *Singapore J. Prim. Indus.*, **24** : 1-4.

Singh, D.S. and Singh, S.P. (1995) *Res. Dev. Reporter*, **12** : 35-39.

Singh, D.S., Singh, S.P. and Maurya, A.N. (1990) *Prog. Hort.*, **22** : 1-5.

Singh, D.S., Singh, S.P. and Maurya, A.N. (1992) *The Hort. J.*, **5** : 15-18.

Singh, G., Pandey, D., Rajan, S. and Singh, A.K. (1996b) *Fruits*, **51** : 241-246.

Singh, G., Rajan, S., Pandey, D. and Singh, A.K. (1997) *Indian J. Agric. Sci.*, **67** : 303-306.

Singh, G., Singh, A.K. and Mishra, D. (2007) *Acta Hortic.*, **735**: 235-241.

Singh, I.P. and Sharma, C.K. (1996) *J. Hill Res.*, **9** : 422-424.

Singh, J. and Singh, M.P. (1970) *Fertilizer News*, **15** : 45.

Singh, J., Sharma, D.P. and Kashyap, R. (1992a) *Adv. Plant Sci.*, **5** : 176-179.

Singh, M. (1988) *Punjab Hort. J.*, **28** : 50.

Singh, M., Patel, Y.R. and Yadava, H.S. (1995) *Adv. Plant Sci.*, **8** : 183-188.

Singh, N.P. and Rajput, C.B.S. (1977) *Indian J. Hort.*, **34** : 120-125.

Singh, R. (1969) *Fruits*, National Book Trust, New Delhi, pp. 86-90.

Singh, R., Singh, S.N., Gupta, M.R., Dhaliwal, G.S. and Kalra, S.K. (1992) *Indian J. Hort.*, **49** : 127-133.

Singh, S. and Krishnamurthi, S. (1967) *Fruit Culture in India*, ICAR, New Delhi.

Singh, U.P., Sharma, R.K. and Hoda, M.N. (1993) *The Hort. J.*, **6** : 1-5.

Singh, U.R., Pandey, I.C., Upadhyay, N.P. and Tripathi, B.M. (1976) *Punjab Hort. J.*, **16** : 121-124.

Singh, U.R., Pandey, I.C., Upadhyay, N.P. and Tripathi, B.M. (1979) *Punjab Hort. J.*, **19** : 111-113.

Singh, V. (1985) *Udyanika*, **5** : 11-16.

Salazar, D.M., Melgarejo, P. Martínez, R. Martínez, J.J. Hernández, F. and Burguera, M. (2006) *Scientia Hort.*, **108** : 157–161.

Saxena, S., Rajan, S., Chandra, R., Srivastava, N. and Bajpai, A. (2007) *Acta Hort.*, **735**: 49-55.

Sehgal, O.P. and Singh, R. (1967) *Indian J. Hort.*, **24**: 118–125.

Seth, J.N. (1959) *Hort Adv.*, **3**: 82-88.

Seth, J.N. (1963) *Agra Uni. J. Res.*, **12**: 193-197.

Shanmugavelu, K.G., Selvaraj, M. and Thamburaj, S. (1987) *South Indian Hort.*, **35**: 1-3.

Sharma, A.K. and Majumder, A. (1957) *Proc. Ind. Sc. Congress* (Ab 5),p. 256.

Sharma, Y.K., Goswami, A.M. and Sharma, R.R. (1992) *Indian J.Hort.*, **49**: 31-36.

Singh D., Gill, M.I.S. Boora, R.S. and Arora, N.K. (2015) *J. Applied Hort.*, **17**(1): 76-78.

Singh, B. and Lal, S.B. (1953) *Agril. An. Hus.*, III : 78-79.

Singh, B.P. and Rana, R.S. (1996) In: *Advances in Horticulture* Vol.1 (Eds. K.L. Chadha and O.P.Pareek), Malhotra Publishing House, New Delhi, pp. 43-66.

Singh, G., Sinha, G.C., Pandey, D. and Rajan, S. (1995) *Indian Fd. Packer*, May-June, pp. 15-20.

Singh, K.K. (2017) *Int. J. Curr. Microbiol. App. Sci.*, **6**(10): 4887-4893.

Singh K.K. (2018) *J. Med. Plants Stud.*, **6**(1): 167-172.

Singh, P., Chandrakar, J., Singh, A.K., Jain, V. and Agrawal, S. (2007) *Acta Hort.*, **735**: 197-200.

Singh, L.B. (1959) *Annual Report*, Fruit Research Station, Saharanpur, pp. 58-60.

Singh, R. and Sehgal, O.P. (1968) *Indian J. Hort.*, **25**: 52-59.

Singh, R.L. (1953) *Annual Report*, Fruit Research Station, Saharanpur, 1950-53.

Singh, S. and Hoda, M.N. (1994) Report on Fruit Research at Sabour, Rajendra Agril. Univeristy, Pusa, India, pp. 74-77.

Singh, U.R., Dhar, L. and Singh, G. (1977) *Haryana J. Hort. Sci.*, **6** : 149-150.

Singh, G., Rajan, S. and Singh, A.K. (2003) In: *Precision Farming in Horticulture* (Eds. H.P. Singh, G. Singh, J.C. Samuel and R.K.Pathak), NCPAH, DAC, MOA, PFDC, CISH, pp.92-113

Sommeijer, M. J., De Rooy, G. A., Punt, W. and De Bruijn, L. L. M. (1983) *Apidologie*, **14** (3): 205-224.

Soubihe Sobrinho, J. (1951) Estudos b´asicos para o melhoramento da goiabeira (*Psidium guajava* L.). S˜ao Paulo: ESALQ. p. 166. Tese de Doutorado.

Srivastava, O P. (1975) *Prog Hort.*, **6**: 71-77.

Subramanyam, M.D. and Iyer, C.P.A. (1982) Report, Fruit Research Workshop, Nagpur, pp. 117-18.

Subramanyam, M.D. and Iyer, C.P.A. (1988) Report, Fruit Research Workshop on Tropical and Subtropical Fruits. Rajendra Agricultural University, Pusa, pp. 81-84.

Subramanyam, M.D. and Iyer, C.P.A. (1993) In: *Advances in Horticulture: Fruit Crops–* Vol. 1, New Delhi, India, Malhotra Publishing House, pp. 337–347.

Syamal, M.M., Singh, R.K. and Chhlonkar, V.S. (1980) *Indian J. Hort.*, **37**(3): 243-49.

Smith, W.S. (1934) *UP Dept. Agr. Fruit. Ser. Bull.*, No. 8.

Sohnika, Rani P., Akash, A., Sharma, V.K., Wali, Bakshi, P. and Shahnawaz, A. (2015) *Indian J. Agric. Sci.*, **85**(9): 1162-1169.

Srivastava, O.P. (1962a) *Prog. Hort.*, **6** : 74-75.

Srivastava, R.P. (1962b) *Sci. Cult.*, **28** : 28.

Stark, J.D. and Vargas, R.I. (1992) *J. Econ. Entomol.*, **85** : 808-812.

Subramanyam, M.D. and Iyer, C.P.A. (1982) *Rep. Fruit. Res. Workshop*, Nagpur, pp. 117-118.

Subramanyam, M.D. and Iyer, C.P.A. (1988) *Rep. Fruit Res. Workshop. on Trop. and Subtrop. Fruits*, RAU, Pusa, pp. 81-84.

Subramanyam, M.D. and Iyer, C.P.A. (1993) In : *Advances in Horticulture*. Vol. I. (Eds : K.L. Chadha and O.P. Pareek), Mulhotra Pub. House, New Delhi. pp. 337-337.

Suhag, L.S. (1976) *Pesticides*, **10**: 42-44.

Sundarajan, S., Shanmugavelu, K.G. and Muthuswamy, S. (1969) *South Indian Hort.*, **17** : 41-43.

Sweeti C, Bisen BP, Pandey SK. (2016) *Intl. J. Agric. Sci.* **8**(57): 3151-3153.

Tamburo, S.E., Lynch, S.J. and Nelson, R.O. (1955) *Proc. Fla. St. Hort. Soc.*, **68** : 321-324.

Tavares, M.S.W., Lucca Filho, O.A. and Kersten, E. (1995) *Ciencia Rural*, **25** : 11-15.

Thonte, G.T. and Chakrawar, V.R. (1981) *National Symposium on Subtropical Fruit Crops*, Bangalore, p. 17.

Teaotia, S.S. and Pandey I.C. (1970) *Prog. Hort.*, **1** : 25-28.

Teaotia, S.S. and Pandey, I.C. (1961) *Sci. and Cult.*, **27** : 442-444.

Teaotia, S.S. and Phogat, K.P.S. (1971) *Prog. Hort.*, **4** : 37-45.

Teaotia, S.S., Pandey, I.C. and Mathur, R.S. (1961) *Curr. Sci.*, 300-312.

Teaotia, S.S., Pandey, I.C. and Agnihotri, B.N. (1962) *Indian Agric.*, **6** : 47-53.

Teaotia, S.S., Pandey, I.C. and Awasthi, P.K. (1968) *Punjab Hort. J.*, **8** : 126-133.

Teaotia, S.S., Pandey, I.C. and Dubey, P.S. (1969) *Punjab Hort. J.*, **9** : 41-43.

Teran, L., Melendez, I., Garcia-Aguilar, L., Acuna, J.G. and Urdaneta, M. (1996) *Revista de la Facultad de Agronomia*, Universidad del Zulia, **13**: 363-370.

Tripathi, S.N., Iyer, B.G., Srivastava, R.C. and Pathak, A.C. (1971) *JNKVV. Res. J.*, **5** : 32-34.

Tsay, J.G. (1991) *Plant Protection Bull.*, **33**: 384-394.

Usman, M.; Shah, M.H.; Badar, A.; Fatima, B.; Sabir, M. and Zaman, Q. (2014) *Pak. J. Agric. Sci.*, **51**(3): 615-623.

Vanaja, L., Swami. D.V., Prasanna, Kumar, B., Subbaramamma, P. (2017) *Int. J Curr. Microbiol. App. Sci.* **6**(10): 771-779.

Valecillos, C. and Fermin, G. (2010) *Acta Hort.*, **849**: 245-250.

Verma, K.S., Kang, M.S. and Sharma, A.K. (1991) *Plant Dis. Res.*, **6**: 140-141.

Verma, R.M., Hajare, V.S., Deshmukh, P.P., Shelke, B.D. and Verma, E.K. (1996) *PKV Res. J.*, **20**: 37-39.

Wagh, A.N. and Mahajan, P.R. (1985) *Curr. Res. Rep.*, **1**: 124-126.

Walker, R.R., Kriedemann, P.E. and Maggs, D.H. (1979) *Australian J. Agric. Res.*, **30**: 477.

Walling, M. and Sanyal, D. (1995) *Crop Res.*, **9**: 272-276.

Watt, G. (1908) *The Commercial Products of India*, John Murray, London.

Welgemoed, C. and Preez, R.D.U. (1986) *Inf. Bull. Citrus and Sub-trop. Fr. Res. Inst.*, No. 162, p. 4.

Widyastutil, R.D., Susanto, S., Melati, M. and Kurniawati, A. (2019) *J. Phys.: Conf. Ser.* 1155 012013.

Wilberg, V.C. and Rodriguez-Amaya, D.B. (1995) *Lebensmittel-Wissenschaft und Technologie*, **28** : 474-480.

Willers, P. (1997a) *Inligtingsbulletin- Inst. vir. Trop. Subtrop. Gewasse*, **293** : 10-12.

Willers, P. (1997b) *Inligtingsbulletin - Inst. vir Trop. Subtrop. Gewasse*, **294** : 19-20.

Wills, R.B.H., Mulholland, E., Brown, B.I. and Scott, K.J. (1983) *Trop. Agric.*, **60** : 175-178.

Wilson, W.C. (1980) In : *Tropical and Subtropical Fruits* (Eds. S. Nagy and P.E. Shaw). Composition, Properties and Uses. AVI Publ. Inc., Westport, Connecticut.

Yadav, E.D., Gaikwad, M.R. and Patil, A.V. (1981) *Abst. Nat. Symp. Trop. & Subtrop. Fruit Crops.*, Bangalore, p. 30.

Yadav, L.B. and Pandey, V. (1994) *Ann. Plant Protec. Sci.*, **2** : 94-95.

Zamir, R., Abdur, Rab., Sajid, M., Khattak, G.S.S., Khalil, S.A. and Shah, S.T. (2017) *The Nucleus*, **54**(1): 46-5.

③

PAPAYA

Maneesh Mishra, I. Irulappan, Rohit Jaswal and V. A. Parthasarathy

1.0 Introduction

Papaya (*Carica papaya*) also called papaw or pawpaw, is a quick growing, typically single-stemmed, short-lived, large perennial herb. Branching may occur with age or if the apical growing point is damaged and can bear fruits for more than 20 years (Malo and Campbell, 1986). Papaya diverged from its closest monoecious relative (*Vasconcellea monoica*) about 27.5 million years ago (Carvalho *et al.*, 2012). Papaya is one of the few tree crops where extensive research on genomics have also been carried out and is designated a model system. Ming *et al.* (2008) led a team to sequence the entire papaya genome which led to the series of functional genomics research. In humor, Wei and Wing (2008) commented "The nice thing about working with some plant genomes is that at the end of the day you can eat the fruits of your work". It is now distributed throughout tropical and subtropical regions of the world. It is a highly problematic, complicated and interesting fruit crop from botanical, genetical, cytogenetical and horticultural points of view (Ram, 1996).

2.0 Composition and Uses

2.1.0 Composition

Papaya is a very wholesome fruit. Aykroyd (1951) ranks it second only to mango as a source of the precursor of vitamin A. While this vitamin is generally associated with carotene, the yellow pigment in the papaya is not carotene but caricaxanthin. Cultivar differences and geographic effects on the carotenoid composition and vitamin A value of papaya have also been reported. β-cryptoxanthin (8.1 μg/g) is the major pigment, representing 62 per cent of the total carotenoid content in the yellow/orange fleshed common cultivars. In red fleshed cultivars, lycopene is the major pigment accounting for 61 per cent of the total in Solo, 56 per cent in Formosa and 66 per cent in Tailandia. β-zeacarotene and cryptoflavin were found only in the common cultivars, while β-carotene was found in Solo fruits. Fruits of cv. Formosa from Bahia (warmer climate) contained higher amount of β-carotene, β-cryptoxanthin and lycopene than fruits from Sau Paulo, Brazil. Lowest vitamin A values were found in Formosa fruits from Sau Paulo and highest values in fruits from Bahia. The vitamin A values did not reflect a marked difference in the carotenoid content of the different cultivars because the principal quantitative variation exists in the inactive lycopene. Geographic effects had a greater influence than cultivar on vitamin A values (Kimura *et al.*, 1991). The vitamin A activecarotenoids (β-carotene, β-cryptoxanthin and β-cryptoxanthin monoepoxide) were extracted from papaya (cv. Formosa) fruits and quantified by Godoy *et al.* (1990) and the structure of the β-cryptoxanthin monoepoxide was confirmed to be as 5,6-epoxy-5, 6-dihydro-β βcarotene-3-ol. The ripe fruits contain vitamin A, B, B_1, B_{12}, C and D and also many biologically active and tonic substances (Kapanadze and Khasaya, 1988). The composition and food value of ripe papaya fruit are given in Table 1.

Table 1: Composition and Food Value of Fruit

Moisture	89.6 per cent	Vitamin A	2020 IU/100 g
Protein	0.5 per cent	Vitamin B2	0.04 mg/100 g
Fat	0.1 per cent	Vitamin C	40mg/100g
Carbohydrate	9.5 per cent	Nicotinic acid	0.2 mg/100 g
Calcium	0.01 per cent	Riboflavin	0.25 mg/100 g
Phosphorus	0.01 per cent	Calorific value	40 per 100 g

The major volatile component in fruits of cv. Solo has been identified as linalool followed by trans-linalool oxide, while in fruits of cv. Formosa, cis-linalool oxide was the major volatile component and linalool was the second most abundant component (Franco and Rodriguez-Amaya, 1993). Several enzymes have been isolated and purified from papaya fruits, *viz.* invertase (Lopez *et al.*, 1988), UDP-glucose phenol-B-D-glucosyltransferase (Keil and Schreier, 1989), proteinases (Redina *et al.*, 1993), pectinesterase (Fayyaz *et al.*, 1994) and cysteine endopeptidase from C. candamarcensis fruits (Moraes *et al.*, 1994).

The leaves contain flavonoids, tannins, organic acids and alkaloids including the valuable carparine (Kapanadze and Khasaya, 1988). Air-dried papaya seeds were found to contain moisture (4.0 per cent), oil (25.6 per cent), ash (17.8 per cent) and protein (26.6 per cent) (Raie *et al.*, 1992). Seed oil from papayas, cultivated in Vietnam, accounted for 28 per cent of the whole dried seeds and 54 per cent of the endosperm. Fatty acid and triacylglycerol composition of papaya seed oil is similar to that of olive oil (Nguyen and Tarandjiska, 1995).

2.2.0 Uses

The ripe fresh fruits of papaya are eaten throughout the tropics and subtropics. They are used in preparation of jam, soft drinks, ice-cream flavouring, crystallized fruits and in syrup. The seeds are also used for their medicinal value. Unripe fruits are commonly used as vegetable for cooking. Ochse (1931) reported that young leaves are eaten in Java as vegetables. Papain, prepared from the dried latex of immature fruits, is a proteolytic enzyme similar in action to pepsin and is used as meat-tenderising preparation; in manufacture of chewing gum and cosmetics; as drug for digestive ailments; in the tanning industry for bating hides; for degumming natural silk and to give shrink resistance to wool (Purseglove, 1968). The dried and powdered latex had a proteolytic activity slightly higher than that of the fresh latex (Narinesingh and Mohammed-Maraj, 1988). Papaya contains many biologically active compounds. Two important compounds are chymopapain and papain which are widely known as being useful for digestive disorders and disturbances of the gastrointestinal tract (Teixeira da Silva *et al.*, 2007). Huet *et al.* (2006) showed that papaya-derived papain, caricain, chymopapain, and glycine endopeptidase can survive acidic pH conditions and pepsin degradation. However, at low pH, a conformational transition that instantaneously converts their native forms into molten globules that are quite unstable and rapidly degraded by pepsin. Thus, they may need to be protected against both acid denaturation and proteolysis for them

Uses of papaya

Green and ripe fruit of papaya

Ripe fresh fruit

Papaya juice

Chunks and powder from papaya

Dried ripe papaya

Uses of papaya

Tutty fruity of papaya used in cake

Papain extracted from fruits

Leaf powder (left) and skin cream (right)

Soap from papaya

to be effective in the gut after oral administration for the control of gastrointestinal nematodes (Teixeira da Silva *et al.*, 2007).

3.0 Origin and Distribution

Fuentes and Santamaria (2014) reviewed some of the arguments for belief that domestication of *C. papaya* L. occurred within the lowland Mayas by at least 1500 AC. They felt that given that no direct archeological evidence is available, it is rather difficult to define a precise origin for *C. papaya* L. However, Vavilov (1987) described three centers of origin of the majority of the species: the Mesopotamic center, the Mesoamerican Center, and the North Chinese center. The Mesoamerican Center is the center of origin of important tropical crops, and it has been suggested as being a good candidate to also be the center of origin of *C. papaya* L. (Harlan, 1971 cited by Fuentes and Santamaria, 2014). It is believed that papaya fruits (and seeds) were distributed from Central America to South America and other parts of the world by the Spaniards during the sixteenth century. The fi rst written report of this specie appears to be that made by Oviedo in 1526, in the coastal areas of Panama (Fuentes and Santamaria, 2014). The papaya is an important fruit of tropical and subtropical regions of the world. It is a native of tropical America (Hofmeyr, 1938) and was introduced in India in the 16th century. The papaya cultivation is confined to 32°N and S latitudes of the equator on the globe (Ram, 1996). It is now grown in all the tropical and subtropical countries like Australia, Hawaii, Taiwan, Puerto Rico, Peru and Florida, Texas, California in the USA, Gold Coast, various parts of Central and South Africa, Pakistan, Bangladesh and India. According to Ram (1996), the cultivated papaya might have originated as a cross between two species of the genus *Carica* native to Mexico. It is closely related to *Carica peltata* which occurs in this area, and may have arisen by hybridization. It was taken by the Spanish to Manila in the mid-16th century and reached Malacca shortly afterwards. From there it was brought to India. The exact date of introduction of papaya to the Hawaiian islands is still obscure. It might have been introduced in between 1800 and 1823 by Don Moris, the Spanish horticulturist who settled in Hawaii or it came to the islands via Asia and the South Sea islands before the Europeans appeared there (Ram, 1996). *C. pubescens* (mountain papaya) probably originated from the South American Andean highlands. It is mainly cultivated in Chile, and to a lesser extent in Colombia, Ecuador, Peru, Venezuela, Hawaii, New Zealand and India (Scheer and Ludders, 1997).

4.0 Species and Cultivars

4.1.0 Species

Papaya belongs to the family Caricaceae – a small, somewhat anomalous family with four genera and 31 species of which three are in tropical and subtropical America and one in Africa (Badillo, 1971). The genera and the number of included species are : *Carica*, 22; *Facaratia*, 6; *Farilla*, 1; and *Cylicomorpha*, 2. The edible fruits are found only in *Carica* of which *C. papaya*, the common papaya, is extensively grown. Other species include *C. chilensis, C. goudotiana, C. monoica and C. pubescens. C. candarmarcensis* [*C. pubescens*], known as mountain papaya, grows to a height of

10 m, produces multiseeded fruits throughout the year which are rich in juice and contain considerable amounts of vitamin C and A (Scheer and Ludders, 1997). *C. erythrocarpa* is similar to *C. papaya*, but the fruit has thin red flesh, while *C. quercifolia* reaches up to 2 m (maximum) with cluster of small ellipsoid fruits and *C. gracilis* is a small, slender and ornamental species (Ram, 1996). *C. monoica* is a monoecious plant susceptible to virus (Ram, 1996) and is found growing in the Amazon basin (Swingle, 1947).

The characteristics of *C. papaya*, the commercially grown *Carica* species, are given below:

1. Small, single-stemmed, ever-green, quick growing plant.
2. Trunk is erect, unbranched (generally) with hollow soft wood.
3. Palm-like leaves are large, deeply lobed with long petiole.
4. Fruits are spherical to oblong in shape, fleshy berry with central cavity.
5. Numerous black seeds are attached to placenta in the cavity.
6. Flesh is edible and on ripening turns yellow, deep orange, pinkish or deep red depending on cultivars.
7. The immature fruits, on scratching, yields white milky latex.
8. Flowers are borne singly (usually female and hermaphrodite) or in large clusters (male) in leaf axils.

4.2.0 Cultivars

A large number of papaya cultivars are found in cultivation. As a matter of fact none of these is a true cultivar since it cannot be relied upon to reproduce the parental characters in all the progenies.

Arka Prabhath

It is from the cross (Surya x Tainung-1) x Local Dwarf. It is gynodioecious in nature. The plants are semi-vigorous and bearing starts at a lower height (60-70 cm). Since the variety is gynodioecious, seed production is easy, as bagging of bisexual flowers ensures true to type plants. The pulp is firm (5.9 kg/cm^2) and pulp colour is deep pink. The fruit weight on an average is 900-1200 g, the TSS is 13-14 & Brix and yield per plant is 90-100 kg. The keeping quality is good.

Arka Surya

It is the offspring of sunrise Solo x Pink Flesh Sweet. It is gynodioecious in nature. The plants are shorter compared to solo. Skin is smooth, become uniformly yellow in colour on ripening. Fruits are medium in size of about 600 – 800 g with a small fruit cavity. Pulp is about 3-3.5 cm thick, deep red in colour and sweet with a TSS of 13.5 -15° brix. Yield per plant is 55-65 kg. The keeping quality is good.

Red Lady

It is one of the commercial papaya hybrid popular across the country due to firmness and red flesh colour of the fruit. The fruit weight ranges from 1500-2500 g with firm fruit. The TSS content ranged 13-16°B.

Species of Papaya

Carica papaya

Carica chilensis

Carica goudotiana

Carica monoica

Species of Papaya

Carica pubescence

Carica candarmarcensis

Carica quercifolia

Cultivars of Papaya

Coorg Honey Dew

CO-7

Red Lady

Pusa Nanha

Sunrise Solo

Chinmay-1

Cultivars of Papaya

RCTP-1

Solo

CO-2

Pusa Delicious

Pusa Majesty

Pusa Dwarf

Pusa Giant

Washington

Ranchi

Washington

Fruit is round to ovate in shape, medium to large in size, over 20 cm long and 40 cm in circumference at its maximum bulge. The weight of fruit is about 1 kg. It is sweet and has desirable flavour.

Honey Dew

The cultivar is best suited for northern India. The bearing is heavy. Fruits have less number of seeds with good flavour and taste.

Coorg Honey Dew

It is a chance seedling of Honey Dew. The plants are mostly hermaphrodite but occasionally pistillate flowers are also observed. The fruits are oblong. It is a dwarf and heavy bearing cultivar. The flesh is thick with good flavour.

Pink Flesh Sweet

Fruit is medium in size, with pink flesh and 12 to 14oBrix. It is recommended for fresh consumption.

Co. 1

It is a selection from the progenies of cv. Ranchi. Plant is dwarf. Fruit is round or oval in shape with golden yellow skin and orange-coloured flesh.

Co. 2

This is a pure line selection from a local type. The plant is medium-tall in height. Fruits are obovate and large in size; skin yellowish green; flesh orange coloured, soft and moderately juicy. It is a good table fruit and also a high papain-yielding cultivar.

Co. 3

This is a hybrid between Co.2 x Sunrise Solo, a tall vigorous plant. The fruit is medium-sized, sweet with good keeping quality.

Co. 4

This was evolved from a cross between Co.1 x Washington. Plant medium-tall. Fruit large; flesh thick, yellow with purple tinge, taste sweet; good keeping quality.

Co. 5

This is a selection from the cv. Washington and yields an average of 15 to 16 q dried papain per hectare.

Co. 6

This dwarf, dioecious cultivar has been evolved by selection from Pusa Majesty. Fruits are large and recommended for both papain and dessert purposes.

Co. 7

This is a gynodioecious cultivar developed through multiple crosses. The parents involved are Pusa Delicious, Co.3, CP.75 and Coorg Honey Dew. This can be grown both in plains and up to an altitude of 1000 m above m.s.l.

Co. 8

This is a dioecious cultivar developed from Co.2. Fruits are obovate and large in size, Red pulp and suitable for dessert. It is a good for processing and also a high papain-yielding cultivar. The fruit yield is 200 to 230 ton per hectare.

The Co. cultivars were evolved and released from the Tamil Nadu Agricultural University, Coimbatore.

Pusa Delicious (Pusa 1-15)

This is a gynodioecious cultivar in which two sex forms, *i.e.*, pistillate and hermaphrodite are found. The fruits are medium in size with deep orange flesh colour having excellent flavour.

Pusa Majesty (Pusa 22-3)

This is also a gynodioecious line. The fruits are medium in size and round in shape. Fruit flesh solid in texture and yellowish in colour having good keeping quality and is less prone to spoilage during transportation and storage. Plants have also tolerance to viral diseases.

Pusa Giant (Pusa 1-45V)

This is a dioecious cultivar and plant bears fruit at one metre height. The plant can withstand strong wind and storm. Fruits have attractive big size and weight ranging from 2.5 to 3.5 kg per fruit. It can also be used for vegetable and canning industry.

Pusa Dwarf (Pusa 1-45D)

This is also a dioecious cultivar having dwarf stature and more precocious in bearing. The plants start bearing from 25 to 30 cm above ground level. Fruit size is medium (1 to 2 kg) and oval in shape. This cultivar is most suitable for high density orcharding, nutrition garden and kitchen garden.

Pusa Nanha (Mutant Dwarf)

This is an extremely dwarf dioecious mutant cultivar suitable for kitchen gardening and pot cultivation and is recommended for high density planting.

The Pusa cultivars were evolved at the Indian Agricultural Research Institute (IARI) Regional Station, Pusa, Bihar.

Pant-1

This dioecious cultivar has been evolved by selection. Fruit is medium in size and is recommended for Terai region in U.P.

The selections made at the G.B. Pant Agricultural University, Pantnagar, *viz.*, Pant Papaya 1,2 and 3 are recommended for commercial cultivation.

HPSC-3

This is a hybrid between Tripura Local and Honey Dew. The hybrid produced higher yield compared to both the parents, performed well for papain production and showed resistance to papaya mosaic potexvirus (Singh and Sharma, 1996a).

Solo

The fruits are pear-shaped weighing 350-500 g and of excellent quality. J.E. Higgins was the first to use the name Solo in his reports when referring to the small-fruited papayas grown in Hawaii. Over the years, major improvements were made possible in many plant and fruit characteristics. In 1948, Line 5 Solo was introduced and this was replaced by Line 8 Solo and Kapoho Solo. Line 10 Solo, Wiaminalo, Higgins and Wilder are also important lines.

Table 2: Plant Height and Fruit Characteristics of some Papaya Cultivars

	Plant Height (cm)	*Weight of Fruit (g)*	*No. of Fruits/ Tree*	*Fruit Length (cm)*	*Fruit Width (cm)*	*T.S.S. (°Brix)*	*Referen-ces*
Arka Prabhath	68	1300	17	21.24	13.2	7.36	Tulasigeri *et al.* (2017)
Arka Surya	101	700	18	16.48	9.4	7.32	
Red Lady	213	2100	24	20.24	17.23	9.6	
Washington-1	145	1000	11	16.5	12.6	10.1	Singh and Sirohi (1977)
Washington-2	233	887	68	18.6	11.0		
Honey Dew-1	122	1112	14	15.8	9.8		
Dew-2	246	1503	27	19.7	9.2		
Coorg Honey Dew-1	127	-	20	-	-	-	
Coorg Honey Dew-2	205	1366	41	16.2	13.3	9.8	
Co1	112	1200-1500	42-45	16.5	15.6	12.5-13.6	Veerannah *et al.* (1982)
Co2	120	1000-1500	30-40	-	-	11.5-12.5	
Co3	Medium	885-1000	35-40	-	-	13.8-14.6	
Co4	250-330	1300-1500	40	-	-	13.2-13.5	
Sunrise Solo	195	250-400	50-60	-	-	13.5-15.0	
Solo Hawaii	237	295	48	10.1	7.6	11.8	Singh and Sirohi (1977)
Pusa Delicious (Pusa 1-15)	216	1005	39	-	-	13.0	
Pusa Majesty (Pusa 22-3)	196	950	40	-	-	9.0	Ram (1982)
Pusa Giant (Pusa 1-45V)	220	1910	16	-	-	7.0	
Pusa Dwarf (Pusa 1-45D)	130	1020	41	-	-	6.5	
Pink Fleshed	-	860	-	15.8	10.0	-	Pal *et al.* (1980)
Sweet	(1980)						
Thailand	-	1920	-	28.6	15.9	-	

Besides the above mentioned cultivars, the other important cultivars are Peradeniya, Red Fleshed, Philippines, Madhubindu, Barwani and Ranchi.

In South Africa, Hortus Gold and Honey Gold are commonly grown. The plant height and fruit characteristics of some papaya cultivars are given in Table 2.

In a varietal trial, Biswas *et al.* (1990) reported that Washington, Ranchi, Pusa 1-15 and Co.1 were high yielding cultivars, while the lowest yielding cultivar was Sunrise Solo but it had the best quality fruits followed by Pusa 1-15, Co.1 and Co.2. Sunrise Solo appeared to be the best adapted cultivar with hermaphrodite and female plants giving yields of 46 and 25 kg/plant, respectively (Rodriguez Pastor *et al.*, 1992). The stand of Exotica papaya (a cross between cultivars Subang-6 and Hawaiian Solo) had 70.9 per cent hermaphrodite plants and 29.1 per cent female plants. Plant height ranged from 1.25 to 2.49 m after 48 weeks with number of fruits/tree and fruit weight ranging from 41 to 119 and 170 to 1020 g, respectively (Chan and Teo, 1992).

5.0 Breeding and Improvement

The genus *Carica* exhibits wide range of genetic variability. It is found that almost all the species under genus *Carica* have the same chromosome number of 2n=18. The related species of papaya are of interest to plant breeders for evolving new varieties. The future advancement of papaya improvement programme lies with collection, conservation, documentation, evaluation and utilization of papaya germplasm (Ram, 1996) and through biotechnology.

5.1.0 Selection

Due to continuous selection several seedling races are found to produce fairly large proportions of progenies true to the parents from seed and of these Washington of Maharashtra and Honey Dew enjoy some popularity in South India (Naik, 1949). Based on the two-season evaluation of papaya germplasm in Nainital, cultivars Barwani Red, Coorg Honey Dew, BR-1,2,3,4, Coimbatore IB and Washington have been found promising for commercial cultivation. Considerable genetical variability was found in many plants and fruit characteristics like fruiting height, size, shape, TSS and taste of fruits, number of fruits per tree and weight per fruit. Selection practised from a mixed population of Co.1 and others resulted in 12 promising selections which were superior to Co.1 (Singh and Sirohi, 1977). Arising as a chance seedling at the Chettali Citrus Research Station, the Coorg Honey Dew papaya has not only spread throughout the length and breadth of Coorg, but is in great demand all over India. Apart from being very delicious, it is a hermaphrodite - a character of economic importance (Aiyappa and Nanjappa, 1959). There are only a few true breeding papaya cultivars in the world, the Solo papaya of Hawaii being the outstanding example. This is because hermaphrodite flowers of this cultivar are 100 per cent bud-pollinated and the Solo papaya has become predominantly homozygous and highly inbred. The Maradol papaya of Cuba is another fairly uniform cultivar when grown from seed of hermaphrodite plants. Many other recognisable papaya types occur throughout the world and have been given names, but are rarely true-breeding enough to be considered legitimate cultivars (Hamilton, 1987). In Florida, the cauliflora-derived from the dioecious lines (K2, K3), bears round fruit with sweet yellow flesh, good aroma and quality. It is dioecious and highly tolerant to papaya ring spot virus (Conover *et al.*,1986). The observations in respect to yield and fruit quality indicated that the cv. Solo (Line 8) appeared to be promising. To develop a cultivar with high papain content, breeding was commenced at the then Agricultural College and Research Institute, Coimbatore and thirteen selections of papaya introduced from different parts of the world were tested. Considerable variation existed among the different selections in papain yielding capacity; among these, Selection 7 consistently yielded the best. Over the years, off-types in this selection were eliminated and the progeny purified by the method of inbreeding until a reasonably pure type was obtained and the same was released as Co.2 in 1973. Individual fruit of this selection yielded 4 to 6 g of papain, compared with about 2.25 g in Sri Lanka (Madhava Rao *et al.*, 1974).

Hamilton and Izuno (1965) introduced 'Line 10 Solo' through selection and breeding as a new strain of export papaya. From a breeding programme, in which widely different strains were used as the original parents, a selection was made by Nakasone and his co-workers in 1972, designated as Wiaminalo and released as a commercial strain. Nakasone *et al.* (1974) introduced two new Hawaiian papaya lines, *viz.*, Higgins and Wilder providing many advantages to the grower.

In South Africa, Hortus Gold type papayas are commonly grown. Allan (1981) reported a new selection of Hortus Gold, *viz.*, Honey Gold which produces attractive golden yellow fruits with high sugar content, pleasant flavour and texture. Of 32 accessions evaluated for virus resistance, 4165 and 4042 (Honey Gold) were moderately resistant. Accessions of Cavite special, Solo and native forms (Comprising 75 per cent of the test materials) were moderately to highly susceptible (Magdalita *et al.*, 1986).

5.2.0 Acclimatization

Heavy or shy bearing is a genetical character in papaya. But some acclimatized cultivars give good yield in a specific zone. The cv. Ranchi is acclimatized to South Bihar condition and has a wide adaptability under North Indian condition also. Coorg Honey Dew performs well in Karnataka but is a poor yielder in North India. Similarly, Nongpoh and Haflong perform well in north eastern hill zone, Barwani in central zone, Washington in western zone, and Co.1 and 2 in a large part of southern zone. Most of these cultivars fail to give satisfactory results when grown at other places. The cultivars Pusa Delicious, Pusa Majesty, Pusa Giant and Pusa Dwarf developed at IARI Regional Station, Pusa were found to have a wide range of adaptability (Ram, 1983).

5.3.0 Hybridization

Hybridization has been used primarily for establishing species relationship and understanding the genetics of sex in papaya. Its use in evolving new cultivars has been limited. Hofmeyr (1942) emphasized the importance of both male and female parents in developing improved cultivars of papaya. Several Solo cultivars in Hawaii have been developed through hybridization. Originally, Solo was a polygamous cultivar which was changed to a gynodioecious cultivar by the use of several back crossing to a harmaphrodite source (Storey, 1958).

Breeding programmes were initiated through intervarietal hybridization between Co.1 x Coorg Honey Dew and Co.1 x Washington. The reciprocal combinations were also tried (Desikan, 1972). The progenies were evaluated over a number of generations through sibmating, and a hybrid-derived CP 60 was identified as a promising selection out of Co.1 x Washington combination. This is characterized by a purple tinge shade throughout the stem, petiole and flower. The fruit flesh is thick and yellow in colour with a purple tinge. The fruit quality is better than both the parents. The hybrid of Co.1 x Coorg Honey Dew possessed certain desirable characters like weight, volume, diameter and firm flesh of fruit. On the other hand, the hybrid Coorg Honey Dew x Co.1 exhibited low height and earliness of fruiting and star-shaped fruit cavity (Shah and Shanmugavelu, 1973). Interspecific hybridization was attempted in the genus Carica. Studies indicated crossability barriers among many species. The barrier was found to be operating at different stages of development (Subramanyam and Iyer, 1982). With a view to produce hermaphrodite lines, hybridization was undertaken using Sunrise Solo as male and Co.2 as female parents and an improved selection, *viz.*, CP 81 with several special attributes was made. This selection is a hermaphrodite type with an attractive red coloured flesh. The fruit is bigger than Sunrise Solo and superior in quality to Co.2. Simultaneously, breeding work among commercially important dioecious cultivars was also taken up to obtain new hybrids with distinct morphological and economical attributes. The hybrids were evaluated and have now been released as Co.3 and Co.4 papaya. The fruits of Co.3 (Co.2 x Sunrise Solo) possess better fruit size with excellent quality of Sunrise Solo, while the fruit quality of Co.4 (Co.1 x Washington) is better than Co.1 and Washington (Veerannah *et al.*,1982). Carica candamarcensis and C. pentagona are said to be cold resistant, while C. cauliflora is known to be resistant to viruses. The crosses between the C. papaya and C. cauliflora did not form mature seed but immature embryos could be germinated and grown by employing embryo culture technique (Ram, 1982).

In Hawaii, interspecific crosses were attempted between Carica papaya and 4 wild Carica species with resistance to papaya ring spot virus. Embryological studies of C. papaya x C. cauliflora revealed no barrier to fertilization, but subsequent development was abnormal. A low percentage of ovules in each of the 4 interspecific crosses yielded F1 plants via embryo rescue. Many of these ovules contained multiple embryos which were of zygotic origin by isozyme analysis. F1 hybrids of C. papaya x C. pubescens and C. papaya x C. quercifolia were vegetatively vigorous, C. papaya x C. cauliflora F1s are slow growing and those of C. papaya x C. stipulata developed an apical necrosis before reaching maturity. Pollen stainability of C. papaya x C. quercifolia F1s was less than 1 per cent, but some pollen tube formation was observed in germination studies and F1 females set fruit when sibbed or back crossed to C. papaya (Manshardt and Wenslaff, 1986).

5.4.0 Heterosis

Studies were also made to assess the extent of heterosis, if any, in intervarietal hybrids in papaya. It was observed that significant heterosis was prevalent for all the vegetative characters as well as yield and its components. Heterosis up to 111 per cent over parents for yield was obtained in certain combinations. From a practical point of view with regard to the absolute values in yield, it was found that the combination Thailand xWashington was the best, followed by Coorg Honey Dew x Thailand, Pink Flesh Sweet x Thailand and Thailand x Solo Yellow Sweet. The yield improvement could be attributed to heterotic effects in major yield components (Iyer and Subramanyam, 1981). It is possible that inbreeding may lead to inbreeding depression which can be restored by the intercrossing of two or more inbreds. Inbreeding Solo strains did not exhibit any loss of vigour as is the case in many cross-pollinated crops. On the contrary, the inbred strains produced more vigorus plants than the commercial strains (Hamilton, 1954). Commercial application of hybrid vigour in papaya is highly feasible. Some phenotypically ambivalent staminate and andromonoecious plants produce a high proportion of self-pollinated fruit from which inbred lines can easily be fixed (Aquilizan, 1986).

5.5.0 Polyploidy and Mutation Breeding

Several workers have been successful in inducing polyploidy in papaya (Hofmeyr, 1945; Singh, 1953; Zerpa, 1957). Tetraploids were found to be better than diploids in many horticultural characteristics. But so far no commercial cultivar has been developed through polyploidy breeding. Similarly, mutation breeding has not resulted in any improved cultivar of papaya.

5.5.1 Mutation Breeding

Irradiation of papaya seeds with gamma-rays has been reported to influence the germination, survival percentage and sex of the plantlets. Exposure of seeds to 15 kR gamma rays resulted into a dwarf mutant. Repeated sibmating among the three dwarf plants (including one male) isolated from M3 population gave rise to a homozygous dwarf papaya variety which was named as Pusa Nanha (Ram *et al.*, 1981).

5.6.0 Genetical Factors of Sex Expression

Sex in papaya is determined by three genetic factors : M1 is dominant for maleness, M_2 is dominant for hermaphroditism and m is recessive for femaleness. Each ovule or pollen grain being haploid can carry only one of these factors. The diploid zygote carr-ying two sex-dominant factors cannot thrive, thus the combination M_1M_1, M_2M_2 and M_1M_2 are eliminated. Of the remaining ones M1m are male plants, M_2m hermaphro-dites and mm females. If a female is crossed with a male (mm x M_1m), mm + M1m is obtained, half the seeds will produce female plants, half male plants. If hermaphrodites are selfed or crossed, the ratio of female : hermaphrodite is 1:2. When a hermaphrodite is crossed with a male the ratio of female : hermaphrodite : male is 1:1:1 (Samson, 1980).

Yaqub *et al.* (1985) stated that sex is controlled by a single major locus with 3 alleles, M_1 (male), M_2 (hermaphrodite) and M (female) and all double dominant contributions being lethal.

The scheme for sex determination has been modified recently (Ram, 1996) as below:

Genes

M_1^{RR} = dominant factor for homozygous sex reversing

M_1^{Rr} = dominant factor for heterozygous sex reversing maleness

M_1^{rr} = dominant factor for pure maleness

M2 = dominant factor for hermaphroditism

m = recessive factor for femaleness

Genetic Constitution

$M_1^{rr}m$ = Sex reversing homozygous male plant

$M_1^{Rr}m$ = Sex reversing heterozygous male plant

$M_1^{rr}m$ = Pure male plant

M_2m = Hermaphrodite

Mm = Female plant

6.0 Propagation

6.1.0 Seed

Papaya is normally propagated by seed. Since it is a cross-pollinated crop, the plants raised from seeds have a mixed inheritance which makes them highly variable in performance. It is important, therefore, for the grower to secure seed from a reliable source, where selection and, perhaps, controlled pollination are done.

6.1.1 Seed Production

Papaya seeds are produced by controlled cross pollination and maintaining isolation distance. Pollens from male plants are collected and hand pollinated on flowers of female plants of the same cultivar. The female flower is then prevented from foreign pollens by bagging. The bag is removed once the fruit is set. Fruits are harvested semi-ripe and ripened in the shade. The seeds are extracted from pulp, washed, dried and stored. Ram and Majumder (1990) stated that the dioecious cvs. Pusa Giant and Pusa Dwarf produced more seeds at lower cost than the gynodioecious cvs. Pusa Delicious and Pusa Majesty (using pollen from hermaphrodite flowers). The cv. Pusa Dwarf produced the highest seed yield (391.7 kg/ha) at lowest cost (Rs.61.10/kg), while Pusa Majesty gave the lowest yield (52.5 kg/ha) at the highest cost (Rs. 416.60/kg). The heaviest fruit produced the highest number of seeds and the highest total weight of viable seeds (Muller *et al.*, 1991). The controlled cross-pollination was reported to be best carried out in August-September to maximize seed production (Ram and Ray, 1992). It is suggested that foundation seed production on a commercial scale may be conducted on isolated fields to meet the increasing demands for papaya seeds in India. However, breeder's seed should be produced under strictly controlled pollination to maintain genetic

Propagation of papaya

Raising of seedlings

Propagation by grafting

Tissue cultured papaya plant

Source : Salgado-Garciglia, Rafael Carreto, Lorena, Chávez-Moctezuma, M.P., Hernández-García, Alejandra, Suarez-Rodriguez, Luis and Lopez, Rodolfo. (2011) *Acta Hort.*, doi: 923. 10.17660/ActaHortic.2011.923.27.

Floral biology and fruiting in papaya

Male (left), female (middle) and hermaphrodite (right) of papaya

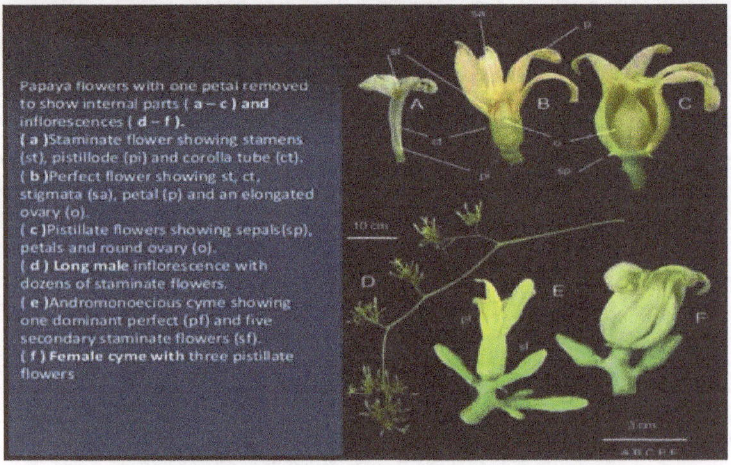

Male (left), female (middle) and hermaphrodite (right) of papaya

Source: Eric *et al*. (2013) In: *Genetics and Genomics of Papaya* (Eds. Ray Ming and Paul H. Moore), pp. 17-33.

purity (Ram, 1995). A 2:1 ratio of bisexual : female plants was recommended for seed production of papaya (Purohit, 1995).

6.1.2 Seed Viability and Storage

The viability of papaya seeds is poor. Germination percentage of seeds declined rapidly with aging and no seed germinated after 3 months (Begum *et al.*, 1988). The viability of seeds decreases when the sarcotesta remains intact and the effect is more pronounced in dry seeds than fresh ones (Arumugam, 1974). Field bulk electrical conductivity test is a promising rapid substitute for a laboratory germination test, but is not a good vigour test for papaya seeds (Shie and Kuo, 1996). Drying of seeds in the shade resulted in higher percentage of germination and took less time to germinate than seeds dried in the sun (Vecchio and Shirwa, 1987). Germination capacity after 3 years was highest (40 per cent) in Coorg Honey Dew seeds stored in N_2 at 5° or 15°C; no seeds stored under ambient conditions germinated after 2 years. Increasing storage temperature reduced the percentage of germination and N2 storage was more effective than CO_2 or vacuum storage (Doijode, 1992). Pre-storage soaking of seeds in tannic acid (10-3M) improved the germination percentage over control after 140 days (Hore and Sen, 1993).

6.1.3 Seed Germination and Seedling Vigour

The position of the seed (top, centre or bottom) in the fruit (cv. Comun) had no effect on germination percentage, sex expression and seedling vigour (Sao Jose and Cunha, 1988). Germination starts 2 to 3 weeks after imbibition. Germination of papaya seeds can be improved by various pre-sowing seed treatments. Soaking of seeds (cv. Maradol Roja) for 24 h in sodium thiosulphate (20 mg/l) resulted in maximum percentage of germination, while treatment with the same biostimulant at 100 mg/l gave the best seedling growth and development (Mederos Olalde and Rodriguez Hernandez, 1988). Pre-soaking of seeds in GA3 (50-200 mg/l) did not increase germination percentage but greatly enhanced vigour (Begum *et al.*, 1988). Pre-sowing seed treatment with tannic acid (10^{-3}M) for 8 h gave maximum percentage of germination after 140 days' storage, followed by GA3 (200 ppm) for 8 h; however, seedling height and leaf number were greatest with 8 h treatment with GA3 (200 ppm) or sodium phosphate (10^{-3}M) as compared to control (Sen *et al.*, 1990). Combining matriconditioning (with Micro- Cel E) with GA4+7 (100 or 200 ?M) could effectively reduce germination time and improve seedling emergence and is recommended as a standard procedure for testing papaya seed germination (Andreoli and Khan, 1993). Treatment of papaya seeds (cv. Co.1) with 2 per cent solution of urea + micronutrient gave the highest germination (Patil *et al.*, 1995a). Germination percentage increased in seeds desiccated in the shade and treated with GA3 (300 ppm), while average germination time decreased as the duration of the desiccation increased in seeds deprived of their sclerotesta and treated with GA3 (Bertocci *et al.*, 1997).

The best time for raising seedlings is the middle of June to the end of October. Cheema and Dani (1930) stated that sowing after this period will not give satisfactory results. In North India, where occurrence of frost is common, seeds are generally sown from February to April (Singh and Dahiya, 1982). Naik (1949) suggested that sowing may be done at any time of the year except during the hottest and the coldest periods and also during heavy rains, but June to November provides the optimum sowing season. In the eastern part of the country, seeds are usually sown in April-May, so that the seedlings are ready for transplanting before the onset of monsoon. About 500 g seed is sufficient to raise seedlings for a hectare of land.

Seeds are sown on raised nursery beds at a depth of 1 cm in rows spaced 15 cm apart with 2.5 cm spacing in the row. Watering with a rose-can is done immediately after sowing and, thereafter, regularly in the morning or evening except during the rains. Seeds start germinating within 15-20 days.

Seedlings can also be raised in polyethylene bags of 22 x15 cm size of 150-200 gauge. Two to three seeds are sown in each bag filled with a mixture of sieved soil, farmyard manure and sand in equal proportions. The seedlings are left to grow in the bags till these are ready for transplanting in the field. It is also possible to raise a plantation by sowing the seeds directly at the orchard site but this method involves higher seed rate and increased cost. Seedlings will be ready for transplanting in the main fields when they are about 60 days old.

6.2.0 Cutting

Papaya can also be propagated by cuttings, but the method is more expensive and laborious, and requires a large number of mother plants; hence, the method is not usually followed. In exceptional cases, where superior types of trees bearing parthenocarpic fruits are met with, it may be of practical use (Shetty, 1953). However, recent findings indicate that rapid propagation of superior papaya clones by cuttings is now possible for individual growers, to ensure the supply of uniform, high quality fruits on the market. Injection of BA/GA3 induced more, evenly sized side shoots all along the stem of 1- or 2-year-old plants and reduces the apical dominance of a few vigorous shoots at the top of the stem. The rooting of cutting was significantly improved and disease problems in the mist bed were minimized when small cuttings were immersed in a solution of fungicides (captan, benomyl, or mancozeb), and growth retardant (paclobutrazol) or IBA. Smaller cuttings (80 to 150 mm long x 8 to 12 mm diameter) gave more rapid and better rooting than the larger cuttings and up to 95 per cent rooting was obtained in 4 weeks with 100 cuttings/m^2 during the summer months (Allan and MacMillan, 1991). Around 24 lateral shoots/year could be taken from a parent tree for cuttings (Chang and Liaw, 1994). Injection or direct application on stems below the leaves of various synthetic cytokinin/gibberellic acid mixturesinduced lateral bud burst on the strong apically dominant stems of 1- to 2-year-old Honey Dew plants. Removal of terminal growing point and fruits, but retention of leaves promoted development of lateral buds. By leaving stubs of the side stems, when cuttings with pencil-sized stems are taken, further growth of new shoots was obtained and cuttings could be harvested monthly provided adequate leafy canopy was left on the parent plant. Cuttings were trimmed to leave 3-4 small leaves, immersed in a fungicidal bath of suitable concentration to prevent phytotoxicity and ensure leaf retention during rooting of cuttings in sterile media under intermittent mist with bottom heat of 30°C, after basal IBA treatment (Allan, 1995). Treatment with IBA (25, 50 or 75 ppm) alone or in combination with ascorbic acid (250, 500 or 750 ppm) gave the best results with regard to rooting (Mazumdar, 1988). Rooting was better when cuttings were inserted horizontally following treatments with IBA (1000, 1500 or 2500 ppm), NAA (1000, 1500 or 2000 ppm) or IAA (1000 ppm) in lanoline. There were no significant differences between these chemical treatments (Yimsawat and Feungchan, 1992).

6.3.0 Budding

Scion buds from vigorous female plants of the cv. Pusa Dwarf were shield or patch budded on seedling rootstocks in mid-July, mid-August, mid-September and mid-October. Patch budding gave higher success rates than shield budding, and mid-September was the best time for budding followed by mid-August (Singh *et al.*, 1986). According to Crabbe (1987) up to 200 plants/year could be produced from a single mother tree and the optimum times for budding were February, March, August and September, when up to 100 per cent success was achieved.

6.4.0 Micropropagation

Rapid multiplication of true-to-type, disease-free desirable genotypes of papaya clones by micropropagation have been attempted successfully (Chang and Liaw, 1994). Explants of apices of nursery and orchard trees of cv. Sun Rise Solo were established in a modified Murashige and Tucker basal medium with half-strength inorganic salts, 0.5 mg/l BA and 0.2 mg/l NAA. Established explants were transferred for proliferation in same medium with 0.5 mg BA and 0.1 mg NAA per litre. Rooting was induced at a higher frequency by subculturing on media with IBA than with NAA (Winnaar, 1988). Transfer of rooted plantlets, regenerated on MS medium with half-strength salts and vitamins supplemented with 0.5 per cent sucrose and 0.5 ppm IBA to hormone-free MS medium for 30 days before transplanting to vermiculite increased the survival rate from 18 per cent to 55 per cent (Chow and Liu, 1987). Small axillary buds were cultured first on solid medium then in liquid medium on roller drum. Apically dominant shoots were produced and grown on Drew and Smith (DS) medium contai-ning NAA and BA. Roots were initiated on a reduced mineral medium and plants were transferred to the field (Drew, 1988). High percentage rooting of shoots, derived from the culture of explants of axillary buds taken from lateral shoots, was obtained on MS medium with half-strength macroelements supplemented with 1.0 mg/l IBA (Reuveni *et al.*, 1990). Rahaman *et al.* (1992) cultured shoot apices of cv. Rajshahi red on MS medium supplemented with different concentrations and combinations of BA and NAA. Successfully established explants were transferred to MS medium supplemented with various concentrations of BA, kinetin or 2iP and NAA with or without GA3 for shoot multiplication. The survival rate of shoot apices increased with increasing concentrations of BA and NAA. Shoots were transferred to half-strength medium containing 1 mg/l IBA for root induction. Rooting of micro-cuttings taken from shoot tip culture was markedly enhanced by the treatment of the base of the cuttings with 250-2500 mg/l IBA (Kataoka and Inoue, 1991). Optimum explant survival and subsequent growth of apices dissected from mature, field-grown plants were obtained on MS medium using 10 mg kinetin + 2 mg NAA per litre. The maximum number of shoots/explant was produced from shoots cultured in 0.5 mg BA + 0.1

mg NAA/l. These shoots rooted well in MS medium supplemented with 1 mg BA/l, and 60 per cent of the shoots developed into plantlets upon transfer to soil (Islam *et al.*, 1993).

Cross-sections of the petioles were induced to form callus with 62 per cent efficiency on MS medium supplemented with 2.5 µM 2,4-D and 0.4 µM BA. Culture of callus in MS medium with 5.0 µM 2,4-D produced numerous embryoids which were able to germinate to form single or multiple shoots on MS medium supplemented with 0.1 µM NAA and 0.8 µM BA. The shoots were then rooted and transplanted to vermiculite (Yang and Ye, 1992). Shoots regenerated from petiole segments produced roots on half-strength MS medium supplemented with 2.5 mg IBA + 5 g glucose/l (Khatoon and Sultana, 1994).

Low temperature below 16°C markedly depressed papaya shoot growth developed *in vitro*. Shoots conserved at 16°C still showed a high survival rate and were able to resume growth even after 12 months of culture (Suksa-Ard *et al.*, 1997). The rooting ability of microcuttings was dependent on the genotype, and those cultured under continuous light intensity of 32.5 and 56.6 mM $m^{-2}s^{-1}$ had rooting percentage of 56.6 per cent (Teo and Chan, 1994).

Ovules, embryos and hypocotyls of papaya cv. Sunrise Solo, used as explants for the induction of callus and somatic embryogenesis, required different concentrations of 2,4-D and sucrose to enable these processes to take place. Mature ovules did not produce callus or somatic embryos, but embryos and hypocotyls were good explants for the production of embryogenic callus. 2,4-D was essential for somatic embryogenesis; hypocotyls required a narrower range of concentrations than mature embryos. Sucrose (>60 g/l) was important for embryogenesis in mature embryos. Hypocotyl callus formed embryos at a high rate, even after subculture on media containing only sucrose (Suksa-Ard *et al.*, 1999).

A transparent plastic chamber has been designed for hardening the *in vitro* generated plantlets (Singh and Pandey, 1988). The success in *in vitro* culture of papaya depends on choice of tissue, season and pre-excision treatments. Explants from decapitated plants showed higher survival percentage than non-decapitated controls and explants of cvs. Pusa Dwarf and Pusa Delicious taken during December-February failed to establish, while highest survival percentages were recorded for June-September, with shoot tip explants performing better than lateral buds (Pandey *et al.*, 1986). The rooting substrates play some role on the morphology of papaya roots formed *in vitro* (Kataoka, 1994).

6.4.1 Embryo and Ovule Culture

In earlier days, one of the chief objectives of papaya improvement programmes was to hybridize *C. papaya* with the sexually incompatible *C. cauliflora* which is resistant to ring spot virus. Incompatibility is mainly due to the failure to endosperm formation. Subsequently, the hybrid embryos resulting from the above-said interspecific crosses have been successfully rescued on White's medium in 30 days by Yang (1986), in 60 days by Khuspe *et al.* (1980) and in 90 days by Rojkind *et al.* (1982). The ninety days old selfed embryos have been successfully rescued on MS (half-strength) medium by Jayalatha (1994) and subsequently by Saraswathi (1996).

Magdalita *et al.* (1996) have also reported an improved protocol for rescuing the embryos of *Carica* interspecific hybrids. Shanthi (1996) germinated 120 days old embryos of Sunrise Solo and Kapoho Solo on MS (half-strength) medium containing additives like caesin hydrolysate (500 mg/l), glutamine (400 mg/l) and coconut water 15 per cent (v/v) in addition to BAP and NAA. While Chen *et al.* (1987 and 1991) and Fitch and Manshardt (1990) demonstrated the induction of somatic embryos from immature embryos of hybrid and selfed papayas, respectively.

Ovules were excised from unpollinated *Carica papaya* ovaries before flowering and cultured on media containing cytokinins, auxins and gibberrellins. Adventitious embryos were induced only in the culture media with mixed gibberellins and gibberellic acid (GA_3) in the dark. No embryos were initiated with cytokinins and auxins, whereas the gibberellin mixture (GA_3 >80 per cent, $GA_1 + GA_5$ about 10 per cent, minute amount of $GA_2 + GA_7$), and GA_3 and GA_5 individually promoted induction, but GA_1, GA_5 and GA_7 were ineffective (Tokumoto *et al.*, 2000).

6.4.2 Anther Culture

Litz and Conover (1978b) recovered low frequency of haploids by culturing anthers excised from 16-20 cm flower buds in liquid MS medium supplemented with 87.6 mM sucrose, 10 g/l activated charcoal, 8.8 µM BA and 2.7µM NAA. They too have excised anthers from 8-16 mm long inflorescence and cultured them in stationary liquid MS medium fortified with 3 per cent sucrose, 1 per cent activated charcoal, 2.0µM BA and 0.5 µM NAA for the recovery of haploids.

6.4.3 Somatic Embryogenesis

Callus derived from cotyledon and hypocotyl segments of papaya cv. Sunrise on MS medium was induced to form embryoids and shoots. Lateral shoots were then rooted on half-strength MS medium with 0.5 per cent sucrose (Yang, 1986). The peroxidase isoenzyme patterns of leaf tissues from 5- to 6-month-old *in vitro*-derived plantlets of hermaphrodite pistillate and staminate plants were identical as those of mature plants and successfully distinguished the sex and cultivar of each plantlets (Sriprasertsak *et al.*,1988). Protoplast isolated from suspension cultures of somatic embryos, induced from immature zygotic embryos of *C. papaya* x *C. cauliflora*, produced somatic embryos on culture in liquid KM8P-S medium. The somatic embryos proliferated rapidly through embryogenesis on MS medium and developed into plantlets (Chen and Chen, 1992). Fresh somatic embryos obtained from suspension cultures of papaya contained 90 per cent water, had poor regeneration capacity and could not be stored. After desiccation to 50-70 per cent water, 82 per cent of somatic embryos produced seedlings on MS medium (Ye *et al.*, 1993a).

An *in vitro* suspension culture system for papaya hybrid Yuan You 1, yielding high frequencies of high quality somatic embryos (artificial seeds) were developed. The survival rate of seedlings reached 72 per cent on transfer to main field and the plants derived from artificial seeds grew normally to flowering and fruit set (Ye *et al.*, 1993b). Embryogenic calli were formed with high frequency on *in vitro*-cultured pieces of stem, epicotyl and petiole from 2-month-old plantlets, which developed into embryos of unicellular origin and produced plantlets (Monmarson *et al.*, 1994). Histocytological study during embryogenic callus culture proved that friable calli capable of producing embryos was of maternal origin, and that the cloning of adult plants was possible (Monmarson *et al.*, 1995).

7.0 Area and Production

The main five countries that have been papaya producers during the last 50 years are India, Nigeria, Brazil, Mexico, and Indonesia. In the 1960s, most of the total papaya production of the world was concentrated in India, Indonesia, and Nigeria; however, in the 1980s and 1990s, Brazil and Mexico increased their production to the extent that they produced equivalent quantities to those of India and Nigeria from 1980 to 1990 (Fuentes and Santamaria, 2014). In the last decade, India and Brazil have maintained their continuous increase in papaya production to rank fi rst and second, after producing almost 4 and 1.8 million t, respectively, in 2009 (FAOSTAT, 2011 cited by Fuentes and Santamaria, 2014).

The production of papaya in different papaya growing countries of the world is presented in Table 3.

Table 3: Production of Papaya in different Countries*

Country	*1961*		*2007*	
	Area (ha)	*Production (MT)*	*Area (ha)*	*Production (MT)*
Brazil	3900	49725	36700	1898000
Mexico	1674	35757	20000	800000
Nigeria	35000	250000	92500	765000
India*	6000	228000	73000	2317200
Indonesia	1800	200000	8000	645000
Ethiopia	NA	NA	11000	230000
Congo	7400	126000	13000	215000
Peru	1580	25000	14000	175000
Philippines	13720	47699	9500	164000
Venezuela	1700	22913	9600	151400
World	114192	1322877	371345	6937097

* Source: FAOSTAT (2008), NHB Indian Horticulture Database (2006).

In India, though it is successfully grown all over the country, the important papaya growing states are Karnataka, Kerala, Assam, West Bengal and Bihar. The area and production of papaya in different states of India are presented in Table 4.

8.0 Soil and Climate

8.1.0 Soil

Papaya is a quick-growing, continuous fruiting, evergreen plant, hence, it requires a good fertile soil. A well-drained sandy loam soil, rich in plant food, is the best for papaya cultivation (Singh and Dahiya, 1982). Papaya thrives well on soil with pH ranging from 5.5 to 7.0 (Maxwell *et al.*, 1984). Plants are highly sensitive to flooding (Maxwell *et al.*, 1984; Malo and Campbell, 1986) and waterlogging may cause death of plants in 3 to 4 days (Samson, 1980). The waterlogging symptoms are defoliation of old leaves and chlorosis of remaining leaves. Reduced leaf gas-exchange is a common response to flooding, even with tolerant species (Schaffer *et al.*, 1992). Complete stomatal closure occurred by day 3 in almost all plants receiving no aeration and by day 9 to 10 in those receiving partial aeration (Marler, 1990).

Table 4: Area, Production and Productivity of Papaya in different States of India*

State	*Area (000 ha)*	*Production (000 tonnes)*	*Productivity (t/ha)*
Assam	7.21	147.40	20.44
Bihar	1.90	42.70	22.45
Gujarat	20.31	1256.51	61.86
Karnataka	8.75	593.69	67.84
Kerala	16.15	95.03	5.88
Maharashtra	10.28	408.30	39.71
Odisha	3.05	70.29	23.05
West Bengal	12.38	365.95	29.56

* Anon (2018).

Reports on tolerance of papaya to salinity stress are contradictory. It has been ranked as not tolerant (Ogden *et al.*, 1981), poorly tolerant (Maxwell *et al.*, 1984), moderately tolerant (Maas, 1984) and extremely sensitive (Raheja, 1966) to salt stress. Papaya seedling growth was not much affected at an EC of 2 $ds.m^{-2}$, but was reduced to about 50 per cent of control plants at 4 $ds.m^{-2}$, while plants receiving NaCl at 6 and 8 $ds.m^{-2}$ died (Makhija and Jindal, 1983). Stomatal conductance of container-grown papaya tended to decline rapidly following irrigation with 4 or 8 $ds.m^{-1}$ water (Marler, 1990). Salinity delays the rate and percentage of papaya seedling emergence.

8.2.0 Climate

Papaya cultivation is extended throughout tropical regions of the world. The optimum temperature for papaya is reported to be 21° to 33°C (Knight, 1980).High temperature tended to induce more maleness in papaya plants that produce flowers with varying degrees of maleness and femaleness (Malo and Campbell, 1986). The net CO_2 assimilation rate of papaya leaves was found to be maximum between 16° and 30°C and declined linearly at above 30°C. It was reduced to 50 per cent at 41°C (Allan and de Jager, 1978). It was suggested that intermittent sprinkling of canopies might be useful to reduce heat stress by evaporative cooling. Low temperature (-1°C or lower) caused severe damage to plants (Malo and Campbell, 1986), and –0.6°C caused freezing damage (Maxwell *et al.*, 1984). In northern India, the occurrence of low temperature and frost restricts its cultivation (Singh and Dahiya, 1982). Low night temperature below 20.5°C caused more carpellodic fruits (Awada, 1958). Ong and Kwok (1983) stated that the fluctuations in the papaya yield was mainly influenced by high day temperature during the time of anthesis – the time when abortions occurred due to dry spell during the second phase of papaya fruits. The incidence of malformation was high on fruits from flowers initiated during low winter temperatures (<16°C) and the cv. Sunrise Solo was highly susceptible to this disorder (Sippel, 1988). In a low temperature regime (day/night temperature of 15/5°C) the decrease in photosynthesis was more pronounced in *C. papaya* than in *C. quercifolia* (15 and 42 per cent of controls respectively). The reduction in stomatal conductance was similar in both species (25-30 per cent of controls), indicating that CO_2 fixation was more affected in *C. papaya*, which showed higher intercellular CO_2 levels under stress (Grau and Halloy, 1997).

Papaya is classified as a shade-avoiding species, as it exhibits considerable morphogenetic plasticity to shade (Grime, 1981). The effects of reduced irradiance and the quantum ratio of red to far-red wave lengths (R : FR) were studied by Buisson and Lee (1993). It was observed that leaf thickness, specific leaf weight, stomatal density, palisade parenchyma cell shape, and the ratio of mesophyll air surface : leaf surface were affected by reductions in irradiance, while some characteristics like reduced chlorophyll a:b ratios, reduced lobing, and greater internode length were reported primarily by low R:FR. The reduced R : FR of foliage shade strongly influenced the morphology and anatomy of papaya leaves.

The quality of papaya fruit has been found to be better in the low humid regions of South Africa (Malan, 1953). However, a relative humidity of 60 per cent was reported to be essential for optimum growth (Anon., 1986). On the other hand, water stress (drought) severely reduced the net CO_2 assimilation rate and it took 3 days following re-watering before net CO_2 assimilation rate returned to pre-stress levels (Allan and de Jager, 1978). For drought stressed plants, stomatal conductance declined to a greter degree following the high-to-low photosynthetic photon fluxes transition and stomatal conductance and net CO_2 asimilation recovered more slowly following the transition from low-to-high photosynthetic photon fluxes than for well-watered plants. It was suggested that papaya stomata were able to track rapid changes in irradiance, and mild drought stress enhanced the tracking response (Clemente and Marler, 1996). Waterstress caused chlorosis of older leaves followed by shedding which was one of the means of reducing future water loss and protecting the young leaves. It, therefore, appeared that papaya plants adapt to drought by dehydration postponement.

Papaya plants are sensitive to strong wind. It can be cultivated up to an elevation of about 1000 m (Madhava Rao, 1974).

9.0 Cultivation

9.1.0 Preparation of Pit and Planting

The orchard site for papaya should be well prepared through repeated ploughing and harrowing. After sufficient weathering, the pits (45x45x45 cm) are filled with top soil and farmyard manure in 3:1 ratio. Cheema and Dani (1930)

Cultivation of papaya

Furrow irrigation in papaya

Drip irrigation in papaya

Intercropping chilli (left) and pineapple (right) with papaya

Intercropping tomato with papaya

Well maintainted papaya orchard

Bearing orchards of papaya

suggested to fill the pits with surface soil mixed with well-rotted farmyard manure at the rate of 80 to 100 lb (36-45 kg) and bone meal at the rate of 3 lb (1.36 kg) per pit. The seedlings raised in polythene bags or in nursery beds are transplanted, preferably in the evening.

In case of dioecious cultivars like Co.1 or Co.2, four seedlings are transplanted per pit whereas it would be sufficient to plant 2 seedlings per pit in the case of hermaphrodite types like Solo, Coorg Honey Dew, *etc.*

There was a positive linear relationship between trench depth and plant height after 6 months, but the number of fruits/plant and fruit weight was not affected by trench depth (Basso-Figuera *et al.*, 1995).

According to Ram (1996) papaya seedlings are planted in three distinct seasons as follows:

Season	*Region*
Spring (February-March)	Where mild climate prevails through-out the year
Monsoon (June-July)	In frost prone areas
Autumn (October-November)	Where rainfall is heavy and virus problem is acute in rainy season

In a trial on the effect of planting season on papaya plants, Singh and Sharma (1996b) observed that planting in January or February resulted in the highest percentage of survival and the highest fruit yields. The lowest yield was obtained with July planting, which was attributed to high rainfall. The percentage of female plants was generally higher when planting was carried out in cool months, being highest for December and January planting and lowest for May planting when the temperature was about 35°C.

9.2.0 Planting Density

Purohit (1981) suggested 1.8 x 1.8 m spacing for cvs. Co.1, Co.2 and Solo, while 2.4 x 2.4 m for Coorg Honey Dew and Washington. Ram (1996) stated that the dwarf mutant cultivar, Pusa Nanha could be planted at a spacing of 1.25 x 1.25 m. The optimum results were obtained by spacing the Coorg Honey Dew papaya plants at 1.33 x 1.33 m *i.e.* 5609 plants/ha (Kohli *et al.*, 1986). Increasing planting density resulted in increase in plant height, decrease in basal girth, plant spread, and petiole length and diameter, delay in flowering and a reduction in the number of fruits/plant, average fruit weight and fruit yield/plant. However, the yield/ha increased with increasing plant density and the highest yield was obtained at 1.85 x 1.85 m (cv. Ranchi) spacing (Biswas *et al.*, 1989).

In Puerto Rico, the cvs. P.R-6-65 and P.R-7-65 gave higher yield when planted at 1.2 x 1.2 m spacing than at 1.8 x 1.8 m spacing (Perez and Varges, 1977). In Cuba, experiment with cv. Maradol Rojo showed that yield per plant decreased from 40.8 to 22.1 kg as planting density was increased from 1250 (4.0 x 4.0 m) to 2500 plants/ha (4.0 x 1.0 m). The maximum yield per hectare (25.1 tonnes) was, however, recorded by using 1666 plants/ha (4.0 x 1.5 m) (Camejo and Alvarez, 1985). Kist and Manica (1995) did not find any effect of spacing on stem diameter, plant height or average fruit weight in the first productive year of papaya cv. Tainung 2 (Formosa group hybrid), but the yields per plant and per hectare were high at 2.0 x 2.0 m, 2.0 x 2.2 m and 2.0 x 4.0 m.

9.3.0 Removal of Male Plants

This operation is necessary only when dioecious cultivars like Co.1, Co.2 are cultivated. The male plants serve only as pollenizer and hence, it would be adequate to leave one male plant for every 20 female plants. At flowering, all the male plants other than those left for pollination should be removed. Similarly, leaving only one robust female plant in each pit, all other female plants are to be removed. It was estimated that the average yield could fall from 37.23 t/ha with a maximum of 6 per cent male plants present to 19.96 t/ha with 50 per cent male plants (Mederos Olalde *et al.*, 1995).

9.4.0 Irrigation

Papaya responds well to better water management. Optimum soil moisture helps in rapid fruit development and also to obtain regular fruit yield. Water deficit significantly reduced number of flowers and fruits by 86 per cent and 58 per cent, respectively and also retarded growth and development of papaya fruit (Masri *et al.*, 1990). The plant is highly sensitive to waterlogged condition and, hence, it is most important to prevent 'wet feet' in papaya irrigation. In well-drained soil, irrigation at shorter intervals during the early crop stages results in good establishment and also encourages better plant development. In general, irrigation to grown up plants is given once in 7-10 days. The ring system of irrigation is better than the bed system, because the ring system prevents irrigation water from coming into direct contact with the stem, thus preventing collar rot (Purohit and Singh, 1978).

According to Awada (1961), proper irrigation and fertilizer application for a particular environment should be practised if maximum productivity is to be achieved. It was observed by him that high moisture level associated with a normal growth rate tended to shift the plants towards the production of commercially desirable Solo fruits. Further, lower moisture levels shifted plants towards sterility and male floral characters, while higher moisture resulted in excessive production of undesirable carpellodic types. Total water requirement decreased from 2191 mm at 20 per cent depletion of available soil moisture (ASM) to 1815 mm at 80 per cent depletion of ASM (Balasubramaniam and Rao, 1988). Awada *et al.* (1979) studied the effects of drip irrigation at 3, 6, 9, 12 or 15 gal/plant daily on vegetative growth, fruit yield andmineral composition of the petiole in Solo papaya and recommended that the irrigtion rate should be 1.3 times of the pan evaporation of water from the previous week. The yield of Solo papaya in terms of number and size of marketable fruits increased with increasing irrigation. It was further reported that petiole concentrations of Ca, B and Na increased and those of Mg and Cl decreased with increasing irrigation. Yields were significantly higher in the drip-irrigated (1 or 2 lateral lines/row) plots than for basin plots (90 x 90 cm or 180 x180 cm). Drip irrigation (both systems) used 50 and 66 per cent less water than the 90 cm and 180 cm check basins respectively (Padmakumari and Sivanappan, 1989).

The effects of drip (trickle) irrigation at 1, 0.8 or 0.6, or conventional irrigation at 1 IW/CPE (ratio of irrigation water:cumulative pan evaporation) on the yield of cv. Honey Dew were investigated by Biswas *et al.* (1999). Higher yields (36,225-40,200 kg/ha) were obtained from drip-irrigated plots compared with those irrigated using a conventional system (31,200 kg/ha). The best treatment in terms of yield and water use efficiency was drip irrigation at an IW/CPE ratio of 0.8.

Papaya plants can tolerate drought to some extent. Fruit yield and quality (TSS, sugar and ascorbic acid contents) in papaya cv. Solo were not significantly affected by irrigation frequency (Jayaprakash *et al.*, 1989). On the contrary, Srinivas and Prabhakar (1993) reported that plants receiving frequent irrigation with replenishment at 75 or 100 per cent evaporation maintained higher relative water content, a higher transpiration rate and a lower diffusive resistance and yielded more than did plants less frequently irrigated. Evapotranspiration increased with irrigation frequency. However, the water use efficiency was highest with 25 per cent evaporation replenishment.

In the subtropical hills of India, papaya is grown as a rainfed crop because these areas receive a well-distributed rainfall from both South-West and North-East monsoons.

9.5.0 Manuring and Fertilization

The nutrition of papaya differs from other fruit crops because of its quick growth, continuous fruiting habit and high fruit yield. Awada and Suehisa (1975) found that the quantity of nutrients removed was in the descending order of K, N, Ca, Mg and P. Nutrient uptake studies conducted at Tamil Nadu Agricultural University, Coimbatore, India showed that the uptake of N, P, K, Ca and Mg was more between flowering and harvesting stages, but uptake was specifically higher between fruit development and harvest stages. The removal of major and minor nutrients per tonne of papaya fruits were as follows : 1170 g N, 200 g P, 2120 g K, 350 g Ca, 180 g Mg, 200 g S, 989 mg B, 300 mg Cu, 3364 mg Fe, 1847 mg Mn, 8 mg Mo and 1385 mg Zn (Cunha and Haag, 1981). Experimental results suggested that liming requirement in papaya grown on acid soil should be considered both on exchangeable Al and Ca as yields were highest on soils with a pH raised to 6.4 or more, and > 8.0 meq. of exchangeable Ca/100 g soil (Villachica and Raven, 1986).

The importance of application of organic manure (FYM, sheep manure, wood ash, bone meal) was stressed upon long back in the year 1930 (Cheema and Dani, 1930; Mowry, 1930; Pope, 1930). Rajput and Sharma (1970) indicated that the application of nutrients to papaya should be 50 per cent as organic and 50 per cent as inorganic manures. Plant height and root growth were not affected by the application rate of 75 or 150 t/ha municipal solid waste (MSW). There was a positive linear correlation between MSW rate and the number of fruits per plant (Basso-Figuera *et al.*, 1995). The percentage of early maturing fruits was positively correlated with MSW rate. Total fruit weight per plant was not affected by MSW rate, but the total number of fruits per plant increased as MSW to the soil can affect yield, increased precocity and improved fruit quality (Basso-Figuera *et al.*, 1994). In an experiment, Hossain *et al.* (1990) reported that N (100 g N/pit) in the form of decomposed cowdung gave highest number of fruits per plant while the highest yield and better quality of fruits were obtained with decomposed cowdung + mustard oil cake (to give 100 g N/pit) in addition to 100 g triple superphosphate and 75 g muriate of potash per pit. Ochse *et al.* (1961) opined that the productiveness of a papaya plant was dependent upon its constant growth for which higher rates of fertilizers and organic matter should be applied at frequent interval. An application of 0.1 kg of 4:8:5 fertilizer mixture to each plant at fortnightly intervals for the first six months and 0.2 kg thereafter resulted in excellent yield in southern Florida. DeGeus (1967) reported that in view of its fast growing character, papaya responded to a continuous supply of nitrogen to ensure good growth and high fruit yield.

From the results of the experiments on fertilization at Coimbatore, it was concluded that application of 200 g each of N, P and K per plant (Co.1) in addition to 25 kg of farmyard manure in split doses during the first, third, fifth and seventh month after planting resulted in higher fruit yield. In another fertilizer trial with Co.2 papaya, it was noted that application of 250 g of N per plant per year in six split doses at bimonthly intervals commencing from the second month after planting was optimum for a good yield of fruits and papain. Experiments conducted at the Indian Institute of Horticultural Research, Bangalore, showed that 240 g N, 500 g P and 500 g K per plant per year gave the best result in Coorg Honey Dew cultivar. Satyanarayana Rao (1971) reported that while phosphorus reduced the acidity, potassium increased the TSS, total sugars and ascorbic acid contents of the fruit. In another nutritional trial conducted at Karnataka, Sulladmath *et al.* (1981) found that application of N and P each at the rate of 250 g/plant/year resulted in the highest yield of 79.9 fruits per plant, weighing 25.76 kg (64.4 t/ha). While K application significantly increased the total soluble solids and skin thickness, seed weight of fruits decreased with increasing levels of N, P and K. Nitrogen and P_2O_5 application resulted in significant increase in number of fruits/plant and fruit yield compared to control (Jayaprakash *et al.*, 1989). Fruit yield was correlated with available nitrate content of the soil (Hossain *et al.*, 1990). Plant height and girth were not affected by N level (Hossain *et al.*, 1990; Reddy *et al.*, 1990). However, the least number of days taken for 50 per cent flowering and highest yield of papaya cv. Coorg Honey Dew were obtained with 250 g N/plant applied in 6-split doses in addition to basal application of 250 g P_2O_5 (as superphosphate) and 500 g K_2O (as muriate of potash) per plant (Reddy *et al.*, 1990). Warner (1993) reported that the yield of papaya cv. Tainung 2 increased linearly with increasing N application from 50 to 150 kg/ha, but application interval did not significantly affect yield. Petiole N content increased with N application rate but fruit quality was not affected. Srinivas and Prabhakar (1993) observed that fruit yield, petiole N content and water use efficiency were higher with 450 g N than with 150 g N/plant. Soil application of 200 g N +100 g P_2O_5/plant, given in 4 split doses at 90-day intervals starting from transplanting, plus foliar spray of 1 per cent urea + 0.2 per cent B and IAA at 50 ppm also applied at 90-day intervals starting from 15 days after transplanting, gave the best results (Lokhande and Moghe, 1991).

Potassium deficiency symptoms included yellowing and necrosis of lower leaves, reduced leaf areas and plant height, poor root and root hair development and reduced leaf water potential. Apart from these, the deficiency of K also caused increases in soluble protein and carbohydrate contents, accumulation of putrescine, increased respiration and reduced pyruvate kinase activity (Pan *et al.*, 1994). The highest yield of good quality fruit was obtained with 350 g N + 600 g K_2O/plant (Biswas *et al.*, 1989). Increasing K application rate from 168 to 504 g/plant, in general, increased the total fruit weight per plant and extended the shelf life, but the proteolytic activity of crude papain tended to decrease. In general, $K_2Mg(SO_4)_2$ form gave better results than KCl in terms of both fruit and latex yields (Harjadi *et al.*, 1995). Best response with respect to number of fruits/plant, yield/plant, and yield/ha was obtained with the highest rates of P (300 g P_2O_5/plant/year) and K (600 g K_2O/plant/year) and the medium rate of N (200 g N/plant/year). Fruit TSS, sugar, acid, ascorbic acid and total carotene contents increased with K and P rates, but were adversely affected by the highest N rate (Ghanta *et al.*, 1995). Nitrogen at 200 g in combination with P at 50 g and K at 100 g/plant was the most effective dose in cv. Pant Papaya-1 for increasing fruit yield and quality (Lavania and Jain, 1995).

Inoculation with vesicular arbuscular mycorrhiza (VAM) has been reported to be beneficial for growth and uptake of nutrients of papaya plants. Plants grown with inoculum (especially that of *Glomus mosseae*) were taller, with greater DW and higher P, N, and Zn contents, even when no P was present. Inoculation was shown to enhance alkaline and acid phosphatase activities on the root surface and also in the enzyme extract from the root surface. The enhancement was greater for *G. mosseae* and for *G. fasciculatum* (Mohandas, 1992). The papaya cv. Solo plants inoculated with *Glomus etunicatum* showed maximum height, shoot and root DW and nutrient content with 20 ppm P_2O_5, whereas with *Entrophospora colombiana*, maximum root DW and Fe content occurred with 20 and 40 ppm P_2O_5 respectively (Weber and Amorim, 1994). In greenhouse-grown papaya cv. Coorg Honey Dew seedlings, inoculation with a mycorrhizal fungus (*G. mosseae* or *G. fasciculatum*) increased plant P uptake and total P content. It was observed that mycorrhizal plants had utilized forms of P not available to non-mycorrhizal plants (Mohandas *et al.*, 1995).

Apart from increasing the growth and yield of papaya plants, the macroelements also play some role on the disease complex of plants. Application of double dose of NPK fertilizers showed great improvement in plant growth when infested with either *Meloidogyne incognita* and/or *Fusarium solani*. Root-rot caused by *F. solani* significantly decreased with the application of different doses of NPK fertilizers. An increase in N reduced the reproduction factor and number of nematode galls whereas increase in the dose of either P or K enhanced the reproduction factor and number of galls (Khan and Khan, 1995). Vallero (1999) reported the highest yield of 66 kg/plant with 366 kg N/ha equivalent to 110 ton fruit/ha for a plant density of 1665 plants/ha. The treatment with 366 kg N/ha showed leaf deficiency of K after twelve growing months and the highest percentage of plants with virus, compared with other treatments. The virus incidence was not significantly different among treatments with N application.

Allan *et al.* (2000) used four levels of each N, P and K combined in an incomplete factorial (San Cristobal Design) to study their effects on the growth of hermaphrodite papaya cv. Solo under glass over 12 months in South Africa. Petiole samples were taken at 5 phenological stages (start of flowering, end of first fruit set, start of first harvest, end of first harvest, and after end of second fruit set period) for analysis of N, P and K contents. There was a significantly linear response to applied N for all 5 stages, with the concentrations increasing up to the last 2 stages when there was luxury consumption. Applied P showed a linear response for the first 2 stages and then quadratic responses. Applied K gave significant quadratic responses for the first 3 stages and then linear responses, with concentrations rising excessively for the fourth stage but decreasing again as the second crop set. There were significant NK, NP, PK interactions. P concentrations in mature seeds showed a distinct quadratic response. For the first crop, total fruit mass was significantly quadratic for N and P and there was a significant negative NK interaction. There was no difference in fruit quality for the different treatments.

Similar to macroelements, the microelements also influenced the growth, yield and fruit quality of papaya. Fruit quality was considered best following treatment with borax at 0.15 per cent (Pant and Lavania, 1989). They (1998) also reported that spraying papaya cv. Pant Papaya-1 with 0.15 per cent $FeSO_4$ + 0.15 per cent $ZnSO_4$ produced the highest percentage of female plants (56.3 per cent). Yields were highest and fruits largest when 0.15 per cent borax was included in the spraying treatment. Ghanta *et al.* (1992) reported that micronutrient treatments *i.e.* 0.1 per cent B, 0.25 per cent Mn and 0.25 per cent Cu, alone or in combination, at 60 and 90 days after transplanting improved plant growth, yield and fruit quality in papaya cv. Ranchi. Highest yield, individual fruit size, weight and quality were obtained with Mn+Cu treatment. The activities of peroxidase, catalase and poly-phenol oxidase (catechol oxidase) increased with rising iron concentrations in the nutrient solution, but decreased at toxic levels. Leaf chlorophyll a and b concentrations increased with Fe concentration of the nutrient solution 7 days after treatment (Lin *et al.*, 1994).

9.5.1 Tissue Analyses and Nutrient Requirement

Awada and Long (1971a) identified the recently matured petiole as the index tissue for analysis of N, P and K in papaya. The maximum fruit yield was recorded in Solo papaya at 1.45 per cent N in the petiole (critical level) and it was suggested that this could serve as a basis for N fertilization to fruiting papaya plants (Awada, 1969; Wada and Long, 1971b). It was also reported by them that the soluble solids content of fruits was not lowered, nor fruit size was reduced under high N fertilization. Further, they observed that the concentration of petiole N and fruit yield increased with N application, particularly when N was applied in the form of ammonium sulphate nitrate at the rate of 0.25 lb (0.11 kg) N per plant at six-week intervals to 10-month-old plant. It was found by Awada and Long (1971b) that the level of K in recently matured petiole was 3.61 per cent when the yield was maximum and recommended that this might serve as the tentative standard for fertilization. Potassium fertilization also brought about improvement in some fruit quality characteristics like TSS.

Subsequently, Awada and Long (1980) reported that the maximum yield of marketable fruits was associated with petiole concentration of 1.44 per cent N and

2.52 per cent K. The N fertilization increased the petiole concentrations of N, Fe, Mn, Cu and Zn and decreased the levels of P and B, whereas K fertilization increased the petiole concentrations of K and Mn and decreased those of N, P, Ca, Mg, Na and B (Table 5).

Table 5: Nitrogen and Potassium Fertilization Effects on Petiole Composition of 24- to 48-month-old Papaya Plants cv. Solo*

Treatments	*N (Per cent)*	*K (Per cent)*	*P (Per cent)*	*Ca (Per cent)*	*Mg (Per cent)*	*Na (Per cent)*	*Fe (ppm)*	*Mn (ppm)*	*Cu (ppm)*	*Zn (ppm)*	*B (ppm)*
Low N	1.13	2.40	0.20	1.10	0.52	0.25	21.7	18.2	4.8	17.8	22.2
Medium N	1.33	2.27	0.18	1.07	0.51	0.24	22.9	17.3	5.1	19.9	22.0
High N	1.53	2.21	0.19	1.10	0.51	0.28	23.6	20.8	5.4	20.6	21.6
Low K	1.45	1.57	0.20	1.18	0.61	0.40	23.8	16.7	5.1	19.7	22.2
Medium K	1.25	2.45	0.18	1.10	0.49	0.19	22.6	19.2	5.1	19.2	21.9
High K	1.29	2.87	0.18	1.00	0.44	0.17	21.7	20.4	5.2	19.4	21.7

Fertilizer rates at 3-month intervals were : low N=0, medium N=0.054 and high=0.108 kg N/tree; low K=0, medium K=0.045 and high K=0.090 kg K/ha.

* Awada and Long (1980).

Regarding P fertilization of papaya, Awada and Long (1969) reported that 0.25 per cent P in the recently matured petiole was the critical level and this would tentatively serve as a basis for P fertilization. It was also reported by Awada and Long (1969) that P fertilisation raised the concentrations of P, N, Mn and Zn and lowered those of Cu and K. In a subsequent study, the critical petiole P level for optimal fruit yield of papaya was found to be 0.186 per cent (Awada and Long, 1977). Papaya exhibited tremendous growth and fruit yield responses to P fertilization at a rate of 202 lb (90.9 kg)/acre applied in two split doses, 102 lb (45.6 kg) during the vegetative stage and 100 lb (45 kg) from flowering through bearing. It was recommended by Awada and Long (1978) that 14 per cent of total P should be applied during the vegetative stage (11 weeks), 54 per cent during the vegetative and bud stages (12 weeks) and 32 per cent during the flowering and fruiting stages (52 weeks) and this rate corresponded to 0.16 per cent P in terms of its petiole concentration, which was slightly below the adequate range of 0.17-0.20 per cent. They observed that an increase in the rate of nitrogen resulted in larger number of fruits and greater yield of marketable fruits, while application of phosphorus was found to increase the number of fruits but did not affect the yield of marketable fruits (Table 6).

An experiment with Co.2 papaya was conducted in Orissa by Das and Sahu (1981). It was reported by them that as the plant was advancing in age, the N and P contents of the leaf showed a decreasing trend, while the leaf K content increased as the plants entered the fruit development stage.

Table 6: Relation of Nitrogen and Phosphorus Fertilization to Fruiting of Solo Papaya*

Treatments	*No. of Fruits per Tree*	*Mean Fruit Weight (kg)*	*Total Marketable Fruit (kg/tree)*	*Total Culls (kg/tree)*
Low N	97	0.479	42.5	3.9
Medium N	189	0.460	65.8	20.8
High N	182	0.450	61.4	20.6
Low P	135	0.457	53.1	8.8
Medium P	155	0.461	57.3	13.8
High P	177	0.465	59.3	22.8

Fertilizer (kg/ha) applied from vegetative through bearing : low N (171) : medium N(686) : high N (1398); low P (41) : medium P (186) and high P (723).

*Awada and Long (1978).

9.6.0 Weed Control

Weeding should be done regularly to keep the field weed free. In the field with grown up plant, the interspace remains covered well with the top growth which helps in checking weeds. Investigations on chemical control of weeds with herbicides in papaya have clearly indicated the effectiveness of the treatments. In Hawaii, application of paraquat at 0.5-1.0 lb/acre (0.23-0.45 kg) and karmex (diuron) at 2.5-5.0 lb/acre (1.13-2.26 kg) was effective in controlling the weeds and increasing yield (Romanowski *et al.*, 1972). Application of glyphosate at 1.1 kg/ha was found to control all annual weed species, while rates of 2.2-4.4 kg/ha were required to control perennials (Nishimoto and Hibbard, 1979). Sauls and Campbell (1980) reported that application of paraquat caused severe bark necrosis when directed on the lower trunks. Soil application of bromacil, bromacil + diuron, terbacil, simazine and ametryne resulted in severe chlorosis and necrosis, stunting and/or death of the plants. Nishimoto (1981) obtained good control of annual weeds by application of oryzalin at 2.2 kg/ha. Pendimethalin and oryzalin showed adequate selectivity on a silty clay soil. Mendoza and Mercado (1983) recommended application of diuron and ametryne at 2 kg/ha at 3 months after planting and repeat every 2 months for the control of weeds in papaya orchards. Papaya cvs. Sunset, Sunrise and Kapoho Solo, tolerated oxyfluorfen at rates up to 4.5 kg if the plants were 120 cm tall when the herbicide was first applied; application to shorter plants often caused injury or death. Oxyfluorfen at 1.1 kg/ha provided control of most weeds except *Bidens pilosa*, which was better controlled by oxyfluorfen + oryzalin combinations (0.56 + 2.2 kg/ha) than either herbicide by itself (Nishimoto, 1993).

9.7.0 Intercropping

During the pre-bearing age, short-duration vegetables like cabbage, cauliflower, onion, chillies, radish, tomatoes, *etc.*, can be grown as intercrops which not only keep the soil free from weeds but also add to the income of the growers (Singh and Dahiya, 1982). However, it has been advised not to grow crops like chillies, tomato, brinjal to avoid virus problem (Ram, 1996). Care should be taken to maintain the intercrops, so that it will not compete with the main crop. Papaya based cropping system (*e.g.*, papaya+tobacco) has been reported to be remunerative (Ram and Pandey, 1990; Singh and Chari, 1991). It was suggested that although intercropping in papaya orchards is beneficial, it should be limited to the early vegetative and late fruiting phases when the leaf area index (LAI) of papaya was low. A relay of okra followed by watermelon or bush greens (*Amaranthus hybridus*), followed by jute (*Corchorus olitorius*) or sweet potato grown for fodder is considered suitable for cropping in the alleys of papaya (Aiyelaagbe and Jolaoso, 1992). Vieira Neto (1995) in Brazil studied the effect of some leguminous plants as green manure to the papaya planting. He planted papaya cv. Sunrise Solo intercropped with cowpea (*Vigna unguiculata*), feijao-de-porco (*Canavalia ensiformis*), mucuna-preta (*Stirolobium aterrinum*), lablab (*Dolichos lablab*) and single papaya as control. The plots intercropped with leguminous plants did not receive any fertilization of nitrogen except foundation fertilization of 1.5 kg of castor bean cake; 300 g of single superphosphate and 50 g of potassium chloride. The result showed good performance of the *C. ensiformis* and *S. aterrinum* in respect of control (which received recommended N dose). It appeared that leguminous crops were a source of N to the main crop and the cost of N fertilization may be reduced (Table 7). Papaya itself is a popular intercrop with fruit trees. In young mango and litchi plantations, one or two rows of papaya between two rows of trees can profitably be grown. As papaya produces fruit for only 2-3 years, it does not interfere with the main crop. A sustainable horti-silvi-pastoral system has been standardized to increase the productivity of rangeland with a variety of multipurpose forest trees, sapota, papaya, guinea grass and various food crops (Nadagoudar *et al.*, 1995).

Table 7: Growth and Yield of Papaya under the Influence of Green Manuring*

Treatments	*Plant Height (cm) (14 months after planting)*	*Yield (t/ha)*	*Average Fruit Weight (g)*
Control	309.5	65.89	530.75
Stirolobium aterrinum	314.0	61.95	511.00
Canavalia ensiformis	304.0	60.47	499.25
Dolichos lablab	297.0	48.18	552.75
Vigna unguiculata	297.0	48.18	499.25
LSD 1 per cent	41.4	38.77	214.26

* Vieira Neto (1995).

10.0 Flowering, Floral Biology, Sex Expression, Fruit Growth and Development

10.1.0 Flowering

Papaya is a polygamous plant with various sex forms. Three sex forms are commonly found in papaya, *i.e.*, pistillate, hermaphrodite and staminate. Out of these three sex forms, pistillate form is stable and sex reversal is not affected by environmental factors. Male flowers are devoid of the functional pistil. Ten stamens, arranged in two groups of five each are attached to the corolla tube of male flower. The hermaphrodite flowers can be classified into three groups, *i.e.*, intermedia, elongata and pentandria (Storey, 1969). The elongata flowers are considered to be the normal bisexual flowers where the corolla tube is formed by the fusion of petals to their 2/3rd length and has ten stamens in two series. Carpels are five in number and fused. There is a short corolla tube in pentandria flower with five stamens and furrowed ovary comprising of five carpels which vary from 5 to 10.

10.2.0 Floral Biology

The different *Carica* species, *viz. C. papaya, C. monoica, C. cauliflora* and *C. goudotiana* were found to flower throughout the year in India. In *C. quercifolia*, periodically the plants remained dormant. The *C. cauliflora* produced flowers straight from the stem while in *C. papaya, C. monoica, C. goudotiana* and *C. quercifolia*, the inflorescence were axillary in position. It took 21 days from bud emergence to anthesis in *C. quercifolia* compared with 36 days for *C. monoica*. Peak anthesis was found between 5 and 6 a.m. in all species except in the pistillate flowers of *C. cauliflora* and staminate flowers of *C. goudotiana*; stigma receptivity in all the species was found to be maximum on the day of anthesis (Subramanyam and Iyer, 1986). Khuspe and Ugale (1977) stated that the inflorescence of Washington papaya emerged 45-48 days after transplanting. The male flower appears in the axil of the 24th leaf and that of female in 18th to 20th leaf. Female and male flowers developed within 32 and 42 days respectively, after bud initiation. The period from bud initiation to anthesis was shorter for male than for female flower buds (Dhaliwal *et al.*, 1991). Most flowers were open between 7.00 and 9.00 h (Mateos Sanchez *et al.*, 1995). Stamen development occurred prior to ovary development in the hermaphrodite flower and stamen differentiation was observed 56-49 days before anthesis (Sippel and Haltzhausen, 1992). Anther dehiscence was completed within 18 to 36 hours before the flowers opened and the stigma became receptive a day before the flowers opened, remaining receptive for 6 days (Khuspe and Ugale, 1977).

10.3.0 Sex Expression

Linnaeus (1753) classified papaya as a dioecious species. Gammie and Patwardhan (1908) recognised several sex types, such as dioecious pistillate, dioecious staminate, andromonoecious, polygamous, staminate and hermaphrodite flowers. Higgins and Holt (1914) recorded 13 sex forms of flowers, while Pope (1930) and Cheema and Dani (1930) recognised 3 and 11 different sex forms respectively. Storey (1937a,b) reported four types of flowers in papaya and classified them as typical female or pistillate, typical male or staminate, hermaphrodite and intermediate. Hofmeyr and Roux (1938) in his work on papaya, reported 9 different sex forms as female, male, elongata, sterile, hermaphrodite, coenomonoecious, pentandria, coexistence of elongata and pentandria, pistillate and hermaphrodite and pistillate and staminate flowers on the same plant. Agnew (1948) described three main types of flowers, namely, pistillate, staminate and hermaphrodite or bisexual flowers. Further, he classified hermaphrodite flowers into 3 forms as pentandria, intermediate and elongata according to the nature of their structural modification. Variations in sex were also reported by Sen (1940) and Kumar (1952). Storey (1958) identified 8 working categories as staminate, teratological staminate, reduced elongata, elongata, carpelloid elongata, pentandria, carpelloid pentandria and pistillate. According to him, *Carica papaya* has 3 basic sex forms, *viz.*, staminate, pistillate and andromonoecious. The pistillate plant is stable, while staminate and andromonoecious plants may be either phenotypically stable or phenotypically ambivalent going through seasonal sex reversals during which they produce varying proportions of staminate, perfect and pistillate flowers.

Many workers have tried sex forecast in papaya before flowering. Singh *et al.* (1977), in their experiment with cv. Honey Dew, opined that dark seed colour was associated with a high proportion of female or hermaphrodite plants. Dark brown seeds of medium size were conducive to pistillate flowers, whereas dark brown seeds of large size produced predominantly hermaphrodite plants. Reproductive tissues of male plants showed greater activities of acid and alkaline phosphatases than those of female plants (Jaiswal *et al.*, 1984). Dhara and Veerannah (1998) working with cvs. Sunrise Solo (hermaphrodite) and Co.5 (dioecious) observed higher germination in black seeds than in brown seeds of both cultivars. Black seeds also produced a higher proportion of females than brown seeds. Larger seeds also germinated more readily. Treating the seeds with concentrated sulphuric acid improved germination and increased the proportion of female plants. Munoz *et al.* (1982) stated that sex of juvenile plants could be identified from the peroxidase composition in various organs. The morphological characters of leaf petiole like angle of orientation, thickness and length of petiole failed to show any relationship with the sex of papaya in seedling plants (Rao *et al.*, 1985).

In an investigation to identify sex-specific differences in papaya, Parasnes *et al.* (1999) employed imformative microsattelite and minisattlite probes. Among these only microsattlite probe (GATA) demonstrated sex-specific differences in all the cultivars analyzed. This study also indicated that the genetic material of the X and Y chromosomes in papaya diverges in a sex-specific manner and hence they are in the process of differentiation.

Levels of auxin-like substances varied significantly between G, E and D plants up to 60 days whereas the levels of GA-like substances were similar in all three forms. Mean levels of both substances increased in G plants on aging. In Eand D plants levels of auxin-like substances increased up to 100 days and decreased thereafter, while the levels of GA-like substances increased progressively as the plants aged. Male plants had the highest levels of these 2 substances while E plants had the lowest (Hasdiseve *et al.*, 1989). The female plants had higher levels of RNA in roots and stems and higher levels of histone in leaves and flowers than male plants, indicating a relatively higher metabolic status of female plants which may provide guidance for identifying the sex of young papaya plants (Nandi and Mazumdar, 1990).

10.3.1 Effect of Chemicals

Sprays of 2,3-dichloroisobutyrate on hermaphrodite forms of dwarf papaya and Solo increased the number of female flowers (Lange, 1961a). Spraying of benzothiazole 2-oxyacetate produced flowers earlier at a lower node and higher yield than the control (Dedolph, 1962). Chacko and Singh (1966) reported that soaking papaya seeds in GA3 solution accelerated germination and increased the height and fresh weight of the stem. Spraying with GA3 and SADH was reported to induce femaleness while application of MH, SADH and phosfon-D reduced the plant height, accelerated flowering and fruiting at lower node, increased fruit set and yield (Padmanabhan, 1970; Alagiamanavalan, 1971). Treatment with TIBA induced femaleness (Singh and Jindal, 1972), while both CCC and TIBA advanced flowering at lower nodes and increased fruit set (Singh and Jindal, 1972; Selvaraj, 1972; Bhattacharya and Madhava Rao, 1982). The quality, yield and size, papain yield and the proteolytic activity of papain were increased by treatment with GA3. Treatment with CCC increased the thickness of pulp and pectin content to an extent of 82 per cent compared with the untreated plants whereas it reduced the plant growth, delayed flowering and lowered the fruit and papain yield. Application of GA3 on Coorg Honey Dew increased the production of hermaphrodite and staminate flowers, caused a two-fold increase in ascorbic acid content, but reduced the seed number, fruit size and pectin content (Shanmugavelu *et al.*, 1973a).

Biswas (1994) reported that application of MH, CCC and ethrel significantly suppre-ssed plant height of papaya cv. Ranchi. A large percentage of female flowers were obtained by spraying with MH, gallic acid, NAA and ethrel, and all growth regulator treatments increased the number of fruits/plant. According to Sharan *et al.* (1994), only GA1, 2,4-D, and NAA had significant effects on sex expression of some of the cultivars. Foliar application of B (0.1 per cent), Mn (0.25 per cent) and Cu (0.25 per cent) singly and in combination at 2 and 3 months after transplanting on cv. Ranchi increased female-ness. Boron increased hermaphroditism and decreased maleness (Ghanta *et al.*, 1994).

Arvind Kumar (1998) recorded that exogenous application of 240-960 ppm ethrel (ethephon) and 20-80 ppm chlorflurenol induced intersexual and female flowers, as well as male flower, on genetically male plants of the papaya cultivar Honey Dew. Fruits developed on the treated flowers. This phenotypic expression of sex reversal was associated with qualitative and quantitative changes in free soluble proteins and peroxidase isozymes.

10.3.2 Influence of Environment

Choudhury (1957) stated that sex in papaya is a physiological phenomenon amenable to change by variation in environment. Storey (1958) reported that season had a deciding effect on shift from female sterility to female fertility, the latter being higher in cool nights or short days. The height of fruiting and also sex in papaya has been reported to be altered by environmental changes and growth regulators (Storey, 1958; Lange, 1961b). Singh *et al.* (1963) found an increase in female sterility during winter months and restoration of female fertility in the same plants during summer. Sex expression has been found to be influenced by planting season (Seemanthini,

1964) indicating the influence of climate on sex differentiation. She observed high percen-tage of females in Coorg Honey Dew, a hermaphrodite type during January planting than in July planting at Coimbatore, but a similar observation could not be recorded in dioecious type like Co.1 and Washington. Teaotia and Singh (1967) studied the effect of seasonal variation on sex expression of papaya cultivar Coorg Honey Dew and reported that the production of female flowers was promoted by long day and high temperature.

Padmanabhan (1970) found that long-day treatment, *i.e.* photoperiod cycle of 16-hour light and 8-hour darkness favoured femaleness in Co.1 papaya. According to Alagia-manavalan (1971), photoperiodic treatment of 16-hour light and 8-hour darkness given at the early growth stages caused considerable reduction in seedling height and number of leaves, but this trend was reversed if the treatment was given at later stages of growth. However, he found that photoperiodic treatments had no significant influence on the height at which first flower was produced, flower production, and on sex in cv. Co.1.

10.4.0 Pollination

Pollens from hermaphrodite flowers are inferior to staminate flowers (Wolfe and Lynch, 1940). Earlier reports say that, papaya is pollinated primarily by wind and small insects (Agnew, 1941; Storey, 1941; Prest, 1957). However, this has been contradicted by Allan (1963). Honey bees and butterflies are thought to be more effective pollinating agents. The time from pollination to first ovule penetration was 25 h at 8°C. No *in vitro* pollen germination was observed at 5°C and at 40°C pollen tube growth was retarded (Cohen *et al.*, 1989). Pollen mother cells entered meiotic division at a mean flower length of 8.5 mm (approx. 35 days before anthesis) and the pollen grain size was larger under warm than in cool climate (Sippel and Holtzhausen, 1992).

10.5.0 Fruit Growth and Development

The pattern of fruit development in papaya followed the double sigmoid type of growth curve. Depending upon the cultivars, the papaya fruit took 145 to 165 days to attain edible ripe stage from the date of flowering. The fruits took 10 days more, if they were allowed to ripe on the plant itself. In general, increasing pulp:peel ratio with increased age of fruit was observed (Table 8). This ratio decreased at ripe stage in cultivars like Coorg Honey Dew, Pink Flesh Sweet, Sunrise, Thailand and Washington and varied between 8.5 and 9.5 (Selvaraj *et al.*, 1982a). In cvs. Coorg Honey Dew, Pink Flesh Sweet and Washington, the fruit pulp contained 87-94 per cent water and 2-12 per cent carbohydrate at different stages. The concentration of total, non-volatile acidity and citric and malic acids decreased to a minimum at the ripe stage. The colour break stage was characterized by an increase in vitamin A and proteins (Selvaraj *et al.*, 1982b). Silva *et al.* (1980) stated that glucose, fructose, and sucrose were the main sugars in fruit of *C. candamarcensis* and there were traces of inositol and sorbitol during ripening. The concentration, however, varied depending on the stage of ripening, with glucose predominating at all stages except at full ripening when fructose accumulated in large quantities.

Table 8: Pulp : Peel Ratio in Developing Papaya Fruits (Open pollinated)*

Days after Anthesis	*Coorg Honey Dew*	*Pink Flesh Sweet*	*Sunrise*	*Thailand*	*Washington*
15	8.98	6.08	3.36	5.44	3.85
30	6.66	6.42	8.02	7.92	7.47
45	7.83	8.64	9.33	9.07	8.58
60	9.44	9.33	11.50	15.02	9.83
75	7.64	7.93	13.82	14.58	13.27
90	14.47	14.51	13.65	11.31	11.64
100	15.50	15.01	13.81	8.74	10.90
110	13.91	15.45	13.84	12.97	10.52
120	17.75	9.56	13.24	13.38	11.63
130	14.24	12.05	12.39	13.45	11.19
140	12.80	13.09	13.20	12.38	10.75
150	9.43	8.86	8.60	9.77	10.26
160	-	-	-	9.01	-
165	-	-	-	-	8.91

* Selvaraj *et al.* (1982a).

The skin colour turning and eating ripe stage of papaya cv. Ranchi occurred between 130-135 days and 155-160 days after anthesis. TSS, total sugar, reducing sugar and sugar : acid ratio were low until about 120 days after anthesis, but thereafter increased sharply up to ripening, while the titratable acidity declined with the advancement of maturity (Ghanta *et al.*, 1994). Fruit weight of cv. Formosa was highest in October, December and April and the weight of placenta+seed was highest in October, March and May. Fruit diameter and firmness were greatest in December and November respectively. Fruits sampled in March-June had the maximum TSS content (Floravanco *et al.*, 1994).

11.0 Pest, Diseases and Disorder

11.1.0 Pests

11.1.1 Mites

Occasionally the plants are attacked by mites Tetranychus spp., Tenuipalpus bioculatus and Hemitarsonemus latus (Purseglove, 1968) and Polyphagotarsonemus latus (Ponte, 1996), which cause considerable damage to the crop. In severe cases the leaves become distorted. Treatment with sulphur or kelthane (0.1 per cent) controls the mite population. Spraying with diluted manipueira (a liquid extract from cassava roots) in water (1:3), 3 times at weekly intervals gave 100 per cent control of papaya white mite (*P. latus*) in Brazil (Ponte, 1996).

Pests of papaya

Mealy bug (left), fruit fly (middle) and mite (right) infestation in papaya

Thrips infestation in papaya

Root knot nematode in papaya

Diseases of papaya

Papaya Meleira Virus (PMV) infection in fruit

Leaf curl infection in papaya

Ring spot virus in papaya leaves and fruits

11.1.2 Fruit Fly [*Dacus cucurbitae* (*Bactrocera cucurbitae*), *D. dorsalis* (*B. dorsalis*)]

The adults lay eggs on fruit and the larvae on hatching bore into the fruits. The fruits left on the ground serve as a major breeding site and thus, as a reservoir of resident (non-immigrant) fly population in papaya growing areas (Liquido, 1991a). Mature green fruits had no infestation and it was observed only in half- and fully-ripe fruits. Proper field sanitation (twice-weekly removal of ripe fruits from trees and the ground) helped in suppressing the fruit fly population (Liquido, 1993). Adult flies can be trapped by the use of methyl eugenol-baited trap. Wrapping the fruits with a semipermeable shrinkwrap film (cryovac, D-955) helped in eradicating the larvae from the infested fruits (Jang, 1990). Two stage hot–water immersion (double dip) treatment is a viable quarantine treatment for controlling the fruit fly (Liquido, 1991b).

11.1.3 Thrips (*Thrips parvispinus*)

Thrips cause damage to both nursery and mature plants through feeding on young developing leaves resulting in slight mottling or streaking of the leaves (Lim, 1989a,b). The feeding on leaves provides the infection sites for invasion by the normally saprophytic fungus *Cladosporium oxysporum* resulting in bunchy and malformed top (Lim, 1989a). Damage by thrips is not serious (Lim, 1989b). Weekly fungicidal sprays with 0.05 per cent benomyl alternating with 0.1 per cent mancozeb provide very good control of the pest as well as disease (Lim, 1989a,b).

Beside these, susceptibility of papaya to *Toxotrypana curvicauda* (Aluja *et al.*, 1994) and *Ceratitis capitata* (Suplicy Filho *et al.*, 1987) has been reported.

11.1.4 Mealybug

The papaya mealybug is believed to be native to Mexico or Central America. It has never been a serious pest there, probably due to the presence of an endemic natural enemy complex (Walker *et al.*, 2003). After invading the Caribbean Islands in 1995, it spread to 24 countries described by Walker *et al.* (2003), Meyerdirk *et al.* (2004), Hue *et al.* (2007), Rich (2008).

The pest was observed to spread rapidly and occurred in large colonies on all aerial plant parts, including leaves and fruits, destroying many plants including papaw and Plumeria (temple trees). On lightly infested plants, papaya mealybugs looked like small pieces of cotton attached to the plant. On papaya trees, this symptom was more prominent, as the insects were attached to the fruits and looked like the oozing of milky sap. On heavily infested plants, the mealybug colonies formed cotton-like masses on the aerial parts. These plants showed symptoms of chlorosis, leaf deformation, and early drop of leaves and fruits; the growth of young plants was stunted. Honeydew excreted by the mealybugs resulted in heavy sooty mould growth, especially on papaw and Plumeria. The sooty mould turned the leaves completely black, blocking out light and air, so interfering with photosynthesis. Heavily attacked plants, mainly papaya, were killed. New flushes of growth on damaged plants were deformed, especially in Plumeria, probably due to toxicity of the saliva injected into the plant by the mealybugs while feeding.

These damage symptoms resembled those caused by papaya mealybug described by Walker *et al.* (2003).Galanihe *et al.* (2010) found that three insecticides, imidocloprid 200SL, thiamethoxam 25 per cent WG and Sparrow oil were found to provide effective chemical control, and were recommended along with an IPM practices for the control of this pest until biological control agents were introduced.

11.1.5 Nematodes

The root-knot nematode (*Meloidogyne incognita*) and reniform nematode (*Pratylenchulus reniformis*) cause severe damage to papaya. Variation in reaction of papaya cultivars to root-knot nematode has been reported. Among the 11 cultivars tested for resistance to *M. incognita* in the greenhouse, the cvs. Pusa Delicious and Washington had the lowest susceptibility index (Darekar and Mhase, 1987). The cvs. Peradeneya, Selection No.7 and Washington were moderately resistant and Co.1 was resistant to root-knot nematode (Khan *et al.*, 1995). In the nursery, to control nematodes, granular insecticide should be applied at the rate of 0.6 g per nursery poly bag. In the main field, 25 g of furadon may be applied per plant. The highest yield per plant was obtained in carbofuran (6 kg a.i./ha) treated plots, followed by that in neem-cake (2.5 t/ha) treated plots (Routary and Das, 1988).

11.2.0 Diseases

Fungal Diseases

11.2.1 Collar Rot

The disease is caused by *Pythium aphanidermatum*, a soil-dwelling fungus. At times, the disease appears to be serious and causes considerable damage to the crop. Water-soaked patches are found on the stem which gradually enlarge and girdle the base of the stem. The affected area turns black and rots. Spraying of one per cent Bordeaux mixture (Nirvan, 1953) or any copper oxychloride at a rate of 2 g per litre of water will check further spread of this disease. Ensuring good drainage and drenching the plants with any one of the above fungicides prior to the commencement of heavy monsoon will help in controlling this disease effectively.

11.2.2 Root Rot

This disease is caused by the fungus *Phytophthora palmivora* and in the nursery by *Rhizoctonia solani* and *Fusarium* spp. Seed treatment with fungicides like captan and ceresan and sterilization of the nursery beds by burning dried leaves, spraying of formaldehyde two weeks before sowing and drenching the soil with copper fungicides will control this disease.

In a screening study conducted by Mosqueda-Vazquez *et al.* (1981) for resistance to root rot caused by Phytophthora palmivora, Wiaminalo-23, Line 8 Solo, Wiaminalo-24, and Line-40 were found resistant, while Line 45-T22 and Kapoho Solo were moderately resistant and Higgins was susceptible. Of the 18 cultivars tested, Ceylon, C.S.C. Dwarf-1, Hybrid Dwarf, Peradenya and Singapore were highly resistant, Improved Sunrise resistant and Mammoth, Philippines and Ranchi moderately resistant to *F. solani* but the presence of *M. incognita* in soil made the cvs. Ceylon, C.S.C. Dwarf-1, Mammoth, Ranchi and Washington susceptible to *F. solani* (Khan *et al.*, 1995). Root rot caused by *F. solani* significantly decreased with the application of different doses of NPK fertilizers (Khan and Khan, 1995).

11.2.3 Black Spot

It is caused by *Asperisporium caricae*. The disease first appears as small speckles on older leaves that increase in size and become velvety black in colour. The initial symptoms on fruits occur as small water-soaked lesions which also become velvety black and may reach 1 cm in diameter. Infected fruits are more prone to post-harvest pathogens, decay more rapidly and have a shorter shelf life. The disease can be controlled by spraying with mancozeb (Schroeder, 1996).

11.2.4 Powdery Mildew

Powdery mildew in papaya is caused by *Oidium caricae*. The disease can be controlled by bayleton W.P. (triadimefon), bayleton E.C. and bavistin (carbendazim) each at 0.1 per cent when applied at 15-day intervals (Rawal, 1986).

Viral Diseases

11.2.5 Bunchy Top

The fungus Cladosporium oxysporum is the causal organism, which is ordinarily a saprophyte. The pathogen invades through the injury caused by thrips. Growth of the plant ceases and the leaves become progressively smaller. The deficiencies of B and Ca may predispose papaya trees to bunchy top disease (Skeete, 1994). Control measure of bunchy top is described in pest section (thrips). The husbandry measures that encourage root development are likely to enhance survival (Skeete, 1994).

11.2.6 Papaya Ringspot Virus

Papaya ring spot virus (PRSV) was discovered by Jensen (1949) who named the virus due to the ring spot symptoms it causes on infected fruit. It is by far the most widespread and damaging virus that infects papaya (Purcifull *et al.*, 1984). The occurrence of this disease has been reported from the USA, Hawaii, West Indies, Cuba, Brazil and India. The affected plants are stunted in growth, show yellow mottling and distortion of leaves, bending down of petiole, followed by death of the plant. Diseased plants yield little or no crop. Electron microscopy and ELISA tests have revealed that it belongs to papaya ringspot virus. This ringspot virus constitutes the main limiting factor to the cultivation of papaya. The disease is found to be spread by several species of aphids, such as Myzus persicae, Aphis gossypii, Aphis malvae, Aphis medicaginis. Careful examination of the papaya plantation and thorough rouging of diseased plants are the most effective methods of minimising the spread of the disease.

The purple coloured papayas and the roselle barrier delayed aphid infestation (Becerra Leov, 1989). Mild strains (PRV HA 5-1 and 6-1) of ring spot virus have been developed and encouraging results have been obtained in cross protection trials. The success of these mild strains in protecting against severe PRV is improved by certain cultural practices including rouging diseased plants in protected orchards, protecting seedlings with a plastic bag after transplanting and intercropping with a high stem barrier crop (Yeh *et al.*, 1988). Transgenic Solo papaya with the coat protein gene of the mild papaya ringspot virus from Hawaii was generated from a cross between a female transgenic Ro line and nontransgenic papaya. The transgenic papaya showed high levels of resistance when inoculated with three severe PRV isolates from Hawaii. However, transgenic papaya inoculated with PRV isolates from different parts of the world showed various degrees of susceptibility (Tennant *et al.*, 1995).

11.2.6.1 Integrated Control of Papaya Ringspot Virus

Papaya plants responded best to integrated management practices (Achicanoy, 1995; Garcia Perez *et al.*, 1995). According to Rezende and Costa (1995) the alternative methods for controlling papaya ringspot potyvirus include planting crops in partially isolated areas, rouging of diseased plants, planting in screen houses, screening of plants resistant to or tolerant of the disease and cross protection with wild strains of the virus. As isolated control measures are ineffective for papaya ringspot potyvirus (PRV), Flores Revilla *et al.* (1995) suggested an integrated crop management strategy that includes : (i) covering seed beds with an insect proof polypropylene mesh, (ii) high density papaya plantings (2222 plants/ha) with rouging of diseased plants, (iii) foliage and soil nutrients to improve plant vigour, (iv) poisoned plant barrier (2 lines of maize and 2 of Hibiscus sabdariffa), (v) 2 plastic strips, 5 cm wide and with a shiny grey-metallic colour above each row of papaya plants, and (vi) biweekly sprays with 1.5 per cent mineral oil. The development of transgenic papayas that are resistant to PRSVprovides a promising approach for controlling this disease, since it can overcome some of the difficulties associated with classical cross protection (Rezende and Costa, 1995).

11.2.7 Papaya Leaf Curl

Papaya leaf curl disease was first reported in India by Thomas and Krishnaswamy in 1939 and was initially suspected to be caused by the tobacco leaf curl virus (Nariani, 1956) which is constituent member of the giminivirus group caused by begomovirus, (Saxena *et al.*, 1998*a*). This disease was discovered in the papaya orchards of southern Taiwan in 2002. The disease is characterize by sever curling, crinkling and distortion of leaves accompanied by vein thickening and reduction in leaf size. The leaf margins are rolled downwards and inward to form inverted cup followed by thickening of veins. The affected leaf becomes leathery and brittle and petioles get twisted in zig-zag manner. The intervened areas are raised on the upper surface due to hypertrophy which gives rugosity to the leaves. The affected plant fails to flower or bear fruits. In advance stage defoliation takes place and the plant growth is arrested (Summanwar and Ram, 1993). The disease has been reported from different part of India *e.g.* Chennai, Coimbatore (Thomas and Krishnaswamy, 1939), Bihar and Karnataka (Govindu, 1964), Lucknow (Uttar Pradesh) (Saxena *et al.*, 1998*c*).

Papaya leaf curl is transmitted by whitefly *Bemisia tabaci* (Haber *et al.*, 1981; Hamilton *et al.*, 1983 and 1984; Stenley, 1983) and *Carica papaya* was the only host on which whiteflies showed a mortality of about 80 per cent within 24 hrs. It was

Diseases of papaya

Mosaic infection in papaya

Anthracnose infection in papaya fruit

Phytoplasma infection in papaya

Fruit and root rot infection in papaya

Diseases of Papaya

Powdery mildew infection in leaf and fruits

Black spot infection in leaf and fruits

***Cercospora* spot infection**

***Phytophthora* Fruit and root rot infection**

thought that the whitefly vector which causes leaf curl is unable to feed continuously on papaya to complete the acquisition, latent and effective inoculation periods. Consequently, a possible role of alternate host in natural spread of papaya leaf curl virus to *Carica papaya* was assumed.

11.2.8 Papaya Mosaic Virus

Papaya Mosaic Virus was first reported in Florida (USA) in 1962 (Conover, 1962), where it seems to be of little economic importance. It is a member of the genus Potexvirus within the Flexiviridae family (Fauquet *et al.*, 2005). papaya mosaic, sap-transmissible, not transmitted by seeds but differ in being aphid-borne and restricted in host range (Rajapakse *et al.*, 1981 a,b) to papaya and cucurbits (Purcifull and Hiebert, 1971; Conover, 1964;), the best known is papaya ringspot, a typical potyvirus (Purcifull *et al.*, 1984; Zettler *et al.*, 1968; Koeing and Lesemann, 1978). In general, potexviruses are naturally transmitted without a vector. They are readily sap transmissible (Purcifull and Hiebert, 1971) spreading through plant contact, vegetative propagation and contaminated tools and hands. PMV retains infectivity in papaya sap for more than six months at room temperature. Potexviruses are not transmitted by seed or pollen. Eliminating *Portulaca* spp. from papaya field may reduce PMV infection.

11.3.0 Disorders

11.3.1 Skin Freckles

The cause and factors that influence this disorder are unknown. Freckle-like blemishes occurred on ripe papaya fruits (cv. Sunset). Young fruits (less than 40 days old) were free from freckle. The freckle diameter increased during the last phase of fruit growth as the fruits approached maturity. More freckles were seen on the exposed side of the fruit away from the stem. Fruit freckle index and freckle diameter were lowest during the late summer-autumn part of the year. Wrapping young fruits in white paper bags significantly reduced freckle incidence (Eloisa *et al.*, 1994).

11.3.2 Dieback

It is a severe problem for the Queensland industry, although the cause of this disorder is still unknown. The plants displayed a brown discoloration of the vascular tissue, developed external symptoms that initiated in the stem zone which supported leaves that developed acropetally into the stem apex and basipetally into the trunk and in part of the root system. The discoloration was autofluorescent under blue or ultraviolet light and was associated with laticifers in xylem and phloem tissue and in ray parenchyma. It is suggested that low calcium content represents a secondary symptom of the disorder (Aleemullah and Walsh, 1996).

12.0 Harvesting and Yield

12.1.0. Harvesting

Mature fruits ripen within a day or so and the colour changes from green to yellowish green. Ripe fruits may also be harvested from the trees but they are often damaged by birds, so it is advisable to harvest the fruits when mature. In an investigation in to the determination of maturity, Greensill and Newman (1999) used transmission spectra for pawpaw fruits at various states of maturity (immature green to skin colour break) over the wavelength range 500-1000 nm using a dual prism spectrometer. An apparent spectral signature of maturity was noted with a distinct change in the relationship of wavelength of peak transmissions at 720 nm to 815 nm between the immature green and ripe fruit respectively.

When the latex of the fruit becomes almost watery, the fruit is considered ready for harvest (Shetty, 1953). Fruits are picked individually by cutting off with a sharp knife. Fruits for local market can be harvested when they are half ripe. In Costa Rica, the number of yellow bands on the fruit surface is used as a maturity index and fruits are harvested between 5.00 and 7.00 a.m. The most commonly used harvesting equipment is a long pole covered with rubber at the top end (Cerdas and Saenz, 1993).

The first crop of fruits is harvested in 12-14 months after transplanting. The cropping will be continuous during the life of the plant. In North India, fruits ripen during spring and summer; in hills it is restricted to three to four months, from February to May, because it requires a warm climate during ripening. It was also observed that high relative humidity and comparatively high temperature during ripening period may not be conducive to development of attractive colour in fruits which may remain pale green even when ripe (Shetty, 1953).

12.2.0 Yield

The yield of papaya is very variable, depending mainly on cultivar, soil and climate, plant density and crop management.

Papaya produces higher yield in the first two years, then declines from the third year. In Punjab, a crop of 15 to 20 fruits per plant, each weighing 1.5 lb (0.68 kg) was observed by Bajwa and Jawanda (1962). Sulladmath *et al.* (1981) recorded an average yield of 79.9 fruits per plant weighing 25.76 kg in cv. Solo. In West Bengal, the cv. Ranchi yielded 40 to 55 fruits, each weighing 1.2 to 3.4 kg (Biswas *et al.*, 1981).

Solo papaya in Hawaii yielded 30 to 80 tonnes per acre (Hamilton, 1987). The cultivar Higgins showed the highest yield of 38489 lb/acre at Maloaa island of Kauai (Yee *et al.*, 1980). In Cuba, the cv. Maradol Rojo yielded 25.1 tonnes/ha at a density of 1666 plants/ha (Camejo and Alvarez, 1985). Spaced at 1.25 x 1.25 m, the dwarf mutant cv. Pusa Nanha yielded 60 to 65 tonnes per hectare (Ram, 1983).

Kist and Manica (1995) reported that yields per plant and per hectare in the first year were 14.74 kg and 36.02 tonnes respectively at a spacing of 2m x 2m, 17.02 kg and 36.02 tonnes respectively at 2m x 2.2m and 17.29 kg and 36.38 tonnes respectively at 2.0 x 2.4 m.

13.0 Post-harvest Management

13.1.0 Packaging and Transport

Commercial trade of papaya between Hawaii and the mainland of the USA is significant. The fruits from Africa are transported to Europe by air and sea. Fruits

Harvesting and post harvest handling of papaya

Hand picking of ripe papaya fruit

Picking of papaya fruit with fruit picker

Sorting and grading of papaya fruits

Papaya fruits in local market

Packaging of papaya for transportation

Post harvest handling of papaya

Mass scale packaging of papaya

Packaging of papaya ready for long distance transport and export

Transportation of papaya for storage

Storage of papaya

Selling of papaya in the market

are packed in bamboo baskets for local market. For long distance transport they are packed stem end down in single-layer in wooden crates or fibreboard containers padded with cushion materials (paper or wood wool) at the bottom, sides and between fruits. For local consumption the fruits after harvest, are stored in a single layer of straw for better colour development. For sea shipment, papayas are held at 8°C and storage for up to four weeks is possible at this temperature for fruit of minimum maturity at harvest (Marriott and Proctor, 1982). Almost 56.5 per cent of the growers in Costa Rica employed bulk transport, usually consisting of 3 or 4 layers of fruit in the truck (Cerdas and Saenz, 1993).

The most common fungal disease caused by *Colletotrichum gloeosporiodes* was found to spoil the fruits during storage and it could be controlled by immersion in hot water (50°C) for 20 minutes (Arriola *et al.*, 1976).

13.2.0 Ripening

The papaya fruit was classified as climacteric (Baile *et al.*, 1954). Ethylene peaks were recorded when fruits were still green (before colour-break), whereas respiratory peaks were associated with colour-break (Jobin-Décor and O'Hare, 1996).

Ripening was characterized by an increase in the activities of invertase, cellulase, pectinesterase, catalase, peroxidase, adenosine triphosphatase, acidphosphatase, polyphenol oxidase and mitochondrial glutamate dehydrogenase and falling levels of amylase, ribonuclease, mitochondrial adenosine triphosphatase, NADH diphorase and malic dehydrogenase, and of the RNA, DNA, protein and mitochondrial protein contents (Pal and Selvaraj, 1987). During ripening, soluble and bound peroxidase activities increased 2.5- and 4.2-fold respectively (Silva *et al.*, 1990). In general, irrespective of enzyme distribution, tissue softening during ripening was more closely related to changes in â-galactosidase activity than to polygalacturonase or pectinesterase activity (Lazan *et al.*, 1995). Polygalacturonase activity increased during ripening concomitantly with an increase in soluble polyuronide level and tissue softness. â -galactosidase activity also increased during ripening (Table 9). However, unlike polygalacturonase, â -galactosidase activity was found consistently higher in the firmer outer than in the softer inner mesocarp (Lazan *et al.*, 1989). Chan and Taniguchi (1985) stated that fatty acid content of phospholipids increased with increasing maturity.

The apparent molecular mass range of different extractable fractions of fruit pectin and hemicellulose during fruit ripening was determined by Paull *et al.* (1999) by analysing fruits of cv. Sunset at the 30, 50, 60 and 100 per cent yellow stages. The pectin molecular mass declined and the solubility of pectin in cyclohexane-diaminotetraacetic acid and Na_2CO_3 solutions increased during ripening. The molecular mass decreased and the solubility of hemicellulose in KOH increased during ripening. Water-soluble uronic acid increased 6-fold during ripening as the yield of cell wall material declined. The loss of high molecular mass pectins decreased throughout ripening while the demethylation rate was greater early in ripening. The pectin fractions were mainly composed of rhamnose, glucose, xylose, galactose, mannose and arabinose, in decreasing order of concentration.

Noncellulosic glucose and xylose were the main neutral sugars in the hemicellulose fraction, followed by mannose and galactose and traces of rhamnose and arabinose.

Table 9: Changes in Polygalacturonase and β-galactosidase Activity in Outer and Inner Mesocarp Tissues during Ripening of cv. Back-cross Solo Papaya (Results are mean of 4-6 fruits ± SE)*

Stage of Ripeness (Per cent yellow)	*Polygalacturonase*		*β-galactosidase*	
	Outer Mesocarp	*Inner Mesocarp*	*Outer Mesocarp*	*Inner Mesocarp*
Green	1.2 ± 0.3	6.2 ± 0.4	3.7 ± 0.1	3.1 ± 0.2
30	6.2 ± 0.2	7.8 ± 0.9	4.9 ± 0.4	4.2 ± 0.1
70	8.0 ± 0.2	10.7 ± 0.7	6.4 ± 0.7	5.9 ± 0.7
100	8.2 ± 0.3	12.6 ± 0.8	7.3 ± 0.2	6.0 ± 0.2

*Lazan *et al.* (1989).

Papaya fruit chloroplasts are mature at the dead green and mature green stage and at these stages resemble chloroplasts in green leaf mesophyll tissue. When internal fruit ripening began (quarter-ripe) photosynthetic and ultrastructural qualities of the exocarp remained relatively unchanged, indicating that external ripening lagged considerably behind internal ripening (Sanxter *et al.*, 1992). Carotenoid loss did not directly accompany chlorophyll loss. The relative degradation rate of chlorophyll b was faster than that of chlorophyll a. The relative degradation rates of violaxanthin and antheraxanthin were faster than those of zeaxanthin. The xanthophyll cycle in the exocarp moved in the direction of de-epoxidation (violaxanthin g antheraxanthin gzeaxanthin) whereas that of the pulp moved in the direction of epoxidation (Hsiu Yu *et al.*, 1996).

Treatment with ethephon 500 to 2000 ppm has been reported to accelerate the ripening process and improve the quality of fruits (Shanmugavelu *et al.*, 1973b; Bal *et al.*, 1993). Ethylene evolution was much greater from ethephon than methionine treated fruits and increased with increasing ethephon concentration (Suzuki *et al.*, 1991). Besides, causing quick ripening, the growth regulator treatment also improved the colour of fruits and quality in terms of TSS, reducing and total sugars, carotenoids and ascorbic acid contents. Rodriguez *et al.* (1974) were able to ripen local Puerto Rican cultivars by treatment with ethylene at 1 ml/l at 25°C and 85-95 per cent relative humidity. A volatile chemical, benzyl isothiocynate, endogenous in papaya was found to delay ripening (Patil *et al.*, 1973). Ripening was delayed when fruits of cv. Solo were vapour-heat treated (47.2°C for 7 h), wrapped and stored (Suzuki *et al.*, 1991).

A significant linear relationship was found between irradiation dose and firmness immediately after irradiation in 10 to 30 per cent yellow fruits, with irradiated fruits declining in firmness with increasing irradiation doses up to 1.5 kGy (Zhao *et al.*, 1996). Fruits treated with 250 Gy gamma-irradiation frequently softened more uniformly than non-irradiated fruits (Paull, 1996). Seal-packaging (wrapping in low density polyethylene film) retarded the development of peel colour, softening and inhibited the increase in titratable acidity during ripening which were partly attributed to decrease in polygalacturonase activity and polyuronide solubilization (Lazan *et al.*, 1990).

13.3.0 Storage

Room temperature (28°-32°C) was found to be practically unsuitable for storage of fruits. A temperature between 13 and 16°C has been reported to be adequate forstoring papaya (Wardlaw *et al.*, 1934). Low temperature (7°C and below) caused chillinginjury to fruits (Lassoudiere, 1969; Thompson and Lee, 1971). Temperature above 20°C made the fruit susceptible to fungal attack, while at lower temperature the onset of the climacteric respiration was delayed and chilling injury was manifested (Broughton *et al.*, 1977). Fruits stored continuously at 17°C, showed delayed ripening over fruits held at 20°C and exhibited reduced development of yellow skin colour (Wills and Widjanarko, 1997). They also quoted that ripe fruit with increasing exposure to sub-ambient (5°-15°C) temperature showed decreasing levels of yellow skin colour and increasing chilling injury symptoms. Untreated fruits stored in air at 16°C were unacceptable after approximately 17 days in storage (Maharaj and Sankat, 1990). Transferring fruits to ambient temperature (25°C) following storage at 10°C resulted in accelerated increase in fruit colour development and firmness loss. Activity of cell wall hydrolases and pectin depolymerization increased rapidly at 25°C. It is possible that polygalacturonase, pectin methylesterase (pectinesterase) and β-galactosidase play significant role in the development of chilling injury in papaya fruits when chill-stored fruits are returned to ambient temperature (Ali *et al.*, 1993).

Controlled atmosphere (CA) storage (16°C, 1.5-2.0 per cent O_2 + 5 per cent CO_2) reduced weight loss during storage, resulted in negligible change in firmness and delayed ripening. CA storage delayed colour development during storage and during ripening under ambient conditions (Sankat and Maharaj, 1989). Fruits wrapped in stretch-film and stored in CA storage had the longest storage life *i.e.* up to 29 days (Maharaj and Sankat, 1990).

The ripening of papaya fruit can be controlled by simply modifying its position with respect to ethylene absorber within the wrapping. The ripening could be completely controlled in cold storage (13°C) by an ethylene absorber at 7 g/kg for 40 days if the absorber is situated on a sheet of micro-perforated PVC over the fruit or at 10 g/kg over 15 days if it is placed at the base (Plaza *et al.*, 1992). Modified atmosphere packaging (MAP) in heat shrinkable film and moderately low temperature (MLT at 15°C) storage delayed and retarded firmness loss. MAP was most effective in retarding texture change when fruits were stored at 15°C. Storage treatments also retarded the initial increase in enzymatic activity during ripening and delayed deploymerization of the cell wall pectins (Lazan *et al.*, 1993).

13.4.0 Post-harvest Losses

The post-harvest losses in papaya have been estimated to be between 5 and 25 per cent in Costa Rica (Cerdas and Saenz, 1993) and up to 75 per cent in Hawaii (Paull *et al.*, 1997). These losses are often blamed on mechanical damage, chilling injury and degree of ripeness. The occurrence of sunken, green patches of skin on fully ripe fruits appeared to be induced by mechanical injury. It has been suggested that the use of liners in the field bins may reduce skin injury caused by abrasion and puncture (Quintana and Paull, 1993). Numerous fungi have been reported to cause post-harvest diseases of papaya, *viz.*, fruit surface rots including anthracnose and chocolate spot caused by *Glomerella cingulata*, dry rot (*Mycosphaerella caricae*), wet rot (*Phomopsis* sp.), alternaria fruit spot (*A. alternata*), stemphylium fruit spot (*S. lycopersici*), fusarium rot (*F. solani*), guignardia spot, stem-end rot caused by *Mycosphaerella* sp., *Botryodiplodia theobromae, S. lycopersici, Phomopsis* sp. and *Rhizopus stolonifer* and internal fruit infections caused by *Cladosporium* sp., *Penicillium* sp., *Fusarium* spp., *Erwinia herbicola* and *Enterobacter cloacae* (Alvarez and Nishijima, 1987). Fruit fly also causes considerable losses during post-harvest ripening of fruits.

Hot water treatments *i.e.*, 42°C for 40 minutes, followed by 49°C for 20 minutes for the disinfestation of fruit flies, or 49°C for 20 minutes for control of decay have been recommended. However, these treatments sometimes resulted in heat injuries which are thought to be due to decreased polygalacturonase activity resulting in delayed fruit softening (Chan, 1986) and an increase in internal lumpiness (Nishijima *et al.*, 1992). Since the polygalacturonases are heat resistant, the deleterious effects of heat are thought to be due to mechanisms controlling the generation of activation of polygalacturonase, such as ethylene. The heat resistant ethylene-forming-enzyme (EFE) comprised 25 per cent of total EFE activity, the remaining 75 per cent being affected by the commercial heat treatments (Chan, 1986). Forced, hot air (48.5°C for 3-4 h) treatment of papaya fruit, a recently developed quarantine treatment for fruit flies, did not significantly reduce incidences of post-harvest diseases. However, when hot-air treatment was combined with thiabendazole (4 g a.i./l) application or hot-water immersion (49°C for 20 min.), the incidence of most post-harvest diseases was reduced (Nishijima *et al.*, 1992). The sensitivity of fruits to injurious treatments was greatest during winter months. The rainfall, evaporation and temperature (min. and max.) in the 3 days before harvest were significantly correlated with fruit heat sensitivity (Paull, 1995). Forced air dry heat (FADH) or vapour heat (VH) treatments are currently used in the US. Both treatments take 6 hours (approx.) of gradually increasing temperature and are completed when the fruits' internal temperature reaches 47.2°C. VH differs from FADH only by the 100 per cent RH during the last phase of treatments. Neither of the treatments was designed for controlling diseases although both provided some level of disease control (Nishijima, 1995).

Coating fruits with either Semperfresh (a sucrose ester preparation, at 0.65, 1.0 or 1.4 per cent) or a polyethylene wax emulsion (1:2, 1:4 or 1:6 wax:total volume of water) and storage at 10°C resulted in reduced moisture loss and delayed colour change, particularly at the mature green, green and quarter ripe stages (Yusof and Salleh, 1992). Cold storage of fruits covered with PVC film or in polyethylene bags delayed weight loss and development of external colour and decay and increased storage life by at least 11 days compared with untreated control (Mosca and Durigan, 1995). Mehta *et al.* (1986) reported that post-harvest treatment of papaya fruits with GA3, vitamin K, silver nitrate, kinetin or cobalt chloride extended the shelf-life appreciably without any adverse effects on palatibility.

14.0 Papain Production

The immature papaya fruit contains milky latex. The dried latex called 'papain' is in great demand in the international markets, particularly in the UK and USA. Papain has several uses in the industry as an essential ingredient in pharmaceutical, textile and tanning industries. Papain is used in tenderising meat; for clearing beer; in the manufacture of cosmetics like snow and face creams and also dental paste; in degumming silk and rayon; in the pre-shrinking of wool. It has also several uses in the medical field in the treatment of necrotic tissues, dyspepsia and other digestive ailments, ring worm and round worm infections, skin lesions and ulcers, eczema and other skin diseases and in kidney disorder. Papain is used in detecting stomach and intestinal cancers and also in correcting diphtheria.

Papain production is influenced by following factors :

14.1.0 Fruit Size

Fruit size plays an important role in tapping of papain. The yield of papain was reported to be high in oblong fruits (Stockdale, 1927). Singh and Tripathi (1957a) found that fruits of 14.3 inches (36.3 cm) in length and 10.9 inches (27.7 cm) in diameter gave the highest papain yield. Seemanthini (1964) stated that the yield of papain was the highest in fruits measuring 1255 to 1544 sq cm in size. She also reported that the yield increased with an increase in the size of fruit, and a significant and positive correlation was established between girth of fruit and papain yield in Co.1 and Selection 7.

14.2.0 Fruit Maturity

Fruit maturity is another important factor influencing the yield of papain. Sen (1931), Hofmeyr and Roux (1938) and Sanders and Robertson (1944) recorded that unripe but fully grown fruits yielded maximum papain. Jayaweera (1957) also recommended fully grown fruits for tapping. Singh and Tripathi (1957b) found 2½-month-old fruits to yield the highest quality of papain. Seemanthini and Balakrishnan (1964) reported that papain yield was more when tapped at 90 days after fruit set. Papain dry weight generally increased from 15 to 90 days and then decreased (Veerannah and Rathnakumari, 1987; Kannan and Muthuswami, 1989).

14.3.0 Season

Becker (1950) reported that the flow of latex was low if the temperature was below 10°C. Studies at the Central Food Technological Research Institute, Mysore, showed that the yield of papain was the highest in the month of October. At Saharanpur, the fruits set during July yielded more latex (Singh and Tripathi, 1957a). According to Muthuswamy *et al.* (1962), season did not show any marked effect on latex production at Coimbatore. But Seemanthini and Balakrishnan (1964) reported that papain yield was higher in the months of July to January. Irulappan (1980) found the yield of papain to fluctuate from May to October, among the different types with the lowest yield in September.

14.4.0 Cultivars

Marked variations were noted among the different cultivars for the yield of papain (Cheema and Dani, 1930; Sen, 1931). In Sri Lanka, an improved strain called 'Botanist's Selection' was recommended for the production of papain by Charavanapavan (1952). Muthuswamy *et al.* (1962) recorded higher yield in the cultivars, Washington, Philippine, Botanist's Selection, Peradeniya and Co.1. A Philippine selection was found to be the most promising by Seemanthini and Balakrishnan (1964), in a study on five inbred selections for papain yield. Khan and Seemanthini (1968) reported that individual plants of some selections made from Peradeniya, Red Fleshed and Philippine types yielded as high as 6 g of papain per fruit. Foyet (1972) stated that cv. Red Panama yielded 21 kg of dry papain per hectare. Madhava Rao (1977a) reported that the cultivar Co.2 evolved from Selection 7, yielded 100-120 kg of papain per acre, and this is one of the recommended cultivars now in India for papain production. The cv. Co.2 recorded the highest papain yield of 686.29 g per plant in a period of six months, among the 70 types evaluated (Irulappan, 1980). Papain dry activity was highest in CP 14, CP 15 and CP 15/2 (Veerannah and Rathnakumari, 1987; Kannan and Muthuswami, 1989).

14.5.0 Effect of Growth Regulators

Chacko *et al.* (1972) reported that spraying papaya plants with ethrel at 200 ppm significantly increased the flow of latex by 172 per cent compared with control in cv. Coorg Honey Dew. In Co.2 papaya the papain yield could be increased four-fold compared with control by the application of ethephon latex stimulant at 37 mM in coconut oil (Shanmugavelu *et al.*, 1976). Highly significant increases in yield and protease activity of the latex over the control were obtained by the application of ethephon (100 or 200 ppm). IBA (25 or 50 ppm) and GA3 (100 or 200 ppm) also increased latex production, but spraying with kinetin or BA solutions had no clear effects (Ali and Mazumdar, 1994).

14.6.0 Papain Extraction

Papaya fruits, which are about 90-100 days old (fully mature but not ripe) are selected for tapping. In the morning hours before 10.00 a.m., four longitudinal incisions are given on the four sides of the selected fruit from the stalk end to the tip. The depth of the incision should be about 0.3 cm. On incising, the latex starts flowing and this is collected in suitable containers (arecanut spathes, aluminium trays or glass vessels). Care should be taken not to use any other container for papain collection, since it will react with papain rendering it unfit for any use. The latex that solidifies in the cuts should also be scrapped carefully and added to the liquid latex. This process of making four incisions in the untapped fruit surface at 3- to 4-day intervals is repeated thrice or four times over a period of 12 to 16 days. The latex thus collected every time should be dried in the sun or in driers at temperatures ranging between 40° and 60°C. The time taken for the fresh latex to dry (solar drying) varied from about 1.5 to 36 hours (Narinesingh and Mohammed-Maraj, 1988). Potassium metabisulphite (0.05 per cent) is added to the liquid latex in small quantities before it is dried, since this helps to extend the storage life of papain. The drying of papain is continued until it comes off in flakes having a porous structure. The dried papain is powdered, sieved in a 10-mesh sieve and stored in polythene bags or in any other suitable container. The latex, whether dripped or congealed, dried as pale yellow flakes which could easily be powdered and had a proteolytic activity slightly higher than that of the fresh latex (Narinesingh and Mohammed-Maraj, 1988).

The biochemical analysis of fruits revealed that latex extraction has no effect on the quality of fruits.

14.7.0 Papain Yield

Sanyal (1931) reported that a papaya tree would give 8 oz (227 g) of papain and Sankaram (1942) stated a yield of 1/3 lb (151.3 g) of crude papain from a plant or 175 lb (79.45 kg) per acre. Sanders and Robertson (1944) estimated 60 to 100 lb (27.24-45.4 kg) per year as the acre yield of papain. Virendrakumar (1952) reported that the yield of papain was about 100 lb (45.4 kg) per acre while Singh and Tripathi (1957b) estimated 60 to 75 lb (27.34-34.0 kg) per acre. Muthuswamy *et al.* (1962) found the papain yield to range from 1.23 to 7.45 g per fruit in 9 cultivars, the cv. Washington recording the highest mean yield of 7.45 g per fruit.

15.0 Biotechnology

Biotechnological research in Papaya has made tremendous progress in the last two decades. Ming *et al.* (2008) led a team of over 85 researchers and presented the draft genome of Papaya. More significant is that this draft genome is from the transgenic papaya which is the first tree transgenic to be fully sequenced.In a one of recent reviews, Dhekney *et al.* (2016) indicated that papaya improvement for stress tolerance and qualitative traits using conventional breeding has been difficult due to the narrow germplasm pool in the *Carica* genus and sexual incompatibility problems encountered during intergeneric hybridization with other genera in the Caricaceae family. Genetic engineering is an important tool in papaya improvement for modifying one or more traits in elite cultivars without altering existing characteristics. Advances in genetic engineering have been facilitated by concerted efforts for genome sequencing of papaya, development of papaya regeneration systems and efficient gene insertion techniques for transfer of desirable traits (Dhekney *et al.*, 2016).

Papaya regeneration via organogenesis and somatic embryogenesis has been refined during the past 3 decades. Early efforts to optimize gene insertion protocols utilized a number of reporter and selectable marker genes, viral- and bacterial-derived regulatory sequences and functional genes for biotic and abiotic stress tolerance (Dhekney *et al.*, 2016). Transgenic plants were routinely produced with several cultivars. One of the best success stories in the commercialization of a genetically modified fruit crop has involved the development of transgenic papaya ring spot virus (PRSV) resistant Rainbow and SunUp cultivars, which saved the Hawaiian papaya industry. Additionally, genetically modified papayas with traits for disease resistance and extended shelf life have been extensively screened in field tests (Dhekney *et al.*, 2016).

15.1.0 Biochemical and Molecular Markers

Manshardt and Drew (1998) reported that phylogenetic analyses based on isozyme and DNA polymorphisms indicate that papaya is a distant relative of the members in the genus.They generated a RAPD map of the papaya genome to identify markers for sex-determination, flowering height and fruit carpellody. The genetic origin of cultivated papaya is not clear. Jobin-Decor *et al.* (1997) studied six wild relatives ofpapaya from central South America using isozyme and RAPD analysis. Seven others pecies (including six from the genus *Carica*) were found to be relatively distant from papaya providing no indication of the genetic origin of papaya. Isozyme and RAPD data gave similar measures of genetic similarity within this group. *C. papaya* was about 70 per cent dissimilar to the other *Carica* species by both methods. The other *Carica* species had average dissimilarities around 50 per cent. Two species, *C. pubescens* and *C. stipulata* were much closer to each other with similarities of 87 per cent by isozyme analysis and 82 per cent by RAPD analysisn although both methods gave similar measures more reliable for analysis of the extremes (*e.g.*, closely related taxa may show no isozyme differences and distant taxa may show no isozyme similarities).The phylogenetic relationships among twelve wild and cultivated species of *Carica* (Caricaceae) were analyzed by Aradhya *et al.* (1999) using restriction fragment length variation in a 3.2-kb PCR amplified intergenic spacer region of the chloroplast DNA. A total of 138 fragments representing 137 restriction sites accounting for 5.8 per cent of the amplified region were examined. Further, ctDNA analyses showed two basic evolutionary lineages within the genus *Carica,* one defined by cultivated *C. papaya* and another consisting of the remaining wild species from South America in a well resolved but poorly supported monophyletic assemblage. This evolutionary split in Carica strongly suggests that *C. papaya* diverged from the rest of the species early in the evolution of the genus and evolved in isolation, probably in Central America.

RFLPs have been used primarily for detection of phytoplasma and to study the genetic relatedness among phytoplasmas associated with various diseases. Guthrie *et al.* (1998) detected phytoplasma DNA and PCR with primers specific for phytoplasmas in general and for the stolbur group of phytoplasms. The dieback-associated phytoplasma was detected one week prior to (4 cases) or the same week (9 cases) as symptom expression, while phytoplasma DNA was detected between 3 and 11 weeks prior to expression of mosaic symptom (6 cases). Lateral shoot regrowth on the lower stem of plants which had suffered dieback disease failed to generate stolbur-specific PCR products in 15 cases. Dual infection with dieback and yellow crinkle or mosaic was diagnosed in a further 2 cases using RFLP digests, and both cases were interpreted as secondary infections by the dieback-associated phytoplasma. Regrowth in 3 of 7 cases of yellow crinkle- and 3 of 9 cases of mosaic-affected plants tested positive for phytoplasma-specific DNA.

15.2.0 Microsatellites

Recently, the functional and evolutionary importance of SSRs has received considerable attention. Wang *et al.* (2008) reported the mining and characterization of the SSRs in papaya genome by analyzing the SSRs from 277.4 Mb of whole genome shotgun (WGS) sequences, 51.2 Mb bacterial artificial chromosome (BAC) end sequences (BES), and 13.4 Mb expressed sequence tag (EST) sequences. The papaya SSR density was one SSR per 0.7 kb of DNA sequence in the WGS, which was higher than that in BES and EST sequences. SSR abundance was dramatically reduced as the repeat length increased. According to SSR motif length, dinucleotide repeats were the most common motif in class I, whereas hexanucleotides were the most copious in class II SSRs. The tri- and hexanucleotide repeats of both classes were greater in EST sequences compared to genomic sequences. In class I SSR, AT and AAT were the most frequent motifs in BES and WGS sequences. By contrast, AG and AAG were the most abundant in EST sequences. For SSR marker development, 9,860 primer pairs were surveyed for amplification and polymorphism. Successful amplification and polymorphic rates were 66.6 per cent and 17.6 per cent, respectively. The highest polymorphic rates were achieved by AT, AG, and ATG motifs. The genome wide analysis of microsatellites revealed their frequency and distribution in papaya genome, which varies among plant genomes. This complete set of SSRs markers throughout the genome will assist diverse genetic studies in papaya and related species (Wang *et al.*, 2008).

15.2.1 Expressed Sequence Tags

The majority of assembled contigs added up to about 271 Mb (73 per cent of the genome) with scaffolds spanning 370 Mb. About 167 Mb of sequence (or 235 Mb of scaffolds) could be anchored to the integrated genetic and physical map of the papaya genome (Wei and Wing, 2008). More than half (52 per cent) of the papaya genome comprises repetitive sequences, mainly long terminal repeat retrotransposons. Cytogenetic studies suggest that the genome is about 65-70 per cent euchromatic and 35-30 per cent heterochromatic. Various measures were used to assess the coverage of the draft genome, such as the percentage of unique genes (unigenes) and genetic markers matching the assembly. Ming *et al.* (2007) estimate that approximately 90 per cent of the euchromatin has been covered, containing 92.1 per cent of the unigenes and 92.4 per cent of the genetic markers (Wei and Wing, 2008).

To identify genes involved in papaya fruit ripening, a total of 1171 expressed sequence tags (ESTs) were generated from randomly selected clones of two independent fruit cDNA libraries derived from yellow and red-fleshed fruit varieties (Devitt *et al.,* 2006). The most abundant sequences encoded: chitinase, 1-aminocyclopropane-1-carboxylic acid (ACC) oxidase, catalase and methionine synthase, respectively. DNA sequence comparisons identified ESTs with significant similarity to genes associated with fruit softening, aroma and colour biosynthesis. Putative cell wall hydrolases, cell membrane hydrolases, and ethylene synthesis and regulation sequences were identified with predicted roles in fruit softening. Expressed papaya genes associated with fruit aroma included isoprenoid biosynthesis and shikimic acid pathway genes and proteins associated with acyl lipid catabolism. Putative fruit colour genes were identified due to their similarity with carotenoid and chlorophyll biosynthesis genes from other plant species (Devitt *et al.,* 2006). The high level of phenotypic diversity seen among papaya cultivars in the field does not correlate with the low levels of genotypic polymorphism thus far elucidated (Eustice *et al.*, 2008). The highly mutable nature of microsatellites or simple sequence repeats (SSRs) make them potentially powerful markers for distinguishing deoxyribonucleic acid (DNA) polymorphisms between closely related genotypes. Genomic research for papaya has resulted in a significant quantity of sequence data. Eustice *et al.* (2008) mined 28.1 Mb of bacterial artificial chromosomes end sequences, 5.8 Mb of complementary DNA, and 1.6 Mb of random genomic sequences for SSRs. They generated 938 SSR markers and tested for polymorphism among seven varieties that had been used to produce five mapping populations. The level of polymorphism was highest for Kaek Dum x 2H94 with 210 markers, followed by UH928 x SunUp with 194, AU9 x SunUp with 189, UH918 x SunUp with 177, and Kapoho x SunUp displaying the lowest level with 97. Variation in levels of polymorphism, motif predominance, and motif length between the genomic and genic fractions indicated differential selection pressures acting on the microsatellites in these two fractions (Eustice *et al.*, 2008).

15.2.2 Genomics

The papaya genome provides basic plant research with an exciting new tool to better understand angiosperm evolution and sex-chromosome biology. It also provides clues to the minimum set of genes that are needed to be a flowering plant. On the practical side, the papaya genome sequence has yielded a vast set of molecular genetic markers that can be used to create higher yielding, more nutritious and hardier papaya varieties.

Papaya (*Carica papaya* L.) is a major tree fruit crop of tropical and subtropical regions with an estimated genome size of 372 Mbp (Lai *et al.*, 2006). The analysis of 4.7 per cent of the papaya genome based on BAC end sequences (BESs) representing 17 million high-quality bases were presented by Lai *et al.* (2006). Microsatellites discovered in 5,452 BESs and flanking primer sequences are available to papaya breeding programs at http://www.genomics. hawaii.edu/papaya/BES. Sixteen percent of BESs contain plant repeat elements, the vast majority (83.3 per cent) of which are class I retrotransposons. Several novel papaya-specific repeats were identified. Approximately 19.1 per cent of the BESs have homology to *Arabidopsis* cDNA. Increasing numbers of completely sequenced plant genomes and BES projects enable novel approaches to comparative plant genomics. Paired BESs of *Carica, Arabidopsis, Populus, Brassica* and *Lycopersicon* were mapped onto the completed genomes of *Arabidopsis* and *Populus*. In general the level of microsynteny was highest between closely related organisms. However, papaya revealed a higher degree of apparent synteny with the more distantly related poplar than with the more closely related *Arabidopsis*. This, as well as significant colinearity observed between peach and poplar genome sequences, support recent observations of frequent genome rearrangements in the *Arabidopsis* lineage and suggest that the poplar genome sequence may be more useful for elucidating the papaya and other rosid genomes. These insights will play a critical role in selecting species and sequencing strategies that will optimally represent crop genomes in sequence databases (Lai *et al.*, 2006).

Ming (2008) reported a 3x draft genome sequence of 'SunUp' papaya, the first commercial virus-resistant transgenic fruit tree to be sequenced. 85 researchers around the world were part of this project. The papaya genome is three times the size of the *Arabidopsis* genome, but contains fewer genes, including significantly fewer disease-resistance gene analogues. Comparison of the five sequenced genomes suggests a minimal angiosperm gene set of 13, 311. A lack of recent genome duplication, atypical of other angiosperm genomes sequenced so far, may account for the smaller papaya gene number in most functional groups. Nonetheless, striking amplifications in gene number within particular functional groups suggest roles in the evolution of tree-like habit, deposition and remobilization of starch reserves, attraction of seed dispersal agents, and adaptation to tropical daylengths. Transgenesis at three locations is closely associated with chloroplast insertions into the nuclear genome, and with topoisomerase I recognition sites. Papaya offers numerous advantages as a system for fruit-tree functional genomics, and this draft genome sequence provides the foundation for revealing the basis of *Carica*'s distinguishing morpho-physiological, medicinal and nutritional properties (Ming *et al.*, 2008).

The papaya genome was sequenced because of its agricultural importance, unique biological features, and successful application of transgenic papaya for resistance to papaya ringspot virus (Ming *et al.*, 2008). As a part of the genome sequencing project, Yu *et al.* (2009) constructed a BAC-based physical map using a high information-content fingerprinting approach to assist whole genome shotgun sequence assembly. A BAC-based physical map of papaya was constructed and integrated with the genetic map and genome sequence. The integrated map facilitated the draft genome assembly, and is a valuable resource for comparative genomics and map-based cloning of agronomically and economically important genes and for sex chromosome research. The physical map consists of 963 contigs, representing 9.4x genome equivalents, and was integrated with the genetic map and genome sequence using BAC end sequences and a sequence-tagged high-density genetic map. The estimated genome coverage of the physical map is about 95.8 per cent, while 72.4 per cent of the genome was aligned to the genetic map. A total of 1,181 high quality overgo (overlapping oligonucleotide) probes representing conserved sequences in *Arabidopsis* and genetically mapped loci in *Brassica* were anchored on the physical map, which provides a foundation for comparative genomics in the Brassicales. The integrated genetic and physical map aligned with the genome sequence revealed recombination hotspots as well as regions suppressed for recombination across the genome, particularly on the recently evolved sex chromosomes. Suppression of recombination spread to the adjacent region of the male specific region of the Y chromosome (MSY), and recombination rates were recovered gradually and then exceeded the genome average. Recombination hotspots were observed at about 10 Mb away on both sides of the MSY, showing 7-fold increase compared with the genome wide average, demonstrating the dynamics of recombination of the sex chromosomes (Yu *et al.*, 2009).

15.2.2.1 Sexuality in Papaya

The phylogenetic positioning of papaya near the major model plant Arabidopsis and within the agronomically significant Brassicales could make it central for comparative structural and evolutional genomic research in this group of dicots

(Chen *et al.*, 2007) from varieties AU9 and SunUp were used for linkage mapping. A total of 707 markers, including 706 microsatellite loci and the morphological marker fruit flesh color, were mapped into nine major and three minor linkage groups. The resulting map spanned 1069.9 cM with an average distance of 1.5 cM between adjacent markers. This sequence-based microsatellite map resolved the very large linkage group 2 (LG 2) of the previous high-density map using amplified fragment length polymorphism markers. The nine major LGs of our map represent papaya's haploid nine chromosomes with LG 1 of the sex chromosome being the largest. This map validates the suppression of recombination at the male-specific region of the Y chromosome (MSY) mapped on LG 1 and at potential centromeric regions of other LGs. Segregation distortion was detected in a large region on LG 1 surrounding the MSY region due to the abortion of the YY genotype and in a region of LG6 due to an unknown cause. This high-density sequence tagged genetic map is being used to integrate genetic and physical maps and to assign genome sequence scaffolds to papaya chromosomes. It provides a framework for comparative structural and evolutional genomic research in the order Brassicales (Chen *et al.*, 2007).

In a brief review on linkage map in their paper, Chen *et al.* (2007) reported that a high-density genetic map of papaya, constructed using 1498 AFLP and three morphological markers revealed severe suppression of recombination at the sex determination locus with 225 cosegregating markers accounting for 66 per cent of the markers on linkage group 1 (LG 1) and 15 per cent of the markers genomewide (Ma *et al.*, 2004 Cited by Chen *et al.*, 2007)). This AFLP map provided a critical piece of evidence for defining the primitive Y chromosome in papaya (Liu *et al.*, 2004 cited by Chen *et al.*, 2007). Sequence comparison of Y-specific fragments from males and hermaphrodites confirmed that a high-density genetic map of papaya (*Carica papaya* L.) was constructed using microsatellite markers derived from BAC end sequences and whole-genome shot gun sequences. Fifty-four F_2 plants derived there are two slightly different Y chromosomes in papaya (Liu *et al.*, 2004). To distinguish these two Y chromosomes, the one controlling male was designated as Y, whereas the other controlling hermaphrodites were designated as Yh (Ming *et al.*, 2007). The sex chromosome genotypes of the three sexes are XY for males, XYh for hermaphrodites, and XX for females. The combinations of YY, YYh, or YhYh are embryonic lethal, indicating the loss of essential genes for embryo development during the Y chromosome degeneration process. This notion was reinforced by the gene paucity in the male specific region of the Y chromosome (MSY) (Yu *et al.*, 2007). The lethal effect of the YY, YYh, or YhYh genotypes resulted in a distorted 2:1 hermaphrodite and female ratio in F2 populations derived from crosses between females and hermaphrodites. The DNA markers at the MSY and its neighboring regions exhibit the 2:1 segregation ratio, matching the observed phenotype (Ma *et al.*, 2004 cited by Chen *et al.*, 2007).

Sex determination in papaya is controlled by a recently evolved XY chromosome pair, with two slightly different Y chromosomes controlling the development of males (Y) and hermaphrodites (Yh). To study the events of early sex chromosome evolution, Wang *et al.* (2012) sequenced the hermaphrodite-specific region of the Yh chromosome (HSY) and its X counterpart, yielding an 8.1-megabase (Mb) HSY pseudomolecule, and a 3.5-Mb sequence for the corresponding X region. The HSY is larger than the X region, mostly due to retrotransposon insertions. The papaya HSY differs from the X region by two large-scale inversions, the first of which likely caused the recombination suppression between the X and Yh chromosomes, followed by numerous additional chromosomal rearrangements. Altogether, including the X and/or HSY regions, 124 transcription units were annotated, including 50 functional pairs present in both the X and HSY. Ten HSY genes had functional homologs elsewhere in the papaya autosomal regions, suggesting movement of genes onto the HSY, whereas the X region had none. Sequence divergence between 70 transcripts shared by the X and HSY revealed two evolutionary strata in the X chromosome, corresponding to the two inversions on the HSY, the older of which evolved about 7.0 million years ago. Gene content differences between the HSY and X are greatest in the older stratum, whereas the gene content and order of the collinear regions are identical. Our findings support theoretical models of early sex chromosome evolution (Wang *et al.*, 2012).

Sexuality in plants is regulated by genetic, epigenetic and physiological mechanisms. The most specialized mechanism of sex determination is sex chromosomes. The sex chromosomes ensure the stable segregation of sexual phenotypes by preventing the recombination of sex determining genes. Despite continuous efforts, sex determining genes of dioecious plants have not yet been cloned (Aryal and Ming, 2014). Concerted efforts with various model systems are necessary to understand the complex mechanism of sex determination in plants. Papaya (*Carica papaya* L.) is a tropical fruit tree with three sex forms, male, hermaphrodite, and female. Sexuality in papaya is determined by an XY chromosome system that is in an early evolutionary stage. The male and hermaphrodite of papaya are controlled by two different types of Y chromosomes: Y and Y^h. Large amounts of information in the area of genetics, genomics, and epigenetics of papaya have been accumulated over the last few decades. Relatively short lifecycle, small genome size, and readily available genetic and genomic resources render papaya an excellent model system to study sex determination and sex chromosomes in flowering plants (Aryal and Ming, 2014).

Aryal *et al.* (2014) provided valuable information toward understanding the papaya sex determination by Analyzing the high throughput sRNAs reads in the Illumina libraries prepared from male, female, and hermaphrodite flowers of papaya. Using the sRNA reads, we identified 29 miRNAs that were not previously reported from papaya. Including this and two previous studies, a total of 90 miRNAs has been identified in papaya. We analyzed the expression of these miRNAs in each sex types. A total of 65 miRNAs, including 31 conserved and 34 novel mirNA, were detected in at least one library. Fourteen of the 65 miRNAs were differentially expressed among different sex types. Most of the miRNA expressed higher in male flowers were related to the auxin signaling pathways, whereas the miRNAs expressed higher in female flowers were the potential regulators of the apical meristem identity genes. Aligning the sRNA reads identified the sRNA hotspots adjacent to the gaps of the X and Y chromosomes. The X and Y chromosomes sRNA hotspots has a 7.8 and 4.4 folds higher expression of sRNA, respectively, relative to the chromosome wide average.Approximately 75 per cent of the reads aligned to the X chromosome hotspot was identical to that of the Y chromosome hotspot (Aryal *et al.*, 2014).

They (Aryal *et al.*, 2014), identified, by analyzing the large-scale sRNA sequences from three sex types, the sRNA hotspots flanking the gaps of papaya X, Y, and Yh chromosome. The sRNAs expression patterns in these regions were reminiscent of the pericentromeric region indicating that the only remaining gap in each of these chromosomes is likely the centromere. They also identified 14 differentially expressed miRNAs in male, female and hermaphrodite flowers of papaya (Aryal *et al.*, 2014).

Previous reports suggest that miRNAs play a role in sex differentiation in plants (Cited by Aryal *et al.*, 2014). Differential expression of miRNAs that regulates the development of carpels and stamens can ultimately lead to the development of unisexual flowers and to the sexual differentiation in plants. Papaya (*Carica papaya*) is a model species to study the plant sex determination. It is a trioecious species with three sex types, hermaphrodite, male, and female. It belongs to the family Caricaceae, which comprises 35 species in six genera including one monoecious, 32 dioecious, and two trioecious species. Sex expression in papaya is controlled by a pair of recently evolved sex chromosomes. The male and hermaphrodite characteristics are determined by two slightly different Y-chromosomes, Y and Yh, respectively (Storey, 1953; Ming *et al.*, 2007; cited by Aryal *et al.*, 2014). The genotype XX determines female, XY determines male, and XYh determines hermaphrodite. All combinations of Y and Yh are embryonic lethal, indicating that the Y and Yh chromosomes have lost some genes necessary for embryo development. The Y and Yh chromosomes show 98.9 per cent sequence similarity on average, differing mainly in intergenic and repetitive regions (Yu *et al.*, 2008).

X chromosomes have long been thought to conserve the structure and gene content of the ancestral autosome from which the sex chromosomes evolved. Gschwend *et al.* (2012) compared the recently evolved papaya sex chromosomes with a homologous autosome of a close relative, the monoecious *Vasconcellea monoica*, to infer changes since recombination stopped between the papaya sex chromosomes. They sequenced 12 V. monoica bacterial artificial chromosomes, 11 corresponding to the papaya X-specific region, and 1 to a papaya autosomal region. The combined V. monoica X-orthologous sequences are much shorter (1.10 Mb) than the corresponding papaya region (2.56 Mb). Given that the *V. monoica* genome is 41 per cent larger than that of papaya, this finding suggests considerable expansion of the papaya X; expansion is supported by a higher repetitive sequence content of the X compared with the papaya autosomal sequence. The alignable regions include 27 transcript encoding sequences, only 6 of which are functional X/*V. monoica* gene pairs. Sequence divergence from the V. monoica orthologs is almost identical for papaya X and Y alleles; the *Carica-Vasconcellea* split therefore occurred before the papaya sex chromosomes stopped recombining, making *V. monoica* a suitable out group for inferring changes in papaya sex chromosomes. The papaya X and the hermaphrodite-specific region of the Yh chromosome and *V.*

monoica have all gained and lost genes, including a surprising amount of changes in the X (Gschwend *et al.,*2012).

A reiterate chromosome walking strategy was applied to construct the two physical maps with three bacterial artificial chromosome (BAC) libraries (Na *et al.,* 2012). The HSY physical map consists of 68 overlapped BACs on the minimum tiling path, and covers all four HSY-specific Knobs. One gap remained in the region of Knob 1, the only knob structure shared between HSY and X, due to the lack of HSY-specific sequences. This gap was filled on the physical map of the HSY corresponding region in the X chromosome. The X physical map consists of 44 BACs on the minimum tiling path with one gap remaining in the middle, due to the nature of highly repetitive sequences. This gap was filled on the HSY physical map. The borders of the non-recombining HSY were defined genetically by fine mapping using 1460 F_2 individuals. The genetically defined HSY spanned approximately 8.5 Mb, whereas its X counterpart extended about 5.4 Mb including a 900 Kb region containing the Knob 1 shared by the HSY and X. The 8.5 Mb HSY corresponds to 4.5 Mb of its X counterpart, showing 4 Mb (89 per cent) DNA sequence expansion. The 89 per cent increase of DNA sequence in HSY indicates rapid expansion of the Yh chromosome after genetic recombination was suppressed 2–3 million years ago. The genetically defined borders coincide with the common BACs on the minimum tiling paths of HSY and X. The minimum tiling paths of HSY and its X counterpart are being used for sequencing these X and Yh-specific regions (Na *et al.,* 2012).

Sex in papaya is controlled by a pair of nascent sex chromosomes. Females are XX, and two slightly different Y chromosomes distinguish males (XY) and hermaphrodites (XYh) (VanBuren *et al.,* 2020). The hermaphrodite-specific region of the Yh chromosome (HSY) and its X chromosome counterpart were sequenced and analyzed previously. We now report the sequence of the entire male-specific region of the Y (MSY). VanBuren *et al.* (2020) used a BAC-by-BAC approach to sequence the MSY and resequence the Y regions of 24 wild males and the Yh regions of 12 cultivated hermaphrodites. The MSY and HSY regions have highly similar gene content and structure, and only 0.4 per cent sequence divergence. The MSY sequences from wild males include three distinct haplotypes, associated with the populations' geographic locations, but gene flow is detected for other genomic regions. The Yh sequence is highly similar to one Y haplotype (MSY3) found only in wild dioecious populations from the north Pacific region of Costa Rica. The low MSY3-Yh divergence supports the hypothesis that hermaphrodite papaya is a product of human domestication. According to VanBuren *et al.* (2020), they estimate that Yh arose only ~4000 yr ago, well after crop plant domestication in Mesoamerica >6200 yr ago but coinciding with the rise of the Maya civilization. The Yh chromosome has lower nucleotide diversity than the Y, or the genome regions that are not fully sex-linked, consistent with a domestication bottleneck. The identification of the ancestral MSY3 haplotype will expedite investigation of the mutation leading to the domestication of the hermaphrodite Yh chromosome. In turn, this mutation should identify the gene that was affected by the carpel-suppressing mutation that was involved in the evolution of males (VanBuren *et al.,* 2020).

15.2.2.2 Fruit Development and Ripening Genes

Auxin and auxin signaling are involved in a series of developmental processes in plants. Auxin Response Factors (ARFs) is reported to modulate the expression of target genes by binding to auxin response elements (AuxREs) and influence the transcriptional activation of down-stream target genes. However, how *ARF* genes function in flower development and fruit ripening of papaya (*Carica papaya* L.) is largely unknown (Liu *et al.*, 2015). A comprehensive characterization and expression profiling analysis of 11 C. papaya *ARF* (*CpARF*) genes was performed using the newly updated papaya reference genome data.Results: We analyzed *CpARF* expression patterns at different developmental stages. *CpARF1, CpARF2, CpARF4, CpARF5,* and *CpARF10* showed the highest expression at the initial stage of flower development, but decreased during the following developmental stages. CpARF6 expression increased during the developmental process and reached its peak level at the final stage of flower development. The expression of *CpARF1* increased significantly during the fruit ripening stages. Many AuxREs were included in the promoters of two ethylene signaling genes (*CpETR1* and *CpETR2*) and three ethylene-synthesis-related genes (*CpACS1, CpACS2,* and *CpACO1*), suggesting that CpARFs might be involved in fruit ripening via the regulation of ethylene signaling. They concluded that the results provided comprehensive information on ARF family in papaya, including gene structures, chromosome locations, phylogenetic relationships, and expression patterns. The involvement of *CpARF* gene expression changes in flower and fruit development allowed us to understand the role of ARF-mediated auxin signaling in the maturation of reproductive organs in papaya (Liu *et al.*, 2015).

Auxin/indole-3-acetic acid (Aux/IAA) family genes encode short-lived nuclear proteins that mediate the responses of auxin-related genes and are involved in several plant developmental and growth processes. However, how *Aux/IAA* genes function in the fruit development and ripening of papaya (Carica papaya L.) is largely unknown (Liu *et al.*, 2017). Hence, a comprehensive identification and a distinctive expression analysis of 18 *C. papaya Aux/IAA* (*CpIAA*) *genes* were performed using newly updated papaya reference genome data. The *Aux/IAA* gene family in papaya is slightly smaller than that in Arabidopsis, but all of the phylogenetic subfamilies are represented. Most of the CpIAA genes are responsive to various phytohormones and expressed in a tissues-specific manner (Liu *et al.*, 2017). To understand the putative biological functions of the *CpIAA* genes involved in fruit development and ripening, quantitative real-time PCR was used to test the expression profiling of *CpIAA* genes at different stages. Furthermore, an IAA treatment significantly delayed the ripening process in papaya fruit at the early stages. The expression changes of *CpIAA* genes in ACC and 1-MCP treatments suggested a crosstalk between auxin and ethylene during the fruit ripening process of papaya. Liu *et al.* (2017) concluded that the results provided comprehensive information on the Aux/IAA family in papaya, including gene structures, phylogenetic relationships and expression profiles. The involvement of *CpIAA* gene expression changes in fruit development and ripening gives us an opportunity to understand the roles of auxin signaling in the maturation of papaya reproductive organs.

15.2.3 Transcriptomics

To identify the candidate genes for sex determination, Urasaki *et al.* (2012) conducted a transcriptome analysis of flower samples from male, female and hermaphrodite plants using high-throughput SuperSAGE for digital gene expression analysis. Among the short sequence tags obtained from the transcripts, 312 unique tags were specifically mapped to the primitive sex chromosome (X or Yh) sequences. An annotation analysis revealed that retroelements are the most abundant sequences observed in the genes corresponding to these tags. The majority of tags on the sex chromosomes were located on the X chromosome, and only 30 tags were commonly mapped to both the X and Yh chromosome, implying a loss of many genes on the Yh chromosome. Nevertheless, candidate Yh chromosome-specific female determination genes, including a MADS-box gene, were identified. Information on these sex chromosome-specific expressed genes will help elucidating sex determination in the papaya (Urasaki *et al.*, 2012).

15.3.0 Genetic Transformation

Though Pang and Sanford (1988) successfully transformed the papaya leaf-disc calli, the first stable transformation was achieved by Fitch *et al.* (1990) through micro-projectile bombardment method. Embryogenic tissues following bombardment were cultured on MS (half-strength) medium supplemented with 10 mg/l 2,4-D, 400 mg/l glutamine and 6 per cent sucrose. Then they were transferred to 2,4-D-free medium contai-ning 15 mg/l kanamycin sulphate where the putative transgenic isolates produced somatic embryos and five regenerated leafy shoots. This served as a base for the successful production of PRSV resistant plants by Fitch *et al.* (1992) using particle bombardment method with zygotic embryos as the initial explant and nptII gene as a selection marker. Similarly, herbicide-resistant transgenic papaya plants have been produced by Ponce *et al.* (1995) through the same particle bombardment method using phosphinothricin (*bar*) and kanamycin (*nptII*) resistant genes as selectable markers and the gus gene as a reporter gene. Here the transformation efficiency reported was as low as 1.24 per cent. Transgenic papaya plants resistant to PRSV could be produced with the help of vectors like C58 strain of *Agrobacterium* as reported by Yang *et al.* (1996). Ponce *et al.* (1996) have also reported the *Agrobacterium rhizogenes*-mediated transformation using somatic embryos as the initial explant. The coat protein (*cp*) gene of a Taiwan strain of papaya ringspot potyvirus (PRSV) was efficiently trans-ferred to papaya by *Agrobacterium tumefaciens*-mediated transformation, assisted by wounding of embryogenic tissue derived from immature zygotic embryos with carborundum in liquid phase. Some 38 transgenic lines were micropropagated by tissue culture and their resistance to PRSV was evaluated by mechanical inoculation with the virus under greenhouse conditions. Among them 9 lines were highly resistant to PRSV and 8 lines were moderately resistant (Yeh *et al.*, 1998). In Hawaii, a genetically transformed papaya cultivar Rainbow was resistant to ringspot virus (Westwood, 1999).

Chemical-based selection for plant transformation is associated with a number of real and perceived problems that might be avoided through visual selection (Zhu *et al.*, 2004). Green fluorescent protein (GFP), as a visual selectable marker, to produce transformed papaya (*Carica papaya*) plants following microprojectile bombardment of embryogenic callus was reported by Zhu *et al.* (2004). GFP selection reduced the selection time from 3 months on a geneticin (G418) antibiotic-containing medium to 3–4 weeks. Moreover, GFP selection increased the number of transformed papaya plants by five-to eightfold compared to selection in the presence of antibiotics. Overall, the use of GFP for selecting transgenic papaya lines improved our throughput for transformation by 15- to 24-fold while avoiding the drawbacks associated with the use of antibiotic resistance-based selection markers. The selectable marker gene phospho-mannose isomerase (pmi), which encodes the enzyme phosphomannose isomerase (PMI) to enable selection of transformed cell lines on media containing mannose (Man), was evaluated for genetic transformation of papaya (Zhu *et al.*, 2005). Papaya embryogenic calli have little or no PMI activity and cannot utilize Man as a carbon source; however, when calli were transformed with a *pmi* gene, the PMI activity was greatly increased and they could utilize Man as efficiently as sucrose Zhu *et al.*, 2005). Plants regenerated from selected callus lines also exhibited PMI activity but at a lower specific activity level. Our transformation efficiency with Man selection was higher than that reported using antibiotic selection or with a visual marker. For papaya, the PMI/Man selection system for producing transgenic plants is a highly efficient addition to previously published methods for selection and may facilitate the stacking of multiple transgenes of interest. Additionally, since the PMI/Man selection system does not involve antibiotic or herbicide resistance genes, its use might reduce environmental concerns about the potential flow of those genes into related plant populations.

Transgenic papaya plants were regenerated from hypocotyls and immature zygotic embryo after cocultivation with *Agrobacterium tumefaciens* LBA-4404 carrying a binary plasmid vector system containing neomycin phosphotransferase (*nptII*) gene as the selectable marker and -glucuronidase (GUS) as the reporter gene Azad *et al.*, 2013). The explants were co-cultivated with *Agrobacterium tumefaciens* on regeneration medium containing 500 mg/L carbenicillin + 200 mg/L cefotaxime for one week. The cocultivated explants were transferred into the final selection medium containing 500 mg/L carbenicillin + 200 mg/L cefotaxime + 50 mg/L kanamycin for callus induction as well as plant regeneration. The callus derived from the hypocotyls of *Carica papaya* cv. Shahi showed the highest positive GUS activities compared to *Carica papaya* cv. Ranchi. The transformed callus grew vigorously and formed embryos followed by transgenic plantlets successfully. The result of this study showed that the hypocotyls of *C. papaya* cv. Shahi and *C. papaya* cv. Ranchi are better explants for genetic transformation compared to immature embryos. The transformed *C. papaya* cv. Shahi also showed the maximum number of plant regeneration compared to that of *C. papaya* cv. Ranchi (Azad *et al.*, 2013).

The adoption of *Agrobacterium*-mediated transformation to agronomically important *Carica papaya* cultivars and genotypes, such as "Kapoho', has been challenging (Carlos-Hilario and Christopher, 2015). To address this problem, an alternative transformation protocol was developed for *papaya* using embryogenic suspension-derived cultures. The ability of the tissues (cultivar 'Kapoho') to regenerate plantlets from these cultures was demonstrated for both transgenic lines and untransformed controls. Suspension-derived cultures at the young globular calli stage contained embryos highly receptive to *Agrobacterium* infection and did not require the problematic wounding treatments inherent in other protocols. The optimum parameters selected to obtain transgenic calli were as follows: 1-d co-cultivation with *Agrobacterium*, the cell density used for transformation was A_{600} of 0.005, and elimination of *Agrobacterium* post-transformation was done using 250 mg/L carbenicillin and cefotaxime. Putative transgenic calli were confirmed positive for the presence of the *eGFP* transgene, *Cp45* promoter sequence, and hygromycin resistance gene (*hptII*) using PCR. The presence of *eGFP* messenger RNA (mRNA) and protein were detected using reverse transcription PCR (RT-PCR) and Western blot analysis, respectively. Visualization of qualitative *eGFP* fluorescence in roots, stems, and leaves further confirmed the expression of *Cp45:eGFP* fusion in the transformed papaya plantlets. This technique serves as an alternative and efficient method to generate transgenic plants in a simple laboratory setup that facilitates *Agrobacterium*-mediated transformation of previously difficult papaya cultivars and genotypes (Carlos-Hilario and Christopher, 2015).

15.3.1 PRSV Resistance

By the late 1990s, the first transgenic line designated as line 55-1 was used to develop PRSV-resistant transgenic cultivars 'Rainbow' and 'SunUp' (Teixeira da Silva *et al.*, 2007). In 1998, two PRSV resistant papaya cultivars, 'SunUp' and 'Rainbow', were released to growers in Hawaii (Fitch *et al.*, 1992; Manshardt, 1998 cited by Teixeira da Silva *et al.*, 2007). The transgenic papayas have offered durable resistance to PRSV and have controlled the virus in Hawaii (Ferreira *et al.*, 2002). According to figures out of the USDA's statistical service, 'Rainbow' makes up 47 per cent of the Big Island's 779 papaya hectares. 'Rainbow' is a yellow-flesh F1 hybrid of a cross between the transgenic cultivar 'SunUp' and nontransgenic cv. 'Kapoho Solo' (Gonsalves, 2002 cited by Teixeira da Silva *et al.*, 2007). Later work showed that the resistance of line 55-1 is RNA-mediated and dependent on the dosage of the cp gene, cp sequence homology of the challenge virus, and plant development stage (Tennant *et al.*, 2001 cited by Teixeira da Silva *et al.*, 2007). Even though 'Rainbow' (RB), a transgenic papaya cultivar hemizygous for PRSV cp gene, exhibited early plant susceptibility (>70 per cent of plants infected) to mechanical inoculation with crude preparations of PRSV isolates from Hawaii, RB plants become highly resistant by approximately 9 weeks after seeding in the greenhouse (2.5 per cent of plants infected) and by 13 weeks in the field (Gaskill *et al.*, 2002). In contrast 'SunUp', a transgenic papaya cultivar homozygous for the cp gene, exhibited complete resistance against all isolates of PRSV from Hawaii, but is susceptible to isolates from outside of Hawaii. Among 18 virus isolates collected in Taiwan, four (5-19, CY4, TD2, and DL1) were able to breakdown the transgenic resistance of papaya lines carrying the cp gene of PRSV and caused symptoms on non-transformed papaya plants different from those induced by the strain YK (Chen *et al.*, 2002 cited by Teixeira da Silva *et al.*,2007)

Papaya ringspot virus (PRSV) is a nonpersistent aphid-transmitted member of Potyvirus that causes one of the most economically important diseases of papaya. PRSV has been reported as a major limiting factor in papaya production around the world (Davis and Ying, 2004). Papaya (Carica papaya) was transformed via *Agrobacterium*-mediated transformation with four constructs containing either the unmodified or modified coat protein (*CP*) gene of Florida isolate H1K of Papaya ringspot virus (PRSV). The CP genes were in the sense orientation (S-CP), antisense orientation (AS-CP), sense orientation with a frame-shift mutation (FS-CP), or sense orientation mutated with three-in-frame stop codons (SC-CP) Davis and Ying, 2004). In all, 256 putative transgenic lines with the CP constructs were inoculated mechanically with PRSV H1K. None of the lines was immune to PRSV; however, highly resistant lines were found in each CP transgene group. For breeding purposes, 21 PRSV-resistant lines representing the four transgene constructs were selected and crossed with six papaya genotypes. The lines from the FS-CP and SC-CP transgene groups were highly fertile, but those from the S-CP and AS-CP transgene groups were practically infertile. Plants derived from 54 crosses and representing 17 transgenic lines were planted in the field. After 1 year in the field, 293 of the 1,258 the plants (23.3 per cent) became naturally infected with PRSV; whereas, 29 of 30 of the nontransgenic control plants (96.7 per cent) became infected. The incidence of PRSV infection varied in the R1 progeny depending on both the transgenic line and the nontransgenic parent (Davis and Ying, 2004).

Lines *et al.* (2002) reported the development of two Australian papaya cultivars that areimmune to infection with PRSV. Papaya somatic embryos were transformed viamicroprojectile bombardment using a construct containing an untranslatable PRSVcoat protein coding region. Immunity was demonstrated by repeated inoculationinthe glasshouse and eighteen months in the field. The immune lines were shown tohave up to five copies of the transgene by Southern hybridisation. Northernhybridisation showed that the coat protein transcript in the immune linesappeared to be degraded; therefore, the mechanism of resistance appears to bepost transcriptional gene silencing via RNA degradation. We have taken aproactive approach to control of PRSV in Australia by developing geneticallyengineered resistance before the Australian papaya industry has been decimatedby the inevitable spread of PRSV (Lines *et al.*, 2002).

Based on observations that viral replicase (RP) gene confers resistance to virus in other plants, Chen *et al.* (2001) designed a pair of primers and cloned the *RP* gene from PRSV by RT-PCR. The 32 -truncated and 52 -extended RP gene fragment was then oriented under the control of the CaMV35 S promoter and nos termination sequence in the mini Ti plasmid vector pRok to construct a plant expression vector, designated pRPTW. Papaya (*C. papaya* L.) cv. Tai-nong-2 embryogenic calli were transformed by *Agrobacterium tumefaciens* LBA4404 harboring the pRPTW vector. After selection on 100 mg/ml kanamycin, 20 putative transgenic papayas were regenerated and confirmed by PCR-Southern blot and Southern blot analyses. PRSV inoculation tests showed that the RP gene conferred resistance to PRSV in transgenic papayas and those offspring carrying the RP gene. The consistency of the presence of the RP gene and PRSV resistance indicates that replicase-mediated resistance against PRSV was attained in papaya. Possible mechanisms include RNA-mediated resistance and protein-mediated resistance, as well as others, although further studies are required (Chen *et al.*, 2001).

Local varieties of papaya grown in the Andean foothills of Mérida, Venezuela, were transformed independently with the coat protein (CP) gene from two different geographical Papaya ringspot virus (PRSV) isolates, designated VE and LA, via *Agrobacterium tumefaciens* (Fermin *et al.,* 2007). The *CP* genes of both PRSV isolates show 92 and 96 per cent nucleotide and amino acid sequence similarity, respectively. Four PRSV-resistant R0 plants were intercrossed or selfed, and the progenies were tested for resistance against the homologous isolates VE and LA, and the heterologous isolates HA (Hawaii) and TH (Thailand) in greenhouse conditions. Resistance was affected by sequence similarity between the transgenes and the challenge viruses: resistance values were higher for plants challenged with the homologous isolates (92 to 100 per cent similarity) than with the Hawaiian (94 per cent similarity) and, lastly, Thailand isolates (88 to 89 per cent similarity). Our results show that *PRSV CP* gene effectively protects local varieties of papaya against homologous and heterologous isolates of PRSV (Fermin *et al.,* 2007).

An efficient transformation method for developing papaya lines with transgenic resistance to these viruses and commercially desirable traits, such as hermaphroditism, is crucial to shorten the breeding program for this fruit crop (Kung *et al.,* 2010). An untranslatable chimeric construct pYP08 containing truncated PRSV coat protein (*CP*) and PLDMV *CP* genes coupled with the 32 untranslational region of PLDMV, was generated. Root segments from different portions of adventitious roots of in vitro multiple shoots of hermaphroditic plants of papaya cultivars 'Tainung No. 2', 'Sunrise', and 'Thailand' were cultured on induction medium for regeneration into somatic embryos. The highest frequency of somatic embryogenesis was from the root-tip segments of adventitious roots developed 2–4 weeks after rooting in perlite medium. After proliferation, embryogenic tissues derived from somatic embryos were wounded in liquid-phase by carborundum and transformed by *Agrobacterium* carrying pYP08. Similarly, another construct pBG-PLDMVstop containing untranslatable *CP* gene of PLDMV was also transferred to 'Sunrise' and 'Thailand', the parental cultivars of 'Tainung No. 2'. Among 107 transgenic lines regenerated from 349 root-tip segments, nine lines of Tainung No. 2 carrying YP08 were highly resistant to PRSV and PLDMV, and 9 lines (8 'Sunrise' and 1 'Thailand') carrying PLDMV *CP* highly resistant to PLDMV, by a mechanism of post-transcriptional gene silencing. The hermaphroditic characteristics of the transgenic lines were confirmed by PCR with sex-linked primers and phenotypes of flower and fruit. This approach has generated transgenic resistance to both PRSV and PLDMV with commercially desirable characters and can significantly shorten the time-consuming breeding programs for the generation of elite cultivars of papaya hybrids (Kung *et al.,* 2010).

Soil samples were collected from an isolated field where transgenic papaya were planted, and the persistences of transgenic genes of 796 bp (located between 35S promoter and coat protein, 35S-P/PRSV-CP), 398 bp (located between plasmid pBI121 and NOS terminator, pBI121/NOS-T), and 200 bp (located between NOS promoter and nptII gene, NOS-P/nptII) were studied (Lo *et al.,* 2007). At the end of planting, the residues of 398 bp in the soil was 0.06 µg g-1 of soil, whereas the residues of 769 and 200 bp were less than 30 pg g^{-1} of soil (detection limit). Kinetics studies on the persistence of these three fragments in sterile distilled water and nonsterile soil microcosms showed that two mechanisms might be involved: an initial fast exponential degradation pattern in the first week and then followed by a slow-release pattern throughout the experiment. Persistence of transgenic DNA in sterile water was longer than in nonsterile soil microcosms, indicating that enzymatic degradation and soil adsorption played important roles on the persistence of DNA in the environment. The reason for the fragment of 398 bp persisted longer than fragments of 769 and 200 bp is not clear, but the guanine plus cytosin (G plus C) content in the DNA fragment might be involved in the stability of DNA in the environment. Biological availability of soil DNA to bacteria conducted by the transformation assay indicated that gene transformation from soil DNA extracts to two Acinetobacter spp. did not occur (Lo *et al.*, 2007).

15.3.2 *Phytophthora* Resistance

Phytophthora spp., some of the more important casual agents of plant diseases, are responsible for heavy economic losses worldwide. Plant defensins have been introduced as transgenes into a range of species to increase host resistance to pathogens to which they were originally susceptible (Zhu *et al.*, 2007). However, the effectiveness and mechanism of interaction of the defensins with *Phytophthora* spp. have not been clearly characterized in planta. In this study, we expressed the Dahlia merckii defensin, DmAMP1, in papaya (*Carica papaya* L.), a plant highly susceptible to a root, stem, and fruit rot disease caused by Phytophthora palmivora. Extracts of total leaf proteins from transformed plants inhibited growth of Phytophthora in vitro and discs cut from the leaves of transformed plants inhibited growth of Phytophthora in a bioassay. Results from greenhouse inoculation experiments demonstrate that expressing the DmAMP1 gene in papaya plants increased resistance against *P. palmivora* and that this increased resistance was associated with reduced hyphae growth of *P. palmivora* at the infection sites. The inhibitory eVects of DmAMP1 expression in papaya suggest this approach has good potential to impart transgenic resistance against *Phytophthora* in papaya (Zhu *et al.*, 2007)

15.3.3 Cold Tolerance

Papaya (*Carica papaya* L.) production is affected by low temperatures that occur periodically in the subtropics (Dhekney *et al.*, 2007). The C-repeat binding factor (CBF) gene family is known to induce the cold acclimation pathway in *Arabidopsis thaliana*. Embryogenic papaya cultures were induced from hypocotyls of "Sunrise Solo" zygotic embryos on semisolid induction medium. The CBF 1/CBF 3 genes along with the neomycin phosphotransferase (NPT II) gene were placed under the control of the CaMV 35 S promoter and introduced into a binary vector pGA 643. Embryogenic cultures were transformed with *Agrobacterium* strain GV 3101 harboring pGA 643. After selection of transformed embryogenic cultures for resistance to 300 mg l^{-1}kanamycin, somatic embryo development was initiated and transgenic plants were regenerated. The presence of the CBF transgenes in regenerated plants was confirmed by Southern blot hybridization. The papaya and the related cold-tolerant *Vasconcella* genomes were probed for the presence of cold inducible sequences using polymerase chain reaction (PCR). Possible cold inducible sequences were present in the *Vasconcella* genome but were absent in the *Carica* genome (Dhekney *et al.*, 2007).

15.4.0 *In vitro* Conservation

Drew (1992) has devised an improved technique for *in vitro* propagation and germplasm storage of papaya. Shoots obtained were dissected and cultured for three days in rooting medium with 10.0 µM IBA and then transferred to hormone-free Drew and Smith medium wherein the growth rate was significantly reduced by substitution of 1 per cent fructose for 2 per cent sucrose. Suksa-Ard *et al.* (1997) made an attempt for the conservation of *C. papaya* shoots *in vitro* by lowering the temperature, applying growth retardants and increasing the osmotic potential of the culture medium. They found that shoots conserved at 16°C showed a high survival rate and were able to resume growth even after 12 months of culture in the medium supplemented with 30 g/l of sucrose, 5 µM ABA and 20 g/l agar.

16.0 References

Achicanoy, H. (1995) *Fitopatologia Colombiana*, **19** : 67-71.

Agnew, G.W.J. (1941) *Qd. Agric. J.*, **56** : 358-375.

Agnew, G.W.J.(1948) *Qd. Agric. J.*, **66** : 208-266.

Aiyappa, K.M. and Nanjappa, P.P. (1959) *Indian Hort.*, **3** : 3.

Aiyelaagbe, I.O.O. and Jolaoso, M.A. (1992) *Agroforestry Systems*, **19** : 1-14.

Alagiamanavalan, R.S. (1971) *M.Sc.Thesis* submitted to the University of Madras.

Ali, S.L. and Mazumdar, B.C. (1994) *J. Hort. Sci.*, **69** : 805-807.

Ali, Z.M., Lazan, H., Ishak, S.N. and Selamat, M.K. (1993) *Acta Horticulturae*, **343** : 230-232.

Allan, P. (1963) *South Afr. J. Agric. Sci.*, **6** : 517-530.

Allan, P. (1981) *Citrus and Subtropical Fruit J.*, **575** : 19-23.

Allan, P. (1995) *Acta Horticulturae*, **370** : 99-102.

Allan, P. and de Jager, J. (1978) *Crop Production*, **7** : 125.

Allan, P., Taylor, N. J. and Dicks, H. M. (2000) *Acta Horticulturae*, **511** : 27-33.

Allan,P. and MacMillan, C.N. (1991) *J. South Afr. Soc. Hort. Sci.*, **1** : 69-72.

Allemullah, M. and Walsh, K.B. (1996) *Australian J. Agric. Res.*, **47** : 371-385.

Aluja, M., Jimenez, A., Camino, M., Aldana, L., Castrejon, V. and Valdes, M.E. (1994) *Folia Entomologica Mexicana*, **90** : 33-42.

Alvarez, A.M. and Nishijima, W.T. (1987) *Plant Disease*, **71** : 681-686.

Andreoli, C. and Khan, A.A. (1993) *HortScience*, **28** : 708-709.

Anonymous (1986) In : *Food and Fruit-bearing Forest Species 3. Examples from Latin America*, FAO Forestry Paper, 44/3, Chap.17.

Aquilizan, F.A. (1986) *Abst. 22nd Int. Hort. Congress*, California, No. 261.

Aradhya, M.K., Manshardt, R.M., Zee, F. and Morden, C.W. (1999) *Genetic Res. Crop Evol.*, **56** : 579-586.

Aravind Kumar (1998) *Acta Horticulturae*, **463** : 251-259.

Arriola, M.C., Menchu, J.F. and Rolz, C. (1976) *Trop. Agric.*, **3** : 14-16.

Arumugam, S. (1974) *M.Sc(Ag.) Thesis* submitted to Tamil Nadu Agricultural University, Coimbatore.

Aryal, R. and R. Ming. (2014) *Plant Sci.*, **217-218**: 56-62.

Aryal,R., G. Jagadeeswaran, Y. Zheng, Q. Yu, R.Sunkar and R. Ming. (2014) *BMC Genomics* **15**: 20. doi: 10.1186/1471-2164-15-20

Awada, M. (1958) *Hawaii Agric. Exp. Sta. Tech. Bull.* No. 38.

Awada, M. (1961) *Hawaii Fm. Sci.*, 7-8.

Awada, M. (1969) *J. Amer. Soc. Hort. Sci.*, **94** : 687-690.

Awada, M. and Long, C. (1969) *J. Amer. Soc. Hort. Sci.*, **94** : 501-504.

Awada, M. and Long, C. (1971a) *J. Amer. Soc. Hort. Sci.*, **96** : 74-77.

Awada, M. and Long, C. (1971b) *J. Amer. Soc. Hort. Sci.*, **96** : 745-749

Awada, M. and Long, C. (1977) *Hawaii Agril. Expt. Sta. Tech. Bull.*, **97** : 25.

Awada, M. and Long, C. (1978) *J. Amer. Soc. Hort. Sci*, **103** : 217-219.

Awada, M. and Long, C. (1980) *J. Amer. Soc. Hort. Sci.*, **105** : 505-507.

Awada, M. and Suehisa, R.H. (1975) *Hawaii Agric. Expt. Sta. Tech. Bull.*, No. 94, p. 20.

Awada, M., Suehisa, R.H. and Padgett, M.M. (1979) *Hawaii Agric. Expt. Sta. Tech. Bull.* No. 103, p. 20.

Aykroyd, W.R. (1951) *The Nutritive Value of Indian Foods and the Planning of Satisfactory Diets*, Govt. of India Res. New Delhi.

Azad,M.A.K., M. G.Rabbani, L. Amin, and N.M. Sidik. (2013) *Int. J.Genomics* https: //doi.org/10.1155/2013/235487

Badillo, V.M. (1971) *Monografia de la familia Caricaceae. Maracay, Venezuela.*

Baile, J.B., Young, R.E. and Olmstead, A.J. (1954) *Plant Physiol.*, **29** : 168-174.

Bajwa, B.S. and Jawanda, J.S. (1962) *Punjab Farmer*, **4** : 204-207.

Bal, J.S., Singh, M.P., Minhas, P.P.S. and Bindra, A.S. (1993) *Acta Horticulturae*, **296** : 119-122.

Balakrishnamurthy, G. and Saraswathi, M.S. (2001) In : *Biotechnology of Horticultural Crops* (Eds. V.A. Parthasarathy, T.K. Bose, S.K. Mitra, P. Das and S. Mohandas), Naya Prokash, Calcutta, pp. 256-276.

Balasubramanium, S. and Rao, V.N.M. (1988) *South Indian Hort.*, **36** : 141-142.

Basso-Figuera, C., Schaffer, B., Crane, J.H., Colls, A.M. and Bryan, H.H. (1994) *Proc. Fla. St. Hort. Soc.*, **107** : 334-347.

Basso-Figuera, C., Schaffer, B., Crane, J.H., Colls, A.M. and Bryan, H.H. (1995) *Trop. Agric.*, **72** : 231-235.

Becerra Leov, E.N. (1989) *Revista Mexicana de Fitopatologia*, **7** : 218-222.

Becker, S. (1950) *Econ. Bot.*, **12** : 62-79.

Begum, H., Lavania, M.L. and Ratna Babu, G.H.V. (1988) *Seed Res.*, **16** : 51-56.

Bertocci, F., Vecchio, V. and Casini, P. (1997) *Advances Hort. Sci.*, **11** : 99-102.

Bhattacharya, R.K. and Madhava Rao, V.N. (1982) *South Indian Hort.*, **30** : 137-138.

Biswas, B., Sen, S.K. and Maiti, S.C. (1981) *Abst. Nat. Symp. Trop. and Subtrop. Fruit Crops*, Bangalore, p.54.

Biswas, B., Sen, S.K. and Maiti, S.C. (1989) *Progressive Hort.*, **21** : 280-284.

Biswas, B., Sen, S.K. and Maiti, S.C. (1990) *The Hort. J.*, **3** : 20-25.

Biswas, N. (1994) *Ann. Agric. Res.*, **15** : 301-305.

Biswas, R. K., Rana, S. K. and Mallick, S. (1999) *Ann. Agric. Res.*, **20** : 116-117.

Broughton, W.J., Hashim, A.W. and Shen, T.C. (1977) *MARDI Research Bulletin*, **5** : 59-72.

Buisson, D.J. and Lee, D.W. (1993) *Amer. J. Bot.*, **80** : 947.

Camejo, B. and Alavarez, P.R. (1985) *Ciencia y Tecnica en la Agricultura*, **6** : 69-79.

Carlos-Hilario, L.R. and Christopher, D.A. (2015) *In Vitro Cell.Dev.Biol.-Plant* **51,** 580–587.

Carvalho F.A. and Renner, S.S. (2012) *Mol Phylogen Evol* **65**(1)**:** 46–53.

Cerdas, M. Del, M. and Saenz, M.V. (1993) *Agronomia Costarricense*, **17** : 49-54.

Chacko, E. K. and Singh, R. N. (1966) *Trop. Agric.*, Trinidad, **43** : 341-346.

Chacko, E.K., Randhawa, G.S., Minon, M.A. and Negi, S.P. (1972) *Curr. Sci.*, **41** : 455.

Chan, H.T. Jr. and Taniguchi, M.H. (1985) *J. Food Sci.*, **55** : 1092-1094.

Chan, H.T.Jr. (1986) *J. Food Sci.*, **51** : 581-583.

Chan, L.K. and Teo, C.K.H. (1992) *Planter*, **68** : 239-242.

Chang, M.T. and Liaw, S.I. (1994) *Proc. Symp. on Practical Aspects of Some Economically Imp. Fruit trees in Taiwan* (Eds. H.S. Lin and L.R. Chang), Sepcial Pub. Taichung Dist. Agric. Improvement Station, **33** : 31-46.

Charavanapavan, C. (1952) *Trop. Agric.*, **108** : 189-190.

Cheema, G.S. and Dani, P.G. (1930) *Bombay Dept. Agric, Bull.*, No. 162.

Chen, G., Ye, C., Huang, J. *et al.* (2001) *Plant Cell Reports* **20,** 272–277.

Chen, L.F.C, H.J. Bau and Yeh, S.D. (2002) *Acta Horticulturae* **575:** 465-474

Chen, M.H. and Chen, C.C. (1992) *Plant Cell Rept.*, **11** : 404-407.

Chen, M.H., Chen, C.C., Wangand, D.N. and Chen, F.C. (1991) *Van. J. Bot.*, **69** : 1913-1918.

Chen, M.H., Wang, P.J. and Maeda, E. (1987) *Plant Cell Rep.*, **6** : 348-351.

Chen, C., Q. Yu,S. Hou, *et al.* (2007) *Genetics* **177:** 2481–2491

Choudhury, R.S. (1957) *Indian J. Hort.*, **14** : 179-185.

Chow, H.T. and Liu, C.S. (1987) *J. College of Sci. Engineering, Nat. Chung Hsing Univ.*, **24** : 1925.

Clemente, H.S. and Marler, T.E. (1996) *J. Amer. Soc. Hort. Sci.*, **121** : 292-295.

Cohen, E., Lavi, U. and Spiegel-Roy, P. (1989) *Scientia Hort.*, **40** : 317-324.

Conover, R. A. (1962) *Phytopathol.*, **52:** 62.

Conover, R.A. (1964) *Proc. Fl. State Hort. Soc.*, **77:** 440-444.

Conover, R.A., Litz, R.E. and Malo, S.E. (1986) *Circular Agric. Exp. Sta.*, Univ. Florida, No. SS-329, 3.

Crabbe, F. (1987) *Information Bull., Citrus and Subtropical Fruits Res. Inst.*, **181** : 1-2.

Cunha, R.J.P. and Haag, H.P. (1981) *Anais da Escola Superior de Agricultura*, **37** : 169-178.

Darekar, K.S. and Mhase, N.L. (1987) *Curr. Res. Reporter, Mahatma Phule Agric. Univ.*, **3** : 96-97.

Das, R.C. and Sahu, A.C. (1981) *Abst. Nat. Symp. Trop. and Subtrop. Fruit Crops*, Bangalore, p. 76.

Davis, M.J. and Z. Ying (2004) *Plant Disease* **88:** 352-358.

Dedolph, P. R. (1962) *Bot. Gaz.*, **124** : 78.

DeGeus, A. (1967) *Fert. Guide for Trop. and Subtropic Fmg.*, Wageningen, pp. 605-606.

Desikan, K.R. (1972) *M.Sc. (Ag.) Thesis* submitted to Tamil Nadu Agricultural University, Coimbatore.

Devitt, L. C, T. Sawbridge, T. A.Holton, K.Mitchelson, and R. G.Dietzgen (2006) *Plant Sci.*, **170:** 356-363.

Dhaliwal, G.S., Gill, M.I.S. and Kalra, S.K. (1991) *J. Res., Punjab Agric. Univ.*, **28** : 354-356.

Dhara, N. and Veerannah, L. (1998) *South Indian Hort.*, **46** : 74-77.

Dhekney, S.A., Litz, R.E., Moraga Amador, D.A. *et al.* (2007) *In vitro Cell. Dev. Biol.-Plant*, **43:** 195–202.

Dhekney, S.A., R.Kandel, D.R.Bergey, V.Sitther, K.Soorianathasundaram and R.E.Litz. (2016) Biocatalysis and Agricultural Biotechnology, **5:** 133-142.

Dinesh, M. R. and Yadav, I. S. (1998) *Indian Hort.*, **43** : 21-33.

Doijode, S.D. (1992) *Acta Horticulturae*, **296** : 123-127.

Drew, R.A. (1988) *HortScience*, **23** : 609-611.

Drew, R.A. (1992) *HortScience*, **27** : 1122-1124.

Eloisa, M., Reyes, A. and Paull, R.E. (1994) *Scientia Hort.*, **58** : 31-39.

Eustice, M., Q. Yu, C.W.Lai, S. Hou *et al.* (2008) *Tree Genetics and Genomes,* **4:** 333–341.

Fauquet, C. M., Mayo, M. A., Maniloff, J., Desselberger, U. and Ball, L. A. (2005) Virus taxonomy: 8th Report of the International Committee on Taxonomy of Viruses, ICTV. Elsevier, San Diego.

Fayyaz, A., Asbi, B.A., Ghazali, H.M., Man, Y.B.C. and Jiap, S. (1994) *Food Chem.,* **49** : 373-378.

Fermin,G., V. Inglessis, C. Garboza, S.Rangel and M. Dagert. (2007) *Plant Disease* **88:** 516 – 522.

Fitch, M.M.M., Manshardt, R.M., Gonsalves, D., Slighton, J.L. and Stanford, J.C. (1990) *Plant Cell Rep.,* **9** : 189-194.

Fitch, M.M.M., Manshardt, R.M., Gonsalves, D., Slighton, J.L. and Stanford, J.C. (1992) *Biotechnol.,* **10** : 1466-1472.

Floravanco, J.C., Paiva, M.C., Carvalho, R.I.N. De and Manica, I. (1994) *Ciencia Rural,* **24** : 519-522.

Flores Revilla, G., Garcia, E., Nieto Angel, D., Teliz Ortiz, D. and Villanueva jimenez, J.A. (1995) *Acta Horticulturae,* **370** : 151-158.

Foyet, M. (1972) *Fruits,* **27** : 305-306.

Franco, M.R.B. and Rodriguez-Amaya, D.B. (1993) *Arquivos de Biologia e Tecnologia,* **36** : 613-632.

Fuentes, G. and J. M. Santamaría. (2014) In: *Genetics and Genomics of Papaya,* Plant Genetics and Genomics (eds. R. Ming and P.H. Moore), Crops and Models 10, DOI 10.1007/978-1-4614-8087-7_1, © Springer Science+Business Media New York 2014

Galanihe, L.D., Jayasundera, M.U.P., Vithana, A., Asselaarachchi, N. and Watson, G.W. (2010) *Tropical Agri. Res. Ext.* **13**(3).

Gammie, G.A. and Patwardhan, G.B. (1908) *Bombay Dept. Agric. Bull.,* No. 30 pp. 54-55.

Garcia Perez, E., Lopez Hernandez, A.Y., Teliz Ortiz, D. and Nieto Angel, D. (1995) *Acta Horticulturae,* **370** : 145-150.

Gaskill,D.A, K.Y. Pitz,S.A. Ferreira, and D. Gonsalves. (2002) *Acta Horticulturae* **575:** 75-83.

Ghanta, P.K., Dhua, R.S. and Mitra, S.K. (1992) *The Hort. J.,* **5** : 43-48.

Ghanta, P.K., Dhua, R.S. and Mitra, S.K. (1994) *J. Res., Birsa Agril. Univ.,* **6** : 155-157.

Ghanta, PK., Dhua, R.S. and Mitra, S.K. (1995) *Ann. Agric. Res.,* **16** : 405-408.

Godoy, H.T., Rodriguez-Amaya, D.B., Connor, A.E. and Britton, G. (1990) *Food Chem.,* **36** : 281-286.

Gonsalves,D. (2002) *Current Topics in Microbiology and Immunology* **266:** 73-83.

Govindu, H.C. (1964) A review on virus disease on crop plant. *Information phamphlet No.2* (*Directorate of Agriculture,* Banglore,13.

Grau, A. and Halloy, S. (1997) *J. Pl. Physiol.*, **150** : 475-480.

Greensill, C.V. and Newman, D.S. (1999) *J. Near Infrared Spectroscopy*, **7** : 109-116.

Grime, J.P. (1981) In : *Plants and the Daylight Spectrum* (Ed. E.H. Smith). Academic Press, London, p. 147.

Gschwend, A.R., Q.Yu, E.J.Tong, *et al.* (2012) *PNAS* **109:** 13716- 13721.

Haber, S., Ikegami, M., Bajet, N.B. and Goodman, R.M. (1981) Evidence for a divided genome in bean golden mosaic virus, a geminivirus. *Nature*, 289-324.

Hamilton, R.A. (1954) *Hawaii Agri. Expt. Sta. Tech. Bull.*, No. 20.

Hamilton, R.A. (1987) *Res. Exten. Series, No. 085, College of Trop. Agric.*, University of Hawaii, pp. 5-6.

Hamilton, R.A. and Izuno, T. (1965) *Hawaii Farm Sci.*, **14** : 1-3.

Hamilton, W.D.O., Bisaro, D.M., Coutts, R.H.A. and Buck, K.W. (1983) Demonstration of the bipartite nature of the genome of a single stranded DNA plant virus by infection with the cloned DNA components. *Nucl. Acids Res.*, **11**: 7387-7396.

Hamilton,W.D.O., Bisaro, D.M., Coutts, R.H.A. and Buck, K.W. (1984) Complete nucleotide sequence of the infectious cloned DNA components of tomato golden mosaic virus: potential coding regions and regulator sequences. *EMBO J.*, **3**: 2197-2205.

Harjadi, S.S., Pribadi, F.I. and Koswara, S. (1995) *Acta Horticulturae*, **379** : 83-88.

Hasdiseve, P., Subhadrabandhu, S., Wasee, S. and Srinivas, P. (1989) *Kasetsart J. Natural Sci.*, **23** : 26-37.

Heu, R.A., Fukada, M.T. and Conant, P, (2007) Papaya mealybug, Paracoccus marginatus Williams and Granara de Willink (Hemiptera: *Pseudococcidae*). *State of Hawaii New Pest Advisory. Dep. of Agri.*

Higgins, J.E. and Holt, V.S. (1914) *Hawaii Agri. Expt. Stat. Bull.*, No. 32.c.

Hofmeyr, J.D.J. (1938) *Report Dept. of Agri. And Forestry*, Union of South Africa.

Hofmeyr, J.D.J. (1942) *South Afr. J. Sci.*, **26** : 286-287.

Hofmeyr, J.D.J. (1945) *South Afr. J. Sci.*, **41** : 225-230.

Hofmeyr, J.D.J. and Roux, J.C.L. (1938) *Fmg. S. Africa*, **14** : 325.

Hore, J.K. and Sen, S.K. (1993) *Environ. Eco.*, **11** : 273-275.

Hossain, M.M., Khan, L.A., Paul, N.K., Hossain, M. and Joarder, O.I. (1990) *Crop Res.*, **3** : 222-233.

Hsiu,Yu, C., YuKuang, L., Changttong, C. and Chiming, Y. (1996) *J. Chinese Agril. Chemical Soc.*, **34** : 460-468.

Huet J, Looze, Y., Bartik, K., Raussens, V., Wintjens, R. and Boussard, P. (2006) *Biochemical and Biophysical Research Communications,* **341**: 620-626.

Irulappan, S. (1980) *M.Sc.* (*Ag.*) *Thesis* submitted to Tamil Nadu Agricultural University, Coimbatore.

Islam, R., Rahman, S.M., Hossain, M. and Joarder, O.I. (1993) *Pakistan J. Botany*, **25** : 189-192.

Iyer, C.P.A. and Subramanyam, M.D. (1981) *Abst. Nat. Symp. Trop. and Subtrop. Fruit Crops*,Bangalore, p. 15.

Jaiswal, V.S., Narayan, P. and Lal, M. (1984) *Biochemic and Physiologie der Pflanzen*, **179** : 799-801.

Jang, E.B. (1990) *Acta Horticulturae*, **269** : 453-458.

Jayalatha, M. (1994) *M.Sc.* (*Hort.*) *Thesis*, Tamil Nadu Agricultural University, Coimbatore, India.

Jayaprakash, R., Bojappa, K.M., Seenappa, K. and Ramanjini, P.H. (1989) *Progressive Hort.*, **21** : 239-243.

Jayaweera, D.M.A. (1957) *Trop. Agric.*, **113** : 55-85.

Jobin-Décor, M.P. and O'Hare, T.J. (1996) *Proc. Australasian Post-harvest Hort. Conf.*, Melbourne, Australia, Dept. of Natural Resources & Environment, pp. 63-66.

Jobin-Decor, M.P., Graham, G.C., Henry, R.J. and Drew, R.A. (1997) *Genet. Res. Crop Evol.*, **44** : 471-477.

Kannan, M. and Muthuswami, S. (1989) *South Indian Hort.*, **37** : 6-9.

Kapanadze, I.S. and Khasaya, G.S. (1988) *Subtropicheski Kultury*, **1** : 136-140.

Kataoka, J. (1994) *Japanese J. Trop. Agric.*, **38** : 251-257.

Kataoka, J. and Inoue, H. (1991) *Japanese J. Trop. Agric.*, **35** : 127-129.

Keil, U. and Schreier, P. (1989) *Phytochemistry*, **28** : 2281-2284.

Khan, K.F. and Seemanthini, B. (1968) *Indian J. Hort.*, **25** : 182-190.

Khan, T.A. and Khan, S.T. (1995) *Pakistan J. Nematology*, **13** : 29-34.

Khan, T.A., Khan, S.T. and Saxena, S.K. (1995) *Annals Plant Protection Sci.*, **3** : 46-51.

Khatoon, K. and Sultana, R. (1994) *Pakistan J. Bot.*, **26** : 191-195.

Khuspe, S.S. and Ugale, S.D. (1977) *J. Maharashtra Agric. Univ.*, **2** : 115-118.

Khuspe, S.S., Hendre, R.R., Mascarenhas, A.F. and Jaganathan, V. (1980) In : *Plant Tissue Culture*, Genetic Manipulation and Somatic Hybridization of Plant Cell (Eds. P.S. Rao, M.R. Heble and M.S. Chandha), BARC, Bombay, pp. 198-205.

Kimura, M., Rodriguez-Amaya, D.B. and Yokoyama, S.M. (1991) *Lebensmittel-Wissenschaft und Technologie*, **24** : 415-418.

Kist, H. and Manica, I. (1995) *Pesquisa Agropecuaria Brasileira*, **30** : 657-666.

Knight, R.J. (1980) In : *Tropical and Subtropical Fruits : Composition, Properties and Uses* (Eds. S. Nagy and P.E. Shaw). AVI Publishing Inc., Westport, CT, Chap.1.

Koenig, R., and Lesemann, D. E. (1978) Potexvirus group. Commonw. *Mycol. Inst. Assoc. Appl. Biol.*, Surrey, UK.

Kohli, R.R., Biswas, S.R., Ramachander, P.R. and Reddy, Y.T.N. (1986) *Indian J. Hort.*, **43** : 88-93.

Kumar, V. (1952) *Indian J. Hort.*, **9** : 20-28.

Kung, Y., Yu, T., Huang, C. *et al.* (2010) *Transgenic Res* **19,** 621–635.

Lai, C.W.J., Q. Yu, S. Hou, R. L. Skelton *et al.*(2006) *Mol Gen Genomics* **276:** 1–12

Lai, Choo Chun, Tes, Tsong Ann, Yeh Shyi Dong and Yand, Jin Sherng (1998) *Plant Cell Tissue Organ Cult.*, **53** : 221-225.

Lange, A.H. (1961a) *Trop. Agric. Trinidad*, **38** : 235-243.

Lange, A.H. (1961b) *Hawaii Fm. Sci.*, **11** : 14.

Lassoudiere, A. (1969) *Fruits*, **11** : 24.

Lavania, M.L. and Jain, S.K. (1995) *Haryana J. Hort. Sci.*, **24** : 79-84.

Lazan, H., Ali, Z.M. and Selamat, M.K. (1993) *Acta Horticulturae*, **343** : 141-147.

Lazan, H., Ali, Z.M. and Sim, W.C. (1990) *Acta Horticulturae*, **269** : 345-358.

Lazan, H., Mohd, Z., Liang, K.S. and Yee, K.L. (1989) *Physiol. Plant.*, **77** : 93-98.

Lazan, H., Selamat, M.K. and Ali, Z.M. (1995) *Physiol. Plant.*, **95** : 106.

Lim, W.H. (1989a) *MADRI Res. J.*, **17** : 200-207.

Lim, W.H. (1989b) *MAPPS Newsletter*, **12** : 32.

Lin, H.C., Lin, H.L. and Lee, K.C. (1994) *Proc. Symp. on Practical Aspects of Some Economically Imp. Fruit trees in Taiwan* (Eds. H.S. Lin and L.R. Chang) Special publication – Taichung Dist. Agric. Improvement Station, 33 : 61-67.

Lines, R.E., Persley, D., Dale, J.L. *et al.* (2002) *Molecular Breeding* **10,** 119–129.

Linnaeus, Carl.Von. (1753) *Species Plantarum*, Lawrentii Salvii Holminae, p. 1200.

Liquido, N.J. (1991a) *Environmental Entomology*, **20** : 620-625.

Liquido, N.J. (1991b) *J. Economic Entomology*, **84** : 948-953.

Liquido, N.J. (1993) *J. Agric. Entomology*, **10** : 163-170.

Litz, R.E. and Conover, R.A. (1980) *J. Amer. Soc. Hort. Sci.*, **106** : 792-794.

Liu, K., C.Yuan, S. Feng *et al.*() *BMC Genomics* **18:** 351. DOI 10.1186/s12864-017-3722-6

Liu, K., C.Yuan,H. Li *et al.* (2015) *BMC Genomics* **16**: 901. DOI 10.1186/s12864-015-2182-0.

Liu, Z., P. H. Moore, H. Ma, C. M. Ackerman, R. Makandar *et al.*(2004) *Nature* **427**: 348–352.

Lo, C.C., Chen, S.C.and Yang, J.Z. (2007) *J. Agric. Food Chem.* **55 :** 7534"7540.

Lokhande, N.M. and Moghe, P.G. (1991) *South Indian Hort.*, **39** : 23-26.

Lopez, M.E., Vattuone, M.A. and Sampietro, A.R. (1988) *Phytochemistry*, **27** : 3077-3081.

Ma, H., P. H. Moore, Z. Liu, M. S. Kim, Q. Yu *et al.* (2004) *Genetics* **166**: 419–436.

Maas, E.V. (1984) In : *Handbook of Plant Science in Agriculture* Vol.II (Ed. B.R. Christie), CRC Press, Boca Raton, F1.

Madhava Rao, V.N. (1974) *Indian Hort.*, **18** : 727.

Madhava Rao, V.N., Balakrishnan, R. and Raman, K.R. (1974) *Indian Hort.*,**18**: 6-7.

Magdalita, P.M., Adkins, S.W., Godwin, I.D. and Drew, R.A. (1996) *Australian J. Bot.*, **44** : 343-353.

Magdalita, P.M., Pimentel, R.B., Rosario, E.E.Del, Bayot, R.G. and Espino, R.R.C. (1986) *Philippine J. Crop Sci.*, **11** : Abst. 3D-9a.

Maharaj, R. and Sankat, C.K. (1990) *Acta Horticulturae*, **269** : 375-385.

Makhija, M. and Jindal, P.C. (1983) *Seed Res.*, **11** : 125.

Malan, E.F. (1953) *Farming in South Africa*, Reprint 57.

Malo, S.E. and Campbell, C.W. (1986) *The Papaya*. Institute of Food and Agricultural Sciences. Fruit Crops Fact Sheet FC-11.

Manshardt, R. and Wenslaff, H. (1986) *Abst. 22nd Int. Hort. Congress*, California, No. 262.

Manshardt, R. and Wenslaff, H. (1998) *Acta Horticulturae*, **461** : 65-73.

Manshardt, R.M. and Drew, R.A. (1998) *Acta Horticulturae*, **461** : 65-73.

Marler, T.E. (1990) Unpublished data

Marriott, J. and Proctor, F.J. (1982) *Outlook in Agriculture*, **9** : 233-239.

Masri, M., Razak, A.S. and Ghazalli, M.Z. (1990) *MARDI Res. J.*, **18** : 191-196.

Mateos Sanchez, M., Perez Flores, J. and Acosta, Z.C. (1995) *Revista Chapingo Serie Horticultura*, **1** : 33-37.

Maxwell, L.S., Krezdorn, A.H., Will, A.A. and Golloy, E.V. (1984) *Florida Fruit*, Maxwell, Tampa, FL.

Mazumdar, B.C. (1988) *Bull. Bot. Soc. Bengal*, **42** : 31-35.

Mederos Olalde, E. and Rodriguez Hernandez, M. (1988) *Centro Agricola*, **15** : 41-49.

Mederos Olalde, E., Carmona Benitez, O., Lopez Perez, A. and Danta Iglesias, C. (1995) *Centro Agricola*, **22** : 5-12.

Mehta, P.M., Raj, S.S. and Raju, P.S.Jr. (1986) *Indian J. Hort.*, **43** : 169-173.

Mendoza, S.P.Jr. and Mercado, B.L. (1983) *Proc. 13th Aniversary and Annual Convention*, Pest Control Council of Philippines.

Meyerdirk DE, Muniappan R, Warkentin R, Bamba J and Reddy GVP (2004) Biological control of the papaya mealybug, Paracoccus marginatus (Hemiptera: *Pseudococcidae*). *Plant Protection Quarterly.*, **19**(3): 110-114.

Miller, R.M. and Drew, R.A. (1990) *Plant Cell Tissue Organ Cult.*, **21** : 39-44.

Ming R, Yu, Q.and Moore, P.H. (2007) *Semin Cell Dev Biol* **18(3):** 401–408.

Ming, R., Hou, S., Feng, Y. *et al.* (2008) *Nature* **452,** 991–996.

Ming, R., Q. Yu and P. H. Moore, (2007) *Semin. Cell Dev. Biol.* **18:** 401–408.

Mohandas, S. (1992) *Fertilizer Res.*, **31** : 263-267.

Mondal, M., Gupta, S. and Mukherjee, B.B. (1990) *Plant Cell Rep.*, **8** : 609-612.

Monmarson, S., Michaux-Ferriere, N. and Teisson, C. (1994) *Fruits*, **46** : 279-288.

Monmarson, S., Michaux-Ferriere, N. and Teisson, C. (1995) *J. Hort. Sci.*, **70** : 57-64.

Moraes, M.G.De, Termignoni, C. and Salas, C. (1994) *Plant Sci.*, **102** : 11-18.

Mosca, J.L. and Durigan, J.F. (1995) *Acta Horticulturae*, **370** : 217-221.

Mosqueda-Vazquez, Aragaki, R.M. and Nakasone, H.Y. (1981) *J. Amer. Soc. Hort. Sci.*, **106** : 484-487.

Mowry, H. (1930) *Press Bull.* 386, Agri. Expt. Stat., Gainesville, University of Florida.

Muller, C.H., Figueiredo, F.J.C. and Carvalho, J.E.U.De (1991) *Circular tecnica – Centro de pesquisa Agropecuaria do Tropico Umido*, No. 61, pp. 15.

Munoz, S., Lima, H., Perez, M. and Rodriguez, O.L. (1982) *Cizecia y Tecnica en la Agricultura*, **5** : 39-48.

Muthuswamy, S., Sriram, T.A. and Balakrishnan, R. (1962) *Madras Agric. J.*, **50**.

Na, J.K., J.Wang, J.E. Murray *et al.* (2012) *BMC Genomics*, **13:** 176. http: //www. biomedcentral.com/1471-2164/13/176

Nadagoudar, B.S., Hanamashetti, S.I., Madiwalar, S.L., Devaranavadagi, S.B., Mutanal, S.M. and Gumaste, S.K. (1995) *Indian Forester*, **121** : 276-279.

Naik, K.C. (1949) *South Indian Fruits and Their Culture*, Varadachary and Co., Madras.

Nakasone, H.Y., Crozier, H.A. and Ikehara, D.K. (1972) *Hawaii Agric. Expt. Sta. Tech. Bull.*, No. 79, p.12.

Nakasone, H.Y., Yee, W., Ikehara, D.K., Doi, M.J. and Ito, P.l. J. (1974) *Hawaii Agric. Expt. Stat. Tech. Bull.*, No. 167, p.24.

Nandi, A.K. and Mazumdar, B.C. (1990) *Indian Biologist*, **22** : 47-50.

Nariani, T.K. (1956) Leaf curl of papaya. *Indian Phytopathol.*, **9:** 151–157.

Narinesingh, D. and Mohammed-Maraj, R. (1988) *J. Sci. Food Agric.*, **46** : 175-186.

Nguyen, H. and Tarandjiiska, R. (1995) *Fett Wissenschaft technologie*, **97** : 20-23.

Nirujogi, Bharathi and Dinesh, M.R. (2012) *Asian J. Hort.*, **7**(2) : 272-275.

Nirvan, R.S. (1953) *Indian Phytopath.*, **6** : 47-48.

Nishijima, K.A., Miura, C.K., Armstrong, J.W., Brown, S.A. and Hu, B.K.S. (1992) *Plant Disease*, **76** : 723-727.

Nishijima, W.T. (1995) *Acta Horticulturae*, **370** : 121-128.

Nishimoto, R.K. (1981) *Proc. Asian Pacific Weed Sci. Soc. Conf.*, pp. 57-61.

Nishimoto, R.K. (1993) *Intern. J. Pest Management*, **39** : 366- 369.

Nishimoto, R.K. and Hibbard, K.L. (1979) *Proc. 7th Asian Pacific Weed Sci. Soc. Conf.*, pp. 71-73.

Ochse, J.J. (1931) *Fruits and Fruit Culture in Dutch East Indies*, G. Kolff and Co., Batavia.

Ochse, J.J., Soule, M.J., Dijkman, M.J. and Wehlburg, C. (1961) *Tropical and Subtropical Agriculture*. The MacMillan Company, New York.

Ogden, M.A.H., Jackson, L.K. and Campbell, C.W. (1981) *Proc. Fla. Sta. Hort. Soc.*, **94** : 222.

Ong, H.T. and Kwok, C.Y. (1983) *Pertanika*, **6** : 7-14.

Padmakumari, O. and Sivanappan, R.K. (1989) *Proc. 11th International Cong. on Agric. Engineering* (Eds. V.A. Dodd and P.M. Grace), Rotterdam, Netherlands, pp. 669-671.

Padmanabhan, V. (1970) *M.Sc.Thesis*, submitted to the University of Madras.

Pal, D.K. and Selvaraj, Y. (1987) *J. Hort. Sci.*, **62** : 117-124.

Pal, D.K., Subramanyam, M.D., Divakar, N.G., Iyer, C.P.A. and Selvaraj, Y. (1980) *J. Fd. Sci. Tech.*, **17** : 254-256.

Pan, H.C., Lin, H.L. and Lee, K.C. (1994) *Proc. Symp. on Practical Aspects of Some Economically Imp., Fruit trees in Taiwan* (Eds. H.S. Lin and L.R. Chang) Special publication – Taichung Dist. Agric. Impovement Station, **33** : 47-60.

Pandey, R.M., Kishore, D.K. and Arulmozhi, K. (1986) *Indian J. Hort.*, **43** : 174-179.

Pang, S.Z. and Sanford, J.C. (1988) *J. Amer. Soc. Hort. Sci.*, **113** : 287-291.

Pant, V. and Lavania, M.L. (1989) *Progressive Hort.*, **21** : 165-167.

Pant, V. and Lavania, M.L. (1998) *South Indian Hort.*, **46** : 5-8.

Parasnis, A.S., Ramakrishna, W., Chowdari, K.V., Gupta, V.S. and Ranjekar, P.K. (1999) *Theor. Appl. Genet.*, **99** : 1047-1052.

Patil, K.B., Patil, B.B. and Patil, M.T. (1995a) *J. Maharashtra Agric. Univ.*, **20** : 447-448.

Patil, S.S., Tang, C.S. and Hunter, J.E. (1973) *Pl. Dis. Reptr.*, **57** : 86.

Paull, R.E. (1995) *Post-harvest Biol. Tech.*, **6** : 167-175.

Paull, R.E. (1996) *Post-harvest Biol. Tech.*, **7** : 359-370.

Paull, R.E., Gross, K. and Qiu YunXia (1999) *Post-harvest Biol. Tech.*, **16** : 79-89.

Paull, R.E., Nishijima, W., Reyes, M. and Cavaletto, C. (1997) *Post-harvest Biol. Tech.*, **11** : 165-179.

Perez, A. and Varges, D. (1977) *J. Agric. Univ. Puerto Rico*, **61** : 68-76.

Plaza, J.L.De La, Alique, R., Calvo, L. and Moure, J. (1992) *Proc. Conf. held in Granada*, Spain, May 3-8, 1992 (Eds. N.C. Prados and J.H. Rodriguez) G-19-29.

Ponce, J.L.C., Garcia, A.V. and Estrella, L.H. (1996) *In vitro Cell Biol. Plant.*, **32** : 86-90.

Ponte, J.J.Da (1996) *Revista de Agricultura* (Piracicaba), **71** : 259-261.

Pope, W.T. (1930) *Hawaii Agril. Expt. Stat. Bull.*, **61**.

Prest, R.L. (1957) *Qd. Agric. J.*, **81** : 144-148.

Purcifull, D. E., and Hiebert, E. (1971) Papaya mosaic virus. Descriptions of Plant viruses. Common wealth. Mycol. *Inst. Assoc. Appl. Biol., Surrey, UK.*

Purcifull, D. E., Edwardson, J., Hiebert, E., and Gonsalves, D. (1984) Papaya ringspot virus. Descriptions of Plant Viruses. *Common wealth. Mycol. Inst. Assoc. Appl. Biol., Surrey, UK.*

Purcifull, D.E. and Hiebert, E. (1971) Papaya mosaic virus. Descriptions of plant viruses. Commonwealth Mycological Institute and Association of Applied Biologists. *Kew, Surrey, England.* No. 56. p4.

Purnima and Bisht, S. (1988) *Curr. Sci.*, **57** : 440-442.

Purohit, A. G. (1981) *Indian Hort.*, **25** : 3-5.

Purohit, A.G. (1995) *J. Maharashtra Agric. Univ.*, **20** : 291-292.

Purohit, A.G. and Singh, H. P. (1978) *Indian Hort.*, **23** : 7-8.

Purseglove, J.W. (1968) *Tropical Crops - Dicotyledons*, The English Language Book Society.

Quintana, M.E.G. and Paull, R.E. (1993) *J. Amer. Soc. Hort. Sci.*, **118** : 618-622.

Rahaman, S.M., Hossain, M., Joarder, O.I. and Islam, R. (1992) *Indian J. Hort.*, **49** : 18-22.

Raheja, P.C. (1966) In : *Salinity and Aridity – New Approaches to Old Problems*, (Ed. H. Boyko and W. Junk), The Hauge, 43.

Raie, M.Y., Kamrah, S., Manzoor, A. and Qureshi, E.E. (1992) *Pakistan J. Scientific and Industrial Res.*, **35** : 43-45.

Rajapakse, R. H., and Herath, H. M. (1981a) Vectors of the papaya virus in Sri Lanka. *Beitr. Trop. Landwirtsch Veterinaermed*, **19**: 359-361.

Rajapakse, R. H., and Herath, H. M. (1981b) Host susceptibility of the papaya mosaic virus in Sri Lanka. *Beitr. Trop. Landwirtsch Veterinaermed*, **19**: 429-432.

Ram, M. (1982) *Punjab Hort. J.*, **22** : 8-14.

Ram, M. (1983) *Indian Hort.*, **28** : 17-20.

Ram, M. (1995) *Seed Res.*, **23** : 98-101.

Ram, M. (1996) In : *A Text Book on Pomology*. Vol. II (Ed. T.K. Chattopadhyay), Kalyani Publishers, Ludhiana, pp. 113-140.

Ram, M. and Majumdar, P.K. (1990) *Seed Res.*, **18** : 117-120.

Ram, M. and Pandey, A.K. (1990) *International Sem. on New Frontiers in Horticulture*, Bangalore, Nov. 25-28, Abst., p. 15.

Ram, M. and Ray, P.K. (1992) *Seed Res.*, **20** : 81-84.

Ram, M., Majumder, P.K. and Singh, R.N. (1981) *Abst. Nat. Symp. Trop. and Subtrop. Fruit Crops*, Bangalore, p. 15.

Rao, O.P., Singh, R.N. and Singh, B.P. (1985) *Progressive Hort.*, **17** : 340-346.

Rawal, R.D. (1986) *Indian J. Hort.*, **43** : 293-298.

Reddy, Y.T.N., Kohli, R.R. and Bhargava, B.S. (1990) *J. Maharashtra Agril. Univ.*, **15** : 146-148.

Redina, E.F., Mezhlumyan, L.G., Kasymova, T.D. and Yuldashev, P.K.H. (1993) *Chemistry of Natural Compounds*, **29** : 781-783.

Reuveni, O., Shlesinger, D.R. and Lavi, U. (1990) *Plant Cell Tissue Organ Cult.*, **20** : 41-46.

Rezende, J.A.M. and Costa, A.S. (1995) *Acta Horticulturae*, **370** : 129-132.

Rodriguez Pastor, C., Galan Sauco, V. and Herrera Rodriguez, G. (1992) *Investigation Agraria, Produccion y Protection Vegetables*, **7** : 49-59.

Rodriguez, A.J., Guadalupe, R. and Iguina, L.M. De George (1974) *J. Agric. Univ. Puerto Rico*, **58** : 184-196.

Rojkind, C., Quezada, N. and Guitierrez, C. (1982) *Proc. 5th Intl. Cong. Plant Tissue and Cell Cult.* (Ed. A. Fujiwara), Maruzen, Tokyo, p. 763.

Romanowski, R.R.Jr., Crozier, J.A.Jr., Ito, P.J. and Tanaka, J. S. (1972) *Res. Rep. Hawaii Agric. Expt. Sta.*, No.181, p.28.

Routary, B.N. and Das, S.N. (1988) *International Nematology Network Newsl.*, **5** : 25-27.

Samson, J.A. (1980) *Tropical Fruits*, Longman, London.

Sanders, F.R. and Robertson, J.K. (1944) *East Africa J.*, **9** : 173-174.

Sankaram, A. (1942) *Madras Agric. J.*, **30** : 247-256.

Sankat, C.K. and Maharaj, R. (1989) *Proc. 5th Intern. CA Res. Conf.*, Wenatchee, Washington, USA (Vol. 2), Pullman Washington, USA, pp. 161-170.

Sanxter, S.S., Yamamoto, H.Y., Fisher, D.G. and Chan, H.T.Jr. (1992) *Canadian J. Bot.*, **70** : 364-373.

Sanyal, P.B. (1931) *Agric. J.*, **16** : 496-507.

Sao Jose, A.R. and Cunha, R.J.P. (1988) *Cientifica*, **16** : 239-243.

Saraswathi, M.S. (1996) *M.Sc. (Ag.) Thesis*, Tamil Nadu Agricultural University, Coimbatore, India.

Satyanarayana Rao, D.V. (1971) *M.Sc. (Ag.) Thesis* submitted to the University of Madras.

Sauls, J. and Campbell, C.W. (1980) *Proc. Trop. Region, Amer. Soc. Hort. Sci.*, **24** : 93-96.

Saxena, S., Hallan, V., Singh, B. P. and sane P.V. (1998a) Leaf curl disease of carica papaya from India may be caused by a bipartite geminiviruses. *Plant Dis.*, **82**: 126.

Saxena, S., Hallan, V., Singh, B. P. and sane P.V. (1998b) *Indian J. Exp. Bot.*, **36(2)**: 229-232.

Saxena, S., Hallan, V., Singh, B. P. and sane P.V. (1998c) *Biochem. Mol. Biol. Int.*, **45**(1): 101-113.

Schaffer, B., Andersen, P.C. and Ploetz, R.C. (1992) *Hortic. Rev.*, **13** : 257.

Scheer, C. and Ludders, P. (1997) *Erwerbsobstbau*, **39** : 54-57.

Schroeder, L. (1996) *Inligtingsbulletin-Instituut vir Tropiese en Subtropiese Gewasse*, **289** : 11-13.

Seemanthini (1964) *Indian J. Hort.*, **24** : 141-154.

Seemanthini and Balakrishnan, R. (1964) *Madras Agril. J.*, **51** : 81.

Selvaraj, P. (1972) *M.Sc.* (*Ag.*) *Thesis* submitted to Tamil Nadu Agricultural University, Coimbatore.

Selvaraj, Y., Pal, D.K., Subramanyam, M.D. and Iyer, C.P.A. (1982a) *Indian J. Hort.*, **39** : 50-56.

Selvaraj, Y., Pal, D.K., Subramanyam, M.D. and Iyer, C.P.A. (1982b) *J. Hort. Sci.*, **57** : 135-143.

Sen, H.D. (1931) *J. Agric. Sci.*, **21** : 209-229.

Sen, P.K. (1940) *Ann. Rept. Fruit Res. Sta.*, Sabour, Bihar for 1939-1940.

Shah, A.H. and Shanmugavelu, K.G. (1973) *South Indian Hort.*, **23** : 100-108.

Shanmugavelu, K.G., Chittiraichelvan, R. and Madhava Rao, V. N. (1976) *J. Hort. Sci.*, **51** : 425-427.

Shanmugavelu, K.G., Madhava Rao, V.N. and Srinivasan, C. (1973a) *South Indian Hort.*, **21** : 19-26.

Shanmugavelu, K.G., Selvaraj, P., Veerannah, L. and Chittiraisedvan, R. (1973b) *Progressive Hort.*, **8** : 89-96.

Sharan, R., Prasad, P.R. and Mohan, K. (1994) *Advances Plant Sci.*, **7** : 273-279.

Shetty, M.H. (1953) *South Indian Hort.*, **1** : 35-46.

Shie, C.H. and Kuo, W.H.J. (1996) *Memoirs College Agric., Nat. Taiwan Univ.*, **36** : 247-258.

Shlesinger, D., Reuveni, O. and Lavi, U. (1987) *Acta Horticulturae*, **212** : 271-273.

Silva, C., Fuentes, F., Cienegos, V. and Lizana, L.A. (1980) *Proc. Trop. Region. Amer. Soc. Hort. Sci.*, **24** : 95-102.

Silva, E.Da, Lourenco, E.J. and Neves, V.A. (1990) *Phytochemistry*, **29** : 1051-1056.

Singh, I.D. and Sirohi, S.C. (1977) In : *Fruit Breeding of India*, (Ed. G.S. Nijjar), pp.187-196.

Singh, I.P. and Sharma, C.K. (1996a) *J. Hill Res.*, **9** : 73-75.

Singh, I.P. and Sharma, C.K. (1996b) *J. Hill Res.*, **9** : 159-160.

Singh, J., Bindra, A.S. and Bhambota, J.R. (1977) *Punjab Hort. J.*, **17** : 139-142.

Singh, K.D. and Chari, M.S. (1991) *XIV Convention, IAUA*. Oct. 29-31, p. 50.

Singh, L.B. and Tripathi, R.D. (1957a) *Hort. Adv.*, **1** : 102-105.

Singh, L.P. and Tripathi, R.D. (1957b) *Indian J. Hort.*, **14** : 77-82.

Singh, R.N. (1953) *Sci. Cult.*, **19** : 44.

Singh, R.N. and Jindal, K.K. (1972) *Abst. 3rd Int. Symp. Trop. and Subtrop. Hort.*, Bangalore, No. 137.

Singh, R.N., Majumdar, P.K. and Sharma, D.K. (1963) *Indian J. Agric. Sci.*, **33** : 261-267.

Singh, R.N., Singh, G. and Rao, O.P. (1986) *Indian J. Hort.*, 43 : 1-8.

Singh, S.P. and Dahiya, S.S. (1982) *Fmr. Parliament*, **17** : 15-16.

Singh, S.P. and Pandey, R.M. (1988) *Indian J. Hort.*, **45** : 271-273.

Sippel, A. (1988) *Information Bull.*, Citrus and Subtropical Fruit Res. Inst., South Africa, **190** : 2,4.

Sippel, A.D. and Holtzhausen, L.C. (1992) *J. South Afr. Soc. Hort. Sci.*, **2** : 89-91.

Skeete, S. (1993) *Proc. 11th Annual Conf., Barbados Soc.* Technologists in Agriculture, Caribbean Dev. Bank, Barbados, pp. 55-56.

Srinivas, K. and Prabhakar, M. (1993) *Singapore J. Prim. Indust.*, **21** : 1-5.

Sriprasertsak, P., Burikam, S., Attathom, S. and Piriyasurawong, S. (1988) *Kasetsart J. Natural Sci.*, **22** : 24-29.

Stanley, J. (1983) Infectivity of the cloned geminivirus genome requires sequences from both DNA. *Nature*, **305**: 643-645.

Stockdale, F.A. (1927) *Trop. Agrist.*, **68** : 3-12.

Storey,W.B. (1953) *J Hered* **44:** 70–78.

Storey, W.B. (1937) *J. Hered.*, **44** : 70-78.

Storey, W.B. (1937) *Proc. Amer. Soc. Hort. Sci.*, **35** : 80-82.

Storey, W.B. (1941) *Hawaii Agr. Expt. Bull.*, **87** : 64.

Storey, W.B. (1958) *Hort. Adv.*, **2** : 49-60.

Storey, W.B. (1969) In : *Outlines of Perennial Crop Breeding in the Tropics* (Eds. F.P. Ferwerda and F. Wit), Landbhogesch, Wageningen, Netherlands, No.4, p. 511.

Subramanyam, M.D. and Iyer, C.P.A. (1982) *South Indian Hort.*, **29** : 167-174.

Subramanyam, M.D. and Iyer, C.P.A. (1986) *Haryana J. Hort. Sci.*, **15** : 179-184.

Suksa-Ard, P., Kataoka, I., Fujime, Y. and Subhadrabandhu, S. (1997) *Japanese J. Trop. Agric.*, **41**: 7-13.

Suksa-Ard, P., Kataoka, I., Fujime, Y. and Subhadrabandhu, S. (1999) *Japanese J. Trop. Agric.*, **43** : 1-4.

Sulladmath, U.V., Narayana Gowda, J.V. and Ravi, S.V. (1981) *Abst. Nat. Symp. Trop. and Subtrop.Fruit Crops*, Bangalore, p. 54.

Summanwar, A.S. and Ram, R.D. (1993) Arrested growth in advance stage defoliation pant. *Adv. in Hort.* **3**(34): 28-30.

Suplicy Filho, N., Raga, A., Calza, R. and Faria, J.T. (1987) *Biologico*, **53** : 37-40.

Suzuki, K., Yosinaga, T., Kaneko, A., Asano, Takano, S. and Hasegawa, T. (1991) *J. Japanese Soc. Food Sci. & Tech.*, **38** : 1057-1062.

Swingle, C.F. (1947) *Proc. Amer. Soc. Hort. Sci.*, **62** : 228-230.

Teaotia, S.S. and Singh, R.N. (1967) *Indian Agric.*, **11** : 45-49.

Teixeira da Silva, J.A., Rashid, Z. and Nhut D. T. *et al.* (2007) *Tree and Forestry Science and Biotechnology* **1(1):** 47 – 73.

Tennant, P., Fitch, M., Lius, S., Gonsalves, C., Manshardt, R., Sanfored, J.L. and Gonsalves, D. (1995) *Acta Hort.*, **370** : 7-11.

Teo, C.K.H. and Chan, L.K. (1994) *J. Hort. Sci.*, **69** : 267-273.

Thomas, K.M. and Krishanaswamy, C.S. (1939) Leaf crinkle – A transmissible disease of Papaya. *Curr. Sci.*, **8:** 316–317.

Thompson, A.K. and Lee, C.R. (1971) *J. Hort. Sci.*, **46** : 511.

Tokumoto, M., Tabel, Y., Kayano, T., Oku, H., Iwasaki, H. and Chinen, I. (2000) *J. Jap. Soc. Hort. Sci.*, **69** : 195-201.

Tulasigeri, G., Prakash, D.P., Sagar, B.S. and Beerappa, J. (2017) *Int.J.Curr.Microbiol. App.Sci.*, **6**(10): 2516-2523.

Urasaki N, K.Tarora K,A. Shudo, H.Ueno, M. Tamaki M, *et al.* (2012) *PLoS ONE* **7(7):** e40904. doi: 10.1371/journal.pone.0040904

Vallero, G.G. (1999) *Facultad Nacional de Agronomia Medellin*, **52** : 515-526.

VanBuren, R., Zeng, F., Chen, C. *et al.* (2020) *Genome Res.*, **25:** 524–533.

Vecchio, V. and Shirwa, A.H. (1987) *Rivista di Agricoltura Subtropicale e Tropicale*, **81** : 175-184.

Veerannah, L. and Rathnakumari, S. (1987) *South Indian Hort.*, **35** : 320-322.

Veerannah, L., Kulasekaran, M. and Muthuswami, S. (1982) *South Indian Hort.*, **30** : 261-263.

Vieira Neto, R.D. (1995) *Acta Hort.*, **370** : 117-119.

Villachica, H. and Raven, K. (1986) *Turrialba*, **36** : 523-531.

Virendrakumar (1952) *Indian J. Hort.*, **9** : 36-47.

Walker A, Hoy M and Meyerdirk DE 2003 Papaya Mealybug (2008) Univ. Florida Featured Creatures.http: //creatures.ifas.ufl.edu/fruit/mealybugs/papaya_mealybug.htm.

Wang, J., Chen, C., Na, J. *et al.* (2008) *Tropical Plant Biol.* **1:** 278–292.

Wang, J., Na, J.K. and Yu, Q. *et al.* (2012) *Proc. Natl. Acad. Sci.* (*US*).**109:** 13710-13715.

Wardlaw, C.W., Leonard, E.R. and Baker, R.F. (1934) *Trop. Agric.*, **11** : 9.

Weber, O.B. and Amorim, S.M.C.De (1994) *Revista Brasileira de Cieneia do Solo*, **18** : 187-191.

Wei, F. and R.A. Wing (2008) *Genome Biology* **9:** 227 (doi: 10.1186/gb-2008-9-6-227)

Werner, H. (1993) *Proc. Interamerican Soc. Trop. Hort.* (Ed. R.J. Campbell), **37** : 94-98.

Westwood, J. (1999) *RIS Biotech. Develop. Rev.*, **2** : 78-80.

Wills, R.B.H. and Widjanarko, S.B. (1997) *Australian J. Expt. Agric.*, **37** : 127-129.

Winnaar, W.De. (1988) *Plant Cell Tissue Organ Cult.*, **12** : 305-310.

Wolfe, H.S. and Lynch, S.J. (1940) *Fla. Agr. Expt. Sta. Bull.*, **350** : 35.

Yang, J.S. (1986) *J. College Sci. & Engineering, Nat. Chung Hsing Univ.*, **25** : 93-108.

Yang, J.S. and Ye, C.A. (1992) *Bot. Bull. Acad. Sinica*, **83** : 375-381.

Yang, J.S., Yu, T.A., Cheng, Y.H. and Yeh, S.D. (1996) *Plant Cell Rep.*, **15** : 456-459

Yaqub, C.M., Ahmad, M.K., Karikari, S.K. and Olerawaju, S.D. (1985) *Pakistan J. Agric. Res.*, **23** : 119-123.

Ye, K.N., Huang, J.C. and Li, B.J. (1993b) In : *Biotechnology in Agriculture* (Eds. C.B.You, Z.L. Chen and Y. Ding) Kluwer Academic publishers, Netherlands, pp. 411-413.

Ye, K.N., Yu, D.Q., Huang, J.C. and Li, B.J. (1993a) *Acta Scientiarum Naturalium Universitatis Sunyatseni*, **32** : 63-69.

Yee, W.J.J., Sekloka, T.T., Neakasone, H.Y., Ikehara, D.K., Ooka, J.J. and Akamine, E.K. (1980) *Res. Ext. Series* No.497, College of Trop. Agric., University of Hawaii, pp. 20.

Yeh, S.D., Gonsalves, D., Wang, H.L., Namba, R. and Chiu, R.J. (1988) *Plant Disease*, **72** : 375-380.

Yimsawat, T. and Feungchan, S. (1992) *Khon Kaen Agric. J.*, **20** : 30-36.

Yu, Q., Tong, E., Skelton, R.L., Bowers, J. E. *et al.* (2009) *BMC Genomics*, **10**: 371 doi: 10.1186/1471-2164-10-371

Yu, Q., Navajas-Pérez, R., Tong, E., Robertson, J., Moore, P.H., Paterson, A.H.and Ming, R. (2008) *Tropical Plant Biology* **1(1):** 49–57.

Yu, Q., Hou, S., Hobza, R., Feltus, F. A., Wang, X.*et al.* (2007) *Mol. Genet. Genomics* **278:** 177–185.

Yusof, S. and Salleh, M. (1992) *Acta Horticulturae*, **292** : 223-230.

Zerpa, D.M.D.F. (1957) *Agronomie Tropical*, **7** : 83-86.

Zettler, F.W., Edwardson, J.R. and Purcifull, D.E. (1968) *Phytopathol.*, **58**: 332.

Zhao, M., Moy, J. and Paull, R.E. (1996) *Post-harvest Biol. Tech.*, **8** : 209-222.

Zhu, Y.J., Agbayani, R. and Moore, P. H. (2007) *Planta*, DOI 10.1007/s00425-006-0471-1

Zhu, Y.J., Agbayani, R., McCafferty, H., Albert, H. H. and Moore P. H. (2005) *Plant Cell Rep.*, **24:** 426–432.

Zhu, Y.J., Agbayani, R. and Moore, P.H. (2004) *Plant Cell Rep.*, **22,** 660–667.

④

BER

V. P. Sharma, S. K. Mitra, Tanusree Saha, Ivi Chakraborty and B.Ghosh

1.0 Introduction

Ber (*Ziziphus mauritiana*) is an ideal fruit tree for arid and semi-arid regions in tropical and subtropical climate where most of the fruit crops cannot be grown either due to lack of irrigation facilities or adverse climatic and soil conditions. The name *Ziziphus* comes from an Arabic word used along the North African coast, zizoufo used for *Z. lotus* (L.) Desf, but also related to the ancient Persian words zizfun and ancient Greeks used the word ziziphon for the Jujube (Williams, 2006). The genus *Ziziphus* belongs to the family Rhamnaceae. Central Asia is believed to be the origin of ber which includes north-west India, Afghanistan, some areas of Tajikistan, Uzbekistan and China (Mitra *et al.*, 2008). It is mainly grown in India, different countries in central Asia, China and Taiwan. Gupta (1983) suggested *Ziziphus* spp. for growing on marginal land in the hot arid region of Rajasthan. Its cultivation has received a great impetus in recent years in the northern part of India, especially in the states of Punjab, Haryana, Rajasthan, Uttar Pradesh and Gujarat.

2.0 Composition and Uses

2.1.0 Composition

The ripe ber fruits are rich in nutritive value. The composition of ber fruits varies in different cultivars and also in different agroclimatic zones.

The major constituents in several cultivars of Indian jujube fruits contain (per 100g): Moisture: 81.6-83.0 g; ash: 0.3-0.59 g; protein: 0.8 g; fiber: 0.60 g; carbohydrate: 17 g; total sugar: 5.4-10.5 g; reducing sugar: 14-6.2 g; calcium: 25.6 mg; phosphorus: 26.8 mg; iron: 0.76-1.8 mg; carotene: 0.021 mg; thiamine: 0.02-0.024 mg; riboflavin: 0.02-0.038 mg; citric acid: 0.2-1.1 mg; ascorbic acid: 65.8-76.0 mg; pectin (per cent dry basis): 2.2-3.4 per cent (Morton, 1987; Pareek and Dhaka, 2008; Pareek *et al.*, 2009, Pareek, 2013) and are given in Table 1.

Table 1: Nutritional Composition of Indian Jujube Fruit (on fresh weight basis)*

Constituents	*Amount (per 100 g)*
Moisture (g)	81.6-83.0
Protein (g)	0.8
Fat (g)	0.07
Fiber (g)	0.60
Carbohydrates (g)	17.0
Total Sugars (g)	5.4-10.5
Reducing Sugars (g)	1.4-6.2
Non-Reducing Sugars (g)	3.2-8.0
Ash (g)	0.3-0.59
Calcium (mg)	25.6
Phosphorus (mg)	26.8
Iron (mg)	0.76-1.8

Constituents	*Amount (per 100 g)*
Carotene (mg)	0.021
Thiamine (mg)	0.02-0.024
Riboflavin (mg)	0.02-0.038
Niacin (mg)	0.7-0.873
Citric Acid (mg)	0.2-1.1
Ascorbic acid (mg)	65.8-76.0
Fluoride (ppm)	0.1-0.2
Pectin (per cent dry basis)	2.2-3.4

*Morton (1987), Pareek and Dhaka (2008), Pareek *et al.* (2009), Pareek (2013).

The proximate composition of five cultivars 'Jinsixiaozao', 'Yazao', 'Jianzao', 'Junzao' and 'Sanbianhong' of Chinese jujube was estimated by Li *et al.* (2007). Significant variations were recorded for moisture (17.38-22.52 per cent), carbohydrates (80.86–85.63 per cent), proteins (4.75–6.86 per cent), lipids (0.37–1.02 per cent), soluble (0.57–2.79 per cent) and insoluble (5.24–7.18 per cent) fibres, reducing sugar (57.61–77.93 per cent) and ash (2.26–3.01 per cent). Fructose and glucose were identified as major soluble sugar in all the five cultivars, while rhamnose, sorbitol and sucrose contribute in lesser amount. Among the minerals K, P, Ca and Mn contributed major portion, while Fe, Na, Zn and Cu were also found in good amount. Chinese jujube is rich in ascorbic acid and it ranged from 192–359 mg/100g among the cultivars studied. Thiamine and riboflavin contents were 0.04–0.08 and 0.05–0.09 mg/100g, respectively. Total phenolic contents ranged from 5.18 to 8.53 mg/g (Table 2).

Table 2: Nutritional Composition of Five Cultivars of Chinese Jujube (on dry weight basis)*

Component	*Cultivars*				
	Jinsixiaozao	*Yazao*	*Jianzao*	*Junzao*	*Sanbianhong*
Carbohydrate (per cent)	81.62	80.86	84.85	82.17	85.63
Reducing sugar (per cent)	57.61	60.24	77.93	58.73	67.32
Fructose (per cent)	19.1	18.6	42.9	25.1	22.8
Glucose (per cent)	22.5	26.3	19.2	19.6	27.2
Rhamnose (per cent)	12.2	12.5	14.1	10.5	14.7
Sorbitol (per cent)	2.1	0.3	1.9	2.6	3.5
Sucrose (per cent)	14.1	11.5	0.21	17.4	11.3
Soluble fiber (per cent)	2.79	1.46	1.51	1.07	0.57
Insoluble fiber (per cent)	6.11	7.18	5.24	5.83	5.56
Lipid (per cent)	0.37	1.02	0.39	0.71	0.65
Protein (per cent)	5.01	6.86	4.75	6.43	6.60
Moisture (per cent)	18.99	20.98	17.38	21.09	22.52
Ash (per cent)	2.26	2.78	2.41	3.01	2.56

Component	Cultivars				
	Jinsixiaozao	*Yazao*	*Jianzao*	*Junzao*	*Sanbianhong*
Potassium (mg/100g)	79.2	458	375	201	344
Phosphorus (mg/100g)	110	59.3	72.3	105	79.7
Calcium (mg/100g)	65.2	91.0	45.6	118	76.9
Manganese (mg/100g)	39.7	36.5	51.2	24.6	42.1
Iron (mg/100g)	4.68	6.93	6.42	7.90	6.01
Sodium (mg/100g)	6.34	7.61	6.21	5.96	3.22
Zinc (mg/100g)	0.55	0.63	0.47	0.42	0.35
Copper (mg/100g)	0.26	0.27	0.42	0.31	0.19
Thiamine (mg/100g)	0.05	0.04	0.09	0.06	0.05
Riboflavin (mg/100g)	0.07	0.07	0.05	0.09	0.05
Vitamin C (mg/100g)	359	192	203	296	315
Phenols (GAE mg/g)	7.42	8.53	8.36	7.01	5.18
Antioxidant capacity (FRAP μMol/g)	1173	1025	794	563	342

*Li *et al.* (2007).

Guil-Guerrero *et al.* (2004) worked with some Spanish cultivars and they found triglycerides of medium chain fatty acids in abundant quantity. On dry weight basis fruits contain 1.33 +/- 0.17 g/100 g saponifiable oil. Carotenes in these Spanish cultivars varied from 4.12 to 5.98 mg/100g on a dry weight basis. Vitamin A value was found in medium range and it was 38 ìg RE/100g on a fresh weight basis.

The cultivar Ponda was found to contain 85.21 per cent moisture, 13.20 per centreducing sugars, 13.61 per cent total sugars, 0.19 per cent acid, 65 mg vitamin C/100 g, and 14.0 per cent TSS. The fruits are large. Raja (1985) observed a maximum TSS of 20 per cent in cv. Ilaichi. A minimum pulp : seed ratio was also observed in cv. Tikadi and maximum pulp:seed ratio in Ilaichi.

Krivencov *et al.* (1970) observed that more vitamin C was found in the fruit flesh near the seed than near the skin of the fruit. Ahmedov and Halmato (1969) observed more vitamin C content in the leaves (1020 mg/100 g). They also recorded succinic and malic acids in the fruits and malic and tartaric acids in the leaves of *Z. jujuba*. The fruits of *Z. mauritiana* is rich in minerals and vitamin-C (Mitra *et al.*, 2008).

Sharma and Kumar (1982) isolated saponin jujubogenin from the ethanolic extract and n-octacosanol and aphitolic acid from the petroleum ether extract of *Z. mauritiana* leaves. The ber seed, roots and stem bark also contain some alkaloids.

2.2.0 Uses

The fruits are mostly eaten fresh but the other forms, such as dried, candied, pickled and other products like squash or juice and ber butter can also be prepared and used.

Various Ayurvedic and Yunani medicines containing ber extract are said to be a blood purifier and also help in digestion. The powder and decoction prepared from the roots are effective in case of fever, ulcers and old wounds. The stem bark is considered to be a remedy for diarrhoea. Besides, the various parts of the plants are being used tradiotionally from ages for the treatment of urinary troubles, diabetes, skin infection, bronchitis, anaemia *etc*. The seed extracts also reported to have antiplasmodial activity (Goyal *et al.*, 2012; Mishra and Bhatia, 2014); whereas the antibacterial activity of stem bark extracts was also proved (Sameera and Mandakini, 2015).

The methnolic extract from leaf powder of *Z. jujuba* may be used to reduce hyperlipidemia (Kumar *et al.*, 2008). The Benzene extract of *Ziziphus jujuba* has more efficient anti-helmenthic activity and also showed various activities like, anti-cancer, anti-pyretic, anti-hepatitis, anti-oxidant and anti-inflamatory properties (Keerti *et al.*, 2016). Jujube possesses neoroprotective activities, including protecting neorone cells against neurotoxin stress.

The medicinal values of various parts of the tree and fruits of *Ziziphus* spp. are many and yet not fully exploited. Sin'ko (1976) also suggested detailed pharmacological investigation of *Z. jujuba*. Phytochemical analysis revealed that the bark extracts is rich in alkaloids, flavonoides, triterpenoids and tannin (Sameera and Mandakini, 2015). In Western India, fruits of *Z. nummularia* are used to prepare churn (Pareek *et al.*, 2009).

Kumar *et al.* (1992) evaluated 4 candy making recipes and found that the use of 75 per cent total soluble solids resulted in the highest organoleptic score. Candy of Umran prepared using this recipe was stored for 10 months at room temperature and its stability determined. The total soluble solids, acidity and browning increased whilst the ascorbic acid content and organoleptic quality score decreased during storage.

The ber plant itself yields many economic products. The leaves are used as a fodder in the dry regions and the tree is a host plant for rearing of the lac insect (*Laccifer lacca*); the stem gives a quality wood which is used for making various agricultural implements. The wood is also considered as valuable timber and is used in the construction of buildings. The ber wood makes excellent charcoal. The thorny branches are commonly used for field fences. The tree is a host plant for rearing of the lac insect (*Laccifer lacca*). Lac rearing is an important practice in the states of Bihar and Madhya Pradesh.

3.0 Origin and Distribution

According to De Candolle (1886), the centre of origin for this fruit crop is Central Asia. This includes north-west India, Afghanistan, areas of Tajikistan, Uzbekistan and China. It is found growing wild, semi-wild and in cultivated forms in almost all parts of India.

Uses of ber

Fresh fruit

Dried ber

Candy from ber

Sour and sweet pickles from ber

Species of ber (*Ziziphus*)

Ziziphus mauritiana

Ziziphus jujuba

Species of ber (*Ziziphus*)

Ziziphus rugosa

Ziziphus mucronata

Ziziphus joazeiro

Ziziphus mistol

Ziziphus spina-christi

Ziziphus rotundifolia

4.0 Species and Cultivars

4.1.0 Species

The genus *Ziziphus* belongs to the family Rhamnaceae and consists of several species (86 species reported by Evreinoff, 1964, Jahnsoton, 1972; 135 species by Bhansali, 1975; 170 species by Liu and Cheng, 1995; more than 100 species by Aktar and Rahaman, 2019) in tropical and subtropical regions of the northern hemisphere. *Ziziphus jujuba* and *Z. mauritiana* are two important species of ber, *Z. mauritiana* is more common in tropical and subtropical regions and *Z. jujuba* is found in temperate parts of the world. *Ziziphus jujuba* commonly known as Chinese jujuba is a small, upright tree (6-8 m high) with bright green leaves which are glabrous on their undersurface. This species is deciduous and has oblong or ovoid fruits. This species together with another 13 species in the genus of *Ziziphus*, with its putative center of origin located in the Yellow River basin in northern China (Liu and Wang, 2009; Zhao *et al.*, 2009). The other species *Z. mauritiana*, commonly known as Indian ber is a vigorous growing, small, spreading tree with almost vinelike drooping branches. The species is evergreen and leaves are densely tomentose on their undersurface. The fruits are round to oval and greenish yellow to reddish brown in colour.

Ziziphus rugosa is found in India, Myanmar and Sri Lanka; *Z. heloia* and *Z. mucronata* in South Africa; *Z. joazeiro* in Brazil; *Z. mistol* in Argentina while *Z. spina-christi* is common in Egypt, Syria and Palestine. Maheswari and Singh (1965) recognized six economically important species of *Ziziphus* in India which are given in Table 3.

Table 3: *Ziziphus* Species of Common Use in India*

Species	*Natural Distribution or Cultivation*	*Economic use*
Ziziphus mauritiana *Z. jujuba*	Native of China, now cultivated in Punjab, UP, Bihar and Rajasthan	Edible fruits, bark for tanning and leaves for fodder
Z. nummularia (syn. *Z. rotundifolia*)	Found in north-west India and Andhra Pradesh	Edible fruits, bark for tanning
Z. oenoplia	Found in north and peninsular India	Edible fruits, bark for tanning
Z. rugosa	Found in UP, Bihar, MP and western peninsula	Edible fruits
Z. sativa	Found in Punjab and Bengal	Edible fruits
Z. xylopyrus	Found in MP	Lac host and for tanning

Other *Ziziphus species are - Z. fumiculosa, Z. spina, Z. glabrata, Z. oxyphylla.*

* Maheswari and Singh (1965).

Flavonoid distribution in 59 *Ziziphus mauritiana* cultivars, and the wild species *Z. nummularia* and *Z. mauritiana* var. *rotundifolia* was determined by Vashishtha *et al.* (1989) in the leaves of 4- to 5-year-old trees. Correlations between flavonoid composition and growth habit in *Z. mauritiana* and its 2 wild types were also suggested.

Liu (1999) reviewed the advances in the study on the distribution and sub-genus division system of genus *Ziziphus*, the taxonomic status of Chinese jujube (*Z. jujuba*) and Chinese acid jujuba and their sub-species division, the cultivar classification of Chinese jujuba, as well as the origin and evolution of Chinese jujuba.

4.2.0 Cultivars

There are more than 125 cultivars grown in India. These cultivars have been developed by selection in different regions. The nomenclature of ber cultivars seems to be very loose and the same cultivar is also known by different names in different regions. The cultivar Umran which normally came from Alwar in Rajasthan (Chadha *et al.*, 1972) is known as Katha, Kotho in Maharashtra. Ajmeri, Chameli, Katha Bombay Madhuri are also similar to Umran. A large number of cultivars similar to Gola have different names, such as Delhi Gola Nazuk, Akrota, Seo, Laddu, *etc.* The cultivars Chhuhara and Rasmi appear to be the same. The important cultivars ber includes Gola, Umran, Nazuk, Chhuhara, Banarasi Karka, Narma, Seo, Thornless, ZG 3 *etc.* (Mitra *et al.*, 2008). The commercial cultivars of different ber-growing states of India are given in Table 4 (Pareek, 1983).

Morphological characteristics of some important ber cultivars are given below:

Banarasi Karaka

Tree tall, vigorous, 8-12 m in height. Leaf green, 6.7 cm long and 4.7 cm broad, elliptic to ovate, young leaves sparsely tomentose. Fruits globose, oblong to long in shape with tapering to pointed styler end, unripe fruits green in colour, turns greenish yellow to light yellow when ripe, skin smooth, glossy, fruit size 5.2 x 3.0 cm.

Chhuhara

Tree semi-tall with spreading branches. Leaf ovate-oblong with short petiole and serrated margin from base to apex, lower surface smooth and upper surface densely tomentose. Fruit ovate-oblong, 2.9 cm long and 2.1 cm thick, weight 6.8 g. Stone subglobose and sharply pointed, surface furrowed, 1.6 cm long and 0.55 cm thick, weight 0.42 g.

Gola

Tree spreading. Leaves cordate, 6.7 cm long and 4.2 cm broad. Fruit almost round, styler end flat, stem end round, skin bright yellow, smooth and glossy, pulp white, fruit size 3.8 cm x 3.5 cm. Early cultivar, fruits ripen during early January to end of January.

Ilaichi

Tree spreading, leaves obovate in shape with tapering base, 9.1 cm long and 6.1 cm in breadth. Fruit oblate to roundish in shape, immature fruits green and golden yellow, turning brown to chocolate-brown when ripe, flesh cavity absent, 2.3 cm in length and 2.3 cm in breadth.

Important cultivars of ber

Apple kool

BAU kool-1

Topa

Umran

Jogia

Dandan

Important cultivars of ber

Madhavpur

Gola

Banarasi karaka

Gomo kirti

Kashmiri apple kool

Jogia

Spreading in habit, leaves oval, dark green, apex acute, slightly cupped or twisted. Fruit oval, styler end slightly tapering to slightly pointed, skin colour green with red pigments, changes to greenish yellow with deeper yellow spots when ripe, fruit size 3.5 x 2.66 cm.

Table 4: Commercial Cultivars in different Ber-Growing States of India

State	*Cultivars*		
	Early	*Mid-season*	*Late*
Haryana	Gola, Safeda, Selected Safeda, Sandhura Narnaul, Seo, Chonchal	Kaithli, Sanaur-5, Muria Mahrara	Umran
Maharashtra	Shamber, Badami, Manuki, Guli	Mehrun, Darakhi, Kharki	—
Punjab	Nazuk, Noki, Seo, Rohataki Gola, Selected Safeda, Sandhura Narnaul	Banarasi, Dandan, Kaithli, Sanaur-2, Walaiti, Thornless	Umran, Ilaichi Pathani, ZG2, ZG3
Rajasthan	Gola, Seb, Seo Bagwadi	Jogia, Mundia, Tikadi	Katha, Maharwali
UP	Narma Varanasi, Delhi Gola, Banarasi Gola	Banarasi Karaka, Muthiya, Muriya, Pewandi	Jogia, Aliganj
Gujarat	—	Meharun	Ajmeri, Chameli, Randeri

Kaithli

Tree medium with bushy growth containing semi-decumbent branches, thorny growth not so pronounced and thorns straight. Leaf ovate with very minutely serrated margins, lower surface smooth and dark green, upper surface slightly pubescent with brownish tint, apex obtuse. Fruit ovate-oblong with broadly mummilate apex, 3.37 cm long and 1.9 cm thick, weight 6.22 g; stone elliptic, oblong and pointed with furrowed surface, 2.7 cm long and 0.68 cm thick, average weight 0.47 g.

Narma

Tree erect, about 6-9 m high. Leaves dark green, 10-12 cm long and 4.7 cm broad, apex is round, upper surface is shining and lower tomentose. Fruit oblong to long with blunt styler end, light yellow when ripe, size 4.4 x 2.85 cm.

Nazuk

Tree semi-tall with round top; branches spreading type and less thorny. Leaf oblong, upper surface green and glossy and lower surface with whitish tomentose, margin wavy and apex obtuse. Fruit elliptic, oblong, 2.19 cm long and 1.37 cm thick, average weight 10.5 g; stone elliptic oblong, size 1.6 cm long and 0.99 cm thick, average weight 0.80 g, surface warty.

Sandhura Narnaul

Tree erect. Leaves light green in colour, 9.5 cm long and 6.5 cm broad, base of leaf round while apex obtuse, slightly curved. Fruits oval, oblong to long, styler end slightly pointed. Fruits greenish yellow to greenish golden yellow when ripe, size 4.5 x 2.18 cm.

Seo

Tree spreading. Leaves ovate to ovate-oblong in shape, leaf base round, obtuse and apex acute, distinctly curved. Fruit round, resembling crab-apple, styler end round, with mild depression in the centre, stem end broad, deeply grooved. Fruits light pinkish yellow with occasional specks at maturity, skin slightly rough, flesh cavity present, size 3.47 x 2.92 cm.

Thornless

Tree erect growing. Leaves oval in shape, 8.0 cm in length and 4.0 cm in breadth, base of leaf tapering and apex subacute, curved, fruit oval-oblong, slightly pointed to bluntly pointed styler end, stem end round, very-slightly ridged, greenish yellow at maturity with glossy skin surface, flesh cavity present, size 4.7 x 3.1 cm.

Tikadi

This is late maturing cultivar, the fruits are small, weing 10g. Mature fruits with large stone, red in colour, flesh is soft creamy with a higher TSS (25 per cent).

Umran

Tree medium-sized with bushy decumbent branches almost touching the ground, less thorny, individual thorn curve. Leaf ovate-oblong, prominently serrated from base to apex, lower surface smooth and upper surface slightly pubescent, apex blunt. Fruit elliptic, 4.21 cm long and 3.22 cm thick, weight 18.7 g; stone oblong and warty and furrowed 2.3 cm long and 0.65 cm thick.

ZG 3

Tree spreading. Leaves oblong, dark green, 7.6 x 5.7 cm, base of leaf round and apex obtuse, flat or curved. The fruits are obovate, roundish or ovate in shape, styler end round with slight depression, stem end round, ridged, broad. Fruits greenish yellow, turning chocolate-brown when ripe, fruit size 4.0 x 3.0 cm.

CAZRI Ber 2018

It is a new high yielding, early maturing variety of ber developed by selection from seedling population of cv. Gola (Meghwal *et al.*, 2019). The plants are vigorous, spreading type which turns into drooping type at fruit maturity due to heavy fruit bearing. Mean plant height 2-3 m, leaves sparsely tomentose on lower surface, dorsal surface-light green, leaf apex-acute, leaf shape cordate, early maturing, mean fruit weight-19.26 g, fruit size 3.19cm x 3.13 cm (L x W), mature fruit colour-greenish yellow, fruit shape round, pulp stone ratio-12.33, TSS-18.83^{o}Brix, pulp acidity 0.51 per cent and vitamin C is 50.73 mg/100g. This variety has about 25 per cent higher yield than Gola variety under rainfed condition (mean fruit yield-61.25 kg/plant) and fruit maturity earlier than Gola by 10-12 days.

In a trial in Punjab, Tomar and Singh (1987) recorded the maximum mean yield of 77.7 kg/tree in cultivar Thornless, followed closely by Sanaur-5, Sanaur-4 and Banarasi Pewandi. Fruit weight, pulp percentage and total soluble solids content were highest in Umran. Sanaur-4 produced the largest fruits, which also had the highest acid content.

Of the 7 cultivars of ber studied by Babu and Kumar (1988) Gola, Banarasi, Kaki and Umran were the tallest and trees of Akola and Mundia were the shortest. Only one flowering season was observed from May to July and its total duration varied from 68 to 94 days. Cultivars Gola, Mundia and Akola flowered early, Umran and Seb flowered in mid-season and Banarasi and Kaki were late flowering.

Vashishtha and Pareek (1989) used data on vegetative and fruit characters in *Ziziphus mauritiana* var. *rotundifolia* to develop a key for identification of 59 cultivars. The key uses 3 tree growth habits (erect, semi-erect and spreading) as primary characters, 6 fruit shapes and 3 leaf shapes as secondary characters, and styler fruit end, pulp cavity, leaf apex and leaf base as tertiary characters. The cultivars were divided into 3 groups based on primary characters and into 18 sub-groups based on secondary characters.

In a study of the ber cultivars Seb, Banarasi Karaka and Gola, Tiwari and Banafar (1995) observed that Banarasi Karaka fruits had the highest nutritive value and yield (99.8 kg fruits/plant). *Ziziphus mauritiana* cultivars Umran, Gola, Kaithli and Sukavani (local control) were evaluated by Chovatia *et al.* (1992) in Gujarat under average rainfall conditions of 625 mm/year. Fruit yield/plant was highest for Gola in all 3 years (mean 47.6 kg), which also had the least fruit drop, followed by Umran (38.4 kg). Gola was also the earliest ripening and recorded the highest acidity (0.45 per cent) and ascorbic acid content (160 mg/100 g pulp). Umran had the highest percentage fruit set (24 per cent).

In a field experiment at Raichur, India 4-year-old trees of 12 popular ber cultivars were compared by Praveen and Patil (1998). Fruit yields were highest in cv. Sanaur-2 and lowest in Rambore. The highest yielding cultivars were those which exhibited intermediate vigour. Aulakh *et al.* (2000) evaluated the cvs. Sanaur-2, Umran and ZG 2 for yield and quality. They suggested cultivation of Sanaur-2 under rainfed conditions in the lower Shiwaliks of Punjab as this cultivar is good yielder and produce quality fruits (Table 5).

Table 5: Fruit Yield and Quality Characteristics of Ber Varieties*

Variety	*Fruit Yield (kg/plant)*	*Fruit Weight (g)*	*Stone Weight (g)*	*Pulp (Per cent)*	*TSS (Per cent)*	*Acidity (Per cent)*	*Total Sugar (Per cent)*
Sanaur-2	32.3	15.3	1.19	92.3	19.4	0.23	13.64
Umran	13.0	18.7	1.26	93.7	15.1	0.26	10.12
ZG 2	11.2	10.1	0.90	91.7	13.1	0.39	10.01
C.D. at 5 per cent	4.4	1.9	0.12	N.S.	0.98	0.08	0.79

* Aulakh *et al.* (2000).

In a survey of wild and cultivated *Ziziphus spinosus* in Hebei province in China by Liu and Wang (1991), 19 promising forms and varieties were identified. Some were cultivated in Zhanhuang and Xiangtai counties, while the variety Laohyyan was cultivated in Beijing; some trees were estimated to be around 1000 years old. Among the selected varieties some were with good fruit quality (for fresh consumption), some with high yields and some with medicinal properties of the stone kernel.

Yu *et al.* (1991) reported that about 30 of the more than 120 varieties of *Ziziphus sativus* in Shanxi province in China are for fresh consumption. They described 7 main varieties – Li Zao, Jixin Mia Zao, Buluoso, Lingling Zao, Xiangfeng Hulu Zao, Yongji Hama Zao which produce high yield of good quality fruits. A jujube cultivar Sihong Dazao reported by Wan (1994) matures in mid to late September. The fruits are large weighing 50-60 g on average, reaching 65 g, crisp, juicy, very sweet flesh with a soluble solids content of 30-36 per cent. Trees begin bearing early and are high yielding, highly adaptable and resistant to drought, waterlogging and diseases.

Kundi *et al.* (1989) in Pakistan reported the physico-chemical characters of the fruits of 7 ber (*Z. mauritiana*) cultivars, *viz.*, LR-9, LR-11, LR-13, Haq Nawaz, Cantonment, Golan and Lal-Wali. There were considerable differences between cultivars in most parameters. Fruit weight was greatest (29.34 g) for Cantonment and was only 9.54 g for Lal-Wali. Fruit length ranged from 3.27 to 4.33 cm, and diameter from 2.04 to 3.30 cm. The percentage of pulp ranged from 94.24 in Lal-Wali to 96.75 in LR-13, which had the highest pulp:peel ratio of 16.22. The content of total sugars (7.98-11.52 per cent) was highest in Haq Nawaz, and that of protein (1.24-2.96 per cent) in Golan.

In Azerbaijan new high-yielding cultivars *viz.*, Akhmedi, Nasimi, Ulduz, Azeri, Irada, Arzu, Khazari and Ordubadi were produced at the Apsheron Experimental Station for Subtropical Crops (Tagiev, 1992).

5.0 Breeding and Improvement

Nehra *et al.* (1983) observed the chromosome numbers of ber cultivars, the basic chromosome number being 12. They found high degree of polyploidy. The chromosome number of Umran was 2n=48 (tetraploid) whereas in some cultivars it was 2n=96 (octaploid). The chromosome number of wild forms, Gola and Boradi was also 2n=96. Another wild form Jhar ber (*Z. rotundifolia*) showed variation in chromosome number. It was either 2n=47 or 72. Khoshoo and Singh (1963) also reported chromosome number of 35 ber cultivars. Out of 35 cultivars, 33 were tetraploid, *i.e.*, 2n=48; one was pentaploid 2n=60 and two were octaploid, *i.e.*, 2n=96. From these studies it seems that high level of ploidy might be responsible for self-incompatibility in ber. Thus the self and cross-incompatibility and long juvenile phase of the crop are the main bottlenecks in carrying out the hybridization in ber.

For efficient and purposeful breeding programme, it is necessary to have knowledge of the morphology and floral biology of a crop. The literature pertaining to floral biology is critically reviewed by Pareek *et al.* (2007) encompassing the events in flowering time and duration, phenology and transmission of buds to flowering up to fruit set and fruit drop in ber. Many researchers observed the phenomenon of cross pollination along with the prevalence of self and cross incompatibility and pollen sterility, fruit set in ber largely depends on timely pollination and its related factors like pollinators and environmental conditions and on the success in fertilization. The life of individual flower is very short and many flowers apparently are not pollinated during their respective period. Consequently, in spite of a very profuse flowering, fruit set in ber is very low even under open pollination.

Very little attempts have been made to evolve superior cultivars of ber by hybridization. Collection of germplasm and selection of promising types are being done at several institutes and agricultural universities. A trial with 40 cultivars was conducted at the Regional Fruit Research Station, Bahadurgarh, and Umran, Thornless, Sanaur-2, Seo, Kala Gold and ZG 3 were found to be promising (Anon., 1966). At the Haryana Agricultural University, Hissar, India cultivars Sandhura Narnaul, Gola, Kaithli, Muria Mahrara, Sanaur-5, Katha Phal and Umran were recommended for commercial cultivation. At the IARI, New Delhi, the cultivars Banarasi Karaka, Katha, Ponda, Narma, Seb, Gola and Ilaichi have been found to be promising.

Keys for classification of ber cultivars were made on the basis of 10 years of observations by Bal (1992). Leaf apex was the most dependable character for classification of cultivars through vegetative characters. Twenty four cultivars were classified with sub-acute to acute apex and eighteen with obtuse apex. Branching habit was another good index and twenty-one cultivars were grouped on spreading branching habit and twenty-one on semi-erect to erect branching habit. Leaf shape, leaf base, leaf colour and petiole length were used as tertiary characters. Fruit characters used in identification are divided into four groups: (i) fruit with a distinctly pointed apex, (ii) fruit with a slightly pointed apex, (iii) fruit with a round apex, and (iv) fruit apex round with a depression. The most commonly found fruit shapes are oval and round. Four cultivars of Sanaur selections, 'Laddu', 'Noki', 'Mirchia' and 'Katha Bombay' have characterstic beak shapes. The obovate shape is found only in the ZG-2 and ZG-3 cultivars. The fruit of 23 cultivars have both stalk and stylar flesh cavities. Both flesh cavities are absent in 'Selected Safeda', 'Illaichi', 'Nalagarh', 'Umran', 'Katha Bombay' and 'Golar' cultivars. Fruit base, fruit colour at maturity, ridges on the fruit base, and apex and fruit surface are used as tertiary characters.

A new variety of *Ziziphus* cultivar Jinsixin 4 was selected by Guo (1999) from open pollinated seedlings of Jinsixua-2. The trees bear early, crop heavily and are highly adaptable to a variety of growing conditions. The fruits mature in early October and have an average fruit weight of 10-12 g. The flesh is crisp and juicy with a soluble solids content 40-45 per cent higher than fruits of most existing cultivars. The fruits are suitable for fresh consumption and for drying.

Fruits of cultivar Daguazao, a chance seedling of *Z. mauritiana* from China matures in early October and are very large weighing up to 50 g (average weight 25.7 g). The skin is thin with a glossy light red-brown colour, the flesh contains 30-40 per cent soluble solids and is crisp with a moderate juice content and very good eating quality (Wang *et al.*, 1999).

Li and Chen (1999) described a new jujuba cultivar Wuqianzhong which produces large, good quality fruits. The trees begin to fruit in the 2nd year after planting, with 2-year-old trees bearing 5-10 kg of fruits and 3-year-old trees bearing 30-40 kg of fruits. The fruits have a crisp and juicy flesh texture with a soluble solids content of 13 per cent and a sweet flavour.

Ma *et al.* (2000) reported that cultivar Juzhou Gongzao was derived from a chance seedling. Trees are very precocious and productive, yield per tree can reach 52.5 kg in fifth year. The flesh is light green, fine and crisp, moderately juicy, with soluble solids content of 31.8 per cent and of very good fresh-eating quality.

Although the two cultivated *Ziziphus species* are in different sections of the Genus *Ziziphus* (*Z. mauritiana* in Section Perdurans and *Z. jujuba* in Section Ziziphus), viable embryos were obtained from *Z. mauritiana* x *Z. jujuba* crosses, suggesting that there is inter-specific cross compatibility (Asatryan and Tel-Zur, 2013). As regards the hybridization work, Vashishtha and Pareek (1983) made an attempt to study the intervarietal crossability in ber. Reciprocal crosses of Seb, Gola, Sanaur-2, Katha and Umran were made. The results indicated that maximum fruit set was recorded in Umran x Katha and minimum in Gola x Sanaur-2. All the cultivars used in crosses were tetraploids and attribute of low pollen fertility is associated with them. It may be one of the reasons of low fruit set amongst the crosses made. The highest fruit set in crosses like Umran x Katha and Katha x Umran seems to be due to their isogenic behaviour. Katha and Umran are believed to be the same cultivars. Srivastava *et al.* (1980) also reported 45.4 per cent fruit set by selfing in Umran. The successful maturity of crossed fruits of Seb x Katha, Seb x Umran and Umran x Seb further supports it.

Godara (1981) showed in cross-compatibility studies that all the ten cultivars studied, *viz.*, Ilaichi, Katha Phal, Reshmi, Sandhura Narnaul, Umran, Banarasi Karaka, Kaithli, Muria Mahrara, Safeda Selected and Kakrola Gola were self-incompatible. Amongst the crosses made by him, Ilaichi x Kakrola Gola resulted in the highest fruit set whereas Katha Phal x Safeda Selected and Kaithli x Kakrola Gola showed complete incom-patibility. Umran showed the best combining ability with other cultivars when used as female or male parent. He also observed that the crossability was considerably increased and premature drop was reduced with colchicine application.

Khan *et al.* (2013) observed increased fruit set due to emasculation of flowers 2 hours before anthesis, followed by pollination by placement of dehisced flowers on stigma in inverted position. Crossing of cultivars Gola and Thar Sevika (early maturing and better fruit quality) with cultivars BS-1 (powdery mildew resistant) and Tikadi (fruit fly and frost tolerant) showed that Thar Sevika is cross compatible with BS-1 and Tikadi; but Gola is cross compatible with BS-1only. Due to crossing the fruit set percentage was increased (2.31-10.92 per cent). The F_1 progeny seddlings were identified as true hybrids using RAPD and ISSR markers.

Three varieties *viz.*, Tikadi, Seb and katha were reported cross compatible and resulted in improved hybrids like F_1 (Seb x Tikadi), F_1 x Seb and Seb x Katha. Ten RAPD primers revealed total 108 bands, out of these 69 bands were polymorphic and showed 33.33-90.9 per cent polymorphism. The dendogram clearly delineated Tikadi and Katha variety from Seb. All the three hybrids formed a single cluster with the variety Seb. Heterozygous positions in ITS sequences clearly showed their hybrid nature. The nucleotide polymorphism in ITS-1 region can serve as a bar code to detect and address genetic diversity within *Z. mauritiana* (Singh *et al.*, 2014).

The cross- pollinating nature and seed propagation are the major reasons of very high intra specific genetic variability in ber may be harnessed for selection and genetic improvement (Singh *et al.*, 2006). Studying genetic variability in ber is prerequisite for efficient use in germplasm conservation. Lack of knowledge about characterization is a big hindrance in managing *Ziziphus* (Soltani *et al.*, 2011). There are several methods to assess variability, but morphological traits provide stronger evidences for agronomic and taxonomic evaluation. Significant morphological variations in ber related to flower, fruit and seed have been revealed in earlier studies (Bal *et al.*, 1992; Gao *et al.*, 2009).

Information on variability, heritability and expected genetic gain (genetic advance as per cent of mean), were reported for twelve growth characters in 30 genotypes of ber (Bisla and Daulta, 1988). High variability was recorded for all the growth characters. In general, heritability estimates were quite high for most of the characters except spread of the tree (22.36 per cent). Maximum heritability was obtained for the days from fruit set to ripening (99.21) followed by per cent fruit set (94.6 per cent). Fruit set per cent itself provides lot of variability, high heritability as well as maximum genetic gain (142.82). It was suggested that for selection of high yielding genotypes attention should be paid upon selection for more fruit set and less fruit drop. Vast genetic variation was found in *Ziziphus* species comprising twenty three varieties forms the basis for systematic tree improvement programme (Ghosh *et al.*, 2015). They observed high heritability coupled with high genetic advance as percent of mean for chlorophyll content index, reducing sugar, soluble protein and scrapedlac yield. Therefore, phenotypic selection for genetic improvement of these traits will be effective as these traits are under the control of additive gene action. Ghosh *et al.* (2015) also suggested that genetic manipulation in ber varieties can be done by making divergent crosses to develop potential varieties of ber for lac production. A total of 10 Zadber (*Ziziphus nummularia* is a branched thorny shrub species in the Indian desert) accessions collected from different agroclimatic regions of India, were evaluated for phenotypic and genotypic variations using morphological and inter-simple sequence repeats (ISSR) markers (Akhtar *et al.*, 2017). Their study revealed that morphological and molecular markers can be successfully utilized for determining genetic diversity and genetic relationship of Zadber cultivars and used in breeding programmes.

Sharif *et al.* (2019) studied the genetic diversity in 60 ber accessions of Pakistan. The measured quantitative traits like leaf length ranged from 1.8 to 9.1 cm, leaf width varied from 1.4 to 6.6 cm, fruit weight lies from 1.88 to 38.45 g, seed weight varied between 0.32–1.84 g, fruit firmness ranges from 0.16 to 1.78 kg, TSS dispersed between 5.9 and 22.96 per cent, while vitamin-C content lies between 60.69 and 165.5 mg/100 g. They observed vast morphological differences for traits like tree shape, branching habit, leaf shape, leaf margins, leaf veins, fruit shape, fruit apex, fruit base, stone shape and for stone apex. Strong positive correlations were detected between leaf, fruit and stone characters. The PCA plot showed high phenotypic diversity and indicated that measured traits can be useful for developing ber plant descriptor and improving taxonomy. Dendrogram generated from agglomeration hierarchical clustering (AHC) successfully grouped wild and domesticated accessions into 3 different clusters. This study appears to be promising for managing genetic resources, in cultivar improvement and for successful breeding programmes in ber.

Rajesh and Mishra (2004) studied the correlation coefficient of 12 growth, fruiting and fruit quality characters of 24 cultivars and obderved that the stem girth was positively correlated with TSS, leaf area with leaf dry weight and titratable acidity, stone weight with acidity and fruit set with fruit drop; whereas stem girth was negatively correlated with fruit set, and stone weight with fruit drop.

Luyuan XiaoZao, a sport of Xiaozao is precocious and high yielding. It is suitable for infertile mountainous soils. The flesh has a soluble solids content of 32.2 per cent and is suitable for fresh consumption and drying. Dry fruit weight is upto 53 per cent of fresh weight (Li *et al.*, 1996).

Hoang and Tuynh (1989) selected mutants following N-methyl-N-nitrosourea treatment (0.02-0.04 per cent for 12 h) of pregerminated seeds of *Z. mauritiana* cultivars. Among those released for cultivation is Ma Hong, which maintains the early maturity of its parental cultivar, Gia Loc, allowing 2 harvests/year. Fruits, however, are round rather than oval, pink rather than yellow and sweet rather than sour. Also released is Dao Tien, which produces round fruits, larger (25 vs. 20 g) and of better flavour than those of its parental cultivar, Thien Phien. It also is one month earlier in maturity, allowing 2 harvests/year rather than the one produced by Thien Phien.

5.1.0 Disease Resistance

A total of 38 diseases/disease complexes have been reported to infect jujube during pre-and post-harvest stages in China (Yao *et al.*, 2013). There are four major diseases/disease complexes that either occur nationwide in China and/or cause the most yield reduction. These include witches' broom (associated with phytoplasma presence), leaf-rust (*Phakopsora zizyphi-vulgaris*), fruit shrink disease (*Coniothyrium olivaceum, Alternaria alternata, Dothiorella gregaria, Macrophoma kawatsukai, Phoma destructiva, Fusicoccum* sp., and Erwinia *jujubovora*) and various fruit rot diseases (Liu *et al.*, 2006; Yao *et al.*, 2013; Mirzaee, 2014). Yield reduction caused by these diseases ranges from 10–30 per cent in average years and yield losses can reach 100 per cent in extreme cases.

Host resistance is expected to be the most efficient way to cope with the above mentioned major diseases/disease complexes in jujube. A substantial amount of research has been carried out in breeding disease resistant jujube cultivars in China (Liu *et al.*, 2006; Yao *et al.*, 2013). Breeding efforts have primarily focused on witch's broom, fruit shrinkage and various fruit rot diseases. The breeding approaches range from germplasm evaluation, selection of farmer cultivars and introgression of host resistance from wild relatives (Wang *et al.*, 2016; Yao *et al.*, 2013). Song *et al.* (2019) reviewed the status of jujube disease resistance breeding and suggests future research needs for developing disease resistant cultivars of Chinese jujube.

6.0 Propagation

In India, seedling trees comprise a large population in old orchards, and the fruits thus produced are inferior in size and quality.

Vegetative propagation is an important method to overcome the constraints and obtain fruits of a desired cultivar.

6.1.0 Vegetative Propagation

Different methods of vegetative propagation, such as cutting, layering and budding have been tried and their success recorded.

6.1.1 Cutting

The leafy cuttings 8-12 cm long, showed high percentage of rooting at a temperature between 26° and 28°C (Shaumarov, 1976). Cuttings taken from the lower part of the shoot and treated with IBA at 50 mg/l for 18 hours showed 88.8 per cent rooting (Jarasenko and Shaumarov, 1977).

Treatment of green wood cuttings of *Z. jujuba* with 250-500 ppm IBA, 250-500 ppm NAA or 250 ppm IBA + 500 ppm NAA resulted in 85 per cent rooting in sand and high percentage of survival (Shen *et al.*, 1992).

In an experiment on rooting, Hatta *et al.* (1996) reported that ten weeks after inocu-lation with *Agrobacterium rhizogenes* (A4 and TR 105) both strain and source of cutting influenced visible symptoms of infection. Jujuba cuttings were more responsive to TR 105 than to A4, and the Li cuttings were more responsive than Contorta cuttings. Strain TR 105 was very effective in increasing rooting percentages and root number. Li cuttings had better rooting percentages and produced more and longer roots than did Contorta cuttings.

6.1.2 Stooling

Ziziphus jujuba suckers were stooled by Kim and Suh (1989) in mounds of soil, vermiculite, 50 per cent vermiculite + 50 per cent perlite, or sand. Some suckers were sprayed with 2500 ppm IBA before stooling. After 2 months, untreated suckers had not formed roots in any medium, while the percentage rooting of suckers treated with IBA was 100 per cent in sand, 66.7 per cent in vermiculite + perlite and 55.6 per cent in soil.

6.1.3 Budding

Budding has been found to be the best and commercial method of ber propagation. Different types of budding, *e.g.*, T or I budding, ring, patch and Forkert budding, *etc.*, have been tried, but the best method in ber is shield budding (T or I budding). The ring budding is a little cumbursome as it requires equal freshness and thickness of stock and scion and hence T budding or shield budding is being preferred and practised all over.

The time of budding depends mainly on temperature, humidity, and availability of budding material. Singh (1952) observed the best result by budding in June, at Saha-ranpur, UP. In Punjab, budding of ber in August showed the best success (Singh *et al.*, 1973). According to Pareek (1978a) the best time for budding in western Rajasthan is in July. In Gujarat, the budding should be completed before the onset of monsoon. In Haryana, Singhrot *et al.* (1980) reported that the sprouting of buds was highest (93.15 per cent) in plants budded in late June. Kaundal *et al.* (1984) at the Punjab Agricultural University, Ludhiana observed high bud-take (84.88 per cent) in June when the humidity was 45-48.5 per cent and temperature was 33.5-34.1°C and in late August and September (80 per cent).

Shield, patch or flute budding was carried out by Mawani and Singh (1992) on cv. Gola plants on 15 or 30 April, 15 or 20 May or 15 or 30 June. Budding success was assessed 105 days after bud break. Flute budding resulted in the longest breaks (average length 92.43 cm), compared with 89.6 and 86.3 cm for patch and shield budding, respectively with the most branches (13.95/plant). Budding on 15 June resulted in the highest number of leaves and branches (62.17 and 14.79/plant, respectively). The best quality bud grafts (in terms of plant health) were obtained with flute budding performed during mid-May to late June.

Nath *et al.* (1998) compared two budding methods (patch or I budding) on the growth of cv. Gola on 1-year-old *Z. mauritiana* var. *rotundifolia* rootstocks. Percentage sprou-ting and shoot lengths were better on patch budded plants than on I budded plants. They (2000) also carried out an experiment to see the effect of cultivars and time of budding on bud sprout for *in situ* establishment of ber orchard under arid ecosystem of Rajasthan. Four cultivars, *viz.*, Gola, Mundia, Kaithli and Umran were patch budded on 15 July, 16 August and 15 September on juvenile boradi (*Z. mauritiana* var. *rotundifolia*) rootstocks. The differences with respect to per cent success among cultivars and time of budding were significant (Table 5). However, interaction effect between both the factors was not significant. The percentage bud sprouts were increased with succeeding date of observations in all the cultivars. Highest success was obtained in Umran budded on 15 September (81.4 per cent), followed by Kaithli budded on 16 August (70.3 per cent), Gola budded on 15 July (67.2 per cent) and Mundia budded on 16 August (64.6 per cent). The observations indicated that even under hot arid ecosystem of Rajasthan, *in-situ* ber orchard can be established successfully by patch budding but success varied significantly with cultivar and time of budding.

Nayak and Sen (2000) standardized the budding method in ber and they reported among the 4 methods (chip, patch, forkert and T budding), patch budding had highest percentage of success (80 per cent) with maximum sprout length (9.1 cm).

The degree of budding success depends on various factors such as time of budding, condition of budding varietal compatibility *etc*. Ghosh (2009) conducted an experiment to know the cultivars compatibility with the local rootstock (*Z. rotundifolia*). Ten cultivars were patch budded on 25th June and the cultivars 'Illaichi' and 'Gola' showed best compatibility with the local rootstock; whereas 'Banarasi Karaka' showed minimum success (70 per cent). Incase of in-situ budding, May,

June and September proved to be the best time as it showed 100 per cent success with maximum plant growth; whereas for polybags grown seedlings in the nursery, the month May and June were best with 75-85 per cent success. In-situ budding (Open field condition) was better as compared to nursery condition in terms of rate of plant growth after budding.

Table 6: *In-situ* Budding Success (Per cent bud sprout) in different Ber Cultivars at different Time Intervals*

Cultivars	*Days after Budding*	*Time of Budding*			*Mean*
		15 July	*16 August*	*15 September*	
Gola	7	18.3 (7.7-27.6)	16.3 (10.0-20.8)	13.0 (5.2-18.9)	15.9
	14	43.4(30.7-58.6)	34.4 (30.5-38.7)	35.8(27.5-47.5)	37.9
	21	57.9 (46.1-70.5)	53.5(46.9-67.2)	54.1(42.0-70.2)	55.2
	28	67.2(56.7-72.3)	58.7(50.0-71.3)	57.3 (48.7-76.4)	61.1
	CV (per cent)	45.6	47.5	62.6	
Mundia	7	15.6 (12.3-29.0)	26.1(20.5-34.4)	22.1(20.0-28.3)	21.3
	14	30.2 (22.9-38.8)	51.1(40.6-59.4)	43.7(35.7-56.0)	41.7
	21	50.2 (40.8-60.3)	58.0(43.7-76.6)5	4.7(43.7-65.7)	54.3
	28	61.2 (50.1-78.5)	64.6(54.5-79.0)	58.0(43.7-76.6)	61.3
	CV (per cent)	5.18	33.7	36.3	
Kaithli	7	17.0 (9.5-25.2)	34.1 (20.0-55.6)	20.8(15.3-30.2)	24.0
	14	28.4 (25.2-33.7)	52.6(28.0-74.1)	47.6(36.7-62.5)	42.9
	21	53.1 (50.0-56.5)	61.6(51.5-80.0)	64.8(51.7-75.3)	59.8
	28	66.8 (56.7-75.3)	70.3(55.3-82.1)	69.1(51.7-81.5)	68.7
	CV (per cent)	54.9	28.3	43.3	
Umran	7	12.4 (5.2-20.3)	9.80 (0.0-18.3)	48.0(20.7-80.0)	23.4
	14	23.5 (20.3-28.6)	37.9(27.5-48.3)	59.5(34.5-80.0)	40.3
	21	55.9 (52.5-60.6)	65.3(50.0-80.5)	79.4(58.6-95.2)	65.2
	28	61.8 (53.0-68.5)	75.3(56.7-90.0)	81.4(62.1-95.2)	72.8
	CV (per cent)	62.9	62.9	24.0	

Figures in parentheses are ranges of variation.

*Nath *et al.* (2000)

6.1.4 Grafting

Grafting of large fruited cultivar of *Z. jujuba* on the young shoots of *Paliurus spina* was not successful. However, the scion grafted onto roots of *P. spina* not less than 2 cm in diameter showed 35 per cent success. After 2 years, the plants reached an average height of 1.5 m and produced fruits similar to those of the original large fruited cultivar (Zaktreger and Solov'ev, 1962).

Omidbaigi and Dadhighi (2003) studied the effect of sucker age and transplanting time on propagation of jujube and they reported the most suitable age for jujube was two-year-old sucker and the best transplanting time was October prior to winter freezing.

6.1.5 Rootstocks

According to Joshi (1960), Chundawat and Srivastava (1978) and Pareek (1978a), the rootstocks are commonly raised from seeds of the wild species. They also suggested the use of *Z. oenoplia*, *Z. rugosa* and *Z. xylocarpa* for rootstock as no graft incompatibility has been reported.

Because of slow growth and spreading habit of *Z. nummularia*, the scion budded on this species remains dwarf compared with other species. Experiments conducted at the Indian Agricultural Research Institute, New Delhi with *Z. nummularia* as a rootstock, the bud graft union recorded delayed incompatibility. However, the species can be used as a rootstock for raising high density orchards keeping in view the other characteristics like precocity, good quality and size of the fruits and above all the easy management of dwarf trees.

Stock and scion compatibility studies were conducted in ber cultivars Banarasi Karaka, Ponda and Gola with 3 rootstocks, *viz.*, Jhar ber (*Z. nummularia*), *Z. mauritiana* ecotype-291 and *Z. mauritiana* ecotype-Assam Gauhati by Verma *et al.* (2000). The rootstock Jhar ber showed incompatibility with inverted bottleneck symptoms budded with Gola cultivar. The all other combinations were compatible with perfect union. The budding success was significantly lesser on rootstock Jhar ber. The histological studies indicated that incompatibility was mainly due to absence of cambium production between the stock and scion which resulted into wide gap in weak and unhealthy plants. There was excessive callus production but no proper formation of union because of lack of differentiation into cambium, phloem degeneration of scion and formation of a thick necrotic layer between the stock and scion which might resulted into the formation of inverted bottleneck symptoms in the later years of the tree growth.

Ziziphus mauritiana cv. Umran was budded on 12 rootstoks (5 *Z. mauritiana* accessions obtained from different regions of India, Chinese *Z. jujuba*, Argentinian *Z. zoaziroo*, Indian *Z. xylopyrus* and 4 accessions of *Z. nummularia* from different regions of India) (Bal *et al.*, 1997). Umran on *Z. mauritiana* rootstocks was the most vigorous and produced the highest yields of fruits (cumulative yield of 115-146 kg/tree).

6.1.5.1 Raising of Seedlings

The germination of ber seeds is slightly difficult due to the hard endocarp and requires about 3-4 weeks to germinate. The germination time can be reduced to one week by cracking the endocarp carefully.

The shortest time taken to start germination (7 days), the highest percentage final germination (46.66 per cent), and the best seedling growth were obtained by Murthy and Reddy (1989) when the kernels were extracted by breaking endocarp + soaking. The corresponding control figures were 24.5 days and 17.5 per cent germination. Mankar *et al.* (1997) found that soaking extracted seeds for 24 h in water promoted early germination, good vegetative growth and higher percentage germination compared with sowing whole seeds (control). Cracked seeds exhibited better germination than control seeds but were not as good as water-treated seeds.

After 30 days from extraction germination was highest (84 per cent) in seeds scarified mechanically; however, after 203 days, it was highest (88 per cent) in seeds treated with 1 per cent potassium dihydrogen phosphate (Ghosh and Sen, 1988).

Kajal (1983) recorded best germination when the seeds were sown at 2 cm depth. He also noted that the germination of freshly harvested seeds was improved by soaking for 6 minutes in sulphuric acid and of one-year-old seed by soaking for 24 hours in water followed by stratification in moist sand for 4 days. Better seedling growth was obtained by sowing in June but March sowing gave seedlings ready for budding in July and it was also the best month for budding.

Studies conducted at various places showed that growth substances promoted germination. Kim and Kim (1984) while studying the germination of *Z. jujuba* seeds at different stages of fruit growth, showed that even immature seeds sown in aseptic media containing 100 ppm BA or GA3 at 50 days after full bloom, germinated in two months after sowing but the untreated seeds did not germinate. Krishnan and Kulasekaran (1984) soaked seeds of *Z. rotundifolia* for 24 hours in solution of GA3 (50-500 ppm) alone or + 5 per cent sucrose or thiourea (250 or 500 ppm). Treatment with GA3 at 500 ppm resulted in the highest germination of 96 per cent followed by 94 and 92 per cent with 200 ppm + 5 per cent sucrose or thiourea at 500 ppm, respectively. The water-soaked seeds gave 68 per cent germination.

Laboratory and field experiments were carried out on seed germination of ber using kernels of cv. Umran (Murthy and Reddy, 1989). Among the treatments with chemicals (GA3, BA, KNO3, thiourea), kernels treated with thiourea showed significantly higher percentage germination compared with the other treatments in both laboratory (78 per cent) and field (67 per cent for 0.5 per cent treatment, 75.66 per cent for 1 per cent treatment). In a subsequent investigation, Murthy and Reddy (1990) reported that treatment of ber seeds cv. Umran with 1 per cent thiourea (for 24 hours) and a temperature of 30°C resulted in highest germination percentage and seedling growth. Lower (20°C or 25°C) or higher (30°C) temperature inhibited germination and seedling growth.

Seeds of ber cv. Gola were treated with GA3, thiourea and cycocel for 24 hours, dried and stored in polythene bags for 1 or 17 months. Hore and Sen (1994) recorded the highest germination rates after 1 and 17 months from seeds treated with 200 ppm GA3 (98.76 and 77.82 per cent, respectively). Incidence of polyembryony was highest with 400 ppm GA3 and 1000 ppm cycocel. Treatment of ber seeds of cv. Umran obtained by breaking open endocarp taken from ripe seeds, with abscisic acid (ABA) at 50 ppm by soaking reduced germination percentage (26.8) compared with control (43.7) (Murthy *et al.*, 1989).

Sin'ko and Chemarin (1979) obtained 6-98 per cent seed germination by treatment with low dose of ã-irradiation. It also promoted seedling growth and development of lateral shoots and length of roots.

Seeds of the *Z. mauritiana* cultivar Umran extracted from ripe fruits and stored in polyethylene bags at room temperature for 2, 4, 6 or 8 months did not show any signi-ficant differences in germination percentages (42-50 per cent germination) (Reddy and Murthy, 1990). Fresh seeds had the lowest percentage germination (33 per cent). Seedling vigour was greatest after seed storage for 8 months.

Seedlings of *Z. mauritiana* do not tolerate transplanting and hence the practice has been to sow the seeds directly in the field and bud them *in situ*. It is cumbursome to carry out the budding operation *in situ*, to take care of the budded plants and the process of raising ber orchards is delayed. Pareek (1978b) developed a method for raising budded plants in polythene tubes. In this method polythene tube of 25 x 10 cm size and 300 gauge thickness are filled with well-prepared pot compost, placed in nursery bed, and the seeds are sown in the first fortnight of April. Ninety-day-old seedlings in the polythene tubes were budded in the second week of July and the budded plants became ready for transplanting in August. Singh *et al.* (1984) transplanted uniform seedlings with 20 x 15 cm soil ball, 95 days after sowing and recorded maximum number of seedlings ready for budding and 91.4 per cent success in bud take.

Srimanthi *et al.* (2002) collected the stones of fruits cv. Umran both from the ground and crown possessing yellow to red pericarp, extracted manually followed by washing with water and air drying. The dried stones were soaked in 15, 18 and 20 per cent salt solution and in water (control). The seeds from fruits collected from the crown showed maximum germination and germination were improved by grading the stones using 20 per cent salt solution

Pre-sowing treatment of medium seeds with H_2SO_4 for 10 minutes increased seed germination percentage (54.22 per cent); while mechanical cracking of seeds resulted in minimum days (23.66 days) to germination (Singh *et al.*, 2004). Karimpour *et al.* (2013) reported germination percentage, vigor index and the rootlet number increased by ascending exposure time with 98 per cent H_2SO_4, while the highest seedling length was obtained in 5 minutes acid exposure.

6.1.6 Top Working

Topworking in wild jujube (*Z. nummularia*) by T-budding of cv. Gola was reported by Yadav (1991).

6.2.0 Micropropagation

Goyal and Arya (1985) described the tissue culture method for clonal multiplication of *Z. mauritiana* cvs Gola and Seb from a single bud culture, in 150 days. BA (1.0 mg/l) and NAA (0.5 mg/l) proved effective for differentiation of shoot. Multiplication of propagules was achieved by subculturing the shoot segments on 1/2 strength Murashige and Skoog's (MS) medium with and without BA and auxin. Shoot segments developed adventitious roots within 20 days in 1/2 strength MS medium containing 0.8 per cent agar, 0.5 mg/l IBA and 0.05 mg/l kinetin.

Success in micropropagation of ber *in vitro* from stem explants of mature tree was reported by Mathur *et al.* (1995) by growing on modified MS medium containing 3800 mg potassium nitrate and 2475 mg ammonium nitrate/l, 11 μM BA and 0.5

μM IAA. During successive subcultures 15-20 shoots per inoculum were produced. Rooting was induced and plantlets grew well in a soil and vermiculite mixture. Proto-cols for clonal propagation of *Z.nummularia* and *Z. mauritiana* by tissue culture were developed by Rathore *et al.* (1992). From a nodal explant of *Z. nummularia,* 5-7 shoots proliferated on MS medium containing 5.0 mg BA + 0.05 mg IAA/l. For *Z. mauritiana,* 4-5 shoots developed from the nodal region of each explant on MS medium containing 7.5 mg BA + 0.1 mg IAA/l. Shoots generated *in vitro* could be further multiplied on fresh medium. The isolated shoots were placed for rooting on a filter paper bridge in White's liquid medium containing 25 mg IBA/l.

Axillary buds of *Z. sativa* cv. Geumsung showed the best shoot and root growth after 8 weeks when 500 mg/l activated charcoal was added to half strength medium with 0.5 mg/l (av. shoot length 6.4 cm, av. number of shoots per explant 10.0), while cv. Bokjo responded best to 1000 mg/l activated charcoal (av. shoot length 7.5 cm, av. number of shoots per explant 12.4) (Kim and Lee, 1988). Shoot tips of jujuba were cultured by Du *et al.* (1997) on half-strength MS medium with 20 g sugar/l and various concentrations of IBA, IAA and NAA. The best results were achieved on media containing 2.0 mg IBA/l or 0.05 mg IAA + 1.0 mg IBA/l. When cultured at 23-29°C, 2000-3000 lx, under a 12-h photoperiod, rooting reached 99-100 per cent, with >5.3 rootlets explant, average rootlet length was >2.2 cm. Before transplanting rooted explants should be acclimatized at 23-29°C, 4000-6000 lx for 20-30 days.

Goyal *et al.* (2006) found MS medium supplemented with BAP at 1.78 μM to be ideal for enhancing shoot regenerative potential of the nodal explant; whereas among the various auxins used, NAA at 2.69 μM proved to be best for induction of healthy roots, which resulted in maximum survival percent (80 per cent) of plants in the field.

Singh *et al.* (2010) carried out an experiment for *in-vitro* clonal propagation using shoot tips, nodal and intermodal segments as explants and they reported shoot tip as a better explant than the other two. For shoot proliferation, media containing BAP at 2 mg L^{-1} + NAA at 0.1 mg L^{-1} + GA_3 at 0.1 mg L^{-1} proved best. IBA at 2 mg L^{-1} resulted in better root induction percent, root regeneration, root number and length of roots; whereas combination of 2,4-D at 2 mg L^{-1} and BAP at 4 mg L^{-1} was successful for callus initiation and root induction from callus.

For *in-vitro* establishment of cv. Comethry and Balady, MS medium enriched with 0.05 mg/L NAA, 2 mg/L N6-(2-isopentyl) adenine proved best, but for cv. Balahy 100 per cent growth was obtained on MS medium containing NAA @ 0.05 mg/L and TDZ @ 1 mg/L. Therefore, MS medium supplemented with BA @ 4 mg/L+ N6-(2-isopentyl) adenine @ 0.5 mg/L significantly affected the shoot multiplication (4 shoots/explants) of cv. Comethry; whereas for the cv. Balady and Balahy both BA @ 3 mg/L+ N6-(2-isopentyl) adenine @ 0.5 mg/L enriched MS medium was useful for inducing maximum shoot multiplication *i.e.* 4.6 shoots/ explants (Soliman and Hegazi, 2013).

Yildirim *et al.* (2015) recorded highest percentage of explants that produced 100 per cent shoots was obtained on MS medium supplemented with TDZ at 0.1 mg/L + BAP at 0.5 mg/L + IBA at 0.1 mg/L + GA_3 at 0.3 mg/L; whereas MS and half strength MS media enriched with 2 mg/L IBA was capable of promoting highest rooting percentage (76.7 per cent).

6.3.0 Micrografting

Micrografting may be a useful technique for improving the rooting ability as well as for the early diagnosis of grafting incompatibility of tissues from mature trees.

Danthu *et al.* (2004) conducted a study which was mainly focused on the 'Gola' variety, multiplied by *in vitro* micrografting technique involving aseptic grafting of miniature scions (5-10 mm long) onto the hypocotyls of seedlings *in vitro* resulted into 80 per cent successful establishment of the micrografts.

7.0 Area and Production

In the north-west parts of India, various cultivars of *Z. mauritiana* are found growing commercially. Species like *Z. nummularia* are also found growing widely. In the eastern part of India, seedlings of *Z. rotundifolia* are found growing abundantly with a very few plantations of grafted cultivars of ber.

In India, ber is being cultivated on an estimated area of 52 thousand hectares producing 639 thousand metric tonnes (Anon, 2019). It is most widely cultivated in Maharashtra, Gujarat, Haryana, the plains of Punjab and Rajasthan. Fairly good cultivation of this arid zone fruit is also found in Madhya Pradesh, Bihar, Uttar Pradesh, Assam and other places. The ber producing states and contribution towards national production of India are given in Table 7.

Table 7: Ber Producing States and its Share Towards National production during the Year 2015-16*

Sl.No.	*State*	*Production (000 tonns)*	*Share (per cent)*
1	Gujarat	120.31	28.34
2	Chattisgarh	70.76	16.67
3	Madhya Pradesh	55.76	13.13
4	Haryana	39.49	9.30
5	Odisha	39.46	9.29
6	Punjab	28.04	6.60
7	Andhra Pradesh	23.76	5.60
8	Karnataka	12.75	3.00
9	Maharashtra	11.04	2.60
10	Jammu & Kashmir	10.80	2.54
11	Rajasthan	6.73	1.59
12	Jharkhand	2.83	0.67
13	Telangana	1.88	0.44
14	Nagaland	0.87	0.20
15	Tamil Nadu	0.09	0.02
	Total	**424.57**	**99.99**

* NHB (2015-16).

8.0 Soil and Climate

8.1.0 Soil

Ber, because of its hardy nature and tap root system, can be grown in a wide range of soil including porous and infertile soils unsuitable for major fruits and other crops.

Studies at the Central Soil Salinity Research Institute, Karnal, India have revealed that ber can also be grown satisfactorily in alkali soils characterized by high pH and sodicity. In such soils, it was suggested to mix five kg gypsum per pit followed by flooding about one week before planting. However, sandy loam soils, with neutral or slightly alkaline reaction, make good growth of ber plants. According to Dhankhar *et al.* (1978) the plants once established continue to grow well even at 21 mmhos/cm sali-nity. Oganesjun (1953) observed that *Z. jujuba* was most tolerant against salinity and also suggested that this species would be used for biological reclamation of saline soils. Dahiya and Dhankhar (1984) studied the salt tolerance of ber seedlings and found adverse effect of salinity on dry shoot weight, plant height, stem diameter, leaf area, root dry weight, nutrient contents in soil above 14.2 mmhos/cm. Patil *et al.* (1981) found decrease in plant height, leaf number, shoot diameter and fresh and dry weight of roots, leaf contents of P, K, Ca, Mg, Zn and Cu with increase in soil ESP from 10 to 50. However, Na content increased in both root and leaves with increasing ESP.

In Azerbaijan, Tagiev (1992) recorded that on the Apsheron peninsula 10-year-old plantations of ber on sandy soil yielded 50 h kg/ha of high quality fruits. In the Shrivanskaya region, with heavy clay soil and hot summers (up to 36°C), plantations irrigated 5-7 times/season yielded upto 100 h kg/ha of high quality fruits.

In an experiment on salt tolerance conducted by Hooda *et al.* (1990) in an orchard at Hissar, India, *Z. mauritiana* cv. Umran was grown in lysimeters treated with NaCl, $CaCl_2$, $MgCl_2$ and $MgSO_4$ to give soil electrical conductivities of 1.45 (control), 5.0, 10.0, 15.0 and 20.0 dS/m. No plants survived at 20.0 dS/m. Tree canopy area and the weight of prunings taken in May 1989 were reduced significantly at >5.0 dS/m, and trunk diameter was affected at a soil electrical conductivity >10.0 dS/m. Leaf tissue Na, Ca, Mg and Cl concentrations increased substantially as salinity increased but leaf K concentration decreased. Yield and fruit set declined with increasing salinity with a 70 per cent yield reduction occurring at a soil electrical conductivity of 15.0 dS/m. Salinity had no effect on fruit quality.

One-year-old plants of 5 cultivars and 1 (rootstock) seedling of ber *Z. mauritiana* were grown in 45 x 45 x 45 cm containers in soil with exchangeable Na contents ranging from 5.0 to 45.3 per cent (Kumar *et al.*, 1990). The leaf Zn, Cu, Mn and Fe contents generally decreased with increasing soil exchangeable Na content. The effect of salinity [(0.80 control), 4.80, 9.68 and 14.70 mmhos/cm)] on the nutrient status of leaves of *Z. mauritiana* var. *rotundifolia* and *Z. mauritiana* cv. Banarasi Karaka plants was investigated in pot experiments (Pandey *et al.*, 1993). Increasing salinity reduced the leaf tissue contents of N, P and K whilst increasing the contents of Mg, Na, Cl and SO_4 and Ca.

In an experiment on the effects of sodicity and salinity level on ber, seedling *Z. mauri-tiana* var. *rotundifolia* (*Z. rotundifolia*) plants and budded plants of *Z. mauritiana* cultivars Banarasi Karaka, Kaithli, Umran and Gola were grown in clay pots in soil containing either exchangeable sodium percentages (ESP) of 9.0 (control), 38.8 or 60.5 ($NaHCO_3$ added), or EC values of 0.82 (control), 14.8 or 20.25 dS/m ($CaCl_2$, $MgCl_2$, $NaCl_2$ and Na_2SO_4 added) (Awasthi *et al.*, 1994). With an ESP of 60.5, the highest percentage plant survival was obtained with *Z. rotundifolia* (62.5 per cent); no Umran and Gola plants survived. *Z. rotundifolia* also showed the best survival under high salinity. The total chlorophyll content influenced the survival percentage with all cultivars. Awasthi *et al.* (1995) also reported that seedlings of wild jujube and budded plants of cv. Banarasi Karaka may be grown successfully with a survival of >50 per cent at upto 60.5 per cent exchangeable sodium and 20.25 dS/m salinity. Cv. Kaithli had 40 per cent survival at the highest salinity level but was susceptible to 60.5 per cent exchangeable sodium. Cultivars Umran and Gola were susceptible, and hence not suitable for cultivation above 38.8 per cent exchangeable sodium and 14.8 dS/m salinity.

In a similar experiment on tolerance to sodicity and alkalinity of ber, Pandey *et al.* (1991) grew seedlings and budded plants of cv. Banarasi Karaka in pots and sub-jected to 4 ESP (exchangeable sodium percentage) levels (control, 15, 30 or 45) or 4 ECe levels(control, 5, 10 or 15 mmhos/cm). Total chlorophyll content decreased and total free amino acid and proline contents increased with increasing levels of sodicity and salinity in both seedlings and Banarasi Karaka plants. Slightly higher total chlorophyll content was observed in plants at an ECe value of 5 mmhos/cm than in the controls, which may be due to higher uptake of N, Ca and Mg. The increase in total free amino acid and proline contents may be due to protein hydrolysis.

Jain *et al.* (1988) observed that salts which accumulated in the soil as a result of saline water irrigation (EC 2.7-9.0 ds/m) of ber cv. Gola were effectively leached by run-off and no significant carryover of salts could be determined after the rainy season during a 7-year study. Irrigation water of EC 2.7-9.0 ds/m was successfully used for establishment of plants. Fruit yield decreased by 10-20 per cent as a result of irri-gation with water of EC 9.0 ds/m. However, no adverse effect on fruit size was observed.

8.2.0 Climate

8.2.1 Temperature

According to Singh *et al.* (1973a), ber is found growing under varying climatic conditions, almost throughout India. However, temperature below freezing is injurious to the fruits as well as to the young plants. In north-western parts of India during the year 1974, ber fruits suffered a heavy loss due to unprecendented frost.

In case of *Z. mauritiana*, the trees come into dormancy during May-June when it is very hot in the northern plains of India. This is considered to be the best time for pruning the plants. The growth, flowering and maturity seem to be more or less correlated with temperature and thus vary accordingly in different regions. The fruits start ripening as early as middle of December in the western part of India whereas in the northern plains the fruits start ripening by mid-February.

The two species, *Z. jujuba* and *Z. mauritiana* are unlike in their climatic requirements. The *Z. jujuba* being deciduous and temperate thrives well under severe cold. Kucherova and Sin'ko (1984) studied the frost tolerance in *Ziziphus* using a refrigerated chamber. They had kept one-year shoots of 22 cultivars and various *Ziziphus* spp. at –27°C and found slight damage in cultivars Da balfszao, Kitaiskii 50, Vakhshskii 30/16, Vakhshskii 40/5 and Kitaiskii. The resistance of *Z. sativa* cultivars to subzero temperatures was studied by Sinko *et al.* (1987) in various climatic zones of the Ukraine. The cultivars Ta-yan-tszao, Suan-tszao and Druzhba showed no frost damage to the buds and shoot tissue after temperatures as low as –21°C. The most frost resistants in the Sudak region were Ta-yan-tszao, Ya-tszao and Vakhsh.

8.2.1 Light

Chiou *et al.* (1994) in an experiment exposed plants of *Z. mauritiana* to light during the night for 15, 30 and 45 days. Flowering of the light-treated plants was significantly advanced compared with the controls. Also, the numbers of flowers and fruits set increased with light treatment. Plants that received lighting for 30 and 45 days cropped 40 and 60 days earlier, respectively than controls.

Effects of night interruption of flowering and yield of *Z. mauritiana* cultivars Kaolangland Telong were reported by Chiou and Weng (1996). Flowering was advanced by 25 days in L3D3 (lighting for 3 hours followed by darkness of the same length) followed by DD (no lighting during night; control). Yields were significantly higher in treatments LL (lighting all through the night), L1D1 (lighting for 1 hour followed by darkness of the same length) and L2D2 (light for 2 hours followed by darkness of the same length) than L3D3 or the control.

9.0 Cultivation

9.1.0 Planting

In Punjab, the best time of planting ber is February-March or in the monsoon during July to September (Bajwa *et al.*, 1972; Bakshi and Singh, 1974). In dry regions of western India, the best time of planting is during July-August. Pits of 90 cm in depth and diameter are dug and kept exposed to the sun for about a month and chlorpyriphos is mixed with soil to protect from the attack of termites, and filled with a mixture of soil and farmyard manure. The budded plants are planted after the soil has settled. After planting, soil around the bud graft is pressed firmly to avoid air-pockets and irrigation is given immediately.

The influences of land treatment (flat bed or basins with a 5 per cent catchment slope) and plant population (278, 185, 139 or 111 plants/ha) on fruit yield of ber (cv. Gola) under rainfed conditions were reported by Reddy *et al.* (1997). In 1990, the highest yields (1989-4863 kg/ha compared with 1676-4004 kg/ha) were produced by plants planted in basins. However, in 1991, plants on flat beds produced the highest yields (mean value of 1913 compared with 1684 kg/ha). The highest yields (kg/ha) were observed at the highest planting density. No interaction was observed between land treatment and plant population.

Sandhu and Dhillon (1983) obtained 100 per cent success in bare-root transplanting of budded ber by defoliation followed by heading back at lifting and wrapping them in moist rice-straw. Sandhu *et al.* (1983) recorded success in transplanting with earthball ranging from 20 to 100 per cent, while in bare-root transplants it varied from 40 to 100 per cent, between January and mid-March.

9.2.0 Planting Distance

Investigations on planting density have revealed higher yield with increase in plant population. Tian *et al.* (1983) tried a high density planting of *Z. jujuba* for high yield at the Changai Institute of Pomology, China. They planted *Z. jujuba* seedlings in double row system at the spacing of 2 x 2 x 4 m. By adopting proper training and pruning and also ringing and knotching practices and by spraying GA_3 at 10 ppm they were able to get high yield.

Thirteen-year-old uniform ber cv. Gola trees were spaced at 4.8 x 4.8 m, 6 x 6 m and 8 x 8 m (S_1, S_2 and S_3, respectively). Fruit drop was lowest in S_1 and highest in S_3 (47.70 and 73.03 per cent, respectively). Fruit weight, yield and total soluble solids were highest in S_1 in both years, whereas acidity was highest in S_3 and lowest in S_1 (Saini *et al.*, 1995).

Plants were grown in Nanpi, China at 2 x 2 m or 1 x 2 m spacing with hedge-row training and were compared with those grown in traditional low density systems (Zhang *et al.*, 1998). The cumulative yields during the investigation were 10 times higher in the plots with higher plant population than those from traditional systems. Spacing of 2 x 2 m was better than 1 x 2 m.

9.3.0 Training and Pruning

9.3.1 Training

Training of ber trees is essential during first 2-3 years to build up a strong framework. The Indian jujube has vine-like growth habit and a tendency to grow horizontally and downwards. If proper training of the trees is not attempted, it will attain a bushy and spreading form with long slender branches. More than one shoot emerge from the base of the stem causing overcrowding of branches and provide a condition suitable for the harbouring of pests and diseases. Such trees are short lived and poor in productivity. The bushy untrained tree also makes the harvesting difficult.

To develop a strong framework, it is necessary to train the trees right from the nursery stage. The young plants transplanted in the field should be supported with bamboo stake to avoid damage of the bud union and to support main stem. As the commercial varieties of ber are spreading in habit, staking is also essential during the first two years to train the tree properly. During the first year, tree puts forth vigorous growth and many side branches emerge from the main shoot. All the secondary branches up to 75 cm from the ground level should be removed, and straight and vigorous growing main trunk be developed (Bajwa and Sarowa, 1977). On the main trunk 3 to 5 laterals/secondary branches most favourably located around the main stem, should be selected to make the proper framework.

Propagation methods of ber

Propagation by T-budding

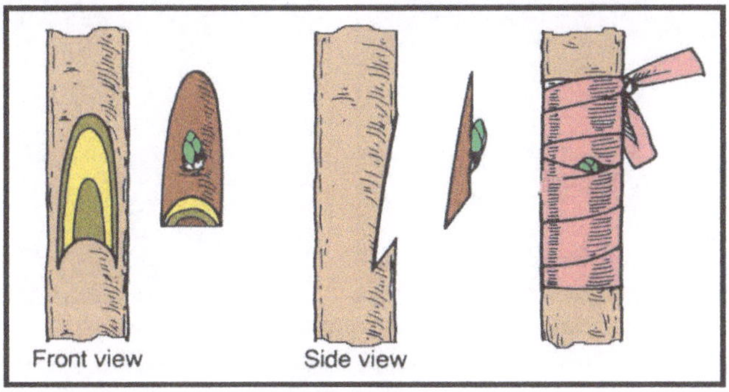

Chip budding

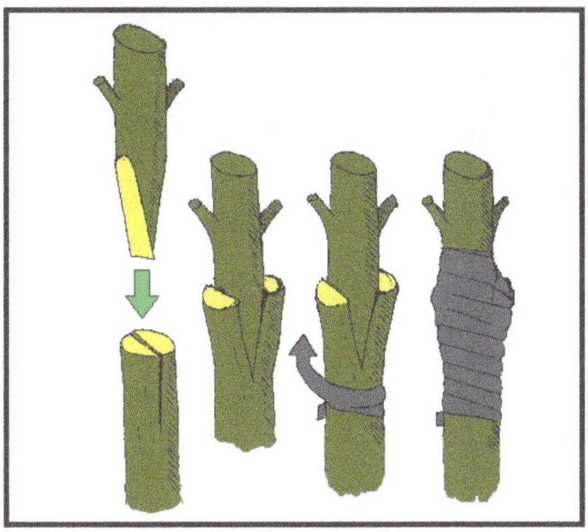

Cleft grafting

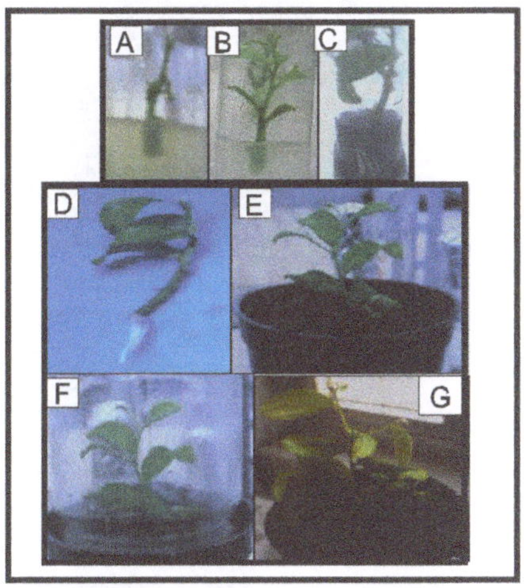

Micropropagation in ber

Source: Abbas, Muayed , Ibrahim, A.H. and Jasim, Abbas (2014) *AACL Bioflux*, 6: 11-15.

Pruning in ber tree

Headed back ber tree

Initiation of shoots on pruned branches

Canopy development in ber

Ber tree with fruits

In the second year, the process should be repeated to develop tertiary branches on the secondary shoots. On each secondary branch 3 to 4 well-distributed tertiary branches should be retained. Efforts to keep the branches and trees upright should be made all the time. Final balancing and correction of framework should be done in the third year so as to obtain a strong framework.

9.3.2 Pruning

The ber fruit is borne in the axils of leaves on the young growing shoots of current season. Therefore, a regular annual pruning in ber is very much necessary to induce a good and healthy growth which will provide the maximum fruit bearing area on the tree. Some thinning out of branches is also necessary to avoid too much crowding which helps in reducing the incidence of pests and diseases.

The best time for pruning a ber tree is during the hot and dry season or when it sheds its leaves and becomes dormant. This state of dormancy is to be observed after the harvesting of fruits. In Delhi, Haryana, Punjab, Rajasthan and Uttar Pradesh a ber tree becomes dormant during May when pruning should be done (Singh and Sandhu, 1984). Dry, dead and diseased wood, criss-cross branches and weak crotches should be removed at the time of pruning. After the establishment of the framework, pruning should be continued every year but drastic thinning and very severe pruning should be avoided; severe pruning adversely affects the growth of plant and gives poor yield. For optimum yield and good quality fruits, 25 per cent of one-year-old shoots should be headed back.

The effect of severity of pruning has been studied by different workers at different locations and no significant difference in yield of fruit has been observed. However, size and weight of fruit and quality in respect of TSS, ascorbic acid, *etc.*, increase with severe pruning (Awasthi and Misra, 1969; Gupta and Singh, 1977; Sharma *et al.*, 1980), while the yield decreases with removal from 25 to 75 per cent of the past season's growth (Gupta and Singh, 1977). In an investigation on the effect of pruning on fruit set, Awasthi and Misra (1969) reported that although the initial fruit set was higher in medium pruning treatment, the final set and yield were more in plants pruned less.

The effect of pruning to 20, 40 or 60 cm was assessed by Bajwa *et al.* (1989) with 13-year-old trees of the cultivar Umran. They reported faster growth rate in pruned plants but fruit set and fruit retention did not differ significantly. Fruit yield decreased and TSS content increased with pruning. Pruning to 60 cm was suggested. Syamal and Rajput (1989) reported that hard pruning (20 cm) of cv. Banarasi Karaka resulted in the most vigorous shoot growth, both in length and diameter, compared with the unpruned control. Fruit set and retention decreased with pruning severity but fruit size and weight increased. The yield was highest in the control (85.32 kg/tree) and in the lightly (60 cm) pruned trees (83.56 kg) and lowest (80.17 kg/tree) in the most severely pruned trees. Fruit quality improved with pruning severity.

In a study carried out by Yadav and Godara (1992) at Haryana, India, 13-year-old trees, at 3 different spacings (4.8 x 4.8 m, 7.2 m x 7.2 m or 9.6 m x 9.6 m), were subjected to light, medium or severe pruning (removal of 20, 40-50 or 70-80 per cent of the previous year's growth, respectively). Flowering, fruit set and fruit maturity was delayed with increasing pruning severity and decreasing planting distance. Percen-tage fruit set increased with increasing severity of pruning and wider planting. Percentage of fruit retention increased with increasing planting distance and decreasing severity of pruning. It is concluded that medium pruning combined with a planting distance of 7.2 m x 7.2 m or 9.6 m x 9.6 m is more appropriate for ber trees.

Pruning of 19-year-old trees of cv. Umran was carried out on 4 dates between 30 April and 10 June and trees were pruned lightly (control), moderately or heavily (Gupta and Godara, 1989). Moderate pruning of 30 May gave the highest yield of 200.33 kg/tree compared with the lowest yield of 66 kg/tree in the control trees pruned on 10 June. Heavy pruning gave the best quality fruits.

In 16-year-old ber trees (cv. Umran) 3 levels of mid-May pruning (heading back to give a shoot diameter of 1, 2 or 3 cm) and 3 spacings (planting distances of 4.8 x 4.8, 7.2 x 7.2 or 9.6 x 9.6 m) were compared by Bisla *et al.* (1991). For most of the para-meters measured (initiation and duration of flowering, fruit set, fruit drop and yield) the least severe pruning (or unpruned controls) and the widest spacing gave the best results. Although fruit set was highest with the most severe pruning, fruit drop was also highest. Kundu *et al.* (1995) pruned ber trees to different severities on 15 or 30 May or 15 June 1991. Highest fruit weight and pulp:stone ratio and lowest stone weight occurred in trees where half the branches were pruned from the base and the remaining half to 15 buds on 30 May. Similarly, highest fruit TSS and ascorbic acid contents and lowest acidity were recorded with this regime. Yields were reduced significantly as the pruning severity was increased; highest yield was recorded by pruning to 25 buds on 30 May (88.33 kg/tree). Pandey *et al.* (1998) in an experiment with cvs. Gola, Banarasi Karaka and Umran evaluated different intensities of pruning. The pruning was done in the last week of May. They observed maximum yield with 50 per cent pruning severity followed by 75 per cent in all the cultivars (Table 8).

Seven-year-old cv. Gola trees on *Z. rotundifolia* rootstock were pruned to 30, 50, 90 or 120 cm of the previous season's growth. Pruning of branches to 90 cm gave the highest yield (28.50 t/ha) and highest fruit TSS content (25.72°Brix), followed by pruning to 60 cm (26.08 t/ha and 24.92°Brix). Control trees produced an average yield of 22.50 t/ha (Hiwale and Ruturi, 1996).

In a trial in Periyakulam, India, on 3-year-old trees of cultivars Banarasi and Umran, Nanthakumar and Balakrisnan (1999) reported that medium pruning increased the average yield of cultivars from 18.7 kg/tree (unpruned controls) to 21.5 kg/tree. Umran responded well to pruning, with higher yields than Banarasi.

Zhao and Deng (2000) in China showed that pruning is an important measure which revives old jujube and increases its yield but pruning intensity should not be heavy taking fruiting branches, for example, branches with 4 buds had a higher yield than branches with 2 or 8 buds or unpruned branches.

The effect of pruning intensity on vegetative growth and yield was studied by Dinesh (2002). The maximum number of primary branches (4.45) was obtained with 25 per cent pruning; whereas 50 per cent pruning was useful for maximizing the secondary (6.70) and tertiary (6.12) branches. The fruit yield was maximum (25.92 kg tree^{-1}) in response to 50 per cent pruning irrespective of cultivars.

Kumar *et al.* (2002) conducted an experiment to determine the effect of different pruning length on the production of high quality fruits. Secondary branches were pruned at 20 cm (light), 40 cm (medium) and 60 cm (severe) length accordingly. Lightly pruned trees had significantly longer fruits as compared to severe pruned trees. On the other hand, pulp percentage and TSS were not significantly influenced by either cultivars or pruning intensities.

Table 8: Effect of Pruning Intensity on Fruit Yield (kg/plant) in Ber Cultivars*

Intensity of Pruning (per cent)	*Gola*	*Banarasi Karaka*	*Umran*	*Mean*
0	23.12	23.33	24.42	23.62
25	26.38	28.10	29.17	27.88
50	35.50	34.33	37.42	35.74
75	27.50	29.33	31.33	29.39
Mean	28.13	28.77	30.59	
Factors	S.Em ±	CD at 5 per cent	–	–
Cultivars	–	1.69	–	–
Pruning intensity	–	1.57	–	–
Interaction (pruning intensity x cultivars)	0.91	N.S.	–	–

* Pandey *et al.* (1998).

9.4.0 Irrigation

Owing to very deep tap root system and xerophytic nature, ber plant is very hardy and once it gets established, needs little care and irrigation. The harvesting of fruit of *Z. mauritiana* is over by April and plants become dormant in May-June and shed their leaves. They do not need any irrigation during the period of fruit maturation, harvesting and dormancy of plants from March to June. With the onset of rainy season tree starts new growth. In some parts of ber growing areas in Punjab and Haryana, there are sufficient rains during July to developing new flush. During September and October, the trees come into blossom. The irrigation should be withheld at the time of flowering, otherwise the flowers will drop down. Fruit setting is complete by the end of first fortnight of November. Ber trees need irrigation from November to February for better fruit development. Trees may bear a crop even if no irrigation is applied during fruit development also. However, it has been observed that fruit size increases and quality improves if irrigation is applied at monthly interval from November to February. Irrigation also reduces fruit drop considerably. Pareek (1983) reported that frequent irrigation between October and February alongwith heavy doses of fertilizers reduced fruit drop and increased fruit size and yield. In trials with the cultivar Nasimi planted at 6 x 6 m, the trees were irrigated at 50-55, 60-65 or 70-75 per cent of soil minimum water content (MWC). The highest yield (66.4 h kg/ha, 60 per cent more than in the control) was obtained from trees irrigated at 60-65 per cent MWC, *i.e.*, with 6 treatments at 350 m^3 of water/ ha. Fruit quality was also best with this treatment.

The treatment combination of 15 cm irrigation + black polythene mulch + 6 per cent kaolin spray at pre-flowering and fruit set stage was found useful for enhanced (13.1 t/ha) fruit yield (Mukherjee *et al.*, 2004).

In the absence of irrigation facilities ber orchards can be raised with a run-off concen-tration system (Yadav *et al.*, 1978; Sharma *et al.*, 1982). In this system the catchment area should be provided either on both sides of tree rows or on all sides of a tree.

The fruit yield was increased by 3.20 times with surface irrigation and 2.86-4.05 times under various drip irrigation treatment over rainfed control (8.0 q ha^{-1}). Scheduling drip irrigation considering 60 per cent wetted area with 25 per cent pan evaporation or 20 per cent wetted area with 75 per cent pan evaporation was found optimum with significantly maximum fruit yield (30.2 q ha^{-1}), saving water (61.5 and 66.3 per cent in two successive years) over surface irrigation; the highest water use efficiency (1.33 and 3.82 q/ha-cm during two successive years) (Yaragattikar and Itnal, 2003).

Ismail and Almarshadi (2013) reported significant increase in branch length, number of branches, soil moisture content as well as fruit quality parameter; while increasing fruit yield by about 78.3 per cent with the use of 8 drippers with a discharge of 4 L ha^{-1}.

Improved fruit quality parameters like TSS (17.1 per cent), ascorbic acid (222 mg/100 g FW), glucose (2431.7 mg/100 g FW), fructose (2633.9 mg/100 g FW), malic acid (392.9 mg/100g FW), catechin (16.5 mg/100g FW) and epicatechin (37.5 mg/100 g FW) can be achieved with the use of drip irrigation at 120 m^3/hm^2 (Gao *et al.*, 2014).

9.5.0 Manuring and Fertilization

Deficiency of N, P and K was found to inhibit shoot length, leaf number and axillary branch production; the flowering was also reduced and the plants failed to set fruits (Sadhu *et al.*, 1978).

Application of 40 to 50 kg farmyard manure in July and 0.75 kg ammonium sulphate per tree in October (after fruit set) was recommended by Singh and Singh (1973). Daulta and Chauhan (1982) recommended application of 50 kg well-rotten farmyard manure during dormant period and one kg of nitrogenous fertilizer per tree in two split doses, once during the rainy season and again at the time of fruit set. Kumar *et al.* (2018) recommended that the use of vermicompost @ 2 kg/plant applied repeatedly for three months at the initial stage of ber, yields maximum fruits and also maintains soil health for sustainable production. Yamdagni *et al.* (1980) observed that irrespective of the levels (250, 500 or 750 g/tree), appli-cation of nitrogen significantly improved fruit set and reduced fruit drop. Plants provided with 500 g N per year showed a yield from 86 to 91 kg per tree compared with 61 to 67 kg under control. Kumar and Tung (1981) reported an increase in vigour of cv.

Umran by fertilizing the plants with 480 g N, 150 g P_2O_5 and 300 g K_2O annually. Increasing doses of N, P and K also increased the leaf N, P and K contents.

A fertilizer trial was undertaken by Dhatt *et al.* (1993) on 3 years old ber cv. Umran using N at 400, 600 or 800 g/tree and P and K each at 100 or 200 g/tree, applied in all combinations. The greatest trunk girth (100.31 cm) was obtained with 400 g N + 100g P + 100 g K/tree, the heaviest fruits (26.5 g) with 400 g N + 200 g P + 100 g K/tree and the highest yield (21.75 kg/tree) with 400 g N + 100 g P + 200 g K/tree. The effect of annual doses of N (0, 250, 500 or 750 g/plant), P_2O_5 (0, 250 or 500 g/plant) and K_2O (0, 50 or 100 g/plant) on the growth and yield of 8-year-old ber cv. Umran trees, spaced at 6 x 6 m was investigated by Supe *et al.* (1997). Pooled results indicated that the yield of fruits/tree increased significantly by 29.5 and 37.12 per cent over the control following application of N and P at 250 g/tree, respectively. K application had no effect on yield. Singh *et al.* (1995) in an experiment with 9-year-old ber cv. Umran applied three levels of N *i.e.*, 250 (N_1), 500 (N_2) and 750 (N_3) g N/tree/year, two levels of P *i.e.*, 250 (P_1) and 500 (P_2) g P_2O_5/tree/year and two levels of K *i.e.*, 50 (K_1) and 100 (K_2) g K_2O/tree/year. They observed an increase in fruit size and weight with increasing levels of N (Table 9) and the maximum yield (1.64 q/tree) was noted with 750 g N, 500 g P_2O_5 and 100 g K_2O/tree/year.

Thirteen-year-old trees *Z. mauritiana* cv. Umran were sprayed with urea at 0, 1.0, 1.5 or 2.0 per cent and zinc sulphate at 0, 0.5 or 1.0 per cent in different combinations in July and November (Singh and Ahlawat, 1995). Foliar application of 1.5 per cent urea and 0.5 per cent zinc sulphate increased fruit total soluble solids and ascorbic acid contents; acidity decreased with increasing urea concentrations. Leaf N concentration was highest with 2.0 per cent urea and 1.0 per cent zinc sulphate spray (2.36 and 2.17 per cent, respectively). Leaf P and K concentrations decreased with increasing concentrations of urea and zinc sulphate.

Prasad and Bankar (2002) conducted an experiment to find out the effect of various concentrations of nitrogen (0, 250, 500 and 750 g $plant^{-1}$) and phosphorus (0 and 250 g $plant^{-1}$) on growth, yield and quality of ber cv. Gola and observed increase in plant height, plant spread and stem girth with increasing concentration of N and P fertilizers. Lal *et al.* (2003) reported N: P_2O_5: K_2O (500:500:50 g $plant^{-1}$) was best for obtaining maximum yield from 10-year-old ber cv. Umran. Yadav and Rathore (2004) reported foliar application of Thiourea (0.1 per cent) + $FeSO_4$ (0.2 per cent) significantly enhanced yield (34.47 kg $tree^{-1}$), while improving the physico-chemical properties of fruits *viz.*, fruit length (2.84 cm), fruit breadth (2.97 cm), fruit weight (12.14 g), TSS (19.10 per cent). Singh and Singh (2009) observed maximum plant height, spread, trunk growth, fruit set and fruit retention in cv. Banarasi Karaka with the application of FYM + 75 per cent NPK + Azotobacter + PSB. The soil application of FYM + 100 per cent NPK + Azotobacter + PSB also improved the soil nutrient status, which was at par with FYM + 75 per cent NPK + Azotobacter + PSB. Combination of organic and inorganic fertilizers such as more potassium and less nitrogen and phosphorus are efficient in improving yield and antioxidant activity (Wu *et al.*, 2013).

An integrated nutrient management was adopted under Malwa plateau conditions of Madhya Pradesh, India with different treatment combinations of organic manures (FYM @ 53, 39.75 and 26.5 kg, vermicompost @ 26.5, 19.86 and 13.25 kg), inorganic fertilizers (full dose of recommended fertilizer, 50 per cent dose of recommended fertilizer and 25 per cent dose of recommended fertilizer) and bio-fertilizers (Azotobacter and PSB) on five-year-old ber trees cv. Gola (Bohane *et al.*, 2016). The results revealed that the application of 50 per cent recommended dose of fertilizer (424 : 315 : 327 g NPK/tree) through vermicompost + 50 per cent RDF through NPK + 50 g Azotobacter + 50 g PSB significantly increased the plant height, canopy volume, number of primary and secondary branches per shoot, fruit set and fruit retention, fruit length, diameter, and volume, pulp weight, stone weight, average fruit weight, yield, TSS, ascorbic acid, reducing sugar, non-reducing sugars, total sugars (9.89 per cent), TSS/acid ratio and chlorophyll content in leaves spad value.

Choudhary *et al.* (2020) recommended that integration of organic manures and inorganic fertilizers was more effective in increasing growth and yield of ber plants than inorganic fertilizers alone under dry tract of Rajasthan, India. Among various combinations, application of vermicompost @ 6 kg/plant with 125 per cent RDF (1100 g urea, 1400 g SSP and 200 g MoP per tree) proved the best treatment combination in terms of number of secondary branches per primary branch, fruit yield and net returns, whereas vermicompost @ 6 kg/plant with 100 per cent RDF proved best in respect to B:C ratio (2.96) as compared to other treatment combinations.

Table 9 : Effect of levels of N, P and K on yield attributes of ber fruit cv. Umran*

Treatments	*Fruit Size (cm)*				*Fruit Weight (g)*		*Yield (q/tree)*	
	Length		*Width*					
	1989	*1990*	*1989*	*1990*	*1989*	*1990*	*1989*	*1990*
N levels								
N1	4.04	4.16	2.94	2.98	21.03	20.77	1.36	1.39
N2	4.30	4.39	3.08	3.14	22.04	22.13	1.61	1.60
N3	4.48	4.49	3.45	3.50	24.25	24.71	1.63	1.63
CD (P=0.05)	0.11	0.14	0.13	0.09	1.41	1.44	0.02	0.03
P levels								
P1	4.23	4.34	3.09	3.16	22.02	22.06	1.53	1.53
P2	4.32	4.41	3.23	3.25	22.86	23.03	1.54	1.54
CD (P=0.05)	NS	NS	0.10	0.08	NS	NS	NS	NS
K levels								
K1	4.26	4.23	3.15	3.19	22.36	22.52	1.53	1.54
K2	4.29	4.37	3.16	3.22	22.46	22.63	1.54	1.54
CD (P=0.05)	NS	NS	NS	NS	NS	NS	NS	NS

NS = Not significant.

* Singh *et al.* (1995).

Lac production on *Zizyphus mauritiana* is a profitable option for farmers in Jharkhand, Chhattisgarh, Odisha, Madhya Pradesh and West Bengal (Ghosal, 2013). The lac insect growth, yield and quality parameters on *Z. mauritiana* improved significantly with the addition of nitrogen, phosphorus and potash (Lal *et al.*, 2003). It was observed that the increase in biomass of *Z. mauritiana* under Kusmi lac production under nutrient management was 4.36 kg per year per tree in comparison to 0.81 kg per year per tree in control (Ghosal, 2013). *Z. mauritiana* plant with basal application of 100g N, 250g P and 75g K per plant had highest lac yield (Paul *et al.*, 2013). Namdev *et al.* (2015) observed that the mean weight of 100 fresh cells of lac at harvest as well as the dry cell weight increased significantly by the basal application of nutrients [Nitrogen(Urea 220g), Phosphorous (SSP 1560g) and Potassium (MoP 125g)] treated *Z. mauritiana* plants as compared to the control. They also recorded the nutrient management of *Z. mauritiana* for lac production significantly increased the yield up to 39.16 per cent.

9.5.1 Tissue Analysis and Nutrient Requirement

The concentrations, distribution and uptake of various nutrients in the exocarp + mesocarp, endocarp and kernel of ber (*Z. mauritiana*) fruits of cultivars Dandan, Banarasi Pewandi, Rashmi, Katha, Banarasi, Banarasi Karaka, Mehrun, Sanaur-5, Sanaur-2 and Aliganj were studied by Varalakshmi *et al.* (1999). The kernel was the major respository for N, P, Ca, Mg, S, Fe, Mn, Zn and Cu, while the exocarp + mesocarp was the major sink for K. The uptake of different nutrients by one tonne of ber fruits was in the order of 5.74-20.39 kg N, 0.98-1.38 kg P, 3.26-7.31 kg K, 0.91-1.60 kg Ca, 0.19-0.41 kg Mg and 0.61-1.06 kg S. Among the micronutrients, Fe was taken up in large quantities (0.24-0.59 kg). Uptake of Mn, Zn and Cu was 13-31, 10-18 and 10-14 g, respectively. Katha contained the greatest amount of protein (8.6 kg per 1 t of fruit), and the P and K contents ranged from 13 to 89 g and 0.3 to 1.8 kg, respectively.

Nitrogen storage and reutilization in *Z. jujuba* trees using 10 per cent ^{15}N-urea alone or with 200, 300 or 400 ppm ethephon was investigated by Hao and Zeng (1991). Spraying with urea + ethephon accelerated the movement of N from the leaves to the rest of the tree and increased the level of stored N, with urea + 200 ppm ethephon being most effective treatment. Treatments increased the length of fruit-bearing deciduous bran-ches, fruit set, average single fruit weight and the amount of reducing sugar in fruits. Urea increased the N contents of shoots and leaves, but decreased fruit sucrose content. The main form of stored N was protein, the N of which was 2-3 times the level of non-protein N.

The influence of the VAM species *Acaulospora morrowae, Gigaspora margarita, Glomus deserticola, G. fasciculatum* and *Scutellospora calospora* on net photosynthesis and transpiration of *Z. mauritiana* was evaluated under greenhouse conditions in sterilized soil (Mathur and Vyas, 1999).VAM inoculation resulted in a significant increase in net photosynthesis, total chlorophyll and carotenoid concentrations, and sugar and starch concentrations. The VA mycorrhizas also increased stomatal resis-tance, thereby reducing the rate of transpiration. Guissou *et al.* (2000) observed that *Z. mauritiana* inoculated with mycorrhizal fungus (*Glomus manihotis*) utilized available P more efficiently than non-inoculated.

Glomus constrictum, one of the commonly occurring VAM fungi of arid and semi-arid regions was multiplied on modified MS medium in association with *Z. mauritiana* under *in vitro* conditions (Mathur and Vyas, 1999). The *in vitro* multiplied VAM strain established efficient symbiosis with *in vitro* raised plantlets. This VAM strain significantly improved the biomass production, nutrient uptake and acclimatization of the *in vitro* produced plantlets into the field.

Bhargava *et al.* (1990) carried out investigation on 4-year-old ber trees (cultivars Gola, Seb, Mundia and Umran budded onto *Z. rotundifolia* seedlings) to determine leaf nutrient composition in relation to its position on the shoot and age following an annual pruning in May. N, P and K concentrations decreased with increasing leaf age. Ca concentration sharply increased with age while the increase in Mg was marginal. Leaf S, Zn and Cu contents decreased with increasing age whereas Fe and Mn contents did not follow a specific pattern. The N, P, K, Ca and Mg concentrations varied among cultivars. The results indicated that the 5th and 6th leaf from the secon-dary shoots, representing the recently matured leaf, should be collected for analysis in nutritional diagnosis. Raturi and Bhargava (1993, 1994) estimated the following ranges of nutrients in the leaves of four different ber cultivars (Table 10).

Table 10: Ranges of Plant Nutrients in Four Cultivars of Ber (g/100 g)*

Nutrients		*Shoot*			
		Gola	*Mundia*	*Seb*	*Umran*
Nitrogen	Mean	2.06	2.28	2.09	2.39
	Range	1.23-2.91	1.58-2.92	1.60-2.75	1.36-3.27
Phosphorus	Mean	0.19	0.27	0.18	0.29
	Range	0.09-0.36	0.14-0.48	0.08-0.34	0.11-0.51
Potassium	Mean	1.34	1.09	1.46	1.67
	Range	0.85-1.86	0.39-1.79	0.83-2.26	0.84-2.52
Calcium	Mean	1.40	1.41	1.94	1.82
	Range	0.60-2.00	0.45-2.08	0.78-2.90	0.99-2.69
Magnesium	Mean	0.34	0.39	0.34	0.41
	Range	0.16-0.87	0.18-0.75	0.21-0.75	0.21-0.98
Sulphur	Mean	0.12	0.10	0.20	0.21
	Range	0.04-0.20	0.05-0.16	0.10-0.32	0.11-0.36

Note : Mean over four types of shoots.

* Raturi and Bhargava (1994).

Li and Wang (1995) analysed leaf nutrient composition of jujuba (*Z. sativa*) during the growing season and found that N, P, K, Fe, Zn and Mg sharply decreased during the early stage, while Ca and Mn increased gradually, then the changes became less obvious. During the late stage of growth, N, P, K, Fe, Zn and Mg decreased again, and Ca and Mn continued to increase. Application of N before flowering and P before or during the flowering period increased production and improved fruit quality.

Soil and leaf samples were analysed by Ahlawat *et al.* (1990) from 20 Z. *mauritiana* orchards, ranging in age from 6 to 20 years, with average growth and productivity and showing no symptoms of poor nutrition. Soil analyses revealed that 75 per cent of orchards were deficient in organic carbon and 90 per cent were deficient in P. However, the K content was optimum. The mean leaf N, P and K concentrations were 2.1-3.55, 0.12-0.20 and 0.94-1.62 per cent, respectively.

9.6.0 Mulching and Weed Control

Effects of mulching with black polythene and rice straw on vegetative growth and mineral element content were compared by Byun *et al.* (1991). The soil water poten-tial in mulched plots was >–0.01 MPa at 12 days after rainfall while the potential in the control plot fell to –0.06 MPa at 8-10 days after rain. Soil temperature at 10 cm depth was always higher with polyethylene than with straw or no mulch. At 10.00 h the soil temperature was the same with straw or no mulch but at 16.00 h the tempe-rature under straw was lower than with no mulch. Growth was best throughout the season with polyethylene, but early in the season growth with straw increased to exceed that of the control. There were no effects of mulching on leaf P, K, Ca or Mg contents but N contents were highest without mulch and lowest with polyethylene.

The effects of diuron, glyphosate and paraquat at 3-4 kg/ha for weed control in ber cv. Umran orchards were evaluated by Bajwa *et al.* (1993). All weed control treatments decreased average weed populations from untreated control values of 14.44-16.41 plants/m^2 to 5.15-11.14 plants, and increased crop yields from control values of 1282 kg/ha to 13590-14990 kg. Glyphosate at 4 kg resulted in the greatest crop yields. In trials on a 12-year-old ber orchard, dalapon and pendimethalin (pre-emergence) and isoproturon and paraquat (post-emergence), each at 3-5 kg/ha, significantly reduced N, P and K uptake losses by weeds. Dalapon followed by para-quat each at 5 kg/ha provided maximum conservation of soil moisture (Bajwa *et al.*, 1990).

Cultural and chemical weed control in ber orchard was studied by Jasmine (2012) and found pre-emergence application of atrazine @ 2 kg/ha followed by glyphosate @ 2 l/ha was effective in controlling weeds in ber.

Kashi *et al.* (2004) reported black polyethylene mulch could increase number and weight of fruits per plant as well as yield by 85 per cent due to suppression of weed growth and conservation of soil moisture. Mukherjee *et al.* (2004) also observed increased growth and productivity by using black polythene mulch. Application of rice straw @ 12.5 t/ha was found effective in weed management in ber orchard (Brar *et al.*, 2020).

9.7.0 Intercropping

After planting ber, 4 to 5 years are required to cover the interspace between the trees. The interspace between the rows of plants can be utilised for growing intercrops. Generally, leguminous crops are preferred as intercrop as they enrich the soil in addition to some income. Under irrigated conditions gram, chillies, leafy vegetables can be grown, while *mung*, cowpea, guar can be grown as rainfed crops. In suitable climate, phalsa and papaya can be grown as intercrop in the first 3 years.

The effect of intercropping in ber cv. Seb with different crops (cluster beans cv. Maru, mung beans cv. S-8 or sesame cv. TC-25) during *kharif* (monsoon) was studied by Singh (1997). Initial fruit yield in the intercropped orchard increased 3-fold (14.8 kg/tree) compared with the control orchard (5.2 kg/tree). Intercropping in a newly planted ber orchard had no adverse effects on plant growth for up to 5 years. The intercrops also exhibited higher yields when planted with ber compared with monocultures under rainfed conditions. The best intercrop was cluster bean.

Ber-based intercropping system recorded 19.6 per cent higher fruit quivalent yield than sole ber. Among different systems, intercropping with green gram gave the highest net monetary return (Rs. 13,099/ha) followed by intercropping with sorghum (Rs. 11,886/ha); while intercropping in ber recorded on an average 10 per cent higher monetary returns (Rs. 11,329/ha) over sole ber cultivation (Rs. 10,287/ha) per annum (Patel *et al.*, 2003).

9.8.0 Rejuvenation of Old Orchards

Rejuvenation of 1.24 million old and weak trees of jujuba (cv. Jinsixiaozao) by rene-wal pruning, enhanced fertilizer application and treatment to improve fruit set was reported by Wang *et al.* (2000). The 5-year average yield of the treated trees was increased by 66.2 per cent. The renewal pruning included light heading back of the skeletal branches (leaving 75 per cent of the length of the branches) and thinning out the weak fruiting branches. Spraying with 10-15 mg GA_3/l + 2.0 per cent borax at flowering increased fruit set.

The 30-year-old trees can be rejuvenated by heading back during the 2nd fortnight of May while retaining 30 cm length of the main limbs. Thinning of the shoots should be done twice- 1st during fourth week of august by retaining 12-18 shoots and 2nd thinning should be conducted before flowering *i.e.* second fortnight of September while retaining 8-12 shoots. This technique is useful for improving the vegetative characters, yield as well as fruit quality parameters (Bal *et al.*, 2004).

10.0 Flowering, Floral Biology, Fruit Growth and Development

10.1.0 Flowering

The flowers in ber are borne in the axil of the leaves of mature as well as current season's shoots and inflorescence is an axillary cyme (Teaotia and Chauhan, 1963; Vashishtha and Pareek, 1979). The time of flowering varies in different parts of India and different regions in other countries and it is influenced by climatic conditions. In Maharashtra, the growth does not cease and hence there is early flowering (May-June), the fruits mature by October (Patil, 1978). In Uttar Pradesh, flowering in cv. Banarasi Karaka starts in the second week of September and continues up to November (Teaotia and Chauhan, 1963). In western Rajasthan, flowering starts in July, peak being in September. At Delhi the flowering in ber starts in early September and continues till mid-November, the longest period being in cv. Ilaichi (61 days) and shortest in Tikadi (47 days) (Raja, 1985). About 12 to 14 flowers are produced per cluster (Vashishtha and Pareek, 1979).

Floral development and pollen studies were carried out on 6 cultivars at the Punjab Agricultural University, Ludhiana (Dhaliwal and Bal, 1998). Development of the bud from its initial appearance up to flowering was divided into 8 stages. Flowering occurred about 21 days from early bud development and length of flowering ranged from 57 to 75 days, depending on cultivar. The mode and time of anthesis was also cultivar specific. Anther dehiscence started about 2 h after anthesis and continued for 2-4 h. Peak receptivity of the stigma appeared to be just as the flower opened.

The flowering response of 27 Chinese jujuba cultivars to photoperiod was investi-gated by Kim and Kim (1984). The cultivars Moodeung, Geumsung, Jg-3 and Jg-10 produced more abundant flowers on the middle part of the reproductive shoot than on the basal or upper part, while, cv. Jc-8 had many flowers on the middle to upper part. The flowering season generally ranged from 10th June to 25th July. Cultivars Moodeung, Geumsung and Jg-3 were in full bloom from mid-June to early July and produced more flowers than Jc-8 and Jg-10 which bloomed from mid to late June.

10.2.0 Floral Biology and Fruit Set

Anthesis is observed to be cultivar dependent. Anthesis in Seb and Sanaur-2 takes place in the morning between 7.30 am and 8.0 am while in Gola, Katha and Umran in afternoon at 1-2 pm (Vashishtha and Pareek, 1983). Teaotia and Chauhan (1963) reported that in Seb, Jogia, Ponda, Aliganj and Ilaichi, the anthesis took place between 7.30 and 8.30 am while in Gola and Mundia it was between 12.0 and 1.0 pm.

In most of the cultivars the dehiscence of anthers starts just after anthesis and is completed within 4 to 5 hours. Vashishtha and Pareek (1979) reported complete dehiscence in an hour after anthesis in cultivars Seb, Gola, Jogia, Mundia, Aliganj, Ponda and Ilaichi. They also observed that stigma was most receptive on the day of anthesis in all these cultivars and pollination 24 hours prior to anthesis failed to set fruits while some fruit set occurred when pollination was done 24 hours after anthesis. In a study on pollen germination in *Z. jujuba,* Yun *et al.* (1994) recorded only a few germinating pollen grains on the stigmas of open-pollinated flowers.

As the pollens of ber are sticky, they are not transferred by wind. During flowering time, lot of insect activities is found in ber orchards. Prominent pollinating insect agents are honey-bees, house flies and yellow wasp.

Godara (1981) studied the floral biology and compatibility behaviour of different ber cultivars and found that Banarasi Karaka, Mundia, Murhara, Reshmik, Sandhura Narnaul, Safeda Selected, Umran, Ilaichi and Kakrola Gola were self-incompatible. Amongst the crosses made, Ilaichi x Kakrola Gola gave the highest fruit set. Umran cultivar showed the best combining ability when used as male or female parent. Fruit set by open pollination ranged between 6.19 per cent in cv. Thornless and 8.16 per cent in cv. Sanaur-5. However, fruit set by controlled cross-pollination (hand pollination) was quite high and varied from 26.66 per cent in Thornless x Umran to

46.66 per cent in Sanaur-2 x Chhuhara and Sanaur-5 x Chhuhara (Josan *et al.*, 1981). Cross incompatibility was also observed in some ber cultivars.

The life of individual flower is very short and in an inflorescence many flowers remain unpollinated during their receptive periods. As such unfertilized flowers drop off. There is also a marked variation in fruit set among the different cultivars which ranged from 5 per cent in Ilaichi to 14.9 per cent in Aliganj in arid conditions of Rajasthan (Vashishtha and Pareek, 1979). It was also observed that branches of eastern side gave maximum set of 8.47 per cent compared with minimum set of 2.04 per cent on the northern side (Yamdagni *et al.*, 1968). At IARI, New Delhi the cultivar Tikadi recorded a maximum fruit set of 27.88 per cent whereas it was minimum in Gola. A maximum fruit drop of 95.63 per cent and a minimum of 79.88 per cent was observed in the cultivars Ilaichi and Tikadi, respectively (Raja, 1985).

An investigation on flowering, fruit set and fruit drop was carried out by Sharma *et al.* (1990) on 8-year-old *Z. mauritiana* trees of cultivars Banarasi Karaka, Ponda, Ilaichi, Gola and Tikadi in New Delhi. The peak period of flowering and fruit set was September-October for all cultivars. Earliest flowering (11 September) and initiation of fruit set (21 September) were observed in Tikadi, which had the shortest duration of flowering (47 days) and fruit set (36 days). The highest number of fruits/branch (239), fruit set (28 per cent), number of fruits reaching maturity/branch (48) and fruit retention (20 per cent) were observed in Tikadi. The lowest number of fruits/branch (73), fruit set (10 per cent) and number of fruits reaching maturity/branch (8) was observed in Gola. The longest duration of flowering (61 days) and fruit set (50 days) and the lowest fruit retention (4 per cent) were observed in Ilaichi.

Singh *et al.* (1991) studied the relation of embryo abortion and fruit drop in 4 cultivars of ber. During the early stages of fruit development (for 30 days after full bloom), yellow fruitlets which were about to abscise and green developing fruitlets were collected and examined for the extent and pattern of embryo degeneration. Cumula-tive fruit drop was highest (85 per cent) in Kaithli and lowest (68.3 per cent) in ZG-2. The bulk of fruit drop occurred during the early stages of fruit development. The yellow fruitlets showed embryo degeneration whereas green fruitlets had normal embryos. The heavy fruit drop during the early stages of fruit development may be attributed to unsuccessful fertilization and/or ovule degeneration.

Bi *et al.* (1996) analysed endogenous GA3, IAA, cytokinin and ABA of flowers of *Z. jujuba*. The last stage of flowering (stigma withering) was a critical period for fruit set, and during this period the levels of IAA and ABA were the main factors affecting setting. High soluble sugar content helped in decreasing physiological fruit drop but had no correlation with fruit set. Spraying with exogenous GA3 not only increased percentage fruit set but it also increased the capacity of the ovary and flower to act as a physiological sink thereby reducing physiological fruit drop.

10.3.0 Fruit Growth and Development

Extensive investigations have been carried out on the physical and biochemical changes in growth and development of ber fruits. Fruit growth of ber followed a double sigmoid curve with 3 distinct phases *viz.*, an early phase of 45 days after fruit set (DAFS), middle phase, 45-90 DAFS and final phase, 90 DAFS to maturity (185 days). The growth patterns of fruits and seeds were similar (Prasad and Babu, 1988).

Pandey *et al.* (1990), in an investigation on physico-chemical changes in fruits of 6-year-old plants of cv. Banarasi Karaka sampled at 15-day intervals between fruit set and 135 days after fruit set (AFS), observed that fruit size increased markedly during 2 phases (0-30 and 60-105 days AFS). Fruit weight and TSS, sugar and ascorbic acid contents increased continuously throughout development. There was a gradual decrease in acid content. Fruit DM content decreased and N content gradually increased. P and K levels markedly increased only in the initial stages of fruit development and then decreased. Ca content showed a decreasing trend from fruit set to harvest, whereas Mg showed the reverse trend. It was shown that 100 kg of fruits removed 43.8 g N, 7.5 g P, 101.2 g K, 4.5 g Ca and 22.7 g Mg.

The ber (*Z. mauritiana*) cultivars Umran, Sanaur-2 and Sanaur-4 were assessed for TSS, acidity, vitamin C, reducing and total sugars, and total phenolics in the skin, stalk and different parts of the fruit (Bal *et al.*, 1988). In all cultivars, vitamin C content was highest in the middle part of the fruit, and acidity was highest in the lower part. Godge and Kale (1991) studied developmental physiology of ber fruit cv. Mehrun and observed that the fruits reached maturity at 110 days after anthesis. Fruit length, diameter, weight and volume increased from fruit set to maturity while specific gravity decreased. Fruits were green in colour at the early stage changing to yellowish red at maturity. TSS, total sugars, reducing sugars, non-reducing sugars and ascorbic acid contents increased from fruit set to maturity while acidity and protein content decreased.

Meel *et al.* (1991) observed that the length, weight and volume of the fruits, the pulp weight, pulp:stone ratio, total soluble solids/acid ratio and chlorophyll content could be used to judge maturity. Ber fruits followed a double sigmoid growth curve. Fruits having a specific gravity of <1 were considered to be ready to harvest; organoleptic tests assisted in judging maturity. It was proposed that in Hissar cv. Sandhura Narnaul ber may be harvested on 12-26 February, 120 days after fruit set when fruits had experienced heat sums of 1840 day-degrees after full bloom.

Abbas *et al.* (1994) studied the physical and chemical constituents of the fruits of *Z. spina-chisti* cultivars Bambawi and Mallacy collected bi-weekly from fruit set until maturity. Fruit growth in terms of length, diameter, volume and FW followed a double sigmoid curve pattern. Stone development proceeded at a fast rate up to 105 days. Specific gravity was low initially, then increased up to 105 days, but decreased as the fruits advanced towards maturity. Total soluble solids, carotenoids, sugars and ascorbic acid contents of fruits of both cultivars increased towards maturity.

Al-Niami *et al.* (1992) studied the physiology of ripening of jujuba fruits of the cultivars Mallacy and Bambawi. The rates of respiration and ethylene production increased as the fruits entered the ripening phase, reached a peak value on days 6-7 and then declined. Total soluble solids, sugars, carotenoids and ascorbic acid contents increased, while titratable acidity, total chlorophylls and total phenolics decreased with fruit maturity. Optimum eating quality was reached at the climacteric peak.

10.3.1 Effect of Growth Regulators

The effects of growth substances on flower drop, fruit set, and fruit growth and deve-lopment have been studied on ber.

Mechanical deblossoming of young plants was found to be time-consuming and uneconomical (Singh *et al.*, 1987). Ethrel or planofix (NAA), each at 200, 400 or 600 ppm were applied at full bloom 9 months after transplanting. Planofix at 400 ppm gave the best results by inducing complete flower abscission but no leaf abscission or damage to buds.

In an investigation, on the effect of NAA, 2,4-D, 2,4,5-T and methyl ester of naphthalene acetic acid (MENA) at 10-30 ppm, Singh and Singh (1976) observed that sprays of 2,4,5-T reduced fruit drop by 14.19 per cent compared with control. All treatments increased fruit size and TSS but NAA at 10 ppm proved most effective. Singh *et al.* (1987) reported marked improvement in yield of ber by treatment with GA_3 at 20, 40, 60 or 80 ppm in October and at the beginning of December. The highest fruit set (20.07 per cent), lowest fruit drop, best fruit retention (47 per cent), best fruit physical and chemical indices, and the highest yield (79.93 kg/tree) were obtained with the highest GA3 rate. The control yield was 49.58 kg/tree. The average fruit weight and size of Umran cultivar were increased appreciably by 10 ppm NAA applied at full bloom (Bal *et al.*, 1988). A general decrease in stone weight was also recorded with NAA treatment, 10 ppm giving the maximum decrease. Fruit quality was improved by 25 ppm NAA.

Eight-year-old trees of ber cv. Gola were sprayed with GA3 and NAA at 10, 20 or 30 ppm, alone or in combination, at flowering and again 15 days later increased fruit set and retention. Fruit weight and length were significantly increased by GA3 or NAA at 30 ppm. TSS content was increased by GA3 but not by NAA (Bankar and Prasad, 1993). In another investigation, Masalkar and Wavhal (1991) reported highest fruit weight, fruit volume, pulp content, non-reducing sugar and ascorbic acid contents and lowest per cent stone with GA3 alone. Ethephon resulted in the highest TSS content and improved the fruit colour to golden yellow.

Bal *et al.* (1982) reported that NAA at 25 ppm applied twice, first application after fruit set and the second one month after, showed a reduction in fruit drop by 9.8 per cent and improved the fruit size and increased TSS (16.2 per cent) content in cv. Sanaur-2. Improve-ment in size and quality of ber was reported by Sandhu and Thind (1988) using 20 ppm NAA at fruit set. Treatment with 2,4,5-T at 25 ppm in cv. Banarasi Karaka also increased fruit size, weight, TSS, sugars and acid contents (Bal *et al.*, 1984). Rajput and Singh (1982) observed the highest yield of 86.8 kg/tree with the application of planofix at 40 ppm. In order to reduce fruit drop which is a serious problem in ber production, Grewal *et al.* (1993) sprayed with 2,4-D, NAA or naphthoxyacetic acid (NOXA) at 10, 20 or 30 ppm on 13-year-old trees of the cultivars ZG 4 and Sandhura Narnaul. The lowest fruit drop for both cultivars (51.97 and 80.97 per cent for ZG 4 and Sandhura Narnaul, respectively) was obtained using NOXA at 10 ppm. This treatment resulted in fruits which were significantly larger, had higher TSS and vitamin A contents and lower acidity, total sugar and reducing sugar contents than most other treatments.

Kale *et al.* (2000) reported enhanced fruit size, weight as well as improved fruit quality with 20 ppm GA_3, which was sprayed twice-first at full bloom stage and second 15 days after flowering. Bhati and Yadav (2003) observed maximum fruit length (3.42 cm), fruit breadth (3.02 cm), fruit weight (1752 g) as well as enhanced yield per tree (56.44 kg) and benefit: cost ratio (2.72) with application of NAA at 20 ppm. Improved fruit retention percentage was obtained with the treatment NAA @ 20 ppm (55.01 per cent) followed by 30 ppm (52.13 per cent). Ghosh *et al.* (2009) found that application of NAA at 25 mg/L gave significantly highest fruit retention (75 per cent) which resulted in highest fruit yield of 120.5 quintals as against 64.7 quintals/ha in control. Katiyar *et al.* (2010) found that foliar sprays of NAA at 25 and 30 ppm alongwith 2 per cent urea produced significantly bigger sized fruits (3.93 cm in length and 2.85 cm in width) with maximum TSS (16.60 °Brix) and sugar content (11.20 per cent). The treatment also improved the fruit yield (147 kg/tree) in comparison to control (55 kg/tree). Gill *et al.* (2011) observed lowest fruit drop (58.52 per cent) in trees sprayed with 30 ppm NAA, thus resulted in highest fruit yield. Arora and Singh (2014) reported huge potential of growth regulators for improvement of ber. They found significant increase in fruit size with the application of NAA at 300 ppm; whereas the texture, colour and taste were improved with the application of GA_3 at 50 ppm. The maximum fruit yield and fruit retention was obtained by GA (75 mg/L) and CPPU (cytofex, 10 mg/L) sprays compared to the control and other treatments (Kassem *et al.*, 2011). Fruit chlorophyll a, b and total content were increased by all chemical treatments, but the fruit carotenoid content was decreased. Majumder *et al.* (2017) recoimmended that twice sprayings (1st after fruit set during 2nd week of October) and 2nd after 30 days from the 1st spray during 2nd week of November) of GA_3 @ 20 mg/L and H_3BO_3 at 0.4 per cent was found effective for optimizing yield components, yield and quality of ber (cv. BAU Kul-1) in trans-Gangetic plains of West Bengal.

11.0 Pests and Diseases

11.1.0 Pests

11.1.1 Fruit Fly (*Carpomyia vesuviana*)

Fruit fly is the most serious pest of ber. Infestation of fruit fly starts in September with the beginning of fruit set. The adult female lays eggs by inserting its ovipositor in the growing fruit. After 2 to 5 days, larvae come out and feed on the pulp. Infested fruits become deformed, rotting may take place and a large number of affected fruits drop down. Singh and Vashishtha (1985) screened some ber cultivars for resistance to fruit fly and found that Dandan, Gola, Gola Cungaon No. 3, Kaithli and Kakrola Gola were most susceptible while Ilaichi was moderately resistant and Tikadi was most resistant. Mann and Bindra (1976) observed that Chinese, Sanaur-1, Safeda Ilaichi, Mirchia, ZG 3 and Umran cultivars were resistant to fruit fly. Pareek *et al.* (2003) observed Tikidi and Mundia were moderately susceptible cultivars while Pramanick *et al.* (2005) identified Tikidi as highly resistant cultivar against fruit fly.

Control of fruit fly is dependent upon the insecticides application of various nature notably among these are dipterex, imidacloprid, triazophos, and neem products (Abbasi *et al.*, 1992; Singh *et al.*, 2000). Baiting and cultural practices for management of fruit flies have been tried in separate studies but not in an integrated manner (Elnkerlin and Mumford, 1997; Alam *et al.*, 1999; Makhmoor and Singh, 1999).

To minimize the damage, cultivation of orchard soil during summer to expose the hibernating pupae to the sun and birds is recommonded. Dropped and infested fruits should be collected and destroyed. Spraying of the trees with 0.17 per cent rogor (Bakshi and Singh, 1974) or 0.12 per cent malathion (Singh *et al.*, 1973) is helpful for the control of this pest.

Patil *et al.* (1989) reported that fenthion at 0.1 per cent applied 3 times was the most effective against the pest, followed by endosulfan at 0.07 per cent, also applied 3 times. Fenthion also resulted in the highest yields, followed by endosulfan and 0.04 per cent malathion (Patil *et al.*, 1989). When 6 insecticides were tested, Raghumoorthi and Arumugam (1992) in Tamil Nadu, India against *Carpomyia vesuviana* on ber, 2 applications (at the pea stage of the fruits and 15 days later) of 0.1 per cent dichlorvos gave the best results (in terms of reduced fruit infestation), followed by 0.036 per cent monocrotophos, 0.05 per cent malathion and 0.07 per cent phosalone.

The integrated management with dipterex 100 g/acre + 5 per cent molasses baiting, hoeing and collection of fallen fruits throughut the season proved to be the best than sigle treatment applied to trees (Ahmed *et al.*, 2005). Application of neem powder and tobacco extracts significantly redeuced the infestation of fruit fly and they are potential candidates for organic cultivation of ber (Mari *et al.*, 2013).

Arora *et al.* (1999) investigated the physico-chemical characteristics of some ber cultivars in relation to fruit fly incidence and reported that fruit fly infestation was positively correlated with fruit weight, pulp-stone ratio, total soluble solids (TSS) and total sugar, whereas it was negatively correlated to acidity, vitamin C (ascorbic acid) and total phenols. The varieties high in pulp content, TSS, total sugar, low acidity, vitamin C and total phenols were highly susceptible to fruit fly attack. The cultivars Umran (49 per cent incidence), Gola (40 per cent) and ZG 2 (33 per cent) were less susceptible.

11.1.2 Bark Eating Caterpillar (*Indarbela tetraonis* Moore and *I. quadrinotata* Walker)

Gallaries of frassy web near the fork of branches indicate the infestation by this pests. The caterpillar eats away the bark of the tree and severe damage reduces the yield considerably.

To control the pest, the exposed holes should be painted with 10 ml monocrotophos 40 EC or 40 ml of trichlorphon 50 EC (Chundawat and Srivastava, 1978). Application of mixture of one litre kerosene and 100 g soap in 19 litres of water has also been found effective. Inject the holes with quinalphos at 0.01 per cent or fenvalerate at 0.05 per cent and alternate spray of carbaryl at 2.5 g/litre of water or quinalphos at 2ml/litre of water or methomyl at 3.5 g/litre was found to be effective in controlling the caterpillar damage (Singh, 2008). Cotton swab soaked in petrol or kerosene also can place in the holes and sealed with mud.Verma and

Pests of ber

Fruit fly infestation

Adult bark eating caterpillar

Symptom of infestation of bark eating caterpillar

Leaf chaffer infestation in ber

Lac insect in ber

Diseases in ber

Powdery mildew infection

Black fruit spot infection

Anthracnose infection in fruits and leaves

Fruit shrinkage disease infection

Singh (1974) screened 25 cultivars and found that Desi Alwar was resistant and Ilaichi, Dandan, Gola and Kaithli were susceptible to this pest.

11.1.3 Hairy Caterpillar (*Euproctis fraterna*)

The caterpillars feed on leaves in initial stage. The older caterpillars also feed on fruits and tender shoots and damage the trees. The caterpillars can be controlled with 0.25 per cent chlorpyriphos.

11.1.4 Mite (*Larvacarus transitans, Eriophyes cernus*)

The incidence of *Larvacarus transitans* was studied by Sharma (1992) on 4 varieties of ber in Rajasthan. Mites were present from the 1st week of October to the 1st week of August. The maximum number of galls and mite population in each gall were observed in the 1st week of June. Seb was the most susceptible variety, followed by Umran, Mundia and Gola. Among the various ber cultivars evaluated against, *E. orientalis*, 'Chomu' local has been found least susceptible while 'Gola' was the most susceptible and 'Jogia', 'Umran', 'Mundia' and 'Seb' were moderately susceptible (Yadav *et al.*, 2000).

The population and control of the gall-forming mite *Eriophyes cernuus* on *Z. mauri-tiana* was investigated by Mukherjee *et al.* (1994). The pest was found throughout the year, except for July-August. It was most numerous in June. Sizes of galls showed similar trends to those of the population of mites. Of 7 acaricide treatments, dicofol (0.04 per cent) was the most effective, followed by dicofol (0.03 per cent) and carbendazim (0.03 per cent). Monocrotophos (0.03 per cent) was found to be effective in reducing this mite followed by phosphamidon (0.03 per cent) and ethion (0.03 per cent) (Yadav *et al.*, 2003).

11.1.5 Leaf Chafer

There are various specis of of chafer beetle among which *Holotichia consanguinea*, *Anomala* sp. and *Adoretus* sp. (Coleoptera: Scarabidae) are the major one which cause considerable defoliation (Singh, 2008; Karuppaiah *et al.*, 2010; Balikai *et al.*, 2013). These are beetles of light brown colour which feed on the leaves during the night. The pest is most active in the monsoon (June to August) when new growth is produced. Seedling treatment with synthetic chemicals or plant origin insecticides protects them from the attack of chafer beetles (Singh, 2008). Application of quinalphos (0.05 per cent), followed by carbaryl (0.2 per cent) and carbosulfan (0.05 per cent) found to be very effective against larvae of *H. consanguinea* (Patel and Patel, 1999).

11.1.6 Lac Insects (*Laccifer lacca; L. Communis; L. indica*)

The shoots are affected with these small insects. They suck the sap from the branches which ultimately dry up. Infested branches are cut off and destroyed. Trees should be sprayed with rogor 1 ml/l of water or dimecron 1 ml in 3 litres of water.

The effect of infestation by lac insects on the quantity and quality of fruit yield was studied in Haryana, India (Lakra and Kher, 1990). An average of 20.8 per cent trees was infested at the 12 localities investigated. An infestation of 5000 nymphs/100 cm twig caused weight loss in fruits of 52.5-58.5 per cent. Of the 3 varieties studied, Gola was the most susceptible to attack, followed by Kaithli and Umran. A large reduction in the total soluble sugar content of infested fruits (average 53.7 per cent) was also recorded.

11.2.0 Diseases

11.2.1 Powdery Mildew (*Oidium* sp.)

Powdery mildew is the most serious disease of ber and causes heavy losses. The disease appears in October-November when the temperature becomes low and weather is cloudy (humid). Initially white spots of fungus appear on the leaves, fruits, *etc*. In case of severe infestation the powdery mass spreads all over the surface of leaves and fruits. The fruits drop prematurely or become sunken, misshapen and corky. Disease severity is maximum in November, when the mean temperature is 19.20°C and relative humidity is 62.64 per cent (Thind *et al.*, 2004). The disease, if not controlled properly may remain in the orchard unless the temperature rises in February-March causing a marked reduction in yield.

The cultivars, Chhuhara, Safeda Selected, Nazuk, Glory, Sanaur-2, ZG 2, ZG 3, ZG 5 and Chinese showed resistance to powdery mildew (Jayarajan and Cheema, 1972). Reddy *et al.* (1997) evaluated ber varieties Gola, Seb, Mundia, Umran and Kaithli for their resistance to powdery mildew. No variety was resistant to the disease. Of the six fungicides tested, triadimefon (0.1 per cent) was the most effective followed by thiopha-nate-methyl (0.1 per cent) and dinocap (0.1 per cent). Highest yield was recorded in triadimefon (0.1 per cent) treated plants followed by dinocap (0.1 per cent) treated plants.

Gupta *et al.* (1977) recommended spraying of karathane (dinocap) 0.2 per cent or morestan (quinomethionate) at 0.05 per cent for the control of powdery mildew. In addition to karathane, bavistin at 0.05 per cent or wettable sulphur (0.2 per cent) at 10-day interval in October and December were recommended by Yadav *et al.* (1980). Dusting the tree with sulphur at 250 g/tree was also found very effective to control the disease.

Das *et al.* (1994) observed that of all the fungicides tested dinocap, fenarimol, carbendazim, chlorothalonil and triadimefon were equally effective as foliar sprays against *Oidium* sp. on *Z. mauritiana*. The efficacy of carbendazim and dinocap was unaffected when used in conjunction with either of the insecticides malathion or dimethoate. Singh *et al.* (1995) reported the best control by using dinocap (0.2 per cent) of powdery mildew followed by carbendazim (as bavistin), and tridemorph (as calixin). Maximum disease control has been found by combined application of Psendomonas fluorescens (1 per cent) and Karathane @ 0.05 per cent (Nallathambi *et al.*, 2003). The application of bayleton 25 WP (0.05 per cent) was found effective against this disease (Thind and Kaur, 2006). The disease can effectively be managed by alternate sprays of triadimefon at 0.1 per cent followed by wettable sulphur at 0.3 per cent at an interval of 12-15 days (Jamadar *et al.*, 2009). Among many genotypes, Jogia and Mundia were found resistant while, popular cultivar Umran was highly susceptible and another cultivar Kadaka was moderately resistant to powdery mildew. The disease can be controlled by three sprays of karathane (0.05 per cent) or wettable sulphur (0.25 per cent) starting from the flowering time in

September followed by a spray in mid October, November (Singh, 2009). Three sprays of karathane (0.1 per cent) followed by difenoconazole (0.05 per cent) at 15 days interval proved 51.0 per cent and 40.3 per cent disease control against the disease of ber (Anonymous, 2016a).

11.2.2 Sooty Mould or Black Leaf Spot (*Isariopsis indica* var. *ziziphi*)

The symptoms of this disease are the appearance of black spots on the lower surface of the leaves and in advanced stages the entire lower surface may be curved by sooty mould and the leaves may drop down.

Use of resistant varieties ZG-3, Seb, Bahadurgarh and Safeda Rohtak against this disease may be effective. The disease is controlled by spraying the trees with dithane Z-78 (zineb) 0.3 per cent or 0.6 per cent copper oxychloride (blitox), or 0.05 per cent difolatan at fortnightly interval (Sharma *et al.*, 1983).

Singh and Andotra (1989) recorded the best control of leaf spot disease caused by *Isariopsis indica* by sprays of bengard (0.1 per cent), closely followed by topsin-M (thiophanate-methyl), in trials with 6 different compounds. The disease was also effec-tively controlled by bavistin (carbendazim) at 0.1 per cent followed by daconil (chlorothalonil) at 0.2 per cent (Rawal and Saxena, 1989). Two sprays of propiconazole, followed by difenoconazole @ 0.1 per cent was found to be effective against this disease and also increased fruit yield (Anonymous, 2016b).

11.2.3 Alternaria Leaf Spot (*Alternaria alternata*)

The disease is characterized by dark grey spots on lower surface of the leaves. It appears in January-February and the affected leaves may fall off. The disease is controlled by spraying zineb (dithane Z-78). *Pseudomonas fluorescens* (CIAH-196 strain) inhibited the maximum mycelia growth (Nallathambi *et al.*, 2006).

11.2.4 Phoma Leaf Spot

The best control of leaf spot disease caused by *Phoma* sp. on ber was given by topsin M (thiophanate-methyl), applied either before or after inoculation (Dang *et al.*, 1986).

11.2.5 Fruit Rot

The various fungi causing this disease are *Phoma* spp., *Colletotrichum* spp., *Alternaria* spp. and *Pestalotiaversicolor*. Light grey-brown spots appear on the lower end of the fruit. Sometime dark brown rings are formed on the fruits. The disease can be controlled by spraying dithane Z-78 or blitox 0.2 per cent at the early stage of infection.

Although eighteen different fungal species caused post-harvest decay of ber fruits, only *Aspergillus flavus* infection was of significance. This post-harvest pathogen not only resulted in maximum loss of ascorbic acid from the fruit tissues but also induced aflatoxin production during pathogenesis (Singh and Sumbali, 2000).

11.2.6 Integrated Control

A 3-year study in China, showed that the major diseases and pests in the growing period of *Z. jujuba* fell into 3 types and an integrated control was suggested by Sun *et al.* (1989). The major disease was *Phacospora zizyphi-vulgaris*, while the major insect pests were *Sucra jujuba, Ancylis sativa, Carposina sasakii* and *Ceroplastes floridensis*. Integrated control measures included chemical treatment, silviculture and a mixed application of fertilizers (urea or KH2PO4) and pesticides (sumicidin, fenvalerate or omethoate). Large scale demonstrations in Xinzheng, Zhongmu and Weishi counties in China showed that spraying a mixed solution of fertilizers and pesticides before flowering secured a control of 96.6-98.7 per cent for *Sucra jujuba*, 85-96 per cent for *Ancylis sativa* and 97.9-98.9 per cent for date ticks. Three sprayings for *Carposina sasakii* produced 90.5-99.5 per cent control. The disease severity index for *Phacospora zizyphi-vulgaris* was reduced by 84.9-97.5 per cent, after application of metalaxyl and Bordeaux mixture. Maheswari and Haldhar (2018) also described the integrated disease management practices of ber.

12.0 Harvesting and Yield

12.1.0 Harvesting

Harvesting of fruits at proper stage of maturity is very important from quality point of view of ripened fruits. Fruits harvested at correct stage of maturity ripen properly giving characteristic colour, flavour and taste. Under-ripe ber fruits when picked do not ripen properly and never attain the desired quality. Hence the market value of such fruits decreases considerably. Over-ripe fruits, on the other hand, loose their attractive colour and become soft. Such fruits have less storage life and deteriorate in quality.

The maturity indices of ber fruits are attainment of full size of a particular cultivar with softening of pulp and development of characteristic yellow or golden yellow colour. Such fruits will have a desired sugar : acid ratio. Bal and Chohan (1981) reported that the specific gravity of Umran fruits at maturity was less than one. The fruits develop golden yellow colour at maturity. In Delhi cv. Gola took only 150 days whereas Tikadi required 173 days to mature (Raja, 1985). In cvs. Gola, Kaithli and Umran, there was an increase in reducing, non-reducing and total sugar, ascorbic acid, TSS, TSS/acid ratio and organoleptic rating throughout the growth and development of fruits. Increase in starch and acidity (Table 11) was observed only during the growth of fruits which declined with the advancement of fruit development (Bhatia and Gupta, 1985).

The time of harvesting varies due to agro-climatic conditions. In south India, the fruits are ready for harvest in October-November. In Gujarat, the harvesting season is December to March while in Rajasthan it is January to March. In north India the harvesting period is between February and April for different ber cultivars.

Siddiqui and Gupta (1989) determined the maturity standard of cultivars Umran, Gola and Kaithli and studied the rate of ethylene evolution and respiration during ripening. Ethylene evolution and respiration increased with ripening and were highest in Umran. The results suggested that fruit of Kaithli could be harvested at the greenish yellow stage for transportation owing to its high wall pressure. Fruit of Gola could be harvested at any stage for storage because of its low ethylene evolution rate.

Table 11: Change in Acidity, Ascorbic Acid and Starch Content in Fruit of Ber Cultivars during Growth and Development*

Date	Acidity (Per cent)			Ascorbic Acid (mg/100 g pulp)			Starch (per cent)		
	Gola	Kaithli	Umran	Gola	Kaithli	Umran	Gola	Kaithli	Umran
15 Nov.	0.037	0.050	0.050	15.5	12.5	16.5	0.42	0.35	0.61
30 Nov.	0.045	0.062	0.052	19.5	15.5	21.7	0.73	0.69	0.75
15 Dec.	0.097	0.102	0.065	52.0	52.2	45.0	0.68	0.62	0.66
30 Dec.	0.102	0.095	0.112	45.0	46.2	44.7	0.63	0.79	0.84
14 Jan.	0.172	0.120	0.131	45.5	54.5	49.9	0.93	0.85	0.97
29 Jan.	0.270	0.172	0.167	65.2	67.5	59.7	0.93	0.82	0.92
13 Feb.	0.287	0.207	0.302	83.7	86.7	66.1	1.08	0.90	0.95
20 Feb.	0.302	-	-	110.7	-	-	0.94	-	-
27 Feb.	0.285	0.217	0.220	134.0	107.7	74.2	0.87	0.27	1.20
6 March	0.271	0.232	-	132.7	126.0	-	0.45	0.77	-
13 March	0.240	0.248	0.228	126.5	249.5	78.5	0.35	0.74	0.79
20 March	0.201	0.228	-	111.2	150.2	86.7	0.20	-	0.72
27 March	-	0.210	0.207	-	144.2	108.2	-	0.33	0.63
3 April	-	0.164	0.191	-	130.7	120.0	-	0.25	0.41
10 April	-	-	0.179	-	-	111.1	-	-	0.32
17 April	-	-	0.162	-	-	98.4	-	-	0.20
CD at 5 per cent	0.017	0.014	0.11	4.02	3.94	3.50	0.03	0.03	0.03

*Bhatia and Gupta (1985).

The harvested fruits are sorted in different groups as over-ripe fruits, unripe fruits, damaged fruits, misshaped fruits, *etc.* The rest of the fruits which are of correct maturity stage are graded as per size, *e.g.*, large, medium and small and grading helps in marketing of the fruits.

12.2.0 Yield

According to Bakshi and Singh (1974) the average yield from 10- to 20-year-old trees of different ber cultivars varies between 80 and 200 kg/tree. Sixteen-year-old Kaithli tree produced a yield of 120 kg and Mundia Murhara produced 125 kg fruits (Godara *et al.*, 1980). Singh *et al.* (1983) reported the highest yield in Umran (142 kg/tree) and the lowest in Chhuhara (65 kg/tree). Shivakumar *et al.* (2017) obtained maximum fruit yield in Madhavpur (50.8 kg/plant) followed by Umran (47.6 kg/plant) and Topa (42.8 kg/plant)in the Gangetic plains of West Bengal, India.

13.0 Post-harvest Management

Effects of various chemicals and biochemical changes during ripening and storage of ber have also been investigated.

13.1.0 Ripening

Jujuba cv. Zaytoni fruits were dipped for 1 min in solutions containing 0, 250, 500, 1000 or 1500 ppm ethephon and stored for 8 days in perforated polyethylene bags (Abbas *et al.*, 1994). Total soluble solids and ascorbic acid contents were significantly increased following treatment with 500 ppm ethephon, whereas titratable acidity was significantly decreased, and ripening was advanced by 6 days compared with controls. In field trials conducted by Bal *et al.* (1993) for 6 years ethephon was applied to 13-year-old ber cv. Umran trees as a foliar spray at 0, 100, 200, 300, 400 or 500 ppm when fruits were just changing colour. Treatment with 400 or 500 ppm ethephon produced fruits of a deep golden-yellow colour with the lowest specific gravity and acidity and the highest TSS, ascorbic acid and total sugar contents; these fruits were also much more palatable and ripened 2 weeks earlier than untreated fruits.

Goel and Siddiqui (1999) studied changes in enzymes and physico-chemical characteristics in ber fruits of different maturity stages. TSS, acidity and total sugars increased whereas ascorbic acid content, starch, proteins and phenols decreased with ripening. Activities of the enzyme amylase decreased, protease and peroxidase increased, whereas activity of the enzymes polyphenol oxidase and catalase increased up to the fully ripe stage but declined at the over-ripe stage.Fruits of ber cv. Kaithli when stored for 6 days showed that fruits with an attached pedicel had significantly less weight loss than those without a pedicel (23.0 and 40.0 per cent, respectively), and fruits with pedicels ripened to a lesser extent (Siddiqui and Gupta, 1996).

13.2.0 Storage

Yadav *et al.* (2009) observed that pre-harvest sprays of $CaCl_2$ up to 1.5 per cent were effective in extending shelf life of the fruits while reducing PLW and maintain the fruit quality. The higher concentration of $CaCl_2$ (2 per cent) had adverse effect on the quality and shelf life of fruits. In an experiment, Gupta *et al.* (1987) treated calcium at 1.7 g/l as chloride, nitrate, sulphate or phosphate as preharvest spray. Fruits of cv. Umran were packed and stored at ambient temperature. Application of Ca compounds reduced the rate of weight and decay losses and decreased ethylene evolution; the fruits retained high TSS and ascorbic acid contents for longer periods than in the controls, and the treatment also slowed the ripening process. Calcium chloride gave the best results, followed by calcium nitrate. Gupta and Mehta (1988) concluded that spraying with CaNO3 at 1 per cent 10 days before harvest was the most effective treatment for increasing fruit shelf life and quality. In another experiment over-ripening and decay losses in fruit were reduced by preharvest spray of $ZnSO_4$ (Siddiqui *et al.*, 1989).

In order to improve the shelf-life of ber, treatment with chemicals, temperature and also wax emulsion has proved effective. Gupta *et al.* (1989) reported that

Harvesting, Marketing and Packaging of Ber

Harvested ber fruits

Marketing of ber

Foam net packing

Nylon net packing

Drying and Selling of Ber

Drying and selling of fruit at supermarket

preharvest spraying with captafol or thiabendazole (TBZ), each at 500 ppm or with 1 per cent calcium nitrate solution improved the shelf-life of ber cv. Kaithli fruits. On the 12th day of storage, the highest reduction in weight loss was observed with captafol; decay loss was reduced most by TBZ. In another experiment, Siddiqui and Gupta (1989) recorded higher weight loss in fruits treated with calcium chloride and calcium nitrate after a storage period of 9 days, than under control.

In an experiment on the influence of post-harvest treatments on shelf-life of ber fruits cv. Umran, Sandbhor and Desai (1991) harvested fruits at the golden yellow colour stage and dipped for 5 min in GA_3 (25, 50 or 75 ppm) or BA (5, 10 or 15 ppm) solutions. Fruits were stored in the open or in 150 gauge polyethylene bags with no perforations. After 7 days fruits stored in the open showed a weight loss of 14 per cent whereas packed fruits showed a weight loss of 0.86 per cent. Vitamin C content and acidity decreased during storage. TSS and sugar contents increased initially and then decreased to below the initial value. These chemical changes were delayed by GA_3 and BA treatments and polyethylene packaging. Fruits treated with 10 ppm BA and packed in polyethylene had lowest weight loss after 8 days of storage (0.33 per cent) and highest TSS, sugar and vitamin C contents and palatability rating.

Among the chemicals (ascorbic acid, KMnO4 and cycocel) used by Siddiqui and Gupta (1995) on extending shelf-life of ber fruits of cv. Umran, cycocel at 500 ppm was considered the most suitable chemical for improving shelf-life. During storage of jujuba fruits ABA content increased to a peak and then declined. Lower storage temperatures reduced the size of this peak and delayed its appearance. 2,3,5-tri-iodo-benzoic acid (TIBA) prevented the formation of ABA and promoted storage life (Zhang *et al.*, 2000).

Jawanda *et al.* (1980b) also studied the storage of Umran and Sanaur-2 in cold storage at 0-3.3°C and 85-90 per cent humidity. They observed that the cultivar Umran could be stored for 30 days while Sanaur-2 stored well for 40 days in cold storage. Wax-coated and polythene-packed fruits kept better than those subjected to other treatments. Banik *et al.* (1988) reported that fruits kept at 10-12°C and 85-90 per cent RH, and fruits coated with paraffin wax and kept at 10-12°C stored well for up to 18 days with minimum spoilage and physiological weight loss when 100 per cent spoilage occurred in untreated fruits (held at 28-32°C and 70-75 per cent RH) on the 9th day of storage. Ripe fruits of cv. Umran and Sanaur-2, were waxed and stored in perforated paper or poly-thene bag at room temperature for 10-12 days. The storage loss (9.32-9.52 per cent) was minimum in fruits of both cultivars dipped in wax formulations for 30 seconds and stored in polythene bags (Jawanda *et al.*, 1980 b). For extending the keeping quality of ber cv. Umran, Naik and Rokhade (1994, 1997) recommended dipping the fruits in 3 or 6 per cent waxol, packing them in polythene bags and storage at room temperature.

Siddiqui *et al.* (1999) in their storage studies of ber cv. Umran, the fruits (harvested at green mature stage) were stored at room temperature after subjecting them to treatments (a) packing in sealed and unperforated 150 guage polythene bags/modified atmosphere packaging (MAP), (b) MAP + $KMnO_4$ impregnated material, and (c) hot water treatment. The fruits of control and hot water treatment were packed in 4.5 per cent perforated polythene bags. They observed that MAP+$KMnO_4$ were the most effective treatment followed by hot water treatment for enhancing the shelf life (Table 12).

Table 12: Effect of Various Treatments on Physiological Loss in Weight, Decay Loss, Ripening, Fruit Firmness, Total Soluble Solids, Acidity, Ascorbic Acid and Organoleptic Score of Ber Fruits during Storage*

Treatments	*Period of Storage (days)*					
	2	*4*	*6*	*2*	*4*	*6*
	PLW			*Dacay Loss (Per cent w/w)*		
Control	2.25	7.94	12.35	1.74	3.00	6.25
MAP	2.00	6.27	13.38	0.00	1.85	4.50
MAP + $KMnO_4$	1.00	4.40	9.62	0.00	0.00	0.00
Hot water	1.75	5.12	10.53	0.00	1.25	2.75
CD at 5 per cent	N.S.	1.30	1.62	-	-	-
	Ripening (Per cent)			*Fruit Firmness (kg/cm_2)*		
Control	22.5	86.0	100.0	3.0	5.5	4.5
MAP	32.5	97.5	96.0	7.0	6.0	4.7
MAP + $KMnO_4$	12.5	33.7	62.0	8.0	7.3	7.0
Hot water	18.7	61.2	80.0	7.5	7.0	6.0
C.D. at 5 per cent	4.3	5.5	7.7	0.2	0.0	0.2
	TSS (Per cent)			*Acidity (Per cent w/w)*		
Control	14.0	16.5	17.5	0.30	0.35	0.36
MAP	13.2	15.5	16.0	0.30	0.34	0.36
MAP + $KMnO_4$	12.5	13.0	13.6	0.29	0.30	0.31
Hot water	13.0	14.5	15.0	0.29	0.32	0.33
C.D. at 5 per cent	0.8	1.4	1.6	0.3	0.3	0.4
	Ascorbic Acid (mg/100 g)			*Organoleptic Score (10 points basis)*		
Control	90.4	80.2	78.3	8.4	6.4	5.0
MAP	90.8	82.1	76.1	8.0	6.0	4.8
MAP + $KMnO_4$	90.9	86.2	81.4	7.8	8.0	7.0
Hot water	90.9	83.7	80.2	7.0	7.5	6.5
C.D. at 5 per cent	N.S.	2.0	2.0	N.S.	1.2	1.0

Initial values of fruits firmness = 9.2 kg/cm^2, TSS = 12 per cent, acidity = 0.27 per cent, ascorbic acid = 82.5 mg/100 g and organoleptic score = 7.5

N.S. = Not significant.

*Siddiqui *et al.* (1999).

Chen *et al.* (2019) found that carnauba wax (CW) and CW containing glycerol monolaurate (CW-GML) coating reduced weight loss, respiration rate and ethylene production, while lowering the activities of various enzymes like polygalacturonase, pectin methylesterase and cellulose to delay the flesh softening. Hoever, these two coatings also delayed the change in skin colour and ensure higher amount of chlorophyll and ascorbic acid. The decay index was 44.3 per cent of the control fruit and 65.6 per cent of the CW-GML coated fruit after 12 days of storage at 20°C. Thus CW-GML coating could be an effective measure to extend the shelf life without deteriorating the fruit quality. The corrugated fiberboard boxes proved best for trabsportation and storage as fruits packed in CFB boxes showed minimum spoilage throughout the storage period while maintaining the fruit quality. These fruits can be stored in CFB boxes up to 6 days without compromising its quality (Lal and Fageria, 2004).

Mature green Z. *mauritiana* cv. Zaytoni fruits were gamma-irradiated at 0, 10, 30 or 50 krad and the fruits were then kept at 20°C and 80-90 per cent RH. Fruits subjected to 30 krad were firmer and greener than unirradiated control fruits after 6 days of storage, and this treatment delayed ripening by 3 days (Abbas *et al.*, 1992). Gamma irradiation treatment up to 1.0kGy significantly increased antioxidant and total flavonoid content compared to fresh ber fruit cv. Gola. On the other hand it brought down the reducing power activity (65 per cent) and total phenolic content (18 per cent) as compared to raw fruit (Kavitha *et al.*, 2014).

13.3.0 Value Addition

Value addition helps to enhance the shelf life of the product along with adding economic value to it. Traditionally ber fruit is dried and pickled in almost every household as a common Grandma technology. Products like jam, candy, preserve, powder, murabba, osmotically dehydrated ber, beverages, wine, pickle *etc.* can be prepared from ber (Khurdia and Sinoh, 1975; Khurdia, 1980 a,b; Chawan *et al.*, 1993; Devi *et al.*, 2015; Dalal *et al.*, 2019; Neeraj and Kaul, 2011; Sharma and Rana, 2016). The formulations of jam, jelly, chutney and pickles were processed from pulp of Jujube and BAU kul (Uddin and Hussain, 2012; Dubey *et al.*, 2014), spray dried Jujube juice powder (Singh *et al.*, 2013) and evaluated the nutritional composition of fresh jujube fruits with an objective to popularize among the farmers and the industry.

14.0 Biotechnology

Ber is an important arid fruit with ample scope for application of biotechnology. The major problems confronting ber cultivation are the attack of fruit fly and black leaf-spot. There is considerable scope for the use of molecular markers to differentiate genotypes and characterize genes. Somatic embryogenesis and plant regeneration need to be perfected and gene transfer has to be attempted for ber improvement (Raj Bhansali, 2001).

14.1.0 Somatic Embryogenesis

A method of direct somatic embryogenesis was developed and conditions for regeneration of 3 cultivars of Chinese date (Z. *jujuba*) *in vitro* were studied by Mitrofanova *et al.* (1997). The highest number of somatic embryos was obtained on half-strength MS medium supplemented with 2.2 μM 2,4-D. In darkness at 26°C, the frequency of embryogenesis was 90 per cent. Plant regeneration was obtained on modified Pierik medium without hormones. Plantlets were acclimatized *in vitro*. Sung and Song (1992) have reported the induction of somatic embryogenesis inChinese jujuba from immature cotyledons and seedling parts on MS medium containing 2.5 mg/l 2,4-D, BAP and 50 per cent sucrose and complete regeneration on 1 mg/l zeatin and 0.01 mg/l 2,4-D. Embryos developed into rooted plantlets on medium supplemented with 2-5 mg/l BAP without auxin under 16-h photoperiod. Toxin resis-tant cell lines against *Fusarium oxysporum* and *Colletotrichum* species were also developed in *Z. rotundifolia* rootstock (Raj Bhansali, 1999).

Regeneration of plantlets *in vitro* from leaf and cotyledon explants of *Z. mauritiana* was reported by Mathur *et al.* (1993) when cultured on MS medium supplemented with BA or kinetin at 2.5 mg/l and silver nitrate (40-50 mg/l). Root and hypocotyl callus regenerated with relatively high concentrations of NAA (2.5 mg/l) and BA (0.6 mg/l), respectively.

Fu *et al.* (2007) reported that immature embryos or tender leaves are suitable for regeneration. To generate embryogenic calli, immature embryos were cultured on MS medium supplemented with TDZ @ 0.05 mg/l + PP333 @ 0.5 mg/l + AgNO3 @ 10 mg/l + Vc @ 100 mg/l + sucrose @ 30 g/l and these calli spontaneously proceeded to high frequency somatic embryogenesis; whereas the tender leaves from adult tree produced embryogenic calli on the same medium but an extra subculture was needed on MS medium containing KT at 4 mg/l, NAA at 0.03 mg/l and sucrose at 50 g/l to start the somatic embryogenesis. Vigorous proliferation was observed from these two kind of somatic embryoids on MS medium supplemented with TDZ @ 0.05 mg/l + PP333 @ 0.5 mg/l+ $AgNO_3$ @ 10 mg/l + Vc @ 100 mg/l + Ac @ 100 mg/l+ sucrose @ 30 g/l. Successful germination followed by conversion to whole plantlets was reported on MS+ KT at 1 mg/l + 6-BA at 1 mg/l + NAA at 0.2 mg/l + GA_3 at 2 mg/l + AC at 2 g/l + sucrose at 30 g/l after 6 days of dehydration.

Sudershan *et al.* (2013) reported production of very compact embryogenic callus after incubation in 2 weeks in darkness at 25°C, followed by 4 weeks in 16hour light at 25°C. Further, proliferation and maintenance of embryogentic callus was on MS medium containing 3 mg/l kinetin with a regular subculture in every 15 days. Hormone free MS medium was suitable for successful somatic embryo multiplication.

14.2.0 Biochemical and Molecular Markers

Work on the use of isozymes (Gupta and Srivastava, 1966; Kumar *et al.*, 1995; Oh *et al.*, 1988) has indicated the use of biochemical markers for classification of ber genotypes. Molecular marker using PCR-RFLP has been used to characterize and identify phytoplasma infecting ber (Han *et al.*, 1996). These techniques can be used to study the phylogenetic linkage and gene tagging.

Singh *et al.* (2006) worked with 33 accessions of *Z. mauritiana* and 8 accessions of its wild relative *Z. nummularia*. Eleven primar pairs used in AFLP analysis detected 952 fragments, of which 783 (83.8 per cent) showed polymorphism. Simmilarity coeffient ranging from 0.14-0.86 suggested that 41 accessions were divergent. The wild relative showed greater diversity than the cultivated ones. Cluster analysis revealed the complete separation of the 33 cultivated accessions and 8 wild relative accessions into 2 distinct groups.

Singh *et al.* (2007) also studied the genetic diversity among 47 ber accessions of *Z. mauritiana* and one wild accession of *Z. nummularia* using ISSR markers. Genetic similarity ranging from 43.07-90.30 per cent indicated that these 48 acessions were divergent.This was the first report on the exploitation of ISSR markers in ber for genetic diversity analysis.

High level of polymorphism of 37 cultivars of *Z. mauritiana* of Indian origin among SCoT (61.6 per cent) and ISSR (61 per cent) markers with higher PIC values of SCoT (63.1-90.4 per cent) and ISSR primers (47.3-88.8 per cent) were recorded. Dendograms revealed similarity coefficient ranging from 0.80-0.92 and 0.79-0.96 for SCoT and ISSR, respectively (Singh *et al.*, 2017).

The genetic relationship among 17 genotypes of *Z. mauritiana* was studies using ISSR markers. A total of 191 scorable bands were detected out of which 162 fragments were polymorphic (84.82 per cent) and remaining 29 showed monomorphism (15.18 per cent). Genetic similarity of these genotypes were 37-76 per cent indicating out of the total genotypes used, 18 genotypes had broad genetic base and considered as divergent. These divergent genotypes were clustered into 3 different clusters using UPGMA method. Cluster 3 was largest among these containing 3 subclusters and 13 genotypes. The study indicated that the cultivar 'Suffon' had maximum genetic diversity as it remained independent during cluster formation (Ahmad *et al.*, 2019).

Liang *et al.* (2019) developed 55 polymorphic nuclear SSR markers based on restriction-site associated DNA sequences ans traanscriptome sequencing. From these they have selected 14 robust nSSR markers to analysis 117 accessions. Cluster analysis showed that the most of the accessions clustered consistently with their geographic region.

15.0 References

Abbas, M.F., Al-Niami, J.H. and Al-Sareh, E.A. (1992) *J. Hort. Sci.*, **67** : 627-629.

Abbas, M.F., Al-Niami, J.H. and Al-Sareh, E.A. (1994) *J. Hort. Sci.*, **69** : 465-466.

Abbas, M.F., Al-Niami, J.H. and Saggar, R.A.M. (1994) *Dirasat. Series B, Pure and Appl. Sci.*, **21** : 171-181.

Abbasi, Q.D., Hussain, T., Talpur, M.A., Ahmed, M., Shakoori, A.R. and Memon, M. (1992) *Proc. Pakistan Congr. Zool.*, **12** : 357-361.

Ahlawat, V.P., Sindhu, S.S., Gupta, O.P. and Thareja, R.K. (1990) *Haryana J. Hort. Sci.*, **19** : 106-111.

Ahmad, B., Anjum, R., Ahmad, A., Yousaf M.M., Hussain, M. and Muhammad, W. (2005) *Pak. Entomol.*, **27** (2): 1-2.

Ahmad, R., Malik, W. and Anjum, M.A. (2019) *Erwerbs-Obstbau.*, **61** : 345-353.

Ahmedov, V.A. and Halmatov, H.H. (1969) *Rast. Resursy*, **5** : 578-581.

Akhtar, Y., Choudhary, R., Sharma, K. Chandra and Sharma, Manju (2017) *Pharmacogn J.*, **9**(3): 417-424.

Akter, A. and Rahman, H. (2019) *Res. Rev.: J. Agric. Sci. Tech.* **8**(1): 11-23.

Alam, M.Z., Akhtaruzzaman, M. and Sardar, M.A. (1999) *Ann. Bangladesh Agric.*, **9**(2): 155-164.

Al-Niami, J.H., Saggar, R.A. and abbas, M.F. (1992) *Sci. Hort.*, **51** : 303-308.

Anonymous (1966) *Ann. Rep. Agric. Univ. Dept. Hort.*, Punjab Agric. Univ., Ludhiana.

Anonymous (2016a) In: *Annual Report, ICAR-All India Co-ordinated Research Project on Arid Zone Fruits*, pp. 106-107.

Anonymous (2016b) In: *Annual Report, ICAR-All India Coordinated Research Project on Arid Zone Fruits*, pp. 104-105.

Anonymous (2019) In: *Indian Horticulture Database* 2019, NHB, MoA, GoI, Gurgaon, Haryana, India.

Arora, P.K., Nirmal, K., Batra, R.C. and Mehrotra, N.K. (1999) *J. Applied Hort. (Lucknow)*, **1** : 101-102.

Arora, R. and Singh, S. (2014) *Agric. Sci. Dig. - A Res. J.*, **34**(2): 102-106.

Asatryan, A. and Tel-Zur, N. (2013) *Flora*, **208** : 390-399.

Aulakh, P.S., Mehrotra, N.K. and Vij, V.K. (2000) *Indian J. Hort.*, **57** : 123-124.

Awasthi, A.N. and Misra, R.S. (1969) *Punjab Hort. J.*, **9** : 54-60.

Awasthi, O.P., Pathak, R.K. and Pandey, S.D. (1994) *Hort. J.*, **7** : 51-54.

Awasthi, O.P., Pathak, R.K. and Pandey, S.D. (1995) *Indian J. Agric. Sci.*, **65** : 363-367.

Babu, R.H. and Kumar, P.S. (1988) *Haryana J. Hort. Sci.*, **17** : 24-27.

Bajwa, G.S., Bal, J.S., Brar, S.S. and Minhas, P.P.S. (1993) *Proc. Indian Soc. Weed Sci. Int. Symp.*, Hisar, India, November 18-20, Vol. III, pp. 225-227.

Bajwa, G.S., Sandhu, H.S. and Bal, J.S. (1989) *Haryana J. Hort. Sci.*, **16** : 209-213.

Bajwa, G.S., Sidhu, M.S. and Shergill, T.S. (1990) *Haryana J. Hort. Sci.*, **19** : 299-301.

Bajwa, M.S. and Sarowa, P.S. (1977) *Punjab Hort. J.*, **17** : 52-53.

Bajwa, M.S., Singh, P. and Singh, R. (1972) *Indian Hort.*, **14** : 7-12.

Bakhshi, J.C. and Singh, P. (1974) *Indian Hort.*, **19** : 27-30.

Bal, J.S. (1992) *Acta Hortic.*, **317**: 245-254.

Bal, J.S. and Chohan, G.S. (1981) *Punjab Hort. J.*, **21** : 70-74.

Bal, J.S., Kahlon, P.S., Jawanda, J.S. and Sandhu, S.S. (1992) *Acta Hortic.*, **1**: 318–325.

Bal, J.S. and Randhawa, J.S. and Jagjit, S. (2004) *J. Res., Punjab Agric. Univ.*, **41**(2) : 210-213.

Bal, J.S., Randhawa, J.S. and Singh, S.N. (1988) *Haryana J. Hort. Sci.*, **17** : 20-23.

Bal, J.S., Singh, M.P. and Sandhu, A.S. (1997) *J. Res. Punjab Agri. Univ.*, **34** : 60-63.

Bal, J.S., Singh, S.N., Randhawa, J.S. and Jawanda, J.S. (1984) *Indian J. Hort.*, **41** : 182-185.

Bal, J.S., Singh, S.N., Randhawa, J.S. and Sharma, S.C. (1982) *Prog. Hort.*, **14** : 148-151.

Balikai, R.A., Kotikal, Y.K. and Prasanna, P.M. (2013) *Acta Hortic.*, **993**: 253-277.

Banik, D., Hore, J.K. and Sen, S.K. (1988) *Haryana J. Hort. Sci.*, **17** : 49-55.

Bankar, G.J. and Prasad, R.N. (1993) *Prog. Hort.*, **22** : 60-62.

Bhansali, A.K. (1975) *Ph.D. thesis*, University of Jodhpur, India

Bhargava, B.S., Raturi, G.B. and Hiwale, S.S. (1990) *Singapore J. Prim. Indus.*, **18** : 85-95.

Bhati, B.S. and Yadav, P.K. (2003) *Progress. Agric.*, **3**(1/2) : 148-149.

Bhatia, S.K. and Gupta, O.P. (1985) *Punjab Hort. J.*, **25** : 62-66.

Bi, P., Niu, Z.M., Wang, X.P. and Lai, F.M. (1996) *Acta Hortic.*, **423** : 8-12.

Bisla, S.S. and Daulta, B.S. (1988) *Indian J. Hort.*, **45** : 29-33.

Bisla, S.S., Dhiman, B.K. and Chharia, A.S. (1991) *Haryana J. Hort. Sci.*, **20** : 26-30.

Bohane, L., Tiwari, R. and Gautam, K.K. (2016) *Indian J. Hort.*, **73**(1): 128-132.

Brar, J.S., Gill, K.S., Kaur, T. and Kaur K. (2020) *Indian J.Weed Sci.*, **52** (1): 58-63.

Byun, J.H. and Chang, K.H. (1991) *J. Korean Soc. Hort. Sci.*, **32** : 81-86.

Chadha, K.L., Gupta, M.R. and Bajwa, M.S. (1972) *Indian J. Hort.*, **29** : 137-150.

Chawan, U.D., Adsule, R.N. and Kadam, S.S. (1993) *Bev. Food World*, **20**(9) : 28-29.

Chen, H., Sun, Z. and Yang, H. (2019) *Sci. Hortic.*, **244** : 157-164.

Chen, J., Liu, X., Li, Z., Qi, A., Yao, P., Zhou, Z., Dong, T.T. and Tsim, K.W. (2017) *Evidence-Based Complementary and Alternative Med.*, pp. 1-10.

Chiou, C.Y. and Huang, M.T. (1994) *Chinese J. Agromet.*, **1** : 115-120.

Chiou, C.Y. and Weng, J.H. (1996) *Chinese J. Agromet.*, **3** : 121-124.

Choudhary, R., Garhwal O.P., Choudhary H.D. and Choudhary I. (2020) *Curr. Hort.*, **8**(1): 44–46.

Chovatia, R.S., Patel, D.S., Patel, A.T. and Patel, G.V. (1992) *Gujarat Agric. Univ. Res. J.*, **17** : 56-60.

Chundawat, B.S. and Srivastava, H.C. (1978) *Haryana Agric. Univ. Publication, Hisar*, p. 38.

Dahiya, S.S. and Dhankhar, O.P. (1984) *Haryana J. Hort. Sci.*, **11** : 53-58.

Dalal, N., Neeraj and Bisht, V. (2019) *Int. J. Curr. Microbiol. App. Sci.*, **8**(1) : 1603-1615

Dang, J.K., Sharma, S. and Singh, D. (1986) *Indian J. Mycol. Plant Pathol.*, **16** : 222-223.

Danthu, P., Toure, M. and Soloviev, P. (2004) *Agrofor. Syst.*, **60** : 247-253.

Das, N.D., Rao, M.S. and Sankar, B.R.M. (1994) *Annals Plant Prot. Sci.*, **2** : 76-78.

Daulta, B.S. and Chauhan, K.S. (1982) *Indian Hort.*, **27** : 7-9.

De Candolle, A. (1886) *Bibliotheque Scientifique Internationale*, 43 Paris, p. 385.

Devi, M.P., Bhowmick, N., Bhanusree, M.R., and Ghosh, S.K. (2015) In: *Value Addition of Horticultural Crops: Recent Trends and Future Directions*, Springer India, pp. 13-41.

Dhaliwal, J.S. and Bal, J.S. (1998) *J. Res. Punjab Agric. Univ.*, **35** : 36-40.

Dhankhar, O.P., Makhija, M. and Singhrot, R.S. (1978) *Intern. Symp. Arid Zone res. Development*, Jodhpur, February 14-16.

Dhatt, A.S., Grewal, G.P.S. and Dhillon, W.S. (1993) *Punjab Hort. J.*, **33** : 70-75.

Dinesh, K. (2002) *Indian J. Agric. Sci.*, **72**(11) : 659-660.

Du, X.M., Guo, H.P., Zhao, Y.J., Hu, X.H. and Zhu, W.R. (1997) *China Fruits*, **4** : 26-27.

Dubey, H., Parihar, P. and Kumar, S. (2014) *JNKVV Res. J.*, **48**(2) : 203-206

Elnkerlin, W. and Mumford, J. (1997) *J. Eco. Entomol.*, **90**(5): 1066-1072.

Evreinoff, V.A. (1964) *J. d' agric. tropicale et de botanique appliqué*, **11**: 177-187.

Fu, J.X., Liu, C.M., Hu, G.B., Tan, J.N. and Zhang, G.D. (2006) *Acta Hortic.*, **764** : 205-212.

Gao, Q.H., Yu, J.G., Wu, C.S., Wang, Z.S., Wang, Y.K., Zhu, D.L. and Wang, M. (2014) *PLoS One.*, **9**(2) : e88912.

Gao, W.H., Liu, X.G. and Wang, C.Z. (2009) *Acta Hortic.*, **840**: 197–202.

Ghosal, S. (2013) Soil fertility management of established Ber (*Zizyphus mauritiana*) plantation under lac cultivation. *The Bioscan*, **8**(3): 787- 790.

Ghosh, S.N. (2009) In : *International Jujube Symposium.*, **840** : 321-326.

Ghosh, S.N., Bera, B., Kundu, A. and Roy, S. (2009) *Acta Hortic.*, **840**: 357-362.

Ghosh, J., Lohot, V.D., Singhal, V., Ghosal, S. and Sharma, K.K. (2015) *The Bioscan*, **10**(4): 1905-1908.

Ghosh, S.N. and Sen, S.K. (1988) *South Indian Hort.*, **36** : 260-261.

Gill, K.S. and Bal, J.S. (2011) In : *II International Jujube Symposium*, 993, pp. 119-124.

Godara, N.R. (1981) *Thesis Abst.*, Haryana Agric. Univ., **7** : 65-66.

Godara, N.R., Chauhan, K.S. and Bisla, S.S. (1980) *Haryana J. Hort. Sci.*, **9** : 101-105.

Godge, P.V. and Kale, P.N. (1991) *J. Maharashtra Agric. Univ.*, **16** : 224-225.

Goel, A. and Siddiqui, S. (1999) *Indian J. Agric. Res.*, **33** : 209-213.

Goyal, D., Bhadauria, S. and Kumar, A. (2006) *Indian J. Plant Physiol.*, **11**(2) : 178-181.

Goyal, M., Nagori, B.P. and Sasmal, D. (2012) *Spatula DD.*, **2**(2) : 107-116.

Goyal, Y. and Arya, H.C. (1985) *J. Pl. Physiol.*, **119** : 399-404.

Grewal, S.S., Gurmail Singh and Dhallon, W.S. (1993) *Punjab Hort. J.*, **33** : 76-83.

Guil-Guerrero, J.L., Diaz Delgado, A., Matallana Gonzalez, M.C. and Torija Isasa, M.E. (2004) *Plant Foods Human Nutr.*, **59**: 23-27.

Guissou, T., Ba, A.M., Guinko, S., Plenchette, C. and Duponois, R. (2000) *Fruits (Paris)*, **55** : 187-194.

Guo, Y.X. (1999) *China fruits*, **1** : 25-26.

Gupta, J.H., Nath, Ram and Srivastava, V.P. (1977) *Prog. Hort.*, **9** : 81-83.

Gupta, M.R. and Singh, S. (1977) *Punjab Hort. J.*, **17** : 54-57.

Gupta, N. and Srivastava, P.S. (1996) *J. Plant Biochem. Biotechnol.*, **5** : 87-90.

Gupta, O.P. and Neema Mehta (1988) *Haryana J. Hort. Sci.*, **17** : 183-189

Gupta, O.P., Siddiqui, S. and Chauhan, K.S. (1987) *Indian J. Agric.*, **1** : 65-70.

Gupta, O.P., Siddiqui, S. and Gupta, A.K. (1989) *Res. Develop. Rep.*, **6** : 35-40.

Gupta, R.B. and Godara, N.R. (1989) *Prog. Hort.*, **21** : 15-20.

Gupta, T. (1983) *Int. Tree Crops J.*, **2** : 155-194.

Han, M.S., Noh, E.W., Lee, J.S., Chung, H.S., Young, Y. and Yun, J.K. (1996) *Res. Rep. Forest Gent. Res. Inst.*, Suwon, **32** : 72-79.

Hao, Z.N. and Zeng, X. (1991) *Acta Hort. Sinica*, **18** : 102-108.

Hatta, M., Beyl, C.A., Garton, S. and Diner, A.M. (1996) *J. Hort. Sci.*, **71** : 881-886.

Hiwale, S.S. and Raturi, G.B. (1996) *Prog. Hort.*, **25** : 161-163.

Hoang, V.T. and Tuynh, N.V. (1989) *Mutation Breed. Newsletter*, **34** : 13.

Hooda, P.S., Singhu, S.S., Mehta, P.K. and Ahlawat, V.P. (1990) *J. Hort. Sci.*, **65** : 589-593.

Hore, J.K. and Sen, S.K. (1994) *Indian J. Agric. Res.*, **28** : 285-289.

Ismail, S.M. and Almarshadi, M.H.S. (2013) *J. Food Agric. Environ.*, **11**(1) : 373-378.

Jain, B.L., Goyal, R.S. and Pareek, O.P. (1988) *Indian J. Hort.*, **45** : 203-207.

Jamadar, M. M., Balikai, R. A. and Sataraddi, A. R. (2009) *Acta Hort.*, **840** : 383-390.

Jarasenko, M.T. and Shaumarov, K.H.B. (1977) *Izvestiya Timirya Zevskoi Skokhoxyaistvennoi Akademi*, **3** : 150-153.

Jasmine, J.A. (2012) *The Asian J. Hort. Res.*, **7**(2): 315-317.

Jawanda, J.S., Bal, J.S., Josan, J.S. and Mann, S.S. (1980b) *Punjab Hort. J.*, **20** : 171-178.

Jayarajan, R. and Cheema, S.S. (1972) *Indian J. Hort.*, **29** : 353.

Johnston, M.C. (1972) In: *Flora of Tropical East Africa* (Eds. E. Milne-Redhead, and R. M. Polhill), Crown Agents, London.

Josan, J.S., Jawanda, J.S., Bal, J.S. and Singh, R. (1981) *Punjab Hort. J.*, **21** : 61-64.

Joshi, N.J. (1960) *Indian For.*, **86** : 169-171.

Kajal, R.S. (1983) *Thesis Abst.*, Haryana Agric. Univ., **9** : 176.

Kale, V.S., Dod, V.N., Adpawar, R.M. and Bharad, S.G. (2000) *Crop Res. (Hisar)*, **20**(2) : 327-333.

Karimpour, S., Gavarynejad, G.H., Rouhbakhsh, H. and Ardakani, E. (2013) *Adv. Environ. Biol.*, **7**(3) : 501-505.

Karuppaiah, V., More, T.A. and Bagle, B.G. (2010) *Karnataka J. Agric. Sci.*, **23**(1): 180-181.

Kashi, A., Hosseinzadeh, S., Babalar, M. and Lessani, H. (2004) *J. Sci. Technol.Agric Nat. Resour.*, **7**(4) : 1-10.

Kassem, H.A., Al-Obeed, R.S., Ahmed, M.A. and Omar, A.K.H. (2011) *Middle-East J. Scientific Res.*, **9** (5) : 628-637.

Katiyar, P.N., Yadav, V. and Singh, J.P. (2010) *Ann. Hortic.*, **3**(1) : 92-94.

Kaundal, G.S., Grewal, S.S. and Bal, J.S. (1984) *J. Res. Punjab Agric. Univ.*, **21** : 372-374.

Kavitha, C., Kuna, A., Supraja, T., Blessy, S.S., Padmavathi, T.V.N. and Prabhakar, N. (2014) *J. Food Sci. Technol.* DOI 10.1007/s13197-014-1359-x

Keerthi, M., Venkateswararao, P., Devi, P.L.A., Laxmi, G.K.M. and Venu, P. (2016) *The Pharma Innovation.*, **5**(6): 107-109.

Khan, H., Sivalingam, P.N., Chauhan, S., Awathi, O.P. and More, T.A. (2013) *Sci. Hortic.*, **150**: 164-171

Khoshoo, T.N. and Singh, N. (1963) *Silvae Genetica*, **12** : 141-180.

Khurdiya, D.S. (1980a) *J. Food Sci. Tech.*,**17** : 158-159.

Khurdiya, D. S. (1980b). *J. Food Sci. Tech.*,**17**: 127-130.

Khurdiya, D.S. and Sinoh, R.N. (1975) *Indian Hort.*, **20**: 5-25.

Kim, D.S. and Lee, S.P. (1988) *J. Korean For. Soc.*, **77** : 445-452.

Kim, Y.K. and Suh, C.K. (1989) *J. Korean Soc. Hort. Sci.*, **7** : 136-137.

Kim, Y.S. and Kim, W.S. (1984) *J. Korean Soc. Hort. Sci.*, **25** : 28-36.

Krishnan, B.M. and Kulasekaran, M. (1984) *South Indian Hort.*, **32** : 153-154.

Krivencov, V.I., Karahanova, S.V. and Savina, G.G. (1970) *Bjull. Gos. Nikitsk. Bot. Sada*, **2** : 57-59.

Kumar, A.V. and Sivakumar, N. (2008) *Natural Prod. An Indian J.*, **4**(2) : 132-139.

Kumar, N., Tikkoo, A. and Mehta, P.K. (1990) *Haryana J. Hort. Sci.*, **19** : 89-95.

Kumar, S., Prasad, R., Shukla, A., Dev, I., Chaturvedi, O.P. and Kumar, A. (2018) *Indian J. Agroforestry*, **20** (2): 58-62.

Kumar, S. and Tung, T.S. (1981) *Proc. Nat. Sem. Trop. Subtrop. Fruit Crops*, Bangalore, India, p. 58.

Kumar, S., Ram, S.N. and Baig, M.J. (2002) *Range Manag. Agrofor.*, **23**(1) : 59-62.

Kumar, S., Sharma, V.P. and Kapoor, H.C. (1995) *J. Hort. Sci.*, **70** : 303-306.

Kundi, A.H.K., Wazir, F.K., Abdul, Ghafoor and Wazer, Z.D.K. (1989) *Sarhad J. Agri.*, **5** : 149-155.

Kundu, S.S., Pareek, O.P. and Gupta, A.K. (1995) *Haryana J. Hort. Sci.*, **24** : 23-30.

Lakra, R.K. and Kher, S. (1990) *Indian J. Plant Prot.*, **18** : 125-127.

Lal, G. and Fageria, M.S. (2004) *Indian J. Plant Physiol.*, **9**(2) : 199-202.

Lal, G., Pareek, C.S., Sen, N.L. and Soni, A.K. (2003) *Indian J. Hortic.*, **60**(2) : 158-162.

Li, B.Y. and Wang, Z.M. (1995) *J. Fruit Sci.*, **12** : 41-43.

Li, D.K., Sun, B.Y., Liu, X.B., Bi, X.M. and Zhang, X.P. (1996) *China Fruits*, **2** : 11-12.

Li, J. W., Fan, L. P., Ding, S. D. and Ding. X. L. (2007) *Food Chem.*, **103**(2): 454-460.

Li, X.Z. and Chen, H.S. (1999) *South China Fruits*, **28** : 48.

Liang, T., Sun, W., Ren, H., Ahmad, I., Vu, N. and Huang, J. (2019) *Planta.*, **246**(6) : 1875-1887.

Liu, M. J. and Cheng, C. Y. (1995) *Acta Hort.*, **390**: 161-165.

Liu, M.J. (1999) *Acta Hort. Sin.*, **26** : 302-308.

Liu, M.J. and Wang, Y.H. (1991) *China Fruits*, **4** : 23-25.

Liu, M.J., Zhou, J.Y., Zhao, J., Wang, J., Liu, P., Dai, L. (2006) *Acta Hortic. Sin.*, **33**: 687.

Ma, Y.K., Wang, S.Y., An, B.G., Sun, J.J. and Yao, F.J. (2000) *China Fruits*, **1** : 22, 38.

Maheswari, P. and Singh Umrao (1965) *Dictionary of Economic Plants in India*, ICAR, New Delhi.

Majumder, Indrani, Sau, Sayan, Ghosh, Bikash, Kundu, Subhasis, Roy, Debjit, and Sarkar, Sukamal (2017) *J. Appl. Nat. Sci.*, **9** (4): 2404-2409.

Makhmoor, H.D. and Singh, S.T. (1999) *Ann. Pl. Protec. Sci.*, **7**(1): 33-36.

Mankar, S.W., Dod, V.N. and Bharad, S.G. (1997) *Crop Res. (Hisar)*, **14** : 437-438.

Mann, G.S. and Bindra, O.S. (1976) *Punjab Hort. J.*, **16** : 64-67.

Mari, J.M., Chachar, Q.I. and Chachar, S.D. (2013) *Intl. J. Agric. Tech.*, **9**(1): 125-133.

Masalkar, S.D. and Wavhal, K.N. (1991) *Maharashtra J. Hort.*, **5** : 37-40.

Mathur, N. and Vyas, A. (1999) *J. Plant Physiology*, **155** : 129-132.

Mathur, N., Ramawat, K.G. and Nandwani, D. (1995) *Plant Cell Tissue Organ Cult.*, **43** : 75-77.

Mathur, N., Ramawat, K.G. and Sonie, K.C. (1993) *Gartenbauwissenschaft*, **58** : 255-260.

Mawani, B. and Singh, S.P. (1992) *Hort. J.*, **5** : 79-83.

Meel, O.P., Chharia, A.S., Ranjeet Kumar and Joon, M.s. (1991) *Haryana J. Hort. Sci.*, **20** : 65-72.

Meghwal, P.R., Vashistha, B.B., Singh, Akath, Prasad, R.N. and Singh, S.K. (2019) *Indian Hort.*, Sept-Oct issue, pp. 11-13.

Maheshwari, S.K. and Haldhar, S.M. (2018) In: *Disease Management in Arid Horticultural Crops*. CIAH/Tech./Pub. No. 68, ICAR-Central Institute for Arid Horticulture, Bikaner, Rajasthan, India, p. 42.

Mirzaee, M.R. (2014) *Arch. Phytopathol. Plant Prot.*, **47**: 82-89.

Mishra, T. and Bhatia, A. (2014) *Int. J. Pharmacol. Res.*, **4**(3) : 111-116.

Mitra, S.K.; Pathak, P.K. and Chakroborty, I. (2008) *Acta Hort.*, **770:** 67-76.

Mitrofanova, I.V., Mitrofanova, O.V. and Pandei, D.K. (1997) *Russian J. Plant Physiol.*, **44** : 108-114.

Morton, J. (1987) In : *Fruits of warm climates, Julia F. Morton*, Miami, FL, pp. 272-275.

Mukherjee, I.N., Singh, R.K. and Singh, J. (1994) *Indian J. Agric. Sci.*, **64** : 343-345.

Mukherjee, S., Paliwal, R. and Pareek, S. (2004) *J. Hortic. Sci. Biotech.*, **79**(6) : 991-994.

Murthy, B.B. and Reddy, Y.N. (1989) *Indian J. Agri. Sci.*, **59** : 269-270.

Murthy, B.N.S. and Reddy, Y.N. (1989) *J. Maharashtra Agric. Univ.*, **14** : 296-298.

Murthy, B.N.S. and Reddy, Y.N. (1990) *Seed Sci. Technol.*, **18** : 621-627.

Murthy, B.N.S., Murti, G.S.R. and Reddy, Y.N. (1989) *Indian J. Plant Physiol.*, **32** : 275-277.

Naik, K.R. and Rokhade, A.K. (1994) *J. Maharashtra Agric. Univ.*, **19** : 180-183.

Naik, K.R. and Rokhade, A.K. (1997) *Karnataka J. Agric. Sci.*, **10** : 388-393.

Nallathambi, P., Umamaheswari, C., Joshi, H. K. and Dhandar, D. G. (2003) In: *Proc. of 6th International Workshop on Plant Growth Rhizobacteria* held during 5-10 October, 2003 at IISR, Calicut, Kerala, pp. 184- 187.

Namdev B.K., Thomas M., Kurmi A., Thakur A.S. and Upadhyaya A. (2015) *The Bioscan*, **10**(3): 1219-1222.

Nanthakumar, S. and Balakrishnan, K. (1999) *Madras Agric. J.*, **85** : 322-324.

Nath, V., Saroj, P.L., Singh, R.S., Bhargava, R.S. and Pareek, O.P. (2000) *Indian J. Hort.*, **57** : 21-26.

Nath, V., Singh, R.S., Bhargava, R. and Pareek, O.P. (1998) *Curr. Agric.*, **22** : 115-116.

Nayak, G. and Sen, S.K. (2000) *Environ. Ecol.*, **18**(1): 241-242.

Neeraj, G. and Kaul, R. K. (2011) *Appl. Biol. Res.*, **13**(2): 94-98.

Nehra, N.S., Sareen, P.K. and Chitkara, S.D. (1983) *Cytologia.*, **48** : 103-108.

Oganesjan, A.P. (1953) *Bot. Zurnal*, **38** : 744-751.

Oh, S.D., Choi, D.G., Kim, J.H. and Kim, Y.S. (1988) *J. Korean Hort. Sci.*, **29** : 304-311.

Omidbaigi, R. and Daghighi, S (2003) In : *III WOCMAP Congress on Medicinal and Aromatic Plants-Volume 2: Conservation, cultivation and Sustainable Use of Medicinal*, 676, pp. 143-146.

Pandey, R.C., Pathak, R.A. and Pathak, R.K. (1990) *Indian J. Hort.*, **47** : 286-290.

Pandey, R.C., Pathak, R.A. and Singh, I.S. (1998) *Indian J. Hort.*, **55** : 306-313.

Pandey, S.D., Parthak, R.K. and Dwivedi, R. (1991) *Hort. J.*, **4** : 33-36.

Pandey, S.D., Pathak, R.K. and Awasthi, O.P. (1993) *Indian J. Hort.*, **50** : 46-48.

Pareek, O.P. (1978a) *CAZRI Leaflet No.* 1, p. 4.

Pareek, O.P. (1978b) *Indian Hort.*, **23** : 5-6.

Pareek, O.P. (1983) *Indian Coun. Agric. Res. Publ.*, p. 8.

Pareek, S. (2013) *Emir. J. Food Agric.*, **25** (6): 463-470.

Pareek, S., Fagera, M.S. and Dhaka, R.S. (2003) *Indian J Plant Protec.*, **31**(2): 89-90.

Pareek, S. and Dhaka, R.S. (2008) *Indian J. Arid Hortic.*, 3 : 77-80.

Pareek, S., Kitinoja, L., Kaushik, R.A. and Paliwal, R. (2009) *Stewart Post-harvest Rev.*, **5**(5) : 1-10.

Pareek, S., Mukherjee, S.and Paliwal, R. (2007) *Agric. Rev.*, **28** (4): 277-282.

Patel, B.D. and Patel, G.M. (1999) *Gujarat Agric. Univ. Res. J.*, **24**(2): 111-113.

Patel, B.M., Patel, S.I., Patel, S.K. and Patel, S.B. (2003) *Agric. Sci Dig.*, **23**(2) : 113-115.

Patil, A.V. (1978) *AGRESCO (Rabi) Mahatma Phule Krishi Vidyapeeth*, Rahuri, pp. 1892-1896.

Patil, B.H., Upadhyay, V.R., Murlidharan, C.M. and Judal, G.S. (1989) *Indian J. Plant Protection*, **17** : 39-45.

Patil, P.K., Patil, V.K. and More, R.M. (1981) *Haryana J. Hort. Sci.*, **10** : 166-170.

Paul, B., Kumar, S. and Das, A. (2013) Lac cultivation and their host trees found in Bastar Forest Division. *Plant Sciences Feed.*, **3**(1): 8-12.

Pramanick, P. K., Sharma V.P. and Singh, S.K. (2005) *Indian J. Hort.*, **62**(4): 389-390.

Prasad, G. and Babu, R.S. (1988) *South Indian Hort.*, **36** : 262-263.

Prasad, R.N. and Bankar, G.J. (2002) *Prog. Hort.*, 34(2) : 190-195.

Praveen, J. and Patil, P.B. (1998) *Karnataka J. Agric. Sci.*, **11** : 143-145.

Purohit, A.K., Rawat, T. and Kumar, A. (2003) *Plant Foods Hum. Nutr.*, **58** : 1-7.

Raghumoorthi, K.N. and Arumugam, R. (1992) *Indian J. Plant Protection*, **20** : 32-36.

Raj Bhansali, R. (1999) In : Research Development *Priorities for Arid Zone beyond 2000 AD*, (Eds. S.K. Verma, R. Raj Bhansali, S.K. Jindal and V. Vir), Scientific Publishers, Jodhpur.

Raj Bhansali, R. (2001) In : *Biotechnology of Horticultural Crops*, Vol. I (Eds. V.A. Pathasarathy, T.K. Bose, P.C. Deka, P. Das, S.K. Mitra and S. Mohandas), Naya Prokash, Calcutta, India, pp. 354-355.

Raja, P.V. (1985) *M.Sc. Thesis* submitted to IARI, New Delhi.

Rajesh, S. and Misra, K.K. (2004) *Sci. Hortic.*, **9** : 37-40.

Rajput, C.B.S. and Singh, C.N. (1982) *Bangladesh Hort.*, **10** : 19-22.

Rathore, D.S., Singh, R.P., Deora, N.S. and Shekhawat, N.S. (1992) *Scientia Hort.*, **51** : 165-168.

Raturi, G.B. and Bhargava, B.S. (1993) *Indian J. Hort.*, **50** : 202-208.

Raturi, G.B. and Bhargava, B.S. (1994) *Indian J. Hort.*, **51** : 346-350.

Rawal, R.D. and Saxena, A.K. (1989) *Indian J. Hort.*, **46** : 413-414.

Reddy, G.S., Reddy, P.K. and Reddy, T.Y. (1997) *Crop Res.* (Hissar), **13** : 99-102.

Reddy, N.M., Reddy, G.S. and Madhusudan, I. (1997) *J. Res. ANGRAI*, **25** : 19-26.

Reddy, Y.N. and Murthy, B.N.S. (1990) *Indian J. Hort.*, **47** : 314-317.

Sadhu, M.K., Ghosh, S.K. and Bose, T.K. (1978) *Mysore J. Agric. Sci.*, **12** : 1-5.

Saini, R.S., Ram Dayai and Rakesh (1995) *Crop Res.* (Hissar), **10** : 324-326.

Sameera, N.S. and Mandakini, B.P. (2015) *Int. Food Res. J.*, **22**(2) : 849-853.

Sandbhor, D.R. and Desai, U.T. (1991) *Maharashtra J. Hort.*, **5** : 24-28.

Sandhu, A.S. and Dhillon, B.S. (1983) *Punjab Hort. J.*, **23** : 57-58.

Sandhu, A.S., Dhillon, B.S. and Singh, Z. (1983) *Punjab Hort. J.*, **23** : 53-56.

Sandhu, S.S. and Thind, S.S. (1988) *Indian J. Hort.*, **45** : 274-282.

Sharif, N., Jaskania M.J., Naqvia, S.A. and Awanb, F. S. (2019) *Sci. Hort.*, **249**: 228–239.

Sharma, A. (1992) *Annals of Arid Zone*, **31** : 231-232.

Sharma, B.D., Cheema, S.S. and Sandhu, A.S. (1983) *Punjab Hort. J.*, **23** : 51-52.

Sharma, K.D., Pareek, O.P. and Singh, H.P. (1982) *Agric. Water Management*, **5** : 73-84.

Sharma, K.K., Jawanda, J.S., Singh, M.P. and Bajwa, M.S. (1980) *Punjab Hort. J.*, **20** : 47-51.

Sharma, R. and Rana, J.C. (2016) *Indian J.Natural Prod. Resources*, **7**(4) : 323-327.

Sharma, S.C. and Kumar, R. (1982) *Pharmazine*, **37** : 809-810.

Sharma, V.P., Raja, P.V. and Kore, V.N. (1990) *Annals Agric. Res.*, **11** : 14-20.

Shaumarov, K.H.B. (1976) *Doklady Mosk-S-Kh Akad in K.A. Timiryazeva*, **216** : 83-89.

Shen, X.D., Gao, F.G., Chen, B.C. and Wang, Y. (1992) *Ningxia J. Agro-Forestry Sci. Technol.*, **2** : 32-34.

Shiva Kumar, S., Kundu, S., Nandi, P., Shekha, R S., Adhikary, R.T., Ghosh, B. and Dutta. P. (2017) *J. Crop Weed*, **13**(1): 189-191.

Siddiqui, S. and Gupta, O.P. (1989) *Haryana Agric. Univ. J. Res.*, **19** : 260-262.

Siddiqui, S. and Gupta, O.P. (1995) *Haryana J. Hort. Sci.*, **24** : 19-23.

Siddiqui, S. and Gupta, O.P. (1996) *Indian J. Plant Physiol.*, **1** : 70-72.

Siddiqui, S., Srivastava, A. and Sharma, R.K. (1999) *Haryana J. Hort. Sci.*, **28** : 179-181.

Sin'ko, L.T. (1976) *Rast. Resursy*, **12** : 303-307.

Sin'ko, L.T. and Chemarin, N.G. (1979) *Subtrop. Kul'tury*, **3** : 26-28.

Singh, A.K., Sharma, R.K., Singh, N.K., Bansal, K.C., Koundal, K.R., Mohapatra, T. (2006) *J. Hort. Sci. Biotech.*, **81**(2): 205-210.

Singh, A.K., Sharma, P., Singh, R., Singh, B., Koundal, K.R. and Singh, N.K. (2007) *J. Plant Biochem. Biot.*, **16**(1) : 35-40.

Singh, A.K., Sharma, R.K., Singh, N.K., Bansal, K.C., Koundal, K.R. and Mohapatra, T. (2006) *J. Hortic. Sci. Biotech.*, **81**(2) : 205-210.

Singh, A.P. (2009) In: *Field Problems of Fruit Crops*. Additional Director of Communication, PAU, Ludhiana, P. 258.

Singh, D. and Andotra, P.S. (1989) *Indian J. Mycol. Plant. Path.*, **19** : 105-106.

Singh, D.K., Barun, S. and Sen, S.K. (2004) *Environ. Ecol.*, **22**(2) : 439-442.

Singh, H.K. and Singh, J.K. (2009) *Asian J. Hort.*, **4**(1) : 230-232.

Singh, K., Singhrot, R.S. and Chauhan, K.S. (1984) *Haryana J. Hort. Sci.*, **11** : 5-12.

Singh, K.K., Chadha, K.L. and Gupta, M.R. (1973a) *Punjab Agric. Univ.*, Ludhiana, Leaflet, p. 23.

Singh, L.B. (1952) *Indian J. Hort.*, **9** : 16-19.

Singh, M. P. (2008) *Indian Hort.*, **53**(1): 31-32.

Singh, M.P. and Vashishtha, B.B. (1985) *J. Pl. Protection*, **12** : 55-56.

Singh, P., Bajwa, M.S. and Singh, R. (1973) *Plant Sci.*, **5** : 137-139.

Singh, R.R., Jain, R.K. and Chauhan, K.S. (1983) *Haryana Agric. Univ. J. Res.*, **13** : 112-114.

Singh, R.S. (1997) *Current Agriculture*, **21** : 117-118.

Singh, S. and Ahlawat, V.P. (1995) *Haryana J. Hort. Sci.*, **24** : 94-97.

Singh, S.B., Maheshwari, S.K. and Singh, P.N. (1995) *Annals Plant Protection Sci.*, **3** : 168-169.

Singh, S., Gupta, R.N., Awasthi, B.J., Verma, R.A. and Singh, S. (2000) *Indian J. Entomol.*, **62**(2): 171-174.

Singh, S.K., Chhajer, S., Pathak, R., Bhatt, R.K. and Kalia, R.K. (2017) *Tree Genet. Genomes.*, **13**(1) : 12

Singh, S.K., Meghwal, P.R., Pathak, R. and Gautam, R. (2014) *Proc. Nat. Academy Sci., India, Section B: Biological Sciences.*, **84**(3) : 603-611.

Singh, S.P., Abhay, M. and Mariya, S. (2010) *Environ. Ecol.*, **28**(1A) : 241-243.

Singh, S.P., Singh, I.B. and Singh, V.N. (1987) *Prog. Hort.*, **19** : 61-64.

Singh, U.R. and Singh, N. (1976) *Haryana J. Hort.*, **5** : 1-8.

Singh, Y.P. and Sumbali, S. (2000) *Indian Phytopath.*, **53** : 38-41.

Singh, Z. and Sandhu, A.S. (1984) *Indian J. Agric. Sci.*, **58** : 688-689.

Singh, Z., Dhillon, B.S. and Sandhu, A.S. (1991) *Indian J. Hort.*, **48** : **277-275**.

Singhrot, R.S., Makhija, M. and Chauhan, K.S. (1980) *Haryana J. Hort. Sci.*, **9** : 40-42.

Sinko, L.T., Kucherova, T.P. and Vilde, E.I. (1987) *Sbornik Nauchnykh Trudov, Gosudarstvennyi Nikitskii Botanicheskii Sad*, **102** : 56-64.

Soliman, H.I. and Hegazi, G.A.E. (2013) *Life Sci.*, **10**(2) : 573-582.

Soltani, A., Gholipoor, M. and Zeinali, E. (2011) *Environ. Exp. Bot.*, **55**: 195–200

Song, L., Meinhardt L. W., Bailey B. and Zhang, D. (2019) *Crop Breed. Genet. Genom.*, **1**: e190015. https: //doi.org/10.20900/cbgg20190015.

Srimanthi, P., Sasthri, G., Venkatasalam, E.P. and Malarkodi, K. (2002) *Prog. Hortic.*, **34**(2) : 179-182.

Srivastava, H.C., Chundawat, B.S. and Bisla, S.S. (1980) *Ann. Arid Zone*, **19** : 365-367.

Sudhersan, C., Jibimanuel, S. and Ashkanani, J. (2013) *Acta Hortic.*, **993** : 83-90.

Sun, J.Z., Qian, Z.G., Liu, F.L., Song, H.W., Wen, T.H., Ma, P.Y., Wang, Y.C. and Li, Y.L. (1989) *Forest Sci. Technol.*, **4** : 1-4.

Sung, D.O. and Song, W.I. (1992) *HortScience*, **27** : 210.

Supe, V.S., Kaulgud, S.N., Shete, M.B. and Karale, A.R. (1997) *J. Maharashtra Agric. Univ.*, **22** : 192-193.

Syamal, M.M. and Rajput, C.B.S. (1989) *Indian J. Hort.*, **46** : 364-367.

Tagiev, T.M. (1992) *Sadovodstvo i Vinogradarstvo*, **1** : 19-20.

Teaotia, S.S. and Chauhan, R.S. (1963) *Punjab Hort. J.*, **3** : 60-70.

Thind, S.K. and Kaur, N. (2006) *Indian J. Hort.*, **63** (3): 267-269.

Thind, S.K., Nirmaljit, K. and Arora, J.K. (2004) *Prog. Hort.*, **36**(1): 164-166.

Tian, S.H., Young, F.N. and Guo, C.Y. (1983) *Forest Sci. Tech.*, No.2, pp. 12-13.

Tiwari, R.J. and Banafar, R.N.S. (1995) *Indian J. Plant Physiol.*, **38** : 88-89.

Tomer, N.S. and Singh, R. (1987) *Haryana J. Hort. Sci.*, **16** : 52-58.

Uddin, M.B. and Hussain, I. (2012) *World J. Dairy and Food Sci.* **7**(1) : 74-78.

Varalakshmi, L.R., Bhargava, B.S. and Raturi, G.B. (1999) *Scientific Hort.*, **6** : 9-17.

Vashishtha, B.B. and Pareek, O.P. (1979) *Ann. Arid. Zone*, **18** : 165-169.

Vashishtha, B.B. and Pareek, O.P. (1983) *Ann. Arid. Zone*, **22** : 43-45.

Vashishtha, B.B. and Pareek, O.P. (1989) *Indian J. Hort.*, **46** : 183-188.

Vashishtha, B.B., Pareek, O.P., Chan-Prasekharan, I. and Ghanim, A. (1989) *Ann. Arid Zone*, **28** : 123-132.

Verma, A.N. and Singh, R. (1974) *Haryana J. Hort. Sci.*, **3** : 62-64.

Verma, M.K., Sharma, V.P. and Saxena, S.K. (2000) *Indian J. Hort.*, **57** : 13-17.

Wan, Z.Z. (1994) *China Fruits*, **1** : 46-47.

Wang, B., Guo, Y.X., Zhou, G.F., Shan, G.H. and Gao, L. (1999) *China Fruits*, **2** : 3-4.

Wang, K.W., Cui, F.X., Liu, H.P. and Li, X.K. (2000) *China Fruit*, **3** : 25-26.

Williamsm J.T. (2006) *Ber and other Jujubes*, pp. 1-17.

Wu, C.S., Gao, Q.H., Kjelgren, R.K., Guo, X.D. and Wang, M (2013) *Molecules*, **18**(10) : 12029-12040.

Yadav, G.R., Nirwan, R.S. and Prasad, B. (1980) *Prog. Hort.*, **12** : 27-32.

Yadav, I., Sharma, R.K., Goyal, R.K. and Siddiqui, S. (2009) *Haryana J. Hort. Sci.*, **38**(3/4) : 243-246.

Yadav, L.S. (1991) *Indian Hort.*, **35** : 132-135.

Yadav, L.S. and Godara, N.R. (1987) *Haryana J. Hort. Sci.*, **16** : 45-51.

Yadav, L.S. and Godara, N.R. (1992) *Prog. Hort.*, **21** : 325-328.

Yadav, P.K. and Rathore, P.S. (2004) *Progress. Agric.*, **4**(1) : 85-86.

Yadav, R.C., Sharma, K.D., Singh, H.P. and Pareek, O.P. (1978) *Proc. Int. Symp. Arid Zone Res. Dev. CAZRI*, February, 1978.

Yadav, S.K., Sharma, A. and Singh, V. (2000) *Curr. Agric.*, **24**(1/2): 97-99.

Yadav, S. K., Sharma, A. and Singh, V. (2003) *Annals of Biol.*, **19**(2): 221-223.

Yamdagni, R., Bajpai, P.N. and Misra, R.S. (1968) *J. Sci. Tech.*, **6b** : 101-103.

Yamdagni, R., Kumar, R. and Jindal, P.C. (1980) *Prog. Hort.*, **12** : 5-7.

Yao, T.M., Zhang, C.H. and Wu, Y.L. (2013) *For. Pest. Dis.*, **1**: 30-33.

Yaragattikar, A.T. and Itnal, C.J. (2003) *Karnataka J. Agril. Sci.*, **16**(1) : 17-21.

Yildirim, A., San, B. and Yildirim, F. (2015) *Erwerbs Obstbau.*, **57** : 135-140.

Yu, J.Z., Duan, L.H. and Wang, L.C. (1991) *Crop Genetic Resources*, No.2, pp. 43-44.

Yun, M.S., Kim, Y.S. and Ahn, C.S. (1994) *J. Korean Soc. Hort. Sci.*, **35** : 466-470.

Zaktreger, N.I. and Solov've, A.K. (1962) *Tr. Turkm Opyt. Stan. Vir.*, **3** : 104-107.

Zhang, X.M., Ji, Q.J., Ren, S.P., Lei, Y.P. and Tian, K.X. (1998) *China Fruits*, **1** : 33-34.

Zhang, Y.L., Chen, J.P., Liu, G.W. and Su, D.H. (2000) *Trans. Chinese Soc. Agri. Engg.*, **16** : 106-109.

Zhao, M.F. and Deng, M.Q. (2000) *Forest Res. Beijing*, **13** : 355-359.

5

ANNONAS

V. A. Parthasarathy, T. Sakthivel and D. S. Rathore

1.0 Introduction

The Annonaceae or custard apple family comprises about 120 genera and more than 2000 species (Leboeuf *et al.*, 1982). The genus Annona is the most economically important one, containing 120 species. The major commercial species throughout the world are: the cherimoya (*Annona cherimola*), sweetsop (*Annona squamosa*) and *atemoya* (a hybrid of *Annona squamosa* and Annona cherimola). The sweetsop species name '*squamosa*' refers to the knobby appearance of the fruit. Sweetsop is a small tropical tree originating in the New World tropics, probably Central America. This species is the most widely grown in the tropical regions of the America, Africa, Asia and the Pacific. Sweetsop is also named sugar apple and has many other regional names, such as Custard apple (India), anon (Spanish), ata (Portuguese), noi-na (Thailand), stis (The Philippines) and fan-li-chi (Taiwan). Sweetsop is the mostly grown Annona species. The fruit is frequently found in village markets but has not shown much potential for large commercial cultivation due to the small fruit size, frequent cracking at maturity and poor shelf life. The perishable nature and supply shortages make marketing localized or air shipment essential.

In different countries, the name custard apple is given to separate species of genus *Annona*. In India, the custard apple is *Annona squamosa,* in the USA, *A. reticulata* or *A. glabra* and in Australia *A. atemoya*. According to Hayes (1957), the term custard apple may be applied to all the *Annona* species producing edible fruits.

2.0 Composition and Uses

2.1.0 Composition

The annonaceous fruits are composed of peel, pulp and seed in varying proportions. The edible portion or pulp is creamy or custard-like, granular with a good blend of sweetness and acidity which vary with the species. The pleasant flavour and mild aroma have a universal liking. The nutritive values of different Annona species are given in Table 1.

Table 1: Composition of Annonaceous Fruits in g/100 g Edible Portion*

Constituents	*Atemoya*	*Cherimoya*	*Soursop*	*Sweetsop*
Moisture	78.7	68.71	80.1	75.97
Protein	1.4	1.54	0.69	1.89
Fat	0.6	0.13	0.39	0.57
Carbohydrate	15.8	28.95	18.23	20.82
Fibre	2.5	-	0.95	1.41
Ash	0.5	0.67	0.58	0.75
Energy (KJ)	310	460	247	360
Calcium	17.0	9.0	9.0	17.0
Magnesium	32.0	-	22.0	22.0
Phosphorus	-	24.0	29.0	54.0

Constituents	*Atemoya*	*Cherimoya*	*Soursop*	*Sweetsop*
Potassium	250.0	-	320.0	142.0
Sodium	4.0	-	22.0	2.0
Iron	0.3	0.25	0.82	0.30
Ascorbic acid	43.0	12.20	16.4	35.9
Thiamine	0.05	0.11	0.07	0.10
Riboflavin	0.08	0.11	0.12	0.06
Niacin	0.8	1.0	1.52	0.89

* Wenkam (1990).

Carotene is not present in the pulp. The carbohydrates mainly constitute sugars, and glucose is the major sugar.

2.2.0 Uses

Annonas are mostly consumed as dessert fruits. They are also used in ice-creams and other milk products and preserved as jam, jelly or other products on a limited scale. Directly baked fruits are commonly eaten in Andhra Pradesh (Rao, 1974).

The immature fruits, seeds, leaves and roots are of considerable medicinal value both in Ayurvedic and Unani systems of medicine (Kirtikar and Basu, 1933). The seeds are abortifacient and roots are drastic purgative. The leaf poultice is used for supperation. The seeds contain about 30 per cent oil which can be used in soap and paint industry (Rao, 1974). Its insecticidal properties, when mixed with neem oil, have been demonstrated against *Nephotettix virescens* and its transmission of rice tungro virus (Mariappan and Saxena, 1984). The seed cake containing about 4 per cent nitrogen is beneficial as manure. At least thirty three compounds are present in the essential oil extracted from the peel.

The fruits of *A. squamosa*, are extensively used to prepare candies, ice creams and beverages (Ma *et al.*, 2017).A wide range of ethno-medicinal uses has been related to different portions of A. squamosa, such as tonic, apophlegmatisant, cool medicine, abortient and heart sedative. Numerous research projects on A. squamosa have found that it has anticancer, anti-oxidant, antidiabetic, antihypertensive, hepatoprotective, antiparasitic, antimalarial, insecticidal, microbicidal and molluscicidal activities. Phytochemistry investigations on *A. squamosa* have considered annonaceous acetogenins (ACGs), diterpenes (DITs), alkaloids (ALKs) and cyclopeptides (CPs) as the main constituents. Until 2016, 33 DITs, 19 ALKs, 88 ACGs and 13 CPs from this species were reported. On the basis of the multiple researches on *A. squamosa*, this review strives to integrate available information on its phytochemicals, folklore uses and bioactivities, hoping to promote a better understanding of its medicinal values (Ma *et al.*, 2017).

The fruits of *A. muricata* are extensively used to prepare syrups, candies, beverages, ice creams and shakes. A wide array of ethnomedicinal activities is contributed to different parts of *A. muricata*, and indigenous communities in Africa and South America extensively use this plant in their folk medicine. Patel and

Patel (2016) extensively reviewed the pharmacological uses of this plant including leaves, fruit and seeds. Numerous investigations have substantiated these activities, including anticancer, anticonvulsant, anti-arthritic, antiparasitic, antimalarial, hepatoprotective and antidiabetic, analgesic hypotensive, anti-inflammatory, and immune enhancing effects. Phytochemical studies reveal that annonaceous acetogenins are the major constituents of *A. muricata*. More than 100 annonaceous acetogenins have been isolated from leaves, barks, seeds, roots and fruits of *A. muricata* (Patel and Patel, 2016).

3.0 Origin and Distribution

All annonaceous fruits are indigenous to tropical America where from the spread to different parts of the world after the discovery of the continent. It is quite possible that a few species may be native of Africa (Hayes, 1957). Philology and phylogenetic studies carried out by DeCandolle have confirmed this viewpoint. Majority of *Annona* species grow wild in South America, and terracotta resembling the shape of cherimoya is also found in Peru. According to Bailey (1949), *A. squamosa* is a native of West Indies and South America while *A. cherimola* has its origin in the mountains of Ecuador and Peru. The distribution of cherimoya probably occurred through Central America and Mexico at an early date and from Mexico to the Caribbean islands and then to the African coast and the Mediterranean region (Nakasone and Paul, 1998). It has been under cultivation on the south coast of Spain for more than one century (Farre *et al.*, 1999).

The occurrence of wild stands in the nature, Sanskrit names, paintings and carvings of the annonaceous fruits and leaves and mention in Ain-i-Akbari have led many to believe that Annonas are of Indian origin. But Watt (1887) in an extensive review concluded that Annonas are introduced plants and none of the *Annona* species is endemic to any Indian forest. It is, however, certain that after being introduced by the Portuguese, the fruits dispersed quickly and got acclimatized throughout South Asia.

Annonas are now cultivated in various countries, such as Australia, Brazil, Chile, Egypt, India, Israel, Mexico, Myanmar, Philippines, Spain, Sri Lanka, South Africa, USA and West Indies. Commercial production of cherimoya occurs in Spain, Bolivia, Chile, Peru and New Zealand (Nakasone and Paull, 1998). The fruits are distributed in the entire tropical region of the globe and are rapidly spreading to the subtropics with the increasing popularity of Annonas hybrids which can tolerate cooler climate.

In India, the fruits are very popular in the Deccan Plateau and are also grown commercially on a smaller scale in Assam, Andhra Pradesh, Bihar, Karnataka, Madhya Pradesh, Maharashtra, Orissa, Rajasthan and Uttar Pradesh.

A. muricata is native to the warmest tropical areas in South and North America and is now widely distributed throughout tropical and subtropical parts of the world, including India, Malaysia and Nigeria, Australia, Africa (Patel and Patel, 2016). Cherimoya originated in Andes' valleys between Ecuador and Peru´, latitude 3–88S, longitude 75–808W and altitude 1500–2000 m where it occurs spontaneously (Cautin and Agust, 2005). The fruit is cultivated mainly in the Mediterranean basin, but some important production occurs in America, China–Taiwan and Australia. By 1998, the estimated production area of Annona species in the world was about 15,000 ha and the total production was about 100,000 t. Spain leads the world in cherimoya production with 3600 ha, which yield 35,000 t, followed by Peru´, 1800 ha and 15,000 t, and Chile, 1200 ha and 12,000 t (Van Dammme and Scheldeman, 1999).

4.0 Species and Cultivars

4.1.0 Species

The annonaceous fruits are the members of the family Annonaceae and genus *Annona*. The family Annonaceae includes about 129 genus and more than 2000 species. Most grown species belong to genus Annona, including *A. muricata* L., soursop or guana´bana, *A. reticulata* L., custard apple, *A. squamosa* L., sugar apple, and *A. cherimola* Mill., cherimoya or chirimoya; and atemoya, a hybrid *between A. cherimola* and *A. squamosa* is also commercially appreciate (Nakasone and Paull, 1998).There are over 100 species of *Annona* and several of these bear edible fruits (Mahdeem, 1990), however, only six species produce the fruits of commercial significance. On the basis of pomological description, a key for identifying these species has been provided by Thakur and Singh (1967).

Annona squamosa

It is the most widely cultivated species in India and probably in the tropics of the world. It is popularly known as custard apple, sugar apple, sweetsop or *ate* and in the northern India *Sharifa*, while in southern part *sitaphal*. It is a deciduous, tall, woody shrub or small tree of about 5 to 6 metres in height with irregularly spreading branches. The alternate double-ranked leaves are lanceolate and acuminate. The greenish yellow flowers arise at an extra axillary position, usually in clusters and rarely solitary. Six petals are in two whorls, the outer petals are thick, linear and rounded at the apex while inner ones are minute, ovate or obovate and keeled on the outside. The flower has numerous stamens and carpels. The fruit is composed of loosely cohering carpels forming a squamose or tuberculated surface. The black seeds are surrounded by white, creamy or custard-like pulp which is very sweet and pleasantly flavoured.

A. cherimola

Cherimoya is called cherimolier or cherimola. Among all the annonaceous fruits, it is considered to be the best fruit because of its acidic sweet flavour, buttery consistency of pulp and low seed content. It is also one of the finest subtropical fruits which are gaining popularity in Australia, California, Hawaii, Brazil, Mexico and India. Though not suitable for tropical climate, yet it has been established well in Telangana (Venkataratnam, 1953).

Cherimoya is deciduous, small and spreading tree with greyish pubescent shoots. The velvety tomentum on the lower surface of leaves is a distinguishing feature of this species. The flowers are usually solitary with narrow outer petals while inner petals are small and scale-like. The fruits are 8-15 cm in diameter, green in colour and round to heart-shaped. As a result of careful selection, some cultivars

Uses of annonas

Annona fruit with pulp

Ice cream prepared from annona

Powder extracted from annona

Milk shake prepared from annona

Oil extracted from annona

Species of annonas

Annona squamosa (green and red)

Annona cherimola

Species of annonas

Annona reticulata

Whole plant

Leaves

Stem

Flowers

Unripe fruit

Ripened fruit

Seeds

Plant parts of annona reticulata

Annona muricata

Species of annonas

Annona diversifolia

Annona atemoya

Annona crassiflora

Different types of *Annona cherimoya*

with fine flavour, fewer seed, and large fruit size have been identified. Bailey (1949) has recognized the following types on the basis of fruit form.

- ☆ **Finger-printed cherimoya**: This is an excellent type having sweet, juicy pulp of good flavour. There are only few seeds in a fruit.
- ☆ **Smooth cherimoya**: The skin of this finest type is smooth. It is largely cultivated in Mexico.
- ☆ **Tuberculate cherimoya** : The fruit is generally sold in Peruvian market.
- ☆ **Mammillate cherimoya** : It is a common form found in the Nilgiri hills of India.
- ☆ **Umbonate cherimoya** : The fruit is oblong-conical having thick skin and numerous seeds. The acidic pulp of pineapple flavour is used in preparing drinks and *sherbets*.

A. reticulata

Custard apple, bullock's-heart, bull's-heart or *ramphal* thrives all over South India up to an elevation of 1,200 m as it is essentially a tropical fruit requiring milder climate than *A. squamosa*. It is a deciduous tree, 6 to 8 m in height with lanceolate, long, acuminate and glabrous leaves. The exterior petals are yellowish, fleshy while the interior ones are scale-like. The heart-shaped fruits are nearly smooth and contain few seeds but they are inferior in quality to sugar apple and cherimoya because of lower sugar content. Flowering occurs from August to October and the fruits ripen in April and May coming later in market than custard apple. Bullock's-heart is not cultivated on a commercial scale anywhere in India but it is widely spread in the Philippines. It differs from *A. glabra* by its elongate, narrow outer petals and small brown seeds.

A. muricata

Annona muricata is a member of the Annonaceae family and is a fruit tree with a long history of traditional use. *A. muricata*, also known as soursop, graviola and guanabana, is an evergreen plant that is mostly distributed in tropical and subtropical regions of the world. *A. muricata* is native to the warmest tropical areas in South and North America and is now widely distributed throughout tropical and subtropical parts of the world, including India, Malaysia and Nigeria, Australia, Africa. The soursop is an evergreen small tree flourishing in tropical climate of Central America and Peru. The fruit is also popular in Cuba and to some extent in Sri Lanka. It grows on a very limited scale in Assam and Andhra Pradesh and rarely in other parts of India. The small obovate-oblong to elliptic leaves have shining upper surface. The exterior petals are fleshy and valvate and interior ones are small and imbricate. The extra large, heart-shaped fruits resemble jackfruit with many soft spines pointing toward apex. The white, juicy pulp has acidic mango-like flavour which is strained and used in preparing drinks, ice-cream, jelly and other preserves. In preparing soup, immature fruits can be used. Though soursop contains a good amount of sugar its sourness is due to the presence of 3 to 4 times more acid than the sugar apple. The soursop as scion is incompatible with the sugar apple and cherimoya. Its low productivity has been attributed to the inadequate pollination.

A. diversifolia

The ilama is a small tree adaptable to subtropical climate and is a popular fruit in Mexico and Central America. The leaves are glabrous and papery thin with leaf-like bracts at the base. The outer petals are linear oblong and inner petals are tiny and scale-like. The fruits resemble sugar apple but are covered by greyish tomentum and pulp is cream- or rose-coloured. Hayes (1957) considered ilama to be one of the best of Annonas which should be introduced in the subtropical regions of India.

A. atemoya

Atemoyas (*Annona atemoya*) are hybrids between sugar apple (*A. squamosa*) and cherimoya (*A. cherimola*) which are commercially grown in many countries, particularly in the coastal and subtropical regions of Australia where African Pride, a cross between cherimoya and sugar apple, has become an important cultivar. The other popular cultivars are Pink's Mammoth and Neilson. Atemoya is a semi-deciduous, subtropical fruit tree being tolerant to frost. Many high yielding cherimoyas in different countries are suspected to be Atemoyas (Oppenheimer, 1947; Prest, 1955). Atemoya fruits are similar to cherimoyas from quality point of view while in climatic requirements, the plant is near to sugar apple (Hayes, 1957). Because of heterozygous parents, Atemoyas differ considerably among themselves in plant and fruit characteristics. The leaves are usually very large like cherimoya but glabrous resembling the leaves of sugar apple. The fruit has a thick coating of glaucous bloom (Thakur and Singh, 1967). The pulp is similar to cherimoya being juicy with an excellent acidic flavour.

The problem of identification of cherimoya and parents of atemoya hybrids has been solved by using isozyme markers by Ellstrand and Lee in 1987. They were able to distinguish the differences among the cultivars of cherimoya and also between cherimoya and atemoya on the basis of isozyme variation patterns at 15 loci influencing 8 enzyme systems.

4.2.0 Cultivars

In India, a number of local cultivars are available in annonaceous fruits but barring few, none of them has got the status of commercial cultivar due to seed propagation which results in considerable variability. Some local cultivars of sugar apple are Balanagar, Barbados, British Guinea, Brandy, Islander, Red Sitaphal, Washington PI 98797 and Washington PI 107005. The cultivars differ from each other in fruit size, shape, colour and eating quality. Red Sitaphal is a distinct cultivar probably originated as a chance seedling (Venkataratnam, 1963). Its fruits are dark pinkish and pulp erythritered (Thakur and Singh, 1967). The fruit quality is similar to other cultivars of sugar apple bearing greenish yellow fruit. Red Sitaphal fruits, however, contain more seeds. In a varietal trial of seven custard apple cultivars at Bangalore, no significant difference was found in yield (Table 2) but the cultivar Balanagar showed better fruit quality (Anon., 1987).

Sanewski (1991) described six atemoya cultivars grown in Australia, where it is grown under the name custard apple.

Pink's Mammoth

The tree bears commercial fruits at the age of 6-7 years. Fruits are large (800 g to 2 kg). It is not as precocious as 'African Pride'. It was probably introduced into Australia from British Guiana.

Table 2: Yield and Physico-chemical Composition of some Cultivars of Custard Apple*

Cultivars	*No. of Fruits/ Tree*	*Fruit Weight (g)*	*Seed Weight (g)*	*Pulp (Per cent)*	*TSS (°Brix)*	*Acidity (Per cent)*	*Reducing Sugar (Per cent)*	*Total Sugar (Per cent)*
Balanagar	48	137	5.7	44.9	20.7	0.20	15.0	17.9
British Guinea	35	151	8.4	37.9	23.3	0.27	15.0	17.9
Barbados Seedling	35	154	8.0	41.2	24.7	0.22	13.7	16.5
Local Sitaphal	31	127	9.2	44.5	20.8	0.26	12.7	16.5
Mammoth	29	122	8.7	44.8	20.0	0.19	15.0	17.9
Red Sitaphal	22	156	5.2	30.5	22.3	0.24	13.7	15.9
Washington PI 98797	21	161	6.9	38.1	20.8	0.28	12.7	17.9
CD at 5 per cent	NS	NS	2.6	3.9	1.5	0.03	-	-

*Anon. (1987).

African Pride

It is reported to have been introduced into Australia from South Africa. It is a precocious cultivar and gives satisfactorily high yield. Fruits are small to medium in size.

Bradley

The tree bears small-sized fruits having relatively smooth and thin skin.

Page

A Florida cultivar produces medium-sized fruit with prominent skin segments.

Gefner

It is introduced into Australia from Israel. Fruits are small to medium in size similar to Page.

Island Gem

It is a high yielding cultivar. Fruits are small and mature early.

Grossberger (1999) listed the most widely grown cultivars of cherimoya in California (Table 3). These cultivars are in cultivation for about 20 years.

Table 3: Cultivars Grown in California*

Principal Cultivars	*Other Cultivars*	*Previous Cultivars*
White	Carmela	Golden Russet
Bays	Knight	Mammilaris
Pierce	Libby	Ryerson
Booth	Ott	Dorothea
Chaffey	Whaley	Wilkerson

* Grossberger (1999).

The fruits of cultivar White are large, somewhat long in shape with irregular shaped skin. Pulp is medium in sweetness. Booth is a medium-sized, very sweet, late maturing cultivar, while Bays is widely produced and has the best shape for packing (Gross-berger, 1999). Chilean cultivars, such as Bronceada and Concha Lisa, possess some favourable characteristics, such as loose seeds and good post-harvest storability, but also have low sugar content and are late-maturing. In terms of organoleptic quality, introduced cultivars Cholan and Loma were superior to the other tested cultivars (Gardiazabal and Cano, 1999).

Anderson and Richardson (1990) selected most promising cultivars of cherimoya in New Zealand and listed them according to the season of maturity. They are (1) early, Smoothey and Bays, (2) mid-season, Jete and Spain, and (3) mid-late, White, Burtons Favourite, Bronceada and Reretai.

Xie *et al.* (1999) described the performance of cherimoya cultivar African Pride introduced in China. The trees were hardy, withstanding temperature of 2.7°C without injury. Yields from 3- to 5-year-old trees were 337.5 per cent higher than that in other cultivars. Liu (2000) also reported that the cultivar African Pride performed very well in the area of Zhanjiang, China. It was highly adapted to the various soils and performed best in sandy loam soil. The trees were much more hardy than other *A. squamosa* cultivars (resistant to 0-5°C). The fruits matured mainly in September to November and were large, weighing 326 g on average, with white, tender flesh having a sugar content of 18.3 per cent.

Arka Sahan

It is a progeny of Island gem (Annona atemoya Hort.) X Mammoth (*A. squomosa* L.). Arka Sahan fruits come to harvest in September-October and mature fruits (210 g) take about 6-7 days to ripe. The skin has a waxy bloom, light green in colour, moderately thick (0.5 cm) with large, flat eyes. The creamy white colour flesh is juicy with mild pleasant aroma and tender with scarce seeds (9/100 g) and large segments. The edible pulp is remarkable for its sweetness with 22.8 per cent total sugars and measures more than 30°B as against 24°B in Mammoth. A 100 g pulp of Arka Sahan contains 2.49 g of crude protein, 42.29 mg P and 225 mg Ca against 1.33 g, 17.05 mg and 159 mg respectively in Mammoth. On an average, 12 tonnes of fruits can be harvested/ha. and resistant to drought (http://www.iihr.res.in/varieties).

Balanagar

Tree grows up to 3 m height. Fruits are spherical, pyramidal or cordate in shape; medium to big in size. Average fruit weight is 200 g but recorded upto 640 g; rarely 8.3 cm long and 3.5 cm in diameter. Areoles tuberculate. Very rough, pitted, forming deep furrows; rind greenish, mesocarp white with coarse and medium granules, pulp white, buttery sweet, with moderate to plenty of juice, flavor excellent; seeds 40–80. Highly productive, fruits remain green when ripe; rich in reducing sugars and proteins.

Barbados

Tree grows upto 2.4 m height. Fruits are spherical to cordate in shape; medium in size. Average fruit weight is 230 g. Fruits are 7.0 cm long and 7.7 cm in diameter; Areoles tuberculate. Very rough; rind fine; mesocarp with medium granules; pulp creamy white, moderately juicy, with agreeable flavor, sweet; seeds 40-60. Flowering late in June to early September. Prolific yielder.

British Guinea

Tree grows upto 1.8 to 2.4 m. Fruits are spheroid to cordate in shape; medium in size. Fruit weight ranges from 170 to 335 g. Fruits are generally 6.8cm long and 6.8 to 7.5 cm in diameter; Areoles tuberculate, acute, rough, rind grayish or grayish green; mesocarp white, soft; pulp white, soft buttery, crisp, moderately juicy, sweet, seeds 30-50. Flowering is late in June to mid September. High yielder and fruits remain green at ripening.

Crimson Custard Apple

Tree grows upto 1.8 to 2.0 m. Fruits are spherical to oval in shape; small to medium in size. Areoles medium-tuberculate, truncate, slightly rough, rind crimson, with grayish blue bloom, mesocarp reddish with medium to big granules, pulp with light red tinge, soft, moderately juicy, sweet, seeds 40-60. Flowering early June to end of August; trunk and branches dark green, new growth pinkish green.

Local Custard Apple

Tree grows upto 1.8 to 2.4 m. Fruits are spherical or cordate or conical in shape; Fruit weight ranges from 140 to 330g and recorded upto 615 g. Areoles small to big. Smooth to very smooth, round or mamillate; rind green, becoming light yellow on ripening. Mesocarp white, with fine, small to medium granules; pulp dull white, juiciness little to plenty, insipid to very sweet, seeds 30-80. It is an average yielder.

Mammoth

Tree grows upto 1.5m. Fruits are irregular, triquetrous to spherical in shape; average in size. Fruit weight ranges from 170 – 250 g, recorded up to 450 g. Areoles medium to big, very smooth, round, rind light russet; mesocarp white with small to medium granule; pulp white and translucent, soft bright, juicy, sweet, seeds 20-40. It has tendency to branch profusely; prolific bearer; rind becomes wrinkled on ripening; possesses low acidity; fruits have a good keeping quality and stand transport better.

Red Sitaphal

Tree grows up to 1.8 m. Fruits are spherical and cordate in shape. Fruit weight ranges from 170-250 g. recorded upto 450 g; areoles medium to big. tuberculate, truncate, rough and pitted; rind pink; grayish blue; mesocarp pinkish with coarse and big granules; pulp with pink tinge, moderately crisp, juicy, slightly acidic, sweet, seeds 40-60. Trunk and stem dark green; section of the stem shows the presence of pink coloured parenchyma in a circular band; mesocarp becomes deep pink on ripening; seeds small as in *Annona reticulata*. It contains maximum reducing sugars.

Red Speckled Custard Apple

Tree grows upto 1.50 to 2.0 m. Fruits are spherical to cordate in shape; small to medium size. Fruit weight ranges from 170 – 335 g. Areoles medium, tuberculate, truncate, very rough, ridged, pitted; rind light russet-green; splashed with red colour; mesocarp white with red colour; with medium to big granules; pulp dull white, moderately juicy, slightly stringy and just sweet, seeds 40–70.

Flowering till the end of August; poor yielder; appears to be a cross between 'Red Custard apple' and Local Custard apple.

Washington 98797

Tree grows upto 1.8 to 3.0 m. Fruits are spheroid to cordate in shape, medium to big size. Average fruit weight is 690 g, areoles medium, truncate, tuberculate, ridged, rough, rind light russet-green; mesocarp with medium to big granules; pulp creamy-white, soft, moderately juicy, pleasant, moderately sweet. Seeds 50–60. Flowering till early September. Heavy yielder.

Washington 107005

Tree grows upto 1.8 to 2.4 m. Fruits are spherical to short conical in shape. Average in size; fruit weight ranges from 170 to 335 g, recorded up to 560 g. Areoles tuberculate, with small pits; rind grayish green, mesocarp white with small to medium granules; pulp soft, bright white, buttery; moderately to very juicy, sweet, flavor, delightful, seeds 30-50. Rich in reducing sugars and also in total acidity and protein.

APK-1

It is a clonal selection from a high yielding type in the State Horticultural farm, Courtallam. It bears higher yield in rainfed vertisol (black soil). It is a drought tolerant variety which bears sweet fruits (TSS of 24.5 Brix with an acidity of 0.2 per cent). The Fruit weight ranges about 207.5 g with a mean of 72 fruits per tree per year. The first bearing commences in a graft/budded plant at 3-3 ½ years. The Optimum productive life is 25 yrs. The best season for planting is May – June/ August – September. It yields about 7300 kg fruits/ha (14.90 kg/tree which are 30.7 percent higher than Balanagar). It is suitable for cultivation in plains of Tamil Nadu especially in semi arid regions and marginal soils of both vertisol and alfisol in dry tracts. It is suitable for both rainfed and irrigated conditions http://www.agritech. tnau.ac.in/horticulture/horti_ TNAU_ varieties_ fruitcrops. html # custard).

Other Custard Apple Cultivars

Some cultivars were identified based on colour of the fruit, its origin and plant part: Yellow custard apple named after the colour of the fruit; Cultivars like Kakarlapahad, Mahaboobnagar, Local Hyderabad and Saharanpur Local named after the place of origin; White stemmed custard apple named after the part of plant.

Cultivars from Alexandria and Egypt

Beni Mazar

Round, large, fruit weight ranges from 150 to 180 g; 56-60 per cent flesh; 15-30 seeds per fruit (Morton, 1987).

Abd El Razik

Light-green or reddish rind; nearly round, large, average fruit weight is 236.3 g; 69.5 per cent flesh; 14 seeds per fruit (Morton, 1987).

Cuban Seedless and Brazilian Seedless

One is a seedless form ('Seedless') with seemingly identical selections from Brazil and Cuba ('Cuban Seedless' and 'Brazilian Seedless'). This form is not likely to be widely grown because the fruit splits very badly as it matures on the tree and fruit quality appears to be slightly inferior to most-seedling fruit. It is not as productive as the seeded types, is nearly identical in quality and flavor. Another "seedless" sugar apple was introduced from Brazil, which appeared to be identical to the Cuban variety (http://university.uog.edu/cals/people/PUBS/Sweetsop/MG33000.pdf).

The list of common varieties of custard apple grown in different countries is given in Table 4.

Table 4: Some Selected Cultivars of Sweetsop*

Country/Region	*Cultivar Name*
Taiwan	Ruan-zhi, Cu-lin, Da-mu, Xi-lin and Tai-nong no.1
India	Balanagar, Mammoth, Arka sahan, Barbados, Washington and Red Sitaphal
Thailand	Fai Kaew, Fai Krung, Nang kaew, nang Sir Krung and Nang Thong
Florida	Lessard, kampong Mauve, Red Sugar, Cuban seedless
Egypt	Abd El-Razik

* Chen and Paull (2008).

5.0 Breeding and Improvement

5.1.0 Cytogenetics

Understanding cytology of a crop is useful in planning breeding programme. The chromosome numbers of cherimoya, Bullock's heart, soursop and sugar apple are 2n = 14 to 16 (Nakasone and Paull, 1998). Although there is some variation in chromosome number, they are all diploids, 2n = 2x. This slight variation in chromosome number may explain the ease or difficulty of interspecific hybridisation and warrants further work to determine if intra-specific variation also exists. *A. glabra*, are known to be tetraploid.

Generally, cross-pollination between Annonas is conducted primarily to determine compatibility for increasing fruit set and occasionally for new hybrid development. The crosses among soursop and other Annonas, such as cherimoya, Bullock's heart custard apple or sugar apple, have not been successful. This may reflect the genetic distance between soursop and the others. However, there is a dearth of information on species' relationships, so this is a topic that needs to be researched.

Cross-pollination between species is conducted primarily to determine compatibility for increasing fruit set. Inter-and intra-specific cross combinations among *Annona squamosa, A. atemoya, A. reticulata, A. glabra, A. cherimola* and *A. muricata* were investigated by Kumar and Jalikop (2000).They reported that *A. squamosa* was the most effective pollen parent of *A. atemoya* and *A. cherimola*. However, *A. squamosa* was not an effective female parent with A. *atemoya*, whereas *A. cherimola* was compatible with both *A. squamosa* and *A. cherimola*. A. *atemoya* can be crossed freely with *A. reticulata, A. glabra* and *A. cherimola*, A. a*temoya* is the only hybrid that has gained importance and it has inherited the glabrate leaf character of *A. squamosa* and a leaf size almost as large as that of *A. cherimola*. Flowering and fruiting seasons are similar to those of sweetsop. Skin, pulp and seed characters of both parents are inherited in varying degrees by each plant. A desirable hybrid would be between the cherimoya and soursop that combines the larger fruit size and acidity of the soursop and the cherimoya's sweetness, flavour and texture. Attempts to cross the soursop with cherimoya, ilama, bullock's heart or sweetsop have not been successful and may reflect a considerable genetic distance of soursop from the other species (Paull and Duarte, 2011). Jalikop (2010a) investigated the inheritance of foliage characteristics in trispecies hybrids of atemoya (*A. cherimola* x *A. squamosa*) and *A. reticulata*. Segregation for leaf colour (green or dark green), leaf angle (erect or drooping), leaf apex shape (rounded or pointed) and time of leaf fall (early or late or intermediate) into discreet phenotypic classes revealed that their inheritance followed simple Mendelian genetics. Duplicate dominant gene interaction governed the leaf colour and leaf position. Individuals with single or both the dominant genes (A-B-, A-bb,aaB-) produce green colour leaves and those with recessive genes (aabb) dark colour leaves. Likewise erect leaf individuals are C-D-, C-dd, ccD- and drooping leaves are ccdd. A single gene determined shape of leaf apex that was either rounded (Rr) or pointed (rr).Segregation of progenies for leaf fall into early, intermediate and late suggested that codominant alleles were responsible for time of leaf fall. The pattern of segregation of tree and fruit traits in trispecies hybrids (*A. cherimola* Mill. x *A. squamosa* L.) and *A. reticulata* L. have been investigated by Jalikop (2010b) with the objective of salvaging useful genes from the three edible Annonas and determining the extent of variation in the progeny. The heterogeneous nature of *Annona* species and simultaneous segregation of three distant genomes resulted in a tremendous heterogeneity in the progenies and thus demonstrated that *A. reticulata* is potentially a valuable source of a variety of novel traits. However, the

Cultivars of Annona

Balangar

African Pride

Bradley

Pink Mammoth

Cultivars of Annona

Arka Sahan

Sugar Apple

F_1 trihybrids exhibited reduced fertility with pronounced undesirable traits of A. *reticulata*, and the survival of F_2 trees was poor, with those that did survive being extremely weak and failing to flower or set fruit. Consequently, backcrossing may be more suitable approach than advancing trihybrid generation for transferring desirable genes from A. *reticulata*. Backcrossing F_1 progeny of [(*A. cherimola* x *A. squamosa*) x *A. reticulata*] to *A. squamosa*, a highly fertile species, is expected to yield productive introgressants in which the contribution of the *A. squamosa* genome will be extended from 25 to 62.5 per cent in F_1BC_1 generation.

5.2.0 Constraints in Breeding

Wester (1913) was the first scientist to realize the possibilities for genetic improvement of Annonas and initiated breeding programmes in Florida and in the Philippines. The long reproductive cycles, higher levels of heterozygosity and the costs associated with evaluating large populations of crosses, limits breeding programmes. Existing commercial cultivars show considerable variation in growth, fruit set, fruit size and quality. No single variety has all the desirable characteristics. The length of the juvenile period varies, with earliest production occurring in 2 years and full production in 5–6 years. This juvenile period is extremely variable with scions on seedling rootstocks. The seedling rootstocks are derived from extremely heterogeneous, open-pollinated seeds; hence it is difficult to fix specific characters in a short period. Breeding programmes have focused on selections from seedling populations. Seedling rootstocks of Annonas are generally derived from heterogeneous open-pollinated plants; hence, it is often difficult to fix specific characters in a short period. Early maturity, better fruit appearance and long post-harvest life for tropical Annonas, and in the subtropics, greater cold tolerance, are objectives for cherimoya breeding (Nakasone and Paull, 1998).

5.3.0 Floral Biology

5.3.1 Floral Biology of *Annona squamosa*

The flowers of sweetsop are hermaphrodite and are produced singly or in small clusters on the current season's growth. The flower buds start appearing on new shoots in the spring, immediately after the commencement of the vegetative growth. The flowering period in sweetsop was from the middle of April to end of June. The average time from visible flower initiation to full flowering was 30.8 days (Nalawadi *et al.*, 1975). Anthesis took place continuously throughout the day and night. Anthesis commenced at 6.00am and continued up to 6.00 pm, with its peak at 6.00 am under Dharwad condition, between 5.30 and 6.30 am under Bhubaneshwar (Sahoo *et al.*, 2000) whereas anthesis occurs at 5.00 am in Brazil (Ribeiro *et al.*,2007). Kiill and Costa (2003) reported from Brazil that the anthesis takes place around 5.00 pm, the flowers last for approximately two days, and show female phase in the first twenty hours and male phase in the following twenty hours, characterizing dichogamy. Anther dehiscence commenced at 12.00 pm and was completed by 4.00 am, with a peak at 2.00 am (Nalawadi *et al.*, 1975).The flower exhibits a protogynous dichogamy nature that limits self-pollination and cross-pollination can increase fruit set. Pollen grains appearing early in a flowering season have lower germination rates than the pollen from late flowers (Chen and Paull, 2008).The highest pollen germination was recorded between 18.9 per cent and 20.4 per cent in 20 per cent sucrose for green and red type, respectively.The stigmas were receptive from one day prior to anthesis till 2-3 days after anthesis indicating a protogynous condition (Thakur and Singh, 1965; Sahoo *et al.*, 2000). Nitidulid beetles, *Carpophilos domidiahus* and *Carpophilos hemipterous* were identified as pollinators in terms of visitation frequency, pollination potential score and pollination efficiency. However, *Carpophilos hemipterus* was relatively effective pollinator than *Carpophilos domidiahus*. Sweetsop clearly exhibit obligate specialization by filtering only *Carpophilos hemipterus* and *Carpophilos domidiahus* as pollinators and thereby explicitly indicating canthrophilous pollination syndrome (Kundan Kishore *et al.*, 2012).Artificial pollination is frequently practiced to ensure pollination and good fruit shape. Hand pollination is normally carried out before 8.00 am using a small brush. Pollen can be collected in the morning between 5.00 and 8.00 am from fully open flowers, when the sacs have turned from white to cream. The collected pollen is used to pollinate half-open flowers whose pistils are already receptive (Jalikop and Ravindra Kumar, 2007).

5.3.2 Flower Biology of *Annona cherimola*

The flowers are hermaphrodite and are produced singly or in clusters on the current season's growth. Cherimoya generally requires 27–35 days for flower bud development from initiation to anthesis. Two major flowering periods occur after periods of vegetative flushes with the second peak coinciding with the onset of monsoon in India. The flowers exhibit both dichogamy and a protogynous nature. This poses a serious problem in obtaining high yields. Cherimoya flowers are receptive, opening around 7–9 a.m. and when pollen is shed at 3.0–4.0 p.m. if the relative humidity is above 80 per cent and temperature is less than 22°C. The flower shows synchrony with that of sweetsop and this together with complementary functional sexes favours cross-pollination leading to natural hybridization. This is attested to by the frequent appearance of hybrid seedlings under the trees of sweetsop and cherimoya when grown in close proximity. Nitidulid beetles (*Carpophilus* and *Uroporus* spp.) are the important pollinators of annona flowers with wind and self-pollination being low (1.5 per cent).Pollen grains appearing early in a flowering season have thick walls, are high in starch, germinate poorly, and give poor fruit set (Saavedra, 1997). Pollen of later flowers shows a high proportion of individual pollen grains without starch grains that germinate well. Hand pollination is frequently practiced to ensure pollination and good fruit shape. Pollen must be collected in the evening from fully open flowers, when the sacs have turned from white to cream. The flowers are held in a paper bag, not a closed container, and should discharge that afternoon. The flowers are shaken over a shallow tray or paper to collect the pollen, which is transferred to a small container and held in the refrigerator for use the next morning. Pollen from 20–30 flowers can give enough pollen to pollinate 50–60 flowers. Hand pollination has shown some variable results and is less successful on very humid, overcast days and on young vigorous trees. About 150 flowers can be pollinated in an hour and a success rate of 80–100 per cent can be achieved.

5.3.3 Floral Biology of *Annona muricata*

Soursop generally requires 27–35 days for flower bud development from initiation to anthesis. The soursop produces fruit throughout the year but peak production in most areas comes during summer and early autumn, sometimes with a secondary peak during early spring. The flowers exhibit both dichogomy and a protogynous nature. Soursop floral anthesis takes place mostly between noon and 8.0 p.m. and 4.0 a.m.–8.0 a.m., with pollen release occurring between 4.0 a.m. and 8.0 a.m. Flower opening and pollen dehiscence are not synchronous and thus very low self-pollination occurs. Cross-pollination may take place early in the morning because at anthesis the flower usually emits a fragrance that attracts insects. In order to increase yield, hand pollination has become an important aspect of cultivation practices.

5.4.0 Breeding Objectives

The first objective is the determination of the specific characteristics that are important for the new cultivar, the ideotype. These characteristics may help a breeder to select parental groups to be used in an Annona breeding programme, in order to obtain desirable progenies. These progenies may not have all desirable characteristics, but at least will have those most important to growers, retailers and consumers. For instance, a sweet sugar apple cultivar with long shelf life fruits (>5 days), which is important from a consumer's viewpoint, may not be selected, if it presents low yield (<20 kg/tree/year) and fruit weight lower than 400 g. Genetic improvement for yield is the most difficult and expensive of all breeding objectives, due to the complex nature of its inheritance and the numerous environmental factors influencing the trait. Quality is another important characteristic for the improvement of Annonas, as important or sometimes more so than yield, since market value is based on the fruit's appearance and its organoleptic characteristics (Pinto *et al.*, 2005).

Breeding for pest and disease resistance is another important part of cultivar development. The soursop cultivar 'Morada' has shown less susceptibility to fruit borer than other cultivars under the conditions of Brazil's Central Region (Pinto and Silva, 1996). Given the stage of domestication of the three principal Annonas, it is not surprising that this type of resistance is rare, especially when most Annona plantations are monoculture orchards.

Quantitative characteristics, like yield, controlled by numerous major and minor genes, are more difficult to manipulate than qualitative ones. In addition, traits whose expression is influenced by environment are also more difficult to select for than those unaffected by environmental factors. As *Annona* breeding programmes have several objectives, it is the responsibility of the breeder to design cost-effective strategies to select genotypes with the maximum number of desired traits (Table 5).

5.4.1 Breeding Strategies

Most *Annona* species and cultivars differ in environmental adaptation, productivity and fruit quality. Therefore, different conventional methods can be used in their breeding. According to Fehr (1987), there are three requirements for the development of an asexually propagated cultivar: a) a suitable source of genetic variability; b) evaluation of individuals from the population; c) asexual multiplication of a new cultivar for commercial use.

Table 5: Main Characteristics of Cherimoya, Soursop and Sugar Apple Ideotypes (Adult plants)

Characteristics	*Cherimoya*	*Soursop*	*Sugar Apple*
Plant			
– Vegetative vigour	Medium to low	Medium to low	Medium to low
– Yield	(>30kg/tree/year)	(>60kg/tree/year)	(>20kg/tree/year)
– Bearing	Regular	Regular	Regular
– Rootstock/Scion compatibility	High Abundant	High Abundant	Abundant
– Flower number	(>150 flowers/tree)	(>100 flowers/tree)	(> 180 flowers/tree)
– Pollen fertility	High (>76 per cent)	High (>76 per cent)	High (>76 per cent)
– Harvest	Out of season	Throughout year	Out of season
Fruit			
– Size/Weight			
For industry	Large/> 650g	Large/> 2.5kg	Large/> 400.0g
For fresh	Medium/300 – 600g	Small/0.8kg	Medium/300 – 400g
Consumption	Heart High (>27 per cent)	Conical High (> 26 per cent)	Rounded to Heart
– Shape	Smooth	Short protuberances	High (>25 per cent)
– Fruit set (naturally)	Sweet, fiberless	Sub-acid, low fibre	Tuberculate
– Skin	Delicate	Sub-acid flavor	Sweet, low fibre
– Pulp	Low or absent/<6seeds	Low/10 -30 seeds	Pleasant acidulous
– Flavour	Long (> 10days)	Long (> 5 days)	Low/absent (<10seeds)
– Seed number per 100g of pulp	–	–	Long (> 5 days)
– Shelf life (15 to 30°C)	–	–	–

Introduction of superior genotypes and/or cultivars to establish a germplasm collection is, basically, the first requirement of any breeding programme. This can be complemented by the introduction into the collection of some wild *Annona* species with useful genes, mainly for resistance to diseases. All accessions require comprehensive characterization and documentation, followed by evaluation and selection. Several types of populations can be developed by hybridization, from which superior clones are selected.

5.4.2 Selection and Cultivar Development

India and Taiwan have selected a few named cultivars of sugar apple from seedling population that are propagated vegetatively. In Cuba, researchers developed 'Cuban Seedless', which is a seedless cultivar with medium-sized fruits, and another cultivar with low fibre content that is very important for the commercial market (Nakasone and Paull, 1998). In the Philippines, there are 3 forms of sugar apple fruits: (a) a green-fruited seedy form, which is grown in most parts of the country; (b) a purple-fruited seedy form, reportedly introduced from India; (c) and the green-fruited seedless form, which is a recent introduction and whose origin is unknown. The selection of superior strains is aimed in the direction of a green-fruited seedy form (Coronel, 1994). The following Tables 6, 7 and 8 list some of the selections in important Annona species.

Table 6: Some Selections and Cultivars of Sugar Apple that is Currently Grown in Various Countries*

Cultivar/Selection	*Country of Origin*	*Cultivar/Selection*	*Country of Origin*
IPA selections	Brazil	Leone	Italy
Libby	USA	Madeira	Portugal
Lisa	USA	Mateus 1	Portugal
Mossman	Australia	Negrito	Spain
Ott	USA	Reretai	New Zealand
Whaley	Australia	White	USA
Molate	Philippines	Cuban Seedless	Cuba
Lobo	Philippines	Noi	Thailand
Red Sugar Apple	USA/Florida	Mammoth	India
Balanagar	India	Red Sitaphal	India

* Pinto *et al.* (2005).

Table 7: Selected Cultivars of Cherimoya

Cultivar(s)	*Country of Origin*
Andrews, Kempsey, Mossman	Australia
Bays, Booth, Libby, Lisa, White	California, USA
Bronceada, Concha Lisa	Chile
Burton's Wonder, Reretai	New Zealand
Campa, Cristalino, Negrito	Spain
E-8	Ecuador

Table 8: Selected Cultivars of Atemoya

Cultivar(s)	*Country of Origin*
Maroochy Gold, KJ Pinks,	
Pink's Mammoth	Australia
African Pride	Southern Africa
Bradley, Page	Florida, USA
Gefner, Kabri, Malalai	Israel

5.4.3 Hybridization

Tri-species hybrids between *A. atemoya* x *A. squamosa* have been attempted by Jalikop and released one hybrid for commercial cultivation during 1997 in India which is a cross *A. atemoya* cv. Island gem x *A. squamosa* cv. Mammoth and named as Arka Sahan (Jalikop and Ravindra Kumar, 2007). Likewise, Girwani *et al.*, 2010 also developed certain sugar apple hybrids at Sangareddy, India.

5.4.3.1 Varietal Evaluation of Sugar Apple/Sweetsop in India

Several varietal evaluation studies have been carried out in various regions of India using different cultivars. Evaluation of sweetsop genotypes (APK-1, Mammoth, Balanagar) under sodic soil at Trichy, Tamil Nadu by Krishnamurthy (2009) showed superiority of APK-1 with regard to yield in nine years old trees which produced 47.78 kg per plant, followed by Balanagar (27.05 kg per plant).Thirty five germplasm of custard apple have been evaluated for scarce rainfall zone under rainfed at Anantapur, Andhra Pradesh by Dhanumjaya Rao and Subramanyam (2011) in 10 years old trees and the maximum yield have been recorded from MMR-10 (15.0 kg per plant). Girwani *et al.* (2011) evaluated sweetsop hybrids under Sangareddy condition and revealed that All the hybrids varied in fruit shape (round, conical and cordate), fruit colour (yellowish green, grayish green, light green and red), pulp colour (creamy white, light pink and white), areole shape (tuberculate, hexagonal) and texture (course, soft and meaty). Time of fruit maturity varied from September 2nd fortnight and lasted till the end of December. Hybrid-1 (17/4 Atemoya x Balanagar) was earliest to mature. Maximum fruit weight (250 g) was recorded in Hybrid-1 (17/4 Atemoya x Balanagar) followed by Hybrid-6 (15/3 Red Sitaphal x Atemoya) (225 g) and Hybrid-4 (1/6 British Guinea x Atemoya) (220 g). Among all the hybrids, maximum number of fruits per tree (94) was recorded in Hybrid-1 (17/4 Atemoya x Balanagar) followed by Hybrid-6 (15/3 Red Sitaphal x Atemoya) (67). The TSS ranged from 22 to 28°B and was found to be maximum in Hybrid-2 (15/2 Red Sitaphal x Pondapple), while the seed content per fruit was also minimum (20) in this hybrid. Based on overall performance the Hybrid-1 (17/4 Atemoya x Balanagar) and Hybrid-3 (15/3 Red Sitaphal x Atemoya) and Hybrid-2 (15/2 Red Sitaphal x Pondapple) were found to be excellent in quality with pleasant aroma, smooth pulp texture with less seed, good sugar acid blend and shelf life.

Since the cultivation of Annonas is increasing and there are very few commercial cultivars, the need for breeding and improvement is immensely felt. The propagation by seeds has given rise to numerous seedlings which vary greatly in productivity, as well as in size, colour and quality of the fruits. A survey of the commercial orchards in Maharashtra revealed a great deal of genetic diversity in sugar apple. The range of variation in fruit weight was 130 to 370 g, in seed number 14 to 66, in pulp percentage 34.4 to 60.6, in total soluble solids 18.0 to 28.2 and titratable acidity 0.20 to 0.92 per cent (Anon., 1987). This variability provides immense possibilities for selection of promising cultivars as well as parents for breeding programmes.

Some superior genotypes of *A. squamosa* have been selected at the Mahatma Phule Krishi Vidyapeeth, Rahuri, India on the basis of yield and size of fruit and areoles, percentage of pulp, number of seeds per fruit and tree vigour. Rahuri

10/4 and 14/3 selections appear outstanding. The type numbers 5, 6, 8, 10 and 15 from Saswad region of Maharashtra are also promising (Anon., 1987). Some of these selections should be propagated vegetatively and released for commercial cultivation after careful evaluation.

Wild Cherimoya accessions were collected in Loja province (southern Ecuador), a centre of biodiversity of cherimoya and probably its origin, between January, 1996 and March, 1998. A total of 448 fruits from 137 trees in 32 sites in 16 parishes of 7 countries in the province of Loja were collected and characterized according to fruit weight, seed percentage, seed index, pulp percentage, soluble solids and skin type. The data indicated a large variability in fruit characteristics, a typical feature for plant species collected in or near their centre of origin, supporting the hypothesis of Loja province being located in the centre of origin (Scheldeman *et al.*, 1999).

Fifteen out of 83 accessions collected in Loja province showed excellent fruit characteristics (Scheldeman and Damme, 1999). Agustin (1999) reported that cherimoya grown in Mexico had wide genetic variability with some genotypes having good agronomic characteristics. This provides an opportunity to collect and conserve superior genotype in the regional germplasm bank. The cherimoya germplasm bank of La Mayora in Malaga, Spain, commenced in 1979 with 29 accessions, reaching 280 in 1998 (Hermoso *et al.*, 1999).

Five ecotypes of cherimoya have been selected in the Pichincha province in Ecuador (E4, E7, E8, E15, and E17) and 8 in Azuay (E4, E14, E16, E27, E41, E64, E66, E69) on the basis of outstanding pomological characteristics (Leon Fuentes, 1999). Silva Junior *et al.* (1999) evaluated genotypes of *A. muricata* in Pernambuco state in Brazil. Apart from morphological characteristics, some of the genotypes showed good adaptation to the specific soil and climatic conditions and great variability was found among them.

Hybridization has tremendous scope in Annonas because of the available genetic variability and comparatively easier intraspecific and interspecific hybridization. Atemoya, an interspecific cross between *A. cherimola* and *A. squamosa* is already popular as commercial fruit in Australia and other countries. A large number of available *Annona* species may be intercrossed in order to evolve not only suitable cultivars but also to develop rootstocks adaptable to various soil and climatic conditions. This will help in extending the cultivation to newer areas. Robust nature of *A. montana* and tropical and swamp loving trait of *A. glabra* may be useful in evolving rootstocks having these characteristics. Improvement can also be effected in *A. montana* and *A. pupurea* by selection (Baily, 1949). All the important cultivated species have the haploid chromosome number as seven except the species *A. glabra* which has fourteen (Singh, 1978). Tetraploid obtained by colchicine treatment had larger and thicker leaves and flowered earlier than diploids (Islam, 1953). With a view to evolving the genotypes bearing fruits of uniform size, better keeping quality, less number of seeds and smooth surface along with high fruit to flower ratio, breeding work is in progress at Bangalore (Anon., 1987). No fruit set occurred by using atemoyas, Island Gem, Pink's Mammoth and Bullock's heart as pollen parent on *A. squamosa* while the reciprocals set fruits freely. Interspecific cross between *A. glabra* and *A. squamosa* was also not successful.

In studies in California on *A. cherimola* both maternal (cultivars White and Booth) and paternal (White, Booth and Chaffey) parents had significant effects on quanti-tative and qualitative fruit and seed characteristics. Kahn *et al.* (1994) found a significant effect of maternal cultivars on mean seed weight, fruit set, surface morphology of carpels and the percentage of carpels with protuberances. Pollen parent cultivar also produced significant differential effects on fruit and seed characteristics or xenia. Pollen parent cultivar had significant effects on fruit characters such as fruit length, fruit weight and percentage of fruit set. Seed characters significantly affected by pollen parent included seed weight per fruit, seed number and mean seed weight.

A major breeding programme, funded by Australian Custard Apple Growers Association, has been launched with the objectives of selecting new varieties of atemoya with smooth skin, symmetrical shape, low seed number, low susceptibility to blemish and high level of self-pollination. A conventional breeding programme using polycrosses of 11 of the best, advanced selections/cultivars (Hillary White, Martin, Paxton, MHRS Gold, MHRS Red, Palethorpe, R11-T3, R11-T4, R11-T6, Ruby Queen, Bullock's Heart) has been initiated. Attempts will be made to evolve seedless cultivars by first producing tetraploids (through gamma irradiation or colchicine) and then crossing to diploids to produce seedless tetraploids (George *et al.*, 1999).

6.0 Propagation

Annonas are propagated both by sexual and asexual methods. Seed propagation is widely prevalent resulting in variability in plant vigour, prolonged juvenility and inferior fruit quality (Campbell and Phillips, 1983). The clonal propagation is, therefore, a necessity to maintain the genetic uniformity for obtaining higher yield of better quality fruits.

6.1.0 Seed

Inspite of a thick impervious seed-coat, very short or no dormancy is observed in sugar apple seeds (Hayat, 1963). Fresh seeds germinate in 20-30 days giving as high as 90 per cent germination (Cantillang, 1976). The seed germination in other annonas, however, varies from 30 to 80 per cent (Campbell and Popenoe, 1968). Seed treatments, such as scarification and soaking in hot water did not increase germination (Hayat, 1963; Casas *et al.*, 1984). Ratan *et al.* (1993) reported that seeds from freshly harvested, fully ripe *A. squamosa* cv. Balanagar soaked for 24 h resulted in the highest germination (71 per cent compared with 15 per cent in unsoaked controls).

Soaking of cherimoya and *A. diversifolia* seeds in gibberellic acid solution (500 ppm) improved germination considerably (Campbell and Popenoe, 1968). Seeds of *A. reticulata* were treated with GA3 at 5000, 7500 or 10,000 ppm. GA3 promoted germination. Germination was significantly better in bags than in beds, but seedling height and diameter were better in beds (Cartagena Valenzuela and Barreto Osorio, 1998). For germination and growth of seedlings of custard apple cocopeat and vermiculite were significantly superior to the standard pot mixture in terms of germination percentage, root number, root length, seedling height and number of leaves per plant (Chopde *et al.*, 1999).

Uneven and irregular germination of cherimoya seeds results in difficulties in propagation. It was found that the cherimoya embryo is small (3-4 mm) and located inside an impermeable testa with the micropyle being the only entrance for water. The effects of soaking in gibberellic acid (GA3) at 500-10000 ppm for 24 or 48 h, soaking in distilled water for 12-72 h and chemical scarification with 75 per cent sulphuric acid were evaluated by Smet *et al.* (1999). GA3 had a positive effect on percentage germination and rate (percentage germination/day), maximum rate being 1.74 after soaking in GA3 at 5000 ppm for 24 h, compared to 0.08-0.16 in the untreated control.Seeds are sown in beds or polythene bags filled with soil, the latter is preferred because of ease and low mortality during transplanting.

6.2.0 Vegetative

Grafting and budding are commercial methods of vegetative propagation of Annonas. Stooling and mist propagation of cuttings are promising methods for clonal propagation while air-layering has not given encouraging result (Rathore, 1976; George and Nissen, 1987b).

6.2.1 Cuttings

With the exception of *A. muricata*, all other Annona species have given poor results from stem or root cuttings (Venkataratnam and Satyanaranswamy, 1956; Bourke, 1976). George and Nissen (1983, 1986a), however, recorded considerable rooting in tip cuttings of Atemoyas under mist at 27°C. The stem cuttings treated with IBA rooted better than the untreated ones and the cultivar African Pride had higher rooting potential than Pink's Mammoth. Etiolation of shoots also improved the rooting and 90 per cent success in cuttings of sugar apple under intermittent mist by etiolating for 15 days and treating with 5000 ppm NAA was reported by Dhua *et al.* (1982). Similar results were also recorded by George and Nissen, (1986a) in cherimoya and *Annona* hybrids. The plants raised from cuttings of African Pride were uniform, vigorous and more productive than from grafts (George and Nissen, 1986b).

Sanewski (1991) reported that atemoya tip cuttings were superior to stem cuttings, with rooting percentage varying between 50 per cent and 60 per cent in comparison to 25 per cent in stem cuttings. Season of taking cutting materials played vital role on rooting success. It was reported that cutting taken at the end of winter season gave better success (Nakasone and Paull, 1998). Median cuttings of atemoya rooted best in Plant max (commercial medium) and treatment with growth regulator failed to improve rooting parameters compared with immersion in water (Ferreira and Cereda, 1999).

6.2.2 Stooling

Rathore (1976) obtained 93.8 per cent rooting in the stools of *A. squamosa* after treating with 20000 ppm IBA. The rooted stools gave 80 per cent survival on planting. Three-year-old seedlings were headed back at ground level in the month of

April. After about three months, a 3-cm wide ring of bark was taken out from the base of each emerged shoot barring one to serve as a nurse shoot, and lanolin paste containing IBA was smeared above the ringed portion followed by mounding of all the shoots. The rooted shoots were detached from the mother plants in the month of September. For clonal propagation, this method appears to be very promising.

6.2.3 Budding and Grafting

Between shield budding and grafting (splice, whip and tongue, cleft, crown) in cherimoya, Moran *et al.* (1972) found no difference in the percentage taken and subsequent growth. Both methods are, therefore, utilized for the propagation of all custard apples. However, Duarte *et al.* (1974) reported that grafting was superior to budding in percentage taken and subsequent growth and side-whip and cleft-graft methods gave the best results. Whip, whip and tongue, cleft, splice, veneer and inarching have been found to be successful in Annonas (Ahmad, 1936a; Choudhry, 1954; Singh, 1959; Campbell, 1970; Fouque, 1972; Bourke, 1976). The operation should be carried out in early spring using one-year-old mature scion and more than one-year-old rootstock. Grafted trees are transplanted in the main field during the following year (Grossberger, 1999). Ledo (1995) found that whip grafted plants of 3 cultivars of *A. muricata* had higher root system than top cleft grafted plants. The thicker rootstocks give better performance (George and Nissen, 1987a). Among the three methods of propagation in *A. reticulata*, the highest percentage of success was achieved through softwood grafting (91.5 per cent), followed by patch budding (89.0 per cent). Shield budding was least successful (80.5 per cent) (Gholap *et al.*, 2000).

Experiments on large scale multiplication of custard apples (*Annona squamosa*, cv. Sindhan Local) were performed by Chauvatia and Singh (1999) in Gujarat, India. The time required for sprouting did not differ significantly between shield budding and wedge grafting, but the time of propagation was significant, bud sprouting occurred earliest (after 13.83 days) when budding or grafting was performed in April. Grafting success (73.53 per cent) was higher than that of budding (69.81 per cent). Irrespective of the method used, greater success was achieved when performed in March (91.11 per cent), or April (85.00 per cent), and success was lowest in July.

Mortality during transplanting in Annonas can be reduced by planting in dormant season, avoiding root damage, over-watering and improper fertilization. Cherimoya is propagated by grafting, with the aim of preserving genetic characteristics. The best method of grafting is terminal scion grafting (Leon Fuentes, 1999).

6.2.4 Rootstock

Seedlings of the scion species are mostly employed as rootstocks in budding and grafting usually giving satisfactory performance. But sometimes a different species of the genus serves better as a rootstock imparting specific characteristics to the plant, *i.e.*, precocity, higher yield, dwarfness, resistance to diseases and tolerance to water logging, drought, cold and high temperatures. When *A. squamosa* was grafted on the seedlings of *A. squamosa, A. reticulata, A. muricata* and *A. palustris,* it survived only on the first two species and started yielding in the fifth year, 41.4 fruits per plant on *A. reticulata* and 29 fruits on *A. squamosa* (Khan and Rao, 1952).

Propagation of Annona

Propagation by seed

Propagation by cutting

Propagation by tissue culture

Pruning, Flowering and Pollination of Annona

Annona tree in orchard

Pruning of annona

Flowers of annona

Annona squamosa

Annona muricata

Annona reticulata

Pollination in annona

Manual pollination in annona

Pollination by bee

Similar results were also reported by Buell (1955) as sugar apple scions died within few months on *A. muricata* and *A. palustris*, showing better tree vigour, fruit quality on *A. reticulata*, than on *A. squamosa* rootstock. Rootstock of *A.reticulata* also imparted longer life to the scion than on *A. squamosa*. The most popular atemoya cultivar, African Pride gave about double the fruit yield on cherimoya than on sugar apple rootstock in the sixth year of planting (George and Nissen, 1986b). They also indicated dwarfing potential of sugar apple rootstock but found it susceptible to *Pseudomonas solanacearum*. In a grafting trial with *A. muricata* scions, *A. muricata* seedlings gave a higher percentage taken in budding and grafting than those on *A. montana*, *A. reticulata* and *A. squamosa* seedlings (Iglesias and Sanchez, 1985).

Seedling rootstocks, however, exhibit wide variability in growth and yield of the scion. Considerable graft incompatibility has been reported among *Annona* and *Rollinia* species and types. Cherimoya was noted to be a vigorous rootstock for Pink's Mammoth, while atemoya was not compatible with *A. glabra, A. montana, A. muricata* and *A. reticulata* as rootstocks (Sanewski, 1991).

The effects of flooding on the net CO_2 assimilation, growth and stem anatomy of grafted and seedling plants of 3 *Annona* species (*A. glabra, A. muricata* and *A. squamosa*) and the scion Gefner, 49-11 and 4-5 were studied by Nunez-Elisea *et al.* (1999). After flooding for up to 60 days, seedlings of *A. glabra* and *A. muricata* survived and grew under flooded conditions and were considered tolerant. Similarly, when the normally flood sensitive *A. reticulata*, Gefner and 49-11 were grafted onto the tolerant rootstocks, they became tolerant. However, *A. squamosa* seedlings and rootstocks and *A. reticulata* rootstocks were flood sensitive, having reduced growth and net CO_2 assimilation. Losses in the sensitive plants were 20-80 per cent.

6.2.5 Topworking

Since large number of Annona plants from seedlings have uncertain yield and quality, their top working with commercial cultivars is a good economical proposition. Cleft or bark grafting using two or three scions on large limbs can be employed successfully at the end of dormant season (Dickinson, 1963). In order to obtain quicker and higher return from topworked trees, frame working is more desirable. A large number of one-year-old shoots emerged out of the beheaded limbs may be grafted or budded.

7.0 Area and Production

The total area under cherimoya cultivation in Spain has increased from 106 hectares in 1953 (Velazquez, 1953) to 3,400 hectares in 1998 (Cabezas Prieto, 1998). The cherimoya industry in California is relatively small (111 hectares) compared with that in Chile and Spain. There are seven times more hectares planted in Chile and some thirty times more hectares planted in Spain as are planted in California (Grossberger, 1999). Cherimoya cultivation in Chile occupies about 1200 hectares, producing for both the internal and export markets, principally the USA, Japan and certain Latin American countries (Gardiazabal and Cano, 1999).

In India, apart from small orchards in semiarid regions, the Annonas groves are found growing here and there in the gardens or home-backyards to meet the demand of local markets. The ripe fruits are perishable and ferment quickly and cannot withstand distant transportation. Maharashtra appears to be the leading state in annona production in India. Recent data on area and production of annona in India are given below (Table 9):

Table 9: Annona Production in India during 2015-16

State	*Production (000tonnes)*	*Share (Per cent)*
Maharashtra	92.32	30.98
Gujarat	61.18	20.53
Madhya Pradesh	56.74	19.04
Chattisgarh	39.58	13.28
Telangana	15.91	5.34
Karnataka	14.45	4.85
Andhra Pradesh	11.35	3.81
Rajasthan	6.12	2.05
Kerala	0.19	0.06
Tamil Nadu	0.10	0.03
Total	297.94	

(http://apeda.in/agriexchange/India per cent 20Production/India_ Productions.aspx?cat =fruit&hscode=1044)

8.0 Soil and Climate

8.1.0 Soil

Annonas grow successfully in varying soil conditions from heavy soils to almost sand (Parsons, 1932). They do well also on rocky soils in many countries. Thus rocky, sandy, marginal and wastelands may be utilized for growing these fruits. But for getting a good crop, a fertile, well-drained sandy loam soil and neutral in chemical reaction (pH) is preferred. Since water logging causes tree decline, heavy soils with poor drainage, sub-soil hardpan or high water table are to be avoided. Proper drainage is essential to prevent root rot diseases. However, *A. glabra* can succeed in shallow ponds owing to its tolerance of wet soils (Nakasone and Paull, 1998). The sugar apple is generally planted in sandy soils with limited nutrients. If moisture deficiencies are not compensated for by frequent irrigations, growth is stunted resulting in low yield. Likewise, fertilizer application is a prerequisite to maintain an optimum level of available nutrients. The plants are shallow-rooted doing away with the necessity of a deep soil. Saline and alkaline soils are not suitable for the cultivation of these crops. However, Oppenheimer (1947) stated that custard apples are capable of growing in as much as 50 per cent lime in the soil and 300 ppm of chlorine in irrigation water.

In Spain, cherimoya is grown in deep, alluvial soil with high lime content or in terraces with soil of metamorphic origin. It can tolerate alkalinity to a certain degree, growing well in soils with pH 7.5/8.5 and with carbonate contents of up to 29-30 per cent without showing any problem of iron chlorosis (Farre *et al.*, 1999). Leaf gas exchange, chlorophyll fluorescence, water relations and mineral nutrient responses of *A. squa-mosa* seedlings to mid salinity were studied by Marler and Zozor (1996) in sand culture. The results indicated that *A. squamosa* is sensitive to salinity stress, and that the responses to salinity were consistent with other salt-sensitive woody perennial species.

8.2.0 Climate

Annonas are mostly subtropical fruits preferring warm climate with moderate winter and humidity for high production. The tree remains dormant during cold season for a short period, yet frost and prolonged cool weather adversely affect its growth. It can tolerate temperature several degrees below freezing (Oppenheimer, 1947), but temperatures beyond 40°C cause heavy flower abscission in northern India. George and Nissen (1987a) reported that high root temperature increased vegetative growth, flower production and dry weight of leaf and root in *A. glabra, A. squamosa, A. cherimola* and *A. atemoya*. In controlled environment the highest day and night temperatures (32-27°C) resulted in strong vegetative growth and adversely affected floral bud production (George and Nissen, 1987b) which led to conclude that tropical conditions might not be conducive to high fruit production. Higuchi and Utsunomiya (1999) studied floral differentiation and development in cherimoya under 20/15°C and 30/25°C day/night temperature. Floral differentiation was more favourable at 20/15°C than at 30/25°C. More flowers were produced at 21/15°C and the flowering period was longer. Flowers were also larger at 20/15°C, however, the growth rate of floral organs was faster at 30/25°C. The temperature effects were more pronounced in the second year. The floral responses to temperature were different depending on the nodal positions, because of the different developmental processes. The soursop is adapted to areas of high humidity and relatively warm winters; temperatures below 5°C (41°F) will cause damage to leaves and small branches, and temperatures below 3°C (37°F) can be fatal (Patel and Patel, 2016).

The phenological growth stages of various species of monocotyledonous and dicotyledonous plants can be uniformly coded using the Biologische Bundesanstalt, Bundessortenamt, und Chemische Industrie (BBCH) scale. In order to determine the duration of different phenological stages and the temperature requirements of the sugar-apple, *Annona squamosa*, during two crop cycles in the semiarid region of Brazil. Mendes *et al.* (2017) divided the phenological stages were divided into eight of ten possible principal stages: (0) bud development, (1) leaf development, (3) shoot/branch development, (5) inflorescence emergence, (6) flowering, (7) fruit development, (8) fruit maturity and (9) senescence and the beginning of dormancy. The phenological cycle of the sugar-apple from having closed leaf buds to the fruit ripening stage lasted 149 and 164 days with temperature requirements of 1684.5 and 1786.7 degree days (DD) for the first and second crop cycles, respectively. The results provided important information that will inform the correct timing for crop management practices.

Detailed crop-specific descriptions and codes are presented for the growth stages of the cherimoya tree, contributing to the standardization of national and international testing systems in fruit growing. Based on the general BBCH-scale, the codes describe growth stages using a two-digit numerical system and differentiate between principal and secondary growth stages to allow separation of developmental steps. Growth stages for bud, leaf and shoot development, inflorescence emergence, flowering and fruit development were described. A feature of the system is that homologous stages of different crops are presented by the same codes (Cautýn and Agust, 2005).

Fruit growth and yield at high temperatures were further affected by soil moisture stress. Hence, warm subtropical conditions showed better fruit growth and yields (George and Nissen, 1988). Poor pollination problem is associated with high temperature (30°C) and low humidity (30 per cent RH), while low temperature (25°C) and high humidity (80 per cent RH) greatly improve pollination (Marler *et al.*, 1994). Progress has been made in understanding the mechanism of pollination in custard apple (*Annona* spp. hybrids) in subtropical Australia. 'KJ Pinks', a bud sport of the custard apple 'Pink's Mammoth' exhibited natural fruit set in excess of 40 per cent compared with its parent which set <3 per cent. The very high fruit set levels of the 'KJ Pinks' will revolutionize the production of custard apple in Australia as it offers major commercial advantages over older cultivars which required expensive hand pollination. Some 30,000 breeding lines have been field planted at the Maroochy Research Station, Nambour, Queensland since 1998 including interspecific crosses between *Annona cherimola*, *Annona squamosa*, *Annona reticulata* and *Annona diversifolia* and intergeneric crosses with *Rollinia deliciosa* (George *et al.*, 2008).

In southern Spain, cherimoyas are grown in the region where the average temperature ranges between 25°C and 13°C, in the warmest and coolest months respectively. Fruit quality is impaired at a temperature below 14°C. Temperatures between 15 and 25°C are considered optimum for growth when water is not a limiting factor (Farre and ermoso, 1985). Leaves, fruit and trunk get damaged and pollen quality diminishes at –2°C and 33°C as well (Farre *et al.*, 1999). The climate in California is acceptable for cultivation of cherimoya only near the coast. Frost damage occurs once in 10 years (Grossberger, 1999).

Annonas thrive from sea level to about 1,000 metres altitude and from few centimetres of rainfall in Rajasthan to 250 centimetres in Sri Lanka (Parsons, 1932). Annual rainfall of 50 to 75 cm is considered adequate for its growth and fruiting (Rao, 1974). They can also tolerate adverse weather conditions, such as extreme temperature, high and low humidity, heavy and scanty rainfall and desiccating winds but at the cost of growth and yield. Inspite of profuse flowering, no fruit set occurs in northern India in summer season due to high temperature and lack of humidity (Hayes, 1957; Kumar *et al.*, 1977; Thakur and Singh, 1965a). High humidity, occasional rains and warm temperature are required for good set and development of fruits in custard apples. *A. cherimoya* performs better at lower temperature than *A. squamosa*, *A. glabra* and *A. atemoya*. Severe summer does not suit *A. reticulata*.

Fruit set is adversely affected by heavy shading. However, photoperiod is reported to have no influence (Nakasone and Paull, 1998). Strong wind causes limb breakage as the wood is relatively soft and the fruit skin gets bruised due to rubbing and exposure to desiccating wind (Marler *et al.*, 1994; Nakasone and Paull, 1998).

9.0 Cultivation

9.1.0 Planting

Commercial orchards of custard apples are rarely found in India. A few uncared seedlings are seen growing here and there in the home gardens, backyards or in orchards of other fruits. For commercial cultivation, the grafted plants should be grown meeting cultural requirements of the crop. The plants raised in nursery beds from seeds suffer heavy mortality due to the damage of roots during lifting and transplanting. It is, therefore, desirable that the seedlings be raised in polythene bags and after budding or grafting they are transplanted in the orchard carefully. In order to promote growth of the planting material, pits (50 x 50 x 50 cm) are dug and filled with a mixture of soil and basal dose of 20 kg farmyard manure and 300 g fertilizer mixture of urea, superphosphate and muriate of potash in equal proportion at least a fortnight before planting. Planting is done in the rainy season at a distance of 5 m adopting square system. In order to secure better pollination, a closer spacing has been suggested by Ahmad (1936a) in desert conditions of Egypt which can be increased up to 6 m in the situations of rich soil and high rainfall. The traditional planting distance in Spain is 8 x 8 m and it is possible to increase the plant population up to 625 trees per hectare or more with the advanced pruning technique (Farre *et al.*, 1999). Ramanan (1995) recommended triangular spacing of 7 m (237 trees/ha) for *A. muricata* in Costa Rica.

9.2.0 Training and Pruning

Generally, training and pruning are not in practice in annonaceous fruits. But leaf shedding in winter season, and production of floral buds on new shoots indicate the necessity of some amount of training and pruning (Nalawadi *et al.*, 1975). Summer pruning has the potential to increase yields with some cultivars of custard apple (Batten, 1983). In an experiment on clonally propagated juvenile trees, George and Nissen (1986) found that in moderate summer and spring dormant pruning resulted in poor yield due to reduction in the laterals and initiation of floral buds. Unpruned trees gave higher yields but were structurally weak and exhibited severe limb breakage. The importance of pruning to develop strong and upright trees was, however, suggested by Stephens (1936). In order to build a strong frame of the plant, only a few scaffold limbs are allowed to develop and other branches are removed. Subsequently, very old and weak branches are thinned out from time to time for giving room for new and young branches. Pruning should, however, be carried out when the plant is to put forth new growth in early spring after dormancy.

The completion of leaf fall is essential for the initiation of new growth in Annonas. Within fifteen days of artificial leaf removal, 100 per cent plants sprouted while partial leaf fall, early initiation of new growth and extent of fruit set determine the yield in custard apples. Chemical defoliants have been tried and two sprays of 0.5 per cent potassium iodide at weekly interval resulted in not only early and prolonged harvest but also gave higher yield. In the cultivar Balanagar, manually and chemically defoliated plants and the controls produced 17.4, 24.4 and 13.0 kg fruits per plant, respectively (Anon., 1981). In the cultivar Red Sitaphal and Mammoth, manual defoliation was, however, better than chemical treatment. George and Nissen (1986) obtained improved bearing precocity by defoliation with 0.2 per cent ethrel and 20 per cent urea during mid-summer. Pre-bud break defoliation of 3-year-old custard apple (*A. cherimola* x *A. squamosa*) cv. African Pride in subtropical Queensland, with ethrel (1 g/l) and urea (250 g/l) hastened, and post-bud break defoliation and summer pruning delayed, shoot growth and flowering, which might be due to a reduction in the number of flowers with defoliation and an increase after summer pruning. The main effect of summer pruning was to increase the number of subpetiolar buds which emerged on new wood (George and Nissen, 1987d). In defoliation studies carried out by Fouad *et al.* (1999) on 3 cherimoya cultivars, urea (2 per cent), 8 per cent and 4 per cent universal (mineral oil + 1.5 per cent dinitro-o-cresol) or 4 per cent dormex sprays resulted in complete or almost complete defoliation and earlier flowering than the untreated plants.

Marler and Crane (1994) observed that pruning of lateral branches of atemoya with narrow crotch angles might be performed during the training of atemoya plants which produced scaffold limbs from supernumerary buds within the same nodes with desirably wide crotch angles.

Atemoya produces first flower of a season on previous season's growth and second and third flushes on current season's growth that bear most crops (Nakasone and Paull, 1998). The first flowering may be avoided because of poor fruit set and fruit splitting on ripening (Sanewski *et al.*, 1991; Nakasone and Paull, 1998). In Queensland, the 1-year-old bearing shoot is shortened to 10 to 15 cm with 4 to 6 buds (Sanewski, 1991). These pruned shoots are retained after fruiting and allowed to become part of the tree architecture, thus extending the canopy by 30-50 cm each year (Nakasone and Paull, 1998).

Farre *et al.* (1999) reported that in Spain, cherimoya trees were traditionally placed in an open vase system with three or four main branches maintained at a height of 1.5 to 2 m above the ground level to facilitate orchard management practices. However, with the introduction of no tillage practice, the tree canopy is kept much lower. Fruit branches and the growing points of main branches are removed each year towards the end of winter season or before the emergence of new growth. Cherimoya trees seemed to remain productive at a reduced size, producing yields similar to those of conventional large trees, but with bigger fruit. In California, pruning is confined to removal of small branches of the previous year's growth (Grossberger, 1999).

9.3.0 Irrigation

Annonas are considered drought-tolerant plants and produce a fair crop even without any irrigation because the fruit set and development occur during the rainy season and thereafter when there is ample humidity in the atmosphere and moisture in the soil. Thakur and Singh (1965a) found no fruit drop in *A. squamosa* trees growing in damp and shady situation and receiving irrigation. The fruit also ripened about two months earlier than usual. George and Nissen (1988) reported reduced flowering and fruit set at high temperature in atemoya due to soil moisture stress. According to them, productivity of cherimoya might be improved by cultural techniques, *viz.*, overhead misting, wind-breaks and efficient irrigation scheduling. Two to three irrigations before the onset of monsoon to promote fruit set and one or two irrigations after the monsoon to increase fruit size are recommended. Heavy soils will obviously need less watering than sandy soils. More frequent irrigation has been advocated in Egypt by Ahmad (1936a). Among four methods of irrigation – furrow, drip with bamboo lateral nozzles, drip with plastic lateral nozzles and sprinkler – drip irrigation with both bamboo and plastic lateral nozzles gave better performance (Rungsimanop *et al.*, 1987).

The water requirement of bearing atemoya has been estimated to be up to 1440 l/tree every 4 weeks during the low-growth phase and 500 to 750 l/tree every 3-5 days during flowering, fruit set and fruit growth (Sanewski, 1991). Traditionally the cherimoya orchards in Spain used to be flood irrigated. However, new plantations are irrigated by drip or spray irrigation methods (Farre *et al.*, 1999). Using spray irrigation along with the appropriate schedule, larger areas can be irrigated (Farre and Hermoso, 1995). Use of 1,800 to 2,000 drippers per hectare are most frequent (*e.g.*, 315 trees/ha at 8 m x 4 m and 6 drippers/tree). Water consumption per hectare and per year is very variable depending on the location of farm and irrigation method adopted. Indicative values for a hectare of cherimoya, reasonably protected from the wind and with a system of fixed irrigation (drip or spray irrigation), range between 5,500 and 6,500 m^3/ha/year (Farre *et al.*, 1999). In California, irrigation in cherimoya orchard is done with mini-sprinklers or drippers (Grossberger, 1999).

Studies were conducted by George and Nissen (1992) to develop reliable methods of measuring and sampling atemoya trees (cv. African Pride) for determination of plant water status. Atemoya leaves were shown to have a low modulus of elasticity (<8.5 MPa) and leaves were shown to adjust osmotically and diurnally (0.14 MPa), and after several slowly developing water stress cycles (0.29 MPa). Bearing trees exhibited higher stress levels than non-bearing ones, and flowers and fruitlets exhibited more stress than leaves, suggesting that reproductive organs may be more appropriate sampling organs than leaves.

9.4.0 Manuring and Fertilization

A balanced nutritional schedule is required to be adopted for getting regular higher yield in a fruit crop. Since the sugar apple is cultivated mostly on poor soils, the need for fertilizer application becomes all the more essential. Ordinarily manures and fertilizers are not applied to this fruit. Guirado (1988) studied the nutrient uptake of cherimoya trees. It was observed that a yield of 14 t/ha extracts, 95 kg/ha N, 4.6kg/ha of P or 11.3 kg/ha of P_2O_5, 38 kg/ha of K or 45 kg of K_2O, 9 kg/ha of Ca and 27.5 kg/ha of Mg.

Sadhu and Ghosh (1976) obtained best growth with the highest rate of nitrogen application in sand culture experiments. In their trial, flowering was advanced by 10 to 15 days with high doses of nitrogen and phosphorus, but low nitrogen delayed flowering by over two months. High nitrogen rates gave higher flower production while high N, P and K enhanced the fruit set. They also described the visual deficiency symptoms of N, P and K in sugar apple. Applications of N, P and K have been recommended in sandy soils (Ahmad, 1936a). According to Sturrock (1940), the sugar apple responds better to organic manures. However, Sanewski (1991) suggested the use of organic fertilizers, with inorganic fertilizers as a supplement to maintain a balance and to regulate cropping.

On the basis of a nutritional trial carried out at Rahuri, India, the fertilizer dose for young custard apple plant was recommended as 250 g N, 125 g P_2O_5 and 125 g K_2O per plant. Further trial on 10-year-old seedlings was continued and results indicated the usefulness of 250 g N and 125 g P_2O_5 per plant (Anon., 1987). The fertilizers should be applied on the commencement of rainy season. Growth and yield of custard apple were evaluated by Mandal and Chattopadhyay (1993) in an experiment with no fertilizer, 120 g N + 120 g P + 120 g K or 240 g N + 240 g P + 240 g K annually for 2 years. The highest NPK rate produced the largest number of flowers and higher fruit set and retention. Number of fruits, yield per plant and per hectare were also highest with the highest NPK rate.

Sanewski (1991) suggested a fertilizer schedule along with time of application for atemoya cv. Pink's Mammoth (Table 4). Field experiments were carried out in Tamil Nadu, India, over 2 seasons, growing 9-year-old cherimoya trees on a sandy clay loam soil. An NPK mixture supplying 375 g N + 202.5 g P^2O_5 + 375 K_2O/tree gave the highest mean yields of 33 fruits/tree and 14.88 kg fruit/tree (50 and 83 per cent higher respectively, than the values for unfertilized trees) (Stalin *et al.*, 1992).

9.4.1 Tissue Analysis and Nutrient Requirement

Leaf analysis has been identified as a valuable diagnostic tool for assessing the nutritional status of atemoya tree. The most recently matured leaf *i.e.*, the 4th or 5th leaf below the growing point can be used as the index tissue (Sanewski, 1991) (Table 10). Seasonal patterns of leaf nutrient concentrations and leaf sampling

Table 10: A Guide to Annual Fertilizer Application in Atemoya cv. Pink's Mammoth Trees and Percentage Distribution of Application of Annual Amounts*

Age of Tree (years)	*Urea*	*Superphosphate g/tree/year*	*Potassium Chloride*
2	400	500	360
4	860	550	930
6	1300	780	1170
8	1600	880	1500
10	1750	880	1650

Fertilizer	*Early Spring*	*Early Summer*	*Early Autumn*	*Late Autumn*
Urea	20 per cent	30 per cent	40 per cent	10 per cent
Super phosphate	-	-	100 per cent	-
Potassium chloride	10 per cent	30 per cent	40 per cent	20 per cent

* Sanewski (1991).

procedures were established by George *et al.* (1989b) for Pink's Mammoth (atemoya) in subtropical Queensland (27°C). One group of nutrients showed a general decline in concentration in the leaves with time (K, 1.4-0.8 per cent, P, 0.3-0.14 per cent) while another group showed an overall increase (Ca, 0.4-1.4 per cent, Mg, 0.3-0.54 per cent; B, 24-66 μg/g). Leaf N levels decreased throughout the growing season (3.5-2.5 per cent), with the pattern of decline strongly influenced by vegetative flushing. Leaf concentration of Ca, Mg, Mn and B generally increased and those of N, P, K, Cu and Zn decreased with leaf age. Movement of the more mobile elements (*i.e.*, N, P, K, Cu and Zn) from the older into the younger leaves was more rapid when trees were flushing compared with trees that had completed flushing.

9.5.0 Weed Control

Annonas are shallow rooted trees and thus limit the possibility of deep cultivation under the tree. The young trees should be protected from weed competition by manual weeding, mulching or use of contact herbicides (Nakasone and Paull, 1998). The most commonly used herbicides in cherimoya orchards in Spain are simazine or terbumeton mixed with terbuthylazine as a pre-emergence herbicide, and glyphosate or a mix of paraquat and diquat as a post-emergence herbicides (Farre *et al.*, 1999). They also reported that a more complex and less contaminating system of soil maintenance was introduced at the end of eighties. It consisted of maintaining the planting line of trees free from weeds by applying pre-emergence herbicides through spray irrigation whereby goal (oxyfluorphene) was used at 12 applications per year. Perennial weeds that survived this treatment were eliminated by glyphosate at 2 or 3 applications per year. The treatment was completed by manual weeding thrice a year.

9.6.0 Intercropping

In the initial years after planting, economic return can be obtained from the unutilized lands between the plants by cultivating intercrops. Groundnut, minor millets and linseed can be grown in the rainy season and pea, oilseeds and gram in the winter months. Besides economic benefit, intercrops also check soil erosion and weed growth. Custard apples in India are mostly grown on uneven sandy or rocky soils which are not fit for clean cultivation.

10.0 Flowering, Pollination, Fruit Growth and Development

10.1.0 Flowering

In comparison with many other tree fruits, custard apple has a short juvenile phase. Seedlings as well as vegetatively propagated plants begin flowering within four years of lanting, the former taking more time to come into bearing than the latter. *Annona* species bear hermaphrodite flowers, either singly or in cluster on the current season's growth and rarely on old wood (Thakur and Singh, 1965a; Nakasone and Paull, 1998). The flowering period of custard apples is very long commencing from March-April, continuing up to July-August (Thakur and Singh, 1965a; Nalawadi *et al.*, 1975; Kumar *et al.*, 1977). The peak flowering, however, takes place in April and May. Variation in flowering pattern in different regions has been reported which might be due to variation in agro-climatic factors and genotype (Kshirsaga *et al.*, 1976). In Australia, the Atemoyas bear flowers twice a year. The first flowering occurs in spring and the second in summer, the latter being more productive (Sanewski, 1991). In the south of Spain, flowering in cherimoya occurs from the end of May until the middle of July, depending on the area (Farre *et al.*, 1999).

From initiation, a floral bud develops into flower in 27 to 35 days depending on the species (Thakur and Singh, 1965a). Flower opening takes place in three phases. In the first phase (so called pre-female stage), the petals begin to separate from the end and it continues for 6 to 15 hours, finishing just after noon (approx.1 pm-3 pm). In this phase stigma is receptive but insects cannot pollinate it. In the second phase (female phase) the petals continue to open, allowing small pollinating insects to visit the stigma. This phase continues for 26-27 hours from 1 pm to 3 pm on day one to 4 pm to 6 pm of the following day. The stigmas remain receptive throughout the period except the last three hours. During the third phase (male phase), the petals separate totally in 20 to 30 minutes coinciding with the ripening of stamens. The step from female to male phase always takes place in the afternoon, from 4 pm to 6 pm. High temperature shortens and low temperature extends the cycle. This mechanism of temporal dioeciousness has probably been developed by the species to favour cross-pollination in the original wild stands (Farre *et al.*, 1999). Flowers of soursop appeared to be protogynous (Worrell *et al.*, 1994).

Although flowers begin to appear in spring, yet no fruit set occurs during the entire spring and summer and it commences only in the rainy season, leaving little period for the late set fruits to develop before the onset of winter season. The setting of fruits early in the season is important because immature fruits instead of developing become inedible in winter season and turn into stone fruits (Hayes, 1957).

Flowering and fruiting were monitored in 4 cherimoya clones growing in Calabria, Italy (Continella *et al.*, 1996). Flowering period depended on weather and peaked between mid-June and early July. Flower density ranged from 29.4 to 58.3 per m of branch and fruit set from 0.5 to 5.1 per cent. Bagged flowers did not set fruit confirming the need for pollinating insects.

10.2.0 Pollination and Fruit Set

Low productivity of annonaceous fruits is the main constraint in expanding their commercial cultivation (Hayes, 1957; George and Nissen, 1986). Enough flowers are borne on a custard apple plant to give a good crop but the poor fruit set causes low yield. Only one to eight per cent fruit set has been reported under natural conditions (Ahmad, 1936a; Venkataratnam, 1963; Thakur and Singh, 1965a; Kumar *et al.*, 1977; George and Nissen, 1988). Insects like, *Apis mellifera*, nitidulid beetles (*Carpophilus* and *Uroporus* spp.) and *Orius* (Anthocoridae) are the important pollinators of annona flowers (Nakasone and Paull, 1998; Farre *et al.*, 1999). Self, wind and insect pollination mechanisms were evaluated by George *et al.* (1989a) in *Annona* spp. and hybrids. Natural fruit set increased linearly with increasing numbers of nitidulid beetles (*Carpophilus hemipterus*) per flower.

The beetle *Notoxus trifasciatus* and the minute pirate bug *Orius laevigatus* were observed on flowers, especially at peak flowering and there was some correlation between the numbers of these insects present and fruit set (Continella *et al.*, 1996). More fruit set and yield of sugar apple and atemoya were obtained by Pena *et al.* (1999), when trees were baited with *Carpophilus dimidiatus*, *C. freemani* and *C. mutilatus* pheromones, and combinations with food odour bait as attractants for pollinators than on untreated trees.

Investigations were carried out to verify the effects of releases of *Orius laevigatus* on the pollination of cherimoya in Sicily (Caleca *et al.*, 1998). The efficacy of the anthocorid was confirmed. Releases of *O. laevigatus* to improved fruit setting and fruit quality of cherimoya could be useful in orchards where the species was present in low numbers and the efficacy of other species was poor. Hand pollination was tested using a small pump. Fruits produced through hand pollinated flowers were significantly more abundant and well proportioned in comparison with those obtained in the control. In *A.squamosa* cultivars Thai Lup and New Caledonian, hand pollination resulted in >90 per cent fruit set, compared with 0 and 3.6 per cent fruit set in the Thai Lup and New Caledonian controls, and mean fruit weights were increased by 17-62 per cent (Cogez and Lyannaz, 1994).

In another study on manual pollination on fruit set and fruit growth in *A. squamosa*, Cogez and Lyannaz (1996) observed that fruit set in control trees was very poor (0 per cent in cv. Thai Lup and 3.6 per cent in New Caledonia trees) and it increased to 100 per cent with manual pollination in cv. Thai Lup and to 90.9 per cent and 92.6 per cent respectively, with intravarietal and intervarietal pollination in New Caledonia trees. Manual pollination increased average fruit weight from 143 g in control Thai Lup trees to 230 g, and from 230 g in New Caledonia trees, to 269 g and 313 g respectively, with intravarietal and intervarietal pollination. Manual pollination also improved the fruit shape.

In cherimoya, a lack of synchrony in maturation between male and female parts of the flower precludes successful natural self-pollination and, in order to achieve uniform fruit shape, a high percentage of the carpels must be fertilized and set seed (Richardson and Anderson, 1990). It was found that using pollen from the same cultivar rather than from a mixture of cultivars resulted in increased fruit set and improved fruit shape.

The low fruit set in custard apples is due to poor pollination which has been attributed to both the external and internal factors, such as very high and low humidity prevailing at the time of flowering, soil moisture stress, competition between vegetative and floral growth, hypogyny, dichogamy, poor pollen germination and dearth of insect pollinators. The peak flowering period of custard apples coincides with the summer season in northern India when the temperature goes beyond 40°C, humidity is extremely low, desiccating wind is prevalent and the soil is in dry condition. Neither the pollen is produced, nor does fruit set occur in this situation (Hayes, 1957; Kumar *et al.*, 1977). Thakur and Singh (1965a) found pollen grains shriveled and sterile in May and June when temperature was very high and relative humidity very low. Hopping (1983) also reported that low humidity and hot days could impair pollen viability. Therefore, fruit set occurs only in rainy season on reversal of environmental conditions. No fruit set occurred from self-pollination of custard apples in Israel at low humidity and high temperatures (Gazit *et al.*, 1982).

On the contrary, in Spain, under high humid conditions, the flowers are mostly self-pollinated as the stigma does not desiccate before anther dehiscence (Popenoe, 1971). Early fruit set also occurs in southern India where humidity is higher and temperature lower than in northern India. In atemoya, high water vapour deficit and soil moisture stress reduced both flowering and fruit set (George and Nissen, 1988). Even the set fruits dried up due to high temperature and low humidity during summer season in northern India (Thakur and Singh, 1965b).

According to George and Nissen (1988), protogynous dichogamy is the main limiting factor for fruit set in atemoya. Poor pollen germination has been reported in all the species of *Annona* (Thakur and Singh, 1965b), and they ascribed it to the compound nature of pollen grains, thick exine and absence of germpores. In Israel, a significant relationship has been demonstrated between the visiting nitidulid beetles and fruit set (Gazit *et al.*, 1982). The number of beetles visiting per flower also affected fruit quality and shape.

Almost all the investigators have reported high fruit set by hand pollination (Ahmad, 1936a; Schroeder, 1943; Thakur and Singh, 1965b; Hopping, 1983; Campbell, 1986; George and Nissen, 1988). Hayes (1957) got as high as 85 per cent fruit set in sugar apple by hand pollination in comparison with 30 per cent of the control. Campbell (1986) obtained 68 and 4 to 24 per cent fruit sets in hand and natural pollination respectively, in atemoya. In *A. muricata* fruits were formed by self-pollination and pollination significantly increased fruit set which was low under natural conditions (Escobar *et al.*, 1986). For hand pollination, flowers should be collected at about 4pm and the number of flowers to be pollinated depends on the vigour of the plant (Ahmad, 1936a; Hopping, 1983). On pollination, pollen tubes reach the embryo sac without any barrier and fertilization takes place normally (Vithanage, 1984). Several workers suggested that for obtaining an economic yield in cherimoya, hand pollination is a must (Rosenberg, 1958; Farre *et al.*, 1976; Farre *et al.*, 1999).

Guirado (1991) noted the following advantages of hand pollination : (i) it guarantees a minimum yield of 10 t/ha; (ii) fruits are better in shape and size; (iii) in areas of precocious ripening, the first flowers can be pollinated in absence of pollinating insects, thus obtaining better market prices; and (iv) it assures early harvesting. Pollen collected during the male phase gave better results after it was refrigerated and could be applied at any moment during the two-day cycle of flower opening (Farre *et al.*, 1999). Pollination is generally done in the afternoon or in the morning, as at mid-day it is either too warm or too dry resulting in less fruit set (Grossberger, 1999).

Schroeder (1943) observed fruit set in cherimoya only after the full development of leaves. The fact that flowers borne on basal nodes are more fruitful, and significant increase in the fruit set near the completion of major vegetative growth indicates that there is a competition between the vegetative growth and flower for photosynthates (George and Nissen, 1988).

Plant growth regulators, *viz.*, NOA, NAA, 2,4-D, 2,4,5-T and GA3 have been tried to increase the fruit set or to reduce the fruit drop (Sundararajan *et al.*, 1968;

Pramanik and Bose, 1974; Campbell, 1986; Keskar *et al.*, 1986). Dipping of freshly opened flowers of the cultivar Mammoth in 50 ppm GA3 gave 70 per cent fruit set and the developed fruits had more weight but less number of seeds. Pramanik and Bose (1974) reported about 18 per cent fruit set in sugar apple by spraying the flowers with 100 ppm GA3 and NOA. Fruit drop was also reduced with the application of 2,4,5-T. In three-year trials on sugar apple, mean numbers of fruits, harvested per 100 m of crown in the trees treated with NAA at 20 and 10 ppm and control, were 94.4, 84.6 and 62 respectively (Keskar *et al.*, 1986). NAA was sprayed four times at 8-day interval during flowering in March-April to increase fruit set. Campbell (1986), however, suggested the GA treatment only in case of little or no natural fruit set. The cultural practices including close planting, regular irrigation in summer months, overhead spraying of water and mulching also help in enhancing the setting of fruits (Ahmad, 1936b; Rao, 1974; George and Nissen, 1988).

Shoot inducers were evaluated in cherimoya trees in Tababela, Ecuador for their regulation of shoot percentage, facilitation cultural practice (pruning, phytosanitation, hand pollination and harvesting) and improvement of productivity (Viteri *et al.*, 1999). Dormex (alzodef) [cyanamide] at 1 and 2 per cent and drop (thidiazuron) at 200 and 300 ppm were applied with and without oil after manual defoliation. At 20 days after treatment (DAT), 200 ppm thidiazuron and 2 per cent cyanamide + 2 per cent oil produced >70 per cent of buds with shoots, compared to the control with 50 per cent. Fruit set was evaluated at 115 DAT and 2 per cent cyanamide caused 34.5 per cent fruit set, compared to 18 per cent in the control. The chemical treatments were concluded to have a generally positive effect on shoot growth and fruit set.

10.3.0 Fruit Growth and Development

Annona fruits usually mature in about four months from the time of anthesis. The period is reduced when the soil moisture is optimum (Thakur and Singh, 1965a). Fruits of *A. reticulata*, however, mature in about six months (Thakur and Singh, 1965b). The increase in diameter of cherimoya and sweetsop follows a typical sigmoidal pattern (Thakur and Singh, 1965a). Worrell *et al.* (1994) reported that fruit growth (size and weight) of soursop was double sigmoidal, comprising initial and final phases of rapid growth interspersed with a period of slow growth. Fruits reached half their final size by the end of the initial rapid growth phase, followed by an intermediate resting phase of 4 weeks and attained the full size and maturity during the final growth phase. The skin colour turned to light green which was due to gradual decline in chlorophyll *a* and *b* concentrations during the last 7 weeks of development.

Aziz and Yusof (1994) studied the physico-chemical characteristics of *A. muricata* fruits and reported that fruit weight and volume increased rapidly during growth and then slowed down as the fruit approached maturity. Titratable acidity and total soluble solids, ascorbic acid and sugar contents increased slowly during growth; then increased rapidly during ripening. Tannin content and pH were initially high but decreased during ripening. Fruit firmness and alcohol insoluble solids and pectin contents increased during the early growth stages; then decreased rapidly during ripening.

Yonemoto *et al.* (2000) studied the effect of GA_3 treatment on fruit growth and quality of *A. cherimoya*. Among the concentrations used (500, 1000 and 1500 ppm), the higher the concentration, the smaller the number of seeds produced in the fruit and the higher the percentage of seedless fruits. Application of GA_3 to the stigma induced fewer seeds than application of GA_3 to the calyx, but it was easier to apply GA_3 to the calyx. Larger fruits were produced at higher concentrations of GA_3, although GA_3 did not affect fruit weight or the total soluble solids content.

11.0 Pests, Diseases and Disorders

Pests and diseases in Annonas have been reported and their control measure suggested.

11.1.0 Pests

Mealy bug (*Planococcus pacificus*) has been noticed harbouring on young shoots and in between the fruit segments. The blemished fruits do not fetch good price in the market. The bug can be controlled by spraying 0.05 per cent of phosphamidon or dichlorvos (Shukla and Tandon, 1984).

The Mediterranean fly (*Ceratitis capitata*) has been reported to be a major pest of cherimoya in the south of Spain. Traps using pheromones combined with field hygiene (removal of fallen fruits) could reduce the pest attack percentage to 4-7 per cent without additional phytosanitary treatments. Immersion of fruits in hot water (45-47°C) for 60 minutes also killed the eggs and larvae of this fly (Farre *et al.*, 1999). Leon Fuentes (1999) recommended the use of toxic bait consisting of 4 per cent hydrolysed protein and 0.15 per cent dimethoate on the entire trees. The highest percentage of healthy fruits (92.0 per cent) was obtained with the application of the bait every 21 days. In studies in Granada, Spain, using traps containing food-based synthetic attractants for females, crop damage caused by *Ceratitis capitata* was significantly reduced in cherimoya compared with untreated areas (Ros *et al.*, 1999).

Damage caused by soft scale pest in *A. muricata* was reported by Cavallazi *et al.* (1998) and *Verticillium lecanii* proved to be an effective biocontrol agent.

11.2.0 Diseases

Leaf spot (*Altenaria* spp.) disease appears in the month of November at the fag end of harvesting causing considerable loss in production. The affected leaves drop down prematurely. Screening of Annona germplasm did not show any resistant genotype but British Guinea, Island Gem and Red Sitaphal cultivars were only moderately susceptible (Anon., 1987). Anthracnose caused by [*Colletotrichum cingulata* (*Glomeralla cingulata*)] has been recorded at Udaipur (Anon., 1981). Both the diseases can be controlled by three fortnightly sprays of benlate (0.05 per cent), bavistin (0.05 per cent) or baycor (0.1 per cent) commencing as soon as the symptoms appear. *Ascochyta* sp. and *Colletotrichum gloeosporioides* (*G. cingulata*) affect cherimoyas, but can be controlled by copper-based products or mancozeb (Leon Fuentes, 1999).

Zarate Reyes (1995) reported leaf anthracnose [*Colletotrichum gloeosporioides* (*G. cingulata*)], dry rot of fruit [(*G. cingulata* and *Fusarium coeruleum* (*F. solani* var.

Fruit Growth and Development of Annona

Growth phenology of Annona

Pests of Annona

Mealy bug infestation in annona

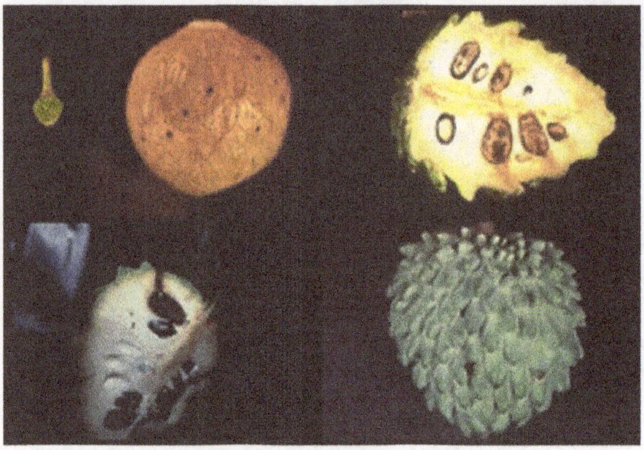

Seed borer infestation in annona

Mealy bug infestation in annona

Diseases of Annona

Black canker infection

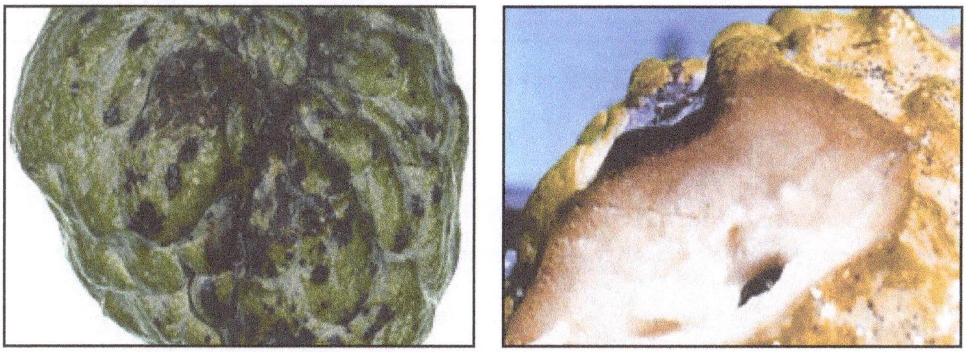

Cylindrocladium leaf spot infection in fruit

Anthracnose infection in leaf and fruit

Diseases of Annona

Symptoms of soft rot caused by *Rhizopus oryzae*

Rot associated with *Bephratelloides cubensis* and *Lasiodeplodia* sp.

coeruleum)], grey rot of fruit (*Rhizopus stolonifer*), white stain of leaf (*Cercosporidium annonae*), root rot (*Armillaria mellea, Rhizoctonia* spp. and *Phytophthora* spp.) and black mildew from *A. muricata* in Colombia. Fruit rot of *A. squamosa* caused by *Phytophthora citrophthora* and *P. nicotianae* was first found in Taiwan in 1990 (Huang *et al.*, 1991). Purplish or brownish black discoloration of the fruits was followed by browning of interior tissue and by complete blackening and mummification in 5-6 days. Ridomil [metalaxyl]-MZ WP at 2.5 g/l and Ridomil-Cupnait WP at 1 g/l protected against infection by *P. citrophthora* but none of the fungicides inhibited disease development when applied after infection.

The causal organism of bacterial wilt of atemoya is *Pseudomonas solanacearum*. The characteristic symptoms are rapid wilting and death of young tree and gradual decline of old trees. Vascular discoloration of woody tissues occurs in the roots and up to the trunk at ground level (Nakasone and Paull, 1998).

11.3.0 Disorders

In sugar apple, some fruits instead of attaining full size, remain very small and become brown. These are known as stone fruits which retain on the tree for a long period, even after the harvest of normal fruits, during the dormancy and initiation of new growth. Some physiological factors, particularly competition among the developing fruits have been suggested to be the causes of stone fruit formation. There was only 7 to 8 per cent stone fruit formation when the fruiting was spread over a longer period due to defoliation treatments compared with 30.7 per cent in the control (Anon., 1981). More stone fruits have been observed in the vicinity of developing fruits.

Sudden and high fluctuations in the water supply to the plants may cause cracking of the fruits. It usually happens from a heavy rainfall or irrigation after a prolonged dry spell. An equitable irrigation schedule will reduce this problem. Water logging in sandy or rocky soils may result in the abrupt decline of the trees. Hence, stagnation of water in custard apple orchards needs to be avoided.

12.0 Harvesting and Yield

Maturity indices in Annona fruits have not been standardized because of variation in the stage of fruit growth and fruit size in the seedling plants which are plucked from the tree while still firm and plumpy. If left on the tree, the fruits get damaged due to splitting instead of ripening properly. Likewise, immature fruits after harvest become soft and rot, though the skin remains intact. It is, therefore, essential to harvest the fruits at the proper stage of maturity. The following three maturity indices are usually adopted in sugar apple (Ezzat *et al.*, 1974) : (a) the light green fruit colour, (b) yellowish white colour between the carpels, (c) initiation of cracking of the skin between the carpels.

The fruits are examined individually every alternate day to determine the stage of maturity for harvesting. The fruits are handpicked. In coastal regions, the fruits ripen earlier than in other regions. Commencing in August, the harvesting of sugar apple continues upto December, the peak period being October and November

(Venkatarat-nam, 1963). The cherimoya fruits reach the market after sugar apples followed by those of *A. reticulata*.

In order to study the effect of harvest time on the storage quality of custard apple, fruits were harvested at normal date or 2 or 5 days earlier and stored at 25-83°C and 85-90 per cent RH. Chen (1999) suggested that for immediate consumption, fruits should be harvested when the TSS reached 15-20 per cent. Otherwise, the harvesting should take place 2-3 days earlier when fruit soluble solids content reached around 10 per cent.

12.1.0 Yield

George *et al.* (1986) studied the factors influencing the productivity of atemoya and established a positive linear relation between yield and tree girth, but cross-sectional areas, canopy volume and number of laterals. Fruit weight was, however, negatively correlated to the number of fruits on a tree. The size of individual fruit is determined by its seed number which depends on the level of pollination. More the number of seeds present in a fruit, higher are its weight.

The yield in Annona fruits varies greatly with climatic and edaphic factors, cultivars, rootstocks, age of the plant, pollination and management of the orchard. Annona plants begin bearing at the age of five years and give as many as fifty fruits while older plants may yield about 100 fruits (Oppenheimer, 1947). He further stated that the productive life of Annona trees is only 10-12 years while Ahmad (1936a) considered it more than 20 years. Hayes (1957) reported 31 to 288 fruits with a mean of 149 fruits per plant from 8-year-old seedlings and estimated a yield of less than 2.5 tonnes/ha. According to Venkataratnam (1963), a sugar apple tree usually produces about 60 to 70 fruits, each weighing from 113 to 227 g. In the 6th year of planting, atemoya cultivar African Pride gave 18.7 tonnes yield of marketable fruits per hectare on cherimoya rootstock and 9.2 tonnes on sugar apple rootstock (George and Nissen, 1986a).

13.0 Post-harvesting Management

13.1.0 Ripening

The cherimoya and sweetsop are climacteric fruits. Lam and Zaipun (1986) did not find any consistent differences in the post-harvest C_2H_4 and CO_2 production in the two maturation stages. On the contrary, Worrell *et al.* (1994) reported that mature fruit produced a biphasic respiratory climacteric, with CO_2 production reaching 100 ml/kg/h and then 350 ml/kg/h at 25-30°C. Peak ethylene production (250-350 ml/kg/h) occurred between the two respiratory maxima. The respiratory climacteric of harvested immature fruit tended to be higher and later than that of mature fruit. Two days after harvest, lower molecular weight (MW) starch fractions were detected in fruits of *A. muricata* (Paull, 1990). The occurrence of these fractions coincided with the preclimacteric respiratory rise. The changes in MW profiles were related to the rise in respiration rate and wall breakdown. The breakdown pattern suggested that an enzyme or enzymes degraded starch to random length glucans. Lugwisha *et al.* (2016) studied the physicochemical composition of sugar apple (*Annona squamosa*

L.) fruits from Kibaha, Coast Region of Tanzania during open air storage ripening process. The ash, titratable acidity, crude fat, crude fiber, moisture and sugars content were determined by proximate analysis. Ascorbic acid contents were determined using the 2, 6-dichlorophenol-indophenol dye method while mineral elements and heavy metals were determined by Flame Atomic Absorption Spectrophotometry (FAAS). The fruits were always harvested at the mature stage and allowed to ripen during open air storage. The determinations were done immediately after arrival at the laboratory and thereafter at intervals of two days from the day of harvest to the 8th day. The results showed that fresh sugar apple fruits had high moisture content range of (64 per cent - 73 per cent), low titratable acidity (<0.28 per cent ca),low crude fat (0.51 g/100 g-fw), high ash content (1.44 g/100 g-fw), low crude fibre content (0.185 g/100 g-fw), high ascorbic acid content (51-34 mg/100 g-fw), high total sugars content (49.7 per cent - 31.1 per cent), moderate reducing sugar content (43.17 per cent -18.57 per cent) and sucrose content (11.8 per cent - 0.9 per cent). Of the mineral elements (K, Ca and Na) determined, the highest content was of Ca (2838.82 mg/100 g-fw.). Heavy metals (Fe, Zn, Cu, Pb and Cd) content was very low in the sugar apple fruits, ranging between <0.0015 mg/100 g-fw for Cd and 1.27 mg/100 g-fw for Fe. Except for acidity and ascorbic acid contents which were decreasing during storage ripening, the moisture content, total sugar content, reducing sugar content and sucrose were all increasing as the fruit was ripening while in storage. There were no significant changes during storage ripening for levels of crude fat, ash, minerals and heavy metals.

During ripening free putrescine levels increased in cherimoya fruit. Escribano *et al.* (1996) reported that the rise in putrescine levels in ripening cherimoya fruit was correlated with greatly enhanced arginine decarboxylase (ADC) activity. No ornithine decarboxylase (ODC) activity was detected under any of the experimental conditions. It is suggested that the ripening related changes in putrescine levels are regulated via ADC activity-mediated synthesis and that no conjugation is involved in the process. The effect of short-term high CO_2 treatment (20 per cent C O_2 + 20 per cent O_2 for 3 days) on ethylene and polyamine synthesis in cherimoya cv. Fino de Jete fruits stored at 20 and 6°C was investigated by Munoz *et al.* (1999). The decrease in putrescine content, with no variation in arginine decarboxylase activity and accumulation of both spermidine and spermine confirmed the preferential transformation of diamenes to polyamines in C O_2-treated fruits. The custard apples ripen within few days after harvest. In an experiment, when the sugar apple fruits were treated with calcium carbide or placed in straw, fruit leaves and open baskets, they ripened in 2,4,5 and 6 days respectively (Prasad and Shukla, 1978). In subsequent storage of ripe fruits at room temperature, the longest shelf-life of four days was found in fruits ripened in straw and fruit leaves. The fruits wrapped in perforated polyethylene bags and ripened in wheat straw suffered the least diameter loss compared with wrapping in paper bags and ripening in sawdust or gunny bags (Chaudhry *et al.*, 1985). Effects of some physiologically active regulating substance on the ripening of *A. squamosa* fruits, stored at 16-24°C were studied by Tsay and Hong (1988). IAA and ACC both enhanced fruit softening and electrolyte leakage; IAA also enhanced respiration. The ethylene biosynthesis inhibitors, AVG and 2,4-dinitrophenol, delayed softening, decreased total soluble solids (TSS) and increased electrolyte leakage relative to the controls. Cycloheximide caused the fruits to remain hard and unripe, with lower TSS and electrolyte leakage than controls. Tsay and Wu (1990) recorded maximum ethylene evolution in fruits of *A. squamosa* during ripening and softening. The abscissic acid content of fruits increased from 16 ng/g FW at harvest to 455 ng/g FW in ripe fruit.

Sugar apple fruits ripen normally at 15 to 30°C in storage, low relative humidity accelerates the process of ripening (Broughton and Tan, 1979). According to them, ripening is not affected by ethylene but is hastened by the removal of CO_2 and addition of O_2 and *vice versa*. Interestingly, fruits dipped in IAA at 10^{-4} to 10^{-2} M ripened faster than undipped ones. The stage of eating ripeness coincides with climacteric.

13.2.0 Storage

Like many other tropical fruits, the mature custard apple fruits get chilled if stored below 15°C; while ripe fruits can be stored at 5°C for six weeks, the pulp remained in good condition but the fruits become unmarketable due to browning of skin. At Udaipur, India, the fruits could be kept for about a week after treating with 8 per cent wax emulsion in combination with 400 ppm 2,4-D or 2,4,5-T (Anon., 1981). The conditions recommended for controlled atmosphere storage of sugar apple are temperatures between 15°C and 20°C, relative humidity 85-90 per cent and low oxygen and ethylene tensions coupled with 10 per cent CO_2 (Broughton and Tan, 1979). The quality of atemoya fruits remained intact for two week storage at 15 to 16°C (Brown and Scott, 1985). Cherimoya fruits could be stored for 12 days without any loss in quality at 9°C, 85 to 90 per cent relative humidity and 2 per cent O_2 + 10 per cent CO_2 in controlled atmosphere storage (Calvo *et al.*, 1983). Plaza *et al.* (1993) recorded that fruits of cherimoya packed in plastic boxes and stored at 8.5°C with 3.5 g ethylene absorbent/kg showed significantly reduced respiration rates and ethylene synthesis; weight losses were reduced during the optimal storage period of 18 days. In an experiment on low temperature storage of cherimoya fruits, Montero *et al.* (1993) concluded that the expression of several fruit ripening-related polypeptides was strongly inhibited at storage temperatures below 10°C and changes in the accumulation of some of these polypeptides appeared to be specifically associated with the rise in ethylene production. A combination of 3 per cent O^2 and 3 or 6 per cent C O^2 extended the storage life of cherimoya cv. Fino de Jete fruit held at 9°C by 2 weeks over that of fruit stored in air. There was no difference between 3 and 6 per cent C O_2 (Alique and Oliveira, 1994). Fruits of cherimoya cv. Concha lisa could be stored in 5 per cent O_2 at 10°C for up to 43 days and still ripened normally after 4 days at room temperature (Palma *et al.*, 1993). Fruits held in 20 per cent CO_2 for 9 days showed unacceptable quality due to bitterness, while 20 per cent CO_2 for 3 days delayed fruit softening, retarded the accumulation of polygalacturonase-related protein, and maintained chlorophyll content of the peel (Del Cura *et al.*, 1996). In an experiment on the post-harvest life of cherimoya, Berger *et al.* (1993) concluded that waxed fruits of cv. Bronceada could be stored at 0 per cent CO_2 + 5 per cent O_2 for 3 weeks without visible deterioration.

Preharvest spray on Balanagar variety of *A.squamosa* with GA_3@ 50 mg/l was the most acceptable treatment considering the physical and quality parameters. For enhanced shelf-life $CaCl_2$ @ 2 per cent treatment was good for ambient storage (Bagul and Masu, 2017). Dipping fruits of custard apple in 500 ppm bavistin and storing in polyethylene bags containing $KMnO_4$ had a storage life of 9 days while untreated fruits could be stored for 5 days (Babu *et al.*, 1990). At ambient conditions (temperature 25-30°C, RH 60-85 per cent) minimum loss in fruit weight of *A.squamosa* fruits was noted when fruits were treated with bavistin and packed in LDPE bags with ethylene absorber. This treatment was proved to be the best on account of slower increase in acidity total soluble solids (Sahu *et al.*, 2016).Enzymatic fruit browning in cherimoya due to the appearance of pigments originated by polyphenoloxidase (PPO) activity, is a considerable problem in the marketing of this fruit. Martinez-Cayuela *et al.* (1988) obtained that PPO from the epicarp catalyzed the hydroxylation of monophenols such as L-tyrosine, tyramine and *p*-cresol and the oxidation of *o*-dihydroxyphenols such as catechol and L-DOPA. According to Patil *et al.* (2011) the shelf life of custard apple fruits was 9 days when wrapped with tissue papers and kept in cardboard boxes when stored in zero energy cool chamber. Gohlani and Bisen, (2012) found the use of packaging films significantly reduced weight loss, preserved firmness and colour of the fruit. Silva *et al.* (2012) reported that the modified atmosphere allowed for a shelf life of 12 days at 10°C. The higher shelf life of fruits *i.e.* 12.28 days obtained with fruits packed in polypropylene bags flushed with air (Venkatram *et al.*, 2013).

13.3.0 Processing

Custard apple being highly perishable fruit, is highly susceptible to spoilage, softens very rapidly during ripening, and becomes squashy and not easy to consume fresh. The fruit has a short shelf-life due to high respiration and ethylene production. Hence there is greater need to preserve the custard apple fruits and to process into suitable products *viz.*, juice, jam, ice cream, milkshake *etc.* to minimize the post-harvest losses which further helps in value addition and in generation of a good income for a needy farmer. Jain *et al.* (2019) presented a comprehensive review on the post-harvest processing of custard apple fruits.They suggested the preparation of various products such as juice, ice cream, vinegar, milk shake besides Jam, fruit-flavoured yoghurt, fruit drinks and syrups *etc.* with fruit extracts. They also reviewed some important aspects such as the effect of packaging material and chemical treatments on the storage and shelf life of fruit pulp including the use of modified and controlled atmosphere, processing of stored pulp into various value-added products.

The fruits should be harvested at the normal time or 2-5 days earlier (when TSS around 10 per cent) and stored at 25-33°C at 85-95 per cent relative humidity. For immediate consumption, fruits should be harvested at total soluble solid content around 15-20 per cent (Chen, 1999). Harvesting of fruits was done at physiological maturity and stored at 26±2°C at 60-70 per cent relative humidity for 4 days (Bolivar *et al.*, 2009). When fruits were harvested at maturity stage, a climacteric peak was noticed in the second day of storage at ambient conditions and respiration peaks of 56.32 mg CO_2/kgh and 27.42 mg CO_2/kgh respectively (Mallikarjuna *et al.*, 2012).

The custard apple fruit harvested at the commercial stage of maturity resulted in an increase in total soluble solids, total sugar and acidity up to 4th day of storage, while on further storage reduction in these parameters and organoleptic characteristics of fruit was observed. The maximum score was observed in case of fruit stored till 4th day, while subsequent storage of fruit decreased in consumer acceptability of fruit (Pawar and Hashmi, 2010).

14.0 Biotechnology

Very little work has been done on biotechnology of custard apple. There is scope for developing varieties resistant to diseases if tissue culture techniques are perfected along with genetic transformation. Molecular markers would aid in studying the phylogenetic relationship among the *Annona* spp. and the genotypes within species (Raj Bhansali, 2001).

14.1.0 Tissue Culture

14.1.1 Micropropagation

Nair *et al.* (1984a and b) were successful in propagating sugar apple (*A. squamosa*) and atemoya (*A. squamosa* x *A. cherimola*) by tissue culture using modified Murashige and Skoog's (MS) medium. On transplanting to soil, 80 per cent plantlets survived in atemoya but only 10 per cent in sugar apple. Sreelata *et al.* (1986) also reported endosperm culture of *A. squamosa*. Despite much research on tissue culture of a wide range of tropical tree species, the Annonaceae has received little attention and only limited success in tissue culture has been reported (Bejoy and Hariharan, 1992). However, the explants had low rooting ability or poor performance during acclimatization.

Adventitious shoot buds developed directly from 1 cm long hypocotyl of *A. muricata* within 30-35 days, without an intervening callus phase, at all auxin and cytokinin concentrations tested. Further growth into shoots with leaves was observed in most cases, when NAA and BA were both present. The maximum regeneration response was obtained on Nitsch and MS media and the greatest number of shoot buds/explant (4.8) was obtained on MS medium. Excised shoots, about 1-2 cm long, were transferred for root formation to a basal MS medium supplemented with IBA at various concentrations and 88 per cent rooting was obtained in presence of 9.8 μM IBA (Bejoy and Hariharan, 1992).

Rasai *et al.* (1995) reviewed the progress in tissue culture of *Annona*. Root and shoot organogenesis was successfully induced in sugar apple anther, endosperm and leaf explants as well as in cherimoya hypocotyl, petiole, internode and nodal cuttings. Hypocotyl and nodal cuttings of atemoya and hypocotyl explants of soursop also proved suitable for *in vitro* culture. Tissue-cultured shoots, especially those obtained from hypocotyl and nodal segments of some *Annona* species were successfully rooted.

Researchers have successfully micropropagated cherimoya, sugar apple, soursop and atemoya. Rasai *et al.* (1994) obtained multiple shoot formation from hypocotyls and nodal cuttings of Atemoya cultivar 'African Pride'. The explants were cultured in MS medium supplemented with BAP, kinetin, biotin and calcium pantothenate. In spite of improved rooting by shoot pre-treatment in liquid MS medium containing IBA, the percentage of rooting was still low (40 *per cent*) and remains a limiting factor for commercial micropropagation of atemoya. Abubacker and Deepalakshmi (2017) achieved *in vitro* callus production in leaf base explants of *A.muricata* on MS + BAP 1.5 mg/l+ 2,4-D 2.0 mg/l, callus cell morphogenesis indicated the establishment of proembryo, perfect dichotomy and shoot apices. The axillary shoot formation from nodal explants was achieved on MS + BAP 1.0 mg/l+ Kin 0.5 mg/l+ IBA 1.5 mg/l and rooting in MS + IBA 2.0 mg/l+ NAA 0.5 mg/l+ BAP 1.0 mg/l. The plantlets were acclimatized and thrived in greenhouse and then in natural environmental conditions.

Although there are numerous experimental protocols for *Annona* tissue multiplication, the final price of the plantlets is still too high for commercial use. Given the potential of this technology, further research is needed to transform experimental protocols into commercial protocols.

14.1.2 Anther Culture

Apart from micropropagation, haploid plants were induced from anther-derived callus on NN medium supplemented with BAP or NAA. Initial exposure to high sucrose and dark period followed by low sucrose and light was noted to be essential for inducing haploid plants (Nair *et al.*, 1984b). The first haploid plants induced by anther culture in fruit trees were reported by Nair *et al.* (1983) with sugar apple. The availability of haploids is very important for fruit-breeding, because of the long generation intervals, the highly heterozygous nature of most fruit species and the presence of parthenocarpy and self-incompatibility. These researchers obtained callus differentiation, and formation of triploid roots and shoots from sugar apple endosperm (Nair *et al.*, 1986). Their aim was development of seedless fruits, but a complete plantlet was not obtained.

14.2.0 Genetic Transformation

Encina *et al.* (1999) have started studying genetic transformation of cherimoya, mainly to optimize the protocol for *Agrobacterium* transformation. The objectives are to control ripening, to change post-harvest characteristics, and to provide pest and disease resistances. For other *Annona* species there are no advanced studies.

14.3.0 Biochemical and Molecular Markers

Samuel *et al.* (1991) suggested the use of allozymes to study diversity and systematics in Annonaceae. They considered these systems to be efficient for investigations of the origin of polyploids for breeding programmes. A preliminary study, using eleven isoenzyme loci, was developed with five *Annona* species. Strangely, soursop, mountain soursop (*A. montana*), and pond apple (*A. glabra*) presented no variation between or within the populations studied.

Some isoenzymes studies with cherimoya have been carried out. Groups at the University of California (USA) and University of Granada (Spain) studied the variation in isoenzyme patterns of cherimoya cultivars from the USA and Spain.

Both groups found sufficient variation to distinguish cultivars and to evaluate cherimoya germplasm (Pascual *et al.*, 1993).

Isoenzyme studies are limited because they are carried out using a relatively small number of loci. RAPDs offer an enormous number of markers covering the whole genome and are a more powerful technique for genotype identification and germplasm evaluation. Ronning *et al.* (1995) estimated variation between cherimoya, sugar apple and atemoya, and determined the inheritance of these markers in the F1. All fifteen primers used, generated repeatable polymorphic patterns, resulting in a very efficient method to distinguish genotypes of *Annona* species.

In short, the recent biotechnological studies, both cellular and molecular, have shown great potential to further Annona development, not only to solve problems of mass propagation of superior cultivars, through micropropagation techniques, but also to identify or fingerprint Annona cultivars, as well as to determine cultivar parentage, through RAPD markers.

Analysis of isozyme patterns indicated that Spanish cherimoya cultivars were distinctly different from cultivars in California (Pascual *et al.*, 1993) and atemoya (Ellstrand and Lee, 1987). The popularly grown Spanish cherimoya cultivars Fino de Jete and Campa exhibited identical banding patterns for 15 enzymes indicating that they might be the same cultivar (Pascual *et al.*, 1993). In the South Coast Research and Extension Centre, California, approximately 30 cherimoya cultivars have been collected and verified by isozyme analysis. Twenty five different isozyme loci have been identified which are useful for determining a cultivar's identity (Grossberger, 1999). For determining diversity and variability in *A. muricata* accessions markers were developed (Brown *et al.*, 2003, Suratman and Sry 2015).

In order to characterize key phenotypic attributes and genetic diversity data of *Annona muricata* from 42 representative accessions of the crop as a precursor to enhancing a systematic varietal improvement and selection programme that would support conservation, Brisibe *et al.* (2017) evaluated phenotypic attributes based on stem, leaf, flower and fruit characteristics in situ to complement molecular tools and data obtained demonstrated that differences amongst the accessions were highly significant ($p < 0.001$) for number of flowers and fruits, with 71.38 per cent of the total variability being accounted for by 3 principal components. Conversely, molecular characterization studies using random amplified polymorphic DNA markers produced a total of 65 reproducible bands. Of these, 59 amplified bands (90.77 per cent) were polymorphic while the remaining 6 were monomorphic loci. Analysis of the data generated from the phenotypic attributes and molecular markers revealed a high degree of divergence with 3 and 4 clusters, respectively, which exhibited continuous variation that was highly distinguishable. The genetic diversity identified here could offer yet-unknown traits of high value that would optimize varietal improvement and use of genetic resources amongst *A. muricata* accessions with great potential as raw material for the food and pharmaceutical industries. Moreira *et al.* (2018) characterized molecularly a population of 60 individuals of *A. muricata* distributed in four counties (Jipijapa, Paján, 24 de Mayo and Olmedo) of this province. Plant DNA was extracted and amplified using 8 ISSR molecular markers. All the markers were informative; however, ISSR 824 and ISSR 873 stood out because they obtained the highest PIC value (0.29). Four groups were formed, with group 2 involving the largest number of individuals, including the entire population of Jipijapa and the highest proportion of the population of 24 de Mayo. Results showed a moderate genetic variability of the study population.

Characterization of genetic diversity within populations of *A.squamosa*, collected from different districts of Rajasthan, based on combined data set showed that percentage of polymorphism bands (PPB) ranged from 29.19 to 50.93. Nei's gene diversity (HE) values varied from 0.1002 to 0.1964 within a total genetic diversity (HT) of 0.2509. The average observed number of alleles per locus (Na) was 1.3602 and Ne was 1.2400. The Shanon's index was 0.2002 (Hpop) and 0.3800 (Hsp) at population and species level, respectively. The moderate gene flow value (Nm=0.59) and significant coefficient of genetic differentiation (GST=0.458) revealed low level of genetic diversity within population and highly significant genetic differentiation among populations of *A. squamosa*. The UPGMA dendrogram resolved all the 21 populations representing 3 districts into various clusters according to their genetic distances. The principal component analysis (PCA) supported the UPGMA results and justified the relationship among populations (Nagori *et al.,* 2018). In order to identify and distinguish the three Annonas varieties namely Abdel Razik of (*A.squamosa* Sugar apple), Senegalensis (*A.Senigalinses*) African Sugar Apple and Balady (*A.squamosa*)-Sugar Apple that are commercially in Egypt, Abdelkawy and El-nawam (2017) used morphological and molecular marker techniques (Random Amplified polymorphic DNA (RAPD) remark unique. Identical fingerprinting for Abdel Razek and Baladi indicate high genetic similarity between these two while *Senegalensis* was separated in an independent cluster.

15.0 Conclusion

Immense prospects exist to develop desirable cultivars of annonaceous fruits by conventional breeding techniques. There is greater scope to exploit inter-specific variation than intra-species variation. Edible Annona species exhibit morphological affinity to one another but each one is unique and distinct for fruit shape, size, skin surface, pulp colour, texture, flavour and taste. Some species like *Annona squamosa* exhibit narrow genetic variation. However, several interesting traits are available in one or the other allied Annona species. Because there are no serious barriers in inter-specific hybridization between *A. squamosa*, *A.cherimola*, and *A.reticulata* gene transfer across species boundaries is possible. Gene exchange among them should result in interesting novel recombinants, at times, most esteemed ones like the popular atemoya.

16.0 References

Abdelkawy M. M. A. and S.M.A. El-nawam (2017) *Alex. J. Agric. Sci.*, **62:** 415-421, 2017

Abubacker, M.N and T. Deepalakshmi (2017) *Biosciences Biotechnology Research Asia*, **14**(1): 123-128.

Agustin, J.A. (1999) *Acta Horticulturae*, **497** : 189-200.

Ahmad, M.S. (1936a) *Ministry Agric. Egypt. Hort. Section Bull.*, No. 157.

Ahmad, M.S. (1936b) *Ministry Agric. Egypt. Hort. Section Bull.*, No. 14.

Alique, R. and Oliveira, G.S. (1994) *J. Agr. Food Chem.*, **42** : 799-803.

Anderson, P. and Richardson, A. (1990) *Orchardist of New Zealand*, **63** : 17-19.

Anonymous (1981) *Ann. Rep. All India Coordinated Fruit Imp. Project.*

Anonymous (1987) *All India Coordinated Res. Project on Arid Zone Fruits Tech. Doc.*, No. 20.

Aziz, P.A. and Yusof, S. (1994) *ASEAN Food Journal*, **9** : 147-150.

Babu, K.H., Zaheeruddin, M. and Prasad, P.K. (1990) *Acta Horticulturae*, **269** : 299.

Bagul, A. A. and Masu, M. M (2017) *International Journal of Agricultural Sciences* **13:** 371-377.

Baily, L.H. (1949) *Manual and Cultivated Plants*, Macmillan Co., New York, pp. 419.

Batten, D.J. (1983) *Tropical Fruit Research Station*, Alstonville, Research Report, 1982-1983, pp. 1-16.

Bejoy, M. and Hariharan, M. (1992) *Plant Cell Tissue Organ Cult.*, **31** : 245-247.

Bejoy, M. and Hariharan, M. (992) *Plant Cell,Tissue and Organ Cult.*, **31**: 245-247.

Berger, H., Galletti, L., Marin, J., Fichet, T. and Lizana, L.A. (1993) *Proc. Interamerican Soc. Trop. Hort.*, **37** : 121-130.

Bolivar, F.N., Solis,P.C., Sauri,D.S. and Saucedo, V.E. (2009) *Agrociencia*, **43:** 133-141.

Bourke, D.U.D. (1976) In : *Propagation of Tropical Fruit Trees* (Eds. R.J. Garner and S.A. Chaudhury), Common Bur. Hort. Plant. Crops Kent, pp. 223-248.

Brisibe, E.A., Ogbonna, N.C. and Chukwurah, P.N. (2017) *Agroforest Syst.*, **91:** 781–793.

Broughton, W.L. and Tan, G. (1979) *Sci. Hort.*, **10** : 73-82.

Brown, J., Laurentin, H., Dávila, M. (2003) *Fruits*, **58**: 255-59.

Brown, B.I. and Scott, K.J. (1985) *Singapore J. Prim. Indust.*, **13** : 23-31.

Buell, E.P. (1955) *Trop. Agriculturist*, **111** : 18-27.

Cabezas Prieto, F. (1998) *Congresos y Jornades*, **47/98** : 49-60.

Caleca, V., Verde, G.Lo, Chiara, S.R.di and Tsolakis, H. (1998) *Bollettino diZoologia Agraria e di Bachicoltura*, **30** : 207-212.

Calvo, M.L., Plaza, I.L. De La, Alique, R. and Rodrigo, M.E. (1983) *Sociaedad Espanola ed-Ciencias Horticolas Primer Congreso Nacional*, II, 1005-1009.

Campbell, C. W. (1970) *Proc. 83rd Annu. Meet. Fla. State Hort. Soc.*, Miami, **83** : 353-356.

Campbell, C. W. (1986) *Proc. Trop. Region Amer. Soc. Hort. Sci.*, **23** : 122-124.

Campbell, C. W. and Phillips, R. L. (1983) *Fruit Crops Factsheet, Fla. Coop. Ext. Serv.*, Univ. of Florida, **3** : 1-3.

Campbell, C. W. and Popenoee, J. (1968) *Proc. Trop. Region Amr. Soc. Hort. Sci.*,**11** : 33-36.

Cantillang, P.L. (1976) *Greenfields*, **6** : 26-27.

Cartagena Valenzuela, J.R. and Borreto Osorio, J.D. (1998) *Revista-Facultad Nacional de Agronomia Medellin*, **5** : 235-244.

Casas, M. H. Victoria, S. M. A., Zarate, R. R. D. (1984) *Acta. Agron.*, **34** : 66-81.

Cautý´n,R. and M. Agustý´(2005) *Sci. Hort.*, **105** : 491–497.

Cavallazzi, G., Prieto, A. and Ariva, R. (1998) *Agronomia Colombiana*, **15** : 106-111.

Chaudhry, A. S., Singh, G. N. and Singh, A. R. (1985) *Indian J. Agric. Res.*, **19** : 90-92.

Chauvatia, R.S. and Singh, S.P. (1999) *Horticultural J.*, **12** : 1-7.

Chen W. (1999) *South China Fruits*. **28**: 27.

Chen,C.C. and Paull,R.E. (2008) In: *The Encyclopedia of fruit and nut*,.(Eds. J. Janick, and R.E. Paull) CABI, Oxfordshire, UK: p: 52.

Chopde, N., Patil, B.N., Pagar, P.C. and Ram Gawande (1999) *J. Soils Crops*, **9** : 69-71.

Choudhry, B. L. (1954) *Indian Farm.*, **4** : 15-21.

Cogez, X. and Lyannaz, J.P. (1994) *Fruits (Paris)*, **49** : 359-360.

Cogez, X. and Lyannaz, J.P. (1996) *Trop. Fruits Newsl.*, **19** : 5-6.

Continella, G., Longo, S., Zappia, R. and Palmer, V. (1996) *Italus Hortus*, **3** : 32-38.

Coronel,R.E. (1994) In: Promising fruits of the Philippines. Coronel,R.E., Zuno,J.C. and Sotto,R.C.(Eds.). College of Agriculture, University of the Philippines, Los Banos, Laguna, The Philippines: pp.1-18

De la Plaza, J.L., Rossi, S. and Calvo, M.L. (1993) *Acta Hort.*, **343** : 181-183.

Dhanumjaya Rao, K. and Subramanyam, K. (2011) *Indian J. Agric. Res.* **45**: 156-160

Dhua, R. S., Mitra, S. K. and Sen S. K. (1982) *Proc. Nat. Sem. on Pl. Propagation*, Bidhan Chandra Krishi Viswavidyalaya, Kalyani, p. 36.

Dickinson, T. A. (1963) *West. Fruit Grower*, **17** : 20-21.

Duarte, O., Villagarcia, J. and Franciosi, R. (1974) *Proc. Amer. Soc. Hort. Sci. of the Trop. Reg.*, **18** : 41-48.

Ellstrand, N. C. and Lee, J. M. (1987) *Sci. Hort.*, **32** : 25-31.

Encina C. L., Padilla I. M. G., Carzola J. M. and Caro E. (1999) *Acta Hort.*, **497** : 289-294.

Escobar, T.W., Zarate, R.R.D. and Bastidas, A. (1986) *Acta Horticulturae*, **36** : 7-20.

Escribano, M.I., Aguado, P., Reguera, R.M. and Merodio, C. (1996) *J. Plant Physiol.*, **147** : 736-742.

Ezzat, A. H., Naguib, M. And Metwalli, S. (1974) *Agric. Res. Rev.*, **52** : 7-17.

Farre, J.M. and Hermoso, J.M. (1985) *Informe del viaje de exploracion sobre la chirimoya realizado al Peru* entre el 22 de abril y el 1 de junio.

Farre, J.M. and Hermoso, J.M. (1995) *Comunicacion I+D Agroalimentaria*, 15/95, Junta de Andalucia.

Farre, J.M., Harmoso, J.M., Guirado, E. and Garcia-Tapia, J. (1999) *Acta Horticulturae.*, **497** : 91-103.

Farre, J.M., Hermoso, J.M. and Gonzalez, M.A. (1976) *Anales INIA*, No. 6.

Fehr,W.R. 1987. Principles of cultivar development - Theory and Technique.Vol.1, Macmillan Publishing Company, New York.

Ferreira, G. and Cereda, E. (1999) *Revista Brasileira de Fruticultura*, **21** : 79-83.

Fouad, M.M., Mahamed, S.M. and Ghonaim, *i.e.* (1999) *Bull. Fac. Agric. Univ. Cairo*, **50** : 245-264.

Fouque, A. (1972) *Fruits*, **27** : 62-72.

Gardiazabal, I.F. and Cano, L.G. (1999) *Acta Hort.*, **497** : 225-253.

Gazit, S., Galon, J. and Podoler, H. (1982) *J. Amer. Soc. Hort. Sci.*, **107** : 849-52.

George, A. P. and Nissen R. J. (1987a) *Scientia Hort.*, **31** : 95-99.

George, A. P. and Nissen R. J. (1988) *Scientia Hort.*, **34** : 183-91.

George, A. P. and Nissen, R. J. (1983) *Australian Hort. Res.Newsletter*. No. 55, 145-6.

George, A. P. and Nissen, R. J. (1986a) *Bienn. Rep. Maroochy Hort. Res. Stn.*, **4** : 46-68.

George, A. P. and Nissen, R. J. (1986b) *Qd. J. Agric. Anim. Sci.*, **43** : 73-81.

George, A. P. and Nissen, R. J. (1987b) *Sci. Hort.*, **32** : 269-274.

George, A. P., Nissen, R. J. and Ko H. L. (1986) *Acta Hort.*, **175** : 37-41.

George, A.P. and Nissen, R.J. (1987c) *Sci. Hort.*, **33** : 75-85.

George, A.P. and Nissen, R.J. (1987d) *Aust. J. Exp. Agric.*, **27** : 915-918.

George, A.P. and Nissen, R.J. (1992) *J. Hort. Sci.*, **67** : 445-455.

George, A.P., Broadley, R.H., Nissen, R.J., Bruun, D., Hormaza, J.I., Escribano, P., Viruel, M.A. and Beppu, K. (2008) *Acta Hort.*, **772:** 215-218.

George, A.P., Broadley, R.H., Nissen, R.J., Hamill, S.D. and Topp, B.L. (1999) *Acta Horticulturae.*, **497** : 255-267.

George, A.P., Nissen, R.J. and Carseldine, M.L. (1989b) *Australian J. Exp. Agric.*, **29** : 587-595.

George, A.P., Nissen, R.J., Ironside, D.A. and Anderson, P. (1989a) *Scientia Hort.*, **39** : 289-299.

Gholap, S.V., Dod, V.N., Bharad, S.G. and Wankar, A.M. (2000) *Crop Research* (*Hisar*), **20** : 158-159.

Girwani, A., Madahvi, A., Kumar,T.S. and Reddy,G.S.(2011) *Acta Hort.*, **890:** 251-254

Gohlani SWATI, Bisen BP.(2012) *The Bioscan*, **7**: 637-640.

Grossberger, D. (1999) *Acta Horticulturae.*, **497** : 119-130.

Guirado, E. (1988) *E.U.I.T.A.*, La Rabida (Huelva).

Guirado, E. (1991) *E.U.I.T.A*, La Rabida (Huelva).

Hayat, M. A. (1963) *Bot Gaz,* **124** : 360-362.

Hayes, W. B. (1957) *Fruit Growing in India,* Kitabistan, Allahabad, pp. 358-387.

Hermoso, G.J.M., Perez de Oteyza, M.A., Ruiz Nieto, A. and Farre, M.J.M. (1999) *Acta Horticulturae.,* **497** : 201-224.

Higuchi, H. and Utsunomiya, N. (1999) *J. Japanese Soc. Hort. Sci.,* **68** : 707-716.

Hopping, M. E. (1983) *Australian Hort.,* **81** : 11-4.

Huang, T.C., Leu, L.H. and Lee, H.I. (1991) *Plant Protection Bulletin (Taipei),* **33** : 102-112.

Iglesias, A. A. ans Sanchez, L. A. (1985) *Acta Agron.,* **35** : 53-58.

Islam, A. S. (1953) *Curr. Sci.,* **22** : 118-120.

Jain,C., Champawat PS, Mudgal VD, Madhu B and Jain SK (2019) *International J. Chemical Studies* **7(3):** 1632-1637

Jalikop, S.H. (2010b) *Euphytica,***173:** 161-171

Jalikop,S.H. (2010a) *Euphytica,***171:** 175-180

Jalikop,S.H.and Ravindra Kumar.(2007) *HortScience,* **42**: 1534-1538

Kahn, T.L., Adams, C.J. and Arpaia, M.I. (1994) *Sci. Hort.,* **59** : 11-25.

Keskar, B. G., Bhore, D. P. and Masalekar, S. R. (1986) *J. Maharashtra Agric Univ.,* **11** : 235-236.

Khan, K. F. and Rao, I. K. S. (1952) *Indian J. Hort.,* **10** : 140-144.

Kiill,L.H.P., Costa,J.G.da.(2003) *Ciencia Rural.,* **33:** 851-856

Kirtikar, K. P. and Basu, P. D. (1933) *Indian Medicinal Plants,* Lalit Mohan Basu, Allahabad, 1, p. 66.

Krishnamoorthy,V. (2009) *Asian J Hort.,* **4**: 149-151

Kshirsaga, S.V., Shinde, N.N., Rana, D.A. and Borikar, S.T. (1976) *South Indian Hort.,* **24** : 6-10.

Kumar, R., Hoda, M. N. and Singh, D. E. (1977) *Indian J. Hort.,* **34** : 252-256.

Kumar,P.S. and Jalikop,S.H. (2000) *Indian J. Hort.,* **57**: 309-313

Kundan Kishore., Shukla, A.K.,Naresh Babu.,Sarangi,D.N.and S.Patanayak (2012) *Sci Hort.*144: 212-217

Lam, P.F. and Zaipun, M.Z. (1986) *MARDI Res. Bull.,* **14** : 231-235.

Leboeuf,M., Cave,A., Bhaumik,P.K., Mukherjee,B. and Mukherjee, R. (1982) *Photochemistry.* **21:** 2783-2813

Ledo, A. da S. (1995) *Fruits (Paris),* **50** : 299-303.

Leon Fuentes, J. (1999) *Acta Hort.,* **497** : 59-69.

Liu, S.B. (2000) *China Fruits,* **3** : 29-30.

Lizana L. A. and Reginato G. (1990) In: *Fruits of Tropical and Subtropical Origin: Composition, Properties and Uses.* (Eds. S. Nagy, P. E. Shaw and W. F. Wardowski), Florida Science Source, Lake Alfred, Florida, USA : pp. 131-148.

Lugwisha,E.H., C. Fabian, and O. C. Othman (2016) *J. Food and Nutrition Sciences.* **4** (3): 41-48. Ma, C., Y. Chen, J. Chen, X. Li and Y. Chen (2017) *The American J. Chinese Medicine,* **45**(5): 1–32

Mahdeem, H. (1990) *Trop. Fruit World,* **1** : 118-120.

Mallikarjuna T, Ranganna B and Dronachari M.(2012) *Journal of Dairying Foods & Home Sciences.* **31**(1).

Mariappan, V. and Sexena, R. C. (1984) *J. Econ. Entom.,* **77** : 519-21.

Marler, J.E., George, A.P., Nissen, R.J. and Andersen, P.J. (1994) In : *Handbook of Environment Physiology of Fruit Crop,* Vol. II, Subtropical and Tropical Crops (Eds. B.C. Schaffer and P.C. Andersen), CRC Press, Boca Raton, Florida, pp. 200-206.

Marler, T.E. and Crane, J.H. (1994) *HortScience,* **29** : 818-820.

Marler, T.E. and Zozor, Y. (1996) *J. Amer. Soc. Hort. Sci.,* **121** : 243-248.

Martinez-Cayuela, M., Sanchez de Medina, L., Faus, M.J. and Gill, A. (1988) *J. Food Sci.,* **53** : 1191-1194.

Mendes, D.S., M.C.T. Pereira, S. Nietsche, J.F. Silva, J. S. Rocha, A. H. Mendes, H.R.A. Xavier and R. C. D. Santos. (2017) *Anais da Academia Brasileira de Ciências* **89**: 2293-2304

Mondal, A. and Chattopadhyay, P.K. (1993) *J. Pot. Res.,* **9** : 375-379.

Montero, L.M., Zamorano, J.P., Plana, J.L. de La and Merodio, V. (1993) *Acta Horticulturae,* **343** : 277-278.

Moran, R. and Others (1972) *An Cient. La Molina,* **10** : 158-176.

Moreira R, J Moreno, J Buitrón, K Orbe, E Hector-Ardisana, F Uguña and W Viera. (2018) *Vegetos,* **31**(3): 13-18.

Morton,J. (1987) Sugar apple.In: fruits of warm climates, Miami, Florida.p.69-72

Munoz, M.T., Escribano, M.I. and Merodio, C. (1999) *Proc. of the EU-TMR-Euroconference*

Symposium Thira (Santorini), Greece, 5-8 September, 1998, Kluwer Academic Publisher, pp. 327-332.

Nagori, R.,P. Sharma and S. D. Purohit. (2018) *Int J Phytocos Nat Ingred.,* **5: 4**. doi: 10.15171/ijpni.2018.04.

Nair, S., Gupta, P. K. and Mascarenhas, A. F. (1984a) *Indian J. Hort.,* **41** : 160-5.

Nair, S., Gupta, P. K., Shirgurkar, M. V. and Mascarenhas, A. F. (1984b) *Pl. Cell Tissue and Organ Culture,* **3** : 29-40.

Nair,S., Gupta,P.K. and Mascarenhas,A.F.(1983) *Plant Cell Reports,* **2:** 198-200

Nair,S., Shirgurkar,M.V. and Mascarenhas,A.F.(1986) *Plant Cell Reports,* **5:** 132-135.

Nakasone, H., Paull, R., 1998. *Tropical Fruits. Crop Production Science in Horticulture* No. 7. CABI Publishing, London, pp.445.

Van Dammme and P., Scheldeman, X. (1999) *Unasylva* **198** (50), 43–46

Nakasone, H.Y. and Paull, R.E. (1998) *Tropical Fruits*, CAB International, Wallingford, UK, pp. 45-76.

Nalawadi, U. G., Sulikeri, G. S. and Singh, C. D. (1975) *Prog. Hort.*, **7** : 15-24.

Nalawadi,U.G., Sulikeri,G.S.and Singh,C.D. (1975) *Prog. Hort.*,**7** (1): 15-24

Nunez-Elisea, R., Schaffer, B., Fisher, J.B., Colls, A.M. and Crane, J.H. (1999) *Annals Bot.*, **84** : 771-780.

Oppenheimer, C. (1947) *Agric. Res. Stn. Rehovot,* Bull. No. 44.

Palma, T., Stanley, D.W., Agnilera, J.M. and Zoffoli, J.P. (1993) *HortScience*, **28** : 647-649.

Parsons, T. H. (1932) *Trop. Agriculturist*, **32** : 71-101.

Pascual L., Perfectti F., Gutierrez M. and Varga A. M. (1993) *HortScience*, **28**(8) : 845-847.

Pascual, L., Perfectti, F., Gutierres, M. and Vargas, A.M. (1993) *HortScience*, **28** : 845-847.

Patel, S. and J. K Patel (2016) *J. Pharmacognosy and Phytochemistry* **5**(1): 137-148.

Patil, S.B., Raut, V.U., Gadekar A.K.and Patil, S.M.(2011) *Asian Journal of Horticulture.* 6(1): 264-268.

Pawar, V.N. and Hashmi, S.I. (2010) *Food Science Research Journal.* **1(2):** 146-149.

Paull, R.E. (1990) *Acta Horticulturae*, **269** : 277-281.

Paull,R.E. and Duarte,O. (2011) Annonas: Cherimoya, Atemoya and sweetsop.In: *Tropical fruits*, Volume 1, CABI, Oxfordshire, UK, pp: 134

Pena, J.E., Castineiras, A., Bartelt, R. and Duncan, R. (1999) *Florida Entomologist*, **82** : 475-480.

Pinto,A.C.de Q., Cordeiro,M.C.R., Andrade,S.R.M.de., Ferreira, F.R., Filgueiras, H.A.de C., Alves, R.E. and D.I.Kinpara. (2005) Annona species.(Eds. Williams, J.T., Smith, R.W., Hughes,A., Haq,N. and Clement,C.R.) International centre for Underutilized Crops, University of Southampton,UK.

Plaza, J.L. de La, Rossi, S. and Calvo, M.L. (1993) *Acta Horticulturae*, **343** : 181-183.

Popenoe, W. (1971) *Calif. Avocado Soc. Year Book*, pp. 133-139.

Pramanik, D. K. and Bose, T. K. (1974) *South Indian Hort.*, **22** : 117-23.

Prasad, A. and Shukla, J. P. (1978) *Plant Sci.*, **10** : 185-186.

Prest, R L. (1955) *Qd. Agric. J.*, **80** : 17-20.

Raj Bhansali, R. (2001) In : *Biotechnology of Horticultural Crops*, Vol. I (Eds. V.A. Parthasarathy, T.K. Bose, P.C. Deka, P. Das, S.K. Mitra and S. Mohandas), Naya Prokash, Calcutta, India, pp. 357-358.

Ramnanan (1995) *Trop. Fruits Newsl.*, **17** : 12-14.

Rao, S. N. (1974) *Indian Hort.*, **19** : 19-21.

Rasai S., Kantharajah A. S. and Dodd W. A.(1994) *Australian J.Botany*, **42 :** 333-340

Rasai, S., George, A.P. and Kantharajan, A.S. (1995) *Scientia Hort.*, **62** : 1-14.

Ratan, P.B., Reddy, S.E. and Reddy, Y.N. (1993) *South Indian Hort.*, **41** : 171-173.

Rathore, D. S. (1976) *Indian J. Hort.*, **33** : 244-245.

Ribeiro,G.S., Jose,A.r.S., Reboucas,T.N.H. and Amaral,C.L.F.(2007) *Acta Sci. Bio. Sci.* **29 (4)** : 369-373

Richardson, A.C. and Anderson, P.A. (1990) *Orchardist of New Zealand*, **63** : 21, 24-25.

Ronning C. M., Schnell R. J., Gazit S. (1995) *J.American Soc.Hort Sci.*, **120** (5): 726-729.

Ros, J.P.O., Escobar, I., Garcia-Tapia, F.J. and Aranda, G. (1999) *Boletin de Sanidad Vegetal, Plagas*, **25** : 395-404.

Rosenberg, G. (1958) *Informe Annual, Seccion Investigacion Fruticola*, Ministerio de Agricultura, Chile.

Rungsimanop, C., Suksri, A. and Srinukul, S. (1987) *Japanese Soc. Agric. Mach.*, pp. 472-478.

Saavedra, E.(1997) *HortScience*, **12:** 117-118

Sadhu, M. K. and Ghosh, S. K. (1976) *Indian Agric.*, **20** : 297-301.

Sahoo,S.C., Panda,J.M. and Mohanty,D.(2000) *Orissa J.Hort.* **28 (1):** 43-45.

Sahu, R. K., Agrawal, A. K. and G.Sinha (2016) *Madras Agric. J.*, **103:** 62-66.

Samuel R., Pinsker W., Balasubhramaniam S. and Morawetz W. (1991) *Plant Systematics And Evolution*, **178 :** 125-134.

Sanewski, G.M. (1991) *Custard Apple – Cultivation and Crop Protection*, Information series Q190031. Qld. Dept. Prim. Industry, Brisbane.

Scheldeman, X. and Damme, P. van (1999) *Acta Hort.*, **497** : 173-188.

Scheldeman, X., Urena, V. and Damme, P. van (1999) *Acta Hort.*, **497** : 153-172.

Schroeder, C. A. (1943) *Proc. Amer. Soc. Hort. Sci.*, **43** : 39-41.

Shukla, R. P. and Tandon, P. L. (1984) *J. Entom.*, **9** : 181-183.

Silva Junior, F.A.Da, Bezerra, J.F.F., Lederman, *i.e.*, Tavares, J.A. and Melo Neto, M.L. De (1999) *Plant Genetic Resources Newsletters*, No. 120, pp. 25-29.

Silva, A.V.C., Muniz, E.N., Rangel, J.H.A., Yaguiu P., Souza, J.P.B. and Carnelossi, M.A.G. (2012) *Acta Hort.*, **934**: 707-712.

Singh, R. N. (1978) *Phal Vigyan* (*Hindi*), Indian Council of Agric. Research, New Delhi.

Singh, S. N. (1959) *Sci. and Cult.*,**25** : 78-79.

Smet, S.de, Damme, P. van, Scheldeman, X. and Romero, J. (1999) *Acta Hort.*, **497** : 269-288.

Sreelata, N., Shirgurkar, M. V. and Mascarenhas, A. F. (1986) *Linn. Pl. Cell. Rep.*, **5** : 132-135.

Stalin, P., Vasudevan, P. and Thamburaj, S. (1992) *South Indian Hort.*, **40** : 248-251.

Stephens, S. E. (1936) *Agric. J.*, **40** : 102-104.

Sturrock, D. (1940) *Arnold Arboretum of Harvard Univ.*, Jamaica Plain, Mass.

Sundarajan, S., Shanmugavelu, K. G. and Muthuswamy (1968) *South Indian Hort.*, **16** : 63-64.

Suratman A. and Sry S (2015) *Biodiversitas* **16**: 247-53.

Thakur, D. R. and Singh, R. N. (1965a) *Indian J. Hort.*, **22** : 10-18.

Thakur, D. R. and Singh, R. N. (1965b) *Indian J. Hort.*, **22** : 238-253.

Thakur, D. R. and Singh, R. N. (1967) *Indian J. Hort.*, **24** : 11-19.

Thakur,D.R. and Singh,R.N.(1965) *Indian J. Hort.* **22:** 238-253.

Tsay, L.M. and Hong, Y.S. (1988) *J. Japanese Soc. Hort. Sci.*, **58** : 538-543.

Tsay, L.M. and Wu, M.C. (1990) *Acta Horticulturae*, **269** : 241-247.

Velazquez, D.J. (1953) *Agriculture*, **22** : 432-434.

Venkataratnam, L. (1953) *Indian J. Hort.*, **10** : 149-53.

Venkataratnam, L. (1963) *Fruit Culture in India*, (Eds. Sham Singh, S. Krishnamutri and S. L. Katyal, Indian Council of Agricultural Research, New Delhi, pp. 217-224.

Venkataratnam, L. and Styanaranaswamy, G. (1956) *Indian J. Hort.*, **13** : 90-100.

Venkatram, A., Bhagwan, A. and Pratap, M. (2013) *J. Research ANGRAU*. **41**: 1-6.

Viteri, P.F., Leon, J.F. and Ortega, C. (1999) *Acta Horticulturae*, **497** : 347-354.

Vithanage, A. I. M. V. (1984) *Ann. Bot.*, **54** : 153-67.

Watt, G. (1887) *Dictionary of Economic Products of India*, Govt. of India.

Wenkam, N.S. (1990) *Foods of Hawaii and the Pacific Basin. Fruits and Fruit Products, Raw, Processed and Prepared*, Vol. 4, Composition, Res. Extn. Series 110, HITAHR, College of Tropical Agriculture and Human Resources, Honolulu.

Wester P. J. (1913) *Philippine Agri.Review*, **6** : 312-321.

Worrell, D.B., Sean Carrington, C.M. and Huber, D.J. (1994) *Sci. Hort.*, **57** : 7-15.

Xie, Z.S., Xie, X.D. and Li, J.Y. (1999) *South China Fruits*, **28** : 24-26.

Yonemoto, Y., Ikeda, M., Higuchi, H. and Tomita, E. (2000) *Japanese J. Trop. Agric.*, **44** : 44-50.

Zarate Reyes, R.D. (1995) *Fitopatologia Colombiana*, **19** : 68-74.

6

SAPOTA

V. A. Parthasarathy, M. A. Narayana Reddy and S. Vinoth

1.0 Introduction

Sapota (*Manilkara zapota* syn. *Achras sapota*) is a delicious fruit introduced from tropical America. It is also known as sapodilla or *chiku* in India. It is mainly cultivated for its fruits, while in south east Mexico, Guatemala, British Honduras and other countries chickle is commercially produced. The unripe fruit and bark yield a milky white latex which solidifies on exposure to air and this forms the base for making chickle. Immature fruits are astringent, while ripe fruits are sweet smelling and delicious. The mature fruits are also used for making mixed jams and they provide a valuable source of raw material for the manufacture of industrial glucose, pectin and natural fruit jellies. They are also canned as slices.

2.0 Composition and Uses

2.1.0 Composition

The sapota fruits are a good source of sugar which ranges between 12 and 14 percent. The constituents in ripe sapota fruits are given in Table 1.

Table 1: Composition of Ripe Sapota Fruit (per 100 g of edible portion)

Constituents	*Amount*
Moisture	73.7 g
Carbohydrate	21.4 g
Protein	0.7 g
Fat	1.1 g
Calcium	28 mg
Phosphorus	27 mg
Iron	2 mg
Ascorbic acid	6 mg

2.2.0 Uses

Sapota when fully ripe is delicious and is eaten as dessert fruit. The pulp is sweet and melting. It is also made into *sherbet* and *halwa* (Singh *et al.,* 1963). The dried sapota pieces could be made into edible powder, while it was found to contain 6.2 per cent protein and 1.39 per cent acidity on dry weight basis. The sapota powder could be utilized for making milk shake and Indian sweets (*burfi*) (Joshi *et al.,* 1993).

In the coastal areas, the fruits are soaked in melted butter for a night and eaten in the morning. In Indonesia, the young leafy shoots are frequently eaten raw or mixed with other vegetables like lablab and consumed as a vegetable after steaming.

It is said to be an excellent preventive against biliousness and febrile attacks. In the West Indies, seeds are known to be aperient and diuretic and the bark is reputed to be tonic and febrifuge. In Guinea, the bark is used as a tonic and antipyretic. In

Kampu-chia, the bark is considered as an astringent and febrifuge. The decoction is given in diarrhoea and in paludism (Kirthikar and Basu, 1975).

3.0 Origin and Distribution

Manilkara zapota (L.) P. Royen of the family Sapotaceae is native to Yucatan region of Mexico. Thompson *et al.* (2015) used genetic approaches to estimate variation and structure in modern populations of this neotropical tree species to discern if genetic patterns were consistent with earlier influences of ancient Maya management or if they could be explained by the natural history of the species. Their work using nine microsatellite markers, consisting of seven novel markers and two markers identified in a related species, were used to characterize the genetic diversity and population genetic structure in three populations of M. zapota collected from reforested, historically urbanized ancient Maya ceremonial centers in Guatemala and Belize, from home gardens in Guatemala, and from a number of cultivars. According to Thompson *et al.* (2015), a high proportion of trees that dominate the forests in the Maya Lowlands of Mesoamerica are proposed to have flourished as a consequence of cultivation and protection by the ancient Maya in agroforestry systems (Folan *et al.*, 1979).

According to Schroeder (1958) it is indigenous to Mexico, south of the Isthmus of Tehuantepec extending into Guatemala, and possibly into parts of Salvador and northern Honduras. It is still found native in the low lands of Tabasco, Chiapas, and the western part of Yucatan in Mexico, and extends into British Honduras. It was introduced into Ceylon about 1802 but did not succeed on any commercial scale. In India it was introduced in Gholrad near Mumbai in 1898. It's cultivation has spread to Maharashtra, Gujarat, Karnataka and Tamil Nadu https://www.indiaagronet.com/indiaagronet/horticulture/CONTENTS/sapota.htm)

Talking about etymology of the word "Sapota/Zapote", Schroeder (1958) opines that the usage of the terms *sapodilla* or chiko for the species *Achras* (*Manilkhara*) *sapote* would be more satisfactory than the general term sapote. The term *sapodilla* is derived from the Spanish word *zapatilla,* which means small sapote. The Mexican name *chicozapote* has been shortened to *chico,* which implies small. The term *naseberry* frequently is used in the West Indies and parts of India. Chiku is also used in Marathi. The name ya, a Maya word, is still used to designate the fruit in Yucatan, Mexico.

Chicle is the latex that was commercially exploited to produce the chewing gum from the Sapota tree. The latex tappers were called Chicleros. Chiclet industry owes its existence and etymology to Sapota. So ancient is the industry and beautifully woven into the Mayan agriculture. A very interesting blog by the hospitality expert Walten (2016) as follows ..."Before the arrival of the Spanish in the mid 1500's, the Maya were farming chicle, a white resin from the Zapote tree found in the Yucatan Peninsula. The Maya had many different uses for the extremely sticky chicle resin but found it to be useful to clean their teeth and mouths. The Chicleros chewed on the white resin once it was processed from a liquid to a solid white resin putty.......... Puerto Morelos, a small town just south of the Cancun airport, was founded during the Chicle industrial era in 1898. Back then it was called Punta Corcha and was the international port where prepared chicle was shipped to buyers. All Chicleros

Uses of sapota

Sapota fruit with slices

Sherbet from sapota

Halwa from sapota

Powder from sapota

Cultivars of Sapota

Kirthibarthi

Pala

Thagarampudi

PKM-4

PKM-5

PKM-1

Cultivars of Sapota

PKM-3

PKM-2

Cricket Ball

Kalipatti

CO-1

CO-3

Propagation of sapota

Propagation by seed

Air layering

Cleft grafting

Approach grafting

brought their bags full of the chicle resin to a jungle village now called Central Vallarta just west of Puerto Morelos. It was here that the sap was cooked down into a white resin/gum and molded into bricks. Workers brought the chicle blocks to Puerto Corcha (Puerto Morelos) by horse and rail wagon on the old rail tracks for exportation to international buyers in the United States".

Lundell (1937) surveyed the vegetation in the Guatemalan lowlands, he found that 75 per cent of the upland forest in the Northern Petén was made up of plant associations dominated by Manilkara zapota (locally known as sapodilla or chicozapote) and concluded that these trees had once been protected by the ancient Maya, who used them as a source of construction timbers and as edible fruit (Thompson *et al.*, 2015).

One of the classical studies using molecular markers was carried out by Kate Armstrong for her doctoral thesis for university of Edinburgh in 2010. According to her, "Resolution along the backbone of the phylogeny is weak and the area of origin is, therefore, difficult to determine. However, all sister taxa to *Manilkara* are African and this suggests that the most likely explanation is an African origin for the genus with subsequent inter-continental dispersal during the Miocene. *Manilkara* spread from Africa to the Neotropics and Asia via at least three separate long distance dispersal events. A single lineage dispersed to the Neotropics ~27-21 Ma and spread across the Isthmus of Panama before its closure. Another lineage dispersed to Southeast Asia ~30-25 Ma from mainland Africa and subsequently diversified throughout the region. A third dispersal from Madagascar to the Sahul Shelf, occurred ~31-16 Ma in the *M. fasciculata/dissecta/udoido* lineage" (Armstron, 2010).

4.0 Species, Types and Cultivars

4.1.0 Species

Sapota belongs to the family Sapotaceae. Its botanical name is *Manilkara zapota, Achras sapota* is a synonym. *Manilkara* is a genus of trees in the Sapotaceae consisting of ~79 species distributed throughout the tropics (30 South and Central American, 35 African and 14 Southeast Asian). According to Armstrong (2010), Sapotaceae is a family of predominantly tropical and subtropical trees and shrubs, which make up a significant component of lowland, wet forest in Africa, Asia and the Neotropics. Aside from its ecological prominence, the family is also economically important, being utilized for its timber, fruit and latex. Within the genus *Manilkara* alone, there are numerous economically important taxa. *M. bidentata, M. huberi, M. obovata* and *M. kauki* are just a few of the many species which are used commercially for timber and are known for their heavy, durable, rot-resistant wood. The latex of *Manilkara zapota* is tapped to form "chicle," (the original chewing gum) and *M. bidentata* latex is the source of "balata," formerly used in the manufacture of golf ball shells and machine belts. *M. zapota* fruit ("sapodilla") is also widely cultivated, while other species, although not in cultivation, are all edible. Additionally, a related species within the tribe Mimusopeae, *Vitellaria paradoxa* (syn = *Butyrospermum parkii*), is known in the cosmetics industry as the source of shea butter (a fat pressed from the seed), and is an important component of many skin creams (Pennington, 1991;

Govaerts *et al.*, 2001; Mathews, 2009; Armstrong, 2010). One of the most classical studies on the classification of Sapotacea was presented by Pennington (1990). According to him, the family sapotaceae includes 53 genera and more than 1,100 species. Pennington recognises five tribes in the Sapotaceae: Chrysophylleae, Isonandreae, Omphalocarpeae, Mimusopeae, and Sideroxyleae, with Mimusopeae subdivided into three subtribes: Mimusopinae, Manilkarinae and Gluemine. *Manilkara* is placed within Pennington's subtribe Manilkarinae (Armstrong, 2010). Armstrong (2010) says that the term Manilkara was first reported by the famous Hortus Malabaricum (Plants of Malabar) written by the Dutch Governor Van Rheede in 1683 much before Linnaeus classification came into vogue. The Plant mentioned was M.kauki. *Manilkara kauki* is not native to India and it is clear from the text that this is a cultivated plant brought to Malabar from China and the Philippines, used mainly for its edible fruits but also to treat ailments ranging from boils to "beriberi." Van Rheede indicates that "Manil" was adapted from the Portuguese "Manilhas Insulas" (for Manila, Philippine Islands) and "kara" refers to the Malayam for edible fruits. The older name for *Manilkara* was *Achras*.

4.2.0 Classification

Sapota is classified into two main types based on fruit shape as 'round' and 'oval'. However, there are several intermediates, as, it is a heterozygous tree crop. Rekha *et al.* (2011) grouped available variability in sapota based on fruit characters for selection of parents for use in further breeding programmes. Cluster analysis revealed four definite clusters. High variability was observed for fresh-fruit weight, fruit length, fruit girth, fruit weight at ripening, pulp weight, peel weight, number of seeds and TSS. Correlation studies among fruit parameters indicated positive relationship between all the parameters studied, except TSS which had negative relationship with the rest of the fruit parameters. Genetic relatedness and diversity among Indian sapota cultivars using random amplification of polymorphic DNA (RAPD) markers were reported by (Meghala *et al.*, 2005) Twenty sapota cultivars widely grown in different regions of India were used. Twenty decamer random primers amplified 232 RAPD markers of which 184 were polymorphic. Cluster analysis based on Ward's minimum variance algorithm showed two major groups. Both cluster analysis and principal components analysis of the RAPD data indicated wide genetic diversity among the cultivars. Despite its recent introduction to India, sapota germplasm appears to have a wide variation, possibly because the open-pollination nature of crop results in high heterozygosity, or because a good number of cultivars were introduced, or both (Meghala *et al.*, 2005). Based on the ratio of the diameter and length of the fruit, sapodilla fruit is classified into three types, *i.e.* round, oval and elliptical (Rozika *et al.*, 2013).

In Indonesia, sapodillas are classed in two main groups: 1) Sawo maneela, normal-size trees having narrow, pointed leaves; and 2) Sawo apel, low, shrublike trees, with oblong leaves broadest above the middle. Belonging to group #1 are the common cultivars **'Sawo betawi'** (fruit large, in clusters of 2-4, popular, perishable, ripening in 3 days from picking); **'Sawo koolon'** (fruit large, solitary, thick skinned, with firm flesh, shipping well); **'Sawo madja'** (large, with persistent scurf, pulp of fine texture, sweet with an acid tang). Belonging to group #2 are **'Sawo apel bener'** (fruits small in clusters of 3-6, thick-skinned); **'Sawo apel klapa'** (fruits medium-size, with persistent scurf). Some others are little grown because the fruits are either very small, too sandy, too gummy, or too dry. (Morton, 1987).

In India they are grouped under four types based on the habit of the tree, *i.e.*, nature of branches and colour of foliage; they are as follows:

4.2.1 Trees with Erect Growing Habit

Branches appear in whorls; foliage deep green, broad and oval. Fruits large in size with smooth and yellow skin. Pulp is butter-like and sweet.

4.2.2 Trees with Drooping Habit

Branches appear in whorls; foliage light green, narrow and elliptical. Fruits small with rough brownish skin and pulp of inferior quality.

4.2.3 Trees with Spreading Habit

Branching irregular; foliage deep green in colour, broad and oval. Fruits with smooth and yellow skin. Pulp is butter-like and sweet.

4.2.4 Trees with Spreading Habit but Fruits Inferior

Another type of sapota tree has spreading habit but the fruits have rough skin and inferior pulp quality.

Each of these groups may further be subdivided into two sections based on shape of fruits, either round or oval.

4.3.0 Cultivars

Based on breeding methods, such as, introduction, selection and hybridization, many varieties have been developed. Details of characters of sapota varieties are given in Table 2 to Table 4.

4.3.1 Introduction

Exotic cultivars introduced in various countries.

Alano

It is grown in United States (Hawaii). The fruit is conical to round, skin light brown, smooth, 115-120g. Pulp is smooth to slightly granular, very good to excellent. Season: November-June.

Betawi (Indonesia)

Fruit is conical, 315g, pulp light amber – yellow, slightly granular, very good, juicy. Season: late December.

Brown Sugar

It is grown in United States. Fruit is round to ovate, skin light brown, moderately scurfy, 133-170g, brown, slightly granular. Season: May –September.

Gonzalez

It is grown in Phillipines. Fruit is round to oval, skin very light brown, slightly scurfy, 260g, pulp is light brown to brown, smooth, very good to excellent. Season: November – April.

Hasya

It is grown in Mexico. Fruit is oval to slightly conical, skin light brown, moderately scurfy, 365 g, pulp is brownish red excellent. Season: November –June.

Makok (Dwarf) Tree

It is grown in Thailand. Fruit is conical, skin light brown, slightly scurfy,140 g, pulp is light brown to slightly greenish-red, smooth. Season: May –November.

Table 2: Sapota Cultivars of Florida, USA*

Cultivar Name	*Country of Origin*	*Fruit Shape and Skin Color*	*Fruit Size*	*Pulp Color and Texture*	*Quality*	*Yield*	*Season*
Alano	United States (Hawaii) smooth	Conical to round, skin light brown,	S to M, 115-250g	Smooth to slightly granular	VG to excellent	G	Nov.-June
Betawi	Indonesia	Conical	ML, 140-315 g	Light amber - yellow, slightly granular	VG, juicy	G	late Dec.
Brown Sugar	United States	Round to ovate, skin light brown, moderately scurfy	S to M, 133-170 g	Brown, slightly granular	VG	G to VG	May-Sept.
Gonzalez	Phili-ppines	Round to oval, skin very light brown, slightly scurfy	M,90-260 g	Light brown to brown, smooth	VG to excellent	VG	Nov.-April
Hasyá	Mexico	Oval to slightly conical, skin light brown, moderately scurfy	ML, 150-365 g	Brownish red	excellent	G	Nov.-June
Makok (dwarf) tree	Thailand	Conical, skin light brown, slightly scurfy	S, 30-140 g	Light brown to slightly greenish- red, smooth	VG	VG	May-Nov.
Modello	United States	Elliptic to ovate, skin light brown, moderately scurfy	M to ML, 227-340g	Whitish to tan, smooth	G	Fair	Feb.-May
Molix	Mexico	Oval	M to M-L, 150-360 g	Brownish red, smooth	VG to excellent	VG	Feb.-April
Morena	Mexico	Oval	M to ML, 170-345 g	Brownish red, smooth	VG to excellent	G to VG	Feb.-April

Cultivar Name	Country of Origin	Fruit Shape and Skin Color	Fruit Size	Pulp Color and Texture	Quality	Yield	Season
Oxkutzcab (Ox)	Mexico	Roundish	L to very L, up to 800 g	Reddish brown	VG	VG	May-Sept.
Prolific	United States	Round to conical, ovate, skin light brown, slightly scurfy	M, 170-225 g	Light tan to reddish tan and smooth	VG	VG	May-Sept
Russell	United States	Round to conical to ovate, skin brown with grayish-brown patches, scurfy	L, 284-454 g	Pinkish tan, granular	Good, mildly fragrant	Poor	May-Sept.
Tikal	United States	Ellipsoid to conical, skin light brown, slightly scurfy	S to ML 80-323 g	Light brown, smooth	Very good, fragrant	VG	Main, Dec. - March; minor, May-Sept.

S=small; M=medium; ML=medium-large; L=large.

*(Balerdi *et al.*, 2013).

Modello

It is grown in United States. Fruit is elliptic to ovate, skin light brown,moderately scurfy, 30 g, pulp is whitish to tan, smooth. Season: February –May.

According to Morton (1987) some introduced cultivars in Florida include: 'Boetzberg', 'Larsen', 'Morning Star', 'Jamaica 8', and 'Jamaica 10'. 'Tikal', a recent seedling selection, seems very promising. It is light-brown, elliptic to conical, much smaller than 'Prolific', but of excellent flavor and comes into season very early. Several cultivars not recommended because of low yield in southern Florida are 'Addley', 'Adelaide', 'Big Pine Key', 'Black', 'Jamaica No. 4', 'Jamaica No. 5', 'Martin' and 'Saunders'.

4.3.2 Selection

Morton (1987) gave details of varieties grown in different countries which are mostly selections as below:

Mexico: 'SCH-02','SCH-03','SCH-07', 'SCH-08', and 'SCH-28'.

Florida, U.S.A.: Seedling selections of high quality have been named and vegetatively reproduced. The first of these was 'Russell' from Islamorada in the Florida Keys, named and propagated by R.H. Fitzpatrick. It is nearly round, up to 4 in (10 cm) in diameter and length, brown-scurfy with gray patches, and luscious, reddish flesh. It is not a dependable bearer. The second, 'Prolific', a seedling grown at the Agricultural Research and Education Center, Homestead, and released in 1941, is round-conical, 2 1/2 to 3 1/2 in (6.25-9 cm) long and broad, with smooth, pinkish-tan flesh. The skin is lighter than that of the 'Russell' and tends to lose much of the scurf as it ripens. The tree bears early, consistently and heavily. Of later selection, 'Modello' is a good quality fruit but not a heavy producer; 'Seedless' yields poorly; 'Brown Sugar' is a good, regular, high yielder; handles and keeps well.

In the Philippines, selected cultivars, 'Ponderosa', 'Java', 'Sao Manila', 'Native', 'Formosa', 'Rangel', and the 'Prolific' from Florida are maintained by the Bureau of Plant Industry for propagation and distribution to farmers. 'Sao Manila' fruits mature in 190 days and ripen 3 to 5 days after picking.

Good number of selections has been released by Tamil Nadu Agricultural University.

PKM-1

It is a clonal selection from Guthi. A mean yield of 3547 fruits (236kg) can be obtained per tree per year. The tree is dwarf and adaptable to Southern and Central districts. Fruits are of two shapes *viz.*, round (rarely) and oval (maximum). The skin is very thin and the pulp has a buttery consistency. High average yielder with medium size fruits. Average fruit weight is 100g. Released in 1981.

PKM (Sa)- 4

It is a open pollinated clone of PKM 1. It is a high yielder with compact canopy and cluster bearing character. It bears spindle shaped fruits suitable for dry flakes production. The fruit contains attractive pulp with light pinkish honey brown colour which are crisp with sweet with TSS of 24 -25° Brix. The fruits are also found to be less seeded (2-3 seeds only). The best season for planting is June – July and October – November. It yields about 20.8 tonnes/hectare (100.4 kg/tree/year) which is 138.29 per cent increase over PKM 1. it is suitable for varied soil types of tropical conditions. Released in 2003.

PKM (Sa)-5

It is a selection from open pollinated seedlings maintained in a private orchard at Virudhunagar. Leaves are waxy, narrow, small, upright and arranged in whorl form in cluster and rich in proline content, relative water content, epicuticular wax and chlorophyll stability index. High yielder, High Total Soluble Solids (25.50 brix) and suitable for preparation of dry flakes, milk shake powder and mixed fruit jam. Suitable for vertisols, water stagnated and drought areas. Oval shaped attractive fruits with smooth, light brown skin which facilitates a good market appeal. Flesh is crisp and retains coppery brown colour while other sapota varieties turn dark brown after ripening. Since the fruits are heavier with coloured flesh than other varieties, it is highly suitable for making flakes, powder and mixed fruit jam. It shows field tolerance to leaf spot and leaf webber incidences. This can be best planted during June – October. This is having a yield potential of 18.70 t/ha which is 19.06 per cent yield increase over (PKM 1) and 17.97 per cent (PKM 4). This is well adapted to the vertisol soils of Southern Tamil Nadu and water stagnated and drought prone areas. Released in 2007.

Table 3: Description of some Cultivars of Sapodilla grown Abroad

Cultivar	Origin	Fruit Size (g)	Yield/Tree	Location
Addley	Bahamas	-	Very poor	Florida
Adelaide	Bahamas	-	150 kg	Venezuela
Big Pine Key	Florida	-	Very poor	Florida
Black	Florida		Very poor	
Brown Sugar	Florida	133-170	901 fruit	Venezuela
Brown Sugar	-	-	125-200 kg	Florida
Jamaica No. 4	Jamaica	-	Very poor	Florida
Jamaica No. 5	Jamaica	-	Very poor	
Jantung	Malaysia	100		Malaysia
Martin	Florida	-	Very poor	Florida
Modello	Florida	227-340	50-100 kg	Florida
Prolific	Florida	170-225	520 fruit	Venezuela
Prolific	-		150-225 kg	Florida
Russell	Florida	284-454	544 fruit	Venezuela
Russell	-	-	25-100 kg	Florida
Saunders	Florida	-	Very poor	
Seedless	Florida	small	<12.5 kg	
Tikal	Mexico	113-170	170-225 kg	Florida
Tikal	-	120	150-175 kg	

CO2

It is a clonal selection from Baramasi. Fruits are obovate to round, medium sized, skin outer surface cinnamon brown in color, inner surface yellowish green, flesh soft, juicy, slightly gritty, light brown in color, aroma slight, taste sweet, seed medium sized, black, obovate in shape, beaked, suture distinct, slightly adhering to flesh and placed centrally in the fruit. The length, diameter and weight of fruits range from 4.0 to 6.5 cm, 5.0 to 7.0 cm and 112-168 g respectively. The number of seeds per fruit varies between two and eight. The mean acidity of fruit is 0.13 with a reducing sugars content of 12.8 per cent, non-reducing sugars of 3.94per cent, total sugars of 16.95 per cent and total soluble solids of 23.10 per cent. Approximately fruits take six months from fruit set to harvest, with two bearing seasons (December – February and May-June). The yield is 175 kg per tree or 11.8 tonnes of fruits per hectare per year at the rate of 67 trees planted at 40 feet spacing. Released in 1974

4.3.3 Hybridization

A large number of hybrids have been released by Tamil Nadu Agricultural University(http://agritech.tnau.ac.in/horticulture/horti_ TNAU_ varieties_ fc_ sapota.html)

CO1

It is a hybrid clone of the cross between Cricket Ball and Oval. The fruit shape is long oval, medium sized with total soluble solids of 18°brix. Each fruit weighs about 125g. A yield of 175 to 200kg per tree can be obtained per year. The fruit is bigger than Oval and sweeter than Cricket Ball. The flesh is granular in texture and reddish brown in color. Released in 1972

CO3

It is a hybrid between Cricket Ball and Vavilavalasa released from Department of Fruit Crops, Horticultural College and Research Institute, Coimbatore. The trees are intermediate stature with compact canopy. Highly suitable for high density planting. It bears fruits all through the year with a peak during February-June and September-October. A well-grown tree yields up to 157kg of fruits as against 109.5 and 101.32kg/tree by CO.2, CO.1 respectively. Under high-density planting system it yields 40-50t/ha. It is suitable for growing in plains up to an elevation of 100m MSL under wide range of soil conditions. Released in 2000.

PKM2

PKM.2 (H-2/4) is a hybrid between Guthi and Kirtibarthi varieties. This hybrid is a high yielder recording a mean yield of 80kg of fruits/tree in during fifth year after planting. The yield increase is 42.76 per cent over PKM.1 variety. The fruits are bigger in size (95g) as compared to PKM.1 (84.0g). They are oblong to oval shaped. The fruits are of good quality with higher TSS (26.3°brix), total sugar (14.25 per cent), reducing sugar (9.46 per cent) with high sugar/acid ratio. Released in 1992.

PKM3

It is a hybrid between Guthi x Cricket Ball. The variety is adaptable to tropical plains of Tami Nadu and yields 14t/ha. The fruits bear in clusters with oval shaped large fruits. The vertical growth habit of tree allows high density planting. The fruits mature earlier than other varieties in this season thus fetching higher return. The variety is tolerant to leaf spot and leaf webber. Released in 1994.

Hybrids released by UAS, Dharwad

DSH-1

It is a cross between Kalipatti and Cricket Ball, developed by UAS, Dharwad. It yields 30 per cent more than both the parents. Fruits are larger (150 g) and oval in shape. Fruit quality is excellent (TSS 26 per cent).

DSH-2

It is a cross between Kalipatti and Cricket Ball, developed by UAS, Dharwad. Fruit yield is more than 30 per cent of both the parents. Fruit is oval and bigger in size (180 g) than Cricket Ball. Fruit TSS is 23 per cent.

Table 4: Important Cultivars of Sapota Released in India

Cultivars	*Methods of Development*	*Description of Cultivars*
Kalipatti	Clonal Selection	✰ Average fruit weight is 98 g and yield up to 160 kg.
Kirtabarti		✰ Average fruit weight is 84 g
Long Oval		✰ Average fruit weight is 140 g
Mohangottee		✰ Average fruit weight is 102 g and yield up to 107 kg
Oval		✰ Average fruit weight is 84 g
Pala		✰ Average fruit weight is 31g
Pilipatti		✰ Average fruit weight is 82 g and yield upto 115 kg.
Badam		✰ Average fruit weight is 45 g.
Baramasi		✰ Average fruit weight is 118g
Calcutta Round		✰ Average fruit weight is 98 g
Tagarampudi		✰ Average fruit weight is 84 g
Cricket Ball		✰ Average fruit weight is 142 g and yield upto 93 kg per fruit
Dwarapudi		✰ Average fruit weight is 90 g
Gavarayya		✰ Average fruit weight is 112g
Guthi		✰ Average fruit weight is 56 g
Vavivalasa		✰ Average fruit weight is 98 g
Zumakhia		✰ Average fruit weight is 57 g and yield upto 55 kg.
CO-1	Hybrid clone of Cricket Ball x Oval (released in 1972)	✰ The fruit shape is long oval, medium sized with total soluble solids of 18°brix.
		✰ Each fruit weighs about 125g with yield of 175 to 200kg.
		✰ The fruit is bigger than Oval and sweeter than Cricket Ball.
		✰ The flesh is granular in texture and reddish brown in color
CO-2	Clonal selection from Baramasi (released in 1974)	✰ Fruits are obovate to round, medium sized, skin outer surface cinnamon brown in color, inner surface yellowish green, flesh soft, juicy, slightly gritty, beaked, suture distinct, slightly adhering to flesh and placed
		✰ The length, diameter and weight of fruits range from 4.0 to 6.5 cm, 5.0 to 7.0 cm and 112-168 g respectively.
		✰ The number of seeds per fruit varies between two and eight.
		✰ The mean acidity of fruit is 0.13 with a reducing sugars content of 12.8 per cent, non-reducing sugars of 3.94 per cent, total sugars of 16.95 per cent and total soluble solids of 23.10 per cent.
		✰ Approximately fruits take six months from fruit set to harvest, with two bearing seasons (December – February and May-June).
		✰ The yield is 175 kg per tree or 11.8 tonnes of fruits per hectare per year at the rate of 67 trees planted at 40 feet spacing

Cultivars	Methods of Development	Description of Cultivars
CO-3	Hybrid of Cricket Ball x Vavilavalasa (released in 2000)	✰ The trees are intermediate stature with compact canopy and highly suitable for HDP.
		✰ It bears fruits all through the year with a peak during February-June and September-October and yield up to 157 kg of fruits as against 109.5 and 101.32 kg/tree by CO-2, CO-1 respectively.
		✰ Under HDP, it yields 40-50 t/ha. It is suitable for growing in plains up to an elevation of 100 m MSL under wide range of soil conditions.
PKM-1	Clonal selection of Guthi (released in 1981)	✰ A mean yield of 3547 fruits (236 kg) of tree which is dwarf
		✰ Fruits are of two shapes *viz.*, round (rarely) and oval (maximum).
		✰ The skin is very thin and the pulp has a buttery consistency with average fruit weight is 100 g.
PKM-2 (H-2/4)	Hybrid of Guthi x Kirtibarthi (released in 1992)	✰ It is a high yielder 80 kg of fruits/tree during fifth year after planting.
		✰ The yield increase is 42.76 per cent over PKM-1 variety.
		✰ The fruits are bigger in size (95g) as compared to PKM.1 (84.0g). They are oblong to oval shaped.
		✰ The fruits are of good quality with higher TSS (26.3°brix), total sugar (14.25 per cent), reducing sugar (9.46 per cent) with high sugar/acid ratio
PKM-3	Hybrid of Guthi x Cricket Ball (released in 1994)	✰ The variety is adaptable to tropical plains of Tami Nadu and yields 14t/ha.
		✰ The fruits bear in clusters with oval shaped large fruits. The vertical growth habit of tree allows HDP.
		✰ The fruits mature earlier than other varieties thus fetching higher return. The variety is tolerant to leaf spot and leaf webber.
PKM-4	Open pollinated clone of PKM-1 (released in 2003)	✰ It is a high yielder with compact canopy and cluster bearing character.
		✰ It bears spindle shaped fruits suitable for dry flakes production.
		✰ The fruit contains attractive pulp with light pinkish honey brown colour which are crisp with sweet with TSS of 24 -25° Brix.
		✰ The fruits are also found to be less seeded (2-3 seeds only). The best season for planting is June-July and October-November.
		✰ It yields about 20.8 tonnes/hectare (100.4 kg/tree/year) which is 138.29 per cent increase over PKM 1.
PKM-5	Selection from open pollinated seedlings (released in 2007)	✰ High yielder, High Total Soluble Solids (25.50 brix) and suitable for preparation of dry flakes, milk shake powder and mixed fruit jam.

Cultivars	Methods of Development	Description of Cultivars
		✰ Suitable for vertisols, water stagnated and drought areas. Oval shaped attractive fruits with smooth, light brown skin which facilitates a good market appeal.
		✰ Flesh is crisp and retains coppery brown colour while other sapota varieties turn dark brown after ripening.
		✰ Since the fruits are heavier with coloured flesh than other varieties, it is highly suitable for making flakes, powder and mixed fruit jam.
		✰ It shows field tolerance to leaf spot and leaf webber incidences having a yield potential of 18.70 t/ha which is 19.06 per cent yield increase over (PKM 1) and 17.97 per cent (PKM 4).
		✰ This is well adapted to the vertisol soils of Southern Tamil Nadu and water stagnated and drought prone areas.
DHS-1	Hybrid of Kalipatti x Cricket Ball	✰ Tree is vigorous, bearing round to slightly oblong fruits with high yield with a TSS of 26°brix.
		✰ The pulp is light orange colour and mean fruit weight is 150g.
DHS-2	Hybrid of Kalipatti x Cricket Ball.	✰ Tree is vigorous, bearing round fruits, high yielder.
		✰ The fruits are sweet with a TSS of 23° brix having a light orange brown pulp, which is soft, granular and mellowing with average fruit weight is 180g.

The chemical compositions of fruits of different sapota cultivars are presented in Table 5.

Table 5: Chemical Composition of Fruits of different Sapota Cultivars*

Cultivars	TSS (°Brix)	Acidity (% tartaric acid)	Ascorbic acid (mg/100 g)	Total Acid (%)	Sugar/Acid ratio
PKM 1	20.05	0.048	2.95	10.68	222.50
Oval	18.50	0.028	1.65	8.22	293.57
Pala	20.46	0.108	3.10	7.23	66.94
Kirtabarthi	23.35	0.038	3.61	8.93	235.00
CO1	15.02	0.020	2.67	8.22	411.00
Cricket Ball	18.48	0.067	1.96	6.94	103.58
Guthi	22.24	0.077	3.15	9.47	122.99
CO_2	23.05	0.046	3.26	8.33	181.07
Gavarayya	19.40	0.046	3.10	8.56	186.09
Thagarampudi	20.60	0.031	2.95	7.10	229.03
Dwarapudi	21.20	0.038	1.97	7.14	187.89
Virudhunagar	19.80	0.041	3.50	9.40	229.27

* Ponnuswami and Irulappan (1987).

In Florida, USA, cultivars Prolific and Brown Sugar have been found to be good producers and used for general planting. Others like Modello and Russel are propagated in the nursery trade but do not bear fruit heavily or regularly. Addley, Adelaide, Big Pine Key, Black, Jamaica No. 4, Jamaica No. 5, Martin, Saunders and Seedless have been found unfruitful in some areas and hence not recommended.

5.0 Breeding and Improvement

Sapota in India has been based on narrow genetic base and most of the varieties are the result of local selection for shape, size and quality of fruits. In sapota, there are a large number of cultivars differing in several fruit characters, such as size, texture of flesh, sweetness and seed number. A dessert sapota of good eating quality should have a pleasant sweet taste and good flavours, with abundant mellow and melting type pulp (Chadha, 1992).

5.1.0 Cytogenetics

Genetic information particularly on the chromosome of sapodilla is still limited. Chromosome number of sapodilla is 26. The karyotype type is "2B" and "1A" respectively, karyotype formula is "18m8sm". (Zhang, 2011). Yuniastuti *et al.* (2018) studied morphological information on its chromosome such as number, size, form, and karyotype of sapodilla. The chromosome number of sapodilla from two locations in Indonesia, Bojonegoro and Wonogiri, had similarity: 2n = 26, metacentric in shape. The chromosome size of sapodilla from Bojonegoro ranged between 1.46 ± 0.38 µM and 3.06 ± 0.48 µM, while that of Sapodilla from Wonogiri ranged between 1.49 ± 0.19 µM and 3.58 ± 0.27 µM. The karyotype formula of Sapodilla from Bojonegoro and Wonogiri was 2n=26m consisting of 13 metacentric chromosome pairs.

5.2.0 Genetic Resources

Even though, Sapota originated in Mexico there has been a lot of variability reported on this crop from many places where it was introduced due to the seedling origin of many of the varieties. Forest fragmentation, habitat loss and isolation may have a strong effect on biodiversity in tropical forests. This can include modification of the genetic diversity and structure of plant populations. In this study, we assessed the genetic diversity and structure of the tree *Manilkara zapota* in 15 naturally formed fragments of semi-evergreen tropical forest, as well as in an adjacent continuous forest for comparison. Forest fragments were scattered within a matrix of wetlands and were highly variable in terms of size and degree of isolation. The naturally fragmented populations of *M. zapota* had slightly less allelic diversity (Ar: 3.4) than those of the continuous forest (Ar: 3.6), when corrected for sample size. However, populations in the fragments and continuous forest had very similar heterozygosity levels (H_E: 0.59 in both cases). Low levels of genetic differentiation were observed among populations (F_{ST}: 0.026) and genetic structure was not consistent with isolation by distance, indicating high levels of gene flow. Genetic diversity was not explained by fragment size or degree of isolation. The relatively high genetic diversity and low inter-population genetic differentiation observed in *M. zapota* may be the result of long-distance pollen and seed dispersal, as well as the high proximity among patches (Martínez-Natarén *et al.*, 2017). Cruz-Rodriguez and

Lopez-Mata (2004) reported an insight into those factors that influence growth and survival of *Manilkara zapota* seedlings, one of the most abundant species in the lowland forests of Mexico and Central America. Eight 10-m^2 rectangular plots 5m x 2m were established, in which all 35-cm high *M. zapota* seedlings were marked, measured and enumerated at bimonthly intervals. In each census seedling height, number of leaves, length of largest leaf, type of damage and seedling death was registered. Seedling survival during 2 yr was high, reaching 82 per cent including newly emerged seedlings. A maximum likelihood regression analysis showed that both number of leaves and length of largest leaf had a direct influence on seedling survival; however, neither type of damage nor seedling density and height had significant effect. Seedling height growth averaged 2.8 cm in the 2 yr of study. The combined effect of high seedling survival, reduced growth, and impact of physical damage mainly due to falling branches and leaves reveals the occurrence of a persistent seedling bank. Such a bank would contribute to recruitment of individuals in the juvenile and eventually in the adult stages. This seedling bank could explain the high density of individuals of *M. zapota* in the tropical forests of Mexico.

In Indonesia, a region of Yogyakarta (DIY) is one of the biggest production centers of sapodilla in Indonesia. It consists of four districts, namely Kulon Progo, Bantul, Gunungkidul, Sleman and the city of Yogyakarta. There is diversity of sapodilla in DIY, either in morphologies or in flavors. Sapodilla accessions spreading in DIY tend to cluster into 2 groups based on their location. The first group was the accessions collected from Bantul, Kulon Progo, and Yogyakarta while the second group consisted of accessions from Gunungkidul and Sleman which were clustered with genetic similarity each of 53 per cent and 37 per cent. Genetic diversity of sapodilla in DIY was high with the value of PLP reaching 98.85 per cent of the 87 loci and h of 0.28±0.14. Genetic diversity within population was found to be greater than that between populations amounted to 53 per cent and 47 per cent, respectively. Due to its highest diversity, Gunungkidul was recommended to be potential site for sapodilla selection to support breeding program (Sari *et al.*, 2018).

5.3.0 Breeding Approaches

Breeding objectives:

a. Varieties with a few seeds, sweet pulp with skin being medium thick and crispy flesh and non-sandy texture
b. Frost resistant cultivars should be screened out for these areas to harvest better fruit quality and productivity.
c. Saline tolerance will be an added trait for rootstock breeding besides Dwarfness.

Hybridization is one of the means of effective improvement in this fruit plant so that desirable traits now found dispersed in different cultivars may be combined in one. Preliminary work on hybridization in sapota was attempted at the then Agricultural College and Research Institute, Coimbatore (Sambamurthi and Ramalingam, 1954). Similar attempts were also made in other parts of the country. At Coimbatore, two commercial parents to combine large size of fruits and superior taste were selected, and they were Cricket Ball and Oval. The seedlings of those crosses were planted in 1954, and they fruited in four years after planting. The fruits were oval and large-sized with an average weight of 125 g. The flesh was granular and reddish brown in colour, very sweet taste with total soluble solids of 18 per cent. The vegetative and floral characters of the hybrids resembled the parental forms. The seeds were slightly longer than those of both the parents with a conspicuous curvature at the tip (Fazlulla Khan *et al.*, 1965).

Similar attempts have also been made at the Agricultural College, Dharwar, Karnataka with four cutivars of sapota, namely Kalipatti, Cricket Ball, Calcutta Round and Oval. The fruit size was large when the pollen sources were Oval, Kalipatti, Calcutta Round and Cricket Ball. Cricket Ball had the largest fruits and the Oval the smallest. The shape of the self-pollinated fruits was oval and the cross-pollinated fruits roundish (Farooqi and Rao, 1976a). The seedlings of the different crosses were planted. In another study by the same workers (1976b) on fruit set through intra- and intervarietal pollination in four cultivars, open pollination, self-pollination and intervarietal cross-pollination were attempted. In case of self pollination, the flower buds were bagged before anthesis or emasculated flowers pollinated with pollen either of the same tree or from a different tree of the same cultivar. In the case of open pollination the percentage of fruit set ranged between 1 and 24, and 0 to 6 per cent in flowers that were bagged. When the flowers were pollinated with pollen of the same tree, fruit set was 10 to 27 per cent and it was 3 to 12 per cent when flowers were pollinated from a different tree. Controlled cross-pollination resulted in the highest fruit set of 78 per cent, but a wide variability existed among the various combinations and their reciprocals. The different cultivars selected for the purpose were Kalipatti, Cricket Ball, Calcutta Round and Oval. The most striking behaviour was in respect of Calcutta Round cultivar which showed one way incompatibility with Kalipatti.

Attempts have been made in India to evolve superior hybrids through systematic research work. At Tamil Nadu Agricultural University, Coimbatore, efforts were made to combine large fruit size of Cricket Ball with the superior taste of Oval, while at Agricultural College, Dharwar, crosses were made among Kalipatti, Cricket Ball, Calcutta Round and Oval. The promising hybrids evolved from the crosses are listed in Table 4.

6.0 Propagation

Sapota is propagated by both seed and vegetative methods.

6.1.0 Seed

In the earlier days, seedlings were used for planting but they had some disadvantages, such as slow growth, very long-pre-bearing age of 8-10 years, growing to a huge size or height and showing too much of variation or not being true-to-type (Cheema *et al.*, 1954; Sayed, 1962).

Seeds are collected from elite trees and sown in pans or beds with light sandy soil as germinating medium and watered at regular intervals. Shanmugavelu (1970) found that GA3 promoted the percentage of germination and increased the growth of seedlings. Soaking of seeds in 400 ppm GA3 resulted into largest seedling (Pampanna and Sulikeri, 1999).

6.2.0 Vegetative Propagation

The vegetative method has many advantages, such as being-true-to type, earliness in bearing, dwarf or easily manageable size of the trees; favourable rootstock may influence and improve bearing. Among the vegetative methods, the important techniques are air-layering or marcottage, ground layering, pot layering, inarching and budding. Recently, attempts have been made to multiply this plant through tissue culture also.

6.2.1 Cutting

Some experiments have been conducted on rooting of sapota stem cutting. Preconditioning with etiolation and ringing followed by treatment with IBA at 2000 ppm gave 80 per cent rooting (Ramsundar and Abdul Khader, 1986). Attempts on rooting of stem cutting with auxin alone did not give any encouraging result (Sarker and Chatterjee, 1993).

6.2.2 Layering

6.2.2.1 Air-layering

Sapota is a difficult-to-root plant. Many factors are known to affect the rooting ability of this plant. Vigour of the shoot has been found to be one of the important factors in this regard. The loss of rooting ability over a period of change from juvenility to maturity is a recognized fact. Air-layers in sapota, from invigorated shoots, produced as a result of pruning of branches, possessed a better rooting ability than mature shoots (Uthaiah *et al.*, 1977).

The best time of air-layering is the beginning of rainy season. High humidity prevailing during the monsoon facilitates rooting. One- to two-year old, 45-60 cm long mature branches of pencil thickness (0.6 to 1.0 cm) with plenty of healthy dark green foliages are selected at a suitable height. Singh *et al.* (1962) reported that rooting percentage in air-layers of sapota increased with a mixture of IBA and NAA at 10000 ppm. Invigoration and etiolation of shoots and application of growth regulator were found to be significantly effective in inducing roots and promoting the root quality of air-layers. Invigorated shoots after etiolation when treated with 10,000 ppm IBA gave 91.7 per cent success and 90.9 per cent survival (Chatterjee *et al.*, 1990).

6.2.2.2 Ground-layering

Ground-layering is not a commercial method but is followed in Maharashtra to some extent. Monsoon (June-July) is the ideal time and can be done upto September. Rooting takes place in about 2 to 3 months and in about another one month they can be separated.

6.2.2.3 Pot-layering

This is followed in Maharashtra and northern parts of Karnataka. A shallow earthen pot with 'v'-shaped cuts on the rim is used to insert the branch to be layered. This pot is known as *'parali'*. It is filled with a moist mixture of equal quantities of farmyard manure and soil. The ringed branch is inserted in between the crotches and covered with moist medium over which a small flat stone is placed to prevent it from being dislodged due to wind. Watering is done regularly. In about two months roots appear and in another 1-2 months the branches are separated and potted.

6.2.3 Grafting

Comparatively longer time for rooting and high percentage of mortality in layer necessitate resorting to grafting which is the most popular and commercial method of propagation of sapota. Some cultivars do not strike roots easily by *gootee* or layering and hence have to be propagated by grafting.

6.2.3.1 Inarch Grafting

It is a commercial method of propagation practised over 40 years. Variousrootstocks, such as sapota seedlings, *khirni* or rayan (*Manilkara hexandra*) and mahua (*Madhuca latifolia*) seedlings are used. The rootstocks are raised in pots. The scion remains attached to the parent tree till the union is complete and if the scion branches are high; the stock plants are placed on bamboo platforms or any such other device. Bhuva *et al.* (1990a,b) reported that survival of graft was significantly influenced by the season of inarching. Maximum success was noted in February (90 per cent) followed by January and March inarching. The grafts could be separated from mother plants 51/2 months after inarching. They further observed that there was negative and significant associations between minimum temperature, relative humidity and rainfall with mean number of grafts survival (Bhuva *et al.*, 1990b).

6.2.3.2 Side Grafting

The scion is prepared in the form of a wedge with one side slightly longer than the other. An incision is made into the stock at an angle of 20-25o and the scion is inserted into this incision. Both are properly tied and sealed at the graft union. After union the top of the stock plant is cut just above the point of union (Singh *et al.*, 1963).

Sapota trees damaged by diseases or rodents near the root system or the crown can be rejuvenated by inarching using Ryan seedlings as rootstock (Sayed, 1962; Gandhi, 1963).

6.2.3.3 Soft Wood Grafting

Several workers reported that soft wood grafting on one-year-old stock plant is simple, economical and can give considerable percentage of success. At Akola, Kumbhar and Desai (1986) obtained more than 80 per cent success in soft wood grafting when performed during July-August, but the percentage of success declined (50 per cent) sharply in September-October. Vasanth *et al.* (1993) reported that soft wood grafting using precured scion during May to July resulted in 100 per cent success on more than 30 cm tall *khirni* rootstocks and 97 per cent on less than 30 cm tall rootstocks. Grafting during August to December gave 72 to 78.6 per cent success while during January to April the success was to the extent of 58.6 to 66.6 per cent only.

6.2.4 Budding

This is a cheaper, easier and more efficient method. A single bud is employed for budding. In Sri Lanka, this method is followed by using rayan rootstock. This method can be adopted all over the coastal regions of India where climate is moist and rainfall heavy. For budding, May is the best month. In this, a flap of bark is eased out in the form of a tongue, and the bud along with a little bark attached to it is inserted underneath the flap and covered keeping the scion bud exposed. After union, the bud starts developing and the top of the seedling is cut a little above the bud union.

6.2.4.1 Rootstock

Use of proper rootstock is important in grafting or budding, since the slow growth of seedling, rootstock becomes a limiting factor for rapid multiplication of the desirable forms in large number. The use of rootstock also poses certain problems, such as incompatibility and undesirable or adverse effects. Often the incompatibility affects fruit quality after several years.

The different rootstocks used are : (i) sapota seedlings (*Manilkara zapota*), (ii) Rayan or *khirni* or *pala* (*Manilkara hexandra* or *Mimusops hexandra*), (iii) Adam's apple (*Manilkara kauki* or *Mimusops kauki*), (iv) *mahua* (*Madhuca latifolia*), (v) Mee tree (*Bassia longifolia*), (vi) star apple (*Chrysophyllum cainito*), (vii) miracular fruit (*Sideroxylon dulicificum*).

Sapota seedlings take a long time to attain a suitable size for grafting. In Andhra Pradesh, Adam's apple is used as a rootstock. In Sri Lanka, both mee tree and miracular fruit are used as rootstocks. Mee tree seedling as rootstock grows quickly and develops deep tap root but exhibits incompatibility, by over growing the scion, *mahua* produces bigger plants but fruits are poor in quality due to the presence of an alkaloid saponin. Of all the rootstocks, rayan has been found to be the most suitable rootstock for sapota (Singh, 1969). It, however, shows variability in the rate of growth of seedlings. Some types are vigorous growing, while others are comparatively slow growing. It is more a compatible rootstock than *Madhuca latifolia*.

Results of rootstock trials on sapota conducted at the Fruit Research Station, Gandevi in Surat district of Gujarat indicated that sapotas on rayan rootstocks were far superior to those raised on sapota seedlings and from air-layers. In terms of yield also the trees grafted on rayan gave 50 per cent more than that on air-layers and twice that on sapota seedling stock. The trees on rayan stock were found to be healthy and strong even after four decades (Cheema *et al.*, 1954; Ram and Sahai, 1962; Sayed, 1962). Further studies would be necessary to confirm the suitability of other species, such as *Chrysophyllum cainito, Manilkara kauki* and others in respect of vigour of seedlings, graft compatibility, fruit quality, and their economic value. Results of the trials carried out in Gujarat, Kerala and Andhra Pradesh revealed that *khirni* was the most vigorous and productive rootstock compared to sapota seedling and *Bassia latifolia* (Chadha, 1992).

6.2.4.2 Raising of Rootstocks

The rayan fruits are small, oval and sweet. They are washed to get rid of the pulp in water and dried under shade and sown. For easy and quick germination, the seeds are soaked in water for a night. They are sown in raised seed beds or pots. The seeds germinate in about 4 weeks and when they are about 15-cm high they are transferred to small pots. Grafting or budding is done on seedlings about 1 cm thick.

6.2.5 Comparison in the Performance of Air-layers and Grafts

Plants from air-layers are shallow rooted and most of the roots concentrate in the top 30-45 cm of soil. These can thrive in both deep and shallow soils. Fruits from such trees have been found to be sweeter with mellow pulp.

Grafts have comparatively deeper root system and most of the roots are concentrated between 60 and 90 cm from the ground surface. The grafts on rootstocks like rayan require at least 90-cm deep soil for good growth. On heavy clay soils grafts have been found to give more yield than air-layers.

In Pune, Maharashtra under dry and light rainfall conditions, the grafts on rayan rootstocks were found to be vigorous in the first two years, but they developed chlorosis later in places where calcareous sub-soil was present, while grafts on sapota seedlings were slow growing compared with layers during that period but later started giving good crops.

In the Telangana area of Andhra Pradesh, grafts on *chalka* soils derived from granite rock which becomes hard on drying show poor growth of plants and low yields. At the Kodur Fruit Research Station, the grafts on rayan rootstocks made better growth and gave more yield than those on *Bassia longifolia* stock.

In Gujarat also, the grafts on rayan rootstocks under low rainfall and deep loamy soil conditions proved better in respect of yield than air-layers and grafts on sapota seedling stocks (Singh *et al.*, 1963). In Maharashtra, air-layers are preferred to grafts because of the reason that grafts bear more granular pulp (Cheema *et al.*, 1954).

A thorough comparative study of propagation of grafting and budding seems to be necessary to study their relative performance over a long period.

6.2.6 Tissue Culture

Sachdeva *et al.* (1986) cultured various explants from sapota seedlings germinated *in vitro* on 3 basal media supplemented with growth regulators. Nitsch's medium (NM) + 15 per cent coconut water (CW) + 4 ppm NAA and 2 ppm kinetin were best for callus induction and growth from root, hypocotyl, stem and shoot-tip explants. After 2 and 4 subcultures of 1 month's duration each, most cells were diploid (2n=26) while some were polyploid or aneuploid.

Purohit *et al.* (1999) developed a complete protocol for micropropagation of sapota using cotyledonary node segments. Multiple shoots were induced *in vitro* from nodal segments through forced axillary branching. Schenk and Hildebrandt's medium supplemented with 2.0 mg/l BAP induced up to 3 shoots per node with an average shoot length of 2.17 cm in 42 days. Further multiplication of shoots required habituation of cultures up to 3 periods of 21 days each on the same medium after establishment of cultures and initiation of shoot buds. Thereafter, a 3-fold multiplication rate was achieved during every subculture.

Purohit and Singhvi (1998) Induced multiple shoots *in vitro* from nodal segments by forced-axillary branching. Up to 3 shoots per node, with an average shoot length of 2.17 cm were induced after 42 days on SH medium supplemented with BA. After further subculturing (up to 3 passages of 21 days) on the same medium, a 3-fold multiplication rate was achieved during each subsequent subculture.

7.0 Area and Production

Presently India is one of the leading producers of Sapota in the world with an estimated area of 98000 ha with an estimated production of 1203000 tonnes of fruits for the year 2017-2018 (http://agricoop.gov.in/sites/default/files/2017-18 per cent 281st per cent 20Adv. per cent 20Est per cent 29.pdf). Among Indian states, Karnataka has the largest area and production while the productivity is highest in Tamilnadu. Sapota is also commercially in a number of countries such as Mexico, Central America, West Indies, Philippines, Pakistan and Malaysia India is considered to be the largest producer of sapota in the world, though it is considered to be a minor crop in India. The area under sapota in India has expanded rapidly from 800 ha in 1954 to 23,824 ha in 1987. The state wise area and production of sapota are given in Table 6.

Table 6: Area and Production of Sapota in India*

Area in '000Ha
Production in '000 MT

Sl. No.	States/Uts	2015-16		2017		2017-18	
		A	P	A	P	A	P
1.	Andhra Pradesh	11.85	148.76	10.59	137.67	10.09	131.46
2	Chhatishgarh	0.22	1.15	0.24	0.90	0.38	1.79
3.	Gujarat	29.56	325.15	30.01	331.54	29.55	326.36
4.	Haryana	1.63	16.02	1.68	16.07	1.69	14.65
5.	Himachal Pradesh	0.05	0.01	0.05	0.00	0.05	0.07
6.	Karnataka	28.29	330.50	25.73	301.44	25.31	315.38
7.	Kerala	0.69	0.70	0.69	1.09	0.16	0.48
8.	Madhya Pradesh	0.13	1.55	0.19	3.25	0.23	4.03
9.	Maharashtra	17.85	159.92	12.55	109.18	15.03	134.78
10.	Mizoram	0.00	0.03	0.00	0.02	0.00	0.02
11.	Odisha	3.35	15.67	3.36	15.72	3.36	15.72
12.	Rajasthan	0.01	0.04	0.01	0.06	0.01	0.03
13.	Tamil Nadu	6.75	225.40	7.83	261.28	5.76	169.96
14.	Telangana	1.45	18.53	0.75	6.84	0.83	9.45
15.	Tripura	0.21	1.81	0.17	1.28	0.14	0.88
16.	West Bengal	4.30	46.73	4.32	46.97	4.38	48.20
	Others	0.35	1.80	0.37	2.28	0.33	2.63
	Total	**106.69**	**1293.76**	**98.53**	**1235.59**	**97.29**	**1175.89**

**Source* : Horticulture Statistics Division, Department of Agriculture, Coopn & Farmers Welfare.

8.0 Soil and Climate

8.1.0 Soil

Sapota is a hardy tree and can be grown on a wide range of soils. Drainage is the most important factor. The most ideal soils are deep alluvium, sandy loams, red laterites and medium black soils. Sapota is sometimes planted in dry river beds with alluvial soil.

In South India, particularly in the Western Ghats of Karnataka, sapota is grown successfully on gravelly laterite soils (Cheema *et al.*, 1954; Choudhury, 1954; Singh *et al.*, 1963; Singh, 1967; Anon., 1975).

There should not be a hard pan in the sub-soil. In sandy soils, the plants are subjected to uprooting by strong winds since they lack strong anchorage. Deep and porous soils are preferred. Sapota trees produced higher root biomass in sub-layer (20-30 cm) than top layer which indicated that this crop could absorb moisture and nutrient from sub-surface at the time of drought (Singh and Sharma, 1997). In soils underlaid with hard gravels in the sub-soil at a depth of 60-90 cm, the plants make a good growth in the first seven to eight years but when the roots touch this hard layer the leaves turn pale in colour and the fruit size diminishes. In black cotton soils or clay soils with poor drainage and aeration during the rainy season, cracks develop in the soil in summer and as a result the smaller roots are damaged. Sapota can tolerate the presence of salts in the soil or in irrigation water to some extent.

The unsuitable soils are those underlaid with hard rock or lime content, yellow-coloured and sticky soils. Presence of high lime content in the soil causes the loss of green matter in the leaves and such plants die within a short time. Sapota gives good return in calcareous soils (pH 6-8) (Nakasone and Paull, 1998).

8.2.0 Climate

Sapota is a tropical fruit crop and can be grown from sea level upto 1200 m. It prefers a warm and moist weather and grows well in both dry and humid areas and coastal climate. In South India, it is grown on the hills upto 1000 m. At higher altitudes the fruit quality and tree health suffer. Areas with an annual rainfall of 125-250 cm are highly suitable. Rain or cloudy weather during any part of the year does not do any harm to fruit set. The optimum temperature is between 11°C and 34°C.

In northern parts of India, young plants undergo frost injury during winter months. Fully grown trees can withstand mild frost for a short period. This is the reason for restricting the sapota growing to only the sheltered pockets of the northern parts of India – in Punjab, Haryana, Uttar Pradesh. Sapota does well on the plains of North India but irrigation and protection in summer months are necessary. A high temperature above 41°C during summer causes drying up of stigmatic fluid, flower drop and fruit scorching. In severe cases, leaves and fruits undergo scorching. Shirsath *et al.* (1998) reported that the maximum and minimum temperatures had a negative corre-lation with the yield of sapota cv. Kalipatti, while relative humidity, rainfall and sunshine hours are positively correlated with yields. Light is reported to influence the growth of juvenile sapota plant. In general, the leaves under rain-shelter conditions were longer and larger than those under the lath-house, probably because the plants could more efficiently use the environmental inputs, especially light (Meza and Bautista, 1999a). Plants under shade (80 per cent light) grew to a height of 40 cm and had 20 leaves, whereas plants grown under a structure with a clear roof grew to a height of 93 cm and had 31 leaves. Internodal lengths were always lower under shade than in full sunlight (Meza and Bautista, 1999b). Dry and strong winds also cause damage to flowers, leaves and fruits (Choudhury, 1954; Narasimham, 1966; Gopalaswamiengar, 1970; Anon., 1975; Puttarudraiah, 1980).

9.0 Cultivation

9.1.0 Selection and Preparation of Land

A location, free from water stagnation should be selected. It is then ploughed two or three times and leveled. Undulated land should be divided into terraces depending on the topography of the land and leveling is done. If the soil is poor, it would be advisable to grow a green manure crop and plough it in so as to improve its physical and chemical conditions before planting operations are taken up.

A strong windbreak should be established by planting tall and thick growing trees on the windward side or on all sides of the orchard. Seedlings of mango, *jamun* and tamarind can serve as windbreaks, but these trees are slow growing and have a spreading habit of growth. While silver oak or casuarina being quick and erect growing makes an ideal windbreak. The plants for windbreak may be planted at 1.5 to 1.8 m apart in the row. The windbreak is helpful in avoiding damages due to dry and strong winds and particularly in sandy soils.

9.2.0 Preparation of Pit and Planting

Pits of the size of 60 cm^3 or 100 cm^3 are prepared at a distance of 8-10 m both ways depending upon the planting material and the soil. In low rainfall areas and soils with low fertility close spacing is followed, while in heavy rainfall tracts and fertile soils a wide spacing is recommended.

In a high density plantation trial in sapota cv. Kalipatti, Patel *et al.* (1993) observed that spacing had significant effect on plant height and yield. Maximum height of plant, number of fruits and fruit weight were obtained at a spacing of 5 m ? 5 m which was mainly due to increased population per unit area. Identification of dwarfing rootstocks and management strategies may help for high density planting of sapota (Chadha, 1992). Rajasekhar and Ratan,(2010) did not observe any difference for plant height cv. Kalipatti planted at different spacing treatments, while plant spread, canopy volume and number of fruits per tree were significant. Maximum yield/ha was recorded in the treatments 7.5 x 7.5 m spacing (19.99 t/ha) and 7.5 x 5 m spacing (19.81 t/ha), followed by 5 x 5 m spacing (15.94 t/ha), while lower yields were recorded in 10 x 10 m spacing (11.14 t/ha). 81.12 t/ha fruit was harvested in 5 x 5 m spacing by 13th year which was 51 per cent higher than the yield of 10 x 10 m spacing (39.79 t/ha). Hence, planting sapota at higher densities can fetch higher early returns to farmers.

The pits are exposed to sun for a month or two. The best time for making pits is during April-May. In Maharashtra, at the time of filling pits 30 kg of sheep manure and 1.2 kg of bone meal are put at the base of each pit and the rest is filled with top soil. One or two showers later will help to settle the soil in the pit.

The best time of planting is during early monsoon. Grafts, budded plants or layers are planted one in each pit in the centre and care should be taken to see that the roots are gently and firmly pressed and stakes are provided to avoid wind damage. The plants are then watered.

9.3.0 Training and Pruning

In sapota, a strong central stem is necessary and in air-layers most of the branches are put forth very near the ground level. In the beginning, the basal branches help in developing a thick central stem and hence care must be taken to maintain proper distribution of branches on all sides. But later, with an advance in age, the lower most branches weigh down to ground level and become unfruitful. Sapota, in general, has a well-balanced distribution of branches and the tree crown assumes a uniform shape. There is no necessity of pruning every year.

All the growths that appear on the rootstock below the graft or bud joint must be removed. After 3 to 4 years of planting, the lower most branches up to a height of 60-90 cm may be removed. Similarly, over shaded and crowded branches are also to be removed. An experiment was conducted to find out the effect of pruning and plant growth regulators,namely, lGA3, IAA, NAA and 2, 4-D on on Cricket Ball. Effect of treatments was significant for the Physico chemical quality like, fruit weight (138.066g), pulp weight (120.076g), TSS (24.6860B), ascorbic acid (14. 63g/100g) and acidity (0.146 per cent) in primary branch pruned trees. Among the growth regulators, NAA@50ppm produced highest ascorbic acid (14.666mg/100g) whereas GA3@20ppm produced highest fruit weight (134.494g), pulp weight (115.974g), TSS (24.9080B) and lowest acidity (0.15 per cent), seed weight (1.116g). It was concluded that the combined effect pruning and growth regulator simultaneously have significant effect on yield and quality.

In sapota, new growth and flowering occur simultaneously and it is a mixed type of bearing habit. Flowers and fruits appear in the leaf axils on the new growth and hence pruning of branches should not be done.

9.4.0 Irrigation

Though sapota can tolerate drought conditions to some extent, yet it responds well to irrigation. Young plants are watered regularly during dry season, and in winter at an interval of 6 to 12 days (Singh *et al.*, 1963). In Maharashtra, young plants are given irrigation once in eight days from October onwards till monsoon starts. Protective irrigation is given during first two seasons in Karnataka for better establishment of plants (Anon., 1975). Insufficient irrigation results in dropping of a large number of flowers leading to a loss upto 40 per cent in yield (Cheema *et al.*, 1954).

In the beginning, small basins are made and watered but as the plant grows in size, the size of the basin is also widened. In coastal Gujarat, there is a practice of raising 0.5-m high mounds of earth around the trees to protect them from strong winds and to conserve moisture (Singh, 1969). Bhuva *et al.* (1990c) suggested that for obtaining economic return from sapota cv. Kalipatti, 10 to 11 irrigations each at 80 mm depth to be applied at 32 days interval in winter and 18 days interval in summer. However, care should be taken not to over-irrigate the plants as it may cause flower and fruit shedding and increase the number of misshapened fruits.

A field trial was conducted with antitranspirants *viz.*, phenyl mercuric acetate, kaolin, salicylic acid, power oil and liquid paraffin on rainfed sapota cv. Kalipatti. It was found that antitranspirants affected the plant height, plant girth, and relative water content (RWC). Power oil (2 per cent) was the best which recorded the maximum RWC compared to control (Table 7). Minimum transpiration and maximum stomatal resistance were recorded with liquid paraffin (1.5 per cent), while salicylic acid (1500 ppm) was found to be best when fruit yield was concerned (Reddy and Khan, 2000).

Table 7: Effect of Antitranspirants on Relative Water Content (per cent) and Fruit Yield of Sapota cv. Kalipatti*

Treatment	*RWC (12 months)*		*RWC (18 months)*		*Fruit yield (kg/plant)*	
	2 Sprays	*4 Sprays*	*2 Sprays*	*4 Sprays*	*2 Sprays*	*4 Sprays*
Phenyl mercuric acetate 10^{-3} M	75.00	82.30	73.82	81.25	55.40	67.43
Phenyl mercuric acetate 10^{-4} M	76.47	81.49	75.33	80.64	55.03	58.16
Kaolin 7.5 per cent	78.63	79.38	77.60	78.41	52.86	50.80
Kaolin 10 per cent	76.96	80.63	75.91	80.05	46.40	65.75
Salicylic acid 1000 ppm	76.43	77.36	75.60	76.50	68.98	74.08
Salicylic acid 1500 ppm	81.36	82.61	80.52	81.95	67.46	69.58
Power oil 2.0 per cent	83.08	84.45	82.26	83.86	67.91	72.25
Power oil 3.0 per cent	78.18	79.90	77.38	79.39	68.63	69.46
Liquid paraffin 1.0 per cent	76.04	78.40	75.33	77.86	60.91	74.93
Liquid paraffin 1.5 per cent	82.93	84.31	82.13	83.78	63.03	64.12
Control (water spray)	73.28	74.38	72.63	73.74	43.56	43.58
Control (no spray)	73.16	73.41	72.41	72.78	42.00	42.86
	CD (at 5 per cent)		*CD (at 5 per cent)*		*CD (at 5 per cent)*	
Main treatments	0.38		0.36		3.19	
Subtreatments	0.43		0.42		14.24	
Main subtreatments	0.61		0.59		NS	

* Reddy and Khan (2000).

9.5.0 Manuring and Fertilization

For healthy growth and good quality fruits, manures and fertilizers should be applied in the required doses. When intercrops, such as bananas are grown in sapota they get heavy fertilization indirectly since bananas are given manures and fertilizers. Beside this, the pseudostems are cut and incorporated into the soil (Singh, 1969). In Andhra Pradesh, on an average, each bearing tree is given 100 kg farmyard manure, 6 kg castor cake and 2 kg superphosphate per year (Narasimham, 1966). In Maharashtra, the sapota trees are given yearly application of manures and fertilizers and a one-year-old tree gets 20 kg farmyard manure and 400 g castor cake and this dosage is increased every year by 4 kg and 400 g, respectively. Thus a ten-year-old tree gets 40 kg farmyard manure and 6 kg of castor cake. For bearing trees of ten years and more, each tree is given 2 kg bone meal also (Singh *et al.*, 1963). Application of super-phosphate to provide phosphorus is said to improve the size of fruit in old orchards. The doses of fertilizer recommended in Karnataka are given in Table 8.

Table 8: Fertilizer Schedule for Sapota*

Age of the Plant	*N*	*P*	*K*
		(in g)	
1-3 years	50	20	75
4-6 years	100	40	150
7-10 years	200	80	300
11 years and more	400	260	450

Farmyard manure is applied at 40 kg per year.

*Anon. (1975)

In a fertilizer trial conducted at Navasari, Gujarat on Kalipatti cultivar with three levels of N, P and K (0, 45 and 80 g for each year of age and stabilized to 0, 450 and 900 g in the tenth year) – the results indicated that during the first two years there was no significant effect on yield but in the subsequent years treatment effects were significant and with an increase in the level of nitrogen a linear increase in respect of both yield and growth of the trees was found, while phosphorus and potash were not effective (Chundawat *et al.*, 1981). Patil *et al.* (1981) in a trial at the Mahatma Phule Agricultural University, Rahuri in Maharashtra found that application of increased level of nitrogen resulted in a significant linear increase in plant growth, size, number and total weight of fruits at harvest, while phosphorus and potash were effective only at lower levels. Nitrogen at 1 kg per tree resulted in a significant improvement in total soluble solids and reducing sugar and the leaf N content increased significantly with increasing levels of applied nitrogen. There is no necessity of application of nitrogen beyond 1.5 kg per tree and optimum economic doses of P_2O_5 and K_2O are at 0.45 and 0.50 kg, respectively. If phosphorus is lacking the fruits become hard and small.

Bafna *et al.* (1983) suggested application of 200 g N, 100 g P_2O_5 and 200 g K_2O per 6-year-old tree, and 275 kg N, 200 g P_2O_5 and 275 g K_2O for 7-year-old tree under rainfed condition of Karnataka. In an experiment at Kaithal, Haryana, Singh *et al.* (2000) observed that increasing N fertilizer rate from 400 to 1200 g/tree increased the tree growth and number of fruits, while the effect of different rates of P and K was not significant. An investigation on sapota carried out by Bhuva *et*

Nutrient deficiency symptoms in sapota

Nitrogen (left), Phosphorus (middle) and Potassium (right) deficiencies

Calcium (left), Sulphur (middle) and Boron (right) deficiencies

Copper (left), Iron (middle) and Manganese (right) deficiencies

Flowering and fruit growth in sapota

Flowering in sapota

Different stages of fruit growth

al. (1991) comprising 4 irrigation schedules and 3 fertilizer rates revealed highest fruit production with 0.50 IC/CPF ratio (162.93 kg/tree) and 1000 g N + 500 g P_2O_5 + 500 g K_2O (151.54 kg/tree). The mean water use efficiency for this treatment was 0.81 kg/tree for each mm of water applied. Tasleema *et al.* (2019a) conducted an experiment with twenty-year-old sapota trees of cv. Kirthabarthi which were planted in a spacing of 8 m x 8 m). The treatments comprised of combinations with FYM, Vermicompost, Recommended Dose of Fertiliser (RDF) and inoculation of effective microorganisms along with a control. From the results, it was observed that the plant height, plant spread and canopy volume was found to be markedly influence by various treatments that were tested. The trees which received the application of RDF plus vermicompost @ 12.5 kg tree^1 plus EM had recorded the tallest plant height (7.29 m), widest plant spread (7.20 m in North –South direction and 7.19 m in East-West direction) and the maximum canopy volume of 96.49 m^3. They (Tasleema *et al.*, 2019) found the application of vermicompost @ 12.5 kg tree^{-1} + RDF (1000 g N + 1000 g P + 1500 g K tree^{-1}) + EM (1:250 dilution) had recorded the maximum positive response with regard to the quality characters in both peak season I and II *viz.*, total soluble solids (TSS) (19.86 p B and 19.80 p B), ascorbic acid (2.85 mg/100 g and 2.69 mg/100 g), total sugar (18.35 per cent and 18.19 per cent), reducing sugar (13.10 per cent and 13.02 per cent) and non-reducing sugar (5.25 per cent and 5.17 per cent) and the minimum values for acidity (0.18 per cent and 0.16 per cent). This was closely followed by the treatment which received 50 kg FYM tree^{-1} + RDF + EM. The lowest values of total soluble solids, ascorbic acid content, total sugar, reducing sugar, nonreducing sugar and the maximum values for acidity were recorded under control.

Leaf sampling techniques in sapota cv. Kalipatti was standardized (Chavan and Patil, 1979) and the best time for sampling of leaves for nitrogen, potassium, magnesium and manganese was suggested as July and for phosphorus, calcium, iron and zinc it was during April-January. Thirupathaiah and Shirol (2017) studied the effects of micronutrients application on soil properties of sapota (*Achras sapota* L.) cv. Kalipatti. They used Zinc and iron sulphates for soil and foliar application, whereas the boron in the form of sodium tetraborate (Jai bore) for soil and solu-bore for foliar application were used. The results revealed that foliar application of 0.5 per cent $ZnSO_4$+ 0.5 per cent $FeSO_4$ + 0.3 per cent B lead to maximum utilization of N, P and K by the plant which resulted into reduced concentration of (123.50 kg ha^{-1}), (11.59 kg ha^{-1}) and (103.50 kg ha^{-1}) as well as exchangeable micronutrient content boron (1.70 mg/kg) in the soil.

9.5.1 Time and Method of Application

For best results, the manures and fertilizers should be applied in two split doses. In Maharashtra and Karnataka, they are first applied in the beginning of monsoon in dry zones and at the end of south-west monsoon in heavy rainfall areas. These are applied in a shallow circular trench, 15 to 22 cm deep, 15 cm wide and 60-90 cm away from the trunk. Split application is followed in the trees that are bearing heavy crops. The fertilizers are incorporated into soil and irrigated.

9.6.0 Intercropping

Sapota has a long pre-bearing age and till the trees cover the entire area with their leaf canopy intercrops can be grown. This period may be as long as 10 years and short duration fruits like banana, papaya or vegetables can be profitably grown. Leguminous vegetables and pulses like cowpea, cluster bean, lima bean, peas and pigeon pea and others benefit in several ways.

9.7.0 Weed Control

Weeds compete for nutrients and moisture and hence should be removed frequently. When intercrops are grown weeds may not be a problem during that period and in such cases the basins only are kept free from weeds.

10.0 Flowering, Floral Biology, Fruit Growth and Development

10.1.0 Flowering

Sapota starts bearing small crops from second or third year of planting but economical yields can be obtained from seventh year onwards. Under tropical conditions, flowers are seen almost throughout the year. However, there are two main seasons of flowering and hence two harvesting seasons. Bhuva and Chundawat (1978) reported that the flowering and fruiting in sapota in south Gujarat is a continuous process with two peak seasons of crop in September-November and February-March.

10.2.0 Floral Biology

Sambamurthi and Ramalingam (1954) studied the blossom biology of sapota at Coimbatore and indicated that pollination occurs by wind and unpollinated flowers failed to set fruits. Madhava Rao and Khader (1960) found that among five different cultivars, studied at the same orchard, one was highly self-sterile and the fertility of the remaining four was very low as indicated by pollination tests. No evidence of bud pollination or parthenocarpy was noticed. The existence of inter-varietal relationship in regard to pollen compatibility was apparent in the cultivars studied.

Hayes (1957) found that in an isolated sapota tree not more than half a dozen fruits were set at a time under Allahabad conditions. Nalawadi *et al.* (1977) studied the floral biology of sapota at the Agricultural College, Dharwad, Karnataka, in Kalipatti, Cricket Ball, Calcutta Round and Oval cultivars. Three seasons of flowering were recorded with maximum in June, while in October and March the flowering was less; peak of anthesis was found at 4 am, in all the cultivars and maximum stigmatic receptivity on the day of opening of flowers. Minhas (1985) in an investigation with 3 cultivars stated that (i) flowers opened at 5 am and flowering reached a peak between 5 am and 7 am, (ii) anther dehiscence started before anthesis, suggesting protandry, (iii) maximum anther dehiscence occurred between 8 am and 11 am, (iv) the cv. Chattri had the largest pollen grains (16 μ), (v) the cv. Kalipatti had the highest pollen viability (95.8 to 97.6 per cent), (vi) pollen germination was maximum (45.5-57.5 per cent) in 15 per cent sucrose solution after 48 hours, (vii) stigmas were receptive from a day before until a day after anthesis.

The shape of pollen grains was roundish and their colour being yellowish, with an average pollen size of 16.25 μ. Acetocarmine stainability was 96.3 per cent in local cultivars and in others 97.6 per cent.

Variations between Round and Oval for the floral characters were reported by Singh (1970). The sapodilla var. Round had 20 per cent of flowers with 8 sepals whereas all flowers of var. Oval had 6 sepals. The style came out of the bud before anthesis in both varieties. Anthers of Round produced fewer pollen grains than Oval. Pollen grain germination in sucrose solution was very poor in both varieties. Artificial pollination increased the percentage of fruit set. Both varieties were self- as well as cross-compatible. There was no self pollination in Round and only 4 per cent fruit set was observed in Oval when the flowers were bagged.

No specific information on pollination has been found, but honeybees collect nectar from the flowers and may contribute to the pollination. Flowers are bisexual; the stigma extends beyond the corolla. The tree flowers and fruits throughout the year; fruit take about 4 months to mature. Seedlings may take 5-8 years to bear fruit, while grafted varieties take only 2-3 years from planting out (Orwa *et al.*, 2009). Sapodilla is an insect pollinated species. Pollinators include *Hermitia* spp., *Oecophylla smaragdina*, thrips (*Thrips hawaiiensis* and *Haplothrips tenuipennis*) (Coronel, 1983; Mickelbart,1996). Honey bees also visit Sapodilla flowers and may be involved in pollination (Pieris, 2007). Reddy (1989) conducted a series of studies in which he showed that thrips (*Thrips hawaiiensis* Morgan and *Haplothrips tenuipennis* Bagnall) are the principle pollinators of sapodilla in India. The thrips apparently take shelter in the flowers and live on the pollen grains, nectar, and stigmatic exudations. They collect pollen grains while feeding on these components and transfer then to other flowers when the food reserves are exhausted. Bees have been observed in flowering sapodilla plantings (Sambamurty and Ramalingam, 1954), although examination of the insects indicated that they were not carrying pollen.

In its place of origin, Yucatan in Mexico, flowering starts at the end of the dry season (June), and flowering peaks during August and September. Sapodilla is self compatible. The artificial pollination experiment defined M. zapota as a self-compatible species, because flowers subject to self-pollination produced an equal proportion of mature fruits compared to cross-pollinated flowers. In nature, the reproductive phenology coincides with the precipitation (Singh, 1970). Generally, plant species with large flower displays (*e.g.*, tropical trees) exhibit high geitonogamy levels because pollinators tend to focus on one individual and visit as many flowers as they can within the same tree in order to minimize foraging costs. This might be another cause that could explain self-compatibility in this species for which large floral displays and high rates of geitonogamy have been observed (Salinas *et al.*, 2007). This is in contrast to the earlier finding of Piatos and Knight (1975) who found sapodilla to be self-incompatible, indicating that cross-pollination might be necessary. Similar cases of incompatibility was reported by many earlier workers (cf. Mickelbert, 1997). Gonzalez and Feliciano (1953) examined flowering and fruit set in 'Ponderosa' sapodilla and found that the trees could not be self-pollinated, either naturally or artificially. The best method of pollination was determined to be cross-pollination by hand, which resulted in a 39.6 per cent fruit set. Open pollination resulted in only 5 per cent fruit set in that study. Similar results were observed in 'Kalipatti' sapodilla by Relekar *et al.* (1991). Mulla and Desle (1990) found that the highest percent fruit set was observed in trees cross-pollinated with pollen of the same cultivar and ranged from 20 to 34 per cent. There was no sign of parthenocarpy and self-pollination resulted in only negligible fruit set in some of the cultivars examined (cf. Mickelbert, 1996).

The time of flowering, morphology of the flowers, time of anthesis, pollen germination percentage and time of maximum receptivity of stigma, were similar in Kalipatti, Cricket Ball, Calcutta Round and Local. Fruit set on selfing ranged from 10 per cent in Calcutta Round to 40 per cent in Kalipatti, and on open pollination from 60 per cent in Calcutta Round to 100 per cent in Kalipatti (Nalawadi *et al.*, 1977). Observations were made on the time and duration of flowering and fruit set by self-, open- and cross-pollination in 22-year-old trees of sapota [sapodilla] cultivars Cricket Ball, Kalipatti and Chhattri (or Chattri?) in North India (Punjab).There were 2 main flowering seasons, June and Nov., and the duration of full bloom ranged from 16 to 19 days. No fruit set was obtained in June with any of the pollination methods. No fruit set was obtained by natural selfing or cross-pollination in Nov. Highest fruit set by open pollination (2.37 per cent) and hand-selfing (1.01 per cent) was obtained with cv. Kalipatti. (Dhaliwal *et al.*, 1990).

Two distinct peak flowering periods were observed in India. While it is June and November in North India, in Eastern India (Oriss) it is February-April and October-November. The crop requires 55.30-60 days from flower bud initiation to anthesis. The percentage of flower buds which developed into flowers ranged from 50.00 in Cricket Ball to 72.72 in Kalipatti. The gap between flowering and fruit set was 50.0-54.5 days. The peak period of anthesis was at 04.00 h and continued up to 08.00 h. Fruit weight was greatest in cv. Co-2 (Lenka *et al.*, 1996). Sapodilla cultivars were compared for fruit set, flower drop and yield in Western India (Karnataka) by Gunaki *et al.* (1999a and b). They found that highest percentage fruit set was observed in the July-August flush; fruit set was lower in the March flush for all the cultivars. Highest flower drop was noticed in Cricket Ball, and lowest flower drop in DHS-1 for all flushes. DHS-1 recorded the highest yield per tree (104.97 kg), and Kalipathi recorded the lowest yield per tree (38.64 kg). All cultivars and hybrids exhibited highest flowering during November-January flush, followed by July-August flush; the lowest flowering was observed in the March flush. Fruit set in all the cultivars and hybrids was highest during July-August (17.42-29.24 per cent) followed by November-January (13.23-25.15 per cent). Fruit set was highest in DHS-1 followed by DHS-2; fruit set in the remaining cultivars was in the range 11.23-20.53 per cent (Gunaki *et al.*, 1999b). Earlier flowering for homegarden trees, as well as two flowering peaks (June–July,and September), which is to some extent evidence of a certain degree of plasticity in reproductive phenology patterns in response to variation in water availability (Salinas Peba *et al.*, 2007).

10.3.0 Fruit Set

The fruit set is the most potent factor which determines the yield. Problems of shedding of flowers and fruits and low fertility in sapota were reported by Cheema *et al.* (1954) and Hayes, (1957). In cultivar Kalipatti, 22 per cent natural fruit set was observed and the maximum fruit drop occurred immediately after fruit setting (Patil and Narwadkar, 1974). The fruit drop during the course of fruit development was, however, meagre. It was also observed that the flowers situated at the base of inflorescence opened and set earlier. Such early set fruits developed rapidly and the remaining, which set relatively late, often dropped down. Pollination studies with 5 sapota cultivars revealed that self pollination by bagging or by hand produced little or no fruit set. Open pollination and hand pollination with pollen from others gave a much higher fruit set, indicating that sapota is cross pollinated (Mulla and Desle, 1990).

10.3.1 Use of Growth Regulators

Self-unfruitfulness is seen in stray plants and pollination becomes difficult (Singh, 1969). Use of growth regulators has been found to be beneficial. In an attempt to get higher fruit set and retention of the set fruits, several growth substances, such as GA_3, ethrel, cycocel, and planofix (á-NAA preparation) and SADH were tried on cultivar Cricket Ball (Das and Mahapatra, 1975). These chemicals were applied in spray solutions before flowering and again at the pea stage. Of all the regulators, SADH at 100 ppm resulted in the highest fruit set and planofix at 300 ppm resulted in the highest fruit retention and largest fruits, followed by 100 ppm GA_3.

In similar studies made at the Gujarat Agricultural University, Navsari, India, of the two growth substances tried at 25 to 100 ppm as spray at flowering and again 15 days later, NAA resulted in a better fruit set than GA_3 (Rathod and Amin, 1981).

10.4.0 Fruit Growth and Development

In an investigation at the Agricultural College, Dharwar, Karnataka on the periodical changes in the fresh weight of developing fruits of cv. Kalipatti, a double sigmoid pattern of growth was noticed with two periods of active growth and one period of retarded growth (Rao, 1978). From the first to twelfth fortnight the fruits showed 23.6 per cent of total weight at maturity, and in the second stage only 14.4 per cent. The third stage of fruit development from the seventeenth to twenty first fortnight had maximum amount of growth of 62 per cent in terms of fresh weight increase. The total time taken from fruit set to maturity was ten months and a half and there was a long lag phase and very slow growth rate at the initial stage of fruit development.

The time taken from flower bud initiation to harvest varies with the climatic and other conditions. Purseglove (1968) reported that fruits take 4 months to mature from the date of flowering, while Sulladmath and Rao (1979) reported that the time taken from fruit set to maturity in Kalipatti cultivar was more than 10 months. Sometimes, the trees bear flowers continuously in several flushes at short intervals throughout the year, as a result, fruits also mature at different times. Fruits that mature during March-May are generally better in quality. Paralkar *et al.* (1987) studied the fruit growth of sapota cv. Kalipatti and noted an increase in weight, volume, length, diameter and flesh to seed ratio from fruit set to maturity. The TSS, sugars, acidity, pH and tannin content were also increased during the course of development.

10.4.1 Influence of Seed on Fruit Growth

The diversity in the shape of fruit in sapota is an interesting feature. Cultivars Calcutta Round and Cricket Ball produce oval or elliptical fruits while cultivars Kalipatti and Chhatri have a special feature of producing both round and elliptical fruits, simultaneously on the same tree during the same season. A correlation was worked out between the shape index and seed number per fruit and between fruit weight and seed number. A significant negative correlation (–0.92) between the shape index and the seed number per fruit was observed (Rao, 1978). The seed number per unit ranged from 1 to 9 and the round fruits had more number of seeds (3.99) than oval (1.52). The shape index was higher for oval fruits (1.53) than round fruits (1.11). The regression analysis suggested that for every decrease in the seed number by a value of one, there was an increase in the shape index by a value of 0.20 and the fruit tended to be oval in shape. The fruit shape may also be governed by the way of distribution of seeds around the placenta. In round-shaped fruits, the seeds were more uniformly distributed around the placenta compared with oval fruits. A positive correlation (0.77) was present between the number of seeds per fruit and fruit weight. Round fruits were heavier (101.0 g) than oval fruits (69.7 g). This indicated the existence of a hormonal influence of seeds on fruit development.

11.0 Pests and Diseases

Sapota is affected by insect pests and diseases and necessary plant protection measures should be taken up in time.

11.1.0 Pests

11.1.1 Stem Borer (*Indarbela tetraonis*)

This is a small beetle. The grub is stout and bores into the bark of the trunk making circular galleries and feeds on the living tissue of the inner bark. The presence of the sect can be detected from the chewed bark thrown out of a hole in the trunk. The borer can be traced by cutting dead bark along the hollow tunnels with a knife. This insect has been reported from Tamil Nadu, Karnataka and other places (Abdul Ravoof, 1964; Puttarudraiah, 1980).

The insect is killed by thrusting a stiff wire into the tunnel. If it is deep inside the wood, the hole is plugged with a wad of cotton wool dipped in kerosene and plastered with wet mud. By this, the borer gets suffocated and dies within the tree.

11.1.2 Scale Insect (*Pulvinaria psidii*)

These are green or brown scales, oval-shaped with a slight twist at the front end. The characteristic feature is the presence of an inverted loop on the body. These infest along the sides of midrib and surface of leaves and on the tender twigs. They suck the sap.

Spraying a suitable insecticide, such as dimethoate or malathion at 30 ml in 18 litres of water effectively controls the pest. Two or three sprays are necessary (Anon., 1975).

Pest of sapota

Stem borer infestation

Leaf webbar

Mealy bug

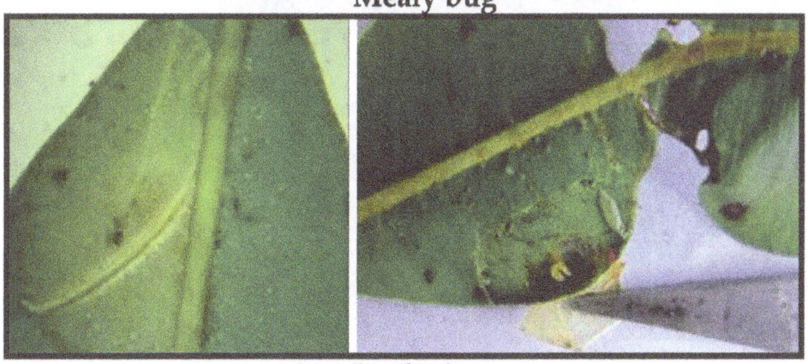

Leaf miner

Pests of sapota

Flower bud borer

AICRP on Fruits (AICRP-Fruits), ICAR, IIHR, 2017

Bark eating catterpillar

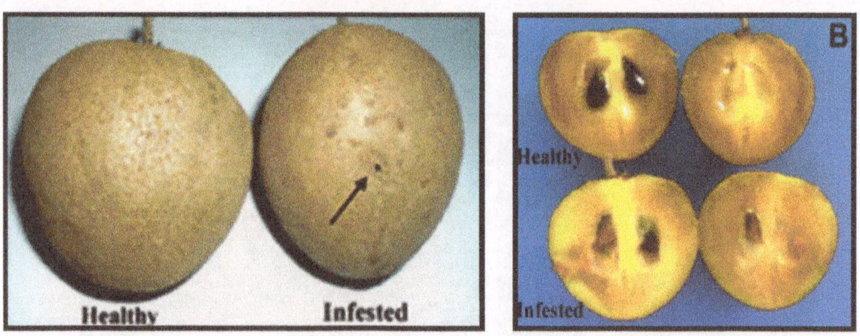

Fruit borer

Diseases of sapota

Leaf spot infection

Sooty mould in sapota

Sapota tree bearing fruits

Sapota tree with fruits

Harvesting, ripening and storage of sapota

Harvested fruits of sapota

Storage of sapota

Ripening of sapota

11.1.3 Leaf Webber (*Nephopteryx eugraphella*)

It feeds on buds, leaves and young fruits. Application of malathion at 0.5 per cent is very effective (Jhala *et al.*, 1985).

11.1.4 Mealy Bugs (*Phenacoccus icerjoides*)

These are small, oval with a cottony white waxy covering on their body. They stick to the under surface of leaves and base of fruit stalks. They suck the sap and secrete large quantities of sugary substance. Spraying of dimethoate at the rate of 30 ml in 18 litres of water can effectively control this pest.

11.1.5 Leaf Miner (*Anarsia gemoniella*)

It is a tiny caterpillar and the adult is a grey-coloured moth. It mines into the surface of young leaves and makes them curl. Infected leaves show glistening irregular galleries or mines on the leaf surface. The tiny caterpillars are seen inside these galleries. Such leaves get distorted, dry up and fall.

One or two sprays of dimethoate (30 ml in 18 litres of water) or malathion (36 ml in 18 litres of water) provide satisfactory control.

11.1.6 Flower Bud Eating Caterpillar

Flowers and flower buds are damaged to a considerable extent by a small caterpillar, resulting in their shedding. In studies at the Agricultural College, Dharwar, Karnataka and in Coimbatore, Tamil Nadu, this insect was found to be responsible for shedding of a large number of flowers. Farooqui *et al.* (1973) recorded a heavy flower drop for a period of three months and it was as high as 88 per cent, at bud, flower or fruit set stage. This can be controlled by the same measures as for leaf miner.

11.1.7 Bark Eating Caterpillar (*Indarbela* sp.)

This bores into the stem bark and the excreta along with fibrous material hang down outside the bark. This can effectively be controlled by giving one or two sprays of dimethoate (30 ml in 18 litres of water).

11.1.8 Fruit Borer (*Virachola isocrates*)

The young fruits are affected. The latex comes out and crystalizes. This insect can also be effectively controlled by the same measures as for leaf miner.

11.1.9 Fruit Fly (*Bactrocera zonata, B. dorsata* and *B. correcta*)

Patel and Patel (1998) were able to catch 49.36 male fruit flies by using a fruit fly trap consisting of a plastic jar, a plastic funnel and a cotton swab impregnated with 5 drops of methyl eugenol as attractant in a glass petridish.

11.2.0 Diseases

11.2.1 Leaf Spot

This disease caused by a fungus *Phaeophleospora indica* was first reported from Dharwar, India by Chinnappa (1968). It is characterized by numerous, small, pinkish to reddish brown spots with whitish centres. In recent years, it is becoming severe. Sohi and Sridhar (1972) and Sohi and Prakash (1972) recommended monthly sprays of 0.2 per cent dithane Z-78 for most effective control, while treatment with 0.5 per cent blitox was less effective under Bangalore conditions. Prasad *et al.* (1979) in a trial on the comparative resistance of several selections and cultivars in sapota at the Agricultural College, Dharwar, Karnataka found that CO1 and Cricket Ball were more resistant to leaf spot disease. The cultivars CO_2 and Kalipatti also showed tolerance, while Calcutta Round was most susceptible. Leaf spot incited by *Pestalotia* sp. (Wilson *et al.*, 1970) and *Glomerella cingulata* (Sohi and Prakash, 1972) has also been reported in sapota.

Topsin M (0.1 per cent) was the most effective fungicide showing 69-79 per cent leaves infected per branch as against 96.93 per cent in control. This was followed by Delan (0.2 per cent) and Bavistin (0.1 per cent) where 71.68 and 72.65 per cent infected leaves respectively, were recorded (Rawal, 1993).

11.2.2 Sooty Mould

It is also a fungal disease developing on the honeydew like excretion by scale insects and mealy bugs. This is a common disease in India, incited by *Capnodium* sp. and adversely affects the photosynthetic function of leaves and disfigures the fruit. This can be controlled by spraying starch solution (100 g in 18 litres of water). On drying this forms thin flakes and drops off. Spraying with 40 g zineb in 18 litres of water checks the disease.

11.2.3 Flattening of Branches

The branches become flattened and this has been attributed to a pathogen *Botryodip-lodia theobromae* (Khurana and Singh, 1972). This has been recorded in South India, coastal areas of Maharashtra and Gujarat and in such trees the fruit set and yields are adversely affected. The affected branches produce small, dry, hard and shrivelled fruits. During summer months these flattened branches may give rise to normal branches. As yet no suitable control measure has been found out.

12.0 Harvesting and Yield

12.1.0 Harvesting

Sapota is a climacteric fruit and it improves in quality after harvesting but immature fruits should not be harvested. The fruits to be harvested must be fully matured and the maturity can be judged by several external symptoms. Sulladmath and Reddy (1990) suggested the following symptoms of maturity.

i) Fruits at full maturity develop a dull orange or potato colour, with a yellowish tinge.
ii) A mature fruit when scratched lightly shows a yellow streak instead of a green streak which is a sign of immature condition.
iii) Brown scaly material disappears from the fruit surface as the fruits approach full maturity.
iv) As the fruit matures, the milky latex content is reduced.

v) The dried spine like stigma at the tip of fruit falls or drops off easily when touched.

Kariyanna *et al.* (1993) recommended harvesting sapota fruits after 245 days of fruit set and morphological parameters like size, external colour, scurf content and colour of skin after scratching could be used to decide the optimum harvesting stage in field. Several workers suggested the fruit weight at maturity as one of the maturity indices and the different varieties showed variations in fruit weight and maturity, which, however, differ in different agro-climatic conditions.

Calcutta Round	48.2 g	
Cricket Ball	140.0 g	Lakshminarayana and Subramanium (1966)
Pala	30.8 g	
Mhangottee	101.43 g	
Kalipatti	98.23 g	Chundawat and Bhuva (1982)
Pilipatti	82.40 g	
Zumakhi	57.47 g	
CO 1	122.22 g	
CO 2	87.41 g	Aravindakshan *et al.* (1987)
Gavarayya	90.48 g	
Guthi	59.62 g	Durairaj *et al.* (1991)

The fully matured fruits are harvested with the stalk intact, individually by giving a twist and collected without bruising. The fruits thus harvested are spread in a thin layer on bamboo mats under shade for an hour or two. To avoid bruising of fruits they are better collected in gunny bags and lowered to the ground carefully. The peak harvest periods are January-February and May-June in the west coast of Maharashtra; March-May and September-October in Andhra Pradesh and Karnataka.

12.2.0 Yield

The yield depends upon several factors, such as the age of the tree, cultivar, agroclimatic conditions of the locality, nutrition and plant protection measures. In round and large-sized cutivars, such as Cricket Ball and Calcutta Round, less number of fruits is obtained, while in others with oval or long fruits more number of fruits is produced but their size is small. In very prolific bearers like Pala, the number of fruits is much more. On an average, a 3-year old tree can yield about 100 fruits; a 5-year old tree, 250 fruits; a 7-year old tree, 700 fruits; a 8-year old tree, 800 fruits; a 10-year old tree, 1000 fruits; a 11-year old tree, 1500 fruits; a 15-year old tree, 2000 fruits; and a 30-year old tree, 2500 to 3000 fruits per year (Gandhi, 1963; Narasimham, 1966 and Singh, 1969; Gopalaswamiengar, 1970; Anon., 1975). Purseglove (1968) also reported similar yield from 30-year-old trees of sapota.

13.0 Post-harvest Management

13.1.0 Ripening

Sapota fruits are extremely perishable in nature. Fruits ripen in about 5 days after harvest at room temperature and the ripening period depends upon the cultivar, stage of maturity and temperature. They ripen well at 12-14°C. Lakshminarayana and Subramanyam (1966) reported a slow ripening in 9-13 days after harvest in fruits of Calcutta Round. Respiration follows a climacteric pattern and the peak was observed on the sixth day in cv. Oblong, eighth day in Cricket Ball (Selvaraj and Pal, 1984) and fifth day in Kalipatti (Rao and Chundawat, 1998). Post-harvest treatment with AVG, ascorbic acid, menadione and hydroquinone retarded the oxidative metabolism, and delayed the occurrence of climacteric by 2-4 days (Rao and Chundawat, 1992).

To reduce the time taken for ripening, a few growth regulators were used and a uniform ripening was possible by treatment with ethephon. Sastry (1970) reported that fruits ripened after 2 days of storage with ethylene, while the untreated ones and those wrapped in gunny bags took 7 days for complete ripening. In a study at the Tamil Nadu Agricultural University, Madhava Rao *et al.* (1971) reported that when fully mature fruits of sapota were dipped in 5000 ppm ethephon before storage at room temperature or kept in airtight chambers in the presence of ethephon at 500-6000 ppm with added NaOH, ripening occurred within 2 days. While there was no change in colour or TSS in the fruits, a marked reduction in the phenolic contents was noticed and there was an increase in total sugars, reducing sugars and titratable acidity due to the treatment. In a similar investigation, Das and Mahapatra (1975) at Bhubaneswar, Orissa found that ethrel was effective in early ripening of sapota fruits without any adverse effect. All fruits dipped in ethrel (250-750 ppm), ripened in 3 days compared with 26 per cent in untreated control. Dipping in 2,4-D (5-15 ppm) and wax emulsion (3-9 per cent) was also effective to hasten ripening.

Soni *et al.* (1981) at Udaipur, Rajasthan were able to enhance the ripening process of Cricket Ball cultivar of sapota fruits through pre- and post-harvest application of ethrel. The fruits were treated at maturity, 15-30 days before expected maturity both on tree and after harvesting. There was no significant effect on the ripening of fruit in the pre-harvest application. However, in the detached fruits application of 1000 and 2000 ppm ethrel at 15 days before and at full maturity and their subsequent storage under ordinary light proved to be the most appropriate treatment for inducing early and uniform ripening. The treated fruits showed an improvement in taste, TSS, ascorbic acid, reducing, non-reducing and total sugars and a reduced acid content. Ethrel resulted in ripening of all the fruits within 71 hours of treatment without any deterioration in organoleptic quality and physico-chemical components. Suryanarayana and Goud (1984) suggested the use of ethrel (5000 ppm) for quick ripening and for improving the quality of fruits.

13.2.0 Storage

The storage period of ripe sapota fruits depends on the respiration, relative humidity, temperature, enzymic activity and carbon dioxide content. The fruits are climacteric and the respiratory peak occurs at the same time or one or two days after peak ethylene production. At the optimum storage temperature of 20°C, the storage life could be increased by removing ethylene and adding 5-10 per cent CO_2 to storage atmosphere. Excess relative humidity or high concentration of C O_2 impaired the quality of stored fruits. Ascorbic acid content decreased with the ripening of fruit (Broughton and Wong, 1979). Under ordinary conditions, fruits kept well for 7-8 days from picking, unripe fruits can be made to ripen slowly and satisfactorily between 12° and 14°C and they keep well for about 5 weeks. Ripe fruits can be kept at 2-3°C and 85-90 per cent relative humidity for 6 weeks and firm fruits at 3 to 5°C for 8 weeks (Singh *et al.*, 1963; Singh, 1969).

Baez *et al.* (1997) observed that fruit weight loss after 6 days at 20°C reached 5.1 per cent, while firmness dropped to 1.0 N; on the other hand, soluble solids content increased to 25.8°Brix. Titratable acidity did not show any significant changes during marketing. Guadaramma and Oriz (2000) investigated the post-harvest changes in 2 sapota cultivars stored at 25 ± 3°C and reported that during ripening, total chlorophyll and tannin decreased and total carotenoids, soluble solids and fresh weight were lost and softening increased, although the post-harvest patterns were different in each cultivar. Both cultivars showed a climacteric respiratory behaviour but suggesting differences in their post-harvest behaviour related to shelf-life.

Sarkar *et al.* (1995) reported that sapota fruits cv. Calcutta Round could be stored in cardboard boxes with waxol (6 per cent) treatment for 16 days without affecting the quality. The shelf-life sapota cv. Kalipatti fruits could be extended to 9 days with minimum weight loss and rotting when packed in polyethylene bags (100 gauge and 1.2 per cent vents) and stored in ambient conditions (Waskar and Nikam, 1998). The shelf-life could also be further extended to 13 and 15 days by packing in polyethylene bags + corrugated fibreboard box or in polyethylene bags, respectively and storing in a cool chamber (Waskar *et al.*, 1999).

Post-harvest dip in GA (300 ppm) was the best treatment followed by kinetin (100 ppm) and silver nitrate (40 ppm) which not only extended the shelf-life but also reduced the post-harvest losses due to moisture loss and degradative metabolism during ripening without adversely affecting the quality (Gautam and Chundawat, 1990). Sapota cv. Victoria fruits were dipped in a 0.75 per cent alum + 0.5g benomyl/l solution (or left untreated) and stored at 10, 15 or 28°C for 20 days. Ripening was non-uniform and was characterized by softening and colour changes from the centre outwards toward the peel. Fruits deteriorated after 6 days' storage at 28°C. Dipping in alum + benomyl prevented storage decay and did not damage the fruits. Storage at 10°C gave the best results. Freshly harvested, well matured sapota fruits of var. Kalipatti were subjected to different post-harvest treatments *viz.*, 50µ LDPE (T1), 75µ LDPE (T2), 100µ LDPE (T3), and control (T4) and stored in ambient condition (Temp: minimum 270C, maximum 38°C, and RH: 85-90 per cent) (Ankalagi *et al.*, 2017). Observations were recorded on physiological loss of weight (per cent), TSS (p Brix), titratable acidity (per cent), reducing sugar (per cent), total sugar (per cent), TSS: Acid ratio (per cent), organoleptic quality and physical parameters at three days intervals. The results revealed that sapota var. Kalipatti fruits packed in 100µ LDPE (T3) proved to be the best treatments followed by 75µ

LDPE (T2) among all the treatments under ambient condition which improves the quality and storability of fruits and also reduced the post–harvest losses without adversely affecting the fruit quality of sapota. As the 100µ LDPE (T3) shows the lowest PLW (per cent), lowest titratable acidity (per cent), and highest TSS (p Brix), reducing sugar (per cent), total sugar (per cent), TSS: Acid ratio (per cent), and Organoleptic quality compare to other treatments. In general, PLW (per cent) and titratable acidity (per cent) continuously decreased during storage while, reducing sugar (per cent) and total sugar (per cent) increased up to 6th day of storage; there after it steadily decreased during subsequent period of storage. 75µ LDPE and 100µ LDPE enhanced shelf life while the control fruits were not acceptable on 9th day.

14.0. Biotechnology

14.1.0 Molecular Markers

According to González-Hernández *et al.* (2012), two earlier studies conducted with molecular markers to identify genetic variation in *M. zapota* (Heaton *et al.*, 1999; Meghala *et al.*, 2005) using RAPDS. Results were contrasting, Heaton *et al.* (1999) detected reduced genetic variation among Mexican populations with different habitats and geographical localities in contrast with the wide morphological diversity observed. Furthermore, Meghala *et al.* (2005) assessed the genetic diversity of cultivars in India and observed extensive genetic diversity. This seems to contradict the probability that it is a crop introduced and the variation depends on the number of genotypes present. We assessed the genetic variation of 20 adult trees from different physiographic regions of the state of Veracruz. The 12 microsatellite loci developed for *M.huberi* were tested in *M. zapota*. Two were monomorphic and three did not amplify. The seven remaining loci were used for genetic analysis. The number of alleles was high (mean 11.85); however, the heterozygosity was low (mean 0.154). Cluster analysis formed a group with specimens from the north and the other, with multiple subgroups, from the south and center of the state. The analysis also showed that the biggest change occurred in conservation areas of the state. The microsatellites employed were useful to learn about the genetic variation of *M. zapota* (González-Hernández *et al.*, 2012).

15.0 Future Thrusts in Breeding

The future of sapodilla appears to be promising, given the attention the crop is receiving from growers and consumers in many countries. Indian production of sapodilla continues to grow and there is an active research program in that country with specific goals toward improving storage, transport, and marketing strategies. Sapodilla has been identified by the Ministry of Agriculture in Malaysia to be promoted under the program for development of its fruit industry (Bakar and Abdul-Karim 1994). The fruit is also gaining popularity as a specialty fruit in restaurants in North America and production of sapodilla as a commercial crop seems to be a possibility in areas where environmental conditions are mild (Mickelbart, 1996)

Sapota has great future as it requires less water compared to many fruit species. Future breeding strategies should embark on the following aspects.

a. Introduction of genetic resources from Americas where it has originated and characterize these genetic resources.
b. Develop varieties suitable for table purposes with less seeds and melting pulp free of sandy texture and thick skin. Flesh should not be hard.
c. Identify genotypes that can withstand cold for cultivation in Northern and eastern part of India.
d. Ideotype breeding should be carried out to develop varieties tolerant to biotic and abiotic stress.

16.0 References

Abdul Ravoof, A. (1964) *South Indian Hort.*, **12** : 1.

Ankalagi, N., Bukya Krishna, A. K. Banik and M.C. Momin (2017) *Int. J. Pure App. Biosci.* **5 (6):** 234-240.

Anonymous (1975) *Cultivation Practices for Fruits*, Dept. of Hort., UAS, and IIHR, Bangalore, pp. 31-34.

Aravindakshan, M., Vatsala Kumari, P.K. and Gopi Kumar, K. (1987) *Agric. Res. J.*, Kerala,**25** : 288-290.

Armstrong, K. (2010) *Ph.D. Thesis. University of Edinburgh & Royal Botanic Garden Edinburgh*. pp.226.

Baez, M.A., Siller, J.H., Heredia, J.B., Portillo, T., Araiza, E., Garcia, R.S. and Muy, M.D. (1997) *Proc. Interamerican Soc. for Trop. Hort.*, **41** : 209-214.

Bafna, A.M., Parikh, N.M., Shah, G.B. and Bhatt, P.M. (1983) *South Indian Hort.*, **31** : 66-69.

Bhuva, H.P. and Chundawat, B.S. (1978) *Udyanika*, **2** : 54-56.

Bhuva, H.P., Katrodia, J.S. and Chundawat, B.S. (1990b) *The Hort. J.*, **3** : 6-9.

Bhuva, H.P., Katrodia, J.S. and Patel, R.G. (1990c) *Haryana J. Hort. Sci.*, **19** : 71-78.

Bhuva, H.P., Katrodia, J.S. and Patel, R.G. (1991) *Indian J. Hort.*, **48** : 291-295.

Bhuva, H.P., Katrodia, J.S. and Raja, K.R.V. (1990a) *Indian J. Hort.*, **47** : 167-169.

Broughton, W.J. and Wong, H.C. (1979) *Scientia Hort.*, **10** : 377-385.

Chadha, K.L. (1992) *Indian J. Hort.*, **49** : 1-17.

Chatterjee, B.K., Ray, D.P. and Chakladar, B.P. (1990) *Orissa J. Hort.*, **18** : 70-74.

Chavan, S.D. and Patil, V.K. (1979) *Haryana J. Hort. Sci.*, **8** : 24-27.

Cheema, G.S., Bhat, S.S. and Naik, K.C. (1954) *Commercial Fruits of India*, Macmillan & Co.

Choudhury, B.L. (1954) *Practical Hints on Fruit Growing*, pp. 217-218.

Chundawat, B.S. and Bhuva, M.P. (1982) *Haryana J. Hort. Sci.*, **11** : 154-158.

Chundawat, B.S., Bhuva, H.P. and Patel, G.L. (1981) *Nat. Symp. Trop. and Subtrop. Fruit Crops*, Bangalore, p. 57.

Cruz-Rodriguez, J. and Lopez-Mata, L. (2004) *Plant Ecology*. **172**: 227–235.

Das, R.C. and Mahapatra, S. (1975) *Plant Sci.*, **7** : 93-94.

Dhaliwal, H. S., Ajmer Singh., Grewal, S. S., Dhillon, D. S. (1990) *Punjab Horticultural J.*, **30**L 150-154.

Durairaj, A., Muthuswami, S. and Irulappan, I. (1991) *Nat. Sem. Optimization of Productivity and Utilization of Sapota*, Gujarat, p. 39.

Farooqi, A.A. and Rao, M.M. (1976a) *Mysore J. Agri. Sci.*, **10** : 413-423.

Farooqi, A.A. and Rao, M.M. (1976b) *Mysore J. Agri. Sci.*, **10** : 28-34.

Farooqi, A.A., Rao, M.M., Bankapur, V.M. and Kololgi, S.D. (1973) *Mysore J. Agri. Sci.*, **7** : 132-134.

Fazlulla Khan, K., Muthuswamy, S. and Ramu, N. (1965) *South Indian Hort.*, **13** : 1-2.

Folan, W.J., Fletcher, L.A.and Kintz, E.R. (1979) *Science*, **204**: 697–701.

Gandhi, S.R. (1963) *Chiku*, Farm Bulletin, Ministry of Agriculture, Govt. of India.

Gautam, S.K. and Chundawat, B.S. (1990) *Indian J. Hort.*, **47** : 264-269.

Gonzalez, L.G. and P.A. Feliciano, Jr. (1953) Philippine Agr. **37**(7): 384-398.

González-Hernández, D., García-Pérez, E. and Guntin-Marey, P. (2012) *Agrociencia*, **46**: 663-675.

Gopalaswamiengar, K.S. (1970) *Complete Gardening in India*, Kalyani Press, Bangalore.

Govaerts, R., Frodin, D. G. and Pennington, T. D. (2001) *World checklist and bibliography of Sapotaceae* London: Royal Botanic Gardens, Kew.

Guadarrama, A. and Oritz, V. (2000) *Acta Horticulturae*, **536** : 363-367.

Gunaki, K. N., Hulamani, N. C., Athani, S. I., Shakargouda Patil. and Swamy, G. S. K. (1999b). *Advances in Agricultural Research in India*, **12**: 57-60.

Gunaki, K. N., Hulamani, N. C., Athani, S. I., Swamy, G. S. K.; Patil, S.(1999a) *Advances in Agricultural Research in India*, **11**: 73-77.

Hayes, W.B. (1957) *Fruit Growing in India*, Kitabistan, Allahabad.

Heaton, H. J., Whitkus. R. and Gómez-Pompa, A. (1999) *Mol. Ecol.* **8**: 627-632.

Jhala, T.C., Shah, A.H., Patel, C.B. and Patel, R.L. (1985) *Gujarat Agric. Univ. Res. J.*, **11** : 27-29.

Joshi, G.D., Sawant, V.S., Relekar, P.P., Pawar, C.D., Salvi, M.J. and Magdum, M.B. (1993) *Golden Jubilee Symp. on Horticultural Research – Changing Scenario*, Bangalore, May 24-28, 1993, Abst. No. 18.5, p. 337.

Kariyanna, Bojappa, K.M. and Reddy, T.V. (1993) *Golden Jubilee Symp. on Horticultural Research – Changing Scenario*, Bangalore, May 24-28, Abst. No. 19.10, p. 347.

Khurana, S.M.P. and Singh, S. (1972) *Zeitschrift fur Pfladzenkrankheitenund Pflanzenschulz*, **79** : 310-312.

Kirthikar, K.R. and Basu, B.D. (1975) *Indian Medicinal Plants*, Bishen Singh Mahendra Pal Singh, Dehra Dun.

Kumbhar, S.S. and Desai, U.T. (1986) *J. Maharashtra Agric. Univ.*, **11** : 184-186.

Lakshminarayana, S. and Subramanyam, H. (1966) *J. Food Sci. and Tech.*, **3** : 151-154.

Lenka, P. C., Das, D. K. and Samal, B. (1996) *Orissa Journal of Horticulture.* 24, (1/2), 42-46.

Lundell, C. L. (1937) *The vegetation of Petén.* Carnegie Institution of Washington, Washington, DC, pp. 46–49.

Madhava Rao, V.N. and Khader, J.B.M.A. (1960) *Acad. Agri. Science*, Coimbatore.

Madhava Rao, V.N., Shanmugavelu, K.G., Srinivasan, C. and Pappaiah, D.R. (1971) *South Indian Hort.*, **19** : 1-8.

Martínez-Natarén, D., Parra-Tabla, V., and Munguía-Rosas, M. (2017) *Journal of Tropical Ecology*, 33(4), 285-294. doi: 10.1017/S0266467417000268.

Mathews, J.P. (2009) *Chicle: The chewing gum of the Americas, from the ancient Maya to William Wrigley.* Tucson: University of Arizona Press.

Meghala, R. K.V. Ravishankar, L. Anand, A. and Rekha, A. (2005) *Plant Genetic Resources Newsletter*, **42**: 43-46.

Meghala, R., K. V. Ravishankar, A. Lalitha, and A. Rekha.(2005) *Plant Genet. Res. Newsletter*, **142**: 43-46.

Meza, N. and Bautista, D. (1999a) *Bioagro*, **11** : 24-28.

Meza, N. and Bautista, D. (1999b) *Agronomia Tropical (Maracay)*, **49** : 187-199.

Meza, N. and Manzano, J. (2000) *Proc. Interamerican Soc. Trop. Hort.*, **42** : 236-245.

Mickelbart, M.V. (1996) In: *Progress in New Crops* (ed. J.Janick), ASHS Press, Alexandria, VA, pp. 439-446.

Minhas, P.P.S. (1985) *J. Res. P.A.U.*, India, **22** : 270-276.

Morton, J. (1987) Sapodilla. In: Julia F. Morton. (ed.) *Fruits of warm climates.* Miami, FL. pp. 393–398.

Mulla, A.L. and G.Y. Desle. (1990) *J. Maharashtra Agr. Univ.* **15**(2), 266-268.

Nakasone, H.Y. and Paull, R.E. (1998) *Tropical Fruits*, CAB International, Wallingford, U.K., pp. 389-400.

Nalawadi, U. G., Dasappa; Sulikeri, G. S. (1977) *Progressive Horticulture.* **9**: 27-32.

Nalawadi, U.G., Dasappa and Sulikeri, G.S. (1977) *Progressive Hort.*, **9** : 13-17.

Narasimham, B. (1966) *Horticulture Part I.*, Agril. College, Bapatla, Andhra Pradesh, pp. 200-201.

Orwa, C. A. Mutua., Kindt. R., Jamnadass, R. and S. Anthony. (2009) Agroforestree Database : a tree reference and selection guide version 4.0 (http: //www. worldagroforestry.org/sites/treedbs/treedatabases.asp).

Pampanna, Y. and Sulikeri, G.S. (1999) *Seed Res.*, **27** : 49-53.

Paralkar, P.S., Joshi, G.D. and Salvi, M.J. (1987) *Indian Fd. Pack.*, **41** : 11-18.

Patel, C.B., Patel, D.B., Patel, B.N. and Katrodia, J.S. (1993) *Golden Jubilee Symp. on Horticultural Research – Changing Scenario*, Bangalore, May 24-28, Abst. No. 8.28, p. 176.

Patel, R.K. and Patel, C.B. (1998) *Insect Environment*, **4** : 52.

Patil, A.V., Warhal, K.N., Kamble, K.B. and Patil, B.S. (1981) *Nat. Symp. Trop. and Subtrop. Fruit Crops*, Bangalore, p. 58.

Patil, V.K. and Narwadkar, P.R. (1974) *Punjab Hort. J.*, **14** : 39-42.

Pennington, T. D. (1990) *Flora Neotropica Monograph 52. Sapotaceae.* Bronx, New York: New York Botanical Garden.

Pieris, K.H.S. (2007) In: Pushpakumara, D.K.N.G., Gunasena H.P.M. and Singh, V.P. (Eds.). *Underutilized fruit trees in Sri Lanka.* World Agroforestry Centre, South Asia Office, New Delhi, India. pp.183-224.

Ponnuswami, V. and Irulappan, I. (1987) *South Indian Hort.*, **35** : 446-447.

Pradumna Singh (1951) *Indian J. Hort.*, **8** : 11-17.

Prasad, K.S.K., Siddaramaiah, A.L. and Jayaramaiah, H. (1979) *Curr. Res.*, **6** : 5-6.

Purohit, S.D. and Singhvi, A. (1998) *Scientia Hort.*, **76** : 219-229.

Purohit, S.D., Singhvi, A., Nagori, R., Kishor, P.B.K. (1999) *Plant Tissue Culture and Biotechnology : Emerging Trends*, Proceedings of a symposium held at Hyderabad, India, 29-31 January, pp. 134-145.

Purseglove, J.W. (1968) *Tropical Crops : Dicotyledons*, Longman Group Ltd., **1**.

Puttarudraiah, M. (1980) *A Guide for Cultivation of Important Crops*, p. 252.

Rajasekhar, M. and Ratan, P. B. (2010) *Research on Crops*.11 : .365-367

Ram, K.B. and Sahai, S.P. (1962) *Fruit Nursery Practices in Bihar*, p. 35.

Ramakrishna Reddy (1976) *Totagaara*, Lalbagh, Bangalore, **4** : 61-62.

Ramsundar, V. and Abdul Khader, J.B.M. Md. (1986) *South Indian Hort.*, **34** : 40-41.

Rao, D.V.R. and Chundawat, B.S. (1988) *Indian J. Plant Physiol.*, **31** : 205-208.

Rao, D.V.R. and Chundawat, B.S. (1992) *Indian J. Plant Physiol.*, **35** : 167-173.

Rao, M.M. (1978) *M.Sc. (Ag.) thesis* submitted to University of Agricultural Science, Bangalore.

Rathod, R.P. and Amin, H.D. (1981) *Nat. Symp. Trop. and Subtrop. Fruit Crops*, Bangalore, p. 115.

Rawal, R.D. (1993) *Golden Jubilee Symp. on Horticultural Research – Changing Scenario*, Bangalore, May 24-28, Abst. No. 14.9, p. 281.

Reddi, E.U.B. (1989) (*Manilkara zapota*). *Proc. Indian Natl. Sci. Acad.* B55 (5-6), 407-410.

Reddy, Y.T.N. and Khan, M.M. (2000) *Indian J. Hort.*, **57** : 125-159.

Rekha, A., M.R. Dinesh., R. Venugopalan. and B.N.S. Murthy. (2011) *J. Hort. Sci.* **6**(2): 101-104.

Relekar, P.P., A.G. Desai, J.C. Rajput, and M.J. Salvi. (1991) *Current Research*, Univ. of Agr. Sci., Bangalore **20**(6): 104-106.

Rozika, Murti, R. H., Purwanti, and Setyastuti. (2013) *Vegetalika*, 2(4), 101–114. https: / /jurnal.ugm.ac.id/jbp/article/viewFile/4009/3281.

Sachdeva, S., Mehra, P.N. and Sachdeva, S. (1986) *Phytomorphology*, **36** : 315-324.

Sahu CK, Patel MK, and Panda CM. (2018) *The Pharma Innovation Journal* **7**: 335-338.

Sambamurty, K. and V. Ramalingham. (1954) *Indian J. Hort.* **11:** 57-62.

Sari, V. K., Wulandari, R. A., and Murti, R. H. (2018) *J. Agric. Sci.*, **40**(2), 295–303. http: / /doi.org/10.17503/agrivita.v40i2.925.

Sarkar, B. and Chatterjee, B.K. (1993) *Golden Jubilee Symp. on Horticultural Research – Changing Scenario*, Bangalore, May 24-28, Abst. No. 26.29, p. 448.

Sarkar, H.N., Hasan, Abu Md. and Chattopadhyay, P.K. (1995) *The Hort. J.*, **8** : 17-24.

Sastry, M.V. (1970) *Indian Fd. Pack.*, **24** : 24-26.

Sayed, I.A. (1962) *Fruit Nursery Practices of India*, pp. 148-153.

Schroeder, C.A. (1958) *The Indian J. Hort.* (Special Symposium), **15**: 116-128.

Selvaraj, Y. and Pal, D.K. (1984) *J. Hort. Sci.*, **59** : 275-281.

Shanmugavelu, K.G. (1970) *Madras Agric. J.*, **57** : 311.

Shanmugavelu, K.G. and Srinivasan, C. (1973) *South Indian Hort.*, **21** : 107-108.

Shirsath, H.K., Chaudhury, S.M., More, T.A. and Patil, B.T. (1998) *J. Maharashtra Agril. Univ.*, **23** : 186-187.

Singh Daljit (1967) *Indian Hort.*, **11** : 51-52.

Singh Ranjit (1969) *Fruits*, National Book Trust, pp. 120-123.

Singh Shyam, Krishnamurthi, S. and Katyal, S.L. (1963) *Fruit Culture in India*, ICAR, New Delhi, pp. 192-198.

Singh, I.P. and Sharma, C.K. (1997) *Haryana J. Hort. Sci.*, **26** : 159-160.

Singh, R. D. (1970) *Indian J. Science and Industry*. **4,** 65-72.

Singh, R., Singh, D. and Ram, T. (2000) *Annals of Biol.*, **16** : 79-82.

Singh, S., Singh, K.K. and Chung, D.V. (1962) *Indian J. Hort.*, **19** : 310-312.

Sohi, H.S. and Prakash, O. (1972) *Progressive Hort.*, **4** : 87-88.

Sohi, H.S. and Sridhar, T.S.S. (1972) *Indian J. Agric. Sci.*, **42** : 1139-1142.

Soni, S.L., Tripati, L.P. and Mohahar, M.S. (1981) *Nat. Symp. Trop. and Subtrop. Fruit Crops*, Bangalore, p. 115.

Srivastava, A.K., Poonam Kashyap., Vijay Singh Meena., Nisha Verma. and S.P. Singh. (2016) In: *Underutilized Fruit Crops: Importance and Cultivation* (Eds. S.N. Ghosh, A.Singh, and A.Thakur), pp.1159- 1194.

Sulladmath, U.V. and Kololgi, S.D. (1969) *South Indian Hort.*, **18** : 9-17.

Sulladmath, U.V. and Rao, M.M. (1979) *J. Maharashtra Agric. Univ.*, **4** : 55-60.

Sulladmath, U.V. and Reddy, N.M.A. (1990) In : *Fruits Tropical and Subtropical* (Eds. T.K. Bose and S.K. Mitra), Naya Prokash, Calcutta, pp. 565-591.

Suryanarayana, V. and Goud, P.V. (1984) *Andhra Agric. J.*, **31** : 308-311.

Tasleema,S.R., Kamalakannan, S., Rajeswari R. and Sudhagar R. (2019a) *Plant Archives* **19** (1): 1086-1088.

Tasleema,S.R., Kamalakannan, S., Rajeswari, R.and Sudhagar, R. (2019b) *Annals of Plant and Soil Research,* **21:** 67 – 70.

Thirupathaiah, G. and Shirol, A. M. (2017) *Asian J. Soil Science and Plant Nutrition* **2:** 1-6.

Thompson, K.M., Culley, T. M., Zumberger, A. M. and Lentz, D. L. (2015) *Tree Genetics & Genomes.* **11:** 40, DOI 10.1007/s11295-015-0867-9.

Uthaiah, B.C., Bhandary, K.R. and Rao, M.M. (1977) *South Indian Hort.*, **25** : 1-5.

Vasanth, K.R., Nache Gowda, V. and Vijaykumar, N. (1993) *Golden Jubilee Symp. on Horticultural Research – Changing Scenario*, Bangalore, May 24-28, Abst. No. 26.29, p. 447.

Walten, Kay. (2016) *Trip Back in Time – Understanding the Chicleros*. https: //www. locogringo.com/rip-time-understanding-chicleros/.

Waskar, D.P. and Nikam, S.K. (1998) *South Indian Hort.*, **46** : 287-290.

Waskar, D.P., Nikam, S.K. and Garande, V.K. (1999) *Indian J. Agric. Res.*, **33** : 240-244.

Wilson, K.I., Balakrishnan, S. and Nair, N.G. (1970) *Sci. and Cult.*, **36** : 109-110.

Yuniastuti, E., Parjanto, E., Yulianingsih and Delfianti, M.N.I. (2018) *Bulgarian Journal of Agricultural Science*, **24**: 421-426.

Zhang, L. (2011) *J. Plant Genetic Resources* **12**: 486-488.

7

AVOCADO

C. Aswath, R. Senthil Kumar and V. A. Parthasarathy

1.0 Introduction

The avocado, *Persea americana*, is a fruit crop of immense importance to Mexican agriculture with an increasing demand worldwide. Avocado lies in the anciently diverged magnoliid clade of angiosperms, which has a controversial phylogenetic position relative to eudicots and monocots (Rendón-Anaya *et al.*, 2019). It is regarded as an important contribution of the New World to the human diet. The name 'avocado' has been derived from the corruption of the Spanish *ahuacate* or *aguacate*, which in turn are adaptations of the Aztec ahuacatl, meaning Avocado (Popenoe, 1920).

2.0 Composition and Uses

2.1.0 Composition

The fruits are rich in fat, proteins and minerals but low in carbohydrate content (Table 1) and can be recommended as a high energy food for diabetics (Samson, 1980). The predominant fatty acids are oleic, palmitic and linolenic acids (Mazliak, 1971; Tango *et al.*, 1972; Swisher, 1984). However, the fatty acid composition of the lipids of avocado fruit varies with cultivar, stage of ripening, anatomical region of the fruit and geographical location (Itoh *et al.*, 1975). The major free amino acids present in avocado fruit cv. Fuerte are aspergine, aspartic acid, glutamine and glutamic acids (Kadam and Salunke, 1995). The immature fruits contain a bitter flavour (Ahmed and Barmore, 1980), which reduces with ripening in some cultivars (Fuerte and Haas) or persists upon ripening in some cultivars like Zutano (Brown, 1972). About 75 per cent of the total volatile compounds in ripe Fuerte avocado are due to the presence of C6 alcohols and aldehyde. Major components are *trans*-hex-3-en-l-ol (25.6 per cent), *trans*-hex-2-en-l-ol (19.1 per cent), hexan-l-ol (17.9 per cent), *cis*-hex-2-enal (8.7 per cent), and hexanal (4.5 per cent) (as quoted by Kadam and Salunke, 1995).

2.2.0 Uses

Ripe fruits are consumed fresh in salads and are used as a sandwich filling or spread. They are also used in ice creams and milk shakes and the pulp may be preserved by freezing. 'Guacamole' (a paste) is prepared from avocado fruit to accompany Mexican cuisine (Whiley and Schaffer, 1994). Three antifungal compounds have been isolated from the peel extract of immature fruits of cv. Green and are identified as 1,2,4-trihydroxyheptadec-16-yne, 1,2,4-trihydroxyheptadec-16-ene and 1-acetoxy-2,4-dihydroxyhepatadec-16-yne (Adikaram *et al.*, 1992).

3.0 Origin and Distribution

The origins of avocado present a particularly challenging case, having been domesticated at least 3 times from geographically distinct populations of the progenitor species (Ashworth and Clegg, 2003). Archeological records suggest that avocado fruits were consumed in Mexico at least 10,000 years ago (Knight, 2002; Galindo-Tovar *et al.*, 2008). Avocado originated in Central America and the

Table 1: Average Nutritional Composition (on a fresh-weight basis) of the Edible Part of Avocado Fruits*

Components	*Fuerte*[a]	*Hass*[a]	*Anaheim*[b]
Water (per cent)	71.2	74.4	79.3
Fat (per cent)	23.4	20.6	15.5
Protein (per cent)	2.0	1.8	1.8
Fibre (per cent)	1.9	1.4	1.7
Ash (per cent)	1.2	1.2	1.0
Sugars (per cent)			
Glucose	0.1	0.3	0.2
Fructose	0.1	0.1	0.2
Sucrose	0.0	0.1	0.1
Starch (per cent)	0.0	0.0	0.1
Organic acids (per cent)			
Malic	0.17	0.32	0.24
Citric	0.13	0.05	0.11
Oxalic	0.0	0.03	0.01
Energy (KJ/100 g)	980	805	956
Vitamins (mg)			
Ascorbic acid	9	11	14
Thiamin	0.07	0.07	0.08
Riboflavin	0.15	0.12	0.21
Pantothenic acid	0.9	1.2	1.11
Nicotinic acid	1.5	1.9	1.56
Vitamin B6	0.61	0.62	0.39
Folic acid	0.03	0.04	0.018
Biotin	0.004	0.006	0.0034
Carotenoids (mg/100 g)			
-Carotene	0.36	0.29	0.24
-Carotene	0.02	0.03	0.03
Cryptoxanthin	0.29	0.16	0.22
Minerals (mg/100 g)			
Potassium	460	480	460
Phosphorus	29	27	29
Calcium	29	14	19
Magnesium	22	23	22
Sodium	2	2	2
Iron	0.6	0.7	0.6
Zinc	0.5	0.5	0.5

a Grown in Australia, b Grown in California,

* Nogalingam (1993).

Uses of avocado (*Persea americana*)

Avocado tree with fruits

Slices of avocado fruit

Chutney and salad from avocado

Seasoning of avocado slices

Bread toast with avocado

Avocado spread in bread

Carrier oil from avocado

Cultivars of avocado

Fuerte

Hass

Ettinger

Brogden

Bacon

Del Rio (Pryor)

early Spanish explorers recorded its cultivation from Mexico to Peru, but not in the West Indies at that time. It was introduced into Jamaica in about 1650 and to Southern Spain in 1601. It was reported in Zanzibar in 1892. Avocado was first recorded in Florida in 1833 and in California in 1856. In India, it may have been introduced in the south and west coast of India from Ceylon (Sri Lanka) about 50 to 75 years ago. Mexico, the USA, Dominican Republic, Brazil, Colombia, Indonesia, Haiti, Venezuela, Chile and South Africa are the leading countries in the world for commercial production of avocado (Anon., 1991). It is also cultivated in Israel, Australia, the New Zealand, the Philippines, Spain, Cyprus, Canary Islands, Ecuador, Greece and the Caribbean area. Berg (1992) feels he exact number of genus *Persea* species is unknown. There have been reports of *Perseas,* or closely related species, in the Japan-Taiwan-southern China region. However, according to Storey *et al.* (1986) that there are over 80 species of *Persea* are recognized as valid. The whole matter needs early, thorough investigation. Nearly all have tiny fruits lacking palatable flesh. Nearly all are native to the Americas. The species are scattered from Mexico north through the southeastern United States, east through the West Indies, south through Central America, Colombia, Venezuela, the Guianas and Brazil, and Ecuador, Peru, Bolivia, and Chile. How one species (*P. indica*) got to the Canary, Madeira, and Azores Islands, and how several that may be *Persea* got to southeastern Asia, is problematic. They (Storey *et al.,* 1986) further stated in the course of time, the primitive avocado, *P. americana,* evidently gave rise to three, and probably more, divergent types which, developing in geographical isolation from one another, became distinct botanical types. These became known in horticultural circles as three races: Mexican, Guatemalan, and West Indian. To systematic botanists, they are botanical varieties. They are designated as *P. americana* var. *drymifolia,* var.*guatemalensis,* and var. *americana,* respectively. Probably the "species" *P. floccosa, P.nubigena,* and perhaps even *P. schiedeana,* are actually botanical varieties of *P.americana,* also. The Mexican race, *P. americana* var. *drymifolia,* is thought to be the primordial race of commercial avocado. Genetic evidence supports the theory that the three races are but variants of a single species. All representatives have the same somatic chromosome number, 2n = 24.They hybridizes readily. In fact, where two races have been brought together in cultivation they soon mongrelize, and racial distinctions disappear.

The geographic races of avocado are each distinguished by a suite of morphological traits; the populations identified in the STRUCTURE and STRUCTURAMA analyses closely correspond to these racial groupings. Indeed, substantial differences in haplotype composition among major racial groups suggest that these populations were separated for appreciable periods of time prior to the advent of human utilization of avocado (Chen *et al.,* 2009). The overall picture is one of modestly sized geographic populations that have retained their genetic integrity. Once, humans entered the picture these different populations appear to have been domesticated more or less in situ and then disseminated more widely. In more recent times, these separate domesticates were admixed and provided the basis for many of the modern cultivated forms. Chen *et al.* (2009), quoting Galindo-Tovar *et al.* (2007) made the suggestion of 2, genetic populations based on elevational and latitudinal differences in the highlands of Central Mexico is new and should stimulate further sampling and genetic analysis of avocado in Mexico and Central America. Avocado is one of the first trees domesticated in the Neotropics and can be used as model of tree domestication in areas of high biological diversity. The ability to unravel the complex hybrid origins of various cultivars should provide useful guidance to genetic resource managers and to breeders. It is also noteworthy that haplotype data can supplement archeological investigations by providing a finer resolution view of the domestication of major tree crops. The avocado occupies an unresolved position near the earliest evolutionary branchings of flowering plants. Nuclear genome sequences of Mexican and Hass variety avocados inform ancient evolutionary relationships and genome doublings and the admixed nature of Hass and provide a look at how pathogen interactions have shaped the avocado's more recent genomic evolutionary history (Rendón-Anaya *et al.*, 2019).

4.0 Species, Races and Cultivars

4.1.0 Species

The avocado (*Persea americana*) (syn. *P. gratissima*, 1807; *P. drymifolia*, 1831; *P. nubigena* 1950), belongs to the family Lauraceae which has about 47 genera and 1900 species including two other species of economic importance, *viz. Cinnamomum zeylanicum* (cinnamon) and *C. camphora* (camphor). All *Persea* species have a chromosome number of 2n = 24 (Bergh, 1975).

The fruit is a berry, pyriform or globose in shape with a thick, fleshy mesocarp surrounding a single large seed and has a yellow-green to maroon or purple skin which can be smooth or warty (Whiley and Schaffer, 1994).

4.2.0 Races

In avocado, though the following three ecological races (subspecies) have been recognized, it is the natural inter-racial hybrids of these races that are of horticultural and commercial importance now. The characteristics of different races are presented in Table 2.

Mexican Race

It is characterised by small fruits weighing less than 250 g and ripening 6-8 months after flowering. Fruits have thin, smooth skin with a large seed fitting loosely in the central cavity and its oil content is up to 30 per cent, the highest of all the three races. Amongst the three races, this is the most resistant to cold temperature.

Guatemalan Race

This race is native to the highlands of Central America. Fruits are fairly large, weighing up to 600 g and borne on long stalks. The fruits ripen 9 to 12 months after flowering. Their skin is thick and brittle and often warty. Its seeds, held tightly in the cavity of the fruit, are small. The oil content ranges between 8 and 15 per cent.

West Indian Race

This group with medium-sized fruit is native to the lowlands of Central America. The fruit skin is smooth but leathery and glossy. Fruits are borne on long stalks and require 6 to 9 months for ripening from the date of flowering. Its seeds are large, fitting loosely in the cavity. The oil content of the fruit is low, ranging between 3 and 10 per cent. This race is the least resistant of the three to cold temperatures.

Table 2: Characteristics of the Avocado Races*

TREE- GENERAL	*Mexican*	*Guatemalan*	*West Indian*
Native Region	Mexican Highlands	Guatemalan Highlands	Tropical lowlands
Climactic Adaptation	Subtropical	Subtropical	Tropical
Cold Tolerance	Most	Intermediate	Least
Salinity Tolerance	Least	Intermediate	Most
Iron Chlorosis Tolerance	Intermediate	Least	Most
Alternate Bearing	Less	More	Less
FORM			
Internodes	Longest	Long	Shortest
Twig Lenticels	Pronounced	Absent	Absent
Bark Roughness	Less	Less	More
Stem Pubescence	More	Less	Less
LEAF			
Size	Smallest	Large	Largest
Color	Green	Green	Pale Green
Flush Color	Greenest	Reddest	Yellowish-green
Anise	Present (usually)	Absent	Absent
Underside waxiness	More	Less	Less
FLOWER			
Season	Early	Late	Early/Intermediate
Bloom to maturity	5-7 months	10-18 months	6-8 months
Perianth Persistence	Greater	Less	Less
FRUIT PEDICEL			
Length	Short	Long	Short
Thickness	Medium	Thick	Thin
Shape	Cylindrical	Conical	Nailhead
FRUIT			
Size	Tiny-Medium	Small-Large	Medium-V. Large
Shape	Mostly elongate	Mostly round	Variable
SKIN			
Color	Usually purple	Black or green	Pale green – maroon
Surface	Waxy coating	Variably rough	Shiny
Thickness	Very thin	Thick	Medium
Stone cells	Absent	Present	Slight
Pliability	Membraneous	Stiff	Leathery
Peeling	No	Variable	Yes

TREE- GENERAL	*Mexican*	*Guatemalan*	*West Indian*
SEED			
Size ratio	Large	Often small	Large
Coats	Thin	Usually thin	Thick
Tightness in cavity	Often loose	Tight	Often loose
Surface	Smooth	Smooth	Rough
PULP			
Flavor	Anise-like, spicy	Often rich	Sweet, mild
Oil Content	Highest	High	Low
Distinct Fibers	Common	Less common	Intermediate
COLD STORE TOLERANCE	More	More	Less

* Berg (1992).

4.3.0 Cultivars

Several cultivars, perpetuated through vegetative means, are under cultivation in the New World. Samson (1980) has listed quite a number of them. The following are, however, the more important ones:

Fuerte

This is the most popular cultivar of avocado. It belongs to group B and is a hybrid of the Mexican and Guatemalan races grown in California. Fruits are pyriform, weigh between 225 and 450 g with 18 to 26 per cent oil. It is fairly resistant to cold; better suited to subtropics than tropical climate.

Hass

This is a group of cultivar also grown in California where it matures much earlier than the Fuerte. It originated as a seedling from the Guatemalan race. Fruits are medium-sized, roundish and turn purple on ripening. This is also more suitable to subtropical climate.

Lula

It is another of Guatemalan origin with pyriform fruit that weighs around 400 to 700 g. Its oil content is only 15 per cent, *i.e.* less than that of Fuerte. This is grown in the tropical climate of Florida in the U.S.A.

Pollock

This cultivar is of West Indian origin with large fruit that weighs up to 1 kg or more but the oil content is less, ranging between 3 and 5 per cent. This is also grown in Florida.

Ardith

Ardith was selected from California material (Blumenfeld and Elimelech, 1986c). It matures in April and the yield is high and consistent. The fruit is green, averages 250 g in weight and has good flavour.

Pinkerton

This is a patented cultivar of California (Blumenfeld and Elimelech, 1986b). The dark green fruit is pear-shaped with a long neck, has a rough rind, and weighs from 250 to 400 g. The fruit remains large in warm regions but sometimes suffers from uneven softening.

Yulia

This cultivar was selected from Californian material (Blumenfeld and Elimelech, 1986a). The fruit has excellent flavour and is similar to Nabal but smaller, averaging 200 g in weight. The fruits tend to be smaller in warm regions. The yield is good and trees yield alternately.

The cultivars that are cultivated in India go by several names, such as Purple, Green, Fuerte, Pollock, Peradeniya Purple Hybrid, Trapp, Round and Long. The Purple and Green cultivars according to Gandhi (1956), were introduced into India from Ceylon in 1941. His descriptions of the two are as follows:

Purple

This cultivar belongs to the West Indian race. The fruit is pear-shaped, about 12 to 15 cm long and 6-9 cm broad, with a long neck and about 250 g in weight. The skin is smooth, shining, deep crimson or maroon in colour, pliable and leathery in texture. It can easily be removed from the pulp. The seed is set loose in the cavity which extends into the neck of the fruit. The flesh is 2 cm thick, firm, smooth and fine in texture, deep yellow, changing to yellowish green close to the skin and with a rich nutty flavour.

Green

The Green avocado belongs to the Guatemalan race. The fruit is large, oval 250-400 g in weight. The surface is slightly rough and yellowish green in colour, the skin being thin but brittle. The seed is large, roundish and sits tight in the cavity. The neck is solid. The flesh is 2-2.5 cm thick, soft, greenish yellow, changing to green close to the skin and has a mild nutty flavour.

Few recent cultivars of avocado are as follows :

Gwen

This has been derived through open pollination in cv. Hass. This is a highly precocious cultivar with semi dwarfing, upright habit. It produces high-quality and medium-sized fruits.

Horshim

It is a new avocado cultivar and produces commercially acceptable yield.

Adi

The cultivar Adi belongs to group A and is derived from a cross between Hass and Horshim. The fruit shape resembles that of Hass with fruit weighing about 230 g. The flesh is pale yellow with a green rim, buttery and having nut-like aroma.

Eden

The avocado cultivar Eden was developed from a cross made between Pinkerton and N 151-2 in Israel (Lavi *et al.,* 1997). Fruit weight of Eden ranges from 250 to 400 g and averages about 260 g, 30 per cent heavier than Hass. Flesh is yellow, buttery and largely free of fibres with a nut-like flavour. The harvest season starts in November and fruit can be kept on the tree until April. Fruits may be stored for at least 24 days at 5°C and 95 per cent RH.

A new cultvar, Lam Haas from South Africa was described by Buzet (1999). It was derived from the cross of a Guatemalan with a Mexican variety. Trees have upright growth habit, fruits have black, pepply skin and good storage traits. Fruit weight is in the range 200-410 g and tree yields average 50 kg (20 t/ha). Lamb Hass has high fibre content and good yields.

Avocado cultivars with low pulp oil contents (6.73-8.07 per cent) from Venezuela (Booth 1, Booth 7, Ceniap 2, Figueroa, Guacara Morado, Luiz de Queiroz, Princesa, and Quebrada Seca. Santa Clara, Santa Cruz, Taylor and Tonnage) were characterized by Gomez-Lopez (1999). Quebrada Seca fruits had the lowest pulp percentage and the highest seed percentage. Princesa fruits had the highest pulp percentage (76 per cent). Seven cultivars had pyriform fruits and 5 had ovate fruits; 7 had rough peel and 5 had smooth peel and 4 had purple peel and the others had green peel. Five cultivars were easy to hand peel. Princesa had the highest yields up to 67 944 kg fruits/ha, 51 675 kg pulp/ha, 3679 kg oil/ha and 3016 kg starch/ha.

Samson (1980) concluded that no single cultivar or group would be recommended for any location without further investigation. If a plantation is to be raised in a new area, the cultivars selected should belong to both groups A and B and their bloom must overlap. The proportion can be 1:1 or 2:1, with Hass, even 4:1. Schroeder (1958) was of the opinion that for testing in a new area, it would always be better to introduce and test avocado seedlings of standard cultivars, rather than their clones, as the latter provide only a very little range of variability, while the former, being more variable and vigorous, would form the best material under such conditions. In a study carried out with this objective at the Tamil Nadu Agricultural University, Coimbatore, Rao and Khader (1978a) noted that the quality and yield performance of seedling progenies were, in general, highly satisfactory. Out of five trees studied, one did not flower, another had poor growth and the rest yielded very well.

Maluma

Ernst *et al.* (2015) outlined in details of evolution of 'Maluma', a new generation 'Hass'-like cultivar, which was discovered, during the early 1990's, by Andries Joubert on his farm Maluma in Levubu, Limpopo, South Africa, as a chance seedling of unknown parentage. An industry dominated by a single cultivar, or that produces too few new cultivars, is vulnerable. Ironically no less than 13 fruiting cultivars presently receives intellectual property protection in various countries, of which less than half has established themselves commercially. Presently 'Hass' is the world's premium avocado cultivar, which originated as a chance seedling planted in 1926 by Rudolph Hass in La Habra Heights, California. A plant patent was granted in 1935 in California, which expired in 1952. 'Hass' only established itself during the 1970's, as the initial acceptance of 'Hass' by packers, distributors, retailers and end users, as a superior commercial cultivar was slow. 'Hass' received resistance because of its fruit being black-skinned, rough and small. 'Maluma', a new generation cultivar, was discovered, during the early 1990's, by Andries Joubert on his farm Maluma in Levubu, South Africa. Plant Breeder's Rights were granted to Allesbeste Nursery on 7 November 2004 and commercially released, after 8 years of intensive semi-commercial research, during November 2007 at the VI World Avocado Congress, Chilé. Successful commercial establishment in the European Union (EU) and United Kingdom (UK) occurred within another 8 years.

This achievement was made possible through the active integration of the following:

- ☆ Horticultural attributes
- ☆ Marketing attributes
- ☆ Competent, loyal participants
- ☆ Vibrant trade relationships
- ☆ Additional economic value for all participants
- ☆ Technical support, able to deal with the critics
- ☆ Branding and promotion
- ☆ Significant marketable volumes
- ☆ Strategic establishment to ensure future 12 months' market presence

This active integration was achieved through Allesbeste's vertical integration in all structures from propagation to exporting/marketing, providing a platform for efficiently applying its inherent skills in all fields.

Gandhi (1956), Chandler (1958) and Biale and Young (1971) recorded that not all avocado cultivars changed their skin colour when they ripened. It was reported by Biale and Young (1971) and Purseglove (1974) that avocado fruits could be stored on the tree even after they attained maturity. The former observed that the fruits of Fuerte and Hass could be retained on the tree even after maturity since they did not soften as long as the stem remained healthy. This characteristic feature was not observed in study at Coimbatore in any of the trees. Almost all the fruits started dropping off from the trees after attaining maturity.

5.0 Breeding and Improvement

5.1.0 Selection

Most of the named avocado cultivars grown throughout the world originated as a chance seedling. An important cultivar Fuerte, which was first introduced in California, originated from a seedling tree in Mexico. Several cultivars are thus found to be originated in Mexico and Central America, which are now grown with popularity in California (Bergh, 1957). The Guatemalan cultivars Benik, Itzamna and Nabal were first introduced as budwood (Janick and Moore, 1975). A number of Mexican race selections have been made in Chile, including Campeon, Choula and the Peuminas (Hodgson, 1959). A number of local selections were also made in Jamaica (David, 1967) of which Elgin, Gimball and Huntley proved worthy. Extensive research programme is under way at the Scientific and Technological Research Centre of Avocado in the state of Mexico to produce new cultivars, including intermediate and dwarf trees. Cultivars Colinmex, Colin V-101 and especially Colia V-23 have been used for breeding and 21 selections with outstanding characteristics of quality and production have been made. Three Fuerte and 2 Rincon dwarf mutants have also been selected. In the selection of dwarfing rootstocks, more than 20 have been selected from Colin V-33 seedlings (Sanchez Colin and Barrentos Priego, 1987).

5.2.0 Hybridization

Phytophthora rot is considered to be a serious problem in avocado cultivation. Extensive attempts to hybridize the avocado with different *Persea* species of sub-genus *Eriodaphne* that are immune to the root rot have failed (Janick and Moore, 1975). According to Zentmyer *et al.* (1967) the Mexican race Duke has some resistance to root rot but crosses with *Persea* species have been unsuccessful; both graft and cross incompatibility appear between sub-genus *Persea* which includes the avocado and sub-genus *Eriodaphne* which includes all the known immune species.

6.0 Propagation

6.1.0 Seed

Samson (1980), reported that the viability period of avocado seed is short (2-3 weeks), which can be improved by storing the seed in dry peat or sand at 5°C. Removal of the seed-coat before sowing helps to speed up the germination. If there is a shortage of rootstocks, the seed can be split lengthwise into four or even six parts, leaving a piece of the embryo on each.

Germination of seed is hypogeal. Seeds are sown in nursery bed with a spacing of 35 and 45 cm between seeds and 60 and 75 cm between rows. The seeds should not be covered with more than 1 cm soil. The number of days to shoot emergence was about 55 to 95 which was positively correlated with the residual dry matter content of the cotyledons. Seeds of varying sizes were of equal value for propagation. The cotyledon DW diminished as shoot size increased and the first two stages of root growth were governed by the cotyledon DW which in turn was controlled by shoot expansion (Flores-Meza *et al.*, 1990). Highest germination rates and least weight loss were recorded in seeds stored in perforated bags (Gordillo *et al.*, 1994).

6.2.0 Vegetative Propagation

6.2.1 Cutting

The rooting capability of avocado cuttings varies greatly depending on the races and cultivars. The Mexican race is relatively easy-to-root whereas the West Indian race is quite difficult. The Guatemalan race is intermediate in rooting ability on cuttings (Reuveni and Raviv, 1976a). Cuttings from young plants root more readily than from old trees (Chandler, 1958). Leaf survival on cuttings under mist has been found to be essential for rooting to take place and leafy cuttings root first. The use of bottom heat (constant 26°C) in the cutting beds plus soaking the basal ends of cuttings in an aqueous solution of IBA greatly improved the rooting of cultivars that normally root slowly (Krezdron *et al.*, 1976). Foliar spraying of a mixture of NAA (1´10-4 M) + cytokinin (5´10-4 M) once a week prevented leaf shedding and improved rooting of cuttings (Reuveni and Raviv, 1976b). Similarly etiolation and use of root promoting substance (IBA) usually promoted rooting on avocado cuttings (Ernst and Holtzhausen, 1978). Rooted cuttings grow slowly at the early stages, but eventually make satisfactory growth (Frolich, 1966). The difficulty in rooting of avocado cuttings was overcome by producing base-etiolated cuttings and such cuttings rooted faster and at a higher percentage (Reuveni and Goren, 1982).

Quantification of endogenous IAA and IAAsp during adventitious root formation in avocado cv. Topa-Topa microcuttings treated with IBA (rooting promoter), TIBA (auxin transport inhibitor) and control (no treatment) revealed that in control cuttings, IAA levels remained constant throughout the rooting process, however, there was a 2-fold increase in IAA level in IBA-treated cuttings during the first 6 days. Application of TIBA resulted in decrease in IAA levels and inhibited rooting. There was a large increase in IAAsp levels in both control and IBA-treated cuttings before root differentiation occurred followed by a decrease as differentiation process progressed.

However, the IAAsp levels did not increase in T1BA-treated cuttings but diminished progressively during root differentiation (Garcia-Gomez *et al.*, 1994).

6.2.2 Budding and Grafting

6.2.2.1 Rootstocks

The avocado industry is based on seedling rootstocks. In most cases seedling rootstocks were chosen according to their ease of propagation and the horticultural value of the rootstock was not studied. The non uniformity of seeded rootstocks and the necessity to overcome various soil problems gave impetus to the search for a clonal propagation method that can produce known uniformity (Ben-Yaacov and Michelson, 1995). The avocado tree (*Persea americana* Mill.) is one of the species most sensitive to salinity. Since the root system of this species is particularly affected by this type of stress, it is crucial to find a rootstock that is tolerant to saline water irrigation. Plant material from the cv. Hass was grafted on five rootstocks to find possible candidates: the clonal 'Duke 7', 'Nabal', 'Degania 117', 'Mexicola', and 'Zutano' seedlings. One year-old plants were placed in 55 L pots; river sand was used as substrate, and the plants were fertilized with a modified Hoagland solution. Vegetative growth and internal nutrient content were compared with control plants with 30 mg L^{-1} of Cl^-, and plants treated mg L^{-1} Cl^- for 130 days using NaCl in the nutrient solution (5.64 mM). No interaction between NaCl and the rootstocks for the vegetative growth variables was detected neither in K^+ internal foliar or root content nor Ca^{+2} root. With regard to the carbon assimilation rate, the Nabal rootstock showed the highest rates under the NaCl treatment by retaining the highest chloride concentration in the roots and greatly limiting the concentration found in the leaves of the cv. Hass. It is therefore a promising rootstock for salt tolerance (Castro *et al.*, 2009).

Menge *et al.* (2012) described three new avocado rootstock cultivars tolerant to *Phytophthora* Root Rot namely, 'Zentmyer', 'Uzi',and 'Steddom'. They have described them as below.

'Zentmyer' (PP4)

This avocado rootstock is named after George Zentmyer who was a pioneer avocado pathologist at UCR and a collector of avocado germplasm worldwide. 'Zentmyer' is an extremely vigorous, highly durable cultivar that is tolerant of PRR under most conditions. It is generally more salt-tolerant than 'Thomas' but less so than 'Dusa' and therefore is not recommended for locations where salt is a problem. This cultivar when grafted with 'Hass' also often does not yield well under non-root rot conditions in comparison with 'Steddom' and 'Uzi'under California conditions. Therefore, it is recommended that the 'Zentmyer' rootstock should be used primarily in replant situations were root rot infested soils are a problem but salinity is not.

'Uzi' (PP14)

This avocado rootstock is named after Uzi Afek, a scientist from the Volcanic Center in Bet-Dagan, Israel. 'Uzi' is an extremely vigorous and fast-growing rootstock that is capable of supporting a 'Hass' tree growing to 15 ft in 2 years. It is highly tolerant to PRR and its yields are generally high and consistent. 'Uzi' is very susceptible to salt based on leaf burn symptoms, but this does not seem to affect either growth or yield of the 'Hass' variety based on field trials to date.

'Steddom' (PP24)

It is an avocado rootstock that has a high tolerance against PRR. It is a moderately fast-growing rootstock, generally larger than 'Thomas' but smaller than 'Dusa'. Steddom' has good salt tolerance, especially compared with 'Thomas', and appears to be an all-around excellent rootstock with moderate stature and excellent yield when grafted to 'Hass'.

6.2.2.2 Grafting and Budding

Avocado should preferably be propagated vegetatively by budding or grafting, Purseglove (1974) reported that in Florida, 2- to 4-month-old rootstock plants are side (veneer) grafted, while in California, somewhat older stocks are shield budded. Pennock (1970) in Puerto Rico obtained 95 per cent success with cleft grafting. At the Kallar Fruit Station in Tamil Nadu, India, layering as well as approach grafting gave up to 75 per cent success, while in West Bengal, chip budding was found successful (Rao and Khader 1978b). Pinheiro and Pinheiro (1971) concluded that cleft, whip and tongue grafting were the most successful methods of avocado propagation. Whip method of grafting was more successful than side grafting and the whip and tongue method gave the lowest percentage of success (Maganjo, 1975). Cleft grafting in avocado under partial shade was very successful and resulted in strong and healthy trees (Uzquiano, 1975).

Propagation of avocado

Saplings of avocado ready for planting

Propagation of avocado by seed

Propagation of avocado by cutting

Propagation of avocado by grafting

Cultivation of avocado

High density planting in avocado

Training and pruning of avocado tree

Drip irrigation in avocado

Brokaw (1987) developed a popular propagation technique. In this method the clonal material is first grafted on a nurse rootstock. A few days later the entire plant is moved to a darkroom, where etiolation takes place and a metal ring is placed above the nurse rootstock. Then the entire etiolated branch and the grafted union are covered by rooting medium. When the etiolated shoot emerges, the plant is moved to the light. This new growth is ready for the second graft, this time with a commercial cultivar. The entire plant grows in the nursery for about 1 year, and during this period the ring induces rooting from the previously etiolated portion. Later the portion below the ring disintegrates and the rooted plant separates from the nurse rootstock. Two other propagation methods will be mentioned briefly. One is the intermediate (sandwich) graft, used experimentally to control tree size by an interstock with a dwarf avocado type (Barrientos *et al.*, 1987).

6.2.3 Micropropagation

In vitro regeneration and multiplication through axillary buds and microcuttings of adult grafted plants (cv. Fuerte) revealed that shoot regeneration was good with 2.0 mg BA/l and 0.5 mg IBA/l in N45K medium. After elongation of the shoot, the axillary buds developed and formed new shoots, which could be excised and subcultured.

The use of an antioxidant (PVP) improved the response of explants in repeated sub-cultures. Depending on the type of explant, differences were observed in the morphology of the regenerated shoots. In both explants the capacity to regenerate shoots varied throughout the year. With microcuttings the elimination of bacterial contaminants was more difficult (Dalsaso and Guevara, 1989). Biasi *et al.* (1994) reported that the nodal segment explant of avocado cv. Ouro Verde produced axillary bud growth on culture medium supplemented with 3 mg BA/1. Explant oxidation was avoided by reducing the concentration of inorganic salts in the medium to half-strength. Shoots rooted at a frequency of 46 per cent after 3 days of growth in a medium containing 25 mg IBA/1 with subsequent change to an auxin-free medium with diluted salts and 1 mg activated charcoal/1. Successful regeneration of plantlet from young shoot by culturing in MS solid medium supplemented with BA at 1.3 mM for bud sprouting and subsequently in liquid medium with Gamborg salt formulation and BA at 1.3 mM, was reported by Barcelo Munoz *et al.* (1999). Micropropagated plantlets of avocado exhibited a very slow rate of growth during the acclimatization phase. Inoculation of plantlets with vesicular arbuscular mycorrhizal fungus (*Glomus fasciculatum*) seemed to be the key factor for subsequent growth and development of micropropagated plants of avocado (Vidal *et al.*, 1992).

6.2.4 Rootstock

According to Bergh (1969), the Mexican race provides nearly all of the rootstocks used in California. The Guatemalan race is more sensitive to cold and has also proved more susceptible to high pH chlorosis and to *Verticillium* wilt. Mexican race rootstocks are used generally in Israel and in South Africa, though in Australia they were found unsatisfactory. *Phytophthora*-resistant rootstocks are probably the single most pressing problem for the industry as a whole. A number of immune or highly resistant *Persea* species are known, but unfortunately all have proved incompatible for hybridisation and grafting with *P. americana. Persea schiedeana* was considered as a possible rootstock to induce vigour in otherwise weak scions or for use where soil conditions are limiting. It was found to be compatible with scions of cultivars Fuerte and Edranol (Schroeder, 1971). The rootstock selections, Duke 6, Duke 7, G-6, Huntalas, Dusa and Latas showed good resistance to *Phytophthora cinnamomi* while Duke 9 was moderately resistant (Zentmyer *et al.*, 1977; Botha, 1991). In South Africa, Duke 7 is the most commonly used avocado rootstock, because of resistance to *P. cinnamomi* (Hannweg, 1997).

The yields of both cvs. Hass and Fuerte were highest on Duke 7 rootstock (Kremer-Kohne and Kohne, 1992). Mexican rootstocks showed a clear dwarfing effect, while trees on Nobal rootstocks produced greater yields than those on one of 2 Mexican rootstocks (Ben-Yaacov *et al.*, 1992a). Rootstocks affected yield, tree size, tree efficiency (yield/unit area) and alternate bearing, and choice of appropriate rootstock may overcome some of the disadvantages associated with the new Israeli avocado cv. Horshim (Ben-Yaacov *et al.*, 1993a). West Indian rootstocks, especially those of Nachlat origin are more lime tolerant than the other types (Kadman, 1985). In an orchard trial on the effects of various rootstocks on avocado cvs. Fuerte and Ettinger grown on calcareous soil and irrigated with saline water, it was observed that the cv. Fuerte grafted on clonal rootstock (Ein Harod) resulted in the biggest trees and the highest yield. Nachlat 2 rootstocks resulted in lower yields than Waldin, Zrifin 99 or Nachlat 3, while trees on Zrifin 67 exhibited chlorosis. In cv. Ettinger, yields were generally much better on Nachlat 3 rootstock (Ben-Yaacov *et al.*, 1992b). Avocado trees (cv. Ettinger) grafted onto the dwarfing West Indian Nachlat-type rootstock (Shiller 10) did not suffer from leaf burn as a result of salinity and were smaller, higher yielding and more uniformly productive than trees on Mexican rootstock (Shiller 1) (Ben-Yaacov *et al.*, 1993b). Three Mexican clonal rootstocks are recommended for heavy soils where non-saline irrigation water is available and 4 West Indian clonal rootstocks are recommended for non-calcareous soils where the irrigation water is saline (Ben-Yaacov, 1998). It is claimed that these clonal rootstocks can confer a 50 per cent increase in yield and greater uniformity (Ben-Yaacov, 1992).

The development and productivity of cv. Ettinger trees were affected by rootstock. The best rootstocks (West Indian avocado seedling) for heavy soils, in the presence of saline conditions and a limited amount of lime, were Degania 62, 113, 115 and 117, Peker 2 and Ashdot 27. Zrifin 99 exhibited superior resistance to lime (Ben-Yaacov, 1995). Mexican seedling rootstocks had significant effects on fruit yields of avocado cv. Ettinger and Schiler 22, 29, 30 and 52 and Gvar-Am 13 showed superior performance (Ben-Yaacov, 1996). When scion cv. Hass was grafted onto Degania 113, 115 and 117 rootstocks, the graft combination showed wide adaptation to different soil conditions. Ashdot 17 showed resistance to lime while Lula and Peker 2 were more sensitive (Ben-Yaacov, 1997). West Indian rootstocks resulted in lower yields than Mexican ones and some of the West Indian rootstocks had a dwarfing effect (Ben-Yaacov *et al.*, 1996).

7.0 Area and Production

Avocado is cultivated commercially in California, Florida and Hawaii as well as in several South American countries, South Africa and Australia. The production of avocado in different parts of the world are presented in Table 3.

Table 3: Production of Avocado in different Parts of the World*

Rank	*Country/Region*	*2018*	*2017*	*2016*
1	Mexico	2,184,663	2,029,886	1,889,354
2	Dominican Republic	644,306	637,688	601,349
3	Peru	504,517	466,758	455,394
4	Indonesia	410,094	363,157	304,938
5	Colombia	326,666	314,275	294,389
6	Brazil	235,788	213,041	196,422
7	Kenya	233,933	194,279	176,045
8	United States	168,528	132,730	124,860
9	Venezuela	139,685	133,922	130,290
10	Israel	131,720	110,000	101,500

Source: Food and Agriculture Organization Corporate Statistical Database.

In India, it is grown in small pockets as scattered trees around Bangalore and in the hill slopes of Tamil Nadu, Kerala and Karnataka and does not find a place among the top 60 avocado producing countries.

8.0 Soil and Climate

8.1.0 Soil

Avocado can be grown on a wide range of soil types from sandy to clay loam provided proper drainage system is ensured to prevent the development of anaerobic conditions. Oxygen diffusion rates of less than 0.18 mg cm^{-1} min^{-1} have been reported to cause 44 to 100 per cent death of 'Mexicola' avocado roots that were healthy prior to flooding (Stolzy *et al.*, 1971). They thrive best in acid soil with a pH range of 4.5-5.5 (Whiley *et al.*, 1984). However, selected West Indian race rootstocks have certain degree of tolerance to calcareous alkaline soils with pH range of 7.6 to 7.9 (Young and Koo, 1976).

The root system of mature cv. Fuerte tree was found to be better distributed horizontally and vertically in a sandy soil than in a clay soil. In both soils, most of the fine roots were found in 0-60 cm depth (Salazar-Garcia and Cortes-Flores, 1986). Durand and Claassens (1987) stated that in a less homogenous soil of bulk density 1.7 g/cm^3, the roots did not effectively penetrate from one soil texture to another and root mass was very low.

8.2.0 Climate

The major avocado growing regions in the world fall into 3 climatic zones : (i) cool, semi-arid climate with winter-dominant rainfall, *e.g.* California, Chile, Israel; (ii) humid, subtropical with summer-dominant rainfall, *e.g.* eastern Australia, Mexico, South Africa; (iii) tropical or semitropical with summer-dominant rainfall, *e.g.* Florida, Indonesia (Whiley and Schaffer, 1994). The ecological races differ widely from temperature adaptation point of view. The Mexican race is most cold-tolerant, the Guatemalan race is intermediate in cold-tolerance, and the West Indian race is the most tropically adapted. Exposure to sub-zero temperature (-2.8°C or below) for 14 consecutive nights caused considerable damage or killed the mature trees of 'Zutano' (Mexican race) while the damage was less severe in 'Haas' trees (Whitney and Arpaia, 1991). Fruits are more sensitive to freezing temperature than leaves and may drop down (Whiley and Schaffer, 1994). Lethal freezing temperature varied from the 5.1° to –9.3°C, depending on time of the year and location (McKellar *et al.*, 1992). Prolonged exposure to non-freezing low temperature may cause physiological damage to avocado leaves and the leaves may suffer from photo inhibition with corresponding photo oxidation of the chlorophyll (Whiley and Schaffer, 1994). The range of mean maximum and minimum temperatures in the Nilgiri hills in Tamil Nadu, India, where, avocado is grown is 27.2°C to 33.9°C and 14.9°C to 22.2°C, respectively.

The photosynthetic characteristics of container grown 'Fuerte' trees exhibited that the optimum temperature for net CO_2 assimilation was between 28 and 31°C and the assimilation rate was 33 per cent of maximum at temperature below 15°C and above 40°C (Scholefield *et al.*, 1980). Root dry matter production was highest between 18° and 25°C and it was reduced at 32°C and at 13°C (Whiley *et al.*, 1990). High day temperature followed by cool nights led to the greatest accumulation of starch (Kaiser and Wolstenholme, 1993). Low temperature stimulates flower initiation in avocado. Container-grown 'Fuerte' produced flowers within 25 to 30 weeks when held at 20°C day and 15°C or lower night temperatures, but flowering did not occur at 25°C/20°C or 30°C/10o or 25°C (day/night) temperatures (Buttrose and Alexander, 1978). It was opined that the absence of temperatures > 20°C rather than low temperature was qualitatively important in promoting floral initiation. On the contrary, flower initiation in Mexican and Guatemalan race cultivars grown in highland tropical Australia, occurred when mean maximum and minimum temperatures varied from 26° to 29°C and 15° to 17°C respectively (Sedgley *et al.*, 1985). Results of the investigation on environmental adaptability in 4 cultivars showed that avocado could not tolerate temperature below 0°C (Azzena and Cossu, 2000).

The dichogamy cycle in avocado was also influenced by temperature. Bringhurst (1951) compared between container-grown type. The flowering trees held in a glasshouse at 33°C day (maximum) and 18°C night (minimum) temperature, and flowering trees in the field where maximum/minimum temperature averaged 19o and 12°C respectively, and observed that the time of opening of female and male flowers was reversed under the cooler field conditions compared with the glasshouse trees. Flowers of 'Fuerte' (a type B cultivar) only opened in the male stage when exposed to maximum day temperatures between 18° and 21°C and minimum night temperatures between 7° and 12°C; however, under the same temperature conditions the dichogamy cycle was completed on type A cultivar 'Hass' (Lesley and Bringhrust, 1951). There is no evidential proof that photoperiod influences floral induction in avocado. The fruits become more elongated with the decreasing mean temperature. However, there are differences in cultivar sensitiveness to this phenomenon (Whiley and Schaffer, 1994). In controlled-temperature studies, pollen germination and pollen tube growth in cvs. Hass and Fuerte were most effective at 25°/20°C, although considerable ovule fertilization in 'Hass' but not in 'Fuerte' occurred at both 17°/12° and 33°/28°C (Sedgley, 1977; Sedgley and Annells, 1981). The fruit production is reported to be maximum in the cool mist-belt zones of the subtropics, where frequent summer rainfall reduces environmental stress on trees (Whiley and Schaffer, 1994). Strong wind may cause damage to the tree as the wood is soft and brittle (Purseglove, 1974); therefore, windbreaks are desirable.

9.0 Cultivation

9.1.0 Planting

The normal planting distance for avocado is 6-12 m depending on vigour of the cultivar and its growth habit; for the cultivars having a spreading type of growth habit (for example, Fuerte) wider spacing is given. If there is any danger of water logging, they should be planted on mounds, as avocado cannot withstand waterlogged condition. They are transplanted with a ball of earth around the roots.

A plant population of 220 plants/ha for the cv. Fuerte and 270 plants/ha for the cvs. Hass and Ettinger appear to be optimum for obtaining higher yield per unit area. The recent trend in avocado planting is high density planting (HDP). The correct choice of cultivar and control of tree size are the prerequisites for obtaining success in HDP. The dwarfing rootstocks exert strong influence in regulating the tree height. Application of paclobutrazol controlled the tree size of avocado cv. Hass grafted on Duke in the HDP (800 trees/ha). The cumulative yield for the first 3 years of cropping was nearly two fold greater in HDP than in standard density planting (400 trees/ha). It was suggested that although the HDP is more expensive to plant and maintain, the point of break even is reached earlier than the standard density planting (Kohne and Kremer-Kohne, 1991).

While reviewing the density of planting of Avocado, Wolstenholme and Sheard (2012) suggested broad generalizations, as many factors affect tree vigour and size. Also, given sufficient time and good management, all avocado trees can eventually become large. Seeing that high vigour and large size are South African features, we will emphasize factors promoting vigour. They further classified selected avocado growing areas for "Subtropical type"avocados (as opposed to the lowland" West Indian" type) into two broad categories, *viz.*:

- ☆ Areas with high tree vigour and large tree size: South Africa (eastern sea-board), Australia (eastern sea-board), Mexico (mid-altitude highland tropics), Guatamela, Colombia, Brazil and New Zealand.

☆ Areas where vigour is reduced and trees are smaller (age basis): California, Israel, Spain, Chile, W. and S. Australia, Peru.

They (Wolstenholme and Sheard, 2012) concluded in seeking the common factors that promote avocado tree vigour, high summer rainfall is probably the most important. This permits root (and therefore shoot) growth, with active roots in the entire moist soil zone throughout the warm growing season. Cool New Zealand therefore has large, vigorous trees – there is sufficient summer rain for this, and the shorter growing season with long days still permits two growth flushes. The second most important factor is probably a mesic, non stressful climate and soil environment. With the other factors in operation to a greater or lesser extent, high density planting (HDP) (500+ tree/ha) will be, to a greater or lesser extent, a horticultural management and economic challenge. It is therefore not surprising that successful HDP orchards are found in semi-arid, or arid, and typically winter rainfall areas. The unique, cool Chilean ultra high-density (>1000/ha) orchards occur in a unique hill slope environment. A highly intensive management package is needed including tree training, pruning, sometimes girdling and intensive use of growth retarding, flower initiating and fruit sizing triazole

9.2.0 Windbreak

As the avocado wood is soft and brittle and liable to wind-damage, windbreaks should be provided. In an orchard of 4-year-old avocado cv. Hass trees (approximately 4.5 m tall), on an east-facing slope, protected by a shelterbelt of *Casuarina cunninghamiana* trees (approximately 9.5 m tall) and planted at 1.5 m apart, it was observed that the average wind reduction efficiency was 72 per cent within the first 45 m leeward of the northern and southern sections of the shelterbelt. The wind reduction decreased towards the middle of the orchards. Daily maximum air temperature and daily maximum and minimum humidity were higher and daily minimum air temperature was lower leeward of the shelterbelt than windward of it. Total monthly evaporation was 3 to 9 per cent less leeward of shelterbelt than windward of it. It was concluded that although shelter-belts effectively reduced wind, their benefits should be balanced against the yield reductions caused by competition effects and the loss of the productive land they occupy (Holmes and Farrell, 1993).

9.3.0 Irrigation

At any point of time, plant water status will reflect the dynamic relationship between the rates of uptake and loss from the soil-plant-atmosphere system. Leaves of avocado develop a waxy cuticle on both adaxial and abaxial surfaces to reduce water loss (Whiley *et al.*, 1988; Blanke and Lovatt, 1993). Avocados are irrigated to improve yield, particularly if there is no rainfall for an extended period. Trees are irrigated for successful cultivation in semi-arid climates such as California, Chile and Israel, while in the humid subtropics of eastern Australia, South Africa and Florida, non-irrigated farming system is practised (Whiley and Schaffer, 1994).

The water requirement of avocados in relation to seasonal demands is yet to be standardized. It has been observed in Israel (Kalmar and Lahav, 1977) and South

Malformation, nutritional deficiencies, mechanical defects, disorders and environmental damage in avocado

(A) malformation of unknown origin; (B) boron deficiency; (C) zinc deficiency; (D) natural variegation; (E) fruit shape; (F) mechanical damage; (G) fruit without pedicel; (H) herbicide damage; (I) ring-neck; (J) impact damage; (K) sunburn damage; (L) hailstone damage; (M–O) overripe fruit; (P) necrotic seed; (Q) lenticel damage; (R) chilling injury.

[Ramírez-Gil, J.G., López, J.H. and Henao-Rojas, J.C. (2020) Agronomy, 10: 8, https://doi.org/10.3390/agronomy10010008].

Flowering and fruiting in avocado

Flowering of avocado

Fruit bearing in avocado

Africa (Rowell, 1979) that during the drying cycles, avocado trees obtain 90 to 95 per cent of their water from a depth of 0 to 60 cm. Therefore, soil monitoring for irrigation scheduling should be located in 0 to 60 cm soil zone (Whiley and Schaffer, 1994). Application of water at 21-day-interval has been reported to be optimum in Israel for higher yield with greater percentage of oil in fruits and advancing the harvest date (Lahav and Kalmar, 1977). Heavy fruit set following high-frequency irrigation (7-day-interval) in a particular year, resulted in a strong alternate cropping pattern in avocado cv. Hass compared to other irrigation regimes which might be attributed to reduced carbohydrate reserve in the following season, thereby limiting yield (Scholefield *et al.*, 1985).

The optimum level of irrigation in avocado has been reported to be at 100 per cent evapotranspiration. Neither the substitution of fertilizer for reduced water use nor a combination of 120 per cent evapotranspiration and high fertilizer application (3.0 lb N and 0.06 lb Zn/tree) was found to maximize returns. The result, however, was inconclusive, as there was considerable variation in yield between years (Takele *et al.*, 1990). Considerable variation in varietal response to water application has also been reported. Trunk cross section in cvs. Ettinger and Hass on West Indian rootstock was greatest at the 100 per cent irrigation rate (8500 and 9500 m^3/ha respectively), but irrigation treatment had no effect on trunk growth in 'Fuerte' grafted on same rootstock. In cv. Ettinger, yields in the exceptionally high yielding years and the overall average yield were significantly higher with 100 per cent irrigation than at other irrigation rates. This difference also occurred in cv. Hass, but was non-significant, while in cv. Fuerte the highest yield was quite low and it was not affected by irrigation treatment. Leaf mineral concentrations were affected by interaction of soil fertility, irrigation and cultivar (Natan *et al.*, 1991). Sprinkler system of irrigation is better than flooding as the latter may promote root rot.

9.4.0 Manuring and Fertilization

Avocado requires adequate fertilization. A crop of 14386 kg/ha removes about 40 kg N, 25 kg P_2O_5, 60 kg K_2O, 11.2 kg CaO and 9.2 kg MgO/ha from the soil (Avilon *et al.*, 1986). Cattle manure alone or combination with lucerne proved more effective to control *Phytophthora cinnamomi* than treatment with metalaxyl, promoted root growth and produced higher yield (Cortes Flores *et al.*, 1993).

Application of nitrogen is very important for the cultivation of avocado. Deficiency of nitrogen caused restricted growth, reduction in leaf size, paleness in leaf colour, early abscission of leaves, smaller and fewer fruits; the affected plants were highly susceptible to frost (Lahav *et al.*, 1985). Nitrogen fertilization significantly increased the yield and foliar sprays of urea were more effective than Ca(NO3)2 added to the soil. Treatment with 250 g N per tree from both sources increased fruit weight and yield (Gazit *et al.*, 1975). In general, young avocado trees should receive N, P_2O_5 and K_2O in the proportion of 1:1:1 and old trees in the proportion of 2:1:2 (Samson, 1980).

Zilakh *et al.* (2000) grew young avocado (cv. Pinkerton on rootstock Schmidt) trees in a coarse sandy soil in Israel and exposed to 15N-labelled KNO3 and (NH4)2SO4 for 10 days during autumn, winter, spring and summer. The uptake efficiency of N-nitrate was significantly higher than N-ammonium in each of the plant organs and during each of the seasonal periods. N absorption efficiency in the whole tree was higher in the summer. During the various seasonal periods, both the N forms were distributed in roots, trunk + branches and leaf organs, in a similar manner. During autumn and winter, N was concentrated in roots and trunk + branches. In the spring, the absorbed N was readily translocated to leaves.

Application of P_2O_5 and N in autumn to an avocado orchard on a light soil increased the yield of cv. Hass by 65 per cent, although no significant effect on yield was observed in cv. Ettinger and the recovery from heat stress as more rapid in treated trees (Reuveni *et al.*, 1990). Nitrogen application alone (limestone ammonium nitrate, 28 per cent N) considerably increased the percentage of fruits with physiological disorders over that in control and increased foliar N content more than the recommended 2 per cent, but not when combined with K (K_2SO_4, 40 per cent K) and Mg ($MgSO_4$, 10.5 per cent Mg). Each application consisted of 2 kg/tree of the fertilizers and used in 5 split doses between September and February (Kremer-Kohne *et al.*, 1993). Nitrogen treatment influenced both nutrient and fatty acid composition of fruits (Arpaia and Eaks, 1990).

The occurrence of Zn deficiency in avocado has been widespread. Reuhle (1940) noted die back of twig and formation of small fruits with dead spots on the skin due to Zn deficiency. The threshold value for Zn deficiency in the leaves was about 10 ppm (Kadman and Cohen, 1977). Koen and Smart (1973) suggested spraying of zinc oxide at 150 g per 100 litres of water when avocado seedlings were grown in nursery soil with a pH above 6.0.

Boron uptake takes place through the underside of the leaf. In general, B uptake was limited in young trees of cv. Hass. The greatest increase in leaf B concentration occurred after spraying with Solubor (soluble boron) at 3 g/dm^3 which caused 3 times increase in fruit set as compared to control (Robbertse *et al.*, 1992). Leaf K and B concentration rose with increasing K and B levels in the nutrient solution; however, higher K levels in the solution resulted in lower leaf B concentrations and vice-versa (Jaime *et al.*, 1992). Foliar spray of urea (1 per cent) + $FeSO_4$ (100 ppm) + $ZnSO_4$ (100 ppm) was the most effective treatment in increasing seedling growth and leaf mineral contents, particularly N, K, Mg, Fe and Zn (Hassan and Atawia, 1995). The physiological response that occurred in the roots during Fe-deficiency influenced the uptake of other metallic micronutrients (Manthey and Crowley, 1997).

9.4.1 Tissue Analyses and Nutrient Requirement

Nitrogen concentrations were significantly lower in N-deficient trees than in healthy trees. Foliar concentrations of Ca, Mg, Mn and Cu were higher in healthy leaves than in N-deficient leaves; concentration of K was higher in N-deficient leaves. In Israel concentrations of most nutrients stabilized when leaves were sampled in autumn (Lahav *et al.*, 1990). The best sample under South African conditions was 6 to 8 months' old leaf collected from the base of a secondary branch in the middle of the spring flush (Koen and Plessis, 1991). The critical leaf nutrient levels given in Table 4 can, however, be used to regulate nutrient application.

Table 4: Critical Leaf Nutrient Levels for Avocado – Two Months Prior to Flowering

Element		*Ranges*	
	Deficient	*Adequate*	*Excess*
Nitrogen	1.6 per cent	1.6-2.0 per cent	2.0 per cent
Phosphorus	0.5 per cent	0.08-0.25 per cent	0.3 per cent
Potassium	0.35 per cent	0.75-2.0 per cent	3.0 per cent
Calcium	0.50 per cent	1.00-3.0 per cent	4.0 per cent
Magnesium	0.15 per cent	0.25-0.80 per cent	1.0 per cent
Sulphur	0.05 per cent	0.20-0.60 per cent	1.0 per cent
Boron	10-20 ppm	50-100 ppm	100-250 ppm
Iron	20-40 ppm	50-200 ppm	-
Manganese	10-15 ppm	30-500 ppm	1000 ppm
Zinc	10-20 ppm	30-150 ppm	300 ppm
Copper	2-3 ppm	5-15 ppm	25 ppm

9.5.0 Pruning and Girdling

Avocado trees are pruned sparingly, mainly by heading back the central shoot in upright growing cultivars, such as Pollock, to develop a spreading habit. On the other hand, in spreading cultivars like Fuerte, branches are thinned and shortened. Old, tall trees may be headed back to facilitate harvesting and other orchard operations. Generally, heavy pruning results in excessive vegetative growth and, consequently, the yield is reduced.

Selective pruning *i.e.* manual removal of vegetative flushes emerging from the centre of the flower panicle weekly for 6 weeks did not increase the yield significantly, but fruit size was enhanced (Bower and Cutting, 1992). Tipping and pruning were successful in limiting the length of the primary shoot axis and increasing axillary branching, without influencing total fruit set (Throp and Sedgley, 1993). Removal of a strip of bark of 5 mm wide (girdling) in avocado cv. Hass on Duke 7 rootstock resulted into 60 per cent increase in yield. Although girdling reduced the percentage of export quality fruits by 6 per cent (due to a greater number of small fruits) this was more than compensated by the increased yield (Kohne, 1992).

Planning of new orchards and tree shaping in the early years of avocado production in South Africa were discussed by Stassen (1999a). Shaping the tree and establishing a branch hierarchy at an early stage could be done with relatively little input. This reduced the development of skewed leaders which reduce light interception. The planting density determined the number of leaders, which should be trained upright to achieve maximum light penetration. The trees should be trained to a pyramid and the tops must not be wider than the base and must not be too dense. A strategy for pruning neglected avocado orchards in South Africa was also reported by Stassen (1999b). Pruning was aimed at correcting overcrowding of trees. Heavy pruning should be done immediately after harvest, followed by pruning vigorous shoots and topping the head of the tree in spring, with summer pruning confined to lightly skimming the shoot tips.

9.6.0 Weed Control and Cover Crop

Avocado orchard is clean cultivated, but deep cultivation should be avoided because of surface roots. Weeds should be controlled and mulch is useful in controlling weeds and conserving moisture. Annual weeds may be controlled by herbicides, such as monouron and simazine. Intercropping with legumes or shallow-rooted crops may be done in young orchards. However, the practice in California was to plant at 5 x 5 m (400 trees/ha) and gradually thining to an ultimate spacing of 10 x 7 m (140 trees/ha) (Storey, 1972).

In an evaluation of three permanent ground covers *viz. Arachis glabrata, Polygonum capitatum* and *Lippia canescens* (*Phyla nodiflora*) in an orchard of avocado cv. Hass on Duke 7, it was observed that the species with greatest potential appeared to be *L. canescens*, owing to its relatively rapid growth rate; however, it competed with the trees for N. In contrast, a cover crop of *A. glabrata* increased the N content of the avocado leaves. The cost of initial establishment was high for all 3 cover species but weed control costs and labour requirements decreased from the second year onwards. Neither incidence of *Phytophthora cinnamomi* in the soil was found before ground cover planting nor one year later (Mans and Hattingh, 1992).

10.0 Flowering, Floral Biology, Fruit Growth and Development

Only a few phenological models have been formulated for rain-fed avocados in warm sub humid environments, and none of them have suggested crop management strategies for improving the trees performance in such conditions. Hence da Silva *et al.* (2017) assessed the occurrence and intensity of major phenological events were in 'Hass' and 'Margarida' avocado trees, in order to properly schedule the main crop practices for optimum management of rain-fed orchards in warm sub-humid growing environments. Shoot growth, root density, number of open flowers, fruit set and fruit dropping rate were evaluated. For both cultivars, two shoot flushes were observed in September-October (spring) and January-February (summer), while a maximum feeder root growth took place in late August, coinciding with flowering. Two other minor root growth flushes occurred after the onset of summer rains. Both cultivars had only one flowering period, followed by an intense fruitlet dropping. A second minor fruit dropping wave occurred in February (midsummer), for both cultivars. By determining the extent of each phenological phase in both avocado cultivars, it is possible to schedule applications of fertilizer, mulch and soil amendments, during the root growth flushes, as well as foliar sprays during shoot flushes, in order to maintain an adequate nutritional and inter cultivation operations (da Silva *et al.*, 2017). Alcaraz *et al.*(2013) standardized a precise description of the different growth stages of this crop is proposed. The BBCH (Biologische Bundesanstalt, Bundessortenamt und Chemische Industrie) code has been used to define the most important phenological growth stages from vegetative bud dormancy to fruit harvest. The code will be an important tool to assist the development and implementation of agronomic management protocols and to standardize observations made in different environments and/or under different experimental conditions in this crop (Table 5).

Table 5: Description of the Phenological Growth Stages of Avocado (*Persea americana* Mill.) According to the BBCH Scale*

BBCH code Description (mesostage 1)	
Principal growth stage	
0	Vegetative bud development
010	Vegetative buds dormant
011	Beginning of bud swell
013	End of bud swell
017	Beginning of bud break
019	End of bud break
Principal growth stage 1: primary leaf expansion	
110	First leaves separating
111	First leaf unfolded
112	More leaves unfolded. First leaf at 20 per cent of its full size
113	More leaves unfolded. First leaf at 30 per cent of its full size
11.	Stages continue until.
119	All leaves unfolded and fully expanded
Principal growth stage 2: axillary (sylleptic) shoot formation	
210	No sylleptic shoots visible
211	First sylleptic shoot visible
212	Two sylleptic shoots visible
213	Three sylleptic shoots visible
21.	Stages continue until.
219	Nine or more sylleptic shoots visible
Principal growth stage 3: primary shoot extension	
310	Beginning of shoot extension
311	10 per cent of final shoot length
312	20 per cent of final shoot length
31.	Stages continue until.
319	90 per cent or more of final shoot length
Principal growth stage 5: reproductive development	
510	Reproductive buds dormant
511	Beginning of reproductive bud swell
512	End of reproductive bud swell
513	Reproductive bud break
515	Inflorescences 50 per cent of final length
517	Inflorescences 70 per cent of final length
519	End of inflorescence extension
Principal growth stage 6: flowering	
610	First flowers open
611	10 per cent of flowers opened

612	20 per cent of flowers opened 61. Stages continue until.
619	90 per cent or more of flowers opened
Principal growth stage 7: fruit development	
710	No ovary growth visible
711	Initial ovary growth
712	First fruitlet abscission
715	50 per cent of final fruit size
71.	Stages continue until.
719	90 per cent or more of final fruit size

* Alcaraz *et al.* (2013).

10.1.0 Flowering

Seedling avocados start bearing in 5-6 years, whereas vegetatively propagated plants usually bear earlier, but it is usual to remove the fruits until the plants are 3-4 years old.

The flowering in avocado has been reviewed extensively by Davenport (1986). Vegetative growth in avocado branches occurs in 2 to 3 distinctive flushes which do not necessarily include the whole tree (Davenport, 1982; Scholefield *et al.*, 1985). However, varietal difference does occur. Reproductive growth occurs following a rest period only once per year in almost all cultivars (Davenport, 1986). The determination of inflorescence morphology is at least in part influenced by environmental conditions (Salomon, 1984). In general, the inflorescence develops laterally with no new leaves evident except the emerging apical bud from the apex (Davenport, 1986). In cv. Topa Topa, however, leafy inflorescences emerge out terminally (Schroeder, 1944).

The last vegetative growth prior to inflorescence development takes place during late September to early October on Florida-grown avocado cultivars and in February on Fuerte, grown in Australia. Inflorescence development becomes apparent during early fall in many cultivars of the Mexican race and Mexican Guatemalan hybrids (Davenport, 1986), whereas the cultivars of West Indian and Guatemalan races and hybrids of West Indian and Guatemalan races produce flower buds from late

November to late January in the northern hemisphere (Davenport, 1982). Floral bud development in Fuerte is reported to begin in April or May in the southern hemisphere. The inflorescence requires 2 months in late-initiating cultivars to about 3.5 months in the earliest-initiating ones for complete development (Davenport, 1982). Full bloom occurs in March and April in the major growing belts throughout the northern hemisphere and in August to October in southern latitudes. Inflorescence and flower development of Hass avocado during 'on' and 'off' years in USA was studied by Salazar-Garcia *et al.* (1998). They observed that the transition from vegetative to reproductive condition occurred from the end of July through August (end of shoot expansion). During this transition, the primary axis meristem changed shape from convex, to flat, to convex. These events were followed by the initiation of additional bracts and their associated secondary axis inflorescence meristems. A period of dormancy was believed not as a prerequisite for inflorescence development. Continued production of secondary axis inflorescence meristems was observed from August to October, followed by anthesis 7 months later. Inflorescence development was correlated with minimum temperature £ 15°C, whereas yield had little effect on the timing of developmental events of individual inflorescence buds. However, the high yield of the 'on' year reduced inflorescence number and increased the number of vegetative shoots. No determinate inflorescences (develop into a terminal flower) were, however, found to develop during the 'on' year. For the 'off' year, 3 and 42 per cent of shoots produced determinate and indeterminate inflorescence (a bud forms on the axis which continues the growth of the shoot) respectively.

Davenport (1982) described the various stages of morphological development of avocado panicles (Table 6).The total time required for panicle development and the rate of development varied among cultivars. Temperature plays an important role in panicle development.

Table 6: Various Stages in Avocado Reproductive Development*

Stage	*Description*
0	Bud in rest, bracts closed with no sign of growth
1	First indication of bud development, bracts show signs of opening
2	Obvious bud growth, bract open showing development in inflorescence bud
3	Bud assumes spherical shape as inflorescence begins enlarging
4	Bud becomes angular, continued enlargement of inflorescence
5	First observable elongation of primary peduncles
6	Continued elongation of primary peduncles. First observable elongation of secondary peduncles
7	Continued elongation of inflorescence and development of flowers
8	First opening of flowers
9	Maximum flower opening

*Davenport (1982).

Flower induction takes place sometime during the period between the last autumn vegetative flush and the time when initiation of inflorescence buds is apparent (Davenport, 1986).Day length and temperature appeared to be involved in the floral induction process. Low day (20°C) and night (5-15°C) temperatures promoted floral induction. Day temperature of 25° to 30°C completely inhibited floral induction. Even a brief exposure (1 hour to 30°C) each day was sufficient to inhibit floral induction. The effect of night temperature was less pronounced. Day length of 9 hour reduced the time to flower compared with 15 hour day length, but the number of flowers produced on plants grown under short day was less than on plants under long day (Buttrose and Alexander, 1978). The inflorescence bud growth was arrested when some Mexican-Guatemalan hybrids were subjected to 33o/23°C day/night temperature (Sedgley *et al.*, 1985). The amount of flower production may be temperature-mediated once induction events are set in motion by short day length (Davenport, 1986).

Inflorescence initiation of 'Hass' avocado in Southern California begins in July to August and anthesis occurs in March to April (Salazar-Garcia *et al.*, 1998). Fruits can be harvested 11 months after flowering (February the following month) (Lovatt, 1990). Mature fruits of 'Hass' avocado is frequently stored on the tree for harvest anytime through August. Excessive fruit load or late harvest contributes to alternate bearing in trees which bear a high yield ('on' year) followed by a bloom of low intensity ('off' bloom) and resulting in low yield ('off' year) (Monselise and Goldschmdt, 1982). In 'Hass' avocado, spring-flush shoots borne on indeterminate inflorescences that set fruit do not flush in summer or fall, nor do they produce inflorescences in the following spring (Salazar-Garcia *et al.*, 1998). Indeterminate inflorescences account for 90 per cent of the total inflorescences produced during spring bloom (Salazar-Garcia and Lovatt, 1998). Thus, when trees are carrying an 'on' year crop, the number of shoots that bear inflorescences is reduced significantly. This results in an 'off' bloom that sets an 'off' year crop with a concomitant increase in both summer and fall vegetative shoot flushes, potential flowering points for the next year. Thus, the cycle of alternate bearing is perpetuated.

Some of the avocado cultivars (*e.g.* Fuerte, Nabal, and Hass) exhibited variation in yields as compared to others (*e.g.* Ettinger, Anaheim) which are partly controlled by genetic factors. An occasional heavy crop causes more carbohydrate depletion resulting in very low stored energy available for flowering and fruiting in the following year. Moreover, the photosynthetic efficiency of avocado leaves is low. Carbohydrate plays an important role in alternate bearing of avocado. The 'off' year trees are reported to have much higher mean starch content than the 'on' year trees in May and July. Leaf chlorophyll content shows a shift in the chlorophyll a:b ratio from 1:1 in 'on' year trees to 3:1 in 'off' year trees (Walt *et al.*, 1993). Pre flowering starch accumulation in shoots were directly correlated with the following season's yield (Whiley *et al.*, 1996 b). Seasonal fluctuations in nonstructural carbohydrates (starch and soluble sugars) were studied in Hass avocado trees on Duke 7 rootstock over a 2-year period in southern California by Liu Xuan *et al.* (1999). On a day weight basis, total soluble sugar (TSS) concentrations ranged from 33.0 to 236.0 mg/g dry weight and were high compared to starch concentration (2.0-109.0 mg/g dry weight) in all measured organs (stems, leaves, trunks and roots). Complementary cycling between stem TSS and starch suggest that a conversion of starch to sugars occurs to support vegetative growth and flowering, while sugars produced photosynthetically may be allocated directly to support flowering and fruit production.

Salazar-Garcia and Lovatt (1998) observed low fruit set potential (0.05 per cent) in the indeterminate inflorescence of 'Hass' avocado. It has been proposed that competition for water, nutrients and photosynthate occurs between the apical vegetative shoot, which normally develops during the period of anthesis and fruit set and setting fruit on indeterminate inflorescences (Whiley, 1990; Cutting and Bower, 1990; Bower and Cutting, 1992).

Recent research provided evidence that GA_3 altered inflorescence phenology and morphology in predictable ways (Salazar-Garcia and Lovatt, 1998). GA_3 applied in November, before inflorescence buds initiated a full complement of secondary inflorescence axes, reduced flowering intensity due to the production of partially formed inflorescences, each bearing fewer flowers, but accelerated inflorescences and vegetative shoot development. The results suggested the possibility of using a November application of GA3 preceding an 'on' bloom to even out alternate bearing by reducing the formation of secondary axes of inflorescence buds. In young trees, injections of 25 or 50 mg GA_3 per tree at bud swelling, prior to bud break caused precocious inflorescence development and anthesis 3 weeks earlier. GA_3 treatment did not affect the number of vegetative shoots produced but increased the number of indeterminate inflorescences at the expense of determinate ones (Salazar-Garcia and Lovatt, 1999). Salazar-Garcia and Lovatt (2000) further studied the effect of canopy spray of GA3 on 'Hass' avocado. In this experiment, GA_3 (25 or 100 mg/l) was applied to separate sets of trees in September (early stage of inflorescence initiation), November (early stage of inflorescence development), January (initial development of the perianth on terminal flowers), March (cauliflower stage of inflorescence development, only 25 mg/l) or monthly from September to January (at 25 mg/l). They found that treatment with GA_3 at 100 mg/l applied in September reduced inflorescence number in both years, but not yield. GA_3 (25 or 100 mg/l) applied in November before the 'on' bloom reduced inflorescence number with a concomitant increase in vegetative shoot number and 47 per cent yield reduction compared to control trees. Treatment with GA_3 at 25 mg/l had higher cumulative yields when applied in September or March than in November or January (Table 7). They concluded that this treatment might provide avocado growers with a tool to break the alternate bearing cycle by reducing yield in an expected 'on' year and to achieve a higher yield in the following year.

Table 7: Effect of GA_3 Applied in the Months (September to March) before both the 'off' and 'on' Bloom (March to April) Years on 2-years Cumulative Yield and Alternate Bearing of 'Hass' Avocado Trees*

Application Date	*GA_3 Concentration (mg/l)*	*Cumulative Yield (kg/tree)*	*ABI (per cent)[z]*
Control	0	98 aby	56.8 a
15 September	25	121 a	76.0 a
	100	90 ab	68.1 a
15 November	25	77 b	27.8 a
	100	86 b	65.0 a
15 January	25	87 b	44.0 a
	100	100 ab	63.8 a
Monthly	25	79 a	66.4 a
15 March	25	123 a	40.7 a

z: ABI-alternate bearing index = (Difference between yield of year 1 and year 2)/(sum of yields of year 1 and year 2) x 100; y: Mean separation in columns by Duncan's multiple range test, P =0.05.

*Salazar-Garcia and Lovatt (2000).

Alternate and irregular bearing (AB) are embedded in the avocado genome through evolutionary adaptation to montane cloud forest environments in Mexico and Central America Current cultivars and rootstocks provide only partial assurance of regular bearing in most environments. Occasional severe climatic events can override the best efforts of management and cause crop failure. Horticulturally, the key is to attempt the maintenance of the ideal vegetative: reproductive balance as determined for the particular environmental conditions and management philosophy. Inevitably, this will swing towards excessive reproduction in "on" crop years, and excessive vegetative growth in "off" crop years - to the detriment of shoot and root flushes in the former, and cropping in the latter. Choice of cultivar, rootstock, environment and management can all aggravate AB; late harvest also. Nevertheless, provided P.c. is under control, there is much that can be done, in a "package deal" that includes orchard intensification through closer spacing, pruning, girdling, provision for cross-pollination, and use of uniconazole, to reduce the intensity of AB. A Chilean management philosophy based on the above, with recycling of "modules" (branches) in an orderly fashion, is an example of a successful adaptive strategy that maintains high yield in a Mediterranean climate (Wolstenholme, 2010).

10.2.0 Floral Biology

The avocado flower is perfect, bearing functional male and female parts, with minor variations among cultivars (*e.g.* presence or absence of trichomes on the pistil and stamens). The open flower is approximately 1 cm in width and 6 to 7 mm in length from the base of the ovary to the stigma. The inflorescence is a compound panicle of raceme. They are protogynous and exhibit a unique mechanism of synchronous dichogamy for alteration of sexes. Each avocado flower opens twice; the first time (stage I) as a functional female and the second time as functional male (stage II). The openings, separated by one overnight period, last up to half-a-day and occur either in the morning or afternoon hours. Flower opening is synchronous, *i.e.* flowers open and close almost uniformly on a tree (Davenport, 1986). The flowers remain closed in between the two stages of opening. The dichogamy is protogynous, *i.e.* the pistil ripens earlier than the stamen. Based on the pattern of opening of flower the avocado cultivars are classified into two types : A and B. In type A, the flower opens first as functional female during the morning hours, closes near midday and reopens as functional male in the afternoon of the following day. Type B cultivars exhibit the first opening of female flowers in the afternoon and the second opening (functional male) occurs in the morning of the following day. Owing to the synchronously dichogamous nature of its flower and existence of type A and B cultivars, the avocado affords interesting opportunities for cross-pollination. Thus, every morning type A pistils can be fertilized by type B pollen and in the afternoon the type B pistils can be crossed by the A pollen. Samson (1980) illustrated diagrammatically the flower opening behaviour of avocado. The synchronously dichogamous nature is extremely sensitive to environmental conditions. The opening of flowers is irregular and retarded during cool, cloudy days. The regularity, continuity and sequence of bloom are disturbed by fog at night and rainy weather. The day-night cycles are important in regulating the floral anthesis (Davenport, 1986).

The avocado cvs. Hass, Nowels and Rincon (all type A), and Bacon, Fuerte, Orotava and Zutano (all type B) showed many variations in the typical flowering cycles, especially in April and early May. The 2-flower-type periods frequently overlapped, even at the end of flowering season when flowering cycles became more regular (Continella *et al.*, 1992).

Abnormalities in avocado flower structures are common such as staminoidy, *i.e.* the conversion of other flower parts to stamens and pistilloidy *i.e.* conversion of other flower parts to pistills, fused stamens, irregular nectaries, modified perianth parts, naked ovules, double pistils with naked ovules, modified perianth number, double embryo sacs, embryo sac degeneration, lack of the embryo sac and compartmentalization, *i.e.* development of voids in the tissue (Kadman *et al.*, 1974; Tomer *et al.*, 1976; Tomer and Gottereich, 1978; Sedgley, 1980). A study of flowers from high and low yielding avocado cv. Fuerte trees revealed that about 50 per cent of the 700 pollen sacs studied contained pollen grains of abnormal shape and diameter, low staining ability and weak resistance to rupturing (Steyn, 1993).

10.3.0 Pollination

According to Gad Ish Am (2005) the avocado cultivars are divided into two complementary flowering groups, according to their daily flowering sequence. 'A type' cultivars, in warm weather, open female-stage flowers from the morning till noon, and the next day reopen these same flowers in the male stage from noon throughout the afternoon (a flower's opening cycle of about 34 hr). 'B type' cultivars, on the other hand, open female-stage flowers from noon throughout the afternoon, and reopen them in the male stage on the following morning till early afternoon (a flower's opening cycle of about 26 hr). Thus, a daily effective overlap occurs between A-type female-stage flowers and B-type pollen releasing male stage flowers during the morning, and vice versa at the afternoon. Furthermore, in most avocado cultivars a daily regular self-overlap phase also exists, of female-stage flowers and pollen releasing male-stage flowers of the same tree, which takes place for a period of 1-3 hr. In cool weather a delay of both female and male opening times occurs, which may result in a complete reversal as to the part of the day female and male-stage flowers are open. At both female and male-stage flower openings nectar is secreted, although by different sets of nectaries. (Gad Ish-Am, 2005).

According to a detailed review by Gad Ish-Am (2005), Self-pollination occurs when pollen grains reach the stigma within the same flower. A clear inability of the male-stage flower to achieve fertilization was demonstrated under Mediterranean-type weather conditions, where mostly Mexican and Guatemalan cultivars, as well as their hybrids, are grown. On the other hand in Florida, where the local avocados are of the West Indian cultivars and their hybrids, highly effective male-flower fertilization was found. Close-pollination, between neighboring flowers within a tree (or a cultivar), occurs regularly during the daily self-overlap period of male and female-stage flowers within the tree. An adult avocado tree, in a normal season, may bear about one million flowers, but only several hundred fruits. Namely, only 0.05

per cent of the flowers (one of 2,000) produce fruit. This effect, which is common in tropical trees, is termed: "mass flowering".

The flowers are entomophilous and dependent on honey bees for pollination. During the early part of its flowering period, the avocado competed with citrus and cruciferous flowers, which are more attractive to the bees. Therefore, the early flowering avocado cultivars may only be visited on 20 to 40 per cent of days during their flowering period, which limits fruit set, while later flowering cultivars are well pollinated and have good rates of fruit set (Ish-Am and Eisikowitch, 1992). The higher the fertility rate and the larger the pollen grain, the better is the fruit yield of the cultivars (Inoue *et al.*, 1992). As a result of synchronous dichogamy, cross-pollination has the best opportunity for fertilizing the egg. Yet, it is plausible that if cross-pollination does not occur, the avocado has evolved a fail-safe system of self-pollination dependent upon environmental conditions (Davenport, 1986). The relative humidity may play some role in prolonging the receptivity of the pistil through the second flower opening of some cultivars. Increased relative humidity helped to sustain the viability of pollen (Loupassaki *et al.*, 1997).

Gad Ish Am (2005) summarized in his detailed review that pollination may be a limiting factor for avocado productivity where:

(a) Self-fertilization within the male flower does not happen;

(b) Pollinator activity is too low (due to a low level of available pollinator population, or to the presence of more attractive competing blooms), and hence total-pollination rates are too low;

(c) Cross-pollination rate is too low, owing to a low mobility of the pollinator insect, or to a large distance from the pollinizer tree;

(d) A significant selection against the self initial-fruits occurs, mainly under stress conditions of a windy, hot and dry spell

10.4.0 Fruit Set

Even though, avocado produces hundreds of flowers in lateral inflorescence yet only a paltry 1 per cent of these flowers set fruits and attain maturity. Two types of fruit-setting habit are common in avocado (Davenport, 1982). The type-I habit is characterized by heavy initial set of fruitlets followed by gradual loss of most of the fruitlets throughout the following months until they reach horticultural maturity. A type-II fruit-setting habit is one in which relatively less fruitlets are initially set, but these generally remain on the tree till maturity. Some cultivars show intermediate fruit-setting habit. Poor yield in avocado may be ascribed to ovule sterility, pollination problems, degeneration of young fruits and defective seed development.

Main period of fruit set in cv. Pinkerton in South Africa was reported to be between August and October and a higher percentage of fruit drop could be expected with fruit set during the second half of the season (Sippel *et al.*, 1992). The dynamics of fruit drop and CO_2 assimilation of fruiting spring shoots of cv. Hass trees were studied for the first 76 days after bud break. Individual leaves showed a net gain in CO_2 assimilation after reaching 80 per cent of full expansion (when about 25 days' old). There was a net CO_2 assimilation loss from developing shoots for the first 27 days after bud break during which 60 per cent of the initial fruits which set dropped. A further 22 per cent of the initial fruit set was lost during the following 15 days when the youngest leaves on the shoots were still strong sinks. Fruit loss stabilized when the shoots reached their maximum CO_2 assimilation capacity (source strength) 60 days after bud break (Whiley, 1990).

Fruit drop can be reduced to some extent by girdling at blossom time and by systematically rubbing off all vegetative growth of branches between March and June. It is recommended that avocado orchards be interplanted with complementary type A and type B cultivars to facilitate cross pollination by insects for obtaining maximum yields.

10.5.0 Fruit Growth and Development

The growth of avocado fruit follows the single sigmoid curve (Valmayor, 1964; Robertson, 1971). The initial phase of fruit growth, irrespective of early or late maturing cultivars, is characterized by rapid cell division (Barmore, 1977). Variations in fruit size of different cultivars maturing at the same time are primarily due to variation in cell division during the initial 6 months of development (Valmayor, 1964). The interesting part of fruit development is that cell division in mesocarp is not restricted to the initial period, rather continues till maturity (van den Dool and Wolstenholme, 1983). Experimental results indicated that avocado fruits contribute to their own carbon requirement by means of CO_2 assimilated in the light (Whiley *et al.*, 1992). The date of fruit set has a marked effect on final fruit size. Fruits of avocado cv. Pinkerton which set late (mid-September to mid-October) had a much faster growth rate than early-set (July to August) fruits, and on an average had the potential to be 40 mm longer at the end of the season (Sippel *et al.*, 1993). Variation in fruit size between locality has been reported. Fruits of avocado cv. Hass grown in warmer climate are smaller than the fruits from the cool mist-belt region (Cutting, 1993).

Several factors, as stated by Bower and Cutting (1988), influence the pattern of fruit growth which are as follows:

10.5.1 Environmental Factors

Photosynthesis is an important factor in Avocado productivity. A detailed review on photosynthesis was presented by Mandemaker (2007). He opines that to increase avocado yield, the maintenance of high photosynthetic capacity for the production of flowering wood each year is required. A failure to achieve good starch reserves going into flowering is considered to contribute to maintaining an alternate bearing cycle. The accumulation of starch is likely to be linked to photosynthetic activity at all times of the year, most especially during winter when low temperatures could be reducing photosynthesis to very low rates. Under the cool temperatures of a New Zealand winter the photosynthetic apparatus of avocado leaves may experience damage from chilling. A greater understanding of the underlying mechanisms of avocado ecophysiology will allow tree management practices to be developed that increase the accumulation of carbohydrate from photosynthesis. Mandemaker

(2007) concludes that potential cause of low production is poor photosynthetic capability during winter that commonly manifests as damage to the photosynthetic apparatus visible as leaf-yellowing. Leaf -yellowing appears to follow periods of cold temperature in winter and is perceived to be an indication of a decline in leaf chlorophyll content, resulting in a large reduction in net photosynthetic rate and consequential decline in starch production. A lack of sufficient starch accumulation during winter may be important in determining a poor fruit set in an 'off' flowering year where a heavy crop of fruit grows in size and increases in oil content over the winter months. Low temperature damage to the photosynthetic apparatus could be an important problem as the damage would restrict leaf longevity and reduce overall plant resources for the coming spring flowering. Schaffer and Whiley (2003), in an earlier review, suggested manipulating and training trees to optimize photosynthesis within the canopy as an important aspect of orchard management. Knowledge of the impact of environmental factors, such as light, temperature, humidity, flooding, salinity and elevated atmospheric CO_2 concentrations on photosynthesis of avocado trees provides information that can be applied to canopy management and plant selection for specific environments.

10.5.1.1 Temperature

Occurrence of frost caused considerable damage to the crop. High temperature forced the tree into vegetative development to the detriment of fruit set and yield (Sedgley *et al.*, 1985) and temperature above 30°C was thought to be damaging to roots (Whitmore, 1986). High temperature also tended to reduce the leaf area rather than leaf number, increasing competition between vegetative growth and fruit set (Lahav and Trochoulias, 1982). Optimum photosynthesis occurred between 19° and 24°C (Bower *et al.*, 1978) and temperature above 25°C probably decreased dry matter production and yield due to reduced photosynthesis (Bower, 1978) or increased photo-respiration (Zelitch, 1971) as fruit growth and development make heavy demands on the carbohydrate reserves of the tree (Wolstenholme, 1985). However, the tree response to temperature changes with age, physiological state, crop size, nutrition and stress status (Bower and Cutting, 1988).

10.5.1.2 Other Climatic Factors

Apart from temperature, rainfall and humidity also play an important role in fruit development. Excess water may adversely affect the yield and quality of fruits (Bower and Cutting, 1987) due to reduction in available root O_2 and promoting root rot (Zentmyer, 1984). High precipitation at certain growth periods hinders fruit set and causes fruit drop (Baxter, 1981). On the other hand, water stress during fruit development causes severe fruit drop and reduces the fruit size. High humidity during and after fruit set is crucial as a sharp decrease in humidity after fruit set is known to cause fruit drop (Bower and Cutting, 1988).

10.5.2 Endogenous Factors

10.5.2.1 Hormonal

Avocado seeds are rich source of auxin activity (Gazit and Blumenfeld, 1972). The IAA concentration reached a peak in avocado cv. Fuerte fruits shortly after set.

Levels declined rapidly thereafter and could not be detected 5 months after fruit set. At all stages of fruit development the mesocarp contained lower levels of IAA than the seed and testa (Cutting *et al.*, 1985). There is strong evidence that ethylene is involved in the abscission of young avocado fruitlets (Davenport and Manners, 1982) even after ovule fertilization (Sedgley, 1980). High levels of cytokinins in both cotyledons and testa of avocado had been reported, which decreased with development (Blumen-feld and Gazit, 1970). The high cytokinin levels detected in young fruitlets may actively assist in increasing the sink strength of the fruit and thereby promote fruit growth (Bower and Cutting, 1988). There were both quantitative and qualitative differences in the cytokinin contents of fruits in cv. Hass from the warm (small fruit) and cool (large fruit) areas. Seed and testa zeatin and dihydrozeatin concentrations were greater in the fruits from the cooler area than warmer area (Cutting, 1993). The ABA levels in developing avocado mesocarp rose dramatically with advancement of age and reached a maximum at maturity (Cutting, 1984; Cutting *et al.*, 1985).

Abscisic acid accumulation down-regulates mesocarp 3-hydroxy-3-methylglutaryl coenzyme–A reductase activity and fruit growth, and that *in situ* cytokinin biosynthesis modulates these effects during phase I of fruit ontogeny, whereas both cytokinins and sterols seem to perform this function during the later phases (Cowan *et al.*, 1997).

The physiological basis for production of phenotypically small fruits of the avocado cultivar Hass, when grown in a warm, low rainfall, subtropical climate was investigated by Bertling *et al.* (1998). Rapidly growing fruits on mature trees were injected through the pedicel with mevastatin, an inhibitor of the key enzyme of isoprenoid biosynthesis, 3-hydroxy-3-methylglutaryl-CoA reductase (HMGR). Growth was reduced significantly. Also mevastatin inhibited HMGR activity in the mesocarp.

Injection of abscisic acid (ABA) affected fruit growth and development similarly, and also inhibited HMGR activity. Injection of the cytokinin, dimethylallylaminopurine (iP) stimulated growth slightly but had no effect on HMGR activity. However, when iP was co-injected with either mevastatin or ABA, the inhibitory effect of these growth regulators was negated.

10.5.2.2 Nutritional

Role of nutrient elements in fruit growth and yield of avocado trees has been well documented (Lahav and Kadman, 1980). Nitrogen and calcium levels have profound influence on yield, fruit size and post-harvest quality and their interactions perhaps control the vegetative/reproductive balance in the avocado trees (Bower and Cutting, 1988). Addition of phosphorus and potassium in a normal fertilizing programme appear to satisfy the plant's requirement.

11.0 Pests, Diseases and Disorders

11.1.0 Pests

Scale insects, mealy bugs and mites (*Oligonichus perseae*) are the important insect pests of avocadoes. Thrips (*Heliothrips haemorrhoidalis*) and (*Selenothrips rubrocinctus*) have been reported to occur sporadically on the leaves and fruits of avocado in South Africa (Steyn, 1991). Acovado cvs. Horshim, 4102 and TX-531 are very susceptible to black vine thrips (*Retithrips syriacus*) and plant insecticide such as 2 per cent extract of Sabadilla (*Schoenocaulon officinale*) was more effective than 2 per cent extract of *Ryania* (Izhar *et al.*, 1992).

Galendromus helveolus and *Neoseiulus californicus* were evaluated as biological control agents against mites (*Oligonychus perseae*) on avocados in California, USA (Kerguelen and Hoddle, 1999). Both predator species, either alone or in combination, significantly reduced numbers of *O. perseae* in comparison to oil-sprayed trees and control trees. Additionally, leaf damage measured as the percentage leaf area damaged by *O. perseae* feeding was reduced on trees treated with *N. californicus* alone or in combination with *G. helveolus*.

11.2.0 Diseases

The most serious disease of avocado is root rot caused by *Phytophthora cinnamomi*, which may kill the tree. Root rot is accentuated by poorly drained soils. Ridomil (mixed with the soil before planting @ 1 g a.i./10 kg soil) or applied as a drench to the soil surface around established plants (1 g a.i./10 litre), controlled root rot for at least 4 months after treatment. Two applications of ridomil, applied at the rate of 5 g a.i./m^2 as a soil drench beneath the canopy of 7-year-old trees affected by root rot, controlled *P. cinnamomi* without affecting its biological antagonists and permitted the trees to recover over a period of 12 months (Allen *et al.*, 1980). Phosphorus acid trunk injections proved successful in Australia (Pegg and Whiley, 1987). Dimethomorph at 10 and 100 mg a.i./l controlled root rot as effectively or better than fosetyl-Al (3000 mg a.i./l) or metalaxyl (150 mg a.i./l) but higher concentration was phytotoxic (Duvenhage and Kohne, 1995). Soil solarization significantly reduced the *P. cinnamomi* incidence (Lopez-Herrera *et al.*, 1997).

Results of investigations carried out on control of *Phytophthora cinnamomi* in Israel were summarized by Zilberstein (1999). The strategies were based on : (a) the selection of resistant rootstocks and (b) providing guidelines to limit disease spread throughout infected plantations. Seventeen resistant rootstocks from different genetic sources were selected including Israeli rootstocks which originated in a gene pool and rootstocks taken from surviving trees in an infected plantation. The rootstocks were planted in commercial plantations and grafted onto commercial varieties to check their yield capacities. The plots were planted in four different regions in Israel. In the field of plantation disease control, in addition to encouraging surface root growth, 'Canon 50' was also shown to be an effective fungicide in the treatment of *P. cinnamomi*. It was shown to be equally effective whether injected into the trunk, sprayed on the foliage, or fertigated. This agro technical management, coupled with stringent nursery standard, contributes considerably to the containment of the disease in Israel.

Phomopsis perseae was isolated from cankerous lesions on the trunks of Martin Grande (G 755) avocado rootstocks and pathogenicity was confirmed. *P. perseae* was not pathogenic on Duke 7 rootstocks *in vivo*. Benomyl (1 ppm) gave complete control of *P. perseae in vitro* (Merwe and Kotze, 1991).

Pests of avocado

Seed moth (left) and shot hole borer (right) infestation in avocado

Source: dos Resi, D.S., Mzachado, L.G., da Silva, E.M. and Fernandes, F.L. (2018) *Florida Entomol.*, 101(2):317-320.

Pests of avocado

Damage caused by pests affecting fruit in Hass avocado: (A) mites (principally associated with *Olygonichus yothersi* Mcgregor); (B) Melolonthidae (principally associated with *Astaena pigydialis* Kirsch); (C) thrips (Thysanoptera); (D) the bug monalonion *Monaloniun velezangeli* Carvalho and Costa); (E,F) ovary gall midge (*Bruggmanniella perseae* Gagné); (G,H) scales (*Hemiptera: Coccoidea*); (I–K) avocado seed moth (*Stenoma catenifer* Walsingham) (L,M) big avocado seed weevil (*Heilipus lauri* Boheman). (N,O) rodent damage.(Photo courtsey: Wilmar Perez)

Source: Ramírez-Gil, J.G., López, J.H. and Henao-Rojas, J.C. (2020) *Agronomy*, **10**(1) 8; doi:10.3390/agronomy10010008

Mealy bug (left), Mite (middle) and Eugnenia psyllid (right) infestation in avocado

Diseases of avocado

Scab (left) and leaf spot (right) infection in avocado

Branch canker and dieback (right) infection

Laurel wilt infection in avocado

Diseases of avocado

Anthracnose infection

Algal leaf spot in avocado

Cercospora infestation in fruit

Avocado fruits are found to be seriously infected with a fruit spot disease caused by *Colletotrichum gloeosporiodes*. Infection results in shedding of young fruits. The remaining unshaded fruits do not attain normal size and become deformed. In some fruits the infection may be latent. The isolated fungus from the fruit does not infect the leaves, but there is another strain of fungus which causes leaf spot disease (Sohi and Sridhar, 1974). Controlled atmospheric storage of fruits in 2 per cent O_2 and 10 per cent CO_2 at 7.2°C for 3-4 weeks prevents development of *C. gloeosporiodes* (Spalding and Reeder, 1975).

Infection of avocado fruits by *Fusarium solani* and *F. sambucinum*, causing accelerated softening of fruits, has been reported by Zauberman and Schiffmann-Nadel (1977).

The fruit rot disease is caused by *Dothiorella aromatica* (*D. gregaria*) (Sridhar, 1978). Brown spots appear on the fruit, covering the entire fruit surface and flesh inside the fruit is discoloured. Stem end rot (*Diplodia natalensis*) usually appears at the stem end and spreads rapidly invading the whole fruit. The pulp of the diseased fruit turns dark brown and soft (Sridhar, 1978).

The post-harvest diseases of avocado in New Zealand are anthracnose, caused by *Colletotrichum* spp., and stem end rot caused by *Botryosphaeria* spp. and *Phomopsis* spp. The rots can be controlled by field sprays and post-harvest dips. The incidence of *Botryosphaeria* spp. was reduced by monthly applications of copper hydroxide, that of *Phomopsis* spp. by pre-harvest application of benomyl and that of *Colletotrichum* spp. by post-harvest dipping in prochloraz (Hartill, 1992). Application of *Bacillus subtilis* in commercial Tag wax at various concentrations as well as in a water dip, significantly reduced anthracnose, *Dothiorella/Colletotrichum* fruit rot complex and stem-end rot (Korsten *et al.*, 1991).

In an investigation on the effect of copper fungicide on non-target microorganisms on avocado (cultivars Hass and Nabal), Stirling *et al.* (1999) found that populations of filamentous fungi, yeasts and bacteria on leaves were at least 10-fold less after a single application of copper hydroxide (as kocide) than on unsprayed leaves. This detrimental effect on copper on microorganisms on the phylloplane was confirmed in random samples collected from 5 commercial avocado orchards in Australia with regular pesticide spray programmes but where isolated unsprayed trees were also available. Fruits were harvested from both orchards, ripened and assessed for anthracnose caused by *Colletotrichum gloeosporioides* (*Glomerella cingulata*).

In both years, there was significantly less disease in unsprayed fruit than in sprayed fruit. These results show that copper fungicide is detrimental to phylloplane micro-organisms and it is suggested that these organisms are providing some natural suppression of *G. cingulata* on avocado.

Other diseases of avocado are *Cercospora* spot (*Pseudocercospora purpurea*), and scab (*Sphaceloma perseae*), which attack the leaves as well as the fruits. *Cercospora* spot of avocado was effectively controlled by pre-harvest applications of flusilazole or copper oxychloride (Lonsdale, 1991). The best control of *Cercospora* could be achieved when spray treatments with Benlate (benomyl) commenced between the third week of October and the second week of November to coincide with peak spore release (Lonsdale and Scott, 1991).

Sooty blotch of avocado is caused by *Akaropeltopsis*. Symptoms include the presence of smoky blotches on the plant surface which gradually become darker and spoil the appearance of fruit without any deterioration. It can be controlled by the application of copper oxychloride and chlorine dips before packing (Schoeman, 1996).

11.2.1 Integrated Pest and Disease Management

In an experiment to control *Phytophthora cinnamomi* on avocado, bovine manure (E), lucerne straw (A) and metalaxyl (M) alone or in combination with straw or manure were applied at each tree basin. Control trees (T), without E, A or M were irrigated by flooding individual basins and chemically fertilized, and double control trees (DT) without E, A or M, according to the traditional management, were irrigated by general flooding, fertilizers rarely applied and without foliar pest control. Control measures reduced the incidence of root rot. Trees with EA were damaged most by the branch weevil and the least damaged were the ones treated with EM. There was a direct relation between damage caused by root rot and injuries caused by the branch weevil. Branch weevil reduced fruit yield (Cabrera Bautista and Salazar Garcia, 1991).

Avocado trees cv. Fuerte (on Duke 7), which had recovered from root rot (*P. cinnamomi*) were supplied with various organic soil amendments while trunk injection of fosetyl-Al was discontinued. The condition (indicated by disease index) of trees in all treatments declined significantly during the 3rd and 4th years of trial, possibly as a result of drought stress. The yield of untreated trees and trees supplied with cattle manure declined significantly after 3 or 4 years, but a leguminous cover crop (*Dolichos lablab*) or a lucerne straw mulch (alone or in combination with cattle manure) seemed to inhibit yield decline, as did the continuation of the fosetyl-Al treatments (Duvenhage *et al.*, 1993).

11.3.0 Disorders

11.3.1 Tip Burn

Occurrence of tip burn is prevalent in spring. It is a chloride toxicity and the affected leaves show necrosis of the tips and edges and fall, resulting in decline in yield. Heavy irrigation before the end of winter, deep ploughing and incorporation of organic matter into the soil alleviate the problem (Sippel and Tomlinson, 1996).

11.3.2 Pulp Spot

The region surrounding the cut vascular bundles of the fruit-stalk becomes black. This is usually localized in nature and the symptom is higher early in the season (Swarts, 1984).

11.3.3 Mesocarp Discoloration

This disorder results in an overall grey to brown flesh discoloration, usually more intense in the distal half of the fruit. The incidence is more predominant towards the end of the season (Bower and Cutting, 1988).Both the disorders of fruits cause browning reactions, and as such implicate in particular enzyme, polyphenol oxidase (Kahn, 1975) and phenolics (Kahn, 1977). Factors affecting polyphenol oxidase appeared to be more important (Bower and Cutting, 1988).

11.3.4 Grey Pulp

The results of a comprehensive study on grey pulp in cv. Pinkerton, a post-harvest disorder indicated that it was most prevalent in fruits grown in warmer areas of the country (Kruger and Kritzinger, 1999). The main problem was due to late marketing. In simulated export conditions it was found that when the moisture content of Pinkerton was about 80 per cent there was no incidence of grey pulp but it developed as the moisture content fell. Fruits were completely unfit for storage when the moisture content approached 70 per cent.

12.0 Harvesting and Yield

12.1.0 Maturity

Identification of horticultural maturity is difficult in the case of avocado as maturation is not accompanied by significant changes in external appearance (Proctor and Caygil, 1985). An avocado of insufficient maturity is likely to shrivel and rot during softening and will in any case lack the desirable flavour found in more mature fruit (Marriott and Proctor, 1984). In Florida, separate maturity standards, based on minimum weight or diameter are defined for each cultivar. California, Israel and Australia use oil content to rate fruit maturity, specifications ranging from 8 to 15 per cent minimum oil content depending on producer country and cultivar grown (Marriott and Proctor, 1984). In Israel, harvesting of Pinkerton avocado began in mid-November when oil content was 9 per cent (19.8 per cent of dry matter) and ceased when uneven ripening of the fruit occurred or when the pericarp did not soften properly (Offer *et al.*, 1986). Cultivars of the three horticultural races differ in oil content, the Mexican race having the highest, Guatemalan race intermediate and West Indian race the lowest. Degnani *et al.* (1986), however, stated that dry weight rather than oil content may be used as the maturity index, because dry weight determination (3 h at 105°C) was more reliable, much simpler, faster and cheaper. Fruit dry weight was closely correlated with oil content indicating that either parameter is suitable as a maturity index (Continella and Rosa, 1992). The optimum maturity for harvest in Japan (detected by a sharp rise in ethylene production) occurred in late November for avocado cvs. Zutano and Bacon and in early December for Fuerte (Inoue and Takahashi, 1991). Nuclear magnetic resonance (NMR) image has been reported to be useful in determining the maturity of avocado fruits (Chen *et al.*, 1993). A harvest delay of 2 months until flesh dry matter reached 30 per cent, reduced average annual yield by 26 per cent and initiated an alternate bearing cycle (Whiley *et al.*, 1996a). Split harvests (early and late) were less detrimental to sustainability of high yields (Whiley *et al.*, 1996b). Delayed harvesting also resulted in decrease in fruit Ca and Mg concentrations and increase in total phenolic compound concentration (Cutting *et al.*, 1992).

Özdemir *et al.* (2009) studied the physical and chemical changes during fruit development and their relationship with optimum harvest maturity for Bacon, Fuerte and Zutano avocado cultivars grown under Dörtyol ecological condition of Turkey. Fruits were harvested from trees grafted on seedlings and planted in a spacing of 5 m x 6 m. Fruit width and length, weight, skin and flesh color (L* and hue), total soluble solid (TSS), titratable acidity (TA), TSS/TA, fruit flesh firmness, juice pH, dry weight and oil content were measured on fruits samples collected from tagged trees from June drop to ripening at a 15 days interval during fruit development. Data indicated that oil and dry matter content, flesh firmness and fruit weight were important criteria in determining optimum harvest maturity. Based on the all criteria, the period from the end of November to middle of December was the optimum harvest time for these cultivars in order to deliver fruit at optimum commercial maturity and quality and to obtain adequate storage time.

12.2.0 Harvesting

Harvesting aovcado fruits is an expensive and, with larger trees, a difficult and hazardous operation (Beck, 1960). In California, harvesting is done with hand clippers or a long pole equipped with a clipper and cloth catching bag (Freistadt, 1958). Eaks (1972) stated that hand snapping (stem removed) instead of clipping (stem in) increased picking rates and reduced cost. Fruit should be cut from the tree and handled with care.

12.3.0 Yield

Average yield is about 100-500 fruits per tree. However, good orchards in California have been found to give a yield of 6,000 to 12,000 fruits per acre, per annum (Purseglove, 1974).

It can be seen from Table 3 that the yield performance of the seedling trees was highly satisfactory. A yield of 20 t ha^{-1} is attainable 6 years after planting, and an average of 22.5 t ha^{-1} is probably sustainable thereafter (Wolstenholme *et al.*, 1991). In an evaluation trial with 98 cultivars in Maracay, Venezuela, Avilan *et al.* (1997) recorded higher fruit yield in the Guatemalan, West-Indian cultivars than in other cultivars. They recorded higher yields in cvs. Santa Clara (296 kg/tree), Princesa (232 kg/tree), CENIAP-13 (216 kg/tree) and CENIAP-4 (185 kg/tree).

13.0 Post-harvest Management

Avocado is an important fruit with a high market value, but with a relatively short storage life. Low temperature storage is utilized to extend its shelf life, but it induces chilling injury (CI) symptoms, expressed as pulp browning, as well as external damage (Zauberman *et al.*, 1985). Electrolyte leakage and disrupted ion balance resulting from ultra structural changes in the membranes have been suggested as leading to the development of CI symptoms. Membrane leakage in avocado fruit, as evaluated from electrical conductivity (EC) measurements, increased during cold storage and ripening (Montoya *et al.*, 1994), and ethylene production is another natural process that occurs during avocado fruit ripening. In the past, strategies such as the use of heat or anaerobiosis, which extend shelf

Harvesting, sorting and packaging of avocado

Harvesting of fruits

Sorting of avocado

Packaging of avocado

Post harvest management in avocado

Washing, grading and storage of avocado

life and reduce CI symptoms in avocado, were based on the control of ethylene activity and maintenance of membrane integrity (Pesis *et al.*, 1994; Woolf, 1997). Of late, the ethylene action inhibitor, 1-methylcyclopropene (1-MCP) was found to be effective in overcoming the effects of ethylene in a range of perishable fruits. Hershkovitz *et al.* (2005) found 1-methylcyclopropene (1-MCP), to be an ethylene action inhibitor, significantly delayed the ripening of the avocado cultivars 'Ettinger', 'Hass' and 'Pinkerton'. Application of 1-MCP at low concentration (300 nl l"1), prior to the climacteric increase, was effective and delayed the onset of the climacteric peaks of CO_2 and ethylene production. The delay was associated with reductions in fruit softening and in electrical conductivity (EC), the latter being an indicator of membrane permeability. The 1-MCP-treated 'Ettinger' and 'Pinkerton' avocado fruit maintained a greener peel color because of their lower levels of chlorophyllase activity and less chlorophyll breakdown. Treatment with 1-MCP at 300 nl l"1 prior to cold storage reduced chilling injury (CI) symptoms that are expressed as mesocarp discoloration and reduced polyphenol oxidase (PPO) and peroxidase (POD) activities in avocadoes stored at 5°C for 3.5 weeks. PPO and POD activities were low in harvested fruit and increased significantly during cold storage and shelf life at 20°C. 1-MCP was effective in reducing pulp browning in all the tested avocado cultivars. Mesocarp discoloration in avocado was found to be correlated with increases in EC values and PPO and POD activities.

13.1.0 Packaging and Transport

The avocado fruits are often packed by hand. The type of box used should be identical. The fruits are placed vertically in the box so as to improve the overall protection and appearance. Fruits should not be packed very tightly (Thompson, 1996).

Fruits are mainly transported by sea freight. Management of transport temperatures between pack house and arrival after shipment overseas has great bearing on the condition of fruits. Bezuidenhout (1992) identified the factors responsible for unsatisfactory condition of fruit such as i) a greater fruit age between picking and arrival overseas; ii) deviation in temperature control on some of the ships; and iii) a higher than normal incidence of black spot (*Pseudocercospora purpurea*) in orchards causing earlier harvest of affected trees than the unaffected ones, resulting in variation in maturity and response to optimum temperature management.

13.2.0 Ripening

Proctor and Caygill (1985) recorded that mature avocado fruits ripened in 6-12 days at 20°C, the ripening time being dependent on physiological maturity. While the climacteric patterns of avocado fruit harvested at various stages of development were similar, the pre climacteric period decreased as a fruit matured (Zauberman and Schiffman-Nadel, 1972). The climacteric peak rate and the peak rate of ethylene production also increased with fruit maturity (Eaks, 1980). Fruits picked from the same tree at the same time were found to differ in ripening time by as much as 10 days (Young, 1979).

A testing technique based on a new non-destructive ultrasonic device was applied to whole avocado fruits to measure internal changes during ripening and to assess fruit quality were reported by Mizrach and Flitsanov (1999). The method was based on local measurements of ultrasonic wave velocity and attenuation, by means of 2 inclined ultrasonic probes, which were moved across the fruit peel. Accepted methods were used for penetration measurements of firmness and physiological tests of dry weight (DW) and oil content of the fruit tissue. Results of the ultrasonic tests were compared with those of destructive penetration tests and physiological tests. Average values of ultrasonic wave attenuation correlated well with storage time and with the ripening process. A linear model represented well the changes in attenuation of the ultrasonic signals in the fruit, while second-order polynomial expressions described the changes in fruit firmness during storage at room temperature.

Treatment with ethylene immediately after harvesting did not accelerate ripening (Gazit and Blumenfeld, 1970), and ripening of fruit was enhanced when ethylene treatment was given 2 days after harvesting (Adato and Gazit, 1974). Rousseau (1981) in South Africa recommended a 24-hour treatment with ethylene at 1 ml/l at 16°C followed by 2 to 4 days at 16°-18°C for Fuerte and 18°-20°C for Hass and Ettinger. It appeared that ethylene did not just trigger the ripening of avocado fruits but was involved in a relatively long process which required its continuous presence (Zauberman *et al.*, 1988). It was been suggested that both synthesis and degradation of proteins were involved in the ripening process which was supported by similar polypeptide profiles (Zamorano and Merodio, 1993a). The delay in ripening at low temperature could be related to a delay in the evolution of the protein profile (Zamorano and Merodio, 1993b).

Treatment with ABA methyl ester at 220 mg/g of fresh weight, immediately after picking, enhanced the soluble polyphenol oxidase activity in the mesocarp, indicating a relation between ABA level and fruit browning potential (Anon., 1986). Fuchs *et al.* (1986) reported that at 20°C, the cellulase and polygalacturonase activities and the reduction in firmness preceded the onset of ethylene production by the fruit. While at 5°C, the ethylene producing process was induced without any significant increase in the activity of these two enzymes. Wounded mature 'Fuerte' fruit ripened faster than non-wounded fruit when stored at 14°C. However, wounded and non-wounded fruits ripened at similar rates when stored after wounding for 10 days at 5°C, and thereafter transferred to either 14° or 20°C (Zauberman and Fuchs, 1981). Tingwa and Young (1974) found that avocadoes with a higher endogenous calcium level took longer time to ripen and had a lower peak production of ethylene. They also found that vacuum infiltration of whole fruits with Ca solutions resulted in inhibition of peak carbon- dioxide and ethylene production, but the time to reach the climacteric was not markedly affected. Wills and Tirmazi (1982) stated that ripening of intact avocado fruit was inhibited by infiltration under reduced pressure with solutions containing calcium chloride. When the fruit did ripen, they were found by organoleptic evaluation to be highly acceptable and of similar quality to fruit that ripened normally. Water stress, both before and after harvest, has a profound effect on avocado fruit-ripening physiology (Bower and Cutting, 1988). Avocado fruits that lost water more rapidly after picking, ripened faster (Adato and Gazit, 1974).

The ethylene peak occurred earlier if fruits were allowed to become dehydrated, whereas the reverse was true if harvested fruits were infiltrated with water. The rate of water loss during storage affected both ripening rate and fruit quality (Bower and Cutting, 1987). More negative fruit water potentials at harvest caused faster ripening after storage (Bower, 1984). Exogenous application of acetaldehyde (AA) vapour to whole or half-peeled fruits (cv. Fuerte), prior to storage inhibited fruit ripening (Pesis *et al.*, 1998). This inhibition was characterized by a delay in fruit softening and a reduction in ethylene production. Activities of the cell wall depolymerizing enzymes polygalacturonase, b-galactosidase and endoglucanase and of ACC oxidase were found to reduce immediately after treatment.

Prusky *et al.* (1999) reported that exposure of whole avocado fruits cv. Fuerte to ethylene (40 ml/l) for 3 h transiently enhanced the level of antifungal 1-acetoxy-2-hydroxy-4-oxo-heneicosa-12, 15-diene (diene) in the pericap without affecting disease resistance to *Colletotrichum gloeosporioides* (*Glomerella cingulata*). Exposure of 1-2 mm slices of pericarp and mesocarp to ethylene enhanced the level of the diene in the mesocarp only. Since most of the diene in the mesocarp was compartmental in idioblasts, the effect of ethylene was tested on isolated idioblasts. Exposure of idioblasts to ethylene increased the level of antifungal diene, 2-fold within 8 h.

13.3.0 Storage

Avocado fruits do not soften while on the tree, but only after it is picked. The period which elapses from picking to softening is a function of the metabolic activity of the fruit, which in turn depends on various factors, one of the most important being storage temperature (Zauberman *et al.*, 1977). Avocados are susceptible to chilling injury, causing symptoms like pitting, browning of pulp near the seed or in the tissue midway between the seed and the skin, as well as failure to soften when transferred to a higher temperature (Thompson, 1996). There is a general impression that West

Indian cultivars are more sensitive to low temperature than Guatemalan and Mexican cultivars are.

Chilling injury has been reported to occur at 10° to 11.1°C for cultivars of the West Indian race and 4.4° to 6.1°C for others (Pantastico, 1975). The recommended storage conditions of avocado fruits are presented in Table 8.

Table 8: Recommended Storage Conditions of Avocado Fruits*

Cultivars	*Temperature (°C)*	*Relative Humidity (per cent)*	*Duration (Weeks)*
Lula	4.5	85-90	4-5
Unripe Fuerte and Hass	5.5-8	85-90	4-5
Ripe Fuerte and Hass	2-5	85-90	1-2
Unripe Ettinger	5.5	85-90	3-4
Unripe Fuchs, Pollock	10-13	85-90	2-3 and Waldrin
Unripe Booth 8, Booth	4.5	85-90	2-4 and Taylor

* Snowdon (1990).

In Israel, cv. Nabal was the most tolerant to chilling injury followed by Ettinger and Fuerte (Zauberman *et al.*, 1973), while in Cyprus, Ettinger was the most sensitive cultivar to chilling injury followed by Fuerte and Hass (Vakis, 1982). Internal discoloration during storage can be reduced by prompt pre-cooling. Irradiation results in excessive flesh browning (Jacobs, 1974). Avocado fruits of cv. Hass do not tolerate exposure to 38°C for longer than 24 h, and even short exposure may cause heat injury (D'Aquino *et al.*, 1996). Dipping of fruits in 1 or 1.5 per cent Vapor Gard (an antitranspirant) for 2 minutes was found to be more effective than gamma-irradiation (with 0.04 or 0.08 kGy) in delaying ripening and extending the shelf life of fruits stored at either room temperature or at 10°C (Awad, 1992). Rates of CO_2 and ethylene production were decreased and softening was delayed by storage at 7°C (Zamorano and Merodio, 1993a). Pretreatment with N2 for 24 hours prior to cold storage of avocado cv. Fuerte fruits reduced chilling injury symptoms and delayed fruit softening which may be ascribed to lower respiration rates and ethelene production during storage and shelf life (Pesis *et al.*, 1993). Parodi and Lerner (1995) reported that storage at 9°C for 35 days improved the fruit quality of avocado cv. Hass. Oil content increased from 15.9 per cent before storage to an average (across storage durations) of 20.2 per cent after storage + the 10 days at ambient temperature.

Fruits wrapped in polythene bags and stored at gradually decreasing temperatures (2 days at 17°, 2 days at 14°, 4 days at 12° and then 8°C) did not soften for 23 days in Fuerte and 46 days in Nabal (Aharoni *et al.*, 1968). Use of ethylene absorber clearly delayed the ripening process, allowing the fruits to be stored longer in sealed polythene bags (Correa and Plaza, 1992). Coating pre-climacteric fruits with Nature Seal, a polysaccharide-based edible film with cellulose as the main component (1 and 2 per cent formulations) delayed ripening at 20°C by 2 days and exhibited a lower respiration rate, lower ethylene production and higher flesh firmness than non-coated fruits (Bender *et al.*, 1993).

Controlled atmosphere storage at 3 to 5 per cent CO_2 and 3 to 5 per cent O_2 delayed fruit softening and 9 per cent CO_2 with 1 per cent O_2 at 10°C kept Lula avocado in an acceptable eating condition and appearance for 60 days (Hardenburg *et al.*, 1990). Long distance sea transport of Hass avocado fruits is feasible in refrigerated CA containers. The best CA storage combination recommended for Hass avocado fruits is 5 per cent CO_2 and 2 per cent O_2 (Lizana and Figueroa, 1997). Another recommendation for storage of California and tropical avocados is 3 to 10 per cent CO_2 with 2 to 5 per cent O_2 (Anon., 1991). Carbondioxide levels > 5 per cent increased the amount of discoloration and the presence of as little as 1 ppm C_2H_4 in the atmosphere could negate the beneficial effects of CA storage with regard to discoloration (Arpaia *et al.*, 1990). Low O_2 levels exert a suppressive effect mainly on *de novo* synthesized proteins of ripening fruits, whereas they induce the synthesis of new proteins irrespective of the developmental stage of fruit (Kanellis *et al.*, 1993). Metzidakis and Sfakiotakis (1993) observed that accumulation and activities of polygalacturonase and cellulase were inhibited in fruits stored in low O_2 atmosphere and the inhibition was mediated via an effect on ethylene action.

Meir *et al.* (1998) reported that Fuerte avocado fruits could be stored for 8 weeks at 5°C, without the appearance of any significant chilling injury symptoms in 40 ´ 70 cm polyethylene (PE) bags of 30 mm thickness with elevated (5-7 per cent) CO_2 concentrations combined with reduced (3-5 per cent) O_2 concentrations within the bags. The modified atmosphere packaging (MAP) significantly reduced fruit weight loss such that at the end of a 6.5-week storage period at 5°C and an additional shelf-life period at 20°C until ripening, fruits sealed in MAP lost only 5 per cent of their initial weight, whereas unpacked fruits lost about 11 per cent of their initial weight.

14.0 Biotechnology

Though avocado is an important fruit crop in the world, research works on various aspects of biotechnology are still inadequate for its application in significant improvement of the crop. Micropropagation has proved successful in avocado but the plants may exhibit a slow rate of growth during acclimatization. Molecular markers have been used for the identification of genotypes and detection of resistance to *Phytophthora*. Report on genetic transformation is scanty.

14.1.0 Tissue Culture

Micropropagation of avocado has been summarized in propagation.

14.1.1 Embryo Culture

Avocado like other subtropical woody species has a high rate of fruit abscission, which is further exaggerated on small potted trees grown in a glasshouse. Skene and Barlass (1983) rescued immature embryos older than 6 weeks at abscission by culturing them in a liquid half-strength MS basal medium supplemented with 0.5 mg/l benzyladenine.

Embryonic tissue explants developed the highest number of multiple adventitious shoots in MS medium containing 30 g sucrose supplemented with 1.0 mg BA/l. From a single piece of embryonic tissue, an average of 10 adventitious multiple shoots were obtained with shoot multiplication medium. WPM medium supplemented with 1.0 mg/l IBA and 10^{-4} M putrescine plus 0.2 per cent activated charcoal was optimal for root development (Ahmed *et al.*, 1998).

14.1.2 Protoplast Culture

Protoplasts were successfully isolated from callus derived from stem explants. Callus cells were plasmolyzed with 0.7 M mannitol and 0.4 per cent PEG (6000). Plasmolyzed cells were then treated with 1 per cent Onozuka cellulase RS and 0.1 per cent pectolyase Y23 which yielded 2-3 x 106 protoplasts/g of callus (Blickle *et al.*, 1986).

14.1.3 Somatic Embryogenesis

Embryogenic avocado callus was developed from immature zygotic embryos on MS medium containing the auxin, picloram (Mooney and VanStaden, 1987; Pliego Alfaro and Murashige, 1988). However, the ability of *in vitro* regeneration of many woody species including avocado, is restricted and seems to be limited to juvenile plant material (Pliego-Alfaro *et al.*, 1987). A procedure of recurrent somatic embryogenesis from cell suspension cultures of avocado was used for genetic transformation (Curz Hernandez *et al.*, 1988). However, regeneration of a complete and normal plant from *in vitro* cultures of avocado on a large scale has not yet been achieved. Only a small percentage of green shoots and normal leafy plantlets was regenerated (Mooney and VanStaden, 1987; Pliego-Alfaro and Murashige, 1988).

Witjaksono and Litz (1999) developed a protocol for somatic embryo formation and maturation from proembryonic masses from suspension culture. Somatic embryos were developed both in liquid and semisolid media but only the latter could develop to maturity. Shoots that developed from somatic embryos were micropropagated.

14.2.0 Biochemical and Molecular Markers

14.2.1 Isozymes

Torres (1983) used 6 enzymes systems specified by 12 genes with 37 alleles. Mesocarp and leaf tissues and later cotyledons were used for analysis (Torres, 1984). The isozyme analysis was used for studying the relationship among cultivars and for identifying hybrids from selfed progenies (Vrecenar-Gadus and Ellstrand, 1985; Goldring *et al.*, 1987), for measuring parentage of a given selection (Ellstrand *et al.*, 1987) and for linkage analysis (Torres *et al.*, 1986).

14.2.2 Molecular Markers

RFLP markers were used to map the avocado genome and linkage analysis (Clegg, 1990; Furnier *et al.*, 1990). More than 15 clones were characterized (Clegg, 1990). Lavi *et al.* (1992) studied the applicability of four DNA fingerprint probes to avocado which resulted in identification of various cultivars, characterization of the three avocado races and genetic analysis of family structure. Some of the probes detected polymorphic sequences in all the species, but others were useful only for one or two species (Sharon *et al.*, 1995). Mhameed *et al.* (1997) studied the genetic relationship within avocado (*Persea americana*) cultivars and between *Persea* species. Ashworth and Clegg (2003) found twenty-five microsatellite markers uniquely differentiated 35 avocado cultivars and two wild relatives. Average heterozygosity was high (60.7 per cent), ranging from 32 per cent in *P. steyermarkii* to 84 per cent in Fuerte and Bacon. In a subset of 15 cultivars, heterozygosity averaged 63.5 per cent for microsatellites, compared to 41.8 per cent for restriction fragment length polymorphisms (RFLPs). A neighbor-joining tree, according to average shared allele distances, consisted of three clusters likely corresponding to the botanical races of avocado and intermediate clusters uniting genotypes of presumably racially hybrid origin. Several results were at odds with existing botanical assignments that are sometimes rendered difficult by incomplete pedigree information, the complexity of the hybrid status (multiple backcrossing), or both. For example, cv. Harvest clustered with the Guatemalan race cultivars, yet it is derived from the Guatemalan 3 Mexican hybrid cv. Gwen. Persea schiedeana grouped with cv. Bacon. The rootstock G875 emerged as the most divergent genotype in our data set. Considerable diversity was found particularly among accessions from Guatemala, including G810 (West Indian race), G6 (Mexican race), G755A (hybrid Guatemalan 3 *P. schiedeana*), and

G875 (probably not *P. americana*). Low bootstrap support, even upon exclusion of (known) hybrid genotypes from the data matrix, suggests the existence of ancient hybridization or that the botanical races originated more recently than previously thought.

Sharon *et al.* (1997) developed a linkage map in avocado. They screened avocado genomic library with various microsatellite repeats. (A/T)/and (TC/AG)/sequences were found to be the most frequent repeats. One hundred and seventy-two positive clones were sequenced successfully of which 113 were found to contain simple sequence repeats (SSR). Polymerase chain reaction primers were designed to the regions flanking the SSR in 62 clones. A GenBank search of avocado DNA sequences revealed 1 sequence containing a (CT)10 repeat. A total of 92 avocado specific SSR markers were screened for polymorphism using 50 offsprings of a cross between the avocado cultivars Pinkerton and Ettinger. Both are standard avocado cultivars which are normally out crossed and highly heterozygous. Fifty polymorphic SSR loci, 17 random amplified polymorphic DNA (RAPD) and 23 minisatellite DNA fingerprint (DFP) bands were used to construct the avocado genetic map. The resulting data were analyzed with various mapping programs in order to assess which program best accommodated data from progeny of heterozygous parents. The analyses resulted in 12 linkage groups with 34 markers (25SSRs, 3 RAPDs and 6 DFP bands) covering 352.6 cm. This initial map can serve as a basis for developing a detailed genomic map and for detection of linkage between markers and quantitative trait loci.

Avocado selection and breeding can be more efficient and less expensive through the development of molecular markers for the estimation of germplasm genetic diversity, marker-assisted selection (MAS), and creation of linkage maps. Kuhn *et al.* (2019) described the development of the first set of avocado genetic markers based on single-nucleotide polymorphism (SNP) variation in expressed genes. Two important breeding resources, the world's two largest avocado mapping populations and an extensive germplasm collection, are housed at the USDA-ARS Subtropical Horticulture Research Station (SHRS) in Miami, Florida. However, to use these resources to their greatest advantage, many thousands of genetic markers are necessary. RNA sequencing was used both to build a reference transcriptome from 'Hass', the most widely grown avocado cultivar worldwide, and to identify SNPs by alignment of RNA sequences from the mapping population parents to the 'Hass' transcriptome. Their study provided a new genomic tool for the avocado community that can be used to assess the genetic diversity of avocado germplasm worldwide and to optimize avocado breeding and selection programs by complementing traditional breeding methods with molecular approaches, thus increasing the efficiency of avocado genetic improvement. They successfully created and used for avocado SNP discovery. Sufficient SNPs that could be used found a leaf and flower transcriptome was for genetic mapping and germplasm evaluation were generated for design of a 6000 SNP Infinium II chip. Genotyping by RNA sequencing and chip array were essentially identical with respect to accuracy. SNP discovery by RNA sequencing when no genomic DNA sequence is available is a successful strategy for marker generation in speciality crops.

Avocado lies in the anciently diverged magnoliid clade of angiosperms, which has a controversial phylogenetic position relative to eudicots and monocots. Recently Rendón-Anaya *et al.* (2019), along with 27 others, sequenced the nuclear genomes of the Mexican avocado race, *P. americana* var. *drymifolia*, and the most commercially popular hybrid cultivar, Hass, and anchored the latter to chromosomes using a genetic map. Resequencing of Guatemalan and West Indian varieties revealed that <"39 per cent of the Hass genome represents Guatemalan source regions introgressed into a Mexican race background. Some introgressed blocks are extremely large, consistent with the recent origin of the cultivar. The avocado lineage experienced 2 lineage-specific polyploidy events during its evolutionary history. Although gene-tree/species-tree phylogenomic results are inconclusive, syntenic ortholog distances to other species place avocado as sister to the enormous monocot and eudicot lineages combined. Duplicate genes descending from polyploidy augmented the transcription factor diversity of avocado, while tandem duplicates enhanced the secondary metabolism of the species. Phenylpropanoid biosynthesis, known to be elicited by *Colletotrichum* (anthracnose) pathogen infection in avocado, is one enriched function among tandems. Furthermore, transcriptome data show that tandem duplicates are significantly up- and down-regulated in response to anthracnose infection, whereas polyploid duplicates are not, supporting the general view that collections of tandem duplicates contribute evolutionarily recent "tuning knobs" in the genome adaptive landscapes of given species.

14.3.0 Genetic Transformation

Ahmed *et al.* (1998) succeeded in gene transfer into avocado cells by co-culture with wild *Agrobacterium tumefaciens* strains Ach 5 and C58, and disarmed vectors AGL1/pBI121 and LBA4404/pCGP904, GV3850/pBI121, GV3850/pCGP904, LBA4404/pBI121 and LBA4404/PcGP90. C58 and disarmed vector GV3850/pBI121 were more effective when the pH of the co-cultivation medium was 5.2 and the medium contained 200 mm acetosyringone. Ahmed *et al.* (2012) compared the strain GV3850/pBI121gave best transformation outcome compared to five other binary vectors (AGL1/pCGP904; AGL1/pBI121; GV3850/pCGP904; LBA4404/pCGP904 and LBA4404/pBI121) under different pH and acetosyringone concentrations. The optimal condition for reliable transformation was by using 200 µm acetosyringone and a pH of 5.2. Transformed embryonic shoots co-cultivated with GV3850/pBI121 were tested using the histochemical x-gluc assay. Further analysis was conducted by polymerase chain reaction using specific primers for the reporter gene (GUS).

15.0 References

Adato, I. and Gazit, S. (1974) *Plant Physiol.*, **53** : 45-46.

Adikaram, N.K.B., Ewing, D.F., Karunaratne, A.M. and Wijeratne, E.M.K. (1992) *Phytochem.*, **31** : 93-96.

Aharoni, Y., Schiffmann-Nadel, M. and Zauberman, G. (1968) *Israel J. Agric. Res.*, **18** : 77-82.

Ahmed, E.M. and Barmore, C.R. (1980)b In : *Tropical and Subtropical Fruits Composition, Properties and Uses* (Eds., S. Nagy and P.B. Shaw) pp. 121-56. AVI. Pub., West Port, CT.

Ahmed, M.F. A. S. Kantharajah and P. (2012) *American Journal of Plant Sciences,* **3**: 1225-1231.

Ahmed, M.F., Kantharajah, A.S., Holford, P. and Drew, R.A. (1998) *Acta Horticulturae,* **461** : 339-347.

Alcaraz, M.L., T.G. Thorp,and J.I. Hormaza (2013) *Scientia Horticulturae* **164**: 434–439.

Allen, R.N., Pegg, K. G., Forsberg, L.T. and Firth D.J. (1980) *Aust. J. Expt. Agri. & Anim. Husb.,* **20** : 119-124.

Anonymous (1986) *Information Bull.* Citrus and Subtrop Fruit Res. Inst. No. 162, p.4.

Anonymous, (1991) *FAO Production year book,* Vol. 45, FAO of the United Nations, Rome, 1992.

Arpaia, M.L. and Eaks, I. (1990) *Calif. Avocado Soc. Yearb.,* **74** : 35-42.

Arpaia, M.L., Faubian, D., Mitchell, F.G. and Mayer, G. (1990) *Calif. Avocado Soc. Yearb.,* **74** : 43-48.

Ashworth, V. E. T. M. and M. T. Clegg. (2003) *J Heredity* **94**: 407–415.

Avilan, L., Rodriguez, M. and Marin, R.C. (1997) *Agronomia Tropical (Maracay),* **47** : 457-474.

Avilon, R.L., Chirinos, A.V. and Figueroa, M. (1986) *Proc. Trop. Reg. Amer. Soc. Hort. Sci.,* **23** : 108-113.

Awad, S.M. (1992) *Ann. Agric. Sci.* (Cairo), **37** : 511-521.

Azit, A.B.A., Desouki, I. and EI-Tanahy, M.M. (1975) *Scientia Hort.,* **3** : 89-94.

Azzena, M. and Cossu, A. (2000) *Rivista di Frutticoltura e di Ortofloricoltura,* **62** : 79-82.

Barcelo-Munoz, A., Encina, C.L., Simon-Perez, E. and Pliego-Alearo, F. (1999) *Plant Cell Tissue Organ Cult.,* **58** : 11-17.

Barmore, C.R. (1977) In : *The Avocado* (Eds : J.W. Sauls, R.L. Phillips and L.K. Jackson) pp. 103-109. *Proc. First International Trop. Fruit short course,* Fla. Univ. Press, Gainesville.

Baxter, P. (1981) *Growing Fruit in Australia.* Thomas Nelson, Melbourne.

Beck, W.R. (1960) *Calif. Avoc. Soc. Yearb.,* **44** : 58-59.

Bender, R.J., Brecht, J.K., Sargent, S.A., Navarro, J.C. and Campbell, C.A. (1993) *Acta Horticulturae,* **343** : 184-186.

Ben-Yaacov, A. (1992) *Alon Hanotea,* **46** : 919-926.

Ben-Yaacov, A. (1995) *Alon Hanotea,* **49** : 574-577.

Ben-Yaacov, A. (1996) *Alon Hanotea,* **50** : 172-175.

Ben-Yaacov, A. (1997) *Alon Hanotea,* **51** : 368-371.

Ben-Yaacov, A. (1998) *Alon Hanotea,* **52** : 34-38.

Ben-Yaacov, A., Michelson, E. (1995) Avocado rootstocks. In: J. Janick (ed.) *Horticultural Reviews*. **17:** 381-429.

Ben-Yaacov, A., Michelson, E. and Sela, I. (1992b) *Alon Hanotea*, **46** : 987-993.

Ben-Yaacov, A., Michelson, E. and Sela, I. (1993a) *Alon Hanotea*, **47** : 424-428.

Ben-Yaacov, A., Michelson, E. and Sela, I. (1993b) *Alon Hanotea*, **47** : 26-30.

Ben-Yaacov, A., Michelson, E., and Sela, I. (1992a) *Alon Hanotea*, **46** : 769-783.

Ben-Yaacov. A., Hanochi, U., Michelson, E. and Sela, I. (1996) *Alon Hanotea*, **50** : 178-185.

Berg, B. (1992) California Avocado Society 1992 Yearbook 76: 61-75.

Bergh, B.O. (1957) *Proc. Fla. St. Hort. Soc.*, **70** : 284-290.

Bergh, B.O. (1969) *Outlines of Perennial Crop Breeding in the Tropics* (Eds. F.P. Ferwerda and F. Wit), Veenman and Zonen, Wageningen.

Bergh, B.O. (1975) In : *Advances in Fruit Breeding* (Eds. J. Janick and J.N. Moore), Purdue Univ. Press, Lafayette, IN, pp. 541-567.

Bertting, I., Cowan, A.K. and Moore-Gordon, C.S. (1998) *Acta Horticulturae*, **463** : 225-230.

Bezuidenhout, J.J. (1992) *Yearb. South Afr. Avocado Growers' Assoc.*, **15** : 39-40.

Biale, J.B. and Young, R.E. (1971) *The Biochemistry of Fruits and Their Products*, II, Academic Press, London.

Biasi, L.A., Koller, U.C. and Kampf, A.N. (1994) *Pesquisa Agropequouria* Brasileria, **29** : 1051-1058.

Blanke, M.M. and Lovatt, C.J. (1993) *Ann. Bot.*, **71** : 543.

Blickle, W., Muhlbach, S.P. and Sanger, H.L. (1986) *VI Intl. Cong. Plant Tissue Cell Cult.*, Minneapolis, P. 357 (Abst.).

Blumenfeld, A. and Elimelech, M. (1986a) *Alon Hanotea*, **40** : 897-900.

Blumenfeld, A. and Elimelech, M. (1986b) *Alon Hanotea*, **40** : 777-780.

Blumenfeld, A. and Elimelech, M. (1986c) *Alon Hanotea*, **40** : 741-743.

Blumenfeld, A. and Gazit, S. (1970) *Plant Physiol.*, **46** : 331-333.

Botha, T. (1991) *Yearb. South Afr. Avocado Growers' Assoc.*, **14** : 87-88.

Bower, J.P. (1978) *M.Sc (Ag) Thesis*, Univ. Natal, Pietermaritzburg.

Bower, J.P. (1984) *Yearb. South Afr. Avocado Growers' Assoc.*, **7** : 55-56.

Bower, J.P. and Cutting, J.C. (1987) *Proc. World Avocado Congress*, Pretoria.

Bower, J.P. and Cutting, J.C. (1988) *Hort. Rev.*, **10** : 229-271.

Bower, J.P. and Cutting, J.G.M. (1992) *Acta Horticulturae*, **296** : 55-58.

Bower, J.P., Wolstenholme, B.N. and Jager, J.M. De. (1978) *Crop Prod.*, **7** : 129-134.

Bringhurst, R.S. (1951) *Calif. Avocado Soc. Yearb.*, **34** : 164-168.

Brokaw, W. H. (1987) *Combined Proc. Int. Plant Prop. Soc.* **37:** 97-103.

Brown, B.I. (1972) *J. Agric. Food Chem.*, **20** : 753.

Buttrose, M.S. and Alexander, D.M. (1978) *Scientia Hort.*, **8** : 213-219.

Buzet, Z. (1999) *Neltropika Bulletin*, **306** : 2-4.

Cabrera Bautista, S. and Salazar Garcia, S. (1991) *Revista Mexicana de Fitopatologia*, **9** : 38-43.

Castro V.M., Iturrieta R.E. E. and Fassio, O.C.(2009) *Chilean J. Agricultural Research* **69**: 316-324.

Chandler, W.H. (1958) *Evergreen Orchards*, Lea and Febiger, Philadelphia.

Chen, P., McCarthy, M.J., Kauten, R., Sarig, Y. and Han, S. (1993) *J. Agric. Engineering Res.*, **55** : 177-187.

Chen, H., Morrell, P.L., Ashworth, V.E.T.M., De La Cruz, M. and Clegg, M.T. (2009) *J. Heredity* 1**00**: 56–65.

Clegg, M.T. (1990) *Annual Rep. Calif. Avocado Soc. Grant*, pp. 3-4.

Continella, G. and Rosa, G. La (1992) *Acta Horticulturae*, **296** : 69-73.

Continella, G., Tribulato, E. and D' Aquino, S. (1992) *Acta Horticulturae*, **296** : 59-67.

Correa, P.C. and Plaza, J.L. De La (1992) In : *XII Proc. Conf. Granada*, Spain (Eds : N.C. Pardos. and J.H. Rodriguez), G 10-18.

Cortes Flores, J.I., Yepez Torres, J.E., *Etc*hevers Barra, J.D. and Teliz Ortiz, D. (1993) *Revista Mexicana de Fitopatologia*, **11** : 107-112.

Cowan, A.K., Moore-Gordon, C.S., Bertling, I. and Wolstenholme, B.N. (1997) *Plant Physiol.*, **114** : 511-518.

Curz-Hernandez, A., Litz, R.E., Witjaksonon and Gomezlim, M. (1988) *Plant Cell Rep.*, 17 : 497-503.

Cutting, J.G.M. (1984) *Ph. D. Thesis*, Univ. Natal, Pletermaritzburg.

Cutting, J.G.M. (1993) *Yearb. South Afr. Avocado Growers' Assoc.*, **16** : 20-21.

Cutting, J.G.M. and Bower, J. P. (1990) *Acta Horticulturae*, **275** : 469-475.

Cutting, J.G.M., Wolstenholme, B.N. and Hardy, J. (1992) *J. Hort. Sci.*, **67** : 761-768.

Cutting, J.G.M., Wolstenholme, B.N., Hofman, P.J. and Lishman, A.W. (1985) *Yearb. South Afr. Avocado Growers' Assoc.*, **8** : 92-96.

D. Sharon,D., P. B. Cregan, S. Mhameed, M. Kusharska, J. Hillel, E. Lahav and U. Lavi. (1997) *Theor. Appl. Genet.*, **95:** 911- 921

D.N.Kuhn, D.N., D.S.Livingstone,J.H.Richard, P.Manosalva N.Van den Berg and A.H.Chambers. (2019) *Scientia Hort.*, **246:** 1-11.

D'Aquino, S., Piga, A., Continella, G. and Agabbio, M. (1996) *Advances in Hort. Sci.*, **10** : 167-172.

da Silva,S.R., T.E. Cantuarias-Avilés, B. Chiavelli, M.A Martins, M.S. Oliveira. (2017) *Pesq. Agropec. Trop., Goiânia*, 47. 321-327.

Dalsaso, L. and Guevara, E. (1989) *Agronomia Costarricense*, **13** : 61-71.

Davenport, T.L. (1982) *Proc. Fla. St. Hort. Soc.*, **95** : 92-96.

Davenport, T.L. (1986) *Hort. Rev.*, **8** : 257-289.

Davenport, T.L. and Manners, M.M. (1982) *J. Expt. Bot.*, pp. 815-825.

David, M.R.L. (1967) *Inf. Bul. Sci. Res. Council*, Jamaica, **8** : 22-31.

Degnani, C., Behor, V., Elbatzri, R. and Blumenfeld, A. (1986) *Alon Hanotea*, **40** : 1017-1022.

Durand, B.J. and Claassens, N.J.F. (1987) *Yearb. South Afr. Avocado Growers' Assoc.*, **10** : 15-19.

Duvenhage, J.A. and Kohne, J.S. (1995) *Yearb. South Afr. Avocado Growers' Assoc.*, **18** : 41-42.

Duvenhage, J.A., Kohne, J.S. and Kirkman, B. (1993) *Yearb. South Afr. Avocado Growers' Assoc.*, **16** : 75-76.

Eaks, I.L. (1972) *J. Amer. Soc. Hort. Sci.*, **98** : 106-108.

Eaks, I.L. (1980) *J. Amer. Soc. Hort. Sci.*, **105** : 744-747.

Ellstrand, N.C., Lee, J.M., Bergh, B.O., Coffey, M.D. and Zentmyer, G.A. (1987) *Calif. Avocado Soc. Yearb.*, **70** : 199-203.

Ernst, A.A., Ernst, Z.R. and Ernst, E.D. (2015) VIII Congreso Mundial De La Palta 13- 18 September 2015, Lima, Peru. *Actas,Proccedings Recursos Genéticos Y Manejo De Viveros • Genetic Resources And Nursery Management* p.53-59.

Ernst. A.A. and Holtzhausen, L. C. (1978) *Citrus and Subtrop. Fruit. J.*, **32** : 6-7.

Flores-Meza, D., Vite-Cisneros, L. and Borys, M.W. (1990) *Calif. Avocado Soc. Yearb.*, **74** : 207-234.

Friestadt, D. (1958) *Calif. Avoc. Soc. Yearb.*, **42** : 39-40.

Frolich, E.F. (1966) *Proc. Int. Plant. Prop. Soc.*, **16** : 51-54.

Fuchs, Y., Zauberman, G. and Akerna, M. (1986) *Acta Horticulturae*, **179** : 801-804.

Furnier, G.R., Cummings, M.P. and Clegg, M.T. (1990) *J. Hered.*, **81** : 183-188.

Gad Ish-Am (2005) *New Zealand and Australia Avocado Grower's Conference* '05. 20-22 September 2005. Tauranga, New Zealand. Session 7. Flowering, fruit set and yield, p. 9.

Galindo-Tovar ME, Ogata-Aguilar N, Arzate-Fernandez AM. (2007) *Genet Resour Crop Evol.* doi: 10.1007/s10722-007-9250-5.

Galindo-Tovar, M.E., Ogata-Aguilar, N., Arzate-Fernandez, A.M. (2008) *Gen. Res. Crop Evol.* **55**: 441–450.

Gandhi, S.R. (1956) *Indian Hort.*, **1** : 11-13.

Garcia-Gomez, M.L., Sanchez Romero, C, Barcelo-Munoz, A., Heredia, A. and Pliego-Alfaro, F. (1994) *J. Exp. Bot.*, **45** : 865-870.

Gazit, S. and Blumenfeld, A. (1970) *J. Amer. Soc. Hort. Sci.*, **95** : 229-231.

Gazit, S. and Blumenfeld, A. (1972) *Physiol. Plant.*, **27** : 77-82.

Goldring, A., Gazit, S. and Degani, C. (1987) *J. Amer. Soc. Hort. Sci.*, **112** : 389-392.

Gomez-Lopez, V.M. (1999) *Food Chemistry*, **47** : 2707-2710.

Gonzalez-Rosas, H., Llano-Agudelo, B.E. and Salazar-Garcia, S. (1990) *Calif. Avocado Soc. Yearb.*, **74** : 201-205.

Gordillo, M., Castro-Lopez, T., Rodriguez, N.N., Estevez, J.A. and Mazaira, M. (1994) *Trop. Fruits Newsl.*, **11** : 5-6.

Hannweg, K. (1997) *Inligtingsbulletin – Instituut vir Tropiese en Subtropiese Gewasse*, No. **293**, p.8.

Hardenburg, R.E., Watada, A.E. and Wang, C.Y. (1990) *The Commercial Storage of Fruits, Vegetables and Florist and Nursery Stocks*. USDA, ARS, Agric. Handbook, 66.

Hartill, W.F.T. (1992) In : *Brighton Crop Protection Conference, Pests and Diseases* - 1992. Vol. 3. Franham, UK, British Crop Protection Council, pp. 1157-1162, NewZealand.

Hassan, A.K. and Atawia, A.R. (1995) *Ann. Agric. Sci.*, **40** : 787-797.

Hershkovit, V., S.I. Saguy and E. Pesis (2005) *Post-harvest Biology and Technology* **37** : 252–264.

Hodgson, R.W. (1959) *Calif. Avoc. Soc. Yearb.*, **43** : 45-49.

Holmes, M. and Farrel, D. (1993) *Yearb. South Afr. Avocado Growers' Assoc.*, **16** : 59-64.

Inoue, H. and Takahashi, B. (1991) *Japanese J. Trop. Agric.*, **35** : 187-194.

Inoue, H., Takahashi, B. and Shirato, K. (1992) *Japanese J. Trop. Agric.*, **36** : 45-50.

Ish-Am, G. and Eisikowitch, D. (1992) *Alon Hanotea*, **46** : 885-902.

Itoh, T., Tamura, T., Matsumato, T. and Dupaigne, P. (1975) *Fruits*, **30** : 687.

Izhar, Y., Yehuda, B., Swirski, E., Wysoki, M. and Dagan, M. (1992) *Alon Hanotea*, **46** : 523-529.

Jacobs, C.J. (1974) *Citrus and Sutrop. Fruit. J.*, **485** : 16-21.

Jaime, S., Subires, M.J., Soria, J.T. and Aguilar, A. (1992) *Acta Horticulturae*, **296** : 75-80.

Janick, J. and Moore, J.N. (1975) *Advances in Fruit Breeding*. Purdue University Press.

Kadam, S.S. and Salunkhe, D.K. (1995) In : *Handbook of Fruit Science and Technology-Production, Composition, Storage and Processing* (Eds. D.S. Salunkhe and S.S. Kadam) pp. 363-375. Marcel Dekker, Inc., New York.

Kadman, A. (1985) *Acta Horticulturae*, **158** : 63-67.

Kadman, A. and Cohen, A. (1977) *Israel J. Bot.*, **26** : 50.

Kadman, A., Tomer, E. and Gottreich, M. (1974) *Calif. Avocado Soc. Yearb.*, **57** : 115.

Kahn, V. (1975) *J. Sci. Food Agr.*, **26** : 1319-1324.

Kahn, V. (1977) *J. Sci. Food Agr.*, **28** : 233-239.

Kaiser, C. and Wolstenholme, B.N. (1993) *Yearb. South Afr. Avocado Growers' Assoc.*, **16** : 46-55.

Kalmar, D. and Lahav, E. (1977) *Australian J. Agric. Res.*, **28** : 859.

Kanellis, A.K., Loulakakis, K.A., Hassan, M.M. and Roubelakis-Angelakis, K.A. (1993) In : *Cellular and Molecular Aspects of the Plant Hormone Ethylene* (Eds. J.C. Pech, A. Latche and C. Balague), Kluwer Academic Pub., Netherlands pp. 117-122.

Kerguelen, V. and Hoddle, M.S. (1999) *Intern. J. Acarology*, **25** : 221-229.

Knight, R.J. (2002) History distribution and uses. In: Whiley, W.A., Scaffer, B.,Wolstenholme, B.N. (Eds.), The Avocado: Botany, Production and Uses. CABInternational, Wallingford, pp. 1–15.

Koen, J.J. and Smart, G. (1973) *Citrus and Subtrop. Fruit J.*, **474** : 4, 5, 9.

Koen, T.J. and Plessis, S.F. Du (1991) *Yearb. South Afr. Avocado Growers' Assoc.*, **14** : 19-21.

Kohne, J.S. (1992) *Yearb. South Afr. Avocado Growers' Assoc.*, **15** : 68.

Kohne, J.S. and Kremer-Kohne, S. (1991) *Yearb. South Afr. Avocado Growers' Assoc.*, **14** : 42-43.

Korsten, L., Villiers, E.E. De, Jager, E.S. De, Cook, N. and Kotze, J.M. (1991) *Yearb. South Afr. Avocado Growers' Assoc.*, **14** : 57-59.

Kremer-Kohne, S. and Kohne, J.S. (1992) *Yearb. South Afr. Avocado Growers' Assoc.*, **15** : 69.

Kremer-Kohne, S., Kohne, J.S. and Schutte, J.M. (1993) *Yearb. South Afr. Avocado Growers' Assoc.*, **16** : 33-36.

Krezdron, A.H., Marte, D. and Marte, R.J. (1976) *Proc. Fla. St. Hort. Soc.*, **89** : 261-263.

Kruger, F. and Kritzinger, M. (1999) *Neltropika Bulletin*, **306** : 5-6.

Lahav, E. and Kadman, A. (1980) *IPI-Bulletin*. No. **6**, Grutter Druck.

Lahav, E. and Kalmar, D. (1977) *Australian J. Agri. Res.*, **28** : 869-877.

Lahav, E. and Trochoulias, T. (1982) *Australian J. Agric. Res.*, **33** : 549-558.

Lahav, E., Bar, Y. and Kalmar, D. (1990) *Commun. Soil Sci. Plant Anal.*, **21** : 1353-1365.

Lahav, E., Kalmer, D. and Bar, Y. (1985) *Alon Hanotea*, **39** : 973-976.

Lavi, U., Lahav, E., Sharon, D., Gazit, S., Desgani, C. and Hillel, J. (1992) *Acta Horticulturae*, **317** : 219-224.

Lavi, U., Sharon, D., Kaufman, D., Sadda, D., Chapnik, A., Zamet, D., Degani, C., Lahav, E. and Gazit, S. (1997) *HortScience*, **32** : 151.

Lesley, J.W. and Bringhurst, R.S. (1951) *Calif. Avocado Soc. Yearb.*, **34** : 169.

Lewis, E. (1992) *Inligtingsbulletin - Instituut vir Tropiese en Subtropiese Gewasse*, **241** : 7-9.

Liu, X., Robinson, P.W., Madore, M.A., Witney, G.W. and Arpaia, M.L. (1999) *J. Amer. Soc. Hort. Sci.*, **124** : 671-675.

Lizana, L.A. and Figueroa, J. (1997) *Proc. Seventh Int. Controlled Atmosphere Research Conf.*, Davis, California, USA. No. 17, pp.219-224.

Lonsdale, J.H. (1991) *Yearb. South Afr. Avocado Grower' Assoc.*, **14** : 61-62.

Lonsdale, J.H. and Scott, M.P. (1991) *Yearb. South Afr. Avocado Growers' Assoc.*, **14** : 63-64.

Lopez-Herrera, C.J., Perez-Jimenez, R.M., Basallote-Ureba, M.J., Bonilla, T.Z. and Meiero-Vara, J.M. (1997) *Plant Pathology*, **46** : 329-340.

Loupassaki, M., Vasilakakis, M. and Androulakis, I. (1997) *Euphytica*, **94** : 247-251.

Lovatt, C. J. (1990) *Calif. Avocado Soc. Yearb.*, **74** : 193-199.

Maganjo, M. (1975) *Acta Horticulturae*, **49** : 91-94.

Mandemaker, A.J. (2007) *New Zealand Avocado Growers' Association Annual Research Report* **7:** *1 – 10.*

Mans, C.C. and Hattingh, D.A. (1992) *Yearb. South Afr. Avocado Growers' Assoc.*, **15** : 94-96.

Manthey, J.A. and Crowley, D.E. (1997) *J. Plant Nutr.*, **20** : 683-693.

Marriot, J. and Proctor, F.J. (1984) *Outlook on Agriculture*, **9** : 233-239.

Mazliak, P. (1971) *Fruits*, **26** : 615.

McKellar, M.A., Buchanan, D.W., Ingram, D.L. and Campbell, C.W. (1992) *HortScience*, **27** : 341-343.

Meir, S., Maiman, D., Hyman, J.Y., Akerman, M., Zauberman, G. and Fuch, Y. (1998) *Acta Horticulturae*, **464** : 397-402.

Menge,J.A., G.W. Douhan, B.McKee, E.Pond, G.S. Bender and B.Faber (2012) *Hortscience*, **47:** 1191–1194.

Merwe, M. De V. van Der and Kotze, J.M. (1991) *Yearb. South Afr. Avocado Growers' Assoc.*, **14** : 85-86.

Metzidakis, J. and Sfakiotakis, E. (1993) In : *Cellular and Molecular Aspects of the Plant Hormone Ethylene* (Eds. J.C. Pech, A.Latche and C. Balague), Kluwer Academic Pub. Netherlands, pp. 129-135.

Mhameed, S., Sharon, D., Kaufman, D., Lahav, E., Hillel, J., Degani, C. and Lavi, U. (1997) *Theor. Appl. Genet.*, **94** : 279-286.

Mizrach, A. and Flitsanov, U. (1999) *J. Food Engineering*, **40** : 139-144.

Monselise, S. P. and Goldschmidt (1982) *Hort. Rev.*, **4** : 128-173.

Montoya, M.M., Plaza, J.L. and Lopez-Rodriguez, V. (1994) *Lebensm. Wiss. Technol.* **27:** 29–33.

Mooney, P.A. and VanStaden, J. (1987) *Can. J. Bot.*, **65** : 622-626.

Natan, R., Guil, I., Klein, I, Spieler, G., Harler, D. and Pedaov, A. (1991) *Alon Hanotea*, **45** : 899-906.

Nogalingam, T. (1993) In : *Encyclopaedia of Food Science, Food Technology and Nutrition* (Eds : R. Marce, R.K. Robinson, and M.J. Sadler) Academic Press, London, p. 293.

Offer, R., Elimelech, M. and Blumenfeld, A. (1986) *Alon Hanotea*, **44** : 77-82.

Özdemir A.E., Çandýr E.E., C. Toplu, M.Kaplankýran, T.K. Demirkeser and E. Yýldýz. (2009) *African J. Biotech.*, **8** : 1878-1886.

Pantastico, Er. B. (1975) In: *Post-harvest Physiology, Handling and Utilization of Tropical and Subtropical Fruits and Vegetables*, AVI Pub. Co.,Westport, Conn.

Parodi, M.G. and Lerner, M. (1995) *Proc. Interamericen. Soc. Trop. Horticulture*, **39** : 86-93.

Pegg, K. G. and Whiley, A.W. (1987) *Yearb. South Afri. Avocado Grower's Assoc.*, **10** : 94-96.

Pennock, W. (1970) *Plant Grafting Techniques for Tropical Horticulture*, Bull. 221, Univ. Puerto Rico.

Pesis, E., Faiman, D. and Dori, S. (1998) *Post-harvest Biol. Technol.*, **13** : 245-253.

Pesis, E., Marinansky, R., Zauberman, G. and Fuchs, Y. (1993) *Acta Hort.*, **343** : 251-255.

Pesis, E., Marinansky, R., Zauberman, G. and Fuchs, Y. (1994) *HortScience*, **29:** 1042–104

Pinheiro, R.V.R. and Pinheiro, F.J.B. (1971) *Revistaceres*, **18** : 351-357.

Pliego-Alfaro, F. (1981) *Ph.D. Thesis*, California University, Riverside, California.

Pliego-Alfaro, F. and Murashige, T. (1988) *South Afr. Avocado Growers' Assoc. Yearb.*, **10** : 36-41.

Pliego-Alfaro, F., Encina, C. and Barcelo-Munoz, A. (1987) *South Afr. Avocado Growers' Assoc. Yearb.*, **10** : 36-41.

Popenoe, W. (1920) In : *Manual of Tropical and Subtropical Fruits*. Hafner Press, New York, p. 9.

Proctor, F.J. and Caygill, J.C. (1985) *Proc. 39th Easter School*, University of Nottingham, UK, pp. 317-322.

Prusky, D., Leikin-Frenkel, A. and Madi, L. (1999) *In Biology and Biotechnology of the Plant Hormone Ethylene II. Proc. of the EU-TMR-Euroconference Symp.* Thira (Santorini) Greece, September 5-8, 1998 (Eds. A.K. Kanellis, C. Chang, H. Klee and A.B. Bleecker) Kluwer Academic Publishers, pp. 277-284.

Purseglove, J.W. (1974) *Tropical Crops – Dicotyledons*, Longman, London, pp. 192-198.

Rao V.N. Madhava and Khader, J.B.M. Md. A. (1978a) *Food Fmg. Agric.*, **9** : 262-264.

Rao V.N. Madhava and Khader, J.B.M. Md. A (1978b) *Indian Hort.*, Jan.-Mar.

Rendón-Anaya, M. *et al.* (2019) *Proceedings of the National Academy of Sciences*, **116**(34): 17081-17089; DOI: 10.1073/pnas.1822129116

Reuhle, G.D. (1940) *roc. Fla. St. Hort. Soc.*, **56** : 150-152.

Reuveni, B., Winer, L. and Shneur, I. (1990) *Alon Hanotea*, **45** : 191-198.

Reuveni, O. and Goren,M. (1982) *Abst. 21st. International Hort. Congress*, Hamburg, Abst. No. 134.

Reuveni, O. and Raviv, M. (1976a) *Special Publication*, Agricultural Research Organisation, Isreal, **65** : 33-35.

Reuveni, O. and Raviv, M. (1976b) *Special Publication*, Agricultural Research Organisation Isreal, **65** : 37-39.

Robbertse, P.J., Coetzer, L.A. and Janse van Vuuren, B.P. (1992) *Yearb. South Afr. Avocado Growers' Assoc.*, **15** : 89-93.

Robertson, B.L. (1971) *Fmg. South Afr. Subtrop. Fruit Series*, **10** : 58.

Rousseau, G.G. (1981) *Yearb. South Afr. Avocado Growers' Assoc.*, **5** : 36-37.

Rowell, A.W.G. (1979) *South Afr. Avocado Growers' Assoc. Res. Rep.*, **3** : 35.

Salazar-Garcia, S. and Cortes-Flores, J.I. (1986) *Calif. Avoc. Soc. Yearb.*, **70** : 165-74.

Salazar-Garcia, S. and Lovatt, C. J. (1998) *J. Amer. Soc. Hort. Sci.*, **123** : 791-797.

Salazar-Garcia, S. and Lovatt, C. J. (1999) *J. Hort. Sci. Biotech.*, **74** : 69-73.

Salazar-Garcia, S. and Lovatt, C.J. (2000) *J. Amer. Soc. Hort. Sci.*, **125** : 25-30.

Salazar-Garcia, S., Lord, E.M. and Lovatt, C.J. (1998) *J. Amer. Soc. Hort. Sci.*, **123** : 537-544.

Salomon, E. (1984) *Acta Hort.*, **149** : 3.

Samson, J.A. (1980) *Tropical Fruits*. Longman, London.

Sanchez Colin, S. and Barrentos Priego, A. (1987)) *Yearb. South Afr. Avocado Growers' Assoc.*, **10** : 24-26.

Sanchez-Romero, C., Garcia-Gomez, M.L., Pliego-Alfaro, F. and Heredia, A. (1993) *J. Plant Growth Regul.*, **12** : 95-100.

Schaffer, B. and A.W. Whiley. (2003) *Proc. V World Avocado Congress* (*Actas V Congreso Mundial del Aguacate*) 2003. pp. 335-342.

Schoeman, M. (1996) *Inligtingsbulletin - Instituut vir Tropiese en Subtropiese Gewasse*, **290** : 4.

Scholefield, P.B., Sedgley, M. and Alexander, D.M. (1985) *Scientia Hort.*, **25** : 99-110.

Scholefield, P.B., Walcott, J.J., Kriedemann, P.E. and Ramadasan, A. (1980) *Calif. Avocado Soc. Yearb.*, **64** : 93.

Schroeder, C.A. (1944) *Calif. Avocado Soc. Yearb.*, **28** : 39-40.

Schroeder, C.A. (1958) *Indian J. Hort.*, **1d** : 116-128.

Schroeder, C.A. (1971) *Calif. Avocado. Soc. Yearb.*, **36** : 18-23.

SeaLand (1991) *Shipping guide to prerishables.* SeaLand Service Inc., P.O. Box 800, Iselim, New Jersey 08830 USA.

Sedgley, M. (1977) *J. Hort. Sci.*, **52** : 135-141.

Sedgley, M. (1980) *Ann. Bot.*, **46** : 771-777.

Sedgley, M. and Annells, D.M. (1981) *Sci. Hort.*, **14** : 27-33.

Sedgley, M., Scholefield, P.B. and Alexander, D.Mc. E (1985) *Sci. Hort.*, **25** : 21-30.

Seligmann, R., Lowengar-Aycicegi, A., Lahav, E., Regev, Y. and Fetterman, T. (1999) *Alon Hanotea*, **53** : 494-497.

Sharon, D., Adato, A., Mohmeed, S.S., Lavi, U., Hillel, J., Gomolka, M., Epplen, C. and Epplen, J.T. (1995) *HortScience*, **30** : 109-112.

Sippel, A.D., Conradie, W. and Claassens, N.J.F. (1992) *Yearb. South Afr. Avocado Growers' Assoc.*, **15** : 72-74.

Sippel, A.D., Holmes, M.A., Claassens, N.J.F. and Claassens, V. (1993) *Yearb. South Afr. Avocado Growers' Assoc.*, **16** : 90-95.

Sippel, I. and Tomlinson, I.R. (1996) *Inligtingsbulletin - Instituut vir. Tropiese en Subtropiese Gewasse*, **286** : 1-3.

Skene, K.G.M. and Barlass, M. (1988) In : *Biotechnology in Agriculure and Forestry*, Vol. 5 (Ed. Y.P.S. Bajaj), Springer-Verlag, Berlin, Heidelberg, New York.

Snowdon, A.L. (1990) *A Colour Atlas of Post-harvest Diseases and Disorders of Fruits and Vegetables*.Vol. I. *General Introduction and Fruits*. Wolfe Scientific Ltd., London.

Sohi, H.S. and Sridhar, T.S. (1974) *FAO Plant Protection Bull.*, **22** : 73-74.

Spalding, D.H. and Reeder, W.F. (1975) *Phytopathology*, **bd**: 458-460.

Sridhar, T.S. (1978) *Curr. Sci.*, **47** : 71.

Stassen, P.J.C. (1999) *Neltropika Bulletin*, **306** : 23-27.

Steyn, E.M.A. (1993) *Yearb. South Afr. Avocado Growers' Assoc.*, **16** : 9-14.

Stirling, A.M., Pegg, K.G., Hayward, A.C. and Stirling, G.R. (1999) *Australian J. Agric. Res.*, **50** : 1459-1468.

Stolzy, L.H., Zentmyer, G.A., Klotz, L.J. and Labanauskas, C.K. (1971) *Proc. Amer. Soc. Hort. Sci.*, **90** : 67.

Storey, W.B. (1972) *Avocado*, Wageningen.

Storey, W.B., Bob Bergh, and G.A. Zentmye (1986) *California Avocado Society 1986 Yearbook* **70**: 127-133.

Swarts, D.H. (1984) *Yearb. South Afr, Avocado Growers' Assoc.*, **7** : 15-19.

Swisher, H.E. (1984) *J. Amer. Oil Chem. Soc.*, **65** : 1704.

Takele, E. Meyer, J.L., Arpaia, M.L., Yates, M.V., Bender, G. and Witney, G. (1990) *Calif. Avocado Soc. Yearb.*, **74** : 85-98.

Tango, J.S., Dacosta, S.T., Antunes, A.J. and Figueriedo, I.B. (1972) *Fruits*, **27** : 143.

Thompson, A.K. (1996) *Post-harvest Technology of Fruits and Vegetables*, Blackwell Sci. Ltd. Great Britain.

Thorp, T.G. and Sedgley, M. (1993) *Sci. Hort.*, **56** : 147-156.

Tingwa, P.O. and Young, R.E. (1974) *J. Amer. Soc. Hort. Sci.*, **99** : 540-542.

Tomer, E. and Gottreich, M. (1978) *Bot. Gaz.*, **139** : 81-86.

Tomer, E., Gottreich, M. and Gazit, S. (1976) *J. Amer. Soc. Hort. Sci.*, **101** : 620-623.

Torres, A.M. (1983) In : *Isozymes in Plant Genetics and Breeding*, Part B (Eds. S.D. Tanksley and T. Orton), Elsevier, Amsterdam, pp. 401-421.

Torres, A.M. (1984) *J. Hered.*, **75** : 300-302.

Torres, A.M., Mau-Lastovicka, T., Vithanage, V. and Sedgley, M. (1986) *J. Hered.*, **77** : 445-450.

Uzquiano, M.C. (1975) *Proc. Fla. St. Hort. Soc.*, **71** : 76-79.

Vakis, N.J. (1982) *J. Hort. Sci.*, **67** : 221-226.

Valmayor, R.V. (1964) *Ph.D. Thesis*, Fruit Crops Dept., Univ. of Fla., Gainesville.

van Den Dool, B. and Wolstenholme, B.N. (1983) *Yearb. South Afr. Avocado Growers' Assoc.*, **6** : 34-40.

Vidal, M.T., Azcon-Agnilar, C., Barea, J.M. and Pliego-Alfaro, F. (1992) *HortScience*, **27** : 785-787.

Vrecenar-Gadus, M. and Ellstrand, N.C. (1985) *Scientia Hort.*, **27** : 215-221.

Walt, M. van der, Davie, S.J. and Smith, D.G. (1993) *Yearb. South Afr. Avocado Growers' Assoc.*, **16** : 82-85.

Whiley, A. W. and Schaffer, B. (1994) In : *Handbook of Environmental Physiology of Fruit Crops*. vol. 2. *Subtropical and Tropical Crops* (Eds. B. Schaffer and P.C. Andersen), CRC Press Inc., USA, pp. 3-35.

Whiley, A. W., Chapman, K.R. and Saranath, J.B. (1988) *Australian J. Agric. Res.*, **39** : 457.

Whiley, A.W. (1990) *Yearb. South Afr. Avocado Growers' Assoc.*, **13** : 28-30.

Whiley, A.W., Pegg, K.G. and Saranah, J.B. (1984) *Calif. Avocado. Soc. Yearb.*, **68** : 179.

Whiley, A.W., Rasmusen, T.S., Saranah, J.B. and Wolstenholme, B.N. (1996b) *Sci. Hort.*, **66** : 35-49.

Whiley, A.W., Rasmussen, T.S., Saranah, J.B. and Wolstenholme, B.N. (1996a) *Sci. Hort.*, **66** : 23-34.

Whiley, A.W., Schaffer, B. and Lara, S.P. (1992) *Tree Physiol.*, 85-94.

Whiley, A.W., Wolstenholme, B.N., Saranah, J.B. and Anderson, P.A. (1990) *Acta Horticulturae*, **275** : 153.

Whitmore, J.S. (1986) *NPWCAR Project*, CSIR, Pretoria.

Whitney, G. and Arpaia, M.L. (1991) *Calif. Avocado Soc. Yearb.*, **75** : 63.

Wills, R.B.H. and Tirmazi, S.I.H. (1982) *Scientia Hort.*, **16** : 323 : 330.

Witjaksono and Litz, R.E. (1999) *Plant Cell Tissue Organ Culture*, 58 : 141-148.

Wolstenholme, B.N. (1985) *Acta Horticulturae*, **175** : 121-126.

Wolstenholme, B.N. (2010) *http: //www.avocadosource.com/papers/southafrica_papers/ wolstenholmenigel2010.pdf.*

Wolstenholme, B.N. and A. Sheard. (2012) https: //www.researchgate.net/ publication/262494704.

Wolstenholme, B.N., Kaiser, C. and Palmer, P. (1991) *Yearb. South Afr. Avocado Growers' Assoc.*, **14** : 15-18.

Woolf, A.B. (1997) *HortScience*, **32:** 1247–1251.

Young, R.E. (1979) *Calif. Avocado Soc. Yearb.*, **63** : 79-80.

Young, T.W. and Koo, R.C.J. (1976) *Proc. Fla. St. Hort. Soc.*, **89** : 238.

Zamorano, J.P. and Merodio, C. (1993a) In : *Cellular and Molecular Aspects of the Plant Hormone Etylene* (Eds. J.C. Pech, A. Latche, and C. Balague) Kluer Academic Pub., Netherlands, pp. 158-59.

Zamorano, J.P. and Merodio, C. (1993b) *Acta Horticulturae*, **343** : 275-276.

Zauberman, G. and Fuchs, Y. (1981) *HortSc*ience, **16** : 496-497.

Zauberman, G. and Schiffmann-Nadel, M. (1972) *J. Amer. Soc. Hort. Sci.*, **97** : 313-315.

Zauberman, G. and Schiffmann-Nadel, M. (1977) *Phytopathologische Zeitschrift*, **89** : 359-365.

Zauberman, G., Fuchs, Y. and Akerman, M. (1985) *Sci. Hort.*, **26:** 261–265.

Zauberman, G., Fuchs, Y., Yanko, U. and Akerman, M. (1988) *HortSc*ience, **23** : 588-589.

Zauberman, G., Schiffmann-Nadel, M. and Yanko, U. (1973) *HortScience*, **8** : 511-513.

Zauberman, G., Schiffmann-Nadel, M. and Yanko, U. (1977) *HortSc*ience, **12** : 353-354.

Zelitch, I. (1971) *Photosynthesis, Photorespiration and Plant Productivity*. Academic Press, New York.

Zentmyer, G.A. (1984) *Trop. Pest Management*, **30** : 388-400.

Zentmyer, G.A., Guillement, F.B., Harjung, M.K. and Zakai, A.I. (1977) *Calif. Avocado. Soc. Yearb.*, **61** : 76-79.

Zentmyer, G.A., Paulus, A.O. and Burns, R.M. (1967) *Calif. Agr. Expt. Sta. Ext. Ser. Cir.*, p. 511.

Zilberstein, M. (1999) *Alon Hanotea*, **53** : 372-375.

Zilkah, S., Yeselson, Y., David, I. and Klein, I. (2000) *Acta Hort.*, **511** : 107-113.

⑧

DRAGON FRUIT

B. Vanlalneihi and B.N. Hazarika

1.0 Introduction

Dragon fruit (*Hylocereus* sp.) is one of the most nutritious and exotic tropical crops. It is a favorite to many, particularly people of Asian origin and is a potential tropical and subtropical epiphytic fruit crop which has high demand in national and international markets. It is commonly known as pitaya, pitayaha, prickly pear, strawberry pear, wonderous fruit of the 21st century, *etc.* The fruit is named as pitaya meaning "the scaly fruit" because of the bracts or scales on the fruit skin of pitaya. The large scented flowers only bloom at night time for a single night; hence referred to as night blooming cereus, queen of the night, moon flower and/or lady of the night. The skin of the dragon fruit is a thin rind and usually covered in scales, and the center of the fruit is made up of a red or white, sweet tasting pulp which reminiscent kiwifruit. The flesh is mildly sweet. Because of its attractive colours, juicy, pleasant taste, exotic and impressive appearance, this fruit enjoys growing popularity in Australia, New Zealand, Israel, Spain, the southwestern US, Colombia, Ecuador, Mexico, Peru, Cambodia, Guatemala, Indonesia, Japan, Laos, Nicaragua, the Philippines, Taiwan, Thailand and Vietnam (Nobel and De la Barrera, 2002). The fruits of Pitayaha is also converted into juice or used to flavour other beverages. It is a remunerative, exotic promising fruit crop introduced in India and the production is gaining marginally and more recognition as a crop within the country.

2.0 Composition and Uses

The fruit is either eaten fresh or made into several processed products. Pitaya is used as food and for its curative properties since prehispanic times in Mexico (Clerck and Negreros-Castillos, 2000). The fruit is consumed raw or transformed into wine, juice, jelly, yogurt, jam, preserves, and other desserts (Herbach *et al.*, 2006; Shetty *et al.*, 2012). Addition of *H. monacanthus* or *H. undatus* pulp in yogurt improves milk fermentation rates, lactic acid content, syneresis percentage, antioxidant activity and total phenolic content (Zainoldin and Baba, 2012). *H. monacanthus* has nutritional properties; fruit pulp is rich in fiber, vitamin C, minerals and phytoalbumin contents, which confer high antioxidant values (Jafaar *et al.*, 2009). Dried juice contains high protein, fat, fiber, ash and antioxidants (Rebecca *et al.*, 2010; Tze *et al.*, 2012). Pitaya peels are a good source of natural phenols and flavonoids (Hua *et al.*, 2018). Phenolic compounds have been associated with colour, sensory qualities and nutritional properties. Phenolics act as antioxidants, due to the free radical scavenging activity of the phenol moiety (hydroxyl substituent on the aromatic ring). Beyond the protective behavior, caffeic acid is known to selectively block the biosynthesis of leukotrienes, components involved in immune regulation diseases, asthma, and allergic reactions (Yasuko *et al.*, 1984). The antioxidant capacity of purple-red pitaya is contributed by betalains while non-betalainic phenolic compounds contributed the least (Esquivel *et al.*, 2007b). Dragon fruit skin and pulp contain a rare organic acid - citramalic acid which can reduce facial skin wrinkle of people and gives a promising usage in the fields of medicine and cosmetics Wu *et al.* (2020); the citramalic acid has been detected in limited plants and the content were much lower than dragon fruit. Young stems *H. undatus* are consider having high nutritional composition and are consumed as vegetable (Castillo, *et al.*, 2005; Juárez-Cruz *et al.*, 2012). The fresh flower buds are also eaten as vegetables, while dried ones are used for homemade medicine (Ortiz-Hernández, 1999). In Taiwan, dry flowers are consumed as vegetable (Mizrahi and Nerd, 1999). The biochemical properties of pitaya differ with differences in the maturation stage and environmental growth variation of the fruits.

Table 1: Nutritional Composition, Minerals and Vitamins Content of Red Pitaya (per 100 g edible portion)*

Moisture	85.1 – 90.0 g	Iron	0.03 – 0.3 mg
Ash	0.5 – 1.2 g	Magnesium	13.7 – 26.4 mg
Carbohydrates	8.4 – 13.0 g	Potassium	158.3 – 437.4 mg
Protein	0.4 – 1.5 g	Sodium	14.3 – 35.6 mg
Fat	ND	Zinc	0.1 – 0.4 mg
Total dietary fiber	2.7 g	Calcium	1.6 – 6.7 mg
Energy	35.4 – 63.0 kcal	Vitamin A	85.2 – 891.0 μg
TSS	8-12°Brix	ß- carotene	1.4 ìg
pH	4.4	Lycopene	3.4 ìg
Titratable acidity (Lactic acid equivalent)	0.20 to 0.30 mg	Vitamin C	25.0 – 30.2 μg
Pulp/peel ratio	1.3 – 1.7	Vitamin E	0.26 mg

* Nurul and Asmah (2014); Mizrahi (2014); Charoensiri *et al.* (2009).

Table 2: Ascorbic Acid (AA) Content of Fruits Pulps of *Hylocereus* sp. (mg/100 g pulp)

Species	*AA content*	*References*
H. polyrhizus	8.0 ± 1.6	Lim *et al.*, 2007
	32.7 ± 1.6	Choo and Yong, 2011
H. undatus	31.1 ± 1.2	Choo and Yong, 2011

Table 3: Phenol Content of Fruits Pulps of *Hylocereus* sp. (mg of gallic acid equivalent/100 g pulp)

Species	*Phenol content*	*References*
H. polyrhizus	42.4 ± 0.04	Wu *et al.*, 2006
	21.0 ± 6.0	Lim *et al.*, 2007
	24.2 ± 1.0	Choo and Yong, 2011
H. undatus	28.7 ± 1.8	Choo and Yong, 2011

Table 4: Betacyanin Contents of Fruits (Peels and pulps) of *Hylocereus* sp. (mg betanin equivalents per 100 g dry pulp/peel)

Species	*Fruit part*	*Betanin Content*	*References*
H. polyrhizus	Pulp	10.3 ± 0.22	Wu *et al.*, 2006
		10.1 ±0.6	Tang and Norziah, 2007
	Peel	13.8 ± 0.85	
		6.7 ± 0.2	Tang and Norziah, 2007
H. undatus	Peel	101.04*	Mello *et al.*, 2015

Table 5: Relationship between Phytoconstituents of Dragon Fruit and Pharmacological Activities*

Pharmacological Activity	*Active Phytoconstituent*	*Class*
Antioxidant	Gallic acid, Vanillic acid, Syringic acid, Protocatechuic acid, p-hydroxybenzoic acid, p-coumaric acid	Polyphenols (Luo *et al.*, 2014)
	Betacyanin (Mahayothee *et al.*, 2019), Lycopene, β-carotene (Charoensiri *et al.*, 2009)	Pigments
Antimicrobial	Betacyanin (Yong *et al.*, 2018)	Pigments
Anticancer	β-amyrin, β-sitosterol, Stigmast - 4 - en-3 -one (Luo *et al.*, 2014)	Terpenoids
Nutraceutical activity	Vitamin C, B1, B2, B3	Polyunsaturated fatty acids, Vitamins, Oligosaccharides (Wichienchot *et al.*, 2010)
Colouring agent	Betacyanin (Priatni and Pradita, 2015)	Pigments

* Joshi and Prabhakar (2020).

2.1.0 Industrial Uses

Hylocereus fruits can be processed by freezing, concentration, dehydration, fermentation, thermal processing and chemical conservation. From the pulp and peel natural colourants and pectin can be extracted using a home artisanal or industrial technology (Esquivel *et al.*, 2007a). Pitahaya peel could potentially be used as a thickening agent and natural colourant (Harivaindaram *et al.*, 2008), in cosmetic hydrating creams (Stintzing *et al.*, 2002) or diet beverages ingredient. The mucilage extracted from fruit peel is highly perishable and requires preservatives or dehydration (Tze *et al.*, 2012), but due to its rheological qualities, its use as an encapsulating agent for active principles under optimized spray drying could be possible (García- Cruz, 2011).

The most important fruit pigments in red dragon fruit are the betalains such as the betacyanins and betaxanthins (Chang *et al.*, 1998; Wybraniec *et al.*, 2007). The production of natural colourant from red or purple pitahaya (*H. monacanthus* and *H. costaricencis*) pigments are potential colourant sources for the food industry

(Wybraniec and Mizrahi, 2002; Esquivel and Araya, 2012). These Betacyanin are very stable during processing and storage (Herbach *et al.*, 2006, 2007). The peel betalins remained stable over a wide pH range (3.2-7.0) and were resistant to heating (100°C) up to 10 minutes at pH range from 3.7 to 5.5. This feature allows the application of pitaya peel in low acidity foods with moderate heat treatment, or else to be added in food after heating (Mello *et al.*, 2015). Several betacyanines have been separated from *H. monacanthus*, such as betanin, isobetanin, phyllocactin, isophyllocactin, betanidin, isobetanidin, bougainvillein-R-I, hylocerenin and isohylocerenine (Wybraniec and Mizrahi, 2002; Wu *et al.*, 2006; Stintzing and Carle, 2007). Pigments identified either in *H. costaricensis* and *H. monacanthus* are similar (Esquivel *et al.*, 2007a); however, some *H. costaricensis* clones have greater betanin and isobetanin content, while others have higher phyllocactin and hylocerenin contents. In addition to that, new betalain types were found in some clones: neobetanin (14, 15-dehydrobetanin) and gromphrenin (Betanidin-6-O-glucoside). The pulp of *H. polyrhizus* is used for the production of red-violet coloured ice cream in Israel (Wybraniec and Mizrahi, 2002)

2.2.0 Medicinal Uses

Hylocereus undatus flowers are consumed as vegetable in China under the named "Bawanghua". It has several health purposes and has been commercialized as a healthy beverage in China (Yi *et al.*, 2012). The fruit possesses medicinal properties: it is known to prevent colon cancer and diabetes, neutralizes toxic substances such as heavy metals, reduce cholesterol and high blood pressure. It has the capacity to prevent cancer and bleeding, control high sugar level and promote dental health. Dragon fruit helps improve digestion and reduce fat. It helps maintain the health of eyes. *H. undatus* and *H. polyrhizus* contain about 50 per cent essential fatty acids: oleic acid (C18:1, 21–24 per cent), linoleic acid (C18:2, 48–50 per cent); linolenic acid (C18:3, 1.0–1.5 per cent), and cis-vaccenic acid (C18:1, 2.8–3 per cent) (Ariffin *et al.*, 2009). The fruits of *H. undatus* are against heart disease, diuretic and hypoglycemia (Argueta, 1994; Ankli *et al.*, 1999), wound disinfectant, tumor dissolution with stem sap (Mendieta and Del Amo, 1981), and dysentery cure (Hopkins and Stepp, 2012). *H. moncanthus* peel and pulp have phytochemically bioactive compounds which are potent nutraceutical for antioxidant and anti-proliferant (Wu *et al.*, 2006; Esquivel *et al.*, 2007b; Nurliyana *et al.*, 2010; Kim *et al.*, 2011; Tenore *et al.*, 2012); this compound is enrich with polyphenols, and peel could inhibit cancer cell growth (melanoma B16F10 and other types) (Kim *et al.*, 2011). *Hylocereus* peel has high antimicrobial activity, particularly against *Salmonella typhi* (Escobar *et al.*, 2010). *H. polyrhizus* extracts have greater antimicrobial activity against gram-positive and gram-negative bacteria than *H. undatus* (Nurmahani *et al.*, 2012). As a result of alteration in blood lipid, the pitaya based diet reduces the chance of cardiovascular diseases particularly coronary disease (Mohd *et al.*, 2009). *H. undatus* is used for abortion prevention (Ankli *et al.*, 1999). In terms of fertility, *H. costaricensis* fruit increased sperm count, sperm viability and production rate (Aziz and Noor, 2010).

Pitahaya seed oil contains phenolics, tocopherols, and sterols and are potent natural antioxidants source (Lim *et al.*, 2010). The seed linoleic acid concentration is greater than in flax seed, canola, sesame or grapevine. However, seed mass relative to the fruit is very low (1: 99) (Ariffin *et al.*, 2009). Nonetheless, with an exceptional

Uses of dragon fruit

Dragon fruit slices and extraction of fruit juice

Dried fruit (left), jam (middle) and concentrate juice (right) from dragon fruit

Syrup (left) and different value added products prepared from dragon fruit

Dragon fruit plant with fruits

Distribution of dragon fruit in the world

level of linoleic acid in *H. megalantu* (660 g/kg); *H. undatus* (540 g/kg) and *H. monacanthus* (480 g/kg); pitahaya seeds have high potential as source of antioxidant and essential fatty acids (Chemah *et al.*, 2010). In addition, oligosaccharides from *H. undatus* and *H. monacanthus* have pre-biotic characteristics, resistance to acid conditions in the stomach, and partial resistance to human á-amylase, and they also promote *Lactobacillus* and *Bifidobacterias* (Wichienchot *et al.*, 2010).

3.0 Origin and Distributions

The first literary record of pitahaya is in "Historia General y Natural de las Indias" (General and Natural History of the Indies), published in 1535 and written by Fernández de Oviedo y Valdés, the first New World relator. The phrase "De los cardones en que nasce la fructa que llaman pitahaya" (Cactus from which the fruit known as pitahaya is born), in chapter XXVI, could refer to the column-like cactus called pitaya in Mexico (Bravo-Hollis, 1978); but Olaya (1991) shows a fruit illustration from General and Natural History of the Indies, which highly resembles a *Hylocereus* fruit. "Pitaya" and "pitahaya" have the similar meaning in Central and South America; however, in Mexico, "pitahaya" is used for fruits from epiphytic cactus like *Hylocereus*, while "pitaya" is used for fruits from column-like cactus (Ortiz-Hernández, 1999). Nine *Hylocereus* species are distributed in Mexico; however only *H. undatus* (Berger) Britton and Rose, *H. purpussi* (Weing.) Britton and Rose, *H. triangularis* (L.) Britton and Rose, and *H. ocamponis* (Salm-Dyck) Britton and Rose are commonly grown. *H. costaricensis* Britton and Rose and *H. megalanthus* Bauer (K Schumann ex Vaupel) are distributed in Nicaragua; Colombia and Peru, respectively (Cálix De Dios, 2004; Cálix de Dios and Castillo, 2005; Hunt *et al.*, 2006). The vine-cacti pitaya of the Cactaceae, subfamily Cactoideae, tribe Hylocereeae, is known to have been used for thousands of years by the indigenous people of the Americas (Hernández and Salazar, 2012).

Most *Hylocereus* species are endemic to America; Mexico, Central and South America (Britton and Rose, 1963; Mizrahi *et al.*, 1997; Barthlott and Hunt, 1993; Pimienta-Barrios and Nobel, 1994). After the discovery of the America, Cactaceae family members were known in Europe. *H. undatus* was introduced to Southeast Asia in XVI Century *via* the Philippines (Casas and Barbera, 2002), or Indochina by Spaniards where it is considered a native species; this introduction was pivotal to dragon fruit cultivation in Southeast Asia region (Mizrahi *et al.*, 1997; Nerd *et al.*, 1999). In 1984, Israel introduced *Hylocereus* accessions from America, and carried out an extensive research in several agronomical, physiological and genetic fields to increase yield and fruit quality (Cohen and Tel-Zur, 2012). In India, *Hylocereus undatus* (Haw.) Britton and Rose is the most cosmopolitan species followed by *Hylocereus polyrhizus* (Web.) Britton and Rose. The French introduced dragon fruit in Vietnam (Mizrahi *et al.*, 1997) and since then Vietnam tremendously increased its Dragon fruit production. Vietnam is now the top exporter of Dragon fruits (Tel Zur, 2013; Mercado-Silva, 2018). In 1995 Vietnam was the first country to sell pitayas in world markets, under the name Dragon Pearl Fruit; however nowadays this crop is grown and marketed in over 20 countries as a new horticultural fruit crop. Vietnam is the leading producer and exporter of this fruit, far ahead of all other countries combined; hundreds of thousands of tons are shipped and sold from Vietnam around the world annually. The Vietnamese clones dominate the world pitaya markets. However, the taste of the Vietnamese fruit is quite bland, which is an obstacle for converting it into a major world fruit crop (Mizrahi, 2015).

Table 6: Center of Origin of *Hylocereus* Species

Species	*Origin*	*References*
H. undatus	Mexico or Colombia	Fouqué, 1972
	South America	Jorge and Ferro,1989
	Martinica or Colombia	Bravo-Hollis, 1978
H. megalanthus	Andean region (Colombia, Peru, Bolivia, Ecuador and Venezuela)	Mizrahi, 2014
H. triangularis or *H. trigonus*	Uruguay, Brazil and Colombia	Fouqué,1972
H. ocamponis	Uruguay, Brazil and Colombia	Fouqué, 1972

4.0 Species and Cultivars

Hylocereus species are tropical and subtropical hemi-epiphytic fruit belonging to the climbing cacti family *Cactaceae*. These cacti bear a glabrous attractive berry with large scales (Fournet, 2002). The vine-cacti belong to three separate genera. The three- ribbed shoots belong to *Hylocereus*, *Epiphyllum* has the two-ribbed shoots while *Selenicereus* genus has four and more ribbed shoots (Hunt *et al.*, 2006, Tel-Zur *et al.*, 2011a). *Selenicereus megalanthus* has combine characteristics of the first two genera (Mizrah, 2014). Recent taxonomic and molecular data suggest that *Selenicereus megalanthus* belongs to the *Hylocereus* genus, hence the new recent name *Hylocereus megalanthus* (Hunt *et al.*, 2006; Tel-Zur *et al.*, 2004a; and Plume *et al.*, 2013). The genus of *Hylocereus* contains 16 species of plants (Britton and Rose, 1963), five of these are planted as fruit trees including *H. purpusii, H. polyrhizus, H. costaricensis, H. undatus, and H. trigonus* (Le Bellec, 2003). Howevwe, Hunt *et al.* (2006) reported 14 *Hylocereus* species, which are endemic to Latin America. Cytological studies reveal that *Hylocereus* sp. are diploid (2n=22) (De Dios, 2004; Lichtenzveig, *et al.*,2000; Tel-Zur *et al.*, 2003; Tel-Zur *et al.*, 2004b) only *H. megalanthus* is tetraploid (Tel-Zur *et al.*, 2011a). The most commonly cultivated species are *H. undatus, H. monacanthus* (previously known as *H. polyrhizus*), *H. megalanthus* (previously known as *Selenecereus megalanthus* – also a vine cactus), and *H. costaricensis* (Bauer, 2003; Tel Zur, 2013). A pollination study by Lichtenzveig *et al.* (2000) confirmed self-incompatibility in *H. polyrhizus* and a weakened incompatibility reaction in *H. undatus* and *S. megalanthus*. Major crossability barriers do not exist between the *Hylocereus* sp. investigated (Lichtenzveig *et al.*, 2000). The successful reciprocal intergeneric crosses between *Hylocereus* sp. and *S. megalanthus*, suggested that *Hylocereus* sp. might be one of the diploid progenitors of the tetraploid *S. megalanthus* (Lichtenzveig *et al.*, 2000)

Species of dragon fruit

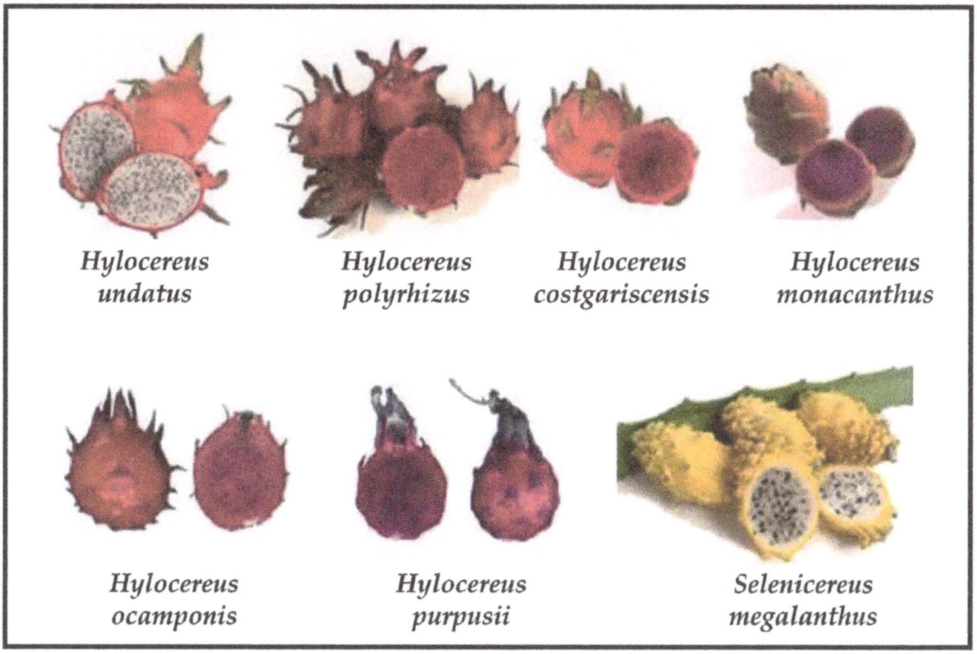

Fruits of different species

Fruits of different colour

Table 7: Classification of *Hylocereus* Species*

Species	*Description*
H. polyrhizus (Web.) Britton and Rose	Long flowers (25–30 cm) with margin; outer reddish perianth segments, especially at the tips; and rather short and yellowish stigma lobes. The fruit is oval shape, large in size, weighing about 300-600 grams, 32– 35 cm diameter and 13 – 15 cm long. The fruit has delicate and sweet flesh with intense red-purple flesh with pink skin. It has a lot of small black seeds which are rich in essential fatty acids.
H. costaricencis	Represented by vigorous vines, perhaps the stoutest of this genus. Flowers are almost similar to *H. polyrhizus* but stems are waxy white. The scarlet fruit is ovoid, weighing 250–600 g, 10–15 cm diameter and covered with scales that vary in size. The fruit has violet red flesh, pink skin, with many small black seeds, pleasant flesh texture and good taste.
H. undatus (Haw.) Britton and Rose	The long green stems are more or less thorny when age. Flowers are very long (up to 29 cm), outer perianth segments are green (or yellow-green) and inner perianth segments pure white. The oblong rosy- red fruit weight about 300–800 g, 15–22 cm long, red and green at the tips and covered with large and long scales; it has a white flesh with pink skin. It has numerous tiny black seeds, pleasant flesh, texture and good taste.
H. trigonus (Haw.) Saff	Stems are slender, green with margins, devoid of thorns. The areoles are located on the top of the rib's undulation. Spines, at first greenish, soon turn dark brown. The red fruit is ovoid or oblong, becoming nearly smooth, 7–9 cm diameter; weighing 120–250 g. The white flesh has many small black seeds, pleasant flesh texture, with mild flavour.
H. megalanthus (*FSelenicereus megalanthus* A. Berger Riccob).	Yellow pitaya plants are self-fruitful; more sun tolerant than most *Hylocereus* species but sensitive to cold. Fruits are medium-sized, weighing 250g and oblong in shape. The peel is yellow with tubercles and spines, which fall off when the fruit is ripe. The white pulp is sweet and tasty, with numerous small edible seeds.

* Britton and Rose (1963); Dag and Mizrahi (2005); Ariffin *et al.* (2009).

Table 8: Differences between *Selenicereus megalanthus* (Real yellow pitaya) and all other *Hylocereus* Clones (Red pitayas)*

Yellow Pitaya Clones	*Hylocereus spp Clones*
Spiny fruit	No spines on fruit
Tiny fruit scales	Big fruit scales
Three years to production	One year to production
Weak vegetative growth	Vigorous vegetative growth
Small weak root system	Well-developed root system
Most sensitive to low temperature	Relative resistance to low temperature
Superior taste	Taste from bland to good
Long shelf-life	Short shelf-life
Long fruit development time	Short fruit development time
Flowering in autumn	Flowering in summer
Tetraploid, self-compatible	Diploid only 10 per cent self-compatible
Fruit size is small-medium	Fruit size is medium-large
Seed number in fruits is low	Seed number in fruits is high

Yellow Pitaya Clones	*Hylocereus spp Clones*
Stomata are big	Stomata are small
Many aborted seeds	All seeds are viable
Parenchyma and chlorenchyma are mixed, as one layer	Parenchyma and chlorencyma are separate layers.
All clones sensitive to nematodes	Many genotypes are resistant to nematodes
Seeds are big	Seeds are small

* Mizrahi (2015).

4.1.0 Cultivars

Variation exists in shape, size and colour of pulp between the varieties. At present, very little knowledge is available on varietal and production aspects (Karunakaran and Arivalagan, 2019). Guanhuahong (red peel with red pulp), Guanhuabai (red peel with white pulp) and Guanhuahongfen' (red peel with pink pulp) are the cultivars from China (Hua *et al.*, 2018). Golden Pitayas (yellow clones of the *Hylocereus undatus*) and Pitaya Amarilla (yellow clones of *Hylocereus costaricensis*) are popular cultivar of Israel (Mizrahi, 2014). For the U.S. market, American Beauty, Armando, Bien Hoa Red, Bien Hoa White, Bloody Mary, Cebra, Colombiana, Costa Rican sunset, Delight, Haley's Comet, Mexicana, Orejona, Physical Graffiti, Purple haze, Rosa, San Ignacio, Seoul Kitchen, Valdivia Roja, Vietnamese Giant, Vietnam White and Yellow Dragon are available (Bozkurt *et al.*, 2020; Khalid, 2020).

5.0 Breeding and Improvement

In a plant breeding programme, the presence of variability and knowledge of genotypic and phenotypic correlation has been considered essential and pre-requisite for hybridization programme in obtaining progenies with desirable characters like higher yield, earliness, good quality and disease resistance. Such study preferred the genetically divergent parents in obtaining desirable combinations among segregating generations. The initial step of Dragon fruit improvement requires collection genotypes from cacti collectors, botanical gardens, private gardens, wild types from countries of origins *etc*. The plant material offers us an opportunity to identify elite lines in terms of yield, fruit quality, resistance to disease, and drought tolerance, enhances our potential to improve breeding efficiency.

Hylocereus undatus yield fruits in both cross-pollination and self-pollination (Valiente-Banuet *et al.*, 2007), although self-incompatible accessions have been reported (Castillo-Martínez *et al.*, 2005); only 10 per cent of *Hylocereus* species *are* self-compatible (Mizrahi, 2015). Cross-pollinated plants produce fruit better than self-pollination (Nerd *et al.*, 1999; Valiente-Banuet *et al.*, 2007 and Weiss *et al.*, 1994). Because the flowers are chiropterophilous, their most common visitors are bats (*Leptonycteris curasoae* and *Choeronycteris mexicana*), sphingid moths (*Sphinx ligustri*) and honeybees (Weiss *et al.*, 1994; Nerd and Mizrahi, 1997; Lichtenzveig *et al.*, 2000; Valiente-Banuet *et al.*, 2007). When cross pollination is performed by bees, fruit weight is greater than with artificial pollination; fruits also tend to have a longer form with cross pollination within the same species (Lichtenzveig *et al.*, 2000).

The major barrier for intra and inter specific hybridization of *Hylocereus* sp. are auto infertility (Weiss *et al.*, 1994, Ramírez, 1999; Castillo *et al.*, 2005) and asynchronicity in the flower aperture. This has resulted in labour intensive manual cross pollination. Yet, fertility between species is evident, since cross pollination among several species promotes 100 per cent fruit set. The diploid *Hylocereus* species bear large (250-800 g) flavourless, red-purple fruits, whereas the tetraploid *H. megalanthus* bears sweeter, but smaller (180-250 g), yellow fruits (Tel-Zur *et al.*, 2003; Tel-Zur *et al.*, 2011a). With the aim of producing elite lines with improved fruit quality, grafting by using *H. megalanthus* rootstock (Mizrahi, 2015) and interspecific-interploid crosses were performed with the aim to combine the size and attractiveness of the diploids with the fruit quality of the tetraploid species (Tel-Zur *et al.*, 2004b). Crosses between *H udatus* and *H. polyrhizus* yield hybrid with increased sugar levels to 8.40 °Brix against uncontrolled pollen flow 6.45 °Brix (Subandi *et al.*, 2018). *H. megalanthus* x *H. undatus* crosses (Tel-Zur *et al.*, 2003: Tel-Zur *et al.*, 2004b), did not yield true hybrids or chimeras but for the reciprocal cross, tetraploids, pentaploids, hexaploids, and 6x-aneuploids, (Tel-Zur *et al.*, 2003), thereby indicating a uni-directional gene flow among these species. The tetraploids exhibited the morphological features of the maternal parent and were therefore not considered true hybrids; rather, they were considered to be either a result of self-pollination due to contamination during the hand-cross pollination process or to be of somatic origin, since polyembryony has been reported in *H. megalanthus* (Cisneros *et al.*, 2011). Among all these polyploids the triploids yielded two commercial hybrids which exhibit good taste, long shelf-life and ripening in late autumn to mid-winter, differing from all other *Hylocereus* genotypes. The fruits are spiny, but to a much lesser extent than the *S. megalanthus*. All these crosses enable us to find dominant, co-dominant and recessive characteristics which will help us in future breeding (Tel-Zur 2004b). The inter generic crossing between *Epiphyllum* fruits that reminiscent passion fruit flavour; the crosses with *Hylocereus* genotypes may be of special interest; on the other *S. megalanthus* have so far yielded un-tasty. Five hybrids were developed by Tel-zur *et al.* (2012); all of which were a cross between the triploid hybrid 'S-75' (*H. monacanthus* x *H. megalanthuus;* 2n=3x=33) and the diploid *Hylocereus undatus;* 2n=2x=22. The hybrids were found to be fully self compatible; they exhibited similar fruit weights and similar total seed numbers following cross or self pollination. Flow cytometric analyses revealed that all the self compatible hybrids were tetraploids, suggesting a possible relation between self compatible and ploidy level (Tel-zur *et al.*, 2012). These interspecific-interploid hybrids performed better than their counterpart parents under extreme high temperature of 45/35 °C (de Oliveira *et al.*, 2020); indicating the hybridization strategy followed by selection is likely to improve *Hylocereus* cultivar.

Doubling the chromosomes also solved this problem by breaking incompatibility (Cohen and Tel-Zur, 2012). By applying colchicines and/or oryzalin, to germinating seeds and to areoles, chromosome doubling was achieved (Tel-Zur *et al.*, 2011b), by crossing between genotypes of various ploidy where some offspring were found to be of different ploidy than the parents (Tel-Zur *et al.*, 2004b), or alternatively - by androgenesis and gynogenesis (Garcia *et al.*, 2009 a,b). Tel-Zur *et al.* (2004b) attempt inter-clonal, inter-specific and inter-generic crosses to produce hybrids. The crosses yield desirable first generation hybrid clones that dominates pitaya growing region in Israel and abroad. The auto polyploidyzation technique resulted hybrids show morphological changes and are auto fertile, thus avoiding artificial pollination. (Cohen and Tel-Zur, 2012).The *Hylocereus* sp hybrids enable staggered harvesting for year round production. Different hybrids mature in several seasons; inter-clonal or interspecific ripen in summer; while the inter-generic hybrids between *Hylocereus* sp and *S. megalanthus* ripen from the autumn to the end of winter; and the real yellow pitaya (*S. megalanthus*) ripen from the end of winter to the late spring (Tel-Zur *et al.*, 2004b). Tel-Zur *et al.* (2020) elucidate the timing and developmental stage of the chromosome doubling of putative hybrids obtained from an embryo rescue procedure following controlled *H. megalanthus* x *H. undatus* crosses. Among different ploidy level only allohexaploids constitute empirical evidence of "hybridization followed by chromosome doubling." All the allohexaploids obtained from the embryo rescue technique where those that regenerated from fertilized ovules 10 days after pollination (at the pro-embryo stage), showing that the chromosome doubling event occurred at a very early development stage, *i.e.*, at the zygote stage or shortly after zygote formation. Therefore the zygotic or postzygotic somatic chromosome doubling constituted the mechanism for the formation of the hexaploid and 6xaneuploid hybrids (Tel-Zur *et al.*, 2003; Tel-Zur *et al.*, 2004b). Chromosome doubling during the very early stage of embryo development probably facilitates the survival of hybrids with specific genomic combinations and maternal/paternal ratios in the endosperm Tel-Zur *et al.* (2020).

6.0 Propagation

6.1.0 Seeds

Hylocereus propagation can be through seed. Seed viability varies among *Hylocereus* species and within the same species (Cisneros and Tel-Zur, 2012). Tripathi *et al.* (2014) reported that the seed propagated seedlings remain smaller with thin stem even after one year of planting. Further the plants produced from seeds are not true to type and there lot of variability among the plants. Therefore, seed propagation is not commercial recommended for multiplication of dragon fruit.

6.2.0 Vegetative

Vegetative propagation is used more often. Dragon fruit plants are easily multiply through stem cutting of 20-25cm long. After the fruiting seasons, the cuttings are taken from elite mother plants. The cutting should be prepared one - two days prior to planting; allowing to dry ooze the latex oozing out of cut end. The cuttings are planted in 12 x 30 cm size polyethylene bags, filled with 1:1:1 ratio of soil, farmyard manure and sand. The bags are kept at a shady place for rooting. For prevention of rotting, excess moisture should be avoided. These cutting roots profusely and become ready for planting within 5-6 months after cutting (Tripathi, *et al.*, 2014). However, *H. undatus* cuttings do not form callus; adventitious roots differentiate in the peryclicle (Cavalcante, 2008). Roots from the cuttings are very sensitive to water salinity, 50 per cent of cuttings die when the electrical conductivity of water is 4.0 dS/m, and organs of the survival cuttings show growth problems

Propagation of dragon fruit

Propagation by seed

Propagation by cutting

Sapling production by vine cutting

Cultivation of dragon fruit

Trellis for planting

Planting of dragon fruit

Dragon fruit plant grown in container

(Cavalcante *et al.*, 2007). The cutting length influences the growth and development of dragon fruit. Kakade *et al.* (2019) found 20–25 cm long cuttings of *Hylocereus undatus* was statistically on par with 35– 40 cm long cuttings which has highest number of 1st order roots. The dependency of shoot upon root and vice versa occurs throughout the plant life (Kozlowski, 1971) and synergetic role between them greatly influence and regulates overall plant growth and development (Davies and Zhang, 1991). Longer cutting lengths has the ability to synthesize higher photosynthates or store more carbohydrates, maintain balance between photosynthesis and transpiration (Leakey and Coutts, 1989) and source-sink interaction (Gifford *et al.*, 1984), which are very much essential for rooting.

7.0 Area and Production

Vietnam is currently the biggest producer and exporter of dragon fruit with a total area of 55,000 ha; production reaching about 1million tones (Ahmad, 2019). Other producers are Thailand, Taiwan, Philippines, Malaysia, Sri Lanka, Australia, Israel, while producers in South America include Mexico, Ecuador, Columbia, Nicaragua and Guatamala (Ahmad, 2019). Malaysia had 1,641 ha in production (Kek Hoe, 2017). In the Philippines, the area planted has been increased to 450 ha (Eusebio and Alaban, 2018). Production in the United States is limited to Florida, California and Hawaii (Merten, 2003: Lobo *et al.*, 2013). Recently, China has embarked on large scale planting adding up to more than 60,000 ha (Ahmad, 2019). Vietnam exports 80 – 85 per cent of its production, mainly to China (Ahmad, 2019). A total production 1,463 MT was recorded in Philippines (Eusebio and Alaban, 2018) and 11,000 MT in Malaysia (Kek Hoe, 2017). The yield depends on planting density, cultivar and crop management. The crop productivity around the world is 10 to 30 t/ha (Le Bellec, 2003 and Anon, 2017). The productivity of dragon fruit averaging to 22 to 35 t/ha (Nguyen *et al.*, 2015).

8.0 Soil and Climate

8.1.0 Soil

Dragon fruit can be grown practically on all soils types provided it is well drained (Daubresse Balayer, 1999). However, slightly acidic with rich in organic matter is preferable for dragon fruit cultivation. Since the area to be used is sub-marginal, an organic fertilizer will be applied to patch up the lacking amount of organic substances in the soil. For better growth and development, an annual rainfall of 1500 to 2000 mm/annum with good drainage soils having 5.5 to 6.5 pH and lower soil slopes of 30 per cent are preferred (Becerra, 1986).

8.2.0 Climate

Hylocereus is considered an epiphyte or hemi-epiphyte, which could turn into a parasite of the host plants by aerial invasion or root introduction into the cambium or root pith, thus causing the death of the host. Pitahaya (*H. undatus*) is a tropical climate cacti, resistant to water stress, and adapted to various climatic conditions. *Hylocereus* is found in regions with greater than 2000 mm per year of rainfall, the extreme temperature range of 11 to 40 °C. Heavy rainfall areas are not suitable for the crop, as excessive rain causes flower drop and fruit drop as a result of rotting. Dragon fruit plants are ideal for growing at a dry tropical climate with an average temperature of 20-29°C, but can withstand temperatures of 38-40°C, and as low as 0°C for short periods (Karunakaran *et al.*, 2014). While either lower and/or higher will decrease or even inhibit induction of flower bud production (Khaimov- Armoza *et al.*, 2012). The temperatures lower than -1.3 °C killed half the chlorenchyma cells in *H. undatus* (Nobel and de la Barrera, 2004). Best results for *H. megalanthus* cultivation are obtained between 1000 to 1750 m above sea level and 18 to 25 °C (Becerra, 1986, 1994). Israel, Northern Africa or the Southeast of United States daily mean temperature exceeds 37 to 38 °C; it causes tissue damage in stems and death (Mizrahi and Nerd, 1999; Nobel and de la Barrera, 2002b; Ben-Asher *et al.*, 2006). At temperatures above 40°C, the plant showed yellowing of the stem (Karunakaran and Arivalagan, 2019), reduction of flower and fruit production (Mizrahi and Nerd, 1999). In Israel, when daytime temperatures reach 45°C; shade clothes ameliorate the effects of high temperatures on *H. undatus* (Raveh *et al.*, 1998).

9.0 Cultivation

9.1.0 Planting of Cutting

An open area with more sunlight is preferred for dragon fruit cultivation over a shady place. Dragon fruit is generally planted in single post system of 30 cm deep and 20 cm wide pit at 3 x 3 m distance Anon, (2017) reported that planting at a distance of 2.5 x 2.5 m with 4 cutting/support can accommodate 6400 plants/ha and also gives good yields and quality of fruits. The pole height is 1.5 m to 2 m at which point they are allowed to branch and hang down. The number of plants per pole maybe 2 to 4 plants depending on the climatic condition. Lateral shoots must be limited and 2-3 main stems are allowed to grow. It is important to arrange a round metal/concrete frame to maintain balanced shrub. The addition of dolomite and organic fertilizer at planting is beneficial. The soil is enriched with organic inputs like farmyard manure, coir compost, vermin-compost and biofertilizers. The growth of dragon fruit vines was so fast that an average growth rate of 8.2 cm was observed per week (Anon, 2017). In about 8 months after planting the dragon fruit forms a thick dense mass of vines on top of the trellis which lies drooping to the ground.

9.2.0 Shade

In tropical areas, dragon fruit plants are grown in open condition but the plant need protected from intense solar radiation and subfreezing temperatures when cultivated under subtropical conditions (Mizrahi and Nobel, 1997). The hemi-epiphytes *Hylocereus* species prefer half-shade conditions provided in nature by trees. The needed degree of shade may vary among cultivar; those having layers of wax on stems might need less shade than those without the wax. In drier and warmer climates, 30 to 60 per cent artificial shade netting is necessary to protect the plants from high solar radiation (Raveh *et al.*, 1998; Mizrahi and Nerd, 1999; Cavalcante *et al.*, 2011). *H. undatus, H. costaricensis, H. purpusii* and *H. monacanthus* are more tolerant to high solar radiation than other species because of a waxy coating that covers their stems (Raveh *et al.*, 1998). Scorch sun with insufficient water supply may lead to sunburn injuries and flower bud drop.

9.3.0 Pruning

Dragon fruit produces dense branches during the initial stage. The lateral buds and branches should be pruned to grow towards stands. Once vines reach up to the top of the stands the branches are then allowed to grow. The removal main stem tip encourages the growth of new shoots. The growth is laterally which climb at the ring and form an umbrella like structure of vines where flowers emanate and develop into fruits. This pruning referred as structural pruning or making a structure on the trellis.

Pruning shoots at the proper times can regulate the flowering in both *S. megalanthus* and its triploid hybrids. When pruning the shoots every week between beginning August to the beginning of September, these genotypes will burst in flower bud production. Consecutive pruning will result in consecutive flowering and ripening. The periods which elapse between the waves of flowering and ripening become longer with time, due to shorter days and lower temperatures. Karunakaran *et al.* (2014) regularly prune the plants to obtain an open and manageable umbrella shape canopy which will induce new shoots for the next cropping season.

9.4.0 Trellising

The Vietnamese system of trellising is followed in major countries where 3-4 cuttings are planted around a support pole which allows the shoots to bend down. This system is inefficient to harvest; in addition, increase in number of cuttings per pole does not correlate with higher yield. In Mexico live trees are used for trellising (Hernández and Salazar, 2012). In Israel the shoots are hang on wires at a height of no more than 160 cm. This allows two "walls" of fruit production on both sides of the system and also provides easy access to growers, to pollinate flowers and harvest fruits. Many trellis designs are used in India. Therefore it is better to go with concrete poles in the initial stage of crop establishment.

9.5.0 Manure and Fertilizers

Pitaya does require mineral fertilization, especially when annual yields are high. The common belief that CAM plants do not respond positively to nutrient enrichment, was proven several times to be a misconception (Weiss *et al.*, 2010). Proper nutrient requirement is needed for better performance and yield of crop. Dragon fruit plant root system is superficial and can rapidly assimilate even the smallest quantity of nutrients. Anon, 2017 applied NPK at the rate of N 450: P_2O_5 350: K_2O 300 for yield and quality. The nutrients were supplied in four split doses to each pillar having four plants @ 10, 10 and 30 per cent of total, before flowering, 20, 40 and 25 per cent at fruit set, 30, 20 and 30 per cent at harvest and finally 40, 30 and 15 per cent of total N, P_2O_5 and K_2O after two months of harvest (Anon, 2017). A combination of organic manure with neem cake and 100g of complete fertilizer (19-19-19) is applied every three to four months.

9.6.0 Irrigation

Hylocereus species are facultative epiphytes that initially roots in the soil and then later become fully epiphytic (Wallace and Gibson, 2002) due to the Crassulacean acid metabolism (CAM) and this makes dragon fruit appealing in arid, semiarid and in other regions with limited water or irrigation (Nerd *et al.*, 2002; Nobel, 2002). Indeed, according to Mizrahi *et al.* (2007), the vine-cacti pitaya, use around 10 per cent of the water used by various C_3 fruit crops such as citrus fruits, peach, avocado and pear. These hemi-epiphytic plants have a shallow root system with depth not more than 40 cm. Hence irrigation should be insured to provide sufficient water during dry season. In well distributed rainfall areas of 1700–2500 mm/annum, pitaya is grown as rainfed crop with proper drainage system. Excess water systematically results in the abscission of flowers and young fruits (Le Bellec, 2004); splitting of older fruits (35 days from flowering) (Le *et al.*, 2000). On the other hand, mulching is necessary to reduce moisture loss. In arid and semi-arid region; irrigation is required only during summer (the dry season). In mid-summer irrigation is given every day with a small amount of water; otherwise the water will slip below the root system. The optimum water requirement for dragon fruit is 120 mm/year is enough (Mizrahi *et al.*, 2007). For production of more fruits, dry period before flowering is required. Irrigation by flooding is not recommended as it wastes water and increases work of weeding. Approximately 2-4 litres of water weekly twice per plant is sufficient during the summer/dry days. Depending upon soil, climate and plant health, water requirement may increase or decrease.

10.0 Flowering, Fruit Growth and Development

Hylocereus species flowering begins at 14 months after planting of the cutting with seven major flowering cycles from May and extended up to 1st week of December in India (Anon, 2017). The funnel shaped flowers starts on small spiral button type attract structures at the stem margins. These develop to flower buds in 10-15 days. The beautiful hermaphrodite nature flowers are 25-30 cm long, white inside and greenish yellow with purple dyes on the outside. They are scented and only blooming at night and last one only night (Karunakaran and M. Arivalagan, 2019). The number of flushes (flowering episodes) depends on the species. Anon, (2017) reported seven to eight flushes for *H. costaricensis* while five to six for *H. undatus*. There is a period of 3 to 4 weeks between flowering flushes (Le Bellec, 2004) which makes it possible to see floral buds, flowers, young fruits and mature fruits on the same plant at the same time. The periods between the appearance of floral buds (lifting of the areole), flowering (stage 1), flower anthesis and fruit harvest (stage 2) are very short duration of 15 to 20 days for the first stage and 30 days for the second stage. Pollen dehisces a few hours prior to complete anthesis. Pollens are yellow colour, abundant and heavy. Anthesis timing of *Hylocereus* species is between 20:00 and 20:30. At the base of a long tube, ovary is located which carry the foliaceous scales to the exterior. There are numerous stamens on a slender anther stalk. Unusually large, tubular style is 20 cm in length and 0.5 cm in diameter. The stigmas are creamy green in colour and have 21 slender lobes.

Cultivation of dragon fruit

Trellising system of cultivation of dragon fruit

Pruning of dragon fruit

Furrow (left) and drip (right) irrigation in orchard

Mulching with black polythene

Flowering in dragon fruit

Dragon fruit plant in flowering stage

Flower buds of different dragon fruit cultivars

Fruit development and maturation

Fruit bearing stage of dragon fruit

Fruit development and maturation

Fruit development and maturation of dragon fruit according to the extended BBCH (Biologische Bundesantalt, Bundessortenamt und Chemische Industrie) scale.

711, A: Fruit set: floral parts drying, flower attached with peri-carpel, development of embryo begins; **713, B**: Fruits about 30% of final size: pericarpel starts growing, endocarp starts developing, seed formation begins; **715,C:** Fruits at 50% of final size: endocarp development continues and occupies about half of the pericarpel, seed maturation begins; **717:** Fruits at 70% of final size; **719:** Fruits at 90% or more of final size; **811:** Beginning of fruit maturation: colour of apical bracts starts changing to light pink, fruits are still firm;**813.** Advance colour development: colour of exocarp (skin) of fruit and basal portion of bracts turns light pink, softening of fruit begins; **815, D:** Advanced maturation: colour of fruit skin along with bracts turns pink, parenchymatous mesocarp light pink in colour, pulp yetto develop fully, fruit ready for commercial picking; **817:** Fruit fully mature for picking: colour changes to red, pulp development completed, pulp attains appropriate consistency and taste; **819:** Fruit over mature: fading of skin colour begins, fruit becomes soft, pulp disintegrates.

Source: Kishore, K. (2016) *Sci. Hort.*,doi.org/10.1016/j.scienta.2016.10.047.

Day length determines the floral growth. In Vietnam, floral induction is often triggered using artificial light to increase day length however water unavailability after bud break will cause death of bud and flower drop. Rainy season is followed by the onset of flowering; the floral buds remain in the latent stage for many weeks (Daubresse Balayer, 1999. The stigma position dominates the stamens which encourages allogamy. The crop is pollinated by hawkmoths and bats (Nerd and Mizrahi, 1997) as well as honey bees (Anon, 2017) in early morning. Before anthesis of the flower (from 4:30 P.M.) until 11:00 A.M. the next day, manual pollination may be carried out.The pinching of bulge flowers parts reveals the stigmata, which are then covered with pollen. Alternatively, the anthers can be directly deposited (with minimal pressure) on the stigmata with the fingers. After removal of pollens from a flower of a different clone (or from another species), it is stored in a box until needed. The pollen collected from two flowers will be enough for around 100 pollinations. It can be stored for from (3 to 9) months at –18°C to – 196°C without risk. Fruits obtained after pollination using pollen stored at 4°C for (3 to 9) months are very small (Metz *et al.*, 2000). However, the fruit quality of manual cross-pollination is superior to the fruits resulting from free pollination (Le Bellec, 2004).

Irrespective of the successful of failed fertilization, flower closed in the subsequent morning after anthesis. Petals become soft and then slowly dry at the following day. The lower part of a fertilized flower remains greenish and increases enormously in volume, indicating fruit set while a non–fertilized flower becomes yellowish drop off 4 to 6 days after anthesis (Anon, 2017). Hernández-Andrade *et al.* (2019) reported the predominance of flowering and fruiting in the rainy season.

As a result of long day length requirements for flowering of pitaya, artificial light is stimulated in the Far East for *Hylocereus* flowering (Jiang *et al.*, 2012); however day length was ineffective in induction of early flowering (Khaimov and Mizrahi, 2006).The researchers in Israel observe that removal of young flower buds delayed the wave of flowering and produced late cropping in *H. undatus and S. megalanthus* without affecting total flower yield (Khaimov and Mizrahi, 2006). In addition to light supplement, Active cytokinins are important inducers of flower bud production in *Hylocereus undatus* (Khaimov–Armoza *et al.*, 2012). The active cytokinins peak just before flower buds appear. Application of CPPU [N-(2-chloro-4-pyridinyl)-N-phenylurea] promotes precocious and early flowering and led to early ripening in *H. undatus* and *S. megalanthus* and increases total flower number, while gibberellic acid (GA_3) application delays flowering, ripening and lowers flower number in these species (Khaimov and Mizrahi, 2006).

Most gas exchange for pitaya occurs at night when air temperatures are lower than daytime temperatures, a characteristic of the Crassulacean acid metabo-lism (CAM) photosynthetic pathway. This physio-logical strategy reduces transpirational loss of water (Taiz and Zeiger, 2002). *H. undatus* has the characteristic potential of CAM plants to respond positively to CO_2 increments. In addition, it has a lower nutrition response, 2 weeks or longer, than most C_3 and C_4 plants, which implies lack of genetic plasticity and low nutrient absorption capacity through root (Nobel and de la Barrera, 2002c and 2004). The first studies on water stress on *H. undatus* were in a growth chamber. The results indicated that two weeks without irrigation had little effect on net CO_2 uptake, but 17 days decreased growth by 78 per cent (Raveh *et al.*, 1995; Nobel and de la Barrera, 2002a, 2004). In another study, when watering was withheld under growth chamber environmental conditions, net daily CO_2 uptake decreased 33 per cent in 7 d and 63 per cent after 12 d (Nobel and de la Barrera, 2002a); however, under field conditions in a desert in Israel, net daily CO_2 decreased 53 per cent after 4 days and 89 per cent after 8 days of water stress. Difference in results between field conditions and growth chamber are explained by the vapor pressure deficit that is three times greater under field conditions (Ben-Asher *et al.*, 2006). The classical metaxenia phenomenon is used for elongating the ripening season of dragon fruit. If pollinated with *Hylocereus* sp, the time which elapses from pollination to ripening is shorter by ~5 to 30 days than if pollinated with of *Selenicereus* sp clone (Mizrahi *et al.*, 2004).

Kundan Kishore (2016) extended the BBCH (Biologische Bundesantalt, Bundessortenamt und Chemische Industrie) scale for seven principal growth stages, *viz.*, bud development (0) shoot development (1) vegetatively propagated organ development (4) reproductive development (5) flowering (6) fruit development (7) and fruit maturation (8) have been described. All the phenophases pertaining to vegetative and reproductive stages and their relative importance in crop management and improvement of dragon fruit are broadly described by extended BBCH scale. The scale will act as a useful tool for adoption of effective crop management practices like nutrient management, pollination, plant propagation, timely harvest of fruits and pest management.

11.0 Pests and Diseases

11.1.0 Pests

Dragon fruit is generally insect free and are not problematic in semi-arid zone which is quite the opposite in tropical and subtropical growing region. The holes and damages left by insects are entrance sites for fungi and bacteria that can eventually kill plants.

Few pests have been recorded on *Hylocereus*. Ants belonging to the genera *Atta* (Barbeau, 1990) and *Solenopsis* (Le Bellec, 2004) are very notorious pest and can cause major damage to the plants as well as to the flowers and fruits. *Leptoglossus zonatus* sucks the sap and *Cotinus mutabilis* perforates the stem, leaving stains and some deformation (Barbeau, 1990).Different species of aphids and scales have also been observed on fruits and flowers. Rats and birds can cause serious damage, mainly to flowers (Le Bellec, 2003). Bees are extremely efficient for pollination but they will have harvested all the pollen after only a few hours of activity. The pollen must thus be collected before the bees arrive and manual pollination carried out the next morning as soon as the bees have left the plantation. *Dasiops saltans* (spear fly) deposits its eggs on the flower buds of *H. megalanthus* and the larvae drill anthers causing 40 to 80 per cent loses. The infected buds turn red and fall easily (Delgado *et al.*, 2010). The fly can be managed by thermal treatment with warm air 46.5 °C for 20 minutes against fruit fly (*Bactrocera* spp) to disinfect *H. undatus* fruits, followed by 2 to 4 weeks storage at 5 °C. Fruits treated as described deteriorate externally, but flavour stays acceptable after some time (Ho *et al.*, 2006). Other important pests includes stem weevil (*Maracayia chlorisalis* Walker) which might be controlled with *Trichogramma; Cotinis mutibales* Gory and Percheron and *Euphoria limatula* Janson beetles, *Metamasiu fareih striatofo* Galli, (NRA-CEE, 1994; Badillo, 2005).

11.2.0 Diseases

Table 9: Fungal, Bacterial, Nematode and Viral Pathogens and their Associated Diseases in Various *Hylocereus* Species

Genus	*Species*	*Disease*	*Hylocereus sp.*	*Infected Country*	*References*
A) Fungus					
Alternaria sp.	*A. alternata*	Post-harvest rot	*H. undatus* and *H. megalanthus*	–	Vilaplana *et al.*, 2018, Castro *et al.*, 2017, Castro *et al.*, 2018
	Alternaria sp.	Alternaria blight	*H. undatus*	US	Patel and Zhang, 2017
Aureo-basidium sp.	*A. pullulans*	Stem and Fruit Spot	*H. undatus* and *H. monacanthus*	China	Wu *et al.*, 2017
Bipolaris sp.	*B. cactivora*	Fruit and stem rot, Post-harvest rot disease	*H. undatus*	Israel, China (Fruit imported from Vietnam), Thailand, Taiwan, US, Korea, Japan	Ben-Ze'ev *et al.*, 2011, He *et al.*, 2012, Oeurn *et al.*, 2015, Wang and Lin, 2005, Tarnowski *et al.*, 2010, Choi *et al.*, 2010, Taba *et al.*, 2007
Botryo-sphaeria sp.	*B. dothidea*	Stem spot	*H. undatus*	Mexico	Vaencia-Botin *et al.*, 2004
Colleto-trichum sp.	*C. aenigma*	Anthracnose	*H. undatus*	Thailand	Meetum *et al.*, 2015a
	C. gloeosop-orioides	Anthracnose	*H. undatus, H. monacanthus,* & *H. megalanthus*	Brazil, China, US, Thailand, Malaysia, Thailand	Takahashi *et al.*, 2008, Masyahit *et al.*, 2009, Ma *et al.*, 2014, Awang *et al.*, 2010, Palmateer *et al.*, 2007, Meetum *et al.*, 2015a, Masyahit *et al.*, 2009
	C. siamense	Anthracnose	*H. undatus* and *H. monacanthus*	India, Thailand, China	Abirami *et al.*, 2019, Meetum *et al.*, 2015a, Zhao *et al.*, 2018
	C. truncatum	Anthracnose	*H. undatus* and *H. monacanthus*	China, Thailand, Malaysia	Guo *et al.*, 2014, Meetum *et al.*, 2015a, Vijaya *et al.*, 2014

Genus	*Species*	*Disease*	*Hylocereus sp.*	*Infected Country*	*References*
	C. karstii	Anthracnose	*H. undatus*	Brazil	Nascimento *et al.*, 2019
Curvularia sp.	*C. lunata*	Spots/lesion on stems	*H. monacanthus*	Malaysia	Hawa *et al.*, 2009
Epicoccum sp.	*E. sorghinum*	Stem mold	*H. undatus*, *H. monacanthus* & *H. megalanthus*	Philippines	Taguiam *et al.*, 2020
Fusarium sp.	*F. fujikuroi*	Stem rot	*H. monacanthus*	Malaysia	Hawa *et al.*, 2017
	F. oxysorum	Basal rot, basal stem rot, stem rot, stem blight	*H. undatus, H. monacanthus* and *H. trigonus*	Argentina, South Korea, Malaysia	Wright *et al.*, 2007, Choi *et al.*, 2007, Mohd Hafifi *et al.*, 2019
	F. proliferatum	Stem rot	*H. monacanthus*	Malaysia	Hawa *et al.*, 2013
	F. semitectum	Stem rot	*H. monacanthus*	Malaysia	Hawa *et al.*, 2010
	F. solani	Stem rot	*Hylocereus* sp.	Indonesia	Rita *et al.*, 2016
Gilbertella sp.	*G. persicaria*	Storage rot	*H. costaricensis*	China	Guo *et al.*, 2012
Monilinia sp.	*M. fructicola*	Brown rot	*H. monacanthus*	-	Awang *et al.*, 2011b
Neocosmospora sp.	*N. rubicola*	Stem rot	*H costaricensis*	China	Zheng *et al.*, 2018a
Neoscytalidium sp.	*N. dimidiatum*	Black rot, stem canker, brown spot, fruit internal browning, fruit canker	*H. undatus and H. monacanthus & H. megalanthus*	Israel, Taiwan, Malaysia, China, Philippines, US	Ezra *et al.*, 2013, Chuang *et al.*, 2012, Hawa *et al.*, 2013, Lu *et al.*, 2015, Lan *et al.*, 2012, Yi *et al.*, 2015, Sanahuja *et al.*, 2016, Xu *et al.*, 2018, Hong *et al.*, 2020, Taguiam *et al.*, 2020
Nigrospora sp.	*N. sphaerica*	Reddish-brown spot	*H. undatus*	China	Liu *et al.*, 2016
Sclerotium sp.	*S. rolfsii*	Southern blight	*H. undatus*	China	Zheng *et al.*, 2018b
B) Bacteria					
Enterobacter sp.	*E. cloacae*	Bacterial soft rot	*Hylocereus* sp.	Malaysia	Masyahit *et al.*, 2009
Paenibacillus sp.	*P. polymyxa*	Bacterial stem rot	*H. undulatus*	China	Zhang *et al.*, 2017

Genus	Species	Disease	Hylocereus sp.	Infected Country	References
C) Virus					
Cactus virus	*CVX*	Viral	*H. undatus* and *H. monacanthus, H. polyrhizus*	US, Taiwan, Korea, China, Malaysia	Gazis *et al.*, 2018, Liou *et al.*, 2001, Liou *et al.*, 2004, Kim *et al.*, 2016, Peng *et al.*, 2016, Masanto *et al.*, 2018
Zygocactus virus X	ZyVX	Viral	*H. polyrhizus*	Taiwan, China	Li *et al.*, 2015, Zheng *et al.*, 2017
Pitaya virus	PiVX	Viral		Taiwan	Li *et al.*, 2015
Pitaya badnavirus	PiBV1	Viral	*H. polyrhizus*	China	Zheng *et al.*, 2020
Tylenchor-hynchus sp.	*T. agri*	Stunting (Growth inhibition and fruit drop)	*H. monacanthus*	China	Zhang *et al.*, 2018
Meloidogyne sp.	*M.icognita, M.javanica*	Gall formation, destroyed the plants	*H. megalanthus*	Israel	Mizrahi, 2015

* Modified from Balendres and Bengoa (2019).

11.2.1 Fungal Diseases

Anthracnose is relatively the most destructive fungal disease of dragon fruit (Balendres and Bengoa, 2019). *Colletotrichum gloeosporioides* is the most widely reported causal organisms of anthracnose (Takahashi *et al.*, 2008; Masyahit *et al.*, 2009; Ma *et al.*, 2014; Awang *et al.*, 2010; Palmateer *et al.*, 2007; Meetum *et al.*, 2015a; Masyahit *et al.*, 2009) and is hosted by all three popular cultivated dragon fruit species (*H. undatus, H. monacanthus and H. megalanthus*). *C. siamense* infection in *H. undatus* starts with reddish-orange spots with severe chlorotic halos (Abirammi *et al.*, 2019) but in *H. monacanthus*, stems showed faint-pink to brown necrotic lesions (Zhao *et al.*, 2018). However, the symptom variation may be attributed to differences in plant varieties, plant age, and *C. siamense* strains. Anthracnose can also reduce yield. In India, 30 per cent of dragon fruit plants were infected and caused yield reduction (Abirammi *et al.*, 2019). In a 15 ha commercial farm in China, 25 per cent of the plants have been affected by anthracnose (*C. siamense*) and 35 per cent of the severely-infected plants died within 40 days from the initial symptom appearance (Zhao *et al.*, 2018). Infection by *C. gloeosporioides* also resulted up to 50 per cent in losses (Ma *et al.*, 2014). In Indonesia, none of the surveyed dragon fruit growing areas in Sumatra were disease-free (Nasir and Nurmansyah, 2016). The abundance of viable spores is a huge challenge with diseases caused by *Colletotrochum* spp. and reducing spore dispersal and production would aid in mitigating disease epidemics.

The dreaded pathogen *Bipolaris cactivora* infected both pre and post-harvest stage of *H. undatus* and reduced the area planted with *H. megalanthus* by 93 per cent (Varela *et al.*, 1995; Bibliowicz and Hernandez, 1998; Nerd *et al.*, 2002). The fungus infects the base of pre-mature fruits and induces yellowing (Varela *et al.*, 1995; Bibliowicz and Hernandez, 1998). The infected fruits showed water-soaked symptoms which later becomes soft rot and thus, affects the quality of fruits (He *et al.*, 2012). The fungus *B. cactivora* causes fruit and stem rotting and remains a problem in several countries (Ben-Ze'ev *et al.*, 2011; He *et al.*, 2012; Oeurn *et al.*, 2015; Wang and Lin, 2005; Tarnowski *et al.*, 2010; Taba *et al.*, 2007). The rapid spread of the pathogen in host tissue is assisted by a large number of destructive spores (Tarnowski *et al.*, 2010).

The potential emerging pathogen *E. sorghinum* infection starts with large irregular-shaped, necrotic, brown lesions developed on rooted stem cuttings and advances to yellowing of stem. (Taguiam *et al.*, 2020) In south Florida, stem and fruit canker, caused by the fungus *N. dimidiatum*, is a major disease of dragon fruit, causing up to 60 per cent –80 per cent losses of sale value due to the symptoms on the fruit (Hong *et al.*, 2020).

Four commercial dragon fruit farms in the Guangdong Province, China have been affected by stem and fruit spot (*A. pullulans*). This infection rapidly expanded when typhoon Ramasoon hit China and more than half of the plants were infected which led to significant fruit-yield loss (Wu *et al.*, 2017). In Malaysia, stem blight (*F. oxysporum*) and spots (*C. lunata*) affected 41 per cent and 20 per cent of dragon fruit production, respectively (Hawa *et al.*, 2009; Mohd Hafifi *et al.*, 2019). In a farm in Florida, US, *Alternaria* blight was responsible for the 10–20 per cent of yield loss (Patel and Zhang, 2017). Plants showing canker (*N. dimidiatum*) also fails to recover as spots that develop into canker subsequently leads to rotting of the stem (Chuang *et al.*, 2012). In Malaysia, canker was identified in 20 dragon fruit plantations. Infection of *N. dimidiatum* in fruits also results in rotting which makes the fruit pulp quality unacceptable (Ezra *et al.*, 2013).

11.2.2 Bacterial Diseases

Bacterial stem rot severely damaged dragon fruit production in China. In 2014, all dragon fruit plants were infected with *P. polymixa*. The severe infection resulted in the decomposition of stems and the infected stems did not produce fruit (Zhang *et al.*, 2017). Soft rot (*Enterobacter cloacae*) and stem rot (*Paenibacillus polymixa*) diseases start with a water-soaked appearance in the infected tissues. Bacterial soft rot-infected stems show yellowish to brownish soft rot symptoms. Soft rotting develops within 24 h after inoculation (hai) of the bacteria in the stems and within 48 hai in the fruits. Stems infected with *P. polymixa*, gradually turn yellow and brown and all fleshy stems decompose (Zhang *et al.*, 2017). Rotting of dragon fruit develops after 48 h from injecting *P. polymixa* into the stems (Zhang *et al.*, 2017). *Klebsiella mobilis, K. oxytoca, Pantoea dispersa, and Rahnella aquatilis* have been also isolated from soft-rot-infected dragon fruit stem, but these bacteria were not pathogenic to dragon fruit (Masyahit *et al.*, 2009). Excessive light makes the plant vulnerable to bacterial rot caused by *Xanthomonas campestris* and *Erwinia carotovora*. According to Nerd *et*

Pests of dragon fruit

Thrips infestation

Mealy bug (left) and leaf footed bug (right)

Aphid infestation

Diseases of dragon fruit

Anthracnose (left) and Botryosphaeria (right) infection

***Bipolaris cactivora* fruit rot**

Bacterial soft rot (*Erwinia caratovora*) in dragon fruit

al. (2002), cactus plant tissues have a relatively high water content and therefore, disease infection rapidly turn into rotting (Nerd *et al.*, 2002).

11.2.3 Viral Diseases

Usually, pitaya plants are vegetatively propagated using mature stem tissues, which results in the accumulation of multiple viruses. A series of virus-like symptoms, such as mottling, yellowing and chlorosis, have been observed previously on the surface of stems of pitaya plants (Liou *et al.*, 2001) In the past decades, a number of viruses belonging to the genus *Potexvirus* in the family *Alphaflexiviridae*, including cactus virus X (CVX), zygocactus virus X (ZyVX), schlumbergera virus X (SchVX), pitaya virus X (PiVX) and pitaya badnavirus 1 (PiBV1) have been shown to infect *Hylocereus* plants, and have been reported in Taiwan, Hainan and Guizhou in China (Liou *et al.*, 2001; Liou *et al.*, 2004; Peng *et al.*, 2016; Zheng *et al.*, 2017, Zheng *et al.*, 2020)) and in other countries, including Brazil, Korea and the USA (Duarte *et al.*, 2008; Kim *et al.*, 2016; Gazis *et al.*, 2018). *Cactus virus X* has been associated with virus disease symptoms in dragon. Necrosis (Gazis *et al.*, 2018; Masanto *et al.*, 2018), mottling (Liou *et al.*, 2001), chlorotic spots (Gazis *et al.*, 2018; Peng *et al.*, 2016), red-brown margins, pale yellow-green mosaics and deformed spines (Gazis *et al.*, 2018) are the associated symptoms in *CVX*-infected plants. *Cactus virus X* can result in reduced growth, such as no new shoots or flowers, enlarged stems, and a darkish-dull green colour. Currently, there is no cure and infected plants should be removed. In Malaysia, secondary infections of fungal pathogens were reported that leads to rotting of the infected tissue (Masanto *et al.*, 2018). CVX from dragon fruit can be sap transmitted to indicator plants *Chenopodium amaranti*colour, *C. quinoa, Celosia argentea,* and *Gomphrena globosa* (Liou *et al.*, 2001; Kim *et al.*, 2016), of which symptoms can be observed 20 days after inoculation (Kim *et al.*, 2016). *C. quinoa* plants manifest chlorotic lesions when infected with *CVX* (Liou *et al.*, 2001; Kim *et al.*, 2016). Small chlorotic spots, then systemic infection, develop in *CVX*-infected *C. argentea* (Liou *et al.*, 2001), while necrotic lesions were observed in *G. globosa* (Liou *et al.*, 2001; Kim *et al.*, 2016). Both necrotic and chlorotic lesions have been associated with *CVX*-infected *C. amaranti*colour (Liou *et al.*, 2001; Kim *et al.*, 2016). Based on the sequences of the replicase and coat protein, CVX is closely related to the *Bamboo mosaic virus (BMV), Cassava common mosaic virus,* and *Papaya mosaic virus (PMV)* (Liou *et al.*, 2001). However, in immunodiffusion tests, CVX did not react with the antisera to *BMV* and *PMV* (Liou *et al.*, 2001). Liou *et al.* (2001) presented the complete nucleotide sequence of the strain *CVX*-Hu and found that the putative product of ORF6 showed no significant similarity to other potexviruses.

11.2.4 Nematode

Only recently Mizrahi (2015) found that two nematode species are causing damage mainly to the *Selenicereus megalanthus* and few other *Hylocereus* genotypes. The nematodes belong to the *Meloidogyne* genus; they are *M. incognita* and *M. javanica,* and cause problems only in sandy soils. The nematode destroyed the plants in years 4 and 5 years. Resistant clones of pitayas are available to be used as rootstocks. It is very easy to graft these plants and they become productive one year after grafting. The other, stunt nematode, *Tylenchorhynchus agri,* reported from

China was the first record of a nematode associated with a cacti disease (Zhang *et al.*, 2018). The glasshouse pot trials conducted by Zhang *et al.* (2018) show that the *H. polyrhizus* plants grown in soil inoculated with 1000 nematodes lost almost 50 per cent of their weight 6 months after the inoculation, stunted growth and fruit drop were observed.

11.3.0 Management of Disease

Temperature management, hot water treatment, and gamma irradiation have been reported to reduce disease incidence and severity of dragon fruit diseases. Storing fruit at 6 °C for 21–26 days can reduce storage rots caused by fungal pathogens *e.g., Alternaria alternata, B. cactivora, N. dimidiatum* and *C. gloeosporioides* (Ngoc *et al.*, 2018). Hot water treatment for 2 min at 50 °C, before storage in 12 °C for 21 days, reduced lesion diameter of *Alternaria alternata* without significant effect on the fruit's sensory qualities (Vilaplana *et al.*, 2017). Irradiating fruits at 800 Gy or less was also effective in inhibiting or reducing fruit decay, although minor changes in fruit quality were observed (Wall and Khan, 2008).

The greatest reduction of brown rot (*M. fructicola*) and anthracnose (*C. gloeosporioides*) disease severity was achieved at 4.0 g CaCl2/L water (30 min treatment) (Awang *et al.*, 2011a,b). Sodium carbonate and potassium sorbate, at varying concentrations have been tested against several fruit-rot-causing fungal pathogens (*C. gloeosporioides, C. capsici,* and *Fusarium* sp.) (Jitareerat *et al.*, 2018). Sodium carbonate completely inhibits C. *gloesporioides* spore germination in media (Jitareerat *et al.*, 2018). Post-harvest black rot management has been achieved by using 2.5 per cent sodium bicarbonate at 21 days at 12 °C (plus 5 days of shelf-life at 20 °C), without negatively affecting the fruits sensory quality (Vilaplana *et al.*, 2018). The quality of fruits treated with sodium bicarbonate also improved. Potassium sorbate completely inhibited the spore germination of all fungal pathogens in culture media. The application of potassium sorbate also had little impact on the plant's respiration rate and ethylene production. Disease severity was reduced in potassium sorbate-treated dragon fruits (at 55 °C in cold water). The use of chitosan in managing post-harvest anthracnose has been also demonstrated. The low molecular weight chitosan at 1 per cent concentration in the form of nanoemulsions (600 nm droplet size) effectively control anthracnose (*C. gloeosporioides*) (Ali *et al.*, 2013; Zahid *et al.*, 2012, 2019). Moreover, Ali *et al.* (2013) found that chitosan maintain the quality of fruits within the 28-day storage period (10 ± 2 °C and 80 ± 5 per cent relative humidity).

The essential oil derived from *Cymbopogon nardus* (Nor Hanis Aifaa *et al.*, 2012), *Cinnamomum zeylanicum, Cymbopogon flexuosus, Eucalyptus globulus, Eugenia caryophyllus* and *Rosmarinus officinalis* (Castro *et al.*, 2017) are effective against post-harvest diseases management of dragon fruit. It was found that *E. caryophyllus* and *C. zeylanicum* both contained high amount (90.50 per cent and 80.70 per cent, respectively) of eugenol (Castro *et al.*, 2017), a chemical with known antimicrobial activity. On fruits, *E. caryophyllus* reduced mycelial growth of *A. alternata* by 31 per cent compared to the non-treated fruits (Castro *et al.*, 2017). Treatment of dragon fruits stored in cold room (10 °C, RH 85–90 per cent for 21 days) with

Cymbopogon nardus essential oil delayed anthracnose incidence without affecting the quality of the fruits at concentrations below 2 per cent oil (Nor Hanis Aifaa *et al.*, 2012). Phytotoxicity effects were observed at above 2 per cent and therefore not recommended for post-harvest treatment. *C. zeylanicum* and *E. caryophyllus* strongly inhibited mycelial growth of *A. alternate* at 250 and 500 ìg/ml, respectively. To control the stem canker, citronella oil is used (Taguiam *et al.*, 2020). Rain tree (*Samanea saman*) leaf extracts at 2.5 per cent have been found to inhibit growth of *F. solani* in culture media (Rita *et al.*, 2016). The use of silicon to reduce the incidence and severity of various fungal diseases in dragon fruit plants has been also reported. Faziha *et al.* (2019) found lower disease incidence and severity of dragon fruit fungal diseases when treated with silicon (5.0 ml/L).

Various biocontrol agents find their way in managing plant diseases. The basal stem rot incidence was reduced by the soil application of spent mushroom sawdust that contains beneficial bacterial and fungal populations (Choi *et al.*, 2007). *Bacillus* sp. (*Bacillus subtilis* GA1-23 and *B. amyloliquefaciens* GA4-4) inhibit the mycelial growth and spore germination of *B. cactivora*; this effect was comparable to the chemical control treatment difenoconazole (Bae *et al.*, 2013). Accordingly, Mixtures of azoxystrobin (200 g/L) and difenoconazole (124 g/L) has been recommended in controlling diseases of dragon fruit, particularly anthracnose and stem canker (Noegrohati *et al.*, 2019). A biopesticide containing *B. subtilis* (2 ml/400 ml), and chemicals isoprothiolane (2.25 ml/400 ml), mancozeb (2 g/400 ml) and pyraclostrobin (1 ml/400 ml) (chemical control) completely inhibited the *in vitro* growth of *E. sorghinum* (Taguiam *et al.*, 2020). The biopesticide is considered safe for the disease management in organic farming.

Plants have an inbuilt beneficial microorganism to fight against various foreign agents. The surface of dragon fruit twigs has *B. methylotrophicus* strains PB182, PB255 and PB257, and *B. subtilis* PB223 that were able to inhibit growth and conidial germination of anthracnose pathogens (Meetum *et al.*, 2015b). *B. methylotrophicus* produces a volatile compound that inhibits mycelial growth and reduced virulence of the pathogen in dragon fruits (Meetum *et al.*, 2015b). Accordingly, *Penicillium oxalicum* T3.3 crude extract is one of the potential biocontrol agents, out of the 126 endophytes isolated from dragon fruit plants that can suppress the growth of *C. gloeosporioides* (Mamat *et al.*, 2018). To control the stem canker *B. subtilis* can be used alternatively or in rotation with synthetic chemicals (Taguiam *et al.*, 2020)

Several pesticides have a fungistatic effect against the stem canker pathogen (*N. dimidiatum*). Hexaconazole, tebuconazole, flusiazole, and pyraclostrobin have had a negative effect on *N. dimidiatum* mycelial growth (Xu *et al.*, 2018). Preventive spray with Chlorothalonil/mancozeb at 2g/L and curative spray with carbendazim at 1g/L is advised for the control of anthracnose (Abirami *et al.*, 2019). Mixtures of azoxystrobin (200 g/L) and difenoconazole (124 g/L) have been used previously to control diseases of dragon fruit, particularly anthracnose and stem canker (Noegrahati *et al.*, 2019). In the field, 250 g/L pyraclostrobin EC controlled stem and fruit canker of up to 80 per cent and 85 per cent, respectively. Azoxystrobin SC (250 g/L) and tebuconazole (430 g/L) also have had a comparable effect to that of pyraclostrobin EC (Xu *et al.*, 2018). Pre-treatment of imazalil fungicide at 0.4 h/L before storage has been also reported to reduce disease severity by 70.5 per cent (Vilaplana *et al.*, 2017). For management of *Neoscytalidium dimidiatum* chemicals *viz.*, mancozeb, pyraclostrobin and isoprothiolane can be used alternatively or in rotation with synthetic chemicals and other biopesticides (Taguiam *et al.*, 2020). Copper sulfate pentahydrate, Extract of *Swinglea glutinosa*, Garlic oil and *Streptomyces lydicus* WYEC 108 are the alternative chemical products registered for managing dragon fruit diseases organic production (Hong *et al.*, 2020).

12.0 Harvesting and Yield

12.1.0 Harvesting

Skin colour is the most frequently used criteria to judge fruit maturity (Nerd and Mizrahi, 1999; Le *et al.*, 2000). Other harvest indices also include: colour, total soluble solids, titratable acidity, and days from flowering; 30-40 days of flowering. As the fruit matures, acidity reaches a peak just as the skin colour change occurs, then declines 25 to 30 days after flowering (Nerd and Mizrahi, 1999; Le *et al.*, 2000). At this stage soluble solids contents increases, to about 14 per cent (Nerd and Mizrahi, 1999; Le *et al.*, 2000). A soluble solids/acidity ratio of 40 has been suggested as a harvest index. Pitaya fruit ripening will occur after. The fruit skin colour varies in its morphological appearance from green to red or rosy-pink depending on the species (Nerd *et al.*, 1999). Harvest time has a much greater effect on quality than varietal differences. The plants yield the fruits 12-15 months after planting yield up to 20 years with long crop cycle from June to September in different flushes. The fruit weight ranged between 300-800 g, and the average yield from the single post is realized about 30 to 35 kg from three years old planting (Karunakaran and Arivalagan, 2019). Final fruit size is dependent upon seed number (Weiss *et al.*, 1994). The clock wise twisting technique of fruit harvesting causes less or no injury to the fruits (Anon, 2017). The first fruits of the season are 50 per cent of the size of the later fruits, probably due to more appropriate temperatures for fruit development and ripening in the late season. Selective harvesting is performing during at full maturity stage.

12.2.0 Yield

The yield varies with the cultivars, the region and management practices, *etc.* In Israel, after three years of planting the plant yields from 32 to 40 t/ha, against 2 to 3 t/ha in drier, warmer and saltier places, aside from reaching productive age much later (Mizrahi and Nerd, 1999). On an average, dragon fruit yields 10 to 30 t/ha in Indian condition (Anon, 2017).

13.0 Post-harvest Management

Dragon fruit are non-climacteric and the post-harvest life is up to 4 to 5 days at room temperature. Pitaya fruits are attractive and fresh after havesting but when it reaches end market, the fruits become shrivel and ultimately decline in flavour. Upon maturation, starch degradation increases in the pulp after fruit maturation, and it is shown partially by soluble sugar accumulation. In *H. undatus*, amylase and invertase activity correlate with increments of fructose and glucose, which

Harvesting and post harvest handling of dragon fruits

Harvested dragon fruits

Manual and mechanical grading of dragon fruits

Packaging, storage, transport and marketing

Packaging of dragon fruit in CFB box

Transport and storage

Marketing of dragon fruits

concentrate around the center of the pulp (Wu and Chen, 1998). *Hylocereus* species contain active stomata concentrated in the scales than other part of the peel. Kaplan-Levy (1999) found that that shriveling occurs more rapidly in *H. undatus* fruit than in the other species as a result of high concentration of stomata in *H. undatus*. Fruits from *Hylocereus* species are identified as not climacteric (Arévalo-Galarza and Ortiz-Hernández, 2004). The main reason that pitaya fruits lose their taste after harvest is a sharp decline in acidity, as malic acid concentration (Nerd *et al.*, 1999). Organic acids play a vital role in the post-harvest management of fruits since they increase the shelf life, stability and microbial safety (Violeta *et al.*, 2010).

Storage temperature and packaging effects the quality of pitaya fruit (Freitas and Mitcham, 2013). Chilling injury, mechanical injury and water loss are the three major disorders observed (Paull, 2014). The fruit should be stored at 10 °C to reduce weight, acidity and sugar losses (Nerd *et al.*, 1999). *H. undatus* and *H. monacanthus* fruits stored at 15 to 20 °C at 85–90 per cent RH have up to10 days shelf life (Arévalo-Galarza y Ortiz-Hernández, 2004, Khalid, 2020). Storing fruits at low temperature exhibit frost damage and this reduces their commercial value. Low temperature inhibits phenol – polyphenol oxidase (PPO) and peroxidase (POD), and temperatures below 11 °C partially degrade fruit cells. Damage is reversible if the cold period is shorter than 7 d (Balois-Morales *et al.*, 2007). High quality dragon fruit should be well-shaped and brightly coloured, with firm flesh and bracts. Mechanical injury (impact and abrasion) leads to the development of sunken areas and loss of scale appearance. More mature fruit has a higher susceptibility to mechanical injury (Le *et al.*, 2000).

Shelf life in *H. monacanthus* fruit is increases by the application of calcium chloride thereby reducing anthracnose (*Colletotrichum gloesosporioides*) lesion size, and increasing calcium content in the rind and firmness without changing pH, soluble solids or fruit tritratable acidity (Awang *et al.*, 2011a). Lum and Norazira (2011) consider that useful life and fruit quality in *H. monacanthus* could increase by fruit immersion in 35 °C water for 60 minutes. The heat treatments for post-harvest rot management affect the flavour and reduce the phytochemical properties. Vargas *et al.* (2007) disinfected *H. undatus* fruits with 1000 ppm $NaCl_2$ stored under controlled atmosphere with 1 per cent oxygen and 12 °C have prolonged shelf life of up to 35 d, without affecting quality. Application of 1-MCP (1-methylcyclopropene) increases shelf life on *H. monacanthus* fruits harvested 30 to 35 days after anthesis and stored for 28 days at 10 °C; under this treatment phenolic, ascorbic acid, and betacyanine contents are unaffected (Novita, 2008). Fruit immersion in a 500 ìg/l of 1-MCP for 4 h at room temperature, delays ethylene production for 14 days of storage because respiration rate is reduced and fruit senescence slowed. 1-MCP application provides greater firmness to the *H. megalanthus* fruit and represents an alternative to reduce undesirable texture changes during storage (Ayala-Aponte and Serna-Cock, 2011). Fruits harvested 28 to 30 days after anthesis can be stored up to 35 d in modified atmosphere chambers (O_2 transmission at 400 ml/m^2/día) at 10 °C (Gross *et al.*, 2002).

Several processed products *viz.*, Ready to serve (RTS Juice,) juice, jam (Sharma, 2016), jelly (Islam *et al.*, 2012; Sharma, 2016; Panchal *et al.*, 2018) and dragon fruit-yoghurt (Jayasinghe *et al.*, 2015) are prepared from pulp of dragon fruit. Sharma (2016) found that 1.5 per cent pectin, 55 per cent sugar and 0.9 per cent citric acid in pulp and juice solution improved the colour as well as other organoleptic characteristics of dragon fruit jam and jelly. In case of dragon fruit RTS, a composition of 14 per cent pulp, 12 per cent sugar and 0.9 per cent was found to be most suitable. The prepared product was organoleptically acceptable. Prepared products can be stored for the period of more than three months at ambient storage condition without microbial spoilage or any considerable loss in quality (Sharma, 2016). For prolong storage of the processed products, dragon jelly was pasteurized at 60°C without affecting antioxidant activity (Glangkarn, 2015). The most favourable sensory score of dragon fruit *i.e.*, pH of 4.2 and 6 per cent sugar supplementation was maintained when nectar was pasteurized at 95°C for 1.0 min. and further preserved at 4 ± 2 °C (Minh *et al.*, 2019).

Till date, dragon fruit has no international grade standards and this can lead to problems in international trade. Fruit are generally graded by size and colour. Size grades suggested for Vietnam are: Extra large fruit >500 g, large 380-500 g, regular 300-380 g, medium 260-300 g, small < 260 g (Le *et al.*, 2000). Fruit exported from Israel to Europe are graded by number of fruits (6, 8, 10, 12, 14 and 16) per 4 kg cardboard box (Paull and Chen, 2019).

14.0 Biotechnology

14.1.0 Tissue Culture

Field grown dragon fruit is labourious, time-consuming, variation in production and susceptible to infections (Guo *et al.*, 2012; Swantara *et al.*, 2013; Oeurn *et al.*, 2015; Mercado-Silva, 2018). Quality pitaya seedlings with high survival rate can be produced efficiently by micropropagation system. Several types of explants *viz.*, anther (Garcia *et al.*, 2009a), ovaries (Garcia *et al.*, 2009a,b), embryo (Tel-Zur *et al.*, 2004b, 2005; Cisneros and Tel-Zur, 2010; Cisneros *et al.*, 2013), dormant buds in spine (Qin *et al.*, 2017), shoot tips (Bozkurt *et al.*, 2020) epicotyl and cotyledon segments (Winson *et al.*, 2020), seeds and leaflets (Kasim *et al.*, 2019) are employed for dragon fruit development. *Hylocereus* species in general are self sterile and only 10 per cent of these species are self-compatible (Mizrahi, 2015). Androgenesis and gynogenesis techniques were employed to developed haploid vine cactus plants. Garcia *et al.* (2009 a,b) produced haploids from both pollen mother cells and ovaries. The success of haploid production through androgenesis is media and species specific (Garcia *et al.*, 2009a). By doubling the chromosomes of these haploids one can obtain homozygous lines from heterozygous donors through a single-step which are easy to work with as parents, for further future breeding.

Novel interspecific hybrids between tetraploid *H. megalanthus* (Vaup.) Bauer as the female parent and either diploid *H. monacanthus* (Lem.) Britton et Rose or *H. undatus* (Haw.) Britton et Rose as the male parent was successfully performed by Tel-Zur *et al.* (2004a, 2005). The genotype differences between the parents can contribute to variations in the *in vitro* responses and influenced the outcomes of rescued embryos (Tonguc and Griffiths, 2004). The embryo rescue technique is an additional stage in *Hylocereus species* breeding program. The main factors influencing the successful development of interspecific-interploid vine cacti hybrids was the rescue timing (Cisneros *et al.*, 2013).The yielded hybrids are of different ploidy level as a result of unbalanced gamete formation (Tel-Zur *et al.*, 2003); the higher than expected ploidy levels encountered may be the product of unreduced gametes or of spontaneous chromosome doubling (Chen *et al.*, 1998; Li and Devaux, 2003; Tel- Zur *et al.*, 2003); unreduced pollen grains and unbalanced gametes due to irregular chromosome disjunction at anaphase (Lichtenzveig *et al.*, 2000). The ploidy level of Tel-Zur *et al.* (2004b, 2005) hybrids yielded only allopentaploids, allohexaploids, and 5x-and 6x-aneuploids instead of the expected allotriploids. The results obtained by Cisneros *et al.* (2013) showed the successful production of viable, interspecific–interploid hybrids with lower ploidy levels, *i.e.*, diploid and triploid, than that of the female parent, an outcome that is impossible to achieve naturally and that provides new avenues for the creation of potentially superior hybrids for commercial applications. The hybrids were further analyzed and confirmed by flow cytometry and amplified fragment length polymorphism (AFLP) analysis. More than 70 per cent of the obtained hybrids grew normally in soil after hardening. Micropropagation technique is suitable for mass production of secondary metabolites. The betalain content is dependent on the developmental and physiological stage of cultivated dragon fruit (Phebe *et al.*, 2009). Therefore, in vitro cultures may provide a controlled condition for the production of desired secondary metabolites without constraints of environmental factors. Winson *et al.* (2020) indicated that *in vitro* protocol of callus and cell suspension culture techniques are efficient system for sustainable biotechnological production of betalain compounds which can contribute to functional food colouring and pharmaceutical industries. Yellow (*Selenicereus megalanthus*) has good taste and long shelf life but slow vegetative growth. Suárez Román *et al.* (2014) studied the reproduction of yellow and red pitaya by *in vitro* tissue culture and observed that the mean of the multiplication coefficient of the red pitaya is higher than the yellow pitaya.

Traditional characterization of genotypes/breeding lines on the basis of morphological descriptors is labourous (Cansian and Echeverrigaray, 2000). Molecular markers provide a simple technique for identification of cultivar and for assigning genotypes to different heterotic groups. The DNA markers have become increasingly recognized tools over other for cultivar identification, where the DNA is not affected by environmental factors or the developmental stage of plants (Astrani *et al.*, 2004). Currently, a range of dominant and co-dominant molecular markers are available for assessing genetic diversity in plants. Dragon fruit diversity was analyzed with several DNA markers *viz.*, random amplified polymorphic DNA (RAPD) (Tel-Zur *et al.*, 2004a, Solano *et al.*, 2005; Rifat *et al.*, 2019), amplified fragment length polymorphism (AFLP) (Pagliaccia *et al.*, 2015), simple sequence repeats (Pan *et al.*, 2017), inter simple sequence repeats (ISSR) (Tao *et al.*, 2014; Hernández-Andrade *et al.*, 2019). The diversity studies will help growers and researchers to choose genetically distinct accessions from the germplasm collection to produce superior hybrids. Detection of genetic differences and elucidation of genetic relationships among genotypes of any plant are of high importance for both proprietary rights protection, conservation, utilization and improvement of its genetic resources (Liu *et al.*, 2011).

14.2.0 Molecular Marker

The protocol for DNA isolation of *Hylocereus* was proposed by Tel-Zur *et al.* (1999); ever since, studies at cytological and chromosomal level of experiment were initiated. Using RAPD technique, the origin of these species is proved to be allopolyploid instead of autopolyploid (Tel-Zur *et al.*,2005). Cisneros and Tel-Zur (2012) mentioned some techniques like fluorescence activated cell sorting (FACS), molecular markers, marker assisted selection (MAS) and quantitative trait loci (QTL) used with phenotypic, genomic and cytological evaluations of *Hylocereus* allowed them to obtain a model that associates agronomical characteristics with the gene that controls their expression and influence inheritance. This model outlines the pathways for polyploidy seed formation and development after interspecific-interploid hybridization.

Tel-Zur *et al.* (2004b) used random amplified polymorphic DNA (RAPD) markers that differentiated 13 species within *Hylocereus* and *Selenicereus*. Moreover, Tel-Zur *et al.* (2004b) were able to differentiate unique genotypes between different accessions of *H. polyrhizus, H. undatus, S. grandiflorus,* and *S. megalanthus* (= *H. megalanthus*). RAPD markers was employed by Solano *et al.* (2005) to evaluate the genetic diversity among 50 accessions of *H. undatus* originating from nine different states in Mexico while one from Colombia was used as a control (unknown accession). In addition to identifying each accession, Solano *et al.* (2005) found three genotypes that were significantly different from the rest and demonstrated substantial genetic diversity of pitahaya in Mexico. Divisity of fifteen germplasms from Bangladesh are studied using random amplified polymorphic DNA (RAPD) markers (Rifat *et al.*, 2019). The clusters suggest that 15 dragon fruit genotypes might come from three different species. Overall heterozygosity or gene diversity and Shannon's information index detected were 0.327 and 0.482, respectively indicating that a substantial level of genetic diversity existed in the studied germplasms collected from Bangladesh (Rifat *et al.*, 2019). To identify the germplasm resources, genetic diversity levels and relationships among 46 accessions collected from China and abroad, Pan *et al.* (2017) investigated using simple sequence repeat markers. Subsequently, based on their geographical distribution, four groups (A, B, C and D) were distinguished by truncating the dendrogram at an 80 per cent similarity level indicating the rich genetic diversity and genetic separation during dispersal. (Pan *et al.*, 2017). Fifty accessions of dragon fruit germplasm in China was differentiated using inter simple sequence repeat markers (Tao *et al.*, 2014). All of the genotypes tested by Tao *et al.* (2014) were unique and clustered into three main groups: one consisting of mostly wild accessions, the red pulp types *i.e H. polyrhizus* accessions occupy the second cluster, and *H. undatus* accessions (white pulp types) represent third cluster. Pagliaccia *et al.* (2015) used Amplified fragment length polymorphism (AFLP) markers to differentiate 230 accessions of pitahaya collection. The tested pitahaya accessions of Pagliaccia *et al.* (2015) yielded seven clades *viz., H. guatemalensis, H. undatus,* Hybrids, Wild-type accessions, *H. megalanthus, H. polyrhizus/H. costaricensis* and *H. ocamponis*. Moreover, the morphologically produced clusters are inconsistency within the cluster analysis compared to molecular data (Tao *et al.*, 2014). This is likely because there are no reproductive barriers among most "species" in pitahaya (Tao *et al.*, 2014; Pagliaccia *et al.*, 2015). Hernández-Andrade *et al.* (2019) collected seventy one dragon fruit specimens to characterize genetic diversity with DNA Inter Simple Sequence Repeats (ISSRs). Two genetic lineages (*K*=2) was identified by Hernández-Andrade *et al.* (2019); one in the West and East regions and another in the West and Central regions. The polymorphism information content (PIC) values were at the most 20 per cent lower than those estimated from an ex situ germplasm collection of *Hylocereus* sp. Li *et al.* (2018) confirmed *H. megalanthus* as an autotetraploid, rather than an allotetraploid, origin of this species using SSR and flow cytometry,

14.3.0 Transcriptomics

q RT-PCR is the most sensitive technology of measuring gene transcript levels depending on stable reference gene(s) for normalization (Nicot *et al.*, 2005). An appropriate reference gene can effectively correct the errors of RNA quantity, reverse transcription efficiency, stability of different reaction channels of PCR instrument and standard operation mode, which can help to obtain the real differential expression of the target genes (Udvardi *et al.*, 2008) Till date, there is no report on the identification of the most reliable reference genes to fit different tissues and growth periods for q RT-PCR normalization. Chen *et al.* (2019) identified *Actin(1)* and *EF1-á(1)* are the most stable candidate reference gene related to betalain biosynthesis at flesh colouration stages. Betalains have high nutritional value and bioactivities. They are water-soluble nitrogen-containing pigments. Among the commercially cultivated fruit, pitaya (*Hylocereus* spp.) contai abundant betalains for consumer. The pitaya genes had the highest number of hits to the *B. vulgaris* (50.63 per cent), and the species distribution with the greatest number of *H. polyrhizus* was *Vitis vinifera* (50.1 per cent) by Illumina RNA-Seq (Qingzhu *et al.*, 2016). Recently few researchers carried out an indepth studies into betalain biosynthesis; their findings provides new information on betalain biosynthesis in *Hylocereus* at the transcriptional level. Wu *et al.* (2020) carried out a new third-generation sequencing (TGS) platform; SMRT sequel by PacBio sequencing. The full-length transcripts of pitaya pulp that are generated from SMRT with Illumina RNA-seq provide an efficient process to the research of genes and facilitate the identification of putative betalain-related genes (Wu *et al.*, 2020). HpCYP76AD4, HpDODA1, HpDODA2, and HpCYP76AD4 involved in betalain biosynthesis were identified in pitaya fruit, and HpNAC and HpGSTs might participate in the regulation of betalain in pitaya fruit (Wu *et al.*, 2020). Furthermore, HpNAC may play a role in the regulation of betalain synthesis in coordination with MYB TF, which provides new insights into betalain synthesis in pitaya (Wu *et al.*, 2020). In addition, Xie *et al.* (2020) indentified twenty-nine candidate genes related to betalain biosynthesis from the transcriptome data. HmB5GT1 and HmHCGT2 were were further analysed. Silencing of HmB5GT1 and HmHCGT2 resulted in a significant reduction in betacyanin and betaxanthin contents Both betalin synthesis gene were conserved in 'PSPG-box' and localized in nucleus (Xie *et al.*, 2020) that suggested that these genes are possibly involved in betalain biosynthesis in *H. megalanthus.*

The long-day pitaya plants are supplemented with lights in short-day winter for induction of flowering. To further explore the molecular mechanisms of supplementary lighting, Xiong *et al.* (2020) used de novo RNA sequencing-based transcriptomic analysis and identified many CONSTANS-LIKE, FLOWERING LOCUS T, and other DEGs (Differentially expressed genes) involved in the direct regulation of flowering including CDF (Cycling DoF factor) and TCP (teosinte branched 1/cycloidea/proliferating cell nuclear antigen factor) which function as typical transcription factor genes in the flowering process. The expression levels of HuTy1P4 retrotransposons in pitaya were associated with stress responses after exposure to hormone and abiotic stress (salt, heat and cold) and may be involved in the occurrence of the somaclonal variation in pitaya (Nie *et al.*, 2019).

15.0 Conclusion

Dragon fruit is one of the super foods of the tropic due to its nutrient richness. It contains many nutraceuticals which today are highly appreciated by consumers. The fruit is extremely visually attractive, besides having great potential for medicine, ornamental and industrial production. Not only pulp, the pitaya peels contain high betalin and citramalic acid which accounts for about 33 per cent of total fruit weight. Hence by-products of pitaya peel and thinned fruit can be considered as a viable resource for betalin and citramalic acid extraction. All these beneficial factors make dragon fruit as the best option for weight loss treatment, control of diabetes, lowering the cholesterol level, *etc*. Apart from nutritional benefits, dragon fruit acclimation to arid and semi-arid conditions and has high Water use efficiency (WUE) - which is characteristics of CAM plants and are tolerant to stresses. The water use efficiency is the highest among all fruit trees. The crop is hardy and can survive in any type of climatic condition favorable for flowering, fruiting and soil condition provided with good drainage. Yields are high; fruits can be produced almost year around which is huge market advantage. Fruits remain fresh under room condition. A dragon fruit production has been demonstrated to provide viable source of food, income and employment to growers locally and abroad. It could be an asset to backyard growers as well as entrepreneurs of medium and large scale plantations. The economic production starts from first year after planting, and full production within three to five years. In current scenario, the marketability of dragon fruit is very high because of limited number of commercial producers of the product. In the future, production is expected to rise; hence, marketing strategies need to be critically examined. In case of any market glut, there is a strong need to explore the avenues of value addition through processing, so that the production will be continue to increase and the surplus produce shall be processed.

Several studies have been taken up to evaluate the genetic diversity of pitahaya in its natural habitat, in particular under ecological niches. The exploring of untapped metabolic profiling in fruits pulp to enhance the fruit quality has been undertaken by many researchers. To keep up the demand, dragon fruit production should be very productive and profitable. Unfortunately, there are challenges that limit the potential production and export of dragon fruit. Making a significant impact on yield reductions are diseases and pests. Several control options are available for select dragon fruit diseases and pest. The use of biological control agents and biopesticides has been considered a promising sustainable disease control approach in the current organic farming scenario. Breeding efforts concerning major diseases and pests, particularly during plant propagation, fruit production and post-harvest fruit life are the challenging research.

16.0 References

Abirami, K., Sakthivel, K., Sheoran, N., Baskaran, V., Gautam, R. K., Jerard, B. A. and Kumar, A. (2019) *Plant Dis.*, **103**(4): 768-768.

Ali, A., Zahid, N., Manickam, S., Siddiqui, Y., Alderson, P. G. and Maqbool, M. (2013) *Post-harvest Biol. Tec.*, **86**: 147-153.

Ankli A., Sticher, O. and Rich, M.H. (1999) *Human Ecol.*, **27**(4): 557-580.

Anonymous Perween, T. (2017) *M.Sc. Thesis*, Bidhan Chandra Krishi Viswavidyalaya, Mohanpur West Bengal, India.

Arévalo-Galarza, M.L. and Ortiz-Hernández, Y.D. (2004) *Cactáceasy Suculentas Mexicanas* XLIX año, **49**(3): 85-90.

Argueta, A.V., Cano, L.M.A. and Rodarte, M.E. (1994) *Instituto Nacional Indigenista* II: 1170– 1171.

Ariffin, A. A., Bakar, J., Tan, C. P., Rahman, R. A., Karim, R. and Loi, C. C. (2009) *Food Chem.*, **114**(2): 561-564.

Astarini, I. A., Plummer, J. A., Lancaster, R. A. and Yan, G. (2004) *Crop Past. Sci.*, **55**(2): 117-124.

Awang, Y.B., Abdul Ghani, M.A., Sijam, K., Mohamad, R. B. and Hafiza, Y. (2011b) In: IInd *International Symposium on Underutilized Plant Species: Crops for the Future-Beyond Food Security*, pp. 763-771.

Awang, Y., Ghani, M. A. A. and Sijam, K. (2010) *Acta Hort.*, **880**: 431-437.

Awang, Y., Ghani, M.A.A., Sijam,K. and Mohamad, R.B. (2011a) *Afr. J. Microbiol. Res.*, **5**(29): 5250-5259.

Ayala-Aponte, A.A. and Serna-Cock, L. (2011) *Revista de la Asociación Colombiana de Ciencia y Tecnología de Alimentos*, **23**: 78-97.

Aziz, F.A. and Noor, M.M. (2010) *Biomed. Res. Int*, **21**(2): 126-130.

Badillo, J. (2005) *Manual Técnico de Fitosanidad en Pitahaya*. MAG-FOR-OIRSA-VIFINEX, Managua, Nicaragua, p. 85.

Bae, S., Kim, S. G. and Kim, Y. H. (2013) *Plant Pathol. J.*, **29**(1): 42.

Balendres, M. A. and Bengoa, J. C. (2019) *Crop Protec.*, **126:** 104920.

Balois-Morales, R., Colinas-León, M. T., Peña- Valdivia, C. B., Chávez-Franco, S. H. and Alia-Tejacal, I. (2007) *Revista Chapingo. Serie Horticultura*, **13**(2): 115-120.

Barbeau, G. (1990) *Fruits*, **45**(2): 141-147.

Barthlott, W. and Hunt, D.R. (1993) In: *The Families and the Genera of Vascular Plants* (Ed. K. Kubitzki), vol. 2. Springer-Verlag, Berlin, Germany, pp. 161–196.

Bauer, R. (2003) *System Initiatives*, **17**: 3–63.

Becerra, L.A. (1986) *El cultivo de la pitaya*. Federación Nacional de Cafeteros de Colombia, Bogotá, Colombia, p. 19.

Becerra, O.L. (1994) In: *Memorias de la primera reunión internacional con demanda nacional e internacional.* (Eds. A. Villegas M *et al.*), *Texcoco, Edo. de México*, Mexico, pp. 123-142.

Ben-Asher, J., Nobel, P. S., Yossov, E. and Mizrahi, Y. (2006) *Photosynthetica*, **44**(2): 181-186.

Ben-Ze'ev, I. S., Assouline, I., Levy, E. and Elkind, G. (2011) *Phytoparasitica*, **39**(2): 195-197.

Bibliowicz, A. and Hernández, S.M. (1998) *Trabajo de grado. Universidad Nacional de Colombia. Bogotá. Facultad de Agronomía.*

Bozkurt, T., Ýnan, S. and Dündar, Ý. (2020) *Intl. J. Agric. Nat. Sci.*, **13**(1): 39-46.

Bravo-Hollis, H. (1978) In: *Las cactáceas de México* Vol. I. UNAM, México, D. F., México, p.743.

Britton N.L and Rose J.N (1963) In: *The Cactaceae: Description and Illustration of Plants of the Cactus Family*. Dover, New York. USA. **1**(2): 183-195.

Cálix de Dios H. (2004) la República Mexicana. *Cactáceas y Suculentas Mexicanas* XLIX: **1**: 4-23.

Cálix de Dios, H. and Castillo, M.R. (2005) *Trópico Rural.*, **1**(2): 31-37.

Cansian, R. L. and Echeverrigaray, S. (2000) *HortScience*, **35**(6): 1155-1158.

Casas, A. and Barbera, G. (2002) In: *Cacti: Biology and Uses* (Ed. P.S. Nobel), University of California Press, Los Angeles, California, pp. 143–162.

Castillo, M.R., Livera, M.M. and Márquez, J.G. (2005) *Agrociencia*, **39**: 183-194.

Castro, J.C., Avincola, A.S., Endo, E.H., Silva, M.V., Dias Filho, B.P., Junior, M.M. and Abreu Filho, B.A. (2018) *Biol. Technol.*, **141:** 71-76.

Castro, J. C., Endo, E. H., de Souza, M. R., Zanqueta, E. B., Polonio, J. C., Pamphile, J. A. and de Abreu Filho, B. A. (2017) *Ind. Crops and Prod.*, **97:** 101-109.

Cavalcante, Í.H.L. (2008) *Ph.D. Thesis*, São Paulo State University, Jaboticabal, Brasil.

Cavalcante, Í.H.L., Beckmann, M.Z., Martins, A.B.G., Galbiatti, J.A. and Cavalcante, L.F. (2007) *Intl. J. Fruit Sci.*, **7**(3): 39-46.

Cavalcante, Í.H.L., Martins, A.B.G., Silva Junior, G.B, Fonseca Rocha L., Neto, R. F. and Cavalcante, L.F. (2011) *Rev. Bras. Frutic.*, **33**(3): 970-982.

Chang, S. C., Lee, M. S., Lin, C. J. and Chen, M. L. (1998) *Asia Pac. J. Clin. Nutr.*, **7**: 206-210.

Charoensiri, R., Kongkachuichai, R., Suknicom, S. and Sungpuag, P. (2009) *Food Chem.*, **113**(1): 202-207.

Chemah, T.C., Aminah, A., Noriham, A. and Wan-Aida, W.M. (2010) *Intl. Fd. Res. J.*, **17**(4): 1003-1010.

Chen, C., Wu, J., Hua, Q., Tel-Zur, N., Xie, F., Zhang, Z. and Qin, Y. (2019) *Plant Methods*, **15**(1): 70.

Chen, Y., Kenaschuk, E. O. and Procunier, J. D. (1998) *Euphytica*, **102**(2): 183-189.

Choi, H.W., Chung, I. M., Sin, M. H., Kim, Y. S., Sim, J. B., Kim, J. W. and Chun, S. C. (2007) *Crop Prot.*, **26**(2): 162-168.

Choi, M. O., Kim, S. G. and Kim, Y. H. (2010) *Plant Pathol. J.*, **26**: 231-237.

Choo, W. S and Yong, W. K. (2011) *Adv. Appl. Sci. Res.*, **2**(3): 418-425.

Chuang, M.F., Ni, H.F., Yang, H.R., Shu, S.L., Lai, S.Y. and Jiang, Y.L. (2012) *Plant Dis.*, **96**(6): 906-906.

Cisneros, A. and Tel-Zur, N. (2010) *Euphytica*, **174**(1): 73-82.

Cisneros, A., Garcia, R. B. and Tel-Zur, N. (2011) *Flora-Morphology, Distribution, Functional Ecology of Plants*, **206**(12): 1076-1084.

Cisneros, A., Garcia, R. B. and Tel-Zur, N. (2013) *Euphytica*, **189**(3): 433-443.

Clerck F.A.J. and Negreros-Castillo, P. (2000) *Agrofor. Syst.*, **48**: 303-317.

Cohen, H. and Tel-Zur, N. (2012) *Euphytica*, **184**(3): 345-354.

Dag, A. and Mizrahi, Y. (2005) *J. Hortic. Sci.Biotech.*, **80**(5): 618-622.

Daubresse Balayer, M. (1999) *Fruits oubliés Reseau*, **1**: 15-17.

Davies, W. J. and Zhang, J. (1991) *Ann Rev. Plant Biol.*, **42**(1): 55-76.

de Oliveira, M.M.T., Shuhua, L., Kumbha, D.S., Zurgil, U., Raveh, E. and Tel-Zur, N. (2020) *Plant Physiol. Bioch.*, doi. 10.1016/j.plaphy.2020.04.044.

Delgado A., Kondo, T., Imbachi, L. K., Quintero, E. M., Belline, M., Burbano, M. and Medina S. J.A. (2010) *Boletín del Museo de Entomología de la Universidad del Valle*, **11**(2): 1-10.

Duarte, L. M. L., Alexandre, M. A. V., Rivas, E. B., Harakava, R., Galleti, S. R. and Barradas, M. M. (2008) *J. Plant Pathol.*, **90**(3): 545-551.

Escobar, B.A., Gómez, G.Y., Bautista, R.E. and Pérez, G.R.S. (2010) In: *Memorias del VII. Encuentro Nacional de Biotecnología del Instituto Politécnico Nacional. Memorias* Mazatlán, México. p.12.

Esquivel, P. and Araya, Y.Q. (2012) *Revista Venezolana de Ciencia y Tecnología de Alimentos*, **3**(1): 113-129.

Esquivel, P., Stintzing F. C. and Carle R. (2007a) *Innov. Food Sci. Emerg. Technol.*, **8** (3): 451-457.

Esquivel, P., Stintzing, F. C.and Carle, R. (2007b). *Zeitschrift für Naturforschung*, **62**(9-10): 636-644.

Eusebio, J.E. and Alaban, M.C.S. (2018) http: //ap.fftc.agnet.org/ap_db.php?id=876.

Ezra, D., Liarzi, O., Gat, T., Hershcovich, M. and Dudai, M. (2013) *Plant Dis.*, **97**(11): 1513-1513.

Faziha, N., Ariff, M., Fahim, M., Ezzry, M. and Suhaizan, L. (2019) *Malay. Appl. Biol. J.*, **48**: 43–49.

Fouqué, A. (1972) *Fruits*, **27**: 200-218.

Fournet J. (2002) *Inra-Cirad-Gondwana*, Paris, France, pp. 224-240.

Freitas, S.T.D. and Mitcham, E.J. (2013) *Sci. Agric.*, **70**(4): 257-262.

Garcia, R. B., Cisneros, A., Schneider, B. and Tel-Zur, N. (2009b) *Plant Cell Rep.*, **28**(5): 719-726.

Garcia, R. B., Schneider, B. and Tel-Zur, N. (2009a) *Plant Cell Tiss. Org.*, **96**(2): 191-199.

García-Cruz, E.E. (2011) *MSc. Thesis*, Instituto Politécnico Nacional, Oaxaca, Mexico.

Gazis, R., Poudel, B., Dey, K. K., Zhang, S., Palmateer, A., Campoverde, E. V and Adkins, S. (2018) *Plant Dis.*, **102**(12): 2666.

Gifford, R.M, Thorne, J.H., Hitz, W.D. and Giaquinta, R.T. (1984) *Science*, **225**: 801–808.

Glangkarn, S. (2015) *Food and Public Health*, **5**(5): 203-206.

Gross, K.C., Wang, CH. Y. and Saltveit, M. (2002) In: *The commercial storage of fruits, vegetables, and florist and nursery stocks*. USDA, Beltsville, USA. (Agricultural Handbook number 66) http: //www. ba.ars.usda.gov/hb66/contents.html.

Guo, L.W., Wu, Y.X., Ho, H.H., Su, Y.Y., Mao, Z.C., He, P.F. and He, Y.Q. (2014) *J. Phytopath.*, **162**(4): 272-275.

Guo, L.W., Wu, Y.X., Mao, Z.C., Ho, H. H. and He, Y.Q. (2012) *Plant Dis.*, **96**(12): 1826-1826.

Harivaindaram, K.V., Rebecca, O.P.S. and Chandran, S. (2008) *Pak. J. Biol Sci.*, **11**(18): 2259-2263.

Hawa, M. M., Salleh, B. and Latiffah, Z. (2013) *Ann. Appl. Biol.*, **163**(2): 269-280.

Hawa, M. M., Salleh, B. and Latiffah, Z. (2009) *Plant Dis.*, **93**(9): 971-971.

Hawa, M. M., Salleh, B. and Latiffah, Z. (2010) *Afr. J. Biotechnol.*, **9**(3).

Hawa, M.M., Nurul Faziha, I., Nik Mohamad Izham, M. N. and Latiffah, Z. (2017) *Ann. Appl. Biol.*, **170**(3): 434-446.

He, P. F., Ho, H., Wu, X. X., Hou, M. S. and He, Y. Q. (2012) *Plant Pathology & Quarantine*, **2**(1): 31-35.

Herbach, K.M., Maier, C., Stintzing, F.C. and Carle, R. (2007) *Eur. Food Res. Technol.*, **224**: 649-658.

Herbach, K.M., Rohe, M., Stintzing, F.C. and Carle, R. (2006) *Int. Food Res. J.*, **39**: 667-677.

Hernández, Y. D. O. and Salazar, J. A. C. (2012) *Comun. Sci.*, **3**(4): 220-237.

Hernández-Andrade, A., Parra-Gómez, L., Ferrer, M. M., Montañez-Escalante, P. I. and Jiménez-Osornio, J. (2019) *J.Ethnobiol.*, **39**(4): 530-548.

Ho, T.T., Clark, C.J., Waddell, B.C. and Woolf, A.B. (2006) *Post-harvest Biol. Tec.*, **41**: 62–69.

Hong, C. F., Zhang, S., Gazis, R., Crane, J. H. and Wasielewski, J. (2020) *Electronic Data Information Source*.

Hopkins, A.L. and Stepp, J.R. (2012) *Econ. Bot.*, **66**(3): 249-254.

Hua, Q., Chen, C., Zur, N. T., Wang, H., Wu, J., Chen, J and Qin, Y. (2018) *Plant Physiol. Biochem.*, **126**: 117-125.

Hunt, D., Taylor, N. P. and Charles, G. (2006) *The new cactus lexicon*, DH Books, Milborne Port.

INRA-CEE. (1994) *Guía tecnológica para la producción de pitahaya*. Instituto Nicaraguense de Reforma Agraria, Managua, Nicaragua, p. 70.

Islam, M. Z., Khan, M. T. H., Hoque, M. M. and Rahman, M. M. (2012) *The Agriculturists*, **10**(2): 29-35.

Jaafar, R.A., Rahman, A.R.B.A., Mahmod, N.Z.C. and Vasudeven, R. (2009) *Am. J. Appl. Sci.*, **6**: 1341-1346.

Jayasinghe, O., Fernando, S., Jayamanne, V. and Hettiarachchi, D. (2015) *Eur. Sci.*, **11**(3).

Jiang, Y.L., Liao, Y.Y., Lin, T. S., Lee, Ch. L., Yen., Ch. R. andYang, W. J. (2012) *Hortic. Sci.*, **47**(8): 1063-1067.

Jitareerat, P., Sripong, K., Masaya, K., Aiamla-or, S. and Uthairatanakij, A. (2018) *Agric. Nat. Resour.*, **52**(6): 543-549.

Jorge, L.I.F and Ferro, V.D O. (1989) *Rev. Farm. Bioquím. Uni. São Paulo*, **25**: 123-136.

Joshi, M. and Prabhakar, B. (2020) *J.Food Biochem.*, e13260.

Juárez-Cruz, A., Livera-Muñoz, M., Sosa-Montes, E., Goytia-Jiménez, M.A., González-Hernández, V.A. and Bárcena-Gama, R. (2012) *Rev. Fitotec. Mex.*, **35**(2): 171-175.

Kakade, V., Dinesh, D. S., Bhatnagar, P. and Kadam, D. (2019) *Indian J. Agric. Sci.*, **89**(11): 143-147.

Kaplan – levy, N.R. (1999) Undergraduate research project. Ben-Gurion: Department of Life Sciences, University of the Negev.

Karunakaran, G. and Arivalagan, M. (2019) *Indian Hortic.*, **63**(1): 8-11.

Karunakaran, G., Tripathi, P. C., Sankar, V., Sakthivel, T. and Senthilkumar, R. (2014) In: *Abstract in proceeding of National Seminar on Strategies for conservation, Improvement and utilization of underutilized fruits* Bengaluru, India, pp. 138-139).

Kasim, D. P., Kishore, N. S., Suneetha, P., Rao, K. B., Kumar, M. N. and Krishna, M. S. R. (2019) *Res. J. Pharm. Technol.*, **12**(4): 1491-1494.

Kek Hoe, T. (2017) *Int. J. Agric. Innov. Res.*, **6**(2): 354-358.

Khaimov, A. and Mizrahi, Y. (2006) *J. Hortic. Sci. Biotech.*, **81**(3): 465-470.

Khaimov-Armoza, A., Novák, O., Strnad, M. and Mizrahi, Y. (2012) *Isr. J. Plant Sci.*, **60**(3): 371-383.

Khalid, S.M.N. (2020) Crop production manual - A guide to fruit and vegetable production in the Federated States of Micronesia, Food and Agriculture Organization pp 62; (http: //www.fao.org/3/ca7556en/CA7556EN.pdf). Accessed on, July 1, 2020.

Kim, H., Choi, H.K., Moon, J.Y., Kim, Y.S., Mosaddik, A. and Cho, S.K. (2011) *J. Food Sci.*, **76**(1): 38-45.

Kim, J. S., Park, C. Y., Nam, M., Lee, J. S., Kim, H. G., and Lee, S. H. (2016) *Plant Dis.*, **100** (12): 2544-2544.

Kishore, K. (2016) *Sci. Hortic- Amsterdam*, **213**: 294-302.

Kozlowski T T. (1971) *Growth and Development of Trees: Cambial Growth, Root Growth, and Reproductive Growth*, Vol 2. Academic Press, New York.

Lan, G. B., He, Z. F., Xi, P. G. and Jiang, Z. D. (2012) *Plant Dis.*, **96**(11): 1702-1702.

Le Bellec F. (2003) La pitaya (*Hylocereus* sp.) enculture de diversification à l'île de la Réunion, Inst. Natl. Hortic. (INH), Mém. Angers, France, p. 55.

Le Bellec, F. (2004) *Fruits*, **59**(6): 411-422.

Le Bellec, F., Vaillant, F., and Imbert, E. (2006) Pitahaya (*Hylocereus* spp.). *Fruits*, **61**(4): 237-250.

Le, V.T., Nguyen, N., Nguyen, D.D., Dang, K.T., Nguyen, T.N.C., Dang, M.V.H., Chau, N.H. and Trink. N.L. (2000) *ACIAR Proceedings*, **100**: 101-114.

Leakey R R B and Coutts M P. (1989) *Tree Physiol.*, **5**: 135–46.

Li, D., Arroyave Martinez, M. F., Shaked, R. and Tel-Zur, N. (2018) *Front. Plant sci.*, **8**: 2142.

Li, H. and Devaux, P. (2003) *Plant Sci.*, **164**(3): 379-386.

Li, Y. S., Mao, C. H., Kuo, T. Y. and Chang, Y. C. (2015) *Mngmt.*, **130**.

Lichtenzveig, J., Abbo, S., Nerd, A., Tel-Zur, N. and Mizrahi, Y. (2000) *Am. J. Bot.*, **87**(7): 1058-1065.

Lichtenzveig, J., Abbo, S., Nerd, A., Tel Zur, N. and Mizrahi, Y. (2000) *Am. J. Bot.*, **87**(7): 1058-1065.

Lim, H.K., Tan, C.P., Karim, R., Ariffin, A.A. and Bakar, J. (2010) *Food Chem.*, **119**: 1326-1331.

Lim, Y. Y., Lim, T. T. and Tee, J. J. (2007) *Food Chem.*, **103**(3): 1003-1008.

Liou, M. R., Chen, Y. R. and Liou, R. F. (2004) *Arch. Virol.*, **149**(5): 1037-1043.

Liou, M. R., Hung, C. L. and Liou, R. F. (2001) *Plant Dis.*, **85**(2): 229-229.

Liu, D., He, X., Liu, G. and Huang, B. (2011) *Genet. Resour. Crop Evol.*, **58**(5): 679-688.

Liu, F., Wu, J. B., Zhan, R. L. and Ou, X. C. (2016) *Plant Dis.*, **100**(8): 1792.

Lobo, R., G. Bender, G. Tanizaki, J. Fernandex de Soto and J. Aguiar. (2013) *Accessed* June 18, 2020.

Lu, Z., Lu, X., Qin, B., Cheng, M., Huang, L., Chen, B. and Liao, Y. (2015) *Journal of Southern Agriculture*, **46**(9): 1606-1612.

Lum M.S. and Norazira, M.A. (2011) *J. Agric. Sci.*, **3**(1): 146-152.

Luo, H., Cai, Y., Peng, Z., Liu, T. and Yang, S. (2014) *Chem. Cent. J.*, **8**(1): 1-7

Ma, W. J., Yang, X., Wang, X. R., Zeng, Y. S., Liao, M. D., Chen, C. J. and Jia, D. M. (2014) *Plant Dis.*, **98**(7): 991-991.

Mahayothee, B., Komonsing, N., Khuwijitjaru, P., Nagle, M. and Müller, J. (2019) *Int. J. Food Sci. Tech.*, **54**(2): 460-470.

Mamat, S., Shah, U. K. M., Remli, N. A. M., Shaari, K., Laboh, R., and Rahman, N. A. A. (2018) *Int. J. Environ. Agric. Res.*, **4**(1).

Masanto, M., Sijam, K., Awang, Y., and Satar, M. G. M. (2018) *Jurnal Perlindungan Tanaman Indonesia*, **22**(1): 1-12.

Masyahit, M., Sijam, K., Awang, Y. and Ghazali, M. (2009) *Int. j. Agric. Biol.*, **11:** 659-666.

Meetum, P., Leksomboon, C. and Kanjanamaneesathian, M. (2015b) In: *International Symposium on Durian and Other Humid Tropical Fruits*, **1186**: 135-142.

Meetum, P., Leksomboon, C. and Kanjanamaneesathian, M. (2015a) *J. Plant Pathol.*, **97**(2): 402-402.

Mello, F. R. D., Bernardo, C., Dias, C. O., Gonzaga, L., Amante, E. R., Fett, R. and Candido, L. M. B. (2015) *Ciênc. Rural*, **45**(2): 323-328.

Mendieta, R.M. and Del Amo, R.S. (1981) *Catálogo de las plantas medicinales del estado de Yucatán*. Instituto Nacional de Investigaciones sobre Recursos Bióticos Xalapa, Veracruz, Mexico, 428p.

Mercado-Silva, E.M. (2018) Pitaya-*Hylocereus undatus* (Haw). In: Exotic Fruits: Reference Guide, (Eds. S. Rodrigues, E.D.O. Silva, E.S.D. Brito), Academic Press, London, pp. 339–349.

Merten, S. (2003) *J. Prof. Assoc. Cactus*, **5**: 98-105.

Metz, C., Nerd, A. and Mizrahi, Y. (2000) *Hortic. Sci.*, **35**(2): 199-201.

Minh, N. P., Nhan, N. P. T., Tha, D. T., Khai, L. Q. and Tu, L. N. (2019) *J. Pharm. Sci. Res.*, **11**(3): 1040-1043.

Mizrahi, Y. (2014) *Rev. Bras. Frutic.*, **36**(1): 124-138.

Mizrahi, Y. (2015) *Improving Pitaya Production and Marketing*, 1-18.

Mizrahi, Y. and A. Nerd (1999) In: J. Janick (eds.), *Perspectives on new crops and new uses*. ASHS Press, Alexandria, VA, USA, Pp. 358-366.

Mizrahi, Y., Mouyal, J., Nerd, A. and Sitrit, Y. (2004) *Ann. Bot.*, **93**(4): 469-472.

Mizrahi, Y., Nerd, A. and Nobel, P. S. (1997) *Hortic. Rev.*, New York **18**: 291-320.

Mizrahi, Y., Raveh, E., Yossov, E., Nerd, A. and Ben-Asher, J. (2007) *Issues in New Crops and New Uses. ASHS Press, Alexandria, VA.* pp. 378-384.

Mohd Adzim Khalili, R., Norhayati, A. H, Rokiah, M. Y., Asmah, R., Siti Muskinah, M. and Abdul Manaf, A. (2009) *Int. Food Res. J.*, **16**: 431-440.

Mohd Hafifi, A. B., Kee, Y. J. and Mohd, M. H. (2019) *Plant Dis.*, **103**(5): 1040-1040.

Nascimento, M. B., Bellé, C., Azambuja, R. M., Maich, S. L. P., Neves, C. G., Souza-Junior, I. T. and Barros, D. R. (2019) *Plant Dis.*, **103**(8): 2137.

Nasir, N. and Nurmansyah. (2016) *Res. J. Pharm. Biol. Chem. Sci.*, **7**(5): 2463-2471.

Nerd, A. and Mizrahi, Y. (1997) *Hortic. Rev.*, **18**: 321-346.

Nerd, A. and Y. Mizrahi. (1999) *Post-harvest Biol. Tec.*, **15**: 99-105.

Nerd, A., Gutman, F. and Mizrahi, Y. (1999) *Post-harvest Biol. Tec.*, **17**: 39-45.

Nerd, A., Tel-Zur, N. and Mizrahi, Y. (2002) Fruits of vine and columnar cacti. In: Nobel, P.S. (Ed.), *Cacti: Biology and Uses*. University of California Press, Los Angeles, California, pp. 185–198.

Ngoc, N. K., Phong Nguyen, N. V., An, P. T. M., Woolf, A. B. and Fullerton, R. A. (2014) In: *III Asia Pacific Symposium on Post-harvest Research, Education and Extension: APS2014 1213*, pp. 453-460.

Nguyen, V. H., Campbell, J., Nguyen, H. H. and Nguyen, M. C. (2015) *Improving pitaya production and marketing. Food and Fertilizer Technology Center, Taipei, Taiwan*, pp. 155-164.

Nicot, N., Hausman, J. F., Hoffmann, L. and Evers, D. (2005) *J. Exp. Bot.*, **56**(421): 2907-2914.

Nie, Q., Qiao, G., Peng, L. and Wen, X. (2019) *Plant Physiol. Biochem.*, **135**: 460-468.

Nobel, P. S. and De la Barrera, E. (2002) *Environ. Exp. Bot.*, **48**(2): 129-137.

Nobel, P. S. and De La Barrera, E. (2004) *Ann. Appl. Biol.*, **144**(1): 1-8.

Nobel, P.S. (2002) *Cacti: Biology and Uses*. University of California Press, Los Angeles, California, p. 291.

Nobel, P.S. and De la Barrera, E. (2002a) *Environ. Exp. Bot.*, **48**(2): 129-137.

Nobel, P.S. and De la Barrera, E. (2002c) *Sci. Hortic-Armsterdam*, **96**: 281-292.

Nobel, P.S. and De la Barrera, E. (2002b) *Biotropica*, **3**(2): 225-231.

Nobel, P.S. andDe la Barrera, E. (2004) *Ann. Appl. Biol.*, **144**: 1-8.

Noegrohati, S., Sulasmi, S., Hernadi, E. and Asviastuti, S. (2019) *Food Quality and Safety*, **3**(2): 99-106.

Nor Hanis Aifaa, Y., Semiah, R., Md. Radzi, M. and Zaulia, O. (2012) In: *VII International Post-harvest Symposium*, **1012**: 735-738.

Nurliyana, R., Syed Zahir, I., Mustapha Suleiman, K., Aisyah, M.R. and Kamarul Rahim, K. (2010) *Int. Food Res. J.*, **17**: 367-375.

Nurmahani, A., Osman, A., Abdul Hamid, F., Mohamad, G. and Pak Dek, M.S. (2012) *Int. Food Res. J.*, **19**: 77–84.

Nurul, S. R. and Asmah, R. (2014) *Int. Food Res. J.*, **21**(4).

Oeurn, S., Jitjak, W. and Sanoamuang, N. (2015) *Asia-Pacific J. Sci. Tech.*, **20**(4): 405-418.

Olaya C., I. (1991) *Frutas de América tropical y subtropical, historia y usos*. Ed. Norma S.A., Bogotá, Colombia, p. 123.

Ortiz-Hernández H., Y.D. (1999) *Pitahaya: un nuevo cultivo para México*. Ed. Limusa-Grupo Noriega Editores, México D.F., Mexico, p. 111.

Pagliaccia, D., Vidalakis, G., Douhan, G. W., Lobo, R. and Tanizaki, G. (2015) *Hortic. Sci.*, **50**(3): 332-336.

Palmateer, A. J., Ploetz, R. C., Van Santen, E. and Correll, J. C. (2007) *Plant Dis.*, **91**(5): 631-631.

Pan, L., Fu, J., Zhang, R., Qin, Y., Lu, F., Jia, L. and Liang, G. (2017) *Sci. Hortic-Armsterdam*, **225**: 171-176.

Panchal, J. B., Gaikwad, R. S., Dhemre, J. K. and Chavan, U. D. (2018) *J. Pharmacogn. Phytochem.*, **7**(4): 2648-2655.

Patel, J. S. and Zhang, S. (2017) *Plant Dis.*, **101**(6): 1046-1046.

Paull, R. E. and Chen, N. J. (2019) (https: //ap.fftc.org.tw/article/1596) Accessed on, June 15, 2020.

Peng, C., Yu, N. T., Luo, Z. W., Fan, H. Y., He, F., Li, X. H. and Liu, Z. X. (2016) *Plant Dis.*, **100**(9): 1956.

Phebe, D., Chew, M. K., Suraini, A. A., Lai, O. M. and Janna, O. A. (2009) *Int. Food Res. J.*, **16**(2).

Pimienta-Barrios, E. and Nobel, P. S. (1994) Pitaya (*Stenocereus* spp., *Cactaceae*): *Econ. Bot.* **48**(1): 76-83.

Plume, O., Straub, S. C., Tel-Zur, N., Cisneros, A., Schneider, B. and Doyle, J. J. (2013) *Syst. Bot.*, **38**(3): 737-751.

Priatni, S. and Pradita, A. (2015) *Procedia Chem.*, **16**: 438-444.

Qin, J., Wang, Y., He, G., Chen, L., He, H., Cheng, X.and Zhang, D. (2017) *Int. J. Agric. Biol.*, **19**(1).

Qingzhu, H., Chengjie, C., Zhe, C., Pengkun, C., Yuewen, M., Jingyu, W. and Yonghua, Q. (2016) *Front. Plant Sci.*, **6**: 1179.

Ramírez, M.F.J. (1999) *Caracterización y compatibilidad en pitahaya* (*Hylocereus spp.*). MSc. Thesis, Universidad Autónoma Chapingo, Chapingo, Mexico.

Raveh, E., Gersani, M. and Nobel, P.S. (1995) *Physiol. Plant.*, **93**: 505-511.

Raveh, E., Nerd, A. and Mizrahi, Y. (1998) *Sci. Hortic-Armsterdam*, **73**(2-3): 151-164.

Rebecca, O.P.S., Boyce, A.N. and Chandran, S. (2010) *Afr. J. Biotechnol.*, **9**(10): 1450-1454.

Rifat, T., Khan, K. and Islam, M. S. (2019) *J. Anim. Plant Sci.*, **29**(3): 809-818.

Rita, W., Suprapta, D., Sudana, I. and Swantara, I. (2016) In: *The Proceeding of International Conference Collaboration Seminar of Chemistry and Industry CoSCI.* October 5-6, Surabaya, Indonesia.

Sanahuja, G., Lopez, P. and Palmateer, A. J. (2016) *Plant Dis.*, **100**(7): 1499-1499.

Sharma, R. K. (2016) *Ph.D Thesis,* Vasantrao Naik Marathwada Krishi Vidyapeeth, Parbhani.

Shetty, A.A., Rana, M.K. and Preetham, S.P. (2012) *J. Food Sci. Tech.*, **49**: 530-536.

Solano, J.P.L., Cano, M.E.A and Hernández, R.G. (2005) *Fitotecnia Mexicana Magazine*, **28**(3): 179-185.

Stintzing, F.C. and Carle, R. (2007) *Trends Food Sci. Tech.*, **18**: 514.

Stintzing, F.C., Schieber, A. and Carle, R. (2002) *Food Chem.*, **77**: 101-106.

Suárez Román, R. S., Caetano, C. M., Ramírez, H. and Morales Osorio, J. G. (2014) *Acta Agron.*, **63**(3): 272-281.

Subandi, M., Mustari, E. and Ari, S. (2018) *Int. J. Eng. Technol.*, **7**(2): 29.

Swantara, I. M. D. and Msi, I. (2013) *J. Biol. Agric. Healthcare*, **3**.

T. K. T. Ngo, and V. H. Nguyen. (2018) In: *Proceedings of Dragon Fruit Regional Network Initiation Workshop*, Taipei, 71–80.

Taba, S., Miyahira, N., Nasu, K., Takushi, T. and Moromizato, Z. I. (2007) *J. Gen. Plant Pathol.*, **73**(5): 374-376.

Taguiam, J. D., Evallo, E., Bengoa, J., Maghirang, R. and Balendres, M. A. (2020) *J. Phytopathol.*, **168**(6): 303-310.

Takahashi, L. M., Rosa, D. D., Basseto, M. A., De Souza, H. G. and Furtado, E. L. (2008) *Australas. Plant Dis. Notes*, **3**(1): 96-97.

Tang, C. S. and Norziah MH. (2007) *Indones. J. Chem.*, **7**(3): 327-331.

Tao, J., Qiao, G., Wen, X. P., Gao, G. L., Liu, T., Peng, Z. J. and Zhang, B. X. (2014) *Sci. Hortic-Armsterdam*, **170**: 82-88.

Tarnowski, T. L. B., Palmateer, A. J. and Crane, J. H. (2010) *Plant Dis.*, **94**(12): 1506-1506.

Tel-Zur, N. (2013) In: *VIII International Congress on Cactus Pear and Cochineal, 1067* (pp. 365-370).

Tel-Zur, N., Abbo, A., Myslabodski, D. and Mizrahi, Y. (1999) *Plant Mol. Biol. Rep.*, **17**: 249-254.

Tel-Zur, N., Abbo, S. and Mizrahi, Y. (2005) *J. Hered.*, **96**(2): 124-131.

Tel-Zur, N., Abbo, S., Bar-Zvi, D. and Mizrahi, Y. (2003) *J. Hered.*, **94**(4): 329-333.

Tel-Zur, N., Abbo, S., Bar-Zvi, D. and Mizrahi, Y. (2004a) *Sci. Hortic- Armsterdam*, **100**: 279-89.

Tel-Zur, N., Abbo, S., Bar-Zvi, D. and Mizrahi, Y. (2004b) *Ann. Bot.*, **94**(4): 527-534.

Tel-Zur, N., Dudai, M., Raveh, E. and Mizrahi, Y. (2011) *Sci. Hortic- Armsterdam*, **129**(4): 570-576.

Tel Zur, N., Dudai, M., Raveh, E. and Mizrahi, Y. (2012) *Plant Breed.*, **131**: 681-685.

Tel-Zur, N., Mizrahi, Y., Cisneros, A., Mouyal, J., Schneider, B. and Doyle, J. J. (2011a) *Genet. Resour. Crop Evol.*, **58**(7): 1075-1085.

Tel-Zur, N., Mouyal, J., Zurgil, U. and Mizrahi, Y. (2020) *Front. Plant Sci.*, **11**: 954.

Tenore, G. C., Novellino, E. and Basile, A. (2012) *J. Funct. Foods*, **4**(1): 129-136.

Tonguç, M. and Griffiths, P. D. (2004) *Euphytica*, **136**(3): 313-318.

Tripathi, P.C., G. Karunakaran, V. Sankar and R. Senthil Kumar. (2014) Technical Bulletin No. 11/2014, Indian Institute of Horticultural Research, Bengaluru, India, pp. 1-9.

Tze, N.L., Han, Ch.P., Yusof, Y.A., Ling, Ch.N., Talib, R.A., Taip, F.S. and Aziz, M.G. (2012) *Food Sci. Biotechnol.*, **21**: 675-682.

Udvardi, M. K., Czechowski, T. and Scheible, W. R. (2008) *Plant Cell*, **20**(7): 1736-1737.

Valencia-Botín, A. J., Kokubu, H. and Ortíz-Hernández, Y. D. (2013) *Australas. Plant Path.*, **42**: 437-440.

Valiente-Banuet, A., Gally, R. S., Arizmendi, M. C. and Casas, A. (2007) *J. Arid Environ.*, **68**: 1-8.

Varela, C. H., Sanchez, D. P. and Bravo, N. (1995) *Fitopatologia Colombiana*, **19**: 40-47.

Vargas, V.M.L., Tamayo Canul., E., Centurión Yah, A., Tamayo Cortez, J., Saucedo Veloz, C. and Sauri Duch, E. (2007) In: V Congreso Iberoamericano de Tecnología Poscosecha y Agroexportaciones (AITEP). *Proceedings of the Universidad Politécnica de Cartagena*, Murcia, España, pp. 1394-1403.

Vijaya, S. I., Anuar, I. S. M. and Zakaria, L. (2015) *J. Phytopathol.*, **163**: 67-71.

Vilaplana, R., Alba, P. and Valencia-Chamorro, S. (2018) *Crop Prot.*, **114**: 90-96.

Vilaplana, R., Páez, D. and Valencia-Chamorro, S. (2017) J. *Food Sci. Technol.*, **82**: 162-169.

Violeta, N. O. U. R., Trandafir, I. and Ionica, M. E. (2010) *Not. Bot. Horti Agrobo.*, **38**(1): 44-48.

Wall, M. M. and Khan, S. A. (2008) *Hortic. Sci.*, **43**: 2115-2119.

Wallace, R.S. and Gibson, A.C. (2002) In: Nobel, P.S. (Ed.), *Cacti: Biology and Uses.* University of California Press, Los Angeles, California, pp. 1–22.

Wang, C. L. and Lin, C. C. (2005) *Plant Pathology Bulletin*, **14**: 269-274.

Weiss, I., Mizrahi, Y. and Raveh, E. (2010) *Sci. Hortic- Armsterdam*, **123**(4): 531-536.

Weiss, J., Nerd, A. and Mizrahi, Y. (1994) *Hortic. Sci.*, **29**: 1487-1492.

Wichienchot, S., Jatupornpipat, M. and Rastall, R. A. (2010) *Food Chem.*, **120**(3): 850-857.

Winson, K. W. S., Chew, B. L., Sathasivam, K. and Subramaniam, S. (2020) *Ind. Crops Prod.*, **155**: 112750.

Wright, E. R., Rivera, M. C., Ghirlanda, A. and Lori, G. A. (2007) *Plant Dis.*, **91**: 323-323.

Wu, J. B., Zhan, R. L., Liu, F. and Cang, J. M. (2017) *Plant Dis.*, **101**: 249.

Wu, L. C., Hsu, H. W., Chen, Y. C., Chiu, C. C., Lin, Y. I. and Ho, J. A. A. (2006) *Food Chem.*, **95**: 319-327.

Wu, M. C. and Chen, C. S. (1998) *In: Proceedings of the annual meeting of the Florida State Horticultural Society.*

Wu, Y., Xu, J., Han, X., Qiao, G., Yang, K., Wen, Z. and Wen, X. (2020) *Int. J. Mol. Sci.*, **21**(9): 3288.

Wu, Y., Xu, J., Shi, M., Han, X., Li, W., Zhang, X. and Wen, X. (2020) *CyTA-J. Food*, **18:** 249-256.

Wybraniec, S. and Mizrahi, Y. (2002) *J. Agr. Food Chem.*, **50**: 6086-6089.

Wybraniec, S., Nowak-Wydra, B., Mitka, K., Kowalski, P. and Mizrahi, Y. (2007) *Phytochemistry*, **68**(2): 251-259.

Xie, F., Hua, Q., Chen, C., Zhang, L., Zhang, Z., Chen, J. and Qin, Y. (2020) *Plant Physiol. Biochem.*

Xiong, R., Liu, C., Xu, M., Wei, S. S., Huang, J. Q. and Tang, H. (2020) *BMC Genom.*, **21:** 1-17.

Xu, M., Peng, Y., Qi, Z., Yan, Z., Yang, L., He, M. D. and Xie, J. (2018) *Australas. Plant Pathol.*, **47**(5): 547-553.

Yacob Ahmad (2019) https: //www.itfnet.org/v1/2019/10/dragonfruit-the-next-potential-major-tropical-fruit/, Accessed on, July 5, 2020.

Yasuko, K., Tomohiro, N., Sei-Itsu, M., Ai-Na, L., Yasuo, F. and Takashi, T. (1984) *BBA-Mol Cell Biol. L.*, **792:** 92-97.

Yi, R. H., Mo, J. J., Wu, F. F.and Chen, J. (2015) *Australas. Plant Dis. Notes*, **10:** 13.

Yi, Y., Zhang, Q. W., Li, S. L., Wang, Y., Ye, W. C., Zhao, J. and Wang, Y. T. (2012) *Food Chem.*, **135**(2): 528-533.

Yong, Y. Y., Dykes, G., Lee, S. M. and Choo, W. S. (2018) *LWT*, **91**: 491-497.

Zahid, N., Alderson, P. G., Ali, A., Maqbool, M. and Manickam, S. (2012) In: *VII International Post-harvest Symposium*, pp. 769-774.

Zahid, N., Maqbool, M., Ali, A., Siddiqui, Y. and Bhatti, Q. A. (2019) *Sci. Hortic-Armsterdam*, **243**: 314-319.

Zainoldin, K. H. and Baba, A.S. (2012) *Int. J. Biol. Life Sci.*, **8**: 93-98.

Zhang, R. Y., Zhao, S. X., Tan, Z. Q. and Zhu, C. H. (2017) *Plant Dis.*, **101**(6): 1031-1031.

Zhang, Y., Lu, X., Huang, J. and Liu, Z. (2018) *Plant Dis.*, **102**(12): 2662.

Zhao, H. J., Chen, S. C., Chen, Y. F., Zou, C. C., Wang, X. L., Wang, Z. H and Ahammed, G. J. (2018) *Plant Dis.*, **102**(6): 1175-1175.

Zheng, F., Chen, L. W., Ma, R., Xu, G., Ding, X. F., Zheng, F. Q. and Xie, C. P. (2018b) *Plant Dis.*, **102**(2): 441-441.

Zheng, F., Xu, G., Zheng, F. Q., Ding, X. F. and Xie, C. P. (2018a) *Plant Dis.*, **102**(12): 2653-2653.

Zheng, Q. M., Wang, Z. and Cai, Y. Q. (2017) *Acta Phytopathol. Sin.*, **47**(3): 364-369.

Zheng, Q., Wang, X., Zhou, J. and Ma, Y. (2020) *Arch. Virol.*, **165**(3): 749-752.

9

MANGOSTEEN

R. Senthil Kumar, Utpala Parthasarathy and V. A. Parthasarathy

1.0 Introduction

Mangosteen (*Garcinia mangostana* L.) is one of the most famous tropical fruits, and is called the Queen of Fruits because of its delicious and much loved taste fruit. In addition, mangosteen is also used as medicine, including as an anti inflammatory and bacterial as treatment for infection and injury. Mangosteen is one of the leading export commodities with a contribution 34.4 per cent of Indonesia total exports [2]. Export destination countries for mangosteen are China, Japan, Singapore, Hong Kong and parts of the Middle East (Suhendra and Mustamu, 2018).

2.0 Composition and Uses

2.1.0 Composition

The fruit is mainly consumed fresh and can be canned, frozen or processed in to jam, sweetened preserved, syrup and candy. The primary active components of the mangosteen fruit are called xanthones which have several benefits including anti- inflammatory properties, anti-allergic and anti-convulants (Arano, 2004). The proximate composition is given in Table 1.

Table 1: Nutritional Composition of the Fruit/100 g Edible Portion*

Components	*Nutritional Value*
Moisture (per cent)	79.7
Energy (calories)	34.0
Protein (g)	0.6
Fat (g)	1.0
Carbohydrates (g)	5.6
Fibre (g)	5.1
Ash (g)	0.1
Calcium (mg)	7.0
Magnesium (mg)	13.0
Phosphorous (mg)	13.0
Sodium (mg)	7.0
Potassium (mg)	45.0
Iron (g)	1.0
Vitamin B1(g)	0.03
Vitamin B2 (g)	0.03
Niacin (g)	0.3
Ascorbic Acid (g)	4.2

* Arano (2004).

2.2.0 Uses

People in mangosteen cultivating countries have used the pericarp (peel, rind, hull or ripe) of GML as a traditional medicine for the treatment of abdominal pain, diarrhea, dysentery, infected wound, suppuration, and chronic ulcer (Pedraza-Chaverri *et al.*, 2008). A detailed medicinal uses have been outline by Morton (1987) as follows. Dried fruits are shipped from Singapore to Calcutta and to China for medicinal use. The sliced and dried rind is powdered and administered to overcome dysentery. Made into an ointment, it is applied on eczema and other skin disorders. The rind decoction is taken to relieve diarrhea and cystitis, gonorrhea and gleet; and is applied externally as an astringent lotion. A portion of the rind is steeped in water overnight and the infusion given as a remedy for chronic diarrhea in adults and children. Filipinos employ a decoction of the leaves and bark as a febrifuge and to treat thrush, diarrhea, dysentery and urinary disorders. In Malaya, an infusion of the leaves, combined with unripe banana and a little benzoin is applied to the wound of circumcision. A root decoction is taken to regulate menstruation. A bark extract called "amibiasine", has been marketed for the treatment of amoebic dysentery (Morton, 1987).

3.0 Origin and Distribution

3.1.0 Origin

Mangosteen (*Garcinia mangostana* L.) has been considered as an obligate agamospermous (formation of fruit without pollination) hybrid, thought to have arisen from two wild species, *G. celebica* L. (syn. *G. hombroniana* Pierre) and *G. malaccensis* Hook. f. Sando *et al.*(2005) has reviewed the origin of mangosteen. According to their review, mangosteen developed by hybridization between the two Malaysian species *G. hombroniana* and *G. malaccensis* (Richards, 1990). Mangosteen is an unusual dioecious species in that it seems to be entirely female. Previously reported male mangosteen trees are now believed confused with closely related species. As males are absent in mangosteen, it has been suggested that all present day mangosteen trees are derived from one female, apomictic clone (Richards, 1990). This theory has been supported by a general lack of phenotypic variation in mangosteen within and between countries (Almeyda and Martin, 1976). However, Sando *et al.* (2005) disproved the theory put forward by some authors (Richards, 1990; Verheij, 1991), suggesting that all present day mangosteen trees may be derived from one original clone. The evidence that genetic diversity exists in mangosteen is promising for potential crop improvement. Further research to facilitate effective germplasm conservation of mangosteen and other Garcinia species is essential for exploration of the genetic resources in Garcinia. Apomictic mangosteen arose from hybridization between two wild species, *G. celebica* L. (syn. *G. hombroniana* Pierre, as the male parent, and *G. malaccensis* as the female. (Nazre, 2010). However, this putative origin was based on a misidentification of *G. malaccensis*, which was confused for *G. penangiana* Pierre. Intensive field studies and molecular investigations based on internal transcribed spacer (ITS) sequence data of 22 samples were conducted, which included six samples of true *G. malaccensis*. Morphological observation shows that mangosteen highly resembles *G. malaccensis*,

Uses of mangosteen

Mangosteen fruit with slices

Ice cream from mangosteen

Juice from mangosteen

Species of mangosteen

Garcinia mangostana

Garcinia prainiana

Garcinia macrophylla

particularly in its vegetative and fruit characters, even sharing similar taste of ripe fruits. ITS data revealed that mangosteen shared more than 99 per cent of its sequence with *G. malaccensis* with a few accessions identical with wild populations in Peninsular Malaysia. Phylogenetic analysis revealed that clades of mangosteen are paraphyletic per se, but monophyletic if both mangosteen and *G. malaccensis* are grouped together. This show that mangosteen and *G. malaccensis* are so closely related that they should be combined together as one species. He proposed two theories on the origin of mangosteen, first, that it is a hybrid of different varieties of *G. malaccensis*, and second, that it may be a product of multiple, superior selections from different populations of female trees of *G. malaccensis* originating in Peninsular Malaysia.

3.2.0 Distribution

According to Brunner and Morales Payan (2010), it is believed that mangosteen originated in the Sunda Islands (currently part of Brunei, East Timor, Indonesia and Malaysia), and the Moluccas or Spice Islands (currently part of Indonesia) in tropical Asia. From Malaysia it arrived in the Philippines, Burma and India, and from in Indonesia praising the mangosteen as a delightful fruit, and in 1770 the famous English explorer Captain Cook described mangosteen fruits he found in Jakarta. The mangosteen plant was introduced in England in 1789, and shortly afterwards, mangosteen and breadfruit (*Artocarpus altilis*) was taken to the Pacific islands and Antilles. Mangosteen was introduced in Sri Lanka in 1800 (Brunner and Morales Payan, 2010)

4.0 Species and Cultivars

Mangosteen, belongs to the large genus Garcinia (family Clusiaceae), which contains about 400 species of evergreen trees or shrubs, distributed mainly in tropical Asia (Fairchild, 1930). Most of the world production is derived from backyard trees in Southeast Asian countries (Thailand, Malaysia, and Indonesia) (Wieble *et al.*, 1992). According to old botanical classification *Garcinia* is placed within the family Guttiferae that includes about 1350 species. Guttiferae is further divided into 42 genera and five subfamilies: Kielmeroideae, Hypericoideae, Calophylloideae, Moronbeioideae and Clusioideae and the last has two tribes, Clusieae and Garcinieae of which the latter has two genera namely *Garcinia* and *Mammea* The mangosteen was first domesticated in Thailand or Myanmar at least 2000 years ago (Corner, 1952). Almeyda and Martin (1976) stated that mangosteen is a native of Indonesia where it is distributed throughout the archipelago, with the main populations in Sumatra and Kalimantan (Sulassih *et al.*, 2013).

4.1.0 Cultivars

Mangosteen (*Garcinia mangostana* L.) is a popular dioecious tropical fruit tree. Unlike other dioecious species which are generally characterised by high levels of genetic diversity, mangosteen does not exhibit much phenotypic differences (Sando *et al.*, 2005). Almost all trees are female with only a few anecdotal reports of male trees reported. Furthermore, seeds from mangosteen are apomictic and consequently produce trees that are clones of the female parent. In the absence of male trees and lack of cross pollination coupled with agamospermy, development of new cultivars is difficult. However, two cultivars have been reported from Philippines (Table 2).

UPLB Sweet

Seedling selection and is the first mangosteen variety registered with National Seed Industry Council of Philippines. It is a semi spreading tree and flowers early and fruits are ready for harvest from July to August. Average yield of fruits is 1148 fruits/tree. Fruit size small (87g). 30 per cent of fruits are seedless. Fruit peel is smooth, 7.5 mm thick and turns purple when ripe. The flesh constitutes 29 per cent fruit weight and snow white aril and with mild aroma. TSS is about 19°Brix. Responds well to cleft grafting (Namuco, 2007).

Roxas Purple

It is a seedling selection. The tree is a semi spreading type. In Laguna area of Philippines, it flowers early and fruits ripen from July to August. Average yield 850 fruits/tree. The fruits are oblong shaped and small (65g.). The apex of the fruit is slightly pointed with a narrow stigmatic lobe (Namuco, 2008).

Table 2: Characteristics of NSIC Registered Varieties of Mangosteen*

Varietal Description	*UPLB Sweet*	*Roxas Purple*
Tree		
Age (years after planting)	55	19
Height (m)	4	7
Growth habit	Semi-spreading	Semi-spreading
Regularity of Bearing	Biennial	Annual
Harvest seasons	July-August	July-August
Yield (Fruits/tree)	1148	850
Other characteristics fruiting	Early	-
Whole Fruit		
Weight (g)	87.0	65.6
Length (mm)	48.6	53.2
Width (mm)	57.0	49.2
Shape	Ovate	Oblong
Peel		
Color	Purple	Purple
Texture	Smooth	Smooth
Thickness (cm)	7.5	6.4
Weight (g)	61	46.1
Flesh		
Color	Snowy White	Snowy White
Texture	Smooth	Smooth
Juiciness	Juicy	Moderate
Aroma	Mild	Mild

Varietal Description	*UPLB Sweet*	*Roxas Purple*
Flavor	Sweet	Sub-acid to sweet
Edible Portion	29 per cent	27.2 per cent
Total Soluble Solids	18.65°Brix	-
Titratable acidity	9	-
Seed		
Number	1	1.1
Size		
Weight		
Total weight (g)	8.73	
Individual weight (g)	0.87	1.30
Length (mm)	1.58	2.03
Width (mm)	1.25	1.29
Thickness (mm)	0.69	-
Shape	Flat	Flat
Other features of the variety	Sweet, High percentage (>25 per cent) of fruits is seedless; other fruits have only 1 or 2 small seeds and high percentage edible portion	Prolific, sub-acid to sweet and high percentage of edible portion

Source: http://bpi.da.gov.ph/bpi/images/Production_guide/pdf MANGOSTEEN.pdf.

5.0 Breeding and Improvement

5.1.0 Botany

Morton (1987) gave the botanical description of mangosteen as follows. The mangosteen tree is very slow-growing, erect, with a pyramidal crown; attains 20 to 82 ft (6-25 m) in height, has dark-brown or nearly black, flaking bark, the inner bark containing much yellow, gummy, bitter latex. The evergreen, opposite, short-stalked leaves are ovate-oblong or elliptic, leathery and thick, dark-green, slightly glossy above, yellowish-green and dull beneath; 3 1/2 to 10 in (9-25 cm) long, 1 3/4 to 4 in (4.5-10 cm) wide, with conspicuous, pale midrib. New leaves are rosy. Flowers, 1 1/2 to 2 in (4-5 cm) wide and fleshy, may be male or hermaphrodite on the same tree. The former are in clusters of 3-9 at the branch tips; there are 4 sepals and 4 ovate, thick, fleshy petals, green with red spots on the outside, yellowish-red inside, and many stamens though the aborted anthers bear no pollen. The hermaphrodite are borne singly or in pairs at the tips of young branchlets; their petals may be yellowish-green edged with red or mostly red, and are quickly shed. Technically, the so-called "seeds" are not true seeds but adventitious embryos, or hypocotyl tubercles, inasmuch as there has been no sexual fertilization (Morton, 1987). When growth begins, a shoot emerges from one end of the seed and a root from the other end. But this root is short-lived and is replaced by roots which develop at the base of the shoot. The process of reproduction being vegetative, there is naturally little variation in the resulting trees and their fruits. Some of the seeds are polyembryonic, producing more than one shoot. The individual nucellar embryos can be separated, if desired, before planting (Morton, 1987).

There are no reports of improvement of mangosteen by conventional breeding. There appears to be no distinct mangosteen cultivars. The different morphological characteristics of the fruit in different regions might be due to climatic and soil factors (Te-chato and Lim,2005). The seeds of mangosteen are formed apomictically and thus all the plants are homogeneous and consist of a single genotype (Horn, 1940). The chromosome number of the mangosteen has been reported to be 2n = 4x = 56–76, 88–90, 96 and 120–130 (Verheij, 1992). Mangosteen is a polyploid,and possibly the allotetraploid derivative of *G. malaccensis* T. Anderson (2n = ?42) and *G. hombroniana* Pierre (2n = ?48) (Richards, 1990).

Te-chato and Lim (2005) listed several breeding constraints as below.

(i) Low seed production;
(ii) Long gestation period;
(iii) Gamboge;
(iv) Sensitivity to drought;
(v) Poor root system;
(vi) Slow growth; and
(vii) Apomixis.

In the absence of a pollen source and because of the apomictic nature of mangosteen seeds, breeding is impossible. Genetic erosion of mangosteen has probably been severe throughout South-east Asia, and there are no truly wild populations. Desirable traits in new mangosteen cultivars include longer seed viability, easier propagation, faster growth rate in nursery, shorter juvenile period, higher tolerance to lower environmental humidity, and higher fruit yield (Brunner and Morales Payan, 2010).

In spite of the reports on lack of diversity in mangosteen, due to agamospermy, efforts are being made to look for diversity in SE Asian countries. Mangosteen clone survey in Eastern Region of Thailand showed diferential morphology as mangosteen phenotype was different and could be distinguished in 6 characters *i.e.* small leave and small fruits trees, oblong shape trees, thin (not prominent) persistent stigma lobe thickness fruit trees, full and partial variegated mature leave color (combination of green and white color) trees, oblong shape leave trees and greenish yellow mature fruit color trees. Generally, rather short shoot, elliptic leaf blade shape, undulate leaf blade margin and thin or with cavity persistent stigma lobe thickness fruits are dominant marker of full seedless fruits that rarely found trees. Survey of mid-sized mangosteen orchards (200-300 trees) showed that over 70 per cent full seedless fruits trees could be found only about 1-3 per cent of all trees (Makhonpas *et al.*, 2015).

6.0 Propagation

6.1.0 Seed Propagation

The process of reproduction in mangosteen being vegetative, there is naturally

Flowering and fruit bearing of mangosteen

Flowers of mangosteen

Female flower development **Male flower development**

[Poovachal, Shameer, Rameshkumar, B. and Nair M.N. (2016) In: *India Diversity of Garcinia species in the Western Ghats*, Jawaharlal Nehru Tropical Botanic Garden and Research Institute (JNTBGRI), Palode, Thiruvananthapuram, Kerala, India].

Fruit bearing of mangosteen

Methods of Propagation

Propagation by seed

Propagation by cleft grafting

Propagation by approach grafting

little variation in the resulting trees and their fruits. Some of the seeds are polyembryonic, producing more than one shoot. The individual nucellar embryos can be separated, if desired, before planting (Morton, 1987).

Seed propagation has many problems, namely, (I) availability of seeds is seasonal during fruiting season. (ii) Each fruit produces hardly 1-2 large seeds suitable for seed germination. (iii) Mangosteen seeds are recalcitrant and the seeds cannot be stored. Suhendra and Mustamu (2018) suggested seed cleavage technique to get adequate mangosteen plant material using the advantage of polyembryony, wherein more than one embryo in one seed is observed. Based on this, it is possible to obtain a large and uniform amount of mangosteen seeds by means of seeds being split into several parts of the same size.

Suhendra and Mustamu (2018) developed the following technique. The mangosteen seeds are cleaned; the seeds are split transversely and longitudinally into 2 parts to see the structure seeds. After the mangosteen seeds are split transversely and longitudinally, the structure of the mangosteen seeds is obtained which consists of testa or the outer part of the mangosteen seeds, then Endosperm namely the food reserves of the mangosteen seeds and Embryo axis namely prospective new plants that surround the mangosteen seeds. So the number of prospective plants in the mangosteen seeds is more than 1 and the embryo is located around the shaft of the embryo. Results showed that mangosteen seeds had polyembryonic properties which had more than one embryo scattered around the embryo axis. The structure of the mangosteen seeds consists of testa, endosperm and embryo axis and after the seeds are planted for 7 days there appears a red line (Micropyle) from the shaft of the embryo to exit the seeds which will become buds. Mangosteen seed germination begins with the appearance of red colored buds that come out randomly around the surface of the mangosteen seeds sourced from the embryo axis, then after a few days red plumules appear and are followed by the appearance of radicles (Suhendra and Mustamu, 2018).

6.2.0 Asexual Propagation

Cleft grafting is widely used methods of asexual propagation. A healthy shoot is taken as the scion from a selected mother plant still in active growth and is then inserted into healthy rootstocks. Healthy rootstocks of about 30-35 cm. in height and are usually obtained from selected plants with consistent fruit size and heavy bearing. These rootstocks are selected from plants which are propagated from seedlings and which are about 2 years of age. At this stage, the stem diameter of the rootstocks should be about the same size as that of scions (1.6-2.0 cm.), so that cambium tissue of each can match closely (Arano, 2004).

According to Morton (1987) that in Florida, approach-grafting has succeeded only by planting a seed of *G. xanthochymus* about 1 1/4 in (3 cm) from the base of a mangosteen seedling in a container and, when the stem of the *G. xanthochymus* seedling has become 1/8 in (3 mm) thick, joining it onto the 3/16 to 1/4 in (5-6 mm) thick stem of the mangosteen at a point about 4 in (10 cm) above the soil. When the graft has healed, the *G. xanthochymus* seedling is beheaded. The mangosteen will make good progress having both root systems to grow on, while the *G. xanthochymus* rootstock will develop very little.

7.0 Area and Production

Information on the area dedicated to mangosteen and its production is scarce. According to Brunner and Morales Payan (2010), quoting Sacramento *et al.* (2007), that Thailand had approximately 11,000 hectares of mangosteen with a production of about 46,000 tons, valued at approximately $46 million in the year 2000; the same year the Philippines reported 1,354 hectares, with a production of 4,700 tons. In 1998, Malaysia had approximately 7,632 hectares, and Indonesia 10,750 hectares. In 1998, the combined commercial production in Thailand, Malaysia, the Philippines, and Indonesia was estimated at more than 150,000 tons. In 2007 there were approximately 350 hectares dedicated to mangosteen production in the Brazilian states of Pará and Bahia, with an annual harvest of 300 tons (Sacramento *et al.*, 2007). There are approximately 50 hectares of commercial production in northern Australia (Brunner and Morales Payan, 2010).

8.0 Soil and Climate

8.1.0 Soil

Mangosteen prefers deep, fertile, well-drained and slightly acidic clay loam rich in organic matter. It is not adapted to limestone soils, sandy alluvial soil or sandy soil low in humus. The plants grow well in soils with a high pH of 5.5—6.8 (Arano, 2004). Deep, porous soils rich in organic matter and nutrients, with good drainage and acidic pH seem to be best for mangosteen. Generally the tree does not grow well in limestone and other alkaline soils. The crop has a high water requirement. The soil must have adequate moisture year-round, provided either by rainfall or irrigation. The most successful orchards receive at least 1200 mm of rain annually. Dry periods should not be longer than 15 days. Drought for 30 days is associated with induction of flowering, and watering may be necessary to prevent undesired flower production. Mangosteen thrives alongside fresh water bodies such as streams and canals. Occasional short dry periods are tolerated, but may have negative impact on yield (Brunner and Morales Payan, 2010). According to Morton (1987) mangosteen is not adapted to limestone and does best in deep, rich organic soil, especially sandy loam or laterite. In India, the most productive specimens are on clay containing much coarse material and a little silt. Sandy alluvial soils are unsuitable and sand low in humus contributes to low yields. The tree needs good drainage and the water table ought to be about 6 ft (1.8 m) below ground level. However, in the Canal Zone, productive mangosteen groves have been established where it is too wet for other fruit trees–in swamps requiring drainage ditches between rows and in situations where the roots were bathed with flowing water most of the year, in spite of the fact that standing water in nursery beds will kill seedlings. The mangosteen must be sheltered from strong winds and salt spray, as well as saline soil or water.

8.2.0 Climate

Mangosteen requires ultra-tropical climates and has a juvenile phase of 10-15 years and is difficult to propagate by traditional vegetative means (Sando *et al.*,

2005). Southern Thailand is a traditional area for tropical fruit plantations. Recently, the phenology in many fruit trees, including mangosteen, has been changed due to climatic variability. Therefore, the impact of climatic variability on phenological change, yield and quality of mangosteen needs to be investigated. In 2008, a study was established in an orchard in Phatthalung province, southern Thailand. Eighteen-year old mangosteen trees with 8 x 8 m spacing were used to investigate flowering during the in-season, off-season and even, alternate bearing during three consecutive years (2008–2010). The 30- year weather data from 1981 to 2010 showed trends of change in the annual rainfall, the number of rainy days, and the maximum and minimum temperature. A marked change in rainfall distribution affected the phenological change in flowering, productivity and fruit quality. In 2008 and 2009, the mangosteen trees experienced the required dry period that regulates fl oral induction before flowering. Thus, flowering was found in both the in-season and off-season. However, in 2010, a prolonged drought in summer followed by rain during July–August caused leaf flushing instead of flowering and this resulted in no off-season fruit production. This indicated that climatic variability resulted in a phenological change of mangosteen in Phatthalung province, southern Thailand where there is usually off-season production. In addition, climatic variability affected the fruit yield and fruit quality of mangosteen (Apiratikorn *et al.*, 2012). In Philippines, The tree is adapted to heavy and well distributed rainfall of 1,200 mm or more without prolonged dry periods. However, the plants require a short dry spell of 15-30 days for flower induction. The optimum temperature is 25-35 C and relative humidity (RH) over 80 per cent. Mangosteen is associated with areas of low elevation *i.e.* less than 500 meters above sea level. It can be cultivated a higher elevation but has a slower growth rate (Arano, 2004).

Prolonged temperatures of 38°C or higher may be lethal to the tree. Mangosteen produces the highest yields at temperatures between 25° and 35°C, with relative air humidity at or above 80 per cent. Growth and yield performance are lower but are still good at 20-25°C. Vegetative growth is slow at temperatures under 20° C, and commercial production is not recommended in places with temperatures of 15° to 20° C. The tree may die at temperatures below 4° C. It commonly grows better at altitudes up to 500 meters above sea level and on slopes of less than 15 per cent, and may even attend 1000 meters providing the minimum temperature remains above tolerable values; growth rates at high altitude are however much slower (Brunner and Morales Payan, 2010).

Morton (1987) described the climatic requirements as follows. The mangosteen is ultra-tropical. It cannot tolerate temperatures below 40° F (4.44° C), nor above 100° F (37.78° C). Nursery seedlings are killed at 45° F (7.22° C). It is limited in Malaya to elevations below 1,500 ft (450 m). In Madras it grows from 250 to 5,000 ft (76-1,500 m) above sea-level. Attempts to establish it north of 20° latitude have all failed. It ordinarily requires high atmospheric humidity and an annual rainfall of at least 50 in (127 cm), and no long periods of drought. In Dominica, mangosteens growing in an area having 80 in (200 cm) of rain yearly required special care, but those in another locality with 105 in (255 cm) and soil with better moisture- holding capacity, flourished.

9.0 Cultivation

9.1.0 Planting and After Care

Mangosteen has a long juvenile period extending beyond 12 years. Good management may bring them to fruiting earlier by a year or two. In Philippines, Arano (2004) recommends a planting distance of 10 and meters between rows and between hills. Planting holes are dig with a minimum size of 12x12x12" and organic matter may be incorporated 1-2 kg one to two months prior to planting. The best time for planting is at the onset of the rainy season. Shade is necessary to the newly transplanted seedlings. Shade is maintained 1-2 years and this is gradually reduced to allow full sun exposure. Morton (1987) recommended a spacing of 35 to 40 ft (10.7-12 m). Planting is preferably done at the beginning of the rainy season. Pits 4 x 4 x 4 1/2 ft (1.2 x 1.2 x 1.3 m) are prepared at least 30 days in advance, enriched with organic matter and topsoil and left to weather. The young tree is put in place very carefully so as not to injure the root and given a heavy watering. Partial shading with palm fronds or by other means should be maintained for 3 to 5 years. Indian growers give each tree regular feeding with well-rotted manure @100 to 200 lbs (45-90 kg) and peanut meal @10 to 15 lbs (4.5-6.8 kg) per year (Morton, 1987).

In Malaysia and Sri Lanka, it is a common practice to spread a mulch of coconut husks or fronds to retain moisture. A 16-inches (40 cm) mulch of grass restored trees that had begun dehydrating in Liberia. It has been suggested that small inner branches be pruned from old, unproductive trees to stimulate bearing. In Thailand, the tree is said to take 12 to 20 years to fruit. In Panama and Puerto Rico trees grown from large seed and given good culture have borne in six years (Morton, 1987).

9.2.0 Manures and Fertilizers

Translucent flesh disorder (TFD) and gamboge disorder (GD) are major problems limiting marketable yield of mangosteen in Thailand (Pechkeo *et al.*, 2007a). The imbalance or deficiency of essential nutrients in soils and plants may cause poor fruit quality. To classify this issues, the pattern of plant nutrient accumulation and nutrient requirement in soils and mangosteen trees during fruit development period were investigated. Pechkeo *et al.* (2007a) conducted an experiment in an established mangosteen orchard with 2 main factors: 1) Fruit positions in the canopy of mangosteen trees (outer and inner canopy fruits), and 2) concentration of plant nutrients in soils and trees. Soil sampling was taken at 2 depths; 0-15 and 15-30 cm from soil surface, around the middle of the tree canopy in blooming and harvesting periods and analyzed for some important chemical and physical properties. Leaves and fruits sampling were taken during fruit development period, every 2 weeks from blooming to harvesting, and analyzed in relation to the changes of soil nutrients. Results indicated that the soil textures varied from sandy loam to sandy clay loam, and the natural soils in the mangosteen orchard was very extremely acid to moderately acid (pH 4.86-5.61, soil: water=1: 5). Most of the soils in the mangosteen orchard had very low level essential nutrients for plant growth. Mangosteen fruit qualities were not significantly different between the outer and inner canopy fruits. Likewise, most of the plant nutrients accumulation in mangosteen leaf, peel and flesh were not significantly different between two fruit positions. It was remarkable that nutrient accumulation in the fruit decreased from blooming to harvesting period. Mangosteen (leaf, peel and flesh) required higher amounts of N, P, K, Ca and Mg for growth in the early stage of fruit development period (from bloom to 6th week after bloom) and S and B in the late stage of fruit development period (from 6th week after bloom) compared with other growth periods. Potassium and Ca contents in the peel of normal fruits (NF) were higher than TFD and GD fruits, whereas P and Mg contents in the flesh of TFD fruits were higher than NF and GD fruits (Pechkeo *et al.*, 2007a).

The mangosteen plant has a positive response to phosphorus fertilizer with increasing crop yields (Kurniadinata *et al.*, 2018). The best phosphorus fertilization recommendations to get the optimum yield (80 per cent) is around 1400 g P_2O^5/ mangosteen tree/year. The threshold value showed that without phosphorus fertilization application, the relative yield of mangosteen plants was 30,889 per cent (Kurniadinata *et al.*, 2018). Under Philippines condition, the following recommendations are given (Table 3).

Table 3: Fertilizer Recommendation in Mangosteen*

Age of Trees (Years)	*Commercial Fertilizer (Kg/tree/year)*		
	45-0-0	*14-14-14*	*0-0-60*
1	0.22	0.35	-
2	0.44	0.70	-
3	0.56	1.40	-
4	-	2.80	-
5	-	3.60	-
6	-	4.20	0.30
7	-	5.70	0.30
8	-	7.10	0.30
9 above	-	8.50	0.50

* Arano (2004).

9.2.1 Organic Manures

The application of growth promoting substances, such as humic acids, during the formation of fruit seedlings can benefit growth and increase the plant nutrient uptake and translocation. Gomes Junior (2019) evaluated the effect of the application of humic acid doses (0 mg L^{-1}, 2 mg L^{-1}, 20 mg L^{-1}, 40 mg L^{-1} and 200 mg L^{-1}) extracted from vermicompost of sugarcane filter cake on the nutrient uptake of seminal mangosteen (*Garcinia mangostana* L.) seedlings. The application of humic acids promoted curvilinear increments with quadratic rates for N, P, K, Ca, Mg and S contents from shoots and roots, as well as quadratic increments in the Zn and Mn uptake by shoots. The application of humic acids increased the chlorophyll a and chlorophyll b contents and reduced the carotenoid content. Overall, the regression analysis demonstrated that a dose of 150 mg L^{-1} is capable of providing the most adequate nutrient uptake by plants, and should be recommended for the production of mangosteen seedlings. Humic acids from vermicompost can provide a better nutrition in mangosteen seedlings at the nursery stage.

9.2.2 Micronutrients

Pechkeo *et al.* (2007b) found the applications of $CaCl_2$ and H_3BO_3 to increase Ca and B concentrations in peel and flesh of mangosteen fruits. Spraying with 10 per cent $CaCl_2$ could also increase the percentages of normal fruits (NF), whereas the percentages of defected fruits (TFD and GD) decreased. Spraying 10 per cent $CaCl_2$+0.5 mg/kg B enhanced the efficiency of Ca to increase the ratio of NF: TFD and GD.

9.3.0 Irrigation

Drip irrigation system in combination with removing weeds under the canopy (A) or removing weeds followed by minimum tillage under the canopy (B) or removing weeds followed by covering with rice hay mulch under the canopy (C), where N, P, K, Ca, Mg fertilizer and yellow fluorescent sticky trap were applied could reduce scarring intensity and percentage of yellow latex on the fruit peel. However, the treatments did not significantly impact fruit diameter or percentage of yellow latex inside the fruit. Nevertheless, treatment C improved mangosteen quality by as much as 67.79 per cent fulfilling export standard requirements.

10.0 Physiological Disorders, Pests and Diseases

10.1.0 Physiological Disorders

A major physiological problem called "gamboge" or Yellow latex is evidenced by the oozing of latex onto the outer surface of the fruits and on the branches during periods of heavy and continuous rains (Morton, 1987). It does not affect eating quality. Fruit-cracking may occur because of excessive absorption of moisture. In cracked fruits the flesh will be swollen and mushy. Bruising caused by the force of storms may be an important factor in both of these abnormalities. Fruits exposed to strong sun may also exude latex. Presence of yellow latex on the mangosteen peel creates a rough, unattractive appearance. In addition, yellow latex contaminating the edible portion of the fruit makes the fruit inedible (Krishnamurthi and Rao, 1962). There is no consensus as to the cause of yellow latex in mangosteen fruit, which makes control impossible. Several experts have summarized that yellow latex is due to damage by insect or mechanical damage, and during harvesting and packaging (Afandi *et al.*, 2011). Morton (1987) argued that the presence of yellow latex on the fruit was the result of a physiological abnormality due to inconsistencies in irrigation or weather patterns. Similarly, Varheij and Cornel, quoted by Afandi *et al.* (2011) believed that yellow latex was caused by over irrigation following by a dry period. Later studies have indicated yellow latex is a physiological response related to turgidity of cells that comprise the endodermis of the fruit peel. Turgidity pressure changes in the cell wall of the fruit peel due to the extreme environmental changes create the yellow latex inside the fruit. Mangosteens produced in Honduras often have crystal-like "stones" in the flesh and they may render the fruit completely inedible (Morton, 1987).

10.2.0 Pests and Diseases

Thrips (*Scirtothrips dorsalis*) is a serious pest of mangosteen. Symptoms of thrips include: silvering of the fruit peel, pale, yellow/brown discoloration, elongated and patchy hardened scars, and an alligator-skin appearance that may cover the entire fruit surface. Heavy scarring can sometimes affect fruit growth. Controlling these most serious constraints in mangosteen cultivation is vital requirement for the export market. Use of drip irrigation may stabilizes physiological activity related to turgidity of cells and installing yellow fluorescent sticky traps can reduces the population of thrips responsible for scarring the fruit peel (Afandi *et al.*, 2011). Work by Affandi and Emilda (2009) revealed that application of yellow fluorescent sticky traps in a mangosteen orchard decreased scarring intensity by 26.28 per cent (year 1) and 10.82 per cent (year 2) compared with the control. In addition, the combined treatment of intensive orchard care (removing weeds using minimum tillage) and use of yellow fluorescent sticky traps (IOC+YST) can reduce scarring intensity (32.14 per cent in year 1 and 15.81 per cent in year 2) compared with the control. Therefore, it is clear that sanitation and use of yellow fluorescent sticky traps reduces thrips population.

Arano (2004) presented a list of pests and diseases under the Philippines condition and their control measures (Table 4).

Table 4: Pests and Diseases of Mangosteen and their Management*

Insect Pests/Diseases	*Plant Parts Affected*	*Control Measures*
Thrips	They prefer to feed on the tender, young plant tissue such as flower buds, opening blooms and	Spray Dimethoate (2ml/l) two to three times at weekly intervals. unopened leaves.
Tussock Caterpillar (*Eupterote favia*)	Larval stage eat on the young leaves	Spray Malathion or Fenvalerate (2ml/l)
Mites	Attack the fruit surface, deface the fruits with small bites and make it unattractive for market	Spray Profenefos or Acrezid
Mealy bugs	Attack on young flushes and fruits	Spray Carbaryl or Dimethoate (1.5ml/lit)
Sooty mold	Attack on young leaves	Improving aeration and sunlight penetration by pruning overlapping branches will reduce infection. Spray with common fungicides such as 3 per cent neem oil, fish oil rosin soap.
Anthracnose	Leaves	Spray with Dithane M-45 (2g/l. of water)
Bacterial Leaf Sheath	Leaves	Spraying antibiotics like Streptocycline (3g/l) at the dose

* Arano (2004).

11.0 Harvesting and Yield

Number of harvests can be many depending the fruit number as they donot come to maturity at the same time. Fruits are ready for harvest when the colour turns from green to purple. The yield ranges from 500 to 1200 depending on the size. According to Morton (1987) Cropping is irregular and the yield varies from tree to tree and from season to season. The first crop may be 200 to 300 fruits. Average yield of a full-grown tree is about 500 fruits. The yield steadily increases up to the 30th year of bearing when crops of 1,000 to 2,000 fruits may be obtained. In Madras, individual trees between the ages of 20 and 45 years have borne 2,000 to 3,000 fruits. Productivity gradually declines thereafter, though the tree will still be fruiting at 100 years of age.

12.0 Post-harvest Management

12.1.0 Grading

Grading is essential for marketing because pricing is tied to produce quality. Quality sorting criteria of fresh mangosteens include size, color and defects. Fruit size is graded differently by various traders. Fruit defects include fruit-cracking, rough surface, pericarp hardening, translucent flesh, gummosis and decay. All defects are typically found at retail level and often reduce consumer demand for the fruit. Dangcham (quoted by Jarimopas *et al.*, 2009) reported symptoms of translucent flesh include flesh changes from white to translucent and textural changes from soft to firm and crisp. The translucency of flesh might occur because of excessive absorption of moisture via the root system, the leaves or the fruit. Grading is done according to following fruit sizes (Arano, 2004).

a. Small (16-18 fruits/kg)
b. Medium (12-14 fruits/kg)
c. Large (8-10 fruits/kg)

12.2.0 Packaging

Yulianti *et al.* (2010) has designed a package made of corrugated board for mangosteen, tested the compression strength of the packaging, analyzed the effect of packaging capacity and fruit packing pattern in the container on some quality parameters after transport, and also analyzed the economy of the packaging used. The research was conducted in three stages, namely: the measurement of physical and mechanical properties of fruit (Stage I); the packaging design *i.e.* determining the packaging dimension (Stage II); and packaging design testing (Stage III). Compression and vibration testing were conducted. For the vibration testing, transport simulations were conducted using the vibration simulator of 3.50 Hz and amplitude of 4.61 cm for 3 hours. The packaging was designed with a capacity of 8 kg and 15 kg using two packing pattern, namely the fcc and Jumble. Based on the physical and mechanical properties, optimal packing pattern of 8 kg container was 8, 4, and 4 pieces of fruit for the long x, y, and z direction, respectively, with the density of 62 per cent. Therefore, they gave packaging dimension of 39.4 x 21 x

Fruit growth of mangosteen

Harvesting, packaging and marketing

Harvested mangosteen

Nylon net packaging

Packaging in wooden box

Marketing

Colour development and grading of mangosteen

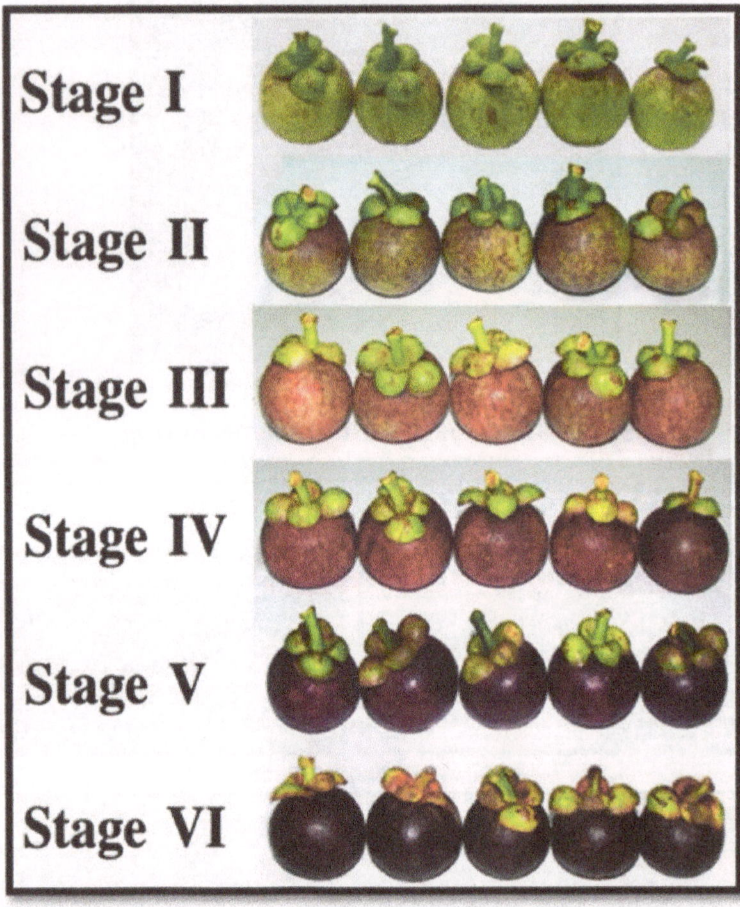

Different stages of colour development in mangosteen

Mechanical grading of mangosteen

21 cm. For the 15 kg container, the optimal packing pattern was 8, 6 and 5 pieces of fruit for the x, y, and z direction, respectively, with the packing density of 65.6 per cent. Thus, gave dimension of 39.4 x 30 x 25 cm. Based on overall test results, it can be stated that the design packaging capacity of 8 kg with dimension of 39.4 cm x 21 cm x 21 cm and the fcc packing pattern gave the optimal setting for mangosteen transportation.

12.3.0 Post-harvest Damage

Mangosteens were found to have a climacteric respiratory pattern indicating that respiration and ethylene production rates increase during maturation and ripening, and these are accompanied by corresponding changes in fruit color from green to pink, red and purple (Jarimopas *et al.*, 2009). It is recommended a skin color-based harvesting index for mangosteens as follows: stage 0 - yellowish white; stage 1 - greenish yellow; stage 2 - pink blush on yellow: background; stage 3 - pink; stage 4 – red; stage 5 - reddish purple; stage 6 – dark purple. Fruit should be picked not earlier than stage 2. At this time, fruits can ripen and sweeten after harvesting. Stage 5 and 6 mangosteens are considered to be ripe (Jarimopas *et al.*, 2009).

Jarimopas *et al.* (2009) investigated post-harvest damage in fresh mangosteens at wholesale level in Thailand from April to October. A total of 37.1 per cent of the production yield was rendered inedible by damage during this period; damages included fruit cracking, hardened rinds, rough surfaces, translucent flesh, gummosis and decay. This study focused on a method of predicting damage based on the color of the skin of the affected mangosteen. As a first step, diameter, height, weight, and volume of large, medium, small, and undersize mangosteens were measured. The term, dimension ratio, was introduced as a sizing parameter identifying conventional trade size. The coefficient of static friction of the glossy- and rough-surface mangosteens on plexiglass, plywood, and galvanized steel sheet varied from 0.31 to 0.46.The color of sound and defective fruits was measured in terms of their tristimulus values X, Y, and Z. The corresponding chromaticity coordinates of a mangosteen, x and z, depended on the maturity stage of the fruit while y depended on the type of fruit surface. A ratio was proposed to test the accuracy of predicting internal defects from the color variation between two spots on the surface of the same fruit. The highest percentage of correct prediction was 67.4 per cent with a color ratio of X1 (pink blush color on yellow ground color) to X2 (pink color) that was greater than 1.25.

12.4.0 Nutraceutical Products

Garrity *et al.* (2004) obtained patent wherein they claim to have discovered a mixture of the mangosteen pericarp and fruit pulp in a single nutraceutical composition yields surprising health benefits. The efficacy of this xanthone-rich mixture of mangosteen pericarp and pulp is enhanced through the addition of selected juice and phytochemical ingredients, which are believed to synergistically react with the natural xanthone compounds. A brief detail of the process is given below.

Whole fruit from mangosteen plant is picked and transported to a production facility. The fresh fruit can kept at ambient air temperatures during transportation or it can be frozen, depending on need. The entire mangosteen fruit, including the fruit pulp and pericarp, is then ground into a pulp and pericarp mixture using commercial grinding or mixing equipment. The resulting mixture of mangosteen fruit pulp and pericarp can then be further processed through the addition of one or more of the selected juice concentrates listed above. In preferred embodiments of the beverage form of the invention, the selected juice concentrates and Water are then added to the mixture in accordance With the amounts, ranges and ratios specified above. The liquid nutraceutical compositions can then be treated, bottled or packaged for distribution to consumers using a variety of methods known to those of ordinary skill in the art, such as pasteurization, ?ash pasteurization, sterilization, UHT sterilization, pressure sealing, freezing, freeze drying, irradiating, *etc*. Dehydrated and other forms of the nutraceutical compositions can also be prepared using standard techniques.

Erlangga *et al.* (2012) conducted a study (1) to determine the value chain of mangosteen at Kiara Pedes Sub district, Purwakarta District, (2) to identify the gap between actual condition at Kiara Pedes and Global GAP standard, (3) to identify internal and external factors that can affect the implementation strategy of Global GAP standards, and (4) to develop alternative strategies that can be applied to improve the system of mangosteen cultivation and post-harvest handling based on Global GAP standards. The analytical tools being used in this study were value chain analysis, gap analysis, internal and external factor evaluation (IFE, EFE, IE matrix), SWOT analysis, and quantitative strategic planning matrix (QSPM). Identified primary actors in mangosteen value chain were farmers, middlemen, suppliers, exporters, and local and overseas retailers. Based on IE Matrix and SWOT analysis, the strategies to implement Global GAP standards were (a) to increase mangosteen productivity and improve its quality by using developed cultivation and post-harvest technology, (b) to increase productivity, and improve quality and transportation network in accordance with Global GAP standard, (c) to improve clean water and post-harvest infrastructure through cooperation with exporters and financial institutions, and (d) to improve warehouse and supporting facilities such as packaging and sanitation according to the Global GAP standard for minimizing the environmental constraints. The most priority strategies from the QSPM analysis were improving clean water and post-harvest infrastructure through cooperation with exporters and financial institutions, followed by using the developed cultivation and post-harvest technology to increase mangosteen productivity and improve its quality.

13.0 Biotechnology

13.1.0 Tissue Culture

Elviana *et al.* (2011) reported that morphologically and histologically, globular somatic embryos of mangosteen resembled closely to the undifferentiated structure of mangosteen seed. They then proposed that mangosteen embryogenesis does not follow the typical somatic embryogenesis stages (globular, heart, torpedo and cotyledon) but resembled the natural anatomy of mangosteen seed itself where there is no differentiated embryo and only procambium structure can be observed (Normah *et al.*, 2016).

Young red leaf segments from 1 to 3 year-old seedlings of mangosteen were incubated on different media, consisting of either B5 or Murashige and Skoog (MS) basal media and supplemented with several plant growth regulator at different levels, including 2, 4-dichlorophenoxyacetic acid (2, 4-D) and kinetin or 2, 4-D and 6-benzyladenine (BA) or thidiazuron (TDZ) and BA. The highest percentage of globular structures (36.0 per cent) was obtained using MS medium with a combination of TDZ and BA, each at a concentration of 0.7 mg L^{-1}. Histological observation of sections of globular structures showed accumulation of dense meristematic cells, a common characteristic of globular formation during somatic embryogenesis. Molecular analysis detected the gene somatic embryogenesis receptor-like kinase 1 (*SERK1*). This gene sequence showed a high similarity (73.0–87.0 per cent) to previously reported plant *SERK1* (Rohani *et al.*, 2012).

An improved procedure for direct shoot bud regeneration from young leaves of mangosteen was reported by Goh *et al.* (1994). Regeneration was achieved by culturing 3-mm transverse sections of about 10-day-old leaves on woody plant medium enriched with 20 µM benzyladenine (BA), 20 g/l sucrose and 2.5 g/l Phytagel. A wound response in the presence of BA at the time of culture was found to be essential for shoot bud induction. Explant size, concentration of BA in the medium, timing of addition of BA to the medium as well as the time of wounding of explant significantly influenced shoot bud regeneration. Leaves from field-grown seedlings produced an average of 45 shoot buds per leaf after 50 days of culture, compared with 8 shoot buds per leaf from in vitro shoots. Regenerated shoots (5–6 mm tall) excised from the explants required BA (5 µM) for further growth. Rooting was induced with indolebutyric acid (IBA) when the shoots reached 10–15 mm and were established in vermiculite/sand mixture in pots.

A four-step procedure was developed by Te-Chato and Lim (2000) for plant regeneration from leaf explants (5–15 mm long) excised from in vitro grown plantlets of mangosteen. Leaf explants formed callus (66.8 per cent) on Murashige and Skoog medium (MS) with 2.22 µM benzyladenine (BA) and 2.25 µM thidiazuron (TDZ) at 3 weeks of culture. The mean number of nodules in the calli on this medium after four serial subcultures was 4.45 nodules per explant. A mean number of 9.3 shoots per explant were induced on woody plant medium (WPM) supplemented with 0.44 µM BA. Shoot elongation (5–6 shoots per explant) at 80–90 per cent was obtained by transferring shoot to the same medium overlaid with half strength liquid MS with 0.32 µM naphthalene acetic acid (NAA) and 0.13 µM BA. Shoots were successfully rooted at 68.2 per cent on WPM with 1.11 µM BA plus 0.25 per cent activated charcoal. The plantlets were successfully established in soil in pots after acclimatization.

Joni *et al.* (2016) used *in vitro* young stems of mangosteen clone Leuwiliang and found the best medium for embryogenic callus induction to be MS supplemented with 0.1 mg L^{-1} TDZ, which resulted semi friable calli. Casein hydrolysate and malt extract could not induce the formation of somatic embryos. After two times subcultures on the same MS medium supplemented with 0.5 mg L^{-1} TDZ and 0.7 mg L^{-1} BA, a total of 33.8 somatic embryos per explant were induced.

Minh (2005) developed the following protocol for micropropagation of mangosteen. MS medium was the most suitable medium for embryogenesis, shoot proliferation and shoot elongation cultures. In embryogenesis cultures, multiple embryos were observed on seed explants cultured on MS medium supplemented with 1mg/l BA, 0.1mg/l NAA and coconut water (10 per cent) after 4-6 weeks of culture. About 80-100 embryos raised on the seed surface and multiple embryos developed to multiple shoots after 2-3 weeks of culture. Secondary embryogenesis continually initiated on the mother embryogenic sample that had 2-3 embryos combined, but did not appear on single embryos. In shoot proliferation cultures, seed with multiple shoots was cut into 4 explants before subculturing on MS medium supplemented with 4mg/l BA, 0.1mg/l NAA, and coconut water (10 per cent). The highest mean number of shoots per explant (18-22) formed after 8 weeks of culture. In shoot elongation cultures, each multiple shoot explant was cut into 4 clumps, having 4-5 shoots per clump, before culturing on MS medium supplemented with 0.1mg/l BA, 0.1 mg/l NAA, 2g/l yeast extract and coconut water (10 per cent). Elongated plants with a height of 20-30mm developed after 4-6 weeks of culture. For rooting, the buds were cultured on WPM medium supplemented with 0.1mg/l BA, 5mg/l IBA, 2g/l yeast extract and coconut water (10 per cent). About 85 per cent of plants produced roots with a root length of 3- 4cm and had 1-2 roots per bud after 8 weeks of culture. The healthy plants were transplanted to pots in ex vitro conditions for further studies on their growth characteristics, flowering and fruit quality characteristics. Micropropagation via embryogenesis culture was the best method for production of mangosteen planting materials.

Field evaluation of tissue cultured plants has been reported by Mohd Asrul *et al.* (2011). A research plot for the field evaluation of tissue culture-derived mangosteen and Mesta was established in 1998 at MARDI Research Station, Bukit Tangga, Kedah. After transplanting to the field, morphological characters and field performance of these plants were observed in comparison with conventional seedlings. Results showed that tissue culture-derived plants started flowering five years after transplanting while the seedlings have not produced any flower in the same period of time. However, there was no fruit set for both mangosteen and mesta. Heavy fruit setting was observed in the seventh year for tissue culture-derived mangosteen, and both tissue culture-derived and seedling Mesta. Fruit quality in terms of total soluble solids (TSS) and total acids (TA) were not much difference between tissue culture-derived and seedlings trees.

13.2.0 Molecular Markers

Unlike other dioecious species which are generally characterised by high levels of genetic diversity, mangosteen does not exhibit much phenotypic differences (Sando *et al.*, 2005). It has been reported by some mangosteen growers in North Queensland that the occasional phenotypic variants observed in their plantations are a result of cross-pollination from wild related species. Whether these differences can be attributed to genetic variation or are purely a consequence of being cultivated at different geographical locations is unclear. To address these questions, we conducted a preliminary survey using a new high-efficiency DNA marker technique, Randomly Amplified DNA Fingerprinting (RAF), to examine the level of genetic diversity in a population of mangosteen trees collected from orchards in Northern Queensland. Additionally, we examined the level of variation among a number of other Garcinia species including mangosteen's closest relative, G. hombroniana, as well as native Australian *Garcinia* species.

The novel molecular marker technique Randomly Amplified DNA Fingerprinting (RAF) was used to survey genetic relationships between 37 accessions of the tropical fruit, mangosteen, and among 11 accessions from eight other *Garcinia* species. Although mangosteen is believed to reproduce exclusively through apomixis, our results show that considerable genetic diversity exists within *G. mangostana* and between other *Garcinia* species. Among the 37 *G. mangostana* accessions examined, nine different genotypes were identified which clustered into three distinct groups based on correspondence analysis (reciprocal averaging). For 26 (70 per cent) of the accessions no marker variation was detected over 530 loci screened. A further eight (22 per cent) accessions exhibited very low levels of variation (0.2–1 per cent) suggesting at least one well conserved mangosteen genotype. The remaining three accessions (8 per cent) showed extensive variation (22–31 per cent) compared with the majority of accessions. The three mangosteen groups were 63–70 per cent dissimilar to the other *Garcinia* species investigated (Ramage *et al.*, 2004).

The phylogenetic relationships among 17 *Garcinia* species including *G. mangostana* (mangosteen) were analyzed by comparing sequences of the internal transcribed spacer (ITS) region of nuclear ribosomal DNA (nrDNA). Both parsimonious and neighbor joining (NJ) analyses revealed that *G. mangostana* is closely related to *G. malaccensis* believed to be a progenitor of mangosteen (Yapwattanaphun *et al.*, 2004). Another suspected progenitor of mangosteen, *G. hombroniana*, was more distant from *G. mangostana* than *G. malaccensis*phylogenetically. *Garcinia hombroniana* formed a cluster with *G. rostrata*, *G, speciosa* and *G. sizygiifolia*, and this cluster was connected with a cluster of *G. mangostana* and *G. malaccensis*. The ITS sequence analysis showed that *G. atroviridis*, *G. cowa*, *G. dulcis*, *G. malaccensis*, *G. mangostana*, *G. rostrata* and *G. vilersiana* have nucleotide additivity (two different nucleotides at the same nucleotide position) at several sites in the ITS region. The occurrence of these species might be related to hybridization with ancestors, but the genomic compositions, even chromosome numbers, of these species are still unknown (Yapwattanaphun *et al.*, 2004).

Sobir *et al.*(2009) employed isoenzyme and AFLP marker on 13 accessions of mangosteen and their close relatives. Isoenzyme marker using four enzyme systems produced 25 bands and 88 per cent out of them were polymorphic and elucidate genetic variability at similarity level ranged between 0.38-0.89. AFLP markers with three primer system produced 220 polymorphic bands and revealed genetic variability at similarity level ranged between 0.38-0.89 successfully produced high polymorphism bands and elucidates genetic variability at similarity coefficient ranged between 0.21-0.77. Both markers exhibited similar clustering pattern, and group successfully *G. mangostana* accessions in one clustering group. Furthermore *G. malaccensis* and *G. porrecta* consistently showed closer genetic relationship to *G. mangostana* clustering group in both markers, in comparison to *G. hombroniana*, which implies the assumption they may be the progenitor of *G. mangostana*, and should be reviewed with more accurate data.

In order to elucidate the genetic diversity of mangosteen and its relatives, morphological and molecular analyses were conducted (Sulassih *et al.*, 2013). The objectives for this study were: (1) to determine the relationships between mangosteen and its relatives; and (2) to confirm the true diversity of allotetraploid mangosteen relatives' G. mangostana. Analysis was conducted using morphological and inter simple sequence repeat (ISSR) between 19 accessions of G. mangostana and their close relatives revealed. Diversity analysis was based on 212 polymorphic characters and 3 groups were formed. Group A consisted of *Garcinia mangostana, Garcinia malaccensis, Garcinia celebica, G. hombroniana* and *Garcinia porrecta*; group B comprised *G. forbesii* and *G. subellptica*; and group C solely with *Calophyllum inophyllum*. The genetic similarity of *Garcinia mangostana, Garcinia malaccensis* and *Garcinia celebica* were 0.78 and 0.63. The epidermis cell observations around stomata cells on the lower surface of leaves revealed that *Garcinia mangostana* has the intermediate shape between *Garcinia celebica* and *Garcinia malaccensis*. It shows that there is a close relationship among *Garcinia celebica, Garcinia malaccensis and Garcinia mangostana*. It was determined that *Garcinia malaccensis* and *Garcinia celebica* were ancestors based on morphological and ISSR markers.

Inter-simple sequence repeat (ISSR) markers were used to examine the level of genetic diversity in mangosteen. Twenty three accessions of the *Garcinia mangostana* collection from Sumatra region were screened for ISSR markers (Mansyah *et al.*, 2010). Eleven random ISSR primers were chosen to differentiate the investigated accessions. The primers generated 72 bands of which 42 (58 per cent) were polymorphic and 30 bands (42 per cent) monomorphic. From the 11 primers tested, two primers were monomorphic. Seven of the nine polymorphic primers produced fingerprint profiles unique to the RT accession from Tembilahan (Riau Province). Cluster analysis divided the accessions into two major groups with genetic similarity coefficient ranging from 0.44 - 0.96. The first group contained only cultivar RT (elliptical stigma lobe) and the second group consisted of 22 accessions (round stigma lobe) which could be divided clearly into six sub-clusters. One of the sub-clusters contained accessions from the same region of Bangka Island and the others contained combination of accessions derived from different locations. The result shows that mangosteen accessions with different genetic background exist in this region. This confirms the general opinion that mangosteen is genetically uniform (Mansyah *et al.*, 2010).

The genetic variation and relationships of the mangosteen (*Garcinia mangostana* L.) were observed across mangosteen populations in Java, Indonesia using newly identified microsatellite loci and morphological characters (Matra *et al.*, 2016). In this study, we developed an improved protocol to isolate microsatellite loci, named Selective Repeats from AFLP Sequence, by using a hybridized membrane. Twenty microsatellite loci were evaluated using 78 individuals from five mangosteen populations, and we successfully amplified four closely related *Garcinia* species, including *G. malaccensis*, *G. hombroniana*, *G. celebica*, and *G. porrecta*. Eight loci were monomorphic and the others were polymorphic. Sixty-nine alleles were found, with 3.491 per locus on an average. Genetic diversity (H^1) was calculated with an average across loci within population (H^1_S) as 0.39, an average loci across many populations (H^1_T) as 0.444, and genetic differentiation (F^1_{ST}) as 0.147. Furthermore, based on morphological characters, mangosteen individuals from four populations including Leuwiliang, Wanayasa, Puspahiang, and Kaligesing also had morphologically distinct fruit weight, rind weight, and rind thickness among populations. The study also elucidated the dispersal pattern of mangosteen in Java; the source of the genotype mangosteen in Java population was the Wanayasa population. In addition, we found evidence of tetraploidy in mangosteen. These results have potential applications in future breeding, conservation studies, and genetic assessment of mangosteen and their closely related species (Matra *et al.*, 2016).

Genetic diversity of mangosteen was developed by using SSR markers (Qosim *et al.*, 2011). An inter-simple sequence repeat (ISSR)- suppression - PCR technique established to SSR marker of plant species. DNA library construction was used restriction enzyme blunt end *Rsa* I and adaptor (consist of 48-mer: 5′-GTAATACGACTCACTATAGG GCACGCGTGGTCGACGGCCCGGGCTGGT-3′ and 8-mer with the 3-end capped by an amino residue: 5′-ACCAGCCC-NH2-3). Primer designed by using compound SSR (AC)10; (TC)6(AC)5 or (AC)6(AG)5 and an adaptor primer AP2 and nested PCR using AP1 primers. The PCR product integrated into the plasmid PGEMT Easy Vector System and competence cell *Escherichia coli* (strain DH5á) and sequence. Eight sequence from sample #G17 with SSR compound (AC)10, (TC)6(AG)5 and (AC)6(AG)5 produced two primer pairs. The primer pairs from sequence positive clone 4 of SSR compound (TC)6(AC)5 were forward primer: 5′-GGCCGTT AAAGTAGCTCAAGAA-3′ and reverse primer:5′-CCGCATAGCATCAGTATCTG TC-3′, while primer pairs from sequence positive clone 10 of SSR compound (AC)6(AG)5 were forward primer: 5′-GTGTTTCCATTTGTTACGCGCT-3′ and reverse primer: 5′TAATGCCGTTGGGCAGTGA-3′.

High-throughput sequencing technology is a powerful tool to generate large amount of sequence data hence efficiently used in comparative transcriptome analysis to study samples under different conditions, treatments or stages. These huge dataset provides a profile of differentially expressed genes, useful in further gene regulation and functional study. Fadryin *et al.* (2019) used this genomic study the functional genomics of somatic embryogenesis related gene in mangosteen. The atypical somatic embryogenesis of mangosteen leads us to enquire what is happening at each stage of somatic embryogenesis (SE) at the level of the molecular regulation controlling transition between embryogenic stages. The expression of selected genes *Somatic embryogenesis receptor-like kinase 1* (*SERK1*), *brassinosteroid insensitive 1-associated kinase 1* (*BAK1*)/*SERK3*, *brassinosteroid insensitive 1* (*BRI1*), *Leafy cotyledon 1* (*LEC1*) *and Wuschel* (*WUS*) were examined during the stages of SE in mangosteen. Somatic embryos of mangosteen were induced on Murashige and Skoog medium with 3.08 µM 6-benzylaminopurine, 3.18 µM thidiazurone and

60 mg/L glucose. Selected genes were isolated from early, mid and late globular stages of somatic embryo development. Quantitative PCR analyses showed that the early globular stage is characterised by low level of expression of *SERK1*, *BRI1* and *WUS* and no expression of *LEC1*. The mid globular stage is characterised by a higher level of expression of the *SERK1* and *BRI1* genes, while the late globular stage is characterised by a high level of expression of LEC1 and WUS genes with low expression of the *BRI1* gene. This demonstrates a clear match between globular stages and the expression of key genes at each stage. Transcriptome data from embryogenic structures showed differentially expressed genes involved in hypothetical signal transduction (brassinosteroid, auxin, cytokinin, jasmonic acid, abscisic acid and glucose) and biosynthetic (glucose and polyamine) pathways. The five selected genes showed interconnection with differentially expressed unigenes (up-regulated) indicating molecular interaction involved in somatic embryogenesis of mangosteen (Fadryin *et al.*, 2019).

Translucent flash disorder (TFD) is one of the important physiological disorders in mangosteen (*Garcinia mangostana* L.). TFD has symptoms such as flesh arils that become firm and appear transparent similar to water core in apple or pear (Matra *et al.*, 2019). Information on the changes of gene expression in TFD-affected tissues remain limited, and investigations into the effects of different water regimes still need to be undertaken. Through an RNA sequencing approach using the Ion Proton, 183,274 contigs with length ranging from 173–13,035 bp were constructed by de novo assembly. Functional annotation was analyzed using various public databases such as non-redundant protein NCBI, SwissProt, and Gene Ontology, and KEGG pathway. Our studies compared different water regimes to incidence and differentially expressed genes of TFD-like physiological disorders. From the differentially expressed gene (DEG) between normal air and TFD-affected aril, we identified DEG-related TFD events, which 6228 DEGs in the control condition and 3327 DEGs in under water stress treatment condition remained, and confirmed these with RT-qPCR, including sucrose synthase (SUSY), endoglucanase (GUN), xyloglucan endotransglucosylase/hydrolase (XTH), and polygalacturonase (PG) showed statistically significant. In addition, transcription factors also indicated changes in MYB, NAC and WRKY between tissues and different water regimes (Matra *et al.*, 2019).

14.0 References

Affandi and Emilda, D. (2009) *J. Fruit Ornam. Plant Res.*, **17**(2): 219-233.

Affandi, Octriana, L., Fatria, D. and Purnama, T. (2011) *AGRIVITA*, **33:** 218-226.

Almeyda, N. and Martin, F.M. (1976) In: *Cultivation of neglected tropical fruits with promise*. Part I. *The Mangosteen*, Agricultural Research Service, USDA, p.18.

Almeyda, N. and Martin, F.W. (1976) In: *Cultivation of neglected tropical fruits with promise. Part 1. The Mangosteen.* ARS-S-155. Agricultural Research Service. U.S. Department of Agriculture, p.55.

Apiratikorn, S., Sdoodee. S., Lerslerwong, L. and Rongsawat, S. (2012) *Kasetsart J. (Nat. Sci.).*, **46**: 1-9.

Araño, B.A. (2004) http: //bpi.da.gov.ph/bpi/images/Production_guide/pdf/MANGOSTEEN.pdf.

Brunner, B.R. and Morales, Payan, J.P. (2010) Mangosteen and Rambutan. *Encyclopedia Of Life Support Systems 2010* - Eolss.Net. Http: //Www.Eolss.Net/Sample-Chapters/C10/E1-05a-52-00.pdf.

Corner, E.J.H. (1952) In: *Wayside trees of Malaya,* Vol. 1, 2nd edn. Government Printing Office, Singapore, pp. 312-320.

Elviana, M., Rohani, E.R., Ismanizan, I., Normah, M.N. (2011) *Biol. Plant.,* **55:** 731-736.

Erlangga, N., Purwadaria, H.K. and Firdaus, M. (2012) *Jurnal Manajemen & Agribisnis.,* **9:** 69-77.

Fadryin, N., Rohania, E.R., Husseina, Z.A.M. and Noora, N. M. (2019) *Plant Gene* **15:** 51-66.

Fairchild, D. (1930) *Exploring for plants.,* The Macmillan Company, New York.

Garrity, A.R., Morton, G.A. and Morton, J.C. (2004) Patent No.: US 6,730,333 B1 dated May 4, 2004.

Goh, C.J., Lakshmanan, P. and Loh, C.S. (1994) *Plant Science.,* **101:** 173-180.

Gomes Júnior, G.A., Pereira, R.A., Sodré, G.A. and Gross, E. (2019) *Pesquisa Agropecuária Tropical,* **49:** e55529. Epub September 26, 2019.https: //doi.org/10.1590/1983-40632019v4955529

Horn, C.L. (1940) Stimulation of growth in juvenile mangosteen plants. *J. Agric. Res.,* **61:** 397-400.

Jarimopas, B., Pushpariksha, P. and Singh, S.P. (2009) *Int. J. Food Prop.,* **12:** 414-426.

Jong, K., Stone, B.C. and Soepadmo, E. (1973) *Proceedings of the Symposium on Biological Research and Natural Development,* pp. 113-121.

Joni, Y.Z., Prihatini, R., Efendi, D. and Roostika, I. (2016) *Indonesian J. Agrl. Sci.,* **17**(1): 9-16.

Krishnamurthi, S., and Rao, N.V.M. (1962) *Indian Horticulture,* **7**(1)**:** 3-8.

Kurniadinata, O.F., Poerwanto, R. and Susila, A.D. (2018) *J. Tropical Horticulture,* **1**(1)**:** 7-9.

Makhonpas, C., Phongsamran, S. and Silasai, A. (2015) *IJAAST,* **11**(2): 227-242.

Mansyah, E., Santosa, S.E. and Poerwanto, R. (2010) *J. Hortic. For.,* **2**(6): 127-134.

Matra, D.D., Kozaki, T., Ishii, K., Poerwanto, R, Inoue, E. (2019) *PLoS ONE* **14** (7): e0219976. https: //doi.org/10.1371/journal. pone.0219976

Matra, D.D., Poerwanto, R., Santosa, E. *et al.* (2016) *Tropical Plant Biol.,* **9:** 29-41. https: //doi.org/10.1007/s12042-016-9161-8

Minh, T.V. (2005) *Acta Hort.,* **692**: 33-36.

Morton, J. (1987) In: *Fruits of warm climates.* Julia F. Morton, Miami, FL, pp. 301-304.

Namuco, L.O. (2007) *Philippines J Crop Sci.,* **32**(3): 81-84.

Namuco, L.O. (2007) *Philippines J Crop Sci.*, **33**(1): 103-106.

Nazre, M. (2014) *Genet Resour Crop Evol.*, DOI 10.1007/s10722-014-0097-2.

Nazre, M. (2010) *Genet Resour Crop Evol.*, **57:** 1249-1259.

Normah, M.N., Aizat, W.M., Hussin, K., Rohani, E.R. (2016) Fruits., **71:** 199-207.

Pechkeo, S., Sdoodee, S. and Nilnond, C. (2007b) *Kasetsart J.* (*Nat. Sci.*)., **41**: 621-632.

Pechkeo, S., Sdoodee, S. and Nilnond, C. (2007a) *Kasetsart J.* (*Nat. Sci.*)., **41:** 61-71.

Pedraza-Chaverri, J., Cárdenas-Rodríguez, N., Orozco-Ibarra, M. and Pérez-Rojas, J.M. (2008) *Food Chem. Toxicol.*, **46**: 3227-3239.

Qosim, W.A., Patarapuwadol, S. and Watanabe, K.N. (2011) *Int. Res. J. Biotechnol.*, **2**(1): 001-008.

Ramage, C.M., Sando, L. and Peace, C.P. *et al.* (2004) *Euphytica.*, **136:** 1-10.

Richards, A.J. (1990) *Bot. J. Linn. Soc.*, **103**: 301-308.

Rohani, E.R., Ismanizan, I. and Mohd Noor, N. (2012) *Plant Cell Tiss Organ Cult.*, **110:** 251-259.

Mohd Asrul, S., Sarip, J., Mohd Shaib, J. and Nor Hanis Aifaa, Y. (2011) *9th Malaysia Genetics Congress,* Kuching, Sarawak, Malaysia, Nov.2011. https: //www. researchgate.net/publication/335977060.

Sandø, L., Peace, C., Ramage, C., Carroll, B.J. and Drew, R. (2005) *Acta Hort.*, **692**: 143-151.

Sobir, S., Sinaga. R., Poerwanto, Rismitasari and Lukman, R. (2009) *Biodiversitas.*, **10**(4): 163-168.

Suhendra, D. and Mustamu, N.E. (2018) *IOP Conf. Series: Earth and Environmental Science.*, **205** (2018) 012015 IOP Publishing doi: 10.1088/1755-1315/205/1/012015.

Sulassih., Sobir, S. and Santosa, E. (2013) *SABRAO J Breed Genet.*, **45**(3): 478-490.

Te-chato, S. and Lim, M. (2000) *Sci. Hort.*, **86:** 291-298.

Te-chato, S. and Lim, M. (2005) In: *Biotechnology of fruit and nut crops* (Biotechnology in Agriculture Series; 29), (Ed. R.E.Litz), CABI publishing, Cambridge, UK. pp. 209-220.

Wieble, J., Chacko, E.K. and Downton, W.J.S. (1992) *Acta Hort.*, **321**: 132-137.

Verheij, E.W.M. (1992) In: *Plant Resources of South-East Asia*, No. 2, *Edible Fruits and Nuts*. (Eds. E.W.M. Verheij, and R.E. Coronel), Pudoc–DLO, Wageningen, pp. 177-181.

Yapwattanaphun, C., Subhadrabandhu, S., Honsho, C. and Yonemori, K. (2004) *J. Amer. Soc. Hort. Sci.*, **129:** 368 -373.

Yulianti, N., Sutrisno. and Darmawati, dan E. (2010) *Jurnal Keteknikan Pertanian*, **24**(1): 55-60.

10

KARONDA

A. R. Karale, Tanushree Saha,
Ivi Chakraborty and B. Ghosh

1.0 Introduction

The karonda (*Carissa carandas*), commonly known as Black Current, Bengal Current 'Christ's thorn', is a non-traditional fruit crop which thrives well as a rainfed crop. Once established, the plant hardly needs any care and gives yield with minimum management. This fruit crop is a monitory reward to tribal communities in various parts of India as it contains several phytochemical constituents belonging to terpenoids category and having tremendous medicinal properties.

2.0 Composition and Uses

2.1.0 Composition

The composition and food value of ripe and dry karonda fruits (Anonymous, 2014) are given in Table 1.

Table 1: Proximate Composition of Karonda Fruit*

Composition	*Nutritional value (100 g)*	
	Fresh	*Dried*
Energy (Calorie)	42	364
Moisture (per cent)	91	18.2
Protein (per cent)	1.1	2.3
Carbohydrate (per cent)	2.9	67.1
Fat (per cent)	2.9	9.6
Mineral (per cent)	-	2.8
Calcium (mg)	2.1	160
Phosphorus (mg)	28	60
Iron (mg)	-	39
Vitamin C (mg)	20.0-50.0	1

* Anon. (2014).

2.2.0 Uses

Ripe fruits are sub-acidic to sweet in taste with peculiar aroma. Whole plant and its parts are used in traditional medicine for treatments of ailments. Unripe fruits could be used to treat liver dysfunction, break fever and counteract putrefaction of blood (Pewlong *et al.*, 2014). They also contain carissone and methyl ester, which contributed to antidiarrhoeal and antianthelmintic properties. Begum *et al.* (2013) reported that the leaf extracts containing carandiol and lot more important constituents were shown to have cytotoxic effects towards HeLa cells (human cervical cancer cells), PC-3 cells (human prostate cancer cells) *etc.* Thus this miraculous plant is reputed for its antioxidant, analgesic, anthelmintic, anti-inflamatory, hypolipidemic, anti-microbial, antidiabetic anticancer, diuretic, nephrotoxicity amelioration, aphrodisiac, appetizing, antipyretic, astringent and hepatoprotective properties (Patel, 2013). Thus, traditionally it can be used to cure various diseases and has been used in the treatment of asthma, skin diseases, cough, cold, sore throat, mouth ulcer, tuberculosis, scabies, intestinal worms, diarrhoea and intermittent fever (Jayakumar and Muthuraman, 2018). Ripened karonda fruit is considered to be antiscorbutic due to the presence of high amount of vitamin C and a remedy for biliousness. Root extracts are used in lumbago, chest complaints and venereal diseases (Achenbach *et al.*, 1985). The alcoholic extract of the roots of *Carissa* had been reported to possess salicylic acid and cardiac glycosides showing antihypertensive and cardio-tonic activity (Arif *et al.*, 2016; Devmurari *et al.*, 2009).

The plant is considered as blessing for the Indian System of medicine particularly Ayurveda. It is used for alleviating vata and pitta disorders. Its fruits and seed latex are used for treating rheumatoid arthritis, anorexia, indigestion, colic, hepatomegaly, splenomegaly, piles, oedema, amenorrhoea, fever and nervine disorder (Pushpangadan., 2003; Sharma *et al.*, 2007).

The fruits may be eaten as a dessert when ripe or used in the preparation of fruit products such as jam, jelly, sauce, *Carissa* cream or jellied salad, preserve, squash, syrup, tarts and chutney, wine (Wani *et al.*, 2013; Yadav *et al.*, 2018) and food colourant. The unripe fruits are sour and astringent and can be used for pickles, sauces and *chutney*. The dried ripe fruits may become a substitute for raisins (Cheema and Cheema, 1971). The fruit can be candied just like cherry, which is known as 'Nakal (duplicate) Cherry' (Mandal *et al.*, 1992).

The unripe fruits yield milky white latex which can also be used in preparing chewing gum and rubber (Monachino, 1946). Fruits can also be used in dyeing and tanning industries. The wood of the karonda plant is white, hard and smooth. It is used for making spoons and combs. The plant makes excellent strong hedge almost impossible to penetrate. This crop is grown for its attractive coloured edible fruits and also used as live fencing around the orchards due to the presence of spines which provide protection against stray and wild animals.

3.0 Origin and Distribution

Karonda is well adapted for arid tropics and sub-tropics since it can withstand high temperatures thrive well as a rainfed crop, and gives yield with the minimum management. It is an evergreen, spiny shrub which is widely grown in South Africa, Australia, India, Malaysia, Pakistan, Sri Lanka and Myanmar (Burma). It is native to India and grows wild in Maharashtra, Rajasthan, Uttar Pradesh, Bihar and South India. It is preferred very much as a protective hedge in Gujarat and Punjab. It is not suited to the hills of Himalayas. Karonda is reported to flourish well in the Western Ghats and commonly found in hills, roadside and riverside.

4.0 Taxonomy and Morphology

Cook (1904) described *C. carandas* as a large, evergreen dichotomously branched, spinous shrub or small tree. Leaves are opposite, small, coriaceous and shining; elliptic or ovate in shape and obtuse (often mucronate) tip. Thorns are forked and 2.5-3.5 cm long. Flowers are white in colour, produced in terminal cyme. Corolla lobe is cylindrical, necked. Anthers included free from stigma; ovary 2-celled. The fruit is

Uses of karonda

Karonda fruits with slices

Karonda juice

Karonda pickle

Karonda chutney

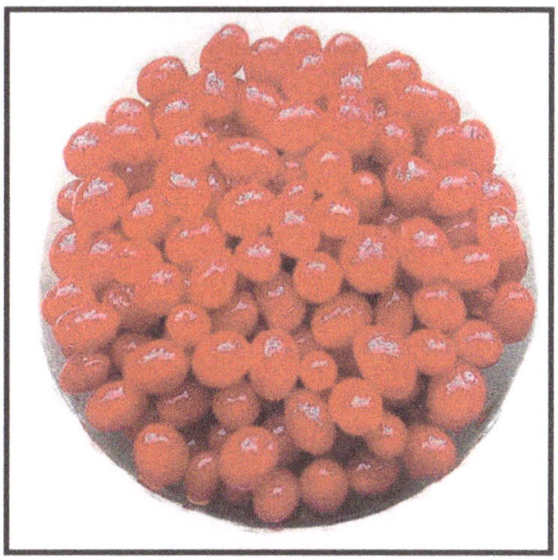

Karonda candy

Uses of karonda

Karonda jelly

Karonda jam

Karonda morobba

Karonda candy

Karonda wine

Karonda achar

a berry, ellipsoid and pointed or globose or oval or round; 1.8 to 2.5 cm long, green in colour when unripe and turn to red and finally black or dark purple on ripening.

5.0 Species and Cultivars

5.1.0 Species

The genus *Carissa* to which karonda belongs is a member of the family Apocynaceae. The genus has nearly 30 species originating in South Africa, Australia, tropical Asia and Malaysia. The five species of this genus of Indian origin including *C. carandas, C. inermis, C. paucinervia, C. Spinarum,* wild in India; the first two in Western peninsula and third in the dry tracts (Singh and Arora, 1978; Anon, 2014). *C. grandiflora* (Natal plum) came to India from South Africa. *C. edulis* has been introduced from the USA (Singh *et al.*, 1982). *Carissa bispinosa, Carissa ovate are two other species available* (Anon, 2014). Different species of *Carissa* are grown for their small berry like edible fruits and closely branched spinous hedges. Some species are also cultivated for ornamental purposes in the gardens. Kanupriya *et al.* (2019) tried to select superior accessions which can be used as parents in breeding programs using morphological, biochemical and molecular markers for successfully categorizing karonda accessions into different groups based on sweetness, colour of fruits and geographical origin.

5.2.0 Cultivars

There are no well established varieties of karonda although the cultivated types are classified according to fruit colour as green-fruited, pink-fruited and white-fruited by Singh (1967). The differences are not much in shape and size of fruits in all the three types.

A few research workers in India have attempted to study the variability in morpho-logical, qualitative and quantitative characters including biochemical ones. These studies have been able to locate some promising types which are not yet commercially exploited.

At Rahuri (Maharashtra) some promising types No. 13, 16, 12 and 3 have been identified (Karale *et al.*, 1989). At Konkan Krishi Vidyapeeth (KKV), Dapoli, three plants namely No. 2, No. 5 and No. 2 (from another location) were reported to be promising by Bhagwat (1984). Similarly Joshi *et al.* (1986) reported that ripe fruits of three genotypes *i.e.*, K-1, K-2, K-3 were highly palatable. Sawant (1998) reported genotype No. 21, 25, 46, 95, 105, 106, 181 and 184 as promising ones. A sweet-type variety, 'Konkan Bold' was developed at Fruit Research Station, Vengurla, Maharashtra, which is large fruited (weighing 16.23g), high in pulp content (92 per cent) and low in acidity (Salvi *et al.*, 2006). At Central Arid Zone Research Institute, Jodhpur, Rajasthan, three cultivars *viz.*, CZK-2011, CZK-2022 and CZK-2031 were selected (Anon, 2011).

Maru Gaurav is an improved high yielding variety of karonda developed by selection from seedling population released by ICAR-Central Arid Zone Research Institute, Jodhpur, Rajasthan (Meghwol *et al.*, 2019). Mature fruit is maroon colored on one side (facing light) while the other side of the fruits remain light green. The peak maturity is attained after middle of August onwards, while fully ripened fruits are available at the end of September. The matured fruits may be harvested for chutney, pickle and candy preparation, while those of ripened fruits may be used for extraction of juice for beverages and for extraction of seeds for multiplication.

Physico-chemical characteristic studies of fruits of five karonda lines *viz.*, PK1, PK4, Pant Manohar, Pant Sudarshan and Pant Suvarna indicated that fruit diameter, shape index, weight, volume, number of seeds per fruit, dry weight of flesh, fruit colour (Table 2), TSS and acidity differed significantly while fruit length, specific gravity, flesh, seed and ascorbic acid contents of fruits did not differ significantly among the selected lines (Misra and Jaiswal, 1999).

Table 2: Physical Characteristics of Fruits of Selected Lines of Karonda*

Selected Lines	*Fruit Length (cm)*	*Fruit Diameter (cm)*	*Fruit Weight (g)*	*Seeds/ Fruit*	*Flesh (Per cent)*	*Seed (Per cent)*	*Fruit Colour*
PK1	2.16	1.70	3.53	4.48	87.30 (69.10)	12.70 (20.90)	White-pink
PK4	2.24	1.71	3.74	5.61	87.85 (69.60)	12.15 (20.40)	Green-purple
Pant Manohar	2.14	1.70	3.50	3.95	88.28 (69.99)	11.72 (20.01)	White-dark pink
Pant Sudarshan	2.17	1.96	3.47	4.68	88.48 (70.14)	11.52 (19.86)	White-pink
Pant Suvarna	2.25	1.67	3.63	5.89	88.25 (69.96)	11.75 (20.04)	Green-purple
S.Em ±	0.02	0.01	0.11	0.20	(0.43)	(0.40)	—
C.D. at 5 per cent	NS	0.03	0.30	0.60	NS	NS	—

Figures in parentheses indicate transformed values.

* Misra and Jaiswal (1999).

Name of some other promising cultivars are Konkanbold developed by Konkan Krishi VidyaPeeth, Dapoli (Maharashtra) in 2004, CHES- K-II-7 and CHES- K- V-6 (identified from the seedling population at CHES (Central Horticultural Experiment Station) Chettalli (Karnataka).

6.0 Breeding and Improvement

Not many research works on improvement of this under-utilized crop has been reported, though improvement in this crop can be done by two methods.

6.1.0 Collection of Superior Types from Exotic Sources

Exotic collections of *Carissa* spp. namely *C. edulis* (EC 35932) and *C. grandiflora* (EC 37515) have been introduced from USA at the Regional Station of NBPGR, Jodhpur. Owing to xerophytic nature of these species, plants have very well adopted in western Rajasthan.

Species of karonda

Carissa carandas

Carissa spinarum

Carissa edulis

Carissa grandiflora

Karonda cultivars

Karonda cultivar "CHES KV 1"

Karonda cultivar "CHES K-V/6"

Karonda hybrid Maru Gaurav

6.2.0 Improvement by Selection

Among the methods of crop improvement, the selection has been known since early time as an easy tool of plant improvement. Selection exploits the natural variability in the population based on economic characters (Kedarnath, 1986). Most of the populations of karonda growing in different parts of the country have adequate genetic variability to make the selection and identify the desirable genotypes. The existing population of this fruit crop is mostly seed originated which shows variability in morphological characters like plant size, canopy, umbel segments, flowers and fruit characters due to heterozygosity. Mehra and Arora (1982) opined that diversity in *Carissa* occurs in the north-western India particularly in Mount Abu, Western Ghats and Khandala in Maharashtra, which offers a great deal of scope for improvement of karonda by seedling selection.

Correlation studies have indicated positive and significant association among the characters which determine the size of berry (berry length, weight and volume). All these three characters had positive and significant association with number of seeds, weight of seed and weight of skin per berry (Dhumal *et al.*, 1993). There existed negative correlation between fruit number/branch and fruit diameter, fruit length, pomace and seed weight per fruit (Sawant, 1998). The estimates of genetic parameters revealed that phenotypic variance was greater than genotypic variance for all the characters. Heritability in broad sense was very high for fruit weight, TSS and acidity and higher genetic advance for pomace, fruit weight and seed weight per fruit. He observed great variability in physico-chemical characters (Sawant *et al.*, 2002). The higher number of umbel segments can be used as selection index for higher yields (Karale *et al.*, 1990). This associationship of characters should aid in the survey work. Genotypic Coefficient of variability, heritability and genetic advance were found lowest for number of seeds per fruit, which indicate that this character could not be improved by selection (Athani *et al.*, 2005).

Recently, genetic variation in karonda population, maintained at IIHR, Bangalore, India, has been analyzed using morphological, molecular and biochemical markers, using multivariate analysis approach. It can be useful in selecting superior accessions that can be used as parents in future breeding programme (Chaturvedi *et al.*, 2019).

7.0 Propagation

7.1.0 Seed

The most common method of propagation of karonda is by seeds. Fresh seeds should be used for raising seedlings since seeds showed good viability upto 10 days and thereafter usually started losing their viability and reached minimum after 60 days of storage (Deepika and Vanajalatha, 2017). The germination of seed starts 40-45 days after sowing and is completed within 60 days. Bankar (1987) obtained early and enhanced germination (68 per cent) with increased vigour of seedlings when seeds were soaked in 100 ppm GA3 for 24 h. Banik *et al.* (2015) observed maximum germination when the seeds were treated with KNO_3 @ 1 per cent. Seedlings are raised in polythene bags during rainy season. The seedlings are slow growing and hence they are ready for transplanting after about two years from sowing of seeds.

7.2.0 Vegetative Propagation

In order to overcome the disadvantages of seed propagation, karonda can also be propagated by vegetative methods. It ensures the perpetuation of promising types. Interspecies grafting in genus *Carissa* has been reported by Shankar (1966).

7.2.1 Cuttings

Karonda is known to be difficult-to-root plant, therefore rooting does not occur easily. Attempts were made by Mehata and Singh (1960) to propagate karonda by stem cuttings. Cuttings from young plants and hedge sources proved to be the best for rooting and shoot growth. Divekar (1984) reported more rooting with 1000 ppm IBA treatment as quick dip. No rooting was obtained without growth regulator treatment. The cuttings are usually taken from mature, about one-year-old shoot and should be 30-35 cm long with six leaves. Highest shooting and rooting of hard-wood stem cutting were obtained with 8000 ppm IBA (Vanajalatha *et al.*, 2015). Rainy season was observed to be the best period for propagation by cuttings. Treatment with 3000 ppm IBA and keeping the cuttings under mist markedly improved root formation. After 6-8 months, the plants attained transplantable height.

7.2.2 Air-layering

Jauhari and Nigam (1958) found high percentage of success with air layering when layers were treated with mixture of 7500 ppm each of IBA and NAA. Mishra and Singh (1990) reported that all the air layered shoots treated with IBA and NAA at 2500 to 10000 ppm in lanolin paste rooted whereas untreated layers failed to root.

The best season for air layering is the beginning of monsoon *i.e.*, June-July, and it can be extended further upto September. Ghosh *et al.* (2011) concluded that air layering in the month of June gave best rooting and field establishment. Though Divekar *et al.* (1984) found 4000 ppm IBA to be the best. But, Raut *et al.* (2015) experimented with different IBA levels and concluded that 5000 ppm IBA could be used for commercial multiplication through air layering. Rooted layers were obtained after 3½ months of layering.

7.2.3 Budding

Propagation of karonda by shield budding is economic and convenient. Natal plum and promising types of karonda could be budded on one-year-old rootstocks of *C. caranda* (Bankar *et al.*, 1994). The best time for budding is July to August in low rainfall areas.

7.2.4 Inarching

Propagation of karonda by inarching also proved successful. However, compared to other methods, inarching is not very popular (Bankar *et al.*, 1994).

7.2.5 Soft Wood Grafting

Soft wood grafting is also successful in karonda and is a very useful technique for *in situ* grafting in arid regions. Nimbalkar (2016) reported that soft wood grafting

Propagation of karonda

Propagation by seed

Propagation by budding

Ripe fruits of karonda

Harvested karonda fruits

with 2 months age scion and polythene bag cover gave significant sprouting (91.96 per cent) and highest survival rate (97.31 per cent).

7.2.6 Micropropagation

To increase the area under cultivation, some alternative way is required for rapid and large scale multiplication. Till now, not much works reported on micro-propagation of this unharnessed crop Karonda.

Rai and Misra (2005) reported that highest shoot proliferation was achieved when the shoot-tips (1.5 cm long) from mature plants of cv. Pant Sudarshan were cultured on MS medium containing 3.0 mg L^{-1} BA. The rooted plants were acclimatized successfully on vermiculite:sand:soil (1:1:1) medium. Hasmah *et al.* (2013) observed effective rooting on MS medium supplemented with 2.0 mg L^{-1} IBA and 84 per cent ±4.3 per cent survival rate of the *in-vitro* grown plantlets.

8.0 Area and Production

Information regarding the area and production of this fruit crop in India is not available because it is not planted in the form of a regular orchard; but preferred as protective hardy hedge and the fruits they bear.

9.0 Soil and Climate

9.1.0 Soil

Karonda can be grown on a wide range of soils including poor, rocky, sandy soils of arid regions. However, on deep loamy or alluvial soils, its growth is better and fruit yield is higher. It is very well suited to waste lands and on field boundaries. It can tolerate saline sodic soils (pH upto 10) to a greater extent (Bankar *et al.*, 1994).

9.2.0 Climate

Owing to the xerophytic nature, plants of karonda are well adopted to the arid and semi-arid climates. Being a hardy plant, it can thrive well under drought conditions.

10.0 Cultivation

10.1.0 Planting

The field is cleaned, thoroughly ploughed and levelled. The pits of 30 to 45 cm^3 are dug and filled with organic manure and soil in the ratio of 1:1. The distance of planting in hedge is recommended at 1 to1.5 m and a regular planting at 1.5 to 2.0 m both ways (Mitra *et al.*, 2008). The best time for planting is the beginning of monsoon.

10.2.0 Training and Pruning

At the time of planting, each plant is supported with the help of wooden stake so that it remains erect. Unwanted laterals are removed from time to time. Bearing tree, usually do not require pruning. However, to give desired shape and making the stem clear, additional twigs are pruned. Twigs which are affected with disease or broken are removed. Old hedge may be headed back to induce new growth.

10.3.0 Irrigation

Karonda is a slow growing plant. Once established, it will not need much water. Moreover, it behaves as a xerophytic plant and therefore, water requirement is very low. However, efforts should be made to collect run-off during rainy season in root zone through land shaping or water harvesting models.

10.4.0 Manuring and Fertilization

About nutrition of karonda, research-based data on the fertilizer doses are not avai-lable. It has been observed that this plant responds better to organic matter and an application of 10-15 kg FYM per bearing plant per year is suggested in the beginning of monsoon. Sharath and Ghosh (2015) reported that cow dung manure at 4kg/plant/year not only improved fruit yield and quality, but continuous use of these organic manures help in improving soil health.

10.5.0 Intercropping

In regular plantation of karonda, many seasonal vegetables can be grown for two to three years during rainy season.

11.0 Flowering, Pollination, Fruit Growth and Development

Hooker (1882) described that the genus *Carissa* produces perfect flowers terminally or axillary. The calyx consists of five sepals and the corolla forms cylindrical tube. The stamens are present at the tip of corolla tube. Rendle (1925) reported that flowers are solitary and present in racemose inflorescence passing into cyme. This species of *Carissa* is divided into two groups, one with short styled pistil and functional anthers and another with long styled pistil and small anthers devoid of pollen. Heterostyly results in low production because of presence of a high proportion of short styled flowers which are unfruitful.

Hains (1910) reported that flowering in karonda starts from April-May and continues upto December under Chhota Nagpur (Bihar) conditions. *Carissa grandiflora* produces flowers and fruits throughout the year but the quantum of flowers was less during winter months. Bhagwat (1984) reported that flowering takes place from

December to March, stigma is receptive upto one day after anthesis and the fruits ripen in about 70 days after fruitset in Konkan region of Maharashtra. Singh and Bajpai (2009) observed that flowering period was from March to April, while full bloom was reported in 1st week of April under Lucknow condition. They also reported occurance of anthesis after 1.00 pm, followed by flower dehiscence within few hours.

According to Misra *et al.* (1968) flower bud differentiation occurred on 7th February, when temperature was 29ºC and it was completed by 14th March. During this period, the average temperature ranged from 30.5 to 33.8ºC. A broadening and flattening of apical meristem with two protuberances, one on either side of it was established as criterion of blossom bud differentiation. Karale *et al.* (1990) reported a large variation in average number of flower buds per umbel (14.6 to

22.6), length of flower bud (1.42 to 2.15 cm) and length of pedicel (0.7 to 1.32 cm), petal number (4 to 6) and flower groups *i.e.*, umbel segments (1 to 6 in number). The fruit set under natural or open pollination was higher (32.02 per cent) than in self-pollination (24.24 per cent).

Fresh pollen grains were fine powdery white mass when observed with naked eyes. Size of pollen grains ranged from 29.1 to 37.5 microns and 15 per cent sucrose was found better medium for pollen germination (Singh *et al.*, 1989).

Studies on flowering and fruiting in *C. edulis* at Jodhpur revealed that it has two frui-ting seasons. One in January-February and another in August-September (Singh *et al.*, 1982). Tiny flower bud takes about 19 days for flowering. The anthesis is between 12.30 to 4.30 pm. Only 36 per cent of the flowers developed fruits to maturity (Banker and Verma, 1987). Flowers develop both terminally and axillary (Mitra *et al.*, 2008 a,b).

12.0 Pests and Diseases

12.1.0 Pests

Karonda is free from any serious pest and disease attack. The leaf eating caterpillar attacks its foliage during spring. Spraying of chlorpyriphos (0.25 percent) is effective to control it.

12.2.0 Diseases

12.2.1 Anthracnose

It is caused by *Colletotrichum inamdarri* (Lal and Singh, 1953). This disease is of wide occurrence, more especially in humid and high rainfall areas. The leaves, shoots, inflorescences and the fruits are affected by it. The affected young shoots finally show dieback symptoms. The diseased young fruits drop down and if the fungus attacks the fruits at maturity, the disease is enhanced by storage. Spraying with copper sulphate (0.1 per cent) or blitox or phytolan (0.3 per cent) or bavistin (0.1 per cent) or benlate (0.1 per cent) thrice a year is recommended (Lal and Singh, 1953).

12.2.2 Bacterial Canker

A bacterial canker of *C. congesta* incited by *Xanthomonas carissa* was described by Moniz *et al.* (1964). Later, Pal and Jalali (2005) and Maheshwari and Haldhar (2018) reported that the crop is seriously affected by cankers caused by different pathovars of *Xanthomonas campestris*. This disease is wide spread in Rajasthan. The symptoms first appear as spots on lower surface of leaves. These spots look like small or round, water soaked, translucent dots gradually expanding and turning dark brown. Twigs show similar spots. Flowers and fruits do not show any symptoms (Kumar and Patel, 1968). At Jobner, Rathore (1970) observed maximum number of leaf lesions in third week of October (24°C and 56 per cent RH). The disease can be checked by removing all the diseased leaves from the shrub pruned to the height of 45 cm from the ground level. Spraying with phytomycin (200 ppm) can also control the disease (Prasada *et al.*, 1971). Removal of diseased leaves or plant parts, and spraying with phytomycin 200 ppm/litre of water at 10 days interval is effective to manage the disease incidence (Maheshwari and Haldhar, 2018).

Post-harvest diseases: *Aspergillus awamori* and *A. niger* were isolated from rotting fruits of *C. carandas* in the market (Garg, and Gupta, 1980).

13.0 Harvesting and Yield

13.1.0 Harvesting

Karonda starts bearing flowers and fruits from the third year of its planting. The plant starts bearing flowers in the month of February and fruits are ready by the second week of June upto August. Both unripe and ripe fruits are harvested. It requires about 120-130 days from fruit set to maturity (Mitra *et al.*, 2008a,b) and complete harvesting for all fruits may not be done at one time. It is done two to three times. Green unripe fruits are available from March to May which is used for preparation of chutney and pickles, while ripe fruits are harvested in May-June. Colour change is a good indicator of maturity of fruits *i.e.*, when the colour of fruits turns to dark purplish black. Pawar (1988) concluded that weight, volume, specific gravity, length, diameter and colour of the fruit can be considered as the physical indices whereas moisture, TSS, sugars, acidity, pH and tannins as chemical indices of maturity of karonda fruits.

13.2.0 Yield

Fruit yield varies with the soil, climate and management of plants. About 4 to 5 kg fruits per bush are obtained. Just after harvesting, fruits are kept in shade. Undesirable fruits are sorted out and rest packed in shallow baskets cushioned with leaves or straw and marketed. Traditionally, the fresh fruits are collected by tribal people from forest areas and marketed in fresh condition. Meghwol *et al.* (2019) reported that mean fruit yield of Maru Gaurav, a regular bearer cultivar during eighth year onwards is 40 kg/plant and the fruits start bearing on 4th year after planting.

14.0 Post-harvest Management

Karonda remained neglected when post-harvest handling and processing technology is concerned. Usually no standard harvesting and handling protocol neither any organized marketing channel is followed for karonda. Generally, both the mature and tree-ripened fruits are harvested depending on their use. Because of its soft flesh with high moisture content, its shelf-life is short (Chauhan *et al.*, 2015). Generally, if harvested at mature stage, fruits may store for 5-7 days and ripe fruit for 2-3 days at room condition. The fruits harvested with stalk intact recorded less shrivelling and microbial spoilage (Pawar, 1988). For reduction of spoilage, the cold storage (13°C and 94 per cent RH) could be considered the best while for reduction of shrivelling, cool chamber could be more effective.

Girdhari Lal *et al.* (1967) described the procedure for making karonda preserve. The wine prepared from ripe fruits contains about 14.5 to 15 per cent alcohol and is very much liked by wine lovers (Nalawadi and Jayasheela, 1975; Mitra *et al.*, 2008). A natural red coloured food colorant (Lalima) can be prepared from the ripe fruits (Arif *et al.*, 2016; Reddy *et al.*, 2018). It was observed that just 1 ml of pigment suspension (Lalima) formulation is sufficient to give excellent red colour to one serving of any colourless beverage (100 ml) such as lemonade. One serve of such supplemented beverage possessed in addition 469.2 ìg anthocyanin, 12.7 mg flavonoids, 14.1 mg phenol, with total antioxidant activities to be 390 µM Trolox Equivalent (http://www.ciah.ernet.in/publication/natural_colorant.pdf.).

The products such as squash, syrup, jam, jelly, pickle, chutney, dried fruits and candy can be successfully prepared from the mature green and ripe fruits. In southern Florida, ripe karonda fruits are utilized for processing into syrup and wine (Sturrock, 1950). Nalawadi (1967) has reported the preparation of attractive pink coloured jelly from ripe fruits. Chaudhary *et al.* (2007) reported that the physicochemical changes in karonda (*Carissa carandas*) jelly during storage were organoleptically acceptable for a period of 4-5 months. Wani *et al.* (2008) observed that karonda jams could be stored for 80 days without deteriorating any physical, chemical and sensory parameters. Singh and Singh (1998) showed that White-yellow fruits are more suitable for pulp based uses like making jam, jelly, chutney, pickles and preserves. Such fruits were less sour than Red-green fruits. Manivasagan *et al.* (2007) reported that two karonda types *i.e.* both pink and green were found suitable for making pickle with acceptable qualitative changes during storage period of four months (Banik *et al.*, 2012). Effects of three different processing methods, sun drying, freezing, microwave drying, and fresh fruits of *Carissa spinarum* in terms of physical, nutritional and phytochemical composition have been studied (Chauhan *et al.*, 2015). The osmo-air dried fruits can be stored at room temperature for 3 months (Duhan *et al.*, 2008). The karonda crush with 35 per cent juice and 40 per cent juice with 55° Brix T.S.S. and 0.8 per cent acidity was found to be the best recipe as a whole with high organoleptic score for colour, flavour and overall acceptability (Divate *et al.*, 2020). However, the workers recommended the crush with 35 per cent juice treatment was economical compared to 40 per cent juice treatment.

15.0 Biotechnology

In karonda, efficient micropropagation protocol has already been developed (Rai and Misra, 2005; Hasmah *et al.*, 2013). Recently, an *in-vitro* organogenesis protocol was also established by Bhandane *et al.* (2018). The highest rate of shoot multiplication was observed on MS medium containing 2.0 mg L^{-1} BAP, 0.5 mg L^{-1} kinetin and 0.75 mg L^{-1} quercetin. However, rooting was successful on medium containing ¼ MS+0.5 mg L^{-1} IAA.

Biotechnological interventions could be the best possible way to be used for the improvement of this valuable crop.

16.0 References

Achenbanch, H., Naibel, R. and Addae Mer Sah (1989) *Phytochem.*, **24** : 2325-2328.

Anonymous (2014) *Bull. IIHR*, Karnataka, India

Anonymous. (2011) Annual Progress Report, CAZRI, Jodhpur.

Arif, M., Kamal, M., Jawaid, T., Khalid, M., Saini, K.S., Kumar, A. and Ahmad, M. (2016) *Asian J. Biomed. Pharm. Sci.*, **6**(58) : 14-19.

Athani, S.I., Patil, B.R., Swamy, G.S.K., Patil, B.R., Kanamadi, V.C., Shirol, A.M. and Dinakara Adiga, J. (2005) *J. Asian Hort.*, **2**(1) : 1-3.

Banik, B.C., Dey, A.N. and Pradhan S. (2015) *Acta Hortic.*, **1074** : 23-26.

Banik, B.C., Ghosh, S.N. and Singh, S.R. (2012) *Acta Hortic.*, **948** : 61-69.

Bankar, G.J. (1987) *Prog. Hort.*, **19**: 90-92.

Bankar, G.J. and Verma, S.K. (1987) *Prog. Hort.*, **19** : 163-166.

Begum, S., Syed, A.A., Siddiqui, B.S. Sattar, S.A. and Choudhary, M.I. (2013) *Phytochem. Letters*, **6** : 91-95.

Bhagwat, N.R. (1984) *M.Sc.(Ag.) thesis* submitted to Konkan Krishi Vidyapeeth, Dapoli.

Bhandane, B.S., Maheshwari, V.L. and Patil, R.H. (2018) *In Vitro cell. Dev. Biol. Plant.*, **54** : 600-605.

Chaturvedi, K., Tripathi, P.C., Venugopalan, R., Singh, P. And Radhika, V. (2019) *Fruits*, **74**(3) : 130-140.

Chaudhary, R., Yadav, M. and Singh, D.B. (2007) *Plant Archives*, **7**(2) : 885-887.

Chauhan, A., Tanwar, B. and Arneja, I. (2015) *Asian J. Pharm. Clin. Res.*, 8(6) : 254-259

Cheema, G.S. and Cheema, B.K. (1971) *Farm J. Calcutta*, **12** : 24.

Cook, T. (1904) In : *Flora of presidency of Bombay*, Vol. II., Botanical Survey of India, Calcutta, p. 186.

Deepika, D. And Vanajalatha, K. (2017) *J. Appl. Nat. Sci.*, **9**(3) : 1767-1771.

Devmurari, V., Shivanand, P. Goyani, M.B., Vagnani, S. and Jivani, N.P. (2009) *Pharmacog. Rev.*, **3** : 375-377.

Dhumal, S.A., Patil, D.S. and Jadhav, M.G. (1993) *J. Maharashtra Agric. Univ.*, **18** : 498-499.

Divate, S.M., Savale, B.G., Patil, N.B. and Relekar, P.P. (2020) *The Pharma Innovation J.*, **9**(1) : 176-179

Divekar, P.B. (1984) *M.Sc.(Ag.) thesis* submitted to Konkan Krishi Vidyapeeth, Dapoli.

Duhan, R., Godara, R.K., Gupta, R.B., Goyal, R.K. and Amarjeet, S. (2008) *Haryana J. Hort. Sci.*, **37**(3/4) : 252-254.

Garg, P. K., and Gupta, M. N. (1980) *Indian Phytopath.*, **33**(3) : 481-482.

Ghosh, S.N., Bera, B. and Roy, S. (2011) *J. Crop Weed.*, **7**(2) : 247-248.

Girdhari Lal, Siddappa, G.S. and Tandon, G.L. (1967) In: *Preservation of fruits and vegetables*, ICAR, New Delhi, p. 182.

Hains, M.H. (1910) In : *A Forest Flora of Chhota Nagpur*, Pub. by M/S. Bishen Singh, Mahindra Pal Singh, Periodical Experts, Delhi, pp. 423-424.

Hasmah, S.N., Bhatt, A. and Kenq, C.L. (2013) *Pertanika J. Trop. Agric.*, **36**(1) : 89-98.

Hooker, J.D. (1882) In : *Flora of British India*, L. Reeve and Co. Ltd., The Oast House, Brook, Nr. Ash Ford, Kent, pp. 630-631.

http: //www.ciah.ernet.in/publication/natural_colorant.pdf.

Jauhari, O.S. and Nigam, U.N. (1958) *Allahabad Fmr.*, **32** : 174.

Jayakumar, K. And Muthuraman, B. (2018) *World Sci. News.*, **96** : 217-224.

Joshi, G.D., Prabhudesai, V.G. and Salvi, M.J. (1986) *Maharashtra J. Hort.*, **3** : 39-44.

Kanupriya, C., Tripathi, P.C., Singh, P., Venugopalan, R. and Radhika, V. (2019) *Fruits*, **74**(3) : 130–140

Karale, A.R., Keskar, B.G., Dhawale, B.C., Kale, P.N. and Choudhari, K.G. (1990) *J. Maharashtra Agric. Univ.*, **15** : 109-110.

Karale, A.R., Keskar, B.G., Shete, M.B., Dhawale, B.C., Kale, P.N. and Choudhari, K.G. (1989) *Maharashtra J. Hort.*, **4** : 125-129.

Kaur, J. and Kaur, A. (2019) *Int. J. Recent Scientific Res.* Res., **10** : 32742-32745

Kedarnath, S. (1986) *Indian J. Genetics*, **46** : 172-186.

Kumar, V. and Patel, P.N. (1968) *Indian Phytopath.*, **2** : 315-317.

Lal, A. and Singh, R.P. (1953) *J. Indian Bot. Soc.*, **32** : 54-63.

Maheshwari, S. K. and Haldhar, S. M. (2018) Disease management in arid horticultural crops. CIAH/Tech./Pub. No. 68, ICAR-Central Institute for Arid Horticulture, Bikaner, Rajasthan, India, p. 42

Mandal, U., Sinha Roy, S. and Mazumdar, B.C. (1992) *Indian J. Landsc. Syst. Ecol. Stud.*, **15**(1) : 100-102.

Manivasagan, S., Rana, G.S., Joon, M.S. and Godara, A.K. (2007) *Haryana J. Hort. Sci.*, **36**(1/2) : 44-45.

Meghwol, P.R., Singh, Akath and Singh, S.K. (2019) *Indian Hort.* (March-April) : 29-31.

Mehata, K.C. and Singh, S.M. (1960) *Indian J. Hort.*, **17** : 31-37.

Mehra, K.C. and Arora, R.K. (1982) *NBPGR. Sci. Monogaph.*, No. 4.

Mishra, K.K. and Jaiswal, H.R. (1999) *Haryana J. Hort. Sci.*, **28** : 176-178.

Mishra, K.K. and Singh, R. (1990) *Annals Agril. Res.*, **11** : 208-210.

Misra, R.S., Yamdagni, R. and Bajpai, P.N. (1968) *Punjab Hort. J.*, **8** : 47-49.

Mitra, S.K., Pathak, P.K. and Chakraborty, I. (2008a) *Acta Hortic.*, 70 : 179-82

Mitra, S. K., Pathak, P. K. and Chakraborty, I. (2008b) In: *New Crops and Uses: Their role in a rapidly changing world*, (Ed. J. Smartt, and N. Haq), Centre for Underutilised

Crops, University of Southampton, Southampton, UK. Monachino, J. (1946) *Lloydia*, **9** : 293-309.

Moniz, L., Sabley, J.E. and More, W.D. (1964) *Indian Phytopath.*, **17** : 256.

Nalawadi, U.G. (1967) *Food Industries J.*, December-January, 1967.

Nalawadi, U.G. and Jayasheela, N. (1975) *Progressive Hort.*, **7**: 37-38.

Nimbalkar, S.D. (2016) *Indian Hort.*, **6**(2) : 211-213.

Pal, V., and Jalali, I. (2005) In: *Challenging problems in horticultural and forest pathology*, Indus Publishing Company, New Delhi, India, pp. 90-114.

Patel, S. (2013) *Environ. Sci. Biotechnol.*, **12** : 201-208.

Pawar, C.D. (1988) *M.Sc.(Ag.) thesis* submitted to Konkan Krishi Vidyapeeth, Dapoli.

Pewlong, W., Sajjabut, S. Eamsiri, J. and Chookaew S. (2014) *Food App. Bioscience*, **13**(1) : 509-517

Prasada, R., Pathak, V.N., Goyal, J.P. and Sharma, H.C. (1971) *Proc. Second Intern. Symp. Plant Pathology*, New Delhi, pp. 145-146.

Pushpangadan P. (2003) In: Karanda: Rural india's rich fruit. *Down to earth*, Thomson Press India, New Delhi : pp. 52-53.

Rai, R. and Misra, K.K. (2005) *Sci. Hortic.*, **103**(2) : 227-232.

Raut, U.A., Jadhav, G.G., Bhogave, A.F. and Deshmukh, M.S. (2015) *Res. Crops.*, **16**(3) : 537-541.

Reddy, S.V.R., Meena, R.K. and Bhargava, R. (2018) *Indian Hort.*, **63**(5) : 101-106.

Rendle, A.B. (1925) In : *Classification of Flowering Plants*, University Press, Cambridge, **2** : 468-474.

Salvi, B.R., Malshe, K.V., Bendale, V.W., Kshirsagar, P.J. and Gurav, S.S. (2006) In : *National symposium on underutilized Horticultural crops*, IIHR, Bangalore, pp. 33.

Sawant, B.R., Desai, U.T., Ranpise, S.A., More, T.A. and Sawant, S.V. (2002) *J. Maharashtra Agric. Univ.*, **27**(3) : 266-268.

Sawant, R.B. (1998) *Ph.D. thesis* submitted to Mahatma Phule Krishi Vidyapeeth, Rahuri.

Shankar, G. (1966) *Allahabad Farmer*, **40** : 169-170.

Sharath, A.A. and Ghosh, S.N. (2014) In : *II International Symposium on Wild relatives of Subtropical and Temperate Fruit and Nut Crops*. **1074** : 73-76.

Sharma, A., Tiwari, R.K., Kaushik, A., Tyagi, L.K., Shankar, K., Virmani, T., Yadav, S. and

Parvez, N. (2007) *Continental J. Pharma. Sci.*, 1 : 9 - 14.

Singh, A.K. and Bajpai, A. (2009) *Progressive Hort.*, **41**(2) : 148-152.

Singh, A.K. and Singh, P. (1998) *Annals Agri. Res.*, **19**(1) : 66-71.

Singh, H.B. and Arora, R.K. (1978) *Wild Edible Plants of India*, ICAR, New Delhi.

Singh, L.B. (1967) *Farm Bulletin*, ICAR, New Delhi.

Singh, M., Singh, G.N. and Singh, H.L. (1989) *Haryana J. Hort. Sci.*, **18** : 4-9.

Singh, R., Chopra, D.P. and Gupta, A.K. (1982) *Indian Hort.*, **23** : 101.

Sturrock, D. (1950) In : *Fruits of Southern Florida. A Handbook of Home Owner*. Horticultural Books, Sturat, Florida, p. 73.

Vanajalatha, K., Sharma, G., Singh, D. And Mishra, G. (2015) *Int. J. Farm Sci.*, **5**(4) : 139-144.

Wani, R.A., Prasad, V.M., Hakeem, S.A., Sheema, S., Angchuk, S. and Dixit, A. (2013) *Afr. J. Agric. Res.*, **8**(21): 2447-2449.

Yadav M., Srilekha, K., Barbhai, M. D. and Maheswari, K. (2018) *J. Pharmacog. Phytochem.*, **7**(5) : 1417-1420

11

DILLENIA

R. K. Bhattacharya, Tanushree Saha, Ivi Chakraborty and B. Ghosh

1.0 Introduction

The *Dillenia* is an important minor fruit grown in subtropical and tropical regions. It is commonly known as elephant apple and also by different names in different Indian languages such as *outenga*, *panchkol* and *chalita* (in Assamese), *chalta* (in Hindi, Bengali and Punjabi) (Table 1). It is also grown for shade on roads and avenues and as windbreaks in plantations.

Table 1: Different Names of Elephant Apple

Language	*Dillenia indica*	*Dillenia pentagyna*
English	Elephant apple	Dog Teak
Sanskrit	Bhavya, ruvya	-
Hindi	Chalta, girnar	Aggai, Kallai
Gujarati and Marathi	Karmbel, Mota Karmal, Mota Karambal	Karmal
Bengali	Chalta, Hargesa	Korkotta
Assam	Chalita, Outenga	Akshi
Oriya	Uvu, Chalota, Ou, Rai	Rai
Nepali	Ramphal, Panchphal, Panchkule	Tatri Chinnakalinga
Telugu	Peddakalinga	Ravudana
Tamil	Akku, Ugakkay, Uva, Uvav,	Naytekku,
Uvatteku	Kanigala.	
Malyalam	Chalitha, Punna, Syalitha, Valapunna	Punna, Kodapunna
Monghyr	Chhilta	-
Trade	Dillenia	-

* Deshmukh *et al.* (2019).

2.0 Composition and Uses

2.1.0 Composition

Dillenia fruit possesses considerable nutritive value. It is a good source of minerals, proteins, mucilage (Sharma and Nath, 2014) and vitamins. The nutritive values of the fruit are presented in Table 2.

The phytochemical constituent of *Dillenia indica* Linn. crude extract include glycoside, steroids, flavonoids, saponings and reducing sugar. The investigation of the phytochemical constituent indicates that the leaves are provided a rich source of triterpenoids and flavonoids. It also contains various chemical constituents like 3, 5,7-trihydroxy-3′, 4′-dimethoxy flavone (dillenetin), betulinic acid, â-sitosterol and stigmasterol. Chief contents of the fleshy sepals are tannins, malic acid, arabinogalactan and glucose. They also contain arabinogalactan, betulin, betulinic acid and flavonoids. Bark and wood contain flavonoids, betulin, betulinic acid, betulinaldehyde, lupeol, â-sitosterol, myricetinhydroxylactone, dihydroisorhamnetin, dillenetin and glucosides. Stem bark contains betulinaldehyde, betulin, lupeol, â-sitosterol, myricetin, a new hydroxylactone, dihydro-isorrhamnetin, dillenetin and glucosides (Talukdar *et al.*, 2012). The exorbitant medicinal property of each and every part of the *Dillenia* is mostly depending on these constituents all over the plant.

Table 2: Nutritive Value of Dillenia Fruits*

Constituent	*Amount*	*Constituent*	*Amount*
Moisture (per cent)	82.3-86.10	Fat (per cent)	0.2-0.34
pH	3.36-3.38	Fibre (per cent)	2.1-2.5
Acidity (per cent)	2.24-2.28	Pectin (per cent)	3.87
Total soluble solids (°Brix)	8.8-9.0	Tannin (per cent)	0.625
Reducing sugar (per cent)	4.59-4.76	Ash (per cent)	3.54
Total sugar (per cent)	5.43-5.44	Phosphorus (mg/100 g)	26
Vitamin C (mg/100 g)	4.04-4.05	Calcium (mg/100 g)	16
Protein (per cent)	0.8	Energy (Kcal/100 g)	59

* Gopalan *et al.* (1971), Neog and Mohan (1993), Saikia and Dutta (1995).

2.2.0 Uses

The ripe fruits are edible, sweet, sour and tasty. The thickened sepals of the ripe fruits are extensively used for garnishing curries due to its excellent flavour. Studies on the processed products from *dillenia* fruits at Assam Agricultural University, Jorhat revealed that good quality ready-to-serve beverages, jam, jelly, pickle and *chutney* can be prepared and the recipes have been standardized (Saikia and Dutta, 1995; Anon., 2001a). The fruit of *dillenia* is an important drug in the indigenous system of medicine and finds mention to some of the oldest *Ayurvedic* records. The leaves and bark also have extensive therapeutic values, acts as astringent and laxative (Kirtikar and Basu, 1933) and induces diarrhoea if indulged in excess. The bruised bark is used externally as a cataplasm in arthritis. It possesses various pharmacological properties such as, anti-inflamatory (Talukdar *et al.*, 2012), antimicrobial (Apu *et al.*, 2010), anti-oxidant (Abdille *et al.*, 2005), analgesic (Alam *et al.*, 2012), anti-diabetic (Kaur *et al.*, 2016; Malviya *et al.*, 2010), antinociceptive (Lim, 2012), central nervous system depressant (Bhakuni *et al.*, 1968), muco-adhesive (Kuotsu and Bandyopadhyay, 2007), anti-cancer (Kumar *et al.*, 2010) antifatigue, cardiotonic (Sharma,1956) properties. Traditionally *D. indica* frequently used as folk medicine as anti-inflammatory purposes (Yeshwante *et al.*, 2009). In Brazil, particularly the fruit extracts are used in healing of psoriasis-like wounds and reduced inflammation via a mechanism associated with protection against oxidative damage in biomolecules (Franco, 2012).

It is known to stop abdominal and joint pains, regulate the heat of the body and tone up the nervous system. The fruit juice mixed with sugar and water is used as a cooling beverage in fevers and against cough (The Wealth of India, 1952). It is used as a remedy in *vata* and *kapha* disorders (Shome *et al.*, 1980).

Uses of dillenia

Dillenia tree with fruits

Sliced fruit of dillenia

Pickle from dillenia

Chutney from dillenia

Species of dillenia

Dillenia pentagyna

Dillenia obovata

Dillenia alata

Dillenia suffruticosa

Dillenia phillipinensis

This beautiful tree is well suited for planting in gardens and parks, as well as along small avenues for shade, particularly in areas along the pools and marshy lands for its elegant appearance, shining green leaves and large white flowers (Bose and Chowdhury, 1991; Bose *et al.*, 1999; Anon., 2001b). The timber made from the tree is considered very well for ridge, poles and very durable under water. The plant is also a good source of firewood. It is a suitable component for making fire retardant treated plywoods (Suh, 1999). A red coloured dye is obtained from the bark of the tree. Dried leaves are used as a substitute for sand paper in polishing ivory and horn. Leaves are used in making leaf cups and plates.

3.0 Origin and Distribution

The place of origin of *Dillenia* is believed to be the South East Asia, more particularly the entire north-eastern region of India and adjacent China. MacMillan (1989) considered Sri Lanka and tropical Asia as the original home of dillenia. According to Singh (1992) and Lim (2012), the origin of dillenia is east India and it occurs in natural forests in Assam and West Bengal. It is undoubtedly indigenous to an area stretching from India to China (Neog and Mohan, 1994).

The distribution of *Dillenia* is in the humid subtropics and tropics. The tree grows in the moist and evergreen forests of the sub-Himalayan terrain stretching from Punjab to Assam, from Kumaon to Garhwal, and southward to central and southern India (Sastri, 1952), Andaman, Gujarat, Mizoram, Tripura and West Bengal (Khanum and Khan, 2007; Khare, 2007). It is very common and often gregarious in damp places and swamps in all the districts of Assam (Neog, 1990).

Dillenia indica and *Dillenia pentagyna* are widely distributed in many Asian countries. *D. indica* plants are mainly distributed in valleys, Bhutan, India, Indonesia, Laos, Malaysia, Myanmar, Nepal, Bangladesh, Philippines, Sri Lanka, Thailand and Vietnam. In India, it is found in Sub Himalayan tract, whole north-east especially Assam, North Bengal, Bihar, Orissa, Madhya Pradesh and Gujarat. Whereas, *D. pentagyna* is distributed in rain forest and hilly areas below 400 m, Hainan, Yunnan in Bhutan, India, Malaysia, Myanmar, Nepal, Thailand and Vietnam. In India, it is widely distributed in Himalayan range.

4.0 Species and Cultivars

4.1.0 Species

Dillenia belongs to the family Dilleniaceae. The genus *Dillenia* has 60 species (Anon., 2001b). *Dillenia indica* (syn. *D. speciosa*) is the common edible species of dillenia which is widely distributed in the orient. *Dillenia pentagyna* is also an edible species of *Dillenia*. The basic differences between these two species are presented in Table 3. Some other less known species of *Dillenia* are *D. parviflora, D. obovata, D. suffruticosa, D. alata, D. philippinensis, D. discolor, D. papuana, D. excels, D. serrata, D. ovate, D. megalantha, D. reifferscheidia, D. turbinate etc*. Among them *D. indica* Linn. and *D.pentagyna* Roxb. are found in India (Dickison, 1979).

Dillenia indica is an evergreen or semi-deciduous tree, 10-25 m in height, with beautifully arranged dense rounded crown. Leaves are oblong, lanceolate, 20-35 cm long, 5-10 cm broad, with pointed apex and toothed margin. The upper part of the leaf as well as the veins beneath, is covered with hairs. The narrow pointed leaves are set on short stalks near the end of the branches.

Flowers are large, 12-20 cm in diameter, white, fragrant and solitary. Sepals five, roundish, concave, thick and fleshy, enlarge and enclose the fruit. Petals are white, five in number, oblong, waved, spreading, which surround numerous yellow stamens and a yellow style with 20 narrow divisions radiating outwards. Inner stamens are larger and arching over the shorter outer ones. The flowers do not fully expand even during anthesis.

Botanically, the fruit of the *Dillenia* is a special type, where the persistent and fleshy calyx forms the prominent and the only edible part of the fruit. Fruits are large weighing about 400-600g, 10-15 cm in diameter, hard, consisting of 5-6 closely fitting imbricate sepals consisting of aggregate of 15-20 carpels, each carpel containing 1-8 small, brown hairy fringed, exarillate, endospermic, compressed, reniform seeds embedded in an edible, glutinous pulp and is made up of the large fleshy imbricate sepals which tightly enclose the true fruit (Lim, 2012). In *Dillenia pentagyna* flowers are yellow, 3 cm across, arising in clusters from the branches (Bose and Chowdhury, 1991).

Table 3: Basic Differences between *Dillenia indica* and *Dillenia pentagyna*

Dillenia indica	*Dillenia pentagyna*
It is a medium sized evergreen trees upto 30 m tall.	It is a deciduous trees upto 15 m tall.
Bark is reddish brown in color, exfoliating. Young branchlet is brown pubescent, glabrescent.	Bark is greyish in color, exfoliating.Branchlets are glabrous and stout,and contains leaf scars.
Leaves are fasciculate at the ends of branches~ veins are close, running into serratures not forking at the margins.	Leaves are petiolate, glabrous, with narrow wings.
Bud is more than 5 cm in diameter.	Buds are less than 2 cm in diameter.
Flowers occur in May-June.	Flowering starts in April-May.
Fruits are aggregate and globose, 10-15 cm.	Fruits are globose, 0.51 cm in diameter.
Seed contains 5 or more per carpel, exarillate, imbedded in glutinous pulp, compressed, with hairy margins	Seeds are exarillate (Metcalfe and Chalk, 1983).

Kirtikar and Basu (1999).

4.2.0 Cultivars

There are no named or standard varieties of this crop under cultivation (Das and Das, 2003). However, two types of fruit are available in the north-eastern part of India: (i) large dusky yellow fruit and (ii) smaller chocolate green fruit.

5.0 Propagation

It is mainly propagated by seeds and cuttings, although propagation through air layering and *in-vitro* shoot multiplication can also be done. This plant is known

Species of dillenia

Dillenia excelsa

Dillenia serrata

Dillenia afloraofindia

Propagation of dillenia

Propagation by seed

Micropropagation in dillenia. (a) Shootlets development that are produced from using MS+2 mg L^{-1} of BA, (b) Rooting of shootlets cultured on liquid MS+NAA at 0.4 mg L^{-1}, (c) Prepared rooted plantlets for the acclimatization stage and(d,e) Acclimatized plants to greenhouse

Source: Taha, L.S., Ibrahim, E.A., Youssef, N.M. and El-Sayed, I.M. (2018) *Amer. J. Plant Physiol.*, **13**: 44-52.

primarily as 'elephant-apple' because the fruits are relished by elephants, which are important seed dispensers for this tree (Apu *et al.*, 2010; Gandhi and Mehta, 2013).

5.1 Seed

Seeds should be collected from ripe fruits. Freshly harvested seeds generally show high percentage of germination. Seeds are shown in the seed bed for raising seedlings. Ideal time for seed sowing is rainy season. Seeds cannot be stored for long period of time as it loses its viability very shortly. The seeds take 30 days to germinate under suitable condition (Thapliyal *et al.*, 2008). Trees grown from seeds are generally hardy and long-lived.

5.2. Cuttings

The hardwood cuttings can be used to obtain true-to-type plants. Angami and Das (2011) suggested that IBA at 250 ppm treated cuttings are superior in length of the primary roots, whereas, IBA @ 2000 ppm treatment leads to superiority in respect of days required for rooting and survival percentage.

5.3 Air Layering

The best time for air layering in *Dillenia* is between July and August (Kishore and Singh, 2015) under hot and humid climate.

5.4 Micropropagation

Micropropagation is most trustworthy approach for rapid and large scale multiplication of plantlets, though it is not being much explored till date especially among the minor crops. An efficient protocol was established by El-kadder and Hammad (2012) for *in vitro* multiplication using nodal explants. The maximum shoot multiplication rate (7.1 shoots per plant) was achieved with MS medium enriched with 2.0 mg L^{-1} BAP (6-Benzylaminopurine). Successful rooting was observed on medium containing ½ MS + 10 mg L^{-1} silver nitrate + IBA (0.5, 0.75 and 1.0 mg L^{-1}) and also in medium with 5 mg L^{-1} + IBA (0.75 and 1.0 mg L^{-1}).

6.0 Area and Production

Information regarding the area and production of this fruit in India is not available, because organized orcharding of *Dillenia* is seldom practised. Scattered trees are commonly found in wild forms in the natural habitat. However, in north-eastern region of India, dillenia is grown in the homestead gardens.

7.0 Soil and Climate

7.1.0 Soil

Dillenia grows well on a wide variety of soil, but for optimum growth and development, it requires deep, fertile, rich, moist but well drained acidic to neutral (pH 5.5-7.0), sandy loam soil. However, it can also thrive in low lying situations.

7.2.0 Climate

Dillenia is successfully grown in subtropical and tropical climate. It requires hot summer with high humidity for flowering and fruiting. The plant can tolerate shade. The recorded temperatures in its natural habitat are 5-10°C (minimum) in winter and 35-40°C (maximum) in summer, while the rainfall ranges from 1500-2000 mm. But, it cannot tolerate frost (Lim, 2012).

8.0 Cultivation

8.1.0 Planting

Dillenia seedlings/cuttings are generally planted in the rainy season (June-August). The saplings are planted with the soil ball intact and irrigation should be provided, in moisture stress, till they get established. *Dillenia* plants can be planted at a distance of 10-12 m in square system in pits of 1 x 1 x 1 m size, dug during January-February and filled with a mixture of surface soil and well rotten farmyard manure.

8.2.0 Training and Pruning

While training the plant, the framework of the branches is allowed to develop above 1 m from the ground level. Pruning in dillenia is not required at regular intervals. In later stages, the dry twigs and criss-crossed branches are to be removed.

8.3.0 Manuring and Fertilization

The young plants should be given 15-20 kg of well rotten farmyard manure annually at the non-bearing stage and subsequently increased after fruiting starts. However, no information is available on the nutritional requirement of dillenia.

8.4.0 Intercropping

Intercropping of *Dillenia* with suitable crop gives additional income and also it improves fertility of the soil. Fruit crops like Assam lemon (*Citrus limon*) as fillers and leguminous crops *viz.*, moong, cowpea, peas can be grown successfully up to 8 years when the plants usually remain in non-bearing stage.

9.0 Flowering, Pollination, Fruit Set and Fruit Growth

9.1.0 Flowering

The *Dillenia* bears the flowers on current season's growth, which appears with the leaves, usually solitary towards the end of each branchlets. The flowers commence opening from the last week of June and the blooming period lasts for three weeks. Anthesis of flower occurs during midnight between 12.30 a.m. to 1.30 a.m. It takes three hours for complete opening of flowers. Anther dehisces after 5 to 6 hours of anthesis (Neog, 1990).

9.2.0 Pollination and Fruit Set

The *Dillenia* is a cross-pollinated crop. The pollination is entomophilous, and is done by honey bees, houseflies, *etc*. Endress (1997) reported that the flowers are adapted to pollination by *Xylocopa* bees. Pollen grains are dark brownish, round to spherical in shape and their size varies from 16.6-21.58 microns. Pollen fertility

Flowering and fruit growth in dillenia

Flower of dillenia

Flower pollinated by bees

Fruit growth in dillenia

Large dusky yellow fruit

Small chocolate green fruit

Sliced fruit of dillenia

Disease of dillenia

Leaf spot disease caused by *Pestalotiopsis* sp.

Harvested fruits

Large scale pickle production from dillenia by self help group (SHG)

has been reported to be 85 per cent (Neog, 1990). After flowering, the petals drop and the persistent calyx again closes and develops to form a thick fleshy covering enveloping the fruit. The fruit, together with its protective covering, measures 10-15 cm in diameter. The many seeded true fruit is greenish yellow and surrounded by the persistent slender, spreading rays of the stigma.

9.3.0 Fruit Growth and Development

Fruit growth pattern in dillenia follows a simple sigmoid curve (Neog, 1990; Dutta, 1992). Increase in fruit weight and volume are pronounced up to 140-160 days after fruit set followed by a decreasing trend. The increase in fruit weight, volume and diameter can be attributed to an increase in the size of the cells and accumulation of carbohydrates and mucilage in the intercellular spaces in fruit (Bollard, 1970). The reduction in physical parameters towards maturity may occur on account of dehydration of fruits and approaching senescence.

The increase in seed weight, volume and diameter follows the same pattern as that of fruit growth. However, it is observed that the increase in volume of the seed is more rapid than that of seed weight.

The pulp is brown, mealy, aromatic, resinous, very slimy and slightly sticky, astringent, acid or sweetish, with numerous small, white seeds scattered through it.

10.0 Pests and Diseases

No major disease has been documented in this crop. Recently, a leaf spot disease caused by *Pestalotiopsis sp.* was reported for the first time from BCKV, India (Banerjee *et al.*, 2018). The disease is characterized by the development of a small, ash-coloured spots that gradually increases from 0.8 to 5.6 cm in diameter, changing from circular to irregular lesions mostly confine to leaf margins. Leaf spots are light brown to ash coloured, surrounded sometimes with a dark brown to black irregular zonate margin.

Infestation of few pests has been observed in the fruit. In *Dillenia*, mite attack is sometimes observed which is caused by *Aculops dilleniae* (Ghosh and Chakrabarty, 1989). The mites suck saps from the leaves causing them to dry up. Besides, in certain *Dillenia* growing areas, fruit and stem borer attack is found causing damage to the fruits (Das and Das, 2003; Deshmukh *et al.*, 2019).

11.0 Harvesting and Yield

11.1.0 Harvesting

The seedling *Dillenia* plants starts bearing after 8-10 years of planting. Fruit ripens in October-January in India when they change to a straw yellow with a greenish tinge (Janick *et al.*, 2008). The fruit reaches harvest maturity at 160 days after fruit set, when it develops straw yellow colour with greenish tinge (Neog and Mohan, 1993) and a pleasant characteristics flavour. Matured and ripe fruits drop off from plants. *Dillenia* fruits are non-climacteric and the fruit should be hand-picked when it is ripe. Care should be taken to avoid all possible damage to the fruits. A 10 year old tree can produce 1000 fruits per year under proper management (Deshmukh *et al.*, 2019).

11.2.0 Yield

On an average, 250-300 fruits can be harvested from a dillenia tree. The average yield of fruit from a full-grown dillenia tree is about 150-200 kg and the average weight of the fruit is 300 g (Das and Das, 2003).

12.0 Post-harvest Management

12.1.0 Storage and Value Addition

The fruit can be stored up to 12 days at ambient temperature of 21.7°C and one month at low temperature of 4.5°C. The chemical changes in fruit at low temperature are slower in comparison to the storage at ambient temperature. Since the fruit is seasonal it may be processed into various food products to make it available round the year. Value added products like jam, jelly, pickle, chutney, beverages and squash can be prepared from *Dillenia indica*, and these products could be shelf stable up to 12 months in ambient conditions with acceptable condition (Burhan Uddin, 2005; Kerrigan *et al.*, 2011).

13.0 References

Abdille, M.H., Singh, R.P., Jayaprakasha, G.K. and Jena, B.S. (2005) *Food Chem.*, **90**(4) : 891-896.

Alam, M.B., Rahman, M.S. Hasan, M., Khan, M.M., Nahar, K. and Sultana, S. (2012) *Int. J. Pharmacol.*, **8**(4) : 243-251.

Angami, T. and Das, R.P. (2011) *Assam J. Hort. Sc.*, **6**(2) : 109-113.

Anonymous (2001a) *Annual report. ICAR ad-hoc scheme on prservation and processing of some important indigenous fruits of North-East India*, Assam Agricultural University, Jorhat.

Anonymous (2001b) *Encyclopaedia Britannica 2001*, Deluxe Edition CD-ROM, Encyclopaedia Britannica Inc.

Apu, A.S., Muhit, M.A., Tareq, S.M., Pathan, A.H., Jamaluddin, A.T.M. and Ahmed, M. (2010) *J. Young Pharm.*, **2**(1) : 50-53.

Banerjee, A., Mandal, R. and Nath, P.S. (2018) *New Dis. Rep.*, **37**: 14.

Bhakuni, D.S., Dhar, M.L., Dhar, M.M., Dhawan, B.N. and Mehrotra, B.N. (1968) *Indian J. Exp. Biol.*, **7** : 250-262.

Bollard, *e.g.* (1970) In : *Biochemistry of Fruits and Their Products*, Vol. 1. (Ed. A.C. Hulme), Academic Press, London, pp. 387-425.

Bose, T.K. and Chowdhury, B. (1991) In : *Tropical Garden Plants in Colour*. Horticulture and Allied Publishers, Calcutta, India.

Bose, T.K., Das, P. and Maity, G.G. (1999) In : *Trees of the World*, Vol. 1. Regional Plant Resource Centre, Bhubaneswar, Orissa.

Burhan Uddin, M. (2005) *Processing of value added food products from elephant apples (Dillenia indica)*. Agricultural Research Management Information System (ARMIS). Project report submitted to the Department of Food Technology and Rural Industries, Bangladesh Agricultural University. Mymensingh 2202.

Das, B.C. and Das, S.N. (2003) In : *Cultivation of minor fruits*, Kalyani publishers, Ludhiana, p. 47-50.

Deshmukh, N.A., Okram, S., Angami, T., Rymbai, H. and Jha, A.K. (2019) In: *Minor Fruits: Nutraceutical Importance and Cultivation*. https: //www.researchgate. net/publication/330351366

Dickison, W.C. (1979) IAWA bulletin, pp. 23.

Dutta, R. (1992) *M.Sc.(Ag.) thesis*, Assam Agricultural University, Jorhat, India.

El-Kadder, E.M.A. and Hammad, H.H. (2012) *Aust. J. Basic Appl. Sci.*, **6**(7): 452-457.

Endress, P.K. (1997) *Plant Systematics and Evolution*, **206** : 99-118.

Franco, I.J. (2012) In: Minhas 500 plantas medicinais. Aparecida, Brasil: Editora Santu_ario.

Gandhi, D. and Mehta, P. (2013) *J. App. Pharm. Sci.*, **3** : 134–142.

Ghosh, N.K. and Chakrabarty, S. (1989) *Entomon.*, **14** : 121-126.

Gopalan, C., Ramsastri, B.V. and Balasubramanian, S.C. (1971) In : *Nutritive value of Indian foods*, National Institute of Nutrition, Indian Council of Medical Research, Hyderabad, India.

Janick, J., and Paull, R.E. (2008) In: The Encyclopaedia of fruits and nuts. *CABI Publishing Series*, pp.321-322.

Kaur, N., Kishore, L. and Singh, R. (2016) *J. Funct. Foods*, **22** : 547-555.

Kerrigan, R.A., Craven, L.A. and Dunlop, C.R. (2011) *Dilleniaceae.* In: Short PS, Cowie ID, eds. Flora of the Darwin Region. Palmerston, Australia: Northern Territory Government : 1-19.

Khanum, A., Khan, I. and Ali, A. (2007) *Ethnomedicine and Human Welfare*, Ukaaz publications, **4** : 52.

Khare, C.P. (2007) In : *Indian Medicinal Plant*, Springer, New York.

Kirtikar, K.R. and Basu, B.D. (1933) In : *Indian Medicinal Plants*, Vol. 1. Bishen Singh Mahendra Pal Singh, Dehradun, India.

Kirtikar, K.R., and Basu, B.D. (1999) In: *Indian Medicinal Plants*, Springer, New York.

Kishore, K. and Singh, H.S. (2015) *Indian Hortic.*, **60**(1) : 30-31.

Kumar, D., Mallick, S., Vedasiromoni, J.R. and Pal, B.C. (2010) *Phytomedicine.*, **17**(6) : 431-435.

Kuotsu, K. and Bandyopadhyay, A.K. (2007) *Sci. Asia.*, **33** : 57-60.

Lim, T.K. (2012) In : *Edible medicinal and non-medicinal plants*, Springer, Dordrecht, p. 410-415.

MacMillan, H.F. (1989) In : *Handbook of Tropical Plants.* Anmol Publication, New Delhi, India, p. 255.

Malviya, N., Jain, S. and Malviya, S. (2010) *Acta Pol. Pharm.*, **67**(2) : 113-118.

Metcalfe, C.R., and Chalk, C. (1983) In: *Anatomy of the Dicotyledons*, Clarendon Press, Oxford, London.

Neog, M. (1990) *M.Sc.(Ag.) thesis*, Assam Agricultural University, Jorhat, India.

Neog, M. and Mohan, N.K. (1993) *South Indian Hort.*, **41** : 115-116.

Neog, M. and Mohan, N.K. (1994) *Indian Hort.*, **39** : 28-31.

Saikia, L. and Dutta, R. (1995) *Indian Food Pack.*, September-October, pp. 25-32.

Sastri, B.N. (1952) *The Wealth of India, Raw Materials*, Vol. III : D-E, CSIR, New Delhi, India, p. 194.

Sharma, P.V. (1956) *Dravya Gun Vigyan*, Varanasi, India.

Sharma, H.K. and Nath, L.K. (2014) *Int. Scholarly Res. Notices.*

Shome, U., Khanna, R.K. and Sharma, H.P. (1980) *Proc. Indian Acad. Sci. (Plant Science)*, **89** : 91-104.

Singh, R. (1992) Fruits, National Book Trust of India, New Delhi, India.

Suh, J.S. (1999) *FRI J. Forest Research*, Seol, No. 62, pp. 41-52.

Talukdar, A., Talukdar, N., Deka, S. and Sahariah, B.J. (2012) *Int. J. phar. Sci. Res.*, **3**(8) : 2482-2486.

Thapliyal, R.C., Phartyal, S.S., Baskin, J.M. and Baskin, C.C. (2008) *Aust. J. Bot.*, **56** : 583-589.

The Wealth of India (1952) CSIR, New Delhi, India, Vol. III, pp. 64-67.

Index

C

D

E

F

G

H

I

J

K

L

M

R

S

Printed at Repro India Ltd.